QUANTUM MECHANICS IN NONLINEAR SYSTEMS

QUANTUM MECHANICS IN NONLINEAR SYSTEMS

Pang Xiao-Feng
University of Electronic Science and Technology of China, China

Feng Yuan-Ping
National University of Singapore, Singapore

NEW JERSEY • LONDON • SINGAPORE • BEIJING • SHANGHAI • HONG KONG • TAIPEI • CHENNAI

Published by

World Scientific Publishing Co. Pte. Ltd.
5 Toh Tuck Link, Singapore 596224
USA office: 27 Warren Street, Suite 401-402, Hackensack, NJ 07601
UK office: 57 Shelton Street, Covent Garden, London WC2H 9HE

Library of Congress Cataloging-in-Publication Data
Pang, Xiao-Feng, 1945-
Quantum mechanics in nonlinear systems / Pang Xiao-Feng, Feng Yuan-Ping.
p. cm.
Includes bibliographical references and index.
ISBN-13 978-981-256-116-9 -- ISBN-10 981-256-116-1 (alk. paper)
ISBN-13 978-981-256-299-9 (pbk) -- ISBN-10 981-256-299-0 (pbk)
1. Nonlinear theories. 2. Quantum theory. I. Feng, Yuang-Ping. II. Title.

QC20.7.N6P36 2005
530.15'5252--dc22

2004060119

British Library Cataloguing-in-Publication Data
A catalogue record for this book is available from the British Library.

Copyright © 2005 by World Scientific Publishing Co. Pte. Ltd.

All rights reserved. This book, or parts thereof, may not be reproduced in any form or by any means, electronic or mechanical, including photocopying, recording or any information storage and retrieval system now known or to be invented, without written permission from the Publisher.

For photocopying of material in this volume, please pay a copying fee through the Copyright Clearance Center, Inc., 222 Rosewood Drive, Danvers, MA 01923, USA. In this case permission to photocopy is not required from the publisher.

Printed in Singapore

Preface

This book discusses the properties of microscopic particles in nonlinear systems, principles of the nonlinear quantum mechanical theory, and its applications in condensed matter, polymers, and biological systems. It is intended for researchers, graduate students, and upper level undergraduate students.

About the Book

Some materials in the book are based on the lecture notes for a graduate course "Problems in nonlinear quantum theory" given by one of the authors (X. F. Pang) in his university in the 1980s, and a book entitled "Theory of Nonlinear Quantum Mechanics" (in Chinese) by the same author in 1994. However, the contents were completely rewritten in this English edition, and in the process, we incorporated recent results related to the nonlinear Schrödinger equations and the nonlinear Klein-Gordon equations based on research of the authors as well as other scientists in the field.

The following topics are covered in 10 chapters in this book, the necessity for constructing a nonlinear quantum mechanical theory; the theoretical and experimental foundations on which the nonlinear quantum mechanical theory is based; the elementary principles and the theory of nonlinear quantum mechanics; the wave-corpuscle duality of particles in the theory; nonlinear interaction and localization of particles; the relations between nonlinear and linear quantum theories; the properties of nonlinear quantum mechanics, including simultaneous determination of position and momentum of particles, self-consistence and completeness of the theory; methods of solving nonlinear quantum mechanical problems; properties of particles in various nonlinear systems and applications to exciton, phonon, polaron, electron, magnon and proton in physical, biological and polymeric systems. In particular, an in-depth discussion on the wave-corpuscle duality of microscopic particles in nonlinear systems is given in this book.

The book is organized as follows. We start with a brief review on the postulates of linear quantum mechanics, its successes and problems encountered by the linear quantum mechanics in Chapter 1. In Chapter 2, we discuss some macroscopic quantum effects which form the experimental foundation for a new nonlinear quantum theory, and the properties of microscopic particles in the macroscopic quantum systems which provide a theoretical base for the establishment of the nonlinear quantum theory. The fundamental principles on which the new theory is based and the theory of nonlinear quantum mechanics as proposed by Pang *et al.* are given in Chapter 3. The close relations among the properties of macroscopic quantum effects; nonlinear interactions and soliton motions of microscopic particles in macroscopic quantum systems play an essential role in the establishment of this theory. In Chapter 4, we examine in details the wave-corpuscle duality of particles in nonlinear systems. In Chapter 5, we look into the mechanisms of nonlinear interactions and their relations to localization of particles. In the next chapter, features of the nonlinear and linear quantum mechanical theories are compared; the self-consistence and completeness of the theory were examined; and finally solutions and properties of the time-independent nonlinear quantum mechanical equations, and their relations to the original quantum mechanics are discussed. We will show that problems existed in the original quantum mechanics can be explained by the new nonlinear quantum mechanical theory. Chapter 7 shows the methods of solving various kinds of nonlinear quantum mechanical problems. The dynamic properties of microscopic particles in different nonlinear systems are discussed in Chapter 8. Finally in Chapters 9 and 10, applications of the theory to exciton, phonon, electron, polaron, proton and magnon in various physical systems, such as condensed matter, polymers, molecules and living systems, are explored.

The book is essentially composed of three parts. The first part consists of Chapters 1 and 2, gives a review on the linear quantum mechanics, and the important experimental and theoretical studies that lead to the establishment of the nonlinear quantum-mechanical theory. The nonlinear theory of quantum mechanics itself as well as its essential features are described in second part (Chapters 3-8). In the third part (Chapters 9 and 10), we look into applications of this theory in physics, biology and polymer, *etc.*

An Overview

Nonlinear quantum mechanics (NLQM) is a theory for studying properties and motion of microscopic particles (MIPs) in nonlinear systems which exhibit quantum features. It was named so in relation to the quantum mechanics established by Bohr, Heisenberg, Schrödinger, and many others. The latter deals with only properties and motion of microscopic particles in linear systems, and will be referred to as the linear quantum mechanics (LQM) in this book.

The concept of nonlinearity in quantum mechanics was first proposed by de Broglie in the 1950s in his book, "Nonlinear wave mechanics". LQM had difficulties explain explaining certain problems right from the start. de Broglie attempted to clarify and solve these problems of LQM using the concept of nonlinearity. Even though a great idea, de Broglie did not succeed because his approach was confined to the framework of the original LQM.

Looking back to the modern history of physics and science, we know that quantum mechanics is really the foundation of modern science. It had great successes in solving many important physical problems, such as the light spectra of hydrogen and hydrogen-like atoms, the Lamb shift in these atoms, and so on. Jargons such as "quantum jump" have their scientific origins and become ever fashionable in our normal life. In this particular case, the phrase "quantum jump" gives a vivid description for major qualitative changes and is almost universally used. However, it was also known that LQM has its problems and difficulties related to the fundamental postulates of the theory, for example, the implications of the uncertainty principle between conjugate dynamical variables, such as position and momentum. Different opinions on how to resolve such issues and further develop quantum mechanics lead to intense arguments and debates which lasted almost a century. The long-time controversy showed that these problems cannot be solved within the framework of LQM. It was also through such debates that the direction to take for improving and further developing quantum mechanics became clear, which was to extend the theory from the linear to the nonlinear regime. Certain fundamental assumptions such as the principle of linear superposition, linearity of the dynamical equation and the independence of the Hamiltonian of a system on its wave function must be abandoned because they are the roots of the problems of LQM. In other words, a new nonlinear quantum theory should be developed.

A series of nonlinear quantum phenomena including the macroscopic quantum effects and motion of solitons or solitary waves have, in recent decades, been discovered one after another from experiments in superconductors, superfluid, ferromagnetic, antiferromagnetic, organic molecular crystals, optical fiber materials and polymer and biological systems, *etc.* These phenomena did underlie nonlinear quantum mechanics because they could not be explained by LQM. Meanwhile, the theories of nonlinear partial differential equations and of solitary wave have been very well established which build the mathematical foundation of nonlinear quantum mechanics. Due to these developments of nonlinear science, a lot of new branches of science, for example, nonlinear vibrational theory, nonlinear Newton mechanics, nonlinear fluid mechanics, nonlinear optics, chaos, synergetics and fractals, have been established or being developed. In such a case, it is necessary to build the nonlinear quantum mechanics described the law of motion of microscopic particles in nonlinear systems.

However, how do we establish such a theory? Experiences in the study of quantum mechanics for several decades tell us that it is impossible to establish such a theory if we followed the direction of de Broglie *et al.* A completely new way of thinking, a new idea and method must be adopted and developed.

According to this idea we will, first of all, study the properties of macroscopic quantum effects, which is a nonlinear quantum effect on macroscopic scale occurred in some matters, for example, superconductors and superfluid. To be more precise, these effects occur in systems with ordered states over a long-range, or, coherent states, or, Bose-like condensed states, which are formed through phase transitions after a spontaneous symmetry breakdown in the systems by means of nonlinear interactions. These results show that the properties of microscopic particles in the macroscopic quantum systems cannot be well represented by LQM. In these systems the microscopic particles are self-localized to become soliton with wave-corpuscle duality. The observed macroscopic quantum effects are just a result produced by soliton motions of the particles in these systems. Therefore, the macroscopic quantum effect is closely related to the nonlinear interaction and to solitary motion of the particles. The close relations among them prompt us to propose and establish the fundamental principles and the theory of NLQM which describes the properties of microscopic particles in the nonlinear systems. We then demonstrate that the NLQM is truely a self-consistent and complete theory. It has so far enjoyed great successes in a wide range of applications in condensed matter, polymers and biological systems. In exploring these applications, we also obtain many important results which are consistent with experimental data. These results confirm the correctness of the NLQM on one hand, and provide further theoretical understanding to many phenomena occurred in these systems on the other hand.

Therefore, we can say that the experimental foundation of the nonlinear quantum mechanics established is the macroscopic quantum effects, and the coherent phenomena. Its theoretical basis is superconducting and superfluidic theories. Its mathematical framework is the theories of nonlinear partial differential equations and of solitary waves. The elementary principles and theory of the NLQM proposed here are established on the basis of results of research on properties of microscopic particles in nonlinear systems and the close relations among the macroscopic quantum effects, nonlinear interactions and soliton motions. The linearity in the LQM is removed and dependence of Hamiltonian of systems on the state wave function of particles is assumed in this theory. Through careful investigations and extensive applications, we demonstrate that this new theory is correct, self-consistent and complete. The new theory solves the problems and difficulties in the LQM.

One of the authors (X. F. Pang) has been studying the NLQM for about 25 years and has published about 100 papers related to this topic. The newly established nonlinear quantum theory has been reported and discussed in many international conferences, for example, International Conference of Nonlinear Physics (ICNP), International Conference of Material Physics (ICMP), Asia Pacific Physics Conference (APPC), International Workshop of Nonlinear Problems in Science and Engineering (IWNPSE), National Quantum Mechanical Conference of China (NQMCC), *etc.*. Pang also published a monograph entitled "The problems for nonlinear quantum theory" in 1985 and a book entitled "The theory of nonlinear quantum mechanics" in 1994 in Chinese. Pang has also lectured in many Universities and Institutes on this subject. Certain materials in this book are based on the above lecture materials and book. It also incorporates many recent results published by Pang and other scientists related to nonlinear Schrödinger equation and non-linear Klein-Gordon equations.

Finally, we should point out that the NLQM presented here is completely different from the LQM. It is intended for studying properties and motion of microscopic particles in nonlinear systems, in which the microscopic particles become self-localized particles, or solitons, under the nonlinear interaction. Sources of such nonlinear interation can be intrinsic nonlinearity or persistent self-interactions through mechanisms such as self-trapping, self-condensation, self-focusing and self-coherence by means of phase transitions, sudden changes and spontaneous breakdown of symmetry of the systems, and so on. In such cases, the particles have exactly wave-corpuscle duality, and obey simultaneously the classical and quantum laws of motion, *i. e.*, the nature and properties of the microscopic particle are essentially changed from that in LQM. For example, the position and momentum of a particle can be determined to a certain degree. Thus, the linear feature of theory and the principles for independences of the Hamiltonian of the systems on the state-wave function of particle are completely removed. However, this is not to deny the validity of LQM. Rather we believe that it is an approximate theory which is only suitable for systems with linear interactions and the nonlinear interaction is small and can be neglected. In other words, LQM is a special case of the NLQM. This relation between the LQM and the NLQM is similar to that between the relativity and Newtonian mechanics. The NLQM established here is a necessary result of development of quantum mechanics in nonlinear systems.

The establishment of the NLQM can certainly advance and facilitate further developments of natural sciences including physics, biology and astronomy. Meanwhile, it is also useful in understanding the properties and limitations of the LQM, and in solving problems and difficulties encountered by the LQM. Therefore, we hope that by publishing this book on quantum mechanics in the non-linear systems would add some value to science and would contribute to our understanding of the wonderful nature.

X. F. Pang and Y. P. Feng
2004

Contents

Chapter 1

Linear Quantum Mechanics: Its Successes and Problems

The quantum mechanics established by Bohr, de Broglie, Schrödinger, Heisenberg and Bohn in 1920s is often referred to as the linear quantum mechanics (LQM). In this chapter, the hypotheses of linear quantum mechanics, the successes of and problems encountered by the linear quantum mechanics are reviewed. The directions for further development of the quantum theory are also discussed.

1.1 The Fundamental Hypotheses of the Linear Quantum Mechanics

At the end of the 19th century, classical mechanics encountered major difficulties in describing motions of microscopic particles (MIPs) with extremely light masses ($\sim 10^{-23} - 10^{-26}$ g) and extremely high velocities, and the physical phenomena related to such motions. This forced scientists to rethink the applicability of classical mechanics and lead to fundamental changes in their traditional understanding of the nature of motions of microscopic objects. The wave-corpuscle duality of microscopic particles was boldly proposed by Bohr, de Broglie and others. On the basis of this revolutionary idea and some fundamental hypotheses, Schrödinger, Heisenberg, etc. established the linear quantum mechanics which provided a unique way of describing quantum systems. In this theory, the states of microscopic particles are described by a wave function which is interpreted based on statistics, and physical quantities are represented by operators and are given in terms of the possible expectation values (or eigenvalues) of these operators in the states (or eigenstates). The time evolution of quantum states are governed by the Schrödinger equation. The hypotheses of the linear quantum mechanics are summarized in the following.

(1) A state of a microscopic particle is represented by a vector in the Hilbert space, $|\psi\rangle$, or a wave function $\psi(\vec{r},t)$ in coordinate space. The wave function uniquely describes the motion of the microscopic particle and reflects the wave nature of microscopic particles. Furthermore, if β is a constant, then both $|\psi\rangle$ and $\beta|\psi\rangle$ describe the same state. Thus, the normalized wave function, which satisfies the condition $\langle\psi|\psi\rangle = 1$, is often used to describe the state of the particle.

(2) A physical quantity, such as the coordinate X, the momentum P and the energy E of a particle, is represented by a linear operator in the Hilbert space, and the eigenvectors of the operator form a basis of the Hilbert space. An observable mechanical quantity is represented by a Hermitian operator whose eigenvalues are real. Therefore, the values a physical quantity can have are the eigenvalues of the corresponding linear operator. The eigenvectors corresponding to different eigenvalues are orthogonal to each other. All eigenstates of a Hermitian operator span an orthogonal and complete set, $\{\psi_L\}$. Any vector of state, $\psi(\vec{r},t)$, can be expanded in terms of the eigenvectors:

$$\psi(\vec{r},t) = \sum_L C_L \psi_L(\vec{r},t), \qquad \text{or} \qquad |\psi(\vec{r},t)\rangle = \sum_L \langle\psi_L|\psi\rangle|\psi_L\rangle \tag{1.1}$$

where $C_L = \langle\psi_L|\psi\rangle$ is the wave function in representation L. If the spectrum of L is continuous, then the summation in (1.1) should be replaced by an integral: $\int dL \cdots$. Equation (1.1) can be regarded as a projection of the wave function $\psi(\vec{r},t)$ of a microscopic particle system on to those of its subsystems and it is the foundation of transformation between different representations in the linear quantum mechanics. In the quantum state described by $\psi(\vec{r},t)$, the probability of getting the value L' in a measurement of L is $|C_{L'}|^2 = |\langle\psi_{L'}|\psi\rangle|^2$ in the case of discrete spectrum, or $|\langle\psi_{L'}|\psi\rangle|^2 dL$ if the spectrum of the system is continuous. In a single measurement of any mechanical quantity, only one of the eigenvalues of the corresponding linear operator can be obtained, and the system is then said to be in the eigenstate belonging to this eigenvalue. This is a fundamental assumption of linear quantum mechanics concerning measurements of physical quantities.

(3) The average $\langle\hat{A}\rangle$ of a physical quantity A in an arbitrary state $|\psi\rangle$ is given by

$$\langle\hat{A}\rangle = \frac{\langle\psi|\hat{A}|\psi\rangle}{\langle\psi|\psi\rangle}, \tag{1.2}$$

or

$$\langle\hat{A}\rangle = \langle\psi|\hat{A}|\psi\rangle,$$

if ψ is normalized. Possible values of A can be obtained through the determination of the above average. In order to obtain these possible values, we must find a wave function in which A has a precise value. In other words, we must find a state such that $\overline{\langle\Delta A\rangle^2} = 0$, where $\overline{(\Delta A)^2} = \langle\hat{A}^2\rangle - \langle\hat{A}\rangle^2$. This leads to the following eigenvalue problem for the operator $\hat{A}$,

$$\hat{A}\psi_L = A\psi_L. \tag{1.3}$$

From the above equation we can determine the spectrum of eigenvalues of the operator $\hat{A}$ and the corresponding eigenfunctions ψ_L. The eigenvalues of $\hat{A}$ are possible values observed from a measurement of the physical quantity. All possible values of A in any other state are nothing but its eigenvalues in its own eigenstates. This hypothesis reflects the statistical nature in the description of motion of microscopic particles in the linear quantum mechanics.

(4) The Hilbert space in which the linear quantum mechanics is defined is a linear space. The operator of a mechanical quantity is a linear operator in this space. The eigenvectors of a linear operator satisfy the linear superposition principle. That is, if two states, $|\psi_1\rangle$ and $|\psi_2\rangle$ are both eigenfunctions of a given linear operator, then their linear combination

$$|\psi\rangle = C_1|\psi_1\rangle + C_2|\psi_2\rangle, \tag{1.4}$$

where C_1 and C_2 are constants, also describes a state of the same particle. The linear superposition principle of quantum states is determined by the linear characteristics of the operators and this is why the quantum theory is referred to as linear quantum mechanics. It is noteworthy to point out that such a superposition is different from that of classical waves, it does not result in changes in probability and intensity.

(5) *The correspondence principle:* If two classical mechanical quantities, A and B, satisfy the Poisson brackets,

$$\{A, B\} = \sum_n \left(\frac{\partial A}{\partial q_n}\frac{\partial B}{\partial p_n} - \frac{\partial A}{\partial p_n}\frac{\partial B}{\partial q_n} \right)$$

where q_n and p_n are generalized coordinate and momentum in the classical system, respectively, then the corresponding operators $\hat{A}$ and $\hat{B}$ in quantum mechanics satisfy the following commutation relation:

$$[\hat{A}, \hat{B}] = (\hat{A}\hat{B} - \hat{B}\hat{A}) = -ih\{\hat{A}, \hat{B}\} \tag{1.5}$$

where $i = \sqrt{-1}$ and h is the Planck's constant. If A and B are substituted by q_n and p_n respectively, we have:

$$[\hat{p}_n, \hat{q}_m] = -ih\delta_{nm}, \quad [\hat{p}_n, \hat{p}_m] = 0, \quad \cdots .$$

This reflects the fact that values allowed for a physical quantity in a microscopic system are quantized, and thus the name "quantum mechanics". Based on this fundamental principle, the Heisenberg uncertainty relation can be obtained as follows,

$$\overline{(\triangle A)^2}\ \overline{(\triangle B)^2} \geq \frac{|C|^2}{4} \tag{1.6}$$

where $iC = [\hat{A}, \hat{B}]$ and $\Delta A = \langle \hat{A} - \langle \hat{A} \rangle \rangle$. For the coordinate and momentum operators, the Heisenberg uncertainty relation takes the usual form

$$|\triangle x||\triangle p \geq \frac{\hbar}{2}.$$

(6) The time dependence of a quantum state $|\psi\rangle$ of a microscopic particle is determined by the following Schrödinger equation:

$$-\frac{\hbar}{i}\frac{\partial}{\partial t}|\psi\rangle = H|\psi\rangle. \tag{1.7}$$

This is a fundamental dynamic equation for microscopic particle in space-time. $\hat{H}$ is the Hamiltonian operator of the system and is given by,

$$\hat{H} = \hat{T} + \hat{V} = -\frac{\hbar^2}{2m}\nabla^2 + \hat{V},$$

where $\hat{T}$ is the kinetic energy operator and $\hat{V}$ the potential energy operator. Thus, the state of a quantum system at any time is determined by the Hamiltonian of the system. As a fundamental equation of linear quantum mechanics, equation (1.7) is a linear equation of the wave function ψ which is another reason why the theory is referred as a linear quantum mechanics.

If the quantum state of a system at time t_0 is $|\psi(t_0)\rangle$, then the wave function and mechanical quantities at time t are associated with those at time t_0 by a unitary operator $\hat{U}(t, t_0)$, *i.e.*

$$|\psi(t)\rangle = \hat{U}(t, t_0)|\psi(t_0)\rangle, \tag{1.8}$$

where $\hat{U}(t_0, t_0) = 1$ and $\hat{U}^+\hat{U} = \hat{U}\hat{U}^+ = I$. If we let $\hat{U}(t, 0) = \hat{U}(t)$, then the equation of motion becomes

$$-\frac{\hbar}{i}\frac{\partial}{\partial t}\hat{U}(t) = \hat{H}\hat{U}(t) \tag{1.9}$$

when $\hat{H}$ does not depend explicitly on time t and $\hat{U}(t) = e^{-i(\hat{H}/\hbar)t}$. If $\hat{H}$ is an explicit function of time t, we then have

$$\hat{U}(t) = 1 + \frac{1}{ih}\int_0^t dt_1\hat{H}(t_1) + \frac{1}{(ih)^2}\int_0^t dt_1\hat{H}(t_1)\int_0^{t_1} dt_2\hat{H}(t_2) + \cdots. \tag{1.10}$$

Obviously, there is an important assumption here: the Hamiltonian operator of the system is independent of its state, or its wave function. This is a fundamental assumption in the linear quantum mechanics.

(7) *Identical particles:* No new physical state should occur when a pair of identical particles is exchanged in a system. In other words, the wave function satisfies $\hat{P}_{kj}|\psi\rangle = \lambda|\psi\rangle$, where $\hat{P}_{kj}$ is an exchange operator and $\lambda = \pm 1$. Therefore, the wave function of a system consisting of identical particles must be either symmetric, ψ_s, ($\lambda = +1$), or antisymmetric, ψ_a, ($\lambda = -1$), and this property remains invariant with time and is determined only by the nature of the particle. The wave function of a boson particle is symmetric and that of a fermion is antisymmetric.

(8) *Measurements of physical quantities:* There was no assumption made about measurements of physical quantities at the beginning of the linear quantum mechanics. It was introduced later to make the linear quantum mechanics complete. However, this is a nontrivial and contraversal topic which has been a focus of scientific debate. This problem will not be discussed here. Interested reader can refer to texts and references given at the end of this chapter.

1.2 Successes and Problems of the Linear Quantum Mechanics

On the basis of the fundamental hypotheses mentioned above, Heisenberg, Schrödinger, Bohn, Dirac, and others established the theory of linear quantum mechanics which describes the properties and motions of microscopic particle systems. This theory states that once the externally applied potential fields and initial states of the particles are given, the states of the particles at any time later and any position can be determined by the linear Schrödinger equation, equations (1.7) and (1.8) in the case of nonrelativistic motion, or equivalently, the Dirac equation and the Klein-Gordon equation in the case of relativistic motion. The quantum states and their occupations of electronic systems, atoms, molecules, and the band structure of solid state matter, and any given atomic configuration are completely determined by the above equations. Macroscopic behaviors of systems such as mechanical, electrical and optical properties may also be determined by these equations. This theory also describes the properties of microscopic particle systems in the presence of external electromagnetic field, optical and acoustic waves, and thermal radiation. Therefore, to a certain degree, the linear quantum mechanics describes the law of motion of microscopic particles of which all physical systems are composed. It is the foundation and pillar of modern physics.

The linear quantum mechanics had great successes in descriptions of motions of microscopic particles, such as electron, phonon, photon, exciton, atom, molecule, atomic nucleus and elementary particles, and in predictions of properties of matter based on the motions of these quasi-particles. For example, energy spectra of atoms (such as hydrogen atom, helium atom), molecules (such as hydrogen molecule) and compounds, electrical, optical and magnetic properties of atoms and condensed matters can be calculated based on linear quantum mechanics and the calculated results are in good agreement with experimental measurements. Being the foundation of modern science, the establishment of the theory of quantum mechanics has revolutionized not only physics, but many other science branches such as chemistry, astronomy, biology, *etc.*, and at the same time created many new branches of science, for example, quantum statistics, quantum field theory, quantum electronics, quantum chemistry, quantum biology, quantum optics, *etc.* One of the great successes of the linear quantum mechanics is the explanation of the fine energy spectra of hydrogen atom, helium atom and hydrogen molecule. The energy spectra predicted by linear quantum mechanics for these atoms and molecules are completely in agreement with experimental data. Furthermore, modern experiments have demonstrated that the results of the Lamb shift and superfine structure of hydrogen atom and the anomalous magnetic moment of the electron predicted by the theory of quantum electrodynamics are in agreement with experimental data within an order of magnitude of 10^{-5}. It is therefore believed that the quantum electrodynamics is one of most successful theories in modern physics.

Despite the great successes of linear quantum mechanics, it nevertheless encountered some problems and difficulties. In order to overcome these difficulties, Einstein had disputed with Bohr and others for the whole of his life and the difficulties still remained up to now. Some of the difficulties will be discussed in the next section. These difficulties of the linear quantum mechanics are well known and have been reviewed by many scientists. When one of the founders of the linear quantum mechanics, Dirac, visited Australia in 1975, he gave a speech on the development of quantum mechanics in New South Wales University. During his talk, Dirac mentioned that at the time, great difficulties existed in the quantum mechanical theory. One of the difficulties referred to by Dirac was about an accurate theory for interaction between charged particles and an electromagnetic field. If the charge of a particle is considered as concentrated at one point, we shall find that the energy of the point charge is infinite. This problem had puzzled physicists for more than 40 years. Even after the establishment of the renormalization theory, no actual progress had been made. Such a situation was similar to the unified field theory for which Einstein had struggled for his whole life. Therefore, Dirac concluded his talk by making the following statements: It is because of these difficulties, I believe that the foundation for the quantum mechanics has not been correctly laid down. As part of the current research based on the existing theory, a great deal of work has been done in the applications of the theory. In this respect, some rules for getting around the infinity were established. Even though results obtained based on such rules agree with experimental measurements, they are artificial rules after all. Therefore, I cannot accept that the present foundation of the quantum mechanics is completely correct.

However, what are the roots of the difficulties of the linear quantum mechanics that evoked these contentions and raised doubts about the theory among physicists? Actually, if we take a closer look at the history of physics, one would know that not so many fundamental assumptions were required for all physical theories but the linear quantum mechanics. Obviously, these assumptions of linear quantum mechanics caused its incompleteness and limited its applicability.

It was generally accepted that the fundamentals of the linear quantum mechanics consist of the Heisenberg matrix mechanics, the Schrödinger wave mechanics, Born's statistical interpretation of the wave function and the Heisenberg uncertainty principle, *etc.* These were also the focal points of debate and controversy. In other words, the debate was about how to interpret quantum mechanics. Some of the questions being debated concern the interpretation of the wave-particle duality, probability explanation of the wave function, the difficulty in controlling interaction between measuring instruments and objects being measured, the Heisenberg uncertainty principle, Bohr's complementary (corresponding) principle, single particle versus many particle systems, the problems of microscopic causality and probability, process of measuring quantum states, *etc.* Meanwhile, the linear quantum mechanics in principle can describe physical systems with many particles, but it is not easy to solve such a system and approximations must be used to obtain approximate solutions. In doing this, certain features of the system which could be important have to be neglected. Therefore, while many enjoyed the successes of the linear quantum mechanics, others were wondering whether the linear quantum mechanics is the right theory of the real microscopic physical world, because of the problems and difficulties it encountered. Modern quantum mechanics was born in 1920s, but these problems were always the topics of heated debates among different views till now. It was quite exceptional in the history of physics that so many prominent physicists from different institutions were involved and the scope of the debate was so wide. The group in Copenhagen School headed by Bohr represented the view of the main stream in these discussions. In as early as 1920s, heated disputes on the statistical explanation and completeness of wave function arose between Bohr and other physicists, including Einstein, de Broglie, Schrödinger, Lorentz, *etc.*

The following is a brief summary of issues being debated and problems encountered by the linear quantum mechanics.

(1) First, the correctness and completeness of the linear quantum mechanics were challenged. Is linear quantum mechanics correct? Is it complete and self-consistent? Can the properties of microscopic particle systems be completely described by the linear quantum mechanics? Do the fundamental hypotheses contradict each other?

(2) Is the linear quantum mechanics a dynamic or a statistical theory? Does it describe the motion of a single particle or a system of particles? The dynamic equation seems an equation for a single particle, but its mechanical quantities are determined based on the concepts of probability and statistical average. This caused confusion about the nature of the theory itself.

(3) How to describe the wave-particle duality of microscopic particles? What is the nature of a particle defined based on the hypotheses of the linear quantum mechanics? The wave-particle duality is established by the de Broglie relations. Can the statistical interpretation of wave function correctly describe such a property? There are also difficulties in using wave package to represent the particle nature of microscopic particles. Thus describing the wave-corpuscle duality was a major challenge to the linear quantum mechanics.

(4) Was the uncertainty principle due to the intrinsic properties of microscopic particles or a result of uncontrollable interaction between the measuring instruments and the system being measured?

(5) A particle appears in space in the form of a wave, and it has certain probability to be at a certain location. However, it is always a whole particle, rather than a fraction of it, being detected in a measurement. How can this be interpreted? Is the explanation of this problem based on wave package contraction in the measurement correct?

Since these are important issues concerning the fundamental hypotheses of the linear quantum mechanics, many scientists were involved in the debate. Unfortunately, after being debated for almost a century, there are still no definite answers to most of these questions. We will introduce and survey some main views of this debate in the following.

As far as the completeness of the linear quantum mechanics was concerned, Von Neumann provided a proof in 1932. According to Von Neumann, if O is a set of observable quantities in the Hilbert space $\mathcal{Q}$ of dimension greater than one, then the self-adjoint of any operator in this set represents an observable quantity in the same set, and its state can be determined by the average $\langle\hat{A}\rangle$ for the operator $\hat{A}$. If this average value satisfies $\langle 1\rangle = 1$, we have $\langle r\hat{A}\rangle = r\langle\hat{A}\rangle$ for any real constant r. If A is non-negative, then $\langle\hat{A}\rangle \geq 0$. If $A, B, C, \cdots$ are arbitrary observable quantities, then, there always exists an observable $A + B + C + \cdots$ such that $\langle\hat{A}+\hat{B}+\hat{C}+\cdots\rangle = \langle\hat{A}\rangle+\langle\hat{B}\rangle+\langle\hat{C}\rangle+\cdots$. Von Neumann proved that there exists a self-adjoint operator $\hat{A}$ in $\mathcal{Q}$ such that $\langle A^a\rangle \neq \langle A\rangle^a$. This implies that there always exists an observable quantity A which is indefinite or does not have an accurate value. In other words, the states as defined by the average value are dispersive and cannot be determined accurately, which further implies that states in which all observable quantities have accurate values simultaneously do not exist. To be more concrete, not all properties of a physical system can possess accurate values. At this stage, this was the best the theory can do. Whether it can be accepted as a complete theory is subjective. It seemed that any further discussion would lead to nowhere.

It was realized later that Von Neumann's theorem was mathematically flawless but ambiguous and vague in physics. In 1957, Gleason made two modifications to Von Neumann's assumptions: $\mathcal{Q}$ should be the Hilbert space of more than two dimensions rather than one; and A, B, C, $\cdots$ should be limited to commutable self-adjoint operators in $\mathcal{Q}$. He verified that Von Neumann's theorem is still valid with these assumptions. Because the operators are commutable, the linear superposition property of average values is, in general, independent of the order in which experiments are performed. Hence, these assumptions seem to be physically acceptable. Furthermore, Von Neumann's conclusion ruled out some nontrivial hidden variable theories in the Hilbert space with dimensions of more than two.

However, in 1966, Bell indicated that Gleason's theorem can essentially only remove the hidden variable theories which are independent of environment and arrangements before and after a measurement. It would be possible to establish hidden variable theories which are dependent on environment and arrangements before and after a measurement. At the same time, Bell argued that since there are more input hidden variables in the hidden variable theory than in quantum mechanics, there should be new results that may be compared with experiments, thus to verify whether the quantum mechanics is complete.

Starting from an ideal experiment based on the localized hidden variables theory and the average value $q(a, b) = \int A_a(\lambda)B_b(\lambda)d\lambda$, Bohm believed that some features of a particle could be obtained once those of another particle which is remotely separated from the first are measured. This indicates that correlation between particles exists which could be described in terms of "hidden parameters". Based on this idea, Bell proposed an inequality which is applicable to any "localized" hidden variables theory. Thus, the natures of correlation in a system of particles predicted by the Bell's inequality and quantum mechanics would differ appreciably which can be used to verify which of the two is correct.

To this end, we discuss a system of spin correlation. We shall first discuss spin correlation from the point of view of quantum mechanics. Assume that there exists a system which consists of two particles A and B, both of spin 1/2, but the total spin of the system is zero. Let A_a be the spin component measured along a direction specified by a unit vector $\hat{a}$, and similarly B_b the spin component measured along a direction specified by a unit vector $\hat{b}$. According to linear quantum mechanics, it is easy to write down the components of the spin operators along directions $\hat{a}$ and $\hat{b}$. They are $(\hat{\sigma}_A \cdot \hat{a})/2$ and $(\hat{\sigma}_B \cdot \hat{b})/2$, respectively, where $\hat{\sigma}_A/2$ and $\hat{\sigma}_B/2$ are the spin operators of particles A and B in terms of the Pauli matrices, respectively. $(\hat{\sigma}_A \cdot \hat{a})/2$ and $(\hat{\sigma}_B \cdot \hat{b})/2$ can be regarded as projections of the spin operators on the unit vectors $\hat{a}$ and $\hat{b}$, respectively. The spin correlation function, $q(a, b)$, may be defined as the average of the product of A_a and B_b, *i.e.* $q(a,b) = 4\overline{A_a \cdot B_b}$, where the factor of 4 is due to "normalization", the horizontal line above $A_a \cdot B_b$ denotes the statistical average of the product of A_a and B_b over all possible results of measurements. According to linear quantum mechanics, we have

$$\overline{A_a \cdot B_b} = \frac{1}{4}\langle 0^+|(\hat{\sigma}_A \cdot \hat{a})(\hat{\sigma}_B \cdot \hat{b})|0^+\rangle$$

where $|0^+\rangle$ represents the spin wave function with zero total spin, of the system consisting of particles A and B of spin 1/2, and can be expressed as

$$|0^+\rangle = \frac{1}{\sqrt{2}}\left[\psi_{+\frac{1}{2}}(A)\psi_{-\frac{1}{2}}(B) - \psi_{-\frac{1}{2}}(A)\psi_{+\frac{1}{2}}(B)\right].$$

$\langle 0^+|$ in the above equations is the Hermitian conjugate of $|0^+\rangle$. Using the above expression and the rules of Pauli matrix, we can obtain

$$q(a,b) = 4\overline{A_a \cdot B_b} = -\hat{a} \cdot \hat{b}.$$

According to this equation, $q(a,b) = -1$ if $\hat{a} = \hat{b}$, which results in "negative" correlation for spin projections measured in the same direction.

On the other hand, if we start from Bell's localized hidden variable theory, we obtain the following Bell's inequality:

$$|q(\hat{a}, \hat{b}) - q(\hat{a}, \hat{c})| \leq 1 + q(\hat{a}, \hat{c}).$$

This involves measurements of the spin components in three directions, specified by unit vectors $\hat{a}$, $\hat{b}$, and $\hat{c}$, respectively, in contrast to the previous case which involves only two directions. If we let $\hat{a} = \hat{b} = \hat{c} = \hat{n}$, then Bell's inequality becomes $q(\hat{n} \cdot \hat{n}) \geq -1$, which is the same as that given by quantum mechanics. Different results can be expected if three directions are really involved in the measurements. For example, if the angles between $\hat{a}$ and $\hat{b}$ and between $\hat{b}$ and $\hat{c}$ are 60° and that between $\hat{a}$ and $\hat{c}$ is 120°, then we have $q(\hat{a}, \hat{b}) = q(\hat{b}, \hat{c}) = 1/2$, and $q(\hat{a}, \hat{c}) = -1/2$ according to quantum mechanics. Substituting these into the Bell's inequality, it is evident that

$$\left|\frac{1}{2} + \frac{1}{2}\right| \leq 1 - \frac{1}{2}$$

which results in $1 \leq 1/2$ that does not make any sense.

It is clearly seen that spin correlation described in linear quantum mechanics contradicts the Bell's inequality. That is to say that all statistical predictions of linear quantum mechanics cannot be obtained from the localized hidden variable theory. In some special cases, if statistical predictions based on linear quantum mechanics are correct, then the localized hidden variable theory does not hold, and vice versa. However, whether the Bell's inequality is correct remained a question.

Since then many physicists, for example Wigner in 1970, had also derived the Bell's inequality using analytical methods which were quite different from Bell's approach. Unfortunately, only single state of particles with zero spin was discussed in an ideal experiment setting. This is equivalent to assume that two particles of spin 1/2 always reach the instrument and therefore the instrument always measures a definite spin along a given axis. Such a measurement is very hard to realize in actual experiments.

This prompted Clayser *et al.* to generalize Bell's inequality by removing the restrictions of single state and spin 1/2, in 1969. The Clayser's generalized inequality

$$|q(\hat{a}, \hat{b}) - q(\hat{a}, \hat{b'})| \leq 2 \pm [q(\hat{a'}, \hat{b'}) + q(\hat{a'}, \hat{b})]$$

is based on some more common and realistic experimental conditions. If $q(\hat{a'}, \hat{b}) = -1$, the Clayser's inequality reduces to the Bell's inequality. Bell himself also obtained the same result in 1971. Since 1972, many experiments, as shown in Table 1.1, have been carried out and results have been reported to verify which theory, the Bell's inequality of localized hidden variable or the linear quantum mechanics, correctly describes the motion of the microscopic particle.

Among the nine experiments listed in Table 1.1, seven of them gave supports to linear quantum mechanics and only two experimental findings are in agreement with the Bell's inequality. It seems that the experimental results are in favor of the linear quantum mechanics than Bell's localized hidden variable theory. This shows that linear quantum mechanics does not satisfy the requirement of localization. The results, however, cannot exclusively confirm its validity either.

1.3 Dispute between Bohr and Einstein

While the view on linear quantum mechanics and its interpretation by Bohr and others in the Copenhagen school dominated the debate, many prominent physicists respected Einstein as the authority who had doubted and continuously criticized

Table 1.1 List of experiments to verify Bell's inequality.

No.	Author(s)	Date	Experiment	Results
1	S. T. Freedman J. F. Clauser	1972	Low-energy photon radiation in transitional process of a calcium atom	Supports linear quantum mechanics
2	R. A. Holt F. M. Pipkin	1973	Low-energy photon radiation in transitional process of mercury-198 atoms	Supports Bell's inequality
3	J. F. Clauser	1976	Low-energy photon radiation in transitional process of mercury-202 atoms	Supports linear quantum mechanics
4	E. S. Firg R. C. Thomson	1976	Low-energy photon radiation in transitional process of mercury-202 atom	Supports linear quantum mechanics
5	G. Fioraci S. Gutkowski S. Natarrigo R. Pennisi	1975	High-energy photon annihilation of electron – positron pair (γ ray)	Supports Bell's inequality
6	J. Kasday J. Ulman Wu Jianxiong	1975	High-energy photon annihilation of electron – positron pair (γ ray)	Supports linear quantum mechanics Supports linear quantum mechanics
7	M. Lamchi-Rachti W. Mitting	1976	Atomic pair in single state	Supports linear quantum mechanics
8	Aspect P. Grangier G. Roger	1981	Cascade photon radiation in transitional process of atoms	Supports linear quantum mechanics
9	P. Grangier P. Grangier G. Roger	1982	Cascade photon radiation in transitional process of ^{46}Ca	Supports linear quantum mechanics

Bohr's interpretation. This resulted in a life-long dispute between Bohr and Einstein, which was unprecedented and went through three stages.

The first stage was during the period from 1924 to 1927 when the theory of quantum mechanics had just been established. Einstein proceeded from his own philosophical belief and his scientific goal for an exact description of causality in the physical world, and expressed his extreme unhappiness with the probability interpretation of linear quantum mechanics. In a letter to Born on December 4, 1926, Einstein said that "Quantum mechanics is certainly imposing. But an inner voice tells me that it is not the real thing (*der Wahre Jakob*). The theory says a lot, but it does not bring us any closer to the secret of the "Old One." I, at any rate, am convinced that *He* is not playing at dice."

The second stage was from 1927 to 1930. After Bohr had put forward his complementary principle and had established his interpretation as the main stream interpretation, Einstein was extremely unhappy. His main criticism was directed at the uncertainty relation on which Bohr's complementary principle was based. At the 5th (1927) and the 6th (1930) International Meetings of Physics at Solway, Einstein proposed two ideal experiments (double slit diffraction and photon box) to prove that the uncertainty relation and formalism of the quantum mechanics contradict each other, and thus to disprove Bohr's complementary principle. But Einstein's idea was demolished each time by Bohr through resourceful analysis. Since then, Einstein had to accept the logical consistency of quantum mechanics and turned his criticism to the completeness of the linear quantum mechanics theory.

The third stage was from 1930 until the death of Einstein. The dispute during this period is reflected in the debate between Einstein and Bohr over the EPR paradox proposed by Einstein together with Podolsky and Rosen. This paradox concerned the fundamental problem of the linear quantum mechanics, *i.e.*, whether it satisfied the deterministic localized theory and the microscopic causality. Since some of the subsequent experiments seem to support the linear quantum mechanics, instead of the Bell inequality, it is necessary to understand the nature of the EPR paradox and results it brought about.

The EPR paradox will be briefly introduced below.

Consider a system consisting of two particles which move in opposite directions. For simplicity but without losing its generality, we assume that the initial relativistic momentum of the pair of particles is $p = 0$. Then there must be $p_1 = -p_2$ after the two particles interact and depart. However, the magnitude and direction of the momentum of each particle are not known. Assume that the momentum of particle 1 is measured, by a detector, and the value $p_1 = +a$ is obtained, then the momentum of the particle 2 is determined and it can only be $p_2 = -a$ according to conservation of momentum in the linear quantum mechanics. However, in the light of the hypothesis of contraction of wave packet in the measuring process, the plane wave with momentum $p_1 = a_1$ is "selected" out by the detector from the wave packet $\psi_1(X_1)$ describing particle 1. In accordance with the traditional linear quantum mechanics, this process of "spectrum resolution" is due to some kind of "uncontrollable interaction" between the instrument and the wave packet. Under the influence of such an "uncontrollable interaction", the momentum of particle 1 could be $p_1 = a$, or $p_1 = b$, $\cdots$. However, what is surprising is that there is always $p_2 = -a$ as long as $p_1 = a$ is measured by the detector. This means that this value should be obtained regardless of the measurement on the wave packet $\psi_2(X_2)$ is made or not. In other words, when the wave packet $\psi_1(X_1)$ is measured and contracted, the wave packet $\psi_2(X_2)$ for particle 2 will also be automatically contracted. A series of questions then arise. For example, what mechanism makes this possible? Does this occur instantaneously, or is it propagating at speed of light according to the special theory of relativity? How can the wave packet contraction caused by measurement automatically guarantee the conservation of momentum? It is very difficult to answer these questions. Only after careful studies by Einstein and others, the following conclusions were obtained: either the description of the linear quantum mechanics was incomplete, or the linear quantum mechanics didn't satisfy the criterion of "localization". Einstein tended to believe that physical phenomena must satisfy the criterion of "localization", *i.e.* physical quantities cannot propagate with speed greater than the speed of light. Thus, he thought that the linear quantum mechanics is an incomplete theory. Due to this remarkable analysis by Einstein, many physicists began to explore the theory of "hidden parameters" of the linear quantum mechanics.

The "queries" to the linear quantum mechanics by Einstein and others had indeed created quite a stir. Bohr had to respond in his own capacity to these queries. In 1935, Bohr published a short essay in Physical Review in which he argued that if a system consists of two local particles 1 and 2, then this system should be described by a wave function $\psi(1,2)$. In such a case, the local particles 1 and 2 are no longer mutually independent entities. Even though they are spatially separated at the instant the system is probed, they cannot be considered as independent entities. Thus, there is no basis for statements such as measurement of subsystem 1 could not influence subsystem 2 within the framework of the linear quantum mechanics, and the idea of Einstein *et al.* cannot be accepted. Essentially, Bohr was not really against the "paradox" proposed by Einstein and others, but only confirmed that linear quantum mechanics might not satisfy the principle of localization. Bohr further commented that in the final decisive steps of measurement in Einstein's ideal experiment, even though there was no mechanical interference to the system being probed, influence on experimental conditions did exist. Thus, Einstein's arguments could not verify their conclusion that the description of quantum mechanics is incomplete.

Many scientists who followed closely the thought of localization and incompleteness of the linear quantum mechanics by Einstein and others believed that there could exist a hidden variables theory behind linear quantum mechanics which might be able to interpret the probability behavior of microscopic particle. The concept of "hidden variables" was proposed soon after linear quantum mechanics was born. However, it was disapproved by Von Neumann in 1932. For a long time since then, no one had mentioned this problem. After the second World War, Einstein repeatedly criticized the linear quantum mechanics and suggested that any actual state should be completely described.

Motivated by this thought, Bohm put forward the first systematic "hidden variable theory" in 1952. He believed that the statistical characteristics of linear quantum mechanics is due to some "background" fluctuations hidden behind the quantum theory. If we can find the hidden function for a microscopic particle, then a deterministic description could be made for a single particle. But how can the existence of such hidden variables be proved? Bohm proposed two experiments, to measure the spin correlation of a single proton and the polarization correlation in annihilating radiation of photons, respectively. It was realized later that in Bohm's theory the single state ψ is essentially a slowly varying state which describes states of a fluid with random fluctuations. Since the wave function itself cannot have such random fluctuation, a hidden variable could not be introduced. Bohm's theory mentioned above was referred to as a random hidden-variables theory.

However, if the motion of particles can also be considered as a stable Markov process. A steady state solution of the Schrödinger equation can then be given from a steady distribution of the Markov chain, and if the Fock-Planck equation was taken as the dynamic equation of microscopic particle, a new "hidden variables theories" of linear quantum mechanics can be set up. After Bell established his inequality on the basis of Bohm's deterministic "localized variables theory" in 1966, various attempts were made to experimentally verify which theory is the right theory and to settle the dispute once and for all. As mentioned earlier, majority of the experiments supported the linear quantum mechanics at that time, and it was clear that not all the predictions by the linear quantum mechanics can be obtained from the localized hidden variables theory. Thus the "hidden variable theory" was abandoned.

To summarize, the long dispute between Bohr and Einstein was focused on three issues. (1) Einstein upheld to the belief that the microscopic world is no different from the macroscopic world, particles in the microscopic world are matters and they exist regardless of the methods of measurements, any theoretical description to it should in principle be deterministic. (2) Einstein always considered that the theory of the linear quantum mechanics was not an ultimate and complete theory. He believed that quantum mechanics is similar to classical optics. Both of them are correct theories based on statistical laws, *i.e.*, when the probability $|\psi(\vec{r},t)|^2$ of a particle at a moment t and location r is known, the average value of an observable quantity can be obtained using statistical method and then compared with experimental results. However, the understanding to processes involving single particle was not satisfactory. Hence, $\psi(\vec{r},t)$ cannot give everything about a microscopic particle system, and the statistical interpretation cannot be ultimate and complete. (3) The third issue concerns the physical interpretation of the linear quantum mechanics. Einstein was not impressed with the attempt to completely describing some single processes using linear quantum mechanics, which he made very clear in a speech at the fifth Selway International Meeting of physics. In an article, "Physics and Reality", published in 1936 in the Journal of the Franklin Institute, Einstein again mentioned that what the wave function ψ describes can only be a many-particle system, or an assemble in terms of statistical mechanics, and under no circumstances, the wave function can describe the state of a single particle. Einstein also believed that the uncertainty relation was a result of incompleteness of the description of a particle by $\psi(\vec{r},t)$, because a complete theory should give precise values for all observable quantities. Einstein also did not accept the statistical interpretation, because he did not believe that an electron possess free will. Thus, Einstein's criticism against the linear quantum mechanics was not directed towards the mathematical formalism of the linear quantum mechanics, but to its fundamental hypotheses and its physical interpretation. He considered that this is due to the incomplete understanding of the microscopic objects. Moreover, the contradiction between the theory of relativity and the fundamental of the linear quantum mechanics was also a central point of dispute. Einstein made effort to unite the theory of relativity and linear quantum mechanics, and attempted to interpret the atomic structure using field theory. The disagreements on several fundamental issues of the linear quantum mechanics by Einstein and Bohr and their followers were deep rooted and worth further study. This brief review on the disputes between the two great physicists given above should be useful to our understanding on the nature and problems of the linear quantum mechanics. It should set the stage for the introduction of nonlinear quantum mechanics.

1.4 Analysis on the Roots of Problems of linear quantum mechanics and Review on Recent Developments

The discussion in the previous section shows that the disputes and disagreement on several fundamental issues of the linear quantum mechanics are deep rooted. Almost all prominent physicists were involved to a certain degree in this dispute which lasted half of a century, which is extraordinary in the history of science. What is even more surprising is that after such a long dispute, there have been no conclusions on these important issues till now. Besides what have been mentioned above, there was another fact which also puzzled physicists. As it is know, the concept of "orbit" has no meaning in quantum mechanics. The state of a particle is described by the wave function ψ which spreads out over a large region in space. Even though this suggests that a particle does not have a precise location, in physical experiments, however, particles are always captured by a detector placed at an exact position. Furthermore, it is always one whole particle, rather than a fraction of it, being detected. How can this be interpreted by the linear quantum mechanics? Given this situation, can we consider that the linear quantum mechanics is complete? Even though the linear quantum mechanics is correct, then it can only be considered as a set of rules describing some experimental results, rather than an ultimate complete theory. In the meantime, the indeterministic nature of the linear quantum mechanics seems against intuition. All these show that it is necessary to improve and further develop the linear quantum mechanics. Attempt to solve these problems within the framework of the linear quantum mechanics seem impossible. Therefore, alternatives that go beyond the linear quantum mechanics must be considered to further develop the quantum mechanics. To do this, one must thoroughly understand the fundamentals and nature of the linear quantum mechanics and seriously consider de Broglie's idea of a nonlinear wave theory.

Looking back to the development and applications of the linear quantum mechanics for almost a century, we notice that the splendidness of the quantum mechanics is the introduction of a wave function to describe the state of particles and the expression of physical quantities by linear Hermitian operators. Such an approach is drastically different from the traditional methods of classical physics and took the development of physics to a completely new stage. This new approach has been successfully applied to some simple atoms and molecules, such as hydrogen atom, helium atom and hydrogen molecule, and the results obtained are in agreement with experimental data. Correctness of this theory is thus established. However, besides being correct, a good theory should also be complete. Successful applications to a subset of problems does not mean perfection of the theory and applicability to any physics system. Physical systems in the world are manifold and every theory has its own applicable scope or domain. No theory is universal.

From the above discussion, we see that the most fundamental features of the linear quantum mechanics are its linearity and the independence of the Hamiltonian of a system on its wave function. These ensure the linearity of the fundamental dynamic equations, *i.e.*, the Schrödinger equation is a linear equation of the wave function, all operators in the linear quantum mechanics are linear Hermitian operators, and the solutions of the dynamic equation satisfy the linear superposition principle. The linearity results in the following limitations of the linear quantum mechanics.

(1) The linear quantum mechanics is a wave theory and it depicts only the wave feature, not their corpuscle feature, of microscopic particles. As a matter of fact, the Schrödinger equation (1.7) is a wave equation and its solution represents a probability wave. To see this clearly, we consider the wave function $\psi = f \cdot \exp(-iEt/\hbar)$ and substitute it into (1.7). If we let $n^2 = (E-U)/(E-C) = k^2/k_0^2$, where C is a constant, and $k_0^2 = 2m(E-C)/\hbar^2$, then (1.7) becomes

$$\frac{\partial^2 f}{\partial t^2} + k^2 n^2 f = 0.$$

This equation is nothing but that of a light wave propagating in a homogeneous medium. Thus, the linear Schrödinger equation (1.7) is only able to describe the wave feature of the microscopic particle. In other words, when a particle moves continuously in the space-time, it follows the law of linear variation and disperses over the space-time in the form of a wave. This wave feature of a microscopic particle is mainly determined by the kinetic energy operator, $\hat{T} = -(\hbar^2/2m)\nabla^2$, in the dynamic equation (1.7). The applied potential field, $V(x,t)$ is imposed on the system by external environment and it can only change the wave form and amplitude, but not the nature of the wave. Such a dispersion feature of the microscopic particle ensures that the microscopic particle can only appear with a definite probability at a given point in the space-time. Therefore, the momentum and coordinate of the microscopic particle cannot be accurately measured simultaneously, which lead to the uncertainty relation in the linear quantum mechanics. Therefore, the uncertainty relation occurs in linear quantum mechanics is an inevitable outcome of the linear quantum mechanics.

(2) Due to this linearity and dispersivity, it is impossible to describe the corpuscle feature of microscopic particles by means of this theory. In other words, the wave-corpuscle duality of microscopic particle cannot be completely described by the dynamic equation in the linear quantum mechanics, because external applied potential fields cannot make a dispersive particle an undispersive, localized particle, and there is no other interaction that can suppress the dispersion effect of the

kinetic energy in the equation. Thus a microscopic particle always exhibits features of a dispersed wave and its corpuscle property can only be described by means of Born's statistical interpretation of the wave function. This not only exposes the incompleteness of the hypotheses of linear quantum mechanics, but also brings out an unsolvable difficulty, namely, whether the linear quantum mechanics describes the state of a single particle or that of an assemble of many particles.

As it is known, in linear quantum mechanics, the corpuscle behavior of a particle is often represented by a wave packet which can be a superposition of plane waves. However, the wave packet always disperses and attenuates with time during the course of propagation. For example, a Gaussian wave packet given by

$$\psi(x,t=0) = e^{-\alpha_0^2 x^2/2}, \quad |\psi|^2 = e^{-\alpha_0^2 x^2} \tag{1.11}$$

at $t = 0$ becomes

$$\begin{aligned}\psi(x,t) &= \frac{1}{\sqrt{2\pi}}\int_{-\infty}^{\infty}\phi(k)e^{i(kx-\hbar k^2 t/2m)}dk \\ &= \frac{1}{\alpha_0\sqrt{1/\alpha_0^2+i\hbar t/m}}e^{-(x^2/2)(1/\alpha_0^2+i\hbar t/m)}, \\ |\psi|^2 &= \frac{1}{\sqrt{1+\left(\alpha_0^2\hbar t/m\right)^2}}\,e^{-x^2/\alpha_t^2}\end{aligned} \tag{1.12}$$

after propagating through a time t, where

$$\phi(k) = \frac{1}{\sqrt{2\pi}}\int_{-\infty}^{\infty}\psi(x,t=0)e^{ikx}dk, \quad \alpha_t = \frac{1}{\alpha_0}\sqrt{1+i\frac{\alpha_0^2\hbar}{m}t}.$$

This indicates clearly that the wave packet is dispersed as time goes by. The uncertainty in its position also increases with time. The corresponding uncertainty relation is

$$\Delta x\Delta p = \frac{\hbar}{2}\sqrt{1+\frac{\alpha_0^4\hbar^2 t^2}{m^2}},$$

where

$$\Delta p = \frac{\hbar\alpha_0}{\sqrt{2}}, \quad \Delta x = \frac{1}{\sqrt{2}\alpha_0}\sqrt{1+\frac{\alpha_0^4\hbar^2 t^2}{m^2}}.$$

Hence, the wave packet cannot be used to describe the corpuscle property of a microscopic particle. How to describe the corpuscle property of microscopic particles has been an unsolved problem in the linear quantum mechanics. This is just an example of intrinsic difficulties of the linear quantum mechanics.

(3) Because of the linearity, the linear quantum mechanics can only be used in the case of linear field and medium. This means that the linear quantum mechanics is suitable for few-body systems, such as the hydrogen atom and the helium atom, *etc.* For many-body systems and condensed matter, it is impossible to solve the

wave equation exactly and only approximate solutions can be obtained in the linear quantum mechanics. However, doing so loses the nonlinear effects due to intrinsic and self-interactions among the particles in these matters. Therefore, the scope of application of the linear quantum mechanics is limited. Moreover, when this theory is applied to deal with features of elementary particles in quantum field theory, the difficulty of infinity cannot always be avoided and this shows another limitation of this theory. Therefore, it is necessary to develop a new quantum theory that can deal with these complex systems.

From the discussion above, we learned that linearity on which linear quantum mechanics is based is the root of all the problems encountered by the linear quantum mechanics. The linearity is closely related to the assumption that the Hamiltonian operator of a system is independent of its wave function, which is true only in simple and uniform physical systems. Thus the linearity greatly limited the applicable scope and domain of the linear quantum mechanics. It cannot be used to study the properties of many-body, many-particle, nonlinear and complex systems in which there exist complicated interaction, the self-interaction, and nonlinear interactions among the particles and between the particles and the environment.

Since the wave feature of microscopic particle can be well described by the wave function, one important issue to be looked into in further development of quantum mechanics is the description of corpuscle feature of microscopic particles, so that the new quantum theory should completely describe the wave-corpuscle duality of microscopic particles. However, this is easily said than done. To this respect, it is useful to review what has already been done by the pioneers in this field, as we can learn from them and get some inspiration from their work.

One can learn from the history of development of the theory of superconductivity. It is known that the mechanism of superconductivity based on electron-phonon interaction was proposed by Fröhlich as early as in 1951. But Fröhlich failed to establish a complete theory of superconductivity because he confined his work to the perturbation theory in the linear quantum mechanics, and superconductivity is a nonlinear phenomenon which cannot be described by the linear quantum mechanics. Of course, this problem was finally solved and the nonlinear BCS theory was established in 1975. This again clearly demonstrated the limitation of the linear quantum mechanics. This problem will be discussed in the next chapter in more details.

In view of this, in order to overcome the difficulties of the linear quantum mechanics and further develop the theory of quantum mechanics, two of the hypotheses of the linear quantum mechanics, *i.e.* linearity of the theory and independence of the Hamiltonian of a system on its wave function must be reconsidered. Further development must be directed toward a nonlinear quantum theory. In other words, nonlinear interaction should be included into the theory and the Hamiltonian of a system should be related to the wave function of the system.

The first attempt of establishing a nonlinear quantum theory was made by de

Broglie, which was described in his book: "the nonlinear wave theory". Through a long period of research, de Broglie concluded that the theory of wave motion cannot interpret the relation between particle and wave because the theory was limited to a linear framework from the start. In 1926, he further emphasized that if $\psi(\vec{r},t)$ is a real field in the physical space, then the particle should always have a definite momentum and position. de Broglie assumed that $\psi(\vec{r},t)$ describes an essential coupling between the particle and the field, and used this concept to explain the phenomena of interference and diffraction.

In 1927, de Broglie put forward a "dual solution theory" in a paper published in J. de Physique. de Broglie proposed that two types of solutions are permitted in the dynamic equation in the linear quantum mechanics. One is a continuous solution, $\psi = Re^{i\theta}$, with only statistical meaning, and this is the Schrödinger wave. This wave can only have statistical interpretation and can be normalized. It does not represent any physical wave. The other type, referred as a u wave, has singularities and is associated with spatial localization of the particle. The corpuscle feature of a microscopic particle is described by the u wave and the position of a particle is determined by a singularity of the u wave. de Broglie generalized the formula of the monochromatic plane wave and stipulated a rule of associating the particle with the propagation of the wave. The particle would move inside its wave according to de Broglie's dual solution theory. This suggests that the motion of the particle inside its wave is influenced by a force which can be derived from a "quantum potential". This quantum potential is proportional to the square of the Planck constant and is dependent on the second derivative of the amplitude of the wave. It can also be given in terms of the change in the rest mass of the particle. In the case of a monochromatic plane wave, the quantum potential is zero. In 1950s, de Broglie further improved his "dual solution theory". He proposed that the u wave satisfies an undetermined nonlinear equation, and this led to his own "nonlinear wave theory". However, de Broglie did not give the exact nonlinear equation that the u wave should satisfy. This theory has serious difficulties in describing multiparticle systems and the s state of a single-particle. The theory also lacked experimental verification. Thus, even though it was supported by Einstein, the theory was not taken seriously by the majority and was gradually forgotten.

Although de Broglie's nonlinear wave theory was incorrect, some of his ideas, such as the quantum potential, the u wave of nonlinear equation which is capable of describing a physical particle, provided inspiration for further development of quantum mechanics.

As mentioned above, de Broglie stated that the quantum potential is related to the second derivative of the amplitude of the wave function $\psi = Re^{i\theta}$. Bohm, who proposed the theory of localized hidden variables in 1952, derived this quantum potential. It is independent of the phase, θ, of the wave function, and is represented in the form of $V = h^2\nabla^2 R/2mR$, where R is the amplitude of the wave, m is the mass of the particle, and h the Planck constant. With such a quantum potential,

Bohm believed that the motions of microscopic particles should follow the Newton's equation, and it is because of the "instantaneous" action of this quantum potential, a measurement process is always disturbed. The latter, however, was less convincing.

Quantum potential and nonlinear equations were again introduced in the Bohm-Bohr theory proposed in 1966. They assumed that, in the dual Hilbert space, there exists a dual vector, $|\psi_1\rangle$ and $|\sigma\rangle$, which satisfy

$$|\psi_1\rangle = \sum_n a_n |A_n\rangle, \qquad |\sigma\rangle = \sum_k \sigma_k |A_k\rangle,$$

where σ is a hidden variable and satisfies the Gaussian distribution in an equilibrium state. They introduced a nonlinear term in the Schrödinger equation, to represent the effect arising from the quantum measurement, and determined the equation containing the nonlinear term based on the relation among the particles, the environment and the hidden variables. Attempts were made to solve the problem concerning the influence of the measuring instruments on the properties of particles being probed. de Broglie pointed out that the quantum potential can be expressed in terms of the change in the rest mass of the particle and tried to interpret Bohm's quantum potential based on the counteraction of the u wave and domain of singularity. Thus, the quantum potential arises from the interaction between particles. It is associated with nonlinear interaction and is able to change the properties of particles. These were encouraging. It seemed promising to make microscopic particles measurable and deterministic by adding a quantum potential with nonlinear effect to the Schrödinger equation, and eventually to have deterministic quantum theory. A delighted Dirac commented that the results will ultimately prove that Einstein's deterministic or physical view is correct.

In summary, we started the chapter with a review on the hypotheses on which the linear quantum mechanics was built, and the successes and problems of theory. We have seen that the linear quantum mechanics is successful and correct, but on the other hand, it is incomplete. Some of its hypotheses are vague and non-intuitive. Moreover, it is a wave theory and cannot completely describe the wave-corpuscle duality of microscopic particles. Therefore, improvement and further development on the linear quantum mechanics are required. The dispute between Einstein and Bohr, the recent work done by de Broglie, Bohm, and Bohr provided positive inspiration for further development of the quantum theory. From the above discussion, the direction for a complete theory seems clear: it should be a nonlinear theory. Two of its fundamental hypotheses of linear quantum mechanics, linearity and independence of the Hamiltonian of a systems on its wave function, must be reconsidered.

However, at what level of theory will the problems of the linear quantum mechanics be solved? What would be a good physical system to start with? What would be the foundation of a new theory? These and many other important questions can only be answered through further research. It is clear that the new theory should not be confined to the scope and framework of the linear quantum mechanics. The work of de Broglie and Bohm gave us some good motivations, but their ideas cannot be indiscriminately borrowed. One must go beyond the framework of the linear quantum mechanics and look into the nonlinear scheme. To establish a new and correct theory, one must start from the phenomena and experiments which the linear quantum mechanics failed, or had difficulty, to explain, and uses completely new concepts and new approaches to study these unique quantum systems. This is the only way to clearly understand the problems in the linear quantum mechanics. For this purpose, we will review some macroscopic quantum effects in the following chapter because these experiments form the foundation of the nonlinear quantum mechanics.

Bibliography

Akert, A. K. (1991). Phys. Rev. Lett. **67** (661) .

Aspect, A., Grangier, P. and Roger, G. (1981). Phys. Rev. Lett. **47** (460) .

Bell, J. S. (1965). Physics **1** (195) .

Bell, J. S. (1987). Speakable and unspeakable in quantum mechanics, Cambridge Univ. Press, Cambridge.

Bell, J. S. (1996). Rev. Mod. Phys. **38** (447) .

Bennett, C. H., Brassard, G., Crépeau, C., Jozsa, R., Peres, R. and Wootters, W. K. (1993). Phys. Rev. Lett. **70** (1895) .

Black, T. D., Nieto, M. M., Pilloff, H. S., Scully, M. O. and Sinclair, R. M. (1992). Foundations of quantum mechanics, World Scientific, Singapore.

Bohm, D. (1951). Quantum theory, Prentice-Hall, Englewood Cliffs, New Jersey.

Bohm, D. (1952). Phys. Rev. **85** (169) .

Bohm, D. and Bub, J. (1966). Rev. Mod. Phys. **38** (453) .

Bohr, N. (1935). Phys. Rev. **48** (696) .

Born, M. and Infeld, L. (1934). Proc. Roy. Soc. **A144** (425) .

Born, M., Heisenberg, W. and Jorden, P. (1926). Z. Phys. **35** (146) .

Czachor, M. and Doebner, H. D. (2002). Phys. Lett. A **301** (139) .

de Broglie, L. (1960). Nonlinear wave mechanics, a causal interpretation, Elsevier, Amsterdam.

Diner, S., Fargue, D., Lochak, G. and Selleri, F. (1984). The wave-particle dualism, Riedel, Dordrecht.

Dirac, P. A. (1927). Proc. Roy. Soc. London **A114** (243) .

Dirac, P. A. (1967). The principles of quantum mechanics, Clarendon Press, Oxford.

Einstein, A., Podolsky, B. and Rosen, N. (1935). Phys. Rev. **47** (777) .

Espaguat, B. D. (1979). Sci. Am. **241** (158) .

Fernandez, F. H. (1992). J. Phys. A **25** (251) .

Ferrero, M. and Van der Merwe, A. (1995). Fundamental problems in quantum physics, Kluwer, Dordrecht, 1995.

Ferrero, M. and Van der Merwe, A. (1997). New developments on fundamental problems in quantum physics, Kluwer, Dordrecht.

Feynman, R. P. (1998). The character of physical law, MIT Press, Cambridge, Mass, p. 129.
French, A. P. (1979). Einstein, A Centenary Volume, Harvard University Press, Cambridge, Mass.
Garola C. and Ross, A. (1995). The foundations of quantum mechanics-historical analysis and open questions, Kluwer, Dordrecht.
Gleason, A. M. (1957). J. Math. Mech. **6** (885) .
Green, H. S. (1962). Nucl. Phys. **33** (297) .
Harut, A. D. and Rusu, P. (1989). Can. J. Phys. **67** (100) .
Heisenberg, W. (1925). Z. Phys. **33** (879) .
Heisenberg, W. and Euler, H. (1936). Z. Phys. **98** (714) .
Hodonov, V. V. and Mizrachi, S. S. (1993). Phys. Lett. A **181** (129) .
Hugajski, S. (1991). Inter. J. Theor. Phys. **30** (961) .
Iskenderov, A. D. and Yaguhov, G. Ya. (1989). Automation and remote control **50** (1631) .
Jammer, M. (1989). The conceptual development of quantum mechanics, Tomash, Los Angeles.
Jordan, T. K. (1990). Phys. Lett. A **15** (215) .
Kobayashi, S., Zawa, H. E., Muvayama, Y. and Nomura, S. (1990). Foundations of quantum mechanics in the light of new technology, Komiyama, Tokyo.
Laudau, L. (1927). Z. Phys. **45** (430) .
Muttuck, M. D. (1981). Phys. Lett. A **81** (331) .
Pang Xiao-feng, (1994). Theory of nonlinear quantum mechanics, Chongqing Press, Chongqing, pp. 1-34.
Polchinki, J. (1991). Phys. Rev. Lett. **66** (397) .
Popova, A. D. (1989). Inter. J. Mod. Phys. **A4** (3228) .
Popova, A. D. and Petrov, A. N. (1993). Inter. J. Mod. Phys. **8** (2683) and 2709.
Reinisch, G. (1994). Physica A **266** (229) .
Roberte, C. D. (1990). Mal. Phys. Lett. A **5** (91) .
Roth, L. R. and Inomata, A. (1986). Fundamental questions in quantum mechanics, Gordon and Breach, New York.

Schrödinger, E. (1928). Collected paper on wave mechanics, Blackie and Son, London.
Schrödinger, E. (1935). Proc. Cambridge Phil. Soc. **31** (555) .
Selleri, R. and Van der Merwe, A. (1990). Quantum paradoxes and physical reality, Kluwer, Dordrecht.
Slater, P. B. (1992). J. Phys. A **25** (L935) and 1359.
Valentini, A. (1990). Phys. Rev. A **42** (639) .
Vitall, D., Allegrini, P. and Grigmlini, P. (1994). Chem. Phys. **180** (297) .
Von Neumann, J. (1955). Mathematical foundation of quantum mechanics, Princeton Univ. Press, Princeton, N.J, Chap. IV, VI.
Walsworth, R. L. and Silvera, I. F. (1990). Phys. Rev. A **42** (63) .
Winger, E. P. (1970). Am. J. Phys. **38** (1005) .
Yourgrau, W. and Van der Merwe, A. (1979). Perspective in quantum theory, Dover, New York.

Chapter 2

Macroscopic Quantum Effects and Motions of Quasi-Pariticles

In this chapter, we review some macroscopic quantum effects and discuss motions of quasi-particles in these macroscopic quantum systems. The macroscopic quantum effects are different from microscopic quantum phenomena. The motions of quasi-particles satisfy nonlinear dynamical equations and exhibit soliton features. In particular, we will review some experiments and theories, such as superconductivity and superfluidity, that played important roles in the establishment of nonlinear quantum theory. The soliton solutions of these equations will be given based on modern soliton and nonlinear theories. They are used to interpret macrosopic quantum effects in superconductors and superfluids.

2.1 Macroscopic Quantum Effects

Macroscopic quantum effects refer to quantum phenomena that occur on the macroscopic scale. Such effects are obviously different from the microscopic quantum effects at the microscopic scale which we are familiar with. It has been experimentally demonstrated that such phenomena can occur in many physical systems. Therefore it is very necessary to understand these systems and the quantum phenomena. In the following, we introduce some of the systems and the related macroscopic quantum effects.

2.1.1 *Macroscopic quantum effect in superconductors*

Superconductivity is a phenomenon in which the resistance of a material suddenly vanishes when its temperature is lower than a certain value, T_c, which is referred to as the critical temperature of the superconducting materials. Superconductors can be pure elements, compounds or alloys. To date, more than 30 single elements and up to a hundred of alloys and compounds have been found to possess such characteristics. When $T \leq T_c$, any electric current in a superconductor will flow forever, without being damped. Such a phenomenon is referred to as perfect conductivity. Moreover, it was observed through experiments that when a material is in the superconducting state, any magnetic flux in the material would be completely repelled, resulting in zero magnetic field inside the superconducting material, and similarly magnetic flux produced by an applied magnetic field also cannot penetrate into supercoducting materials. Such a phenomenon is called the perfect anti-magnetism or Maissner effect. There are also other features associated with superconductivity.

How can these phenomena be explained? After more than 40 years' effort, Bardeen, Cooper and Schreiffier established the microscopic theory of superconductivity, the BCS theory, in 1957. According to this theory, electrons with opposite momenta and antiparallel spins form pairs when their attraction due to the electron and phonon interaction in these materials overcomes the Coulomb repulsion between them. The so-called Cooper pairs are condensed to a minimum energy state, resulting in quantmn states which are highly ordered and coherent over a long range, in which there is essentially no energy exchange between the electron-pairs and lattice. Thus, the electron pairs are no longer scattered by the lattice and flow freely, resulting in superconductivity.

The electron-pair in a superconductive state is somewhat similar to a diatomic molecule, but it is not as tightly bound as a molecule. The size of an electron pair, which gives the coherent length, is approximately 10^{-4} cm. A simple calculation will show that there can be up to 10^6 electron pairs in a sphere of 10^{-4} cm in diameter. There must be mutual overlap and correlation when so many electron pairs are brought together. Therefore, perturbation to any of the electron pairs would certainly affect all others. Thus, various macroscopic quantum effects can be expected in a material with such coherent and long range ordered states. Magnetic flux quantization, vortex structure in the type-II superconductors, and Josephson effect in superconductive junctions, are just some examples of such macroscopic quantum phenomena.

2.1.1.1 *Quantization of magnetic flux*

Consider a superconductive ring. Assume that a magnetic field is applied at $T > T_c$, then the magnetic flux lines produced by the external field pass through and penetrate into the body of the ring. We now lower the temperature to a value below T_c, and then remove the external magnetic field. The magnetic induction inside the body of circular ring equals to zero ($\vec{B} = 0$) because the ring is in the superconductive state and the magnetic field produced by the superconductive current cancels the magnetic field which was in the ring. However, part of the magnetic fluxes in the hole of the ring ramains because the induced current in the ring vanishes. This residual magnetic flux is referred to as frozen magnetic flux. It was observed experimentally that the frozen magnetic flux is discrete, or quantized. Using the macroscopic quantum wave function in the theory of superconductivity, it can be shown that the magnetic flux is given by $\Phi' = n\phi_0$ $(n = 1, 2, 3\cdots)$, where $\phi_0 = hc/2e = 2.07 \times 10^{-15}$ Wb is the flux quantum, representing the flux of one magnetic flux line. This means that the magnetic fluxes passing through the hole of the ring can only be multiple of ϕ_0. In other words, the magnetic field lines are discrete. What does this imply? If the applied magnetic field is exactly $n\phi_0$, then the magnetic flux through the hole is $n\phi_0$ which is not difficult to understand. What if the applied magnetic field is $(n + 1/4)\phi_0$? According to the above, the magnetic flux through the hole cannot be $(n + 1/4)\phi_0$. As a matter of fact, it should be $n\phi_0$. Similarly, if the applied magnetic field is $(n + 3/4)\phi_0$, the magnetic flux passing through the hole is not $(n + 3/4)\phi_0$, but rather $(n + 1)\phi_0$. The magnetic fluxes passing through the hole of the circular ring is always quantized.

An experiment conducted in 1961 surely proved so. It indicated that magnetic flux does exhibit discrete or quantized characteristics on the macroscopic scale. The above experiment was the first demonstration of macroscopic quantum effect. Based on quantization of magnetic flux, we can build a "quantum magnetometer" which can be used to measure weak magnetic field with a sensitivity of 3×10^{-7} Oersted. A slight modification of this device would allow us to measure electric current as low as 2.5×10^{-9} A.

2.1.1.2 *Structure of vortex lines in type-II superconductors*

The superconductors discussed above are referred to as type-I superconductors. This type of superconductors exhibit perfect Maissner effect when the external applied field is higher than a critical magnetic value $\tilde{H}_c$. There exists another type of materials such as the NbTi alloy and Nb_3Sn compound in which the magnetic field partially penetrates inside the material when the external field $\tilde{H}$ is greater than the lower critical magnetic field $\tilde{H}_{c1}$ but less than the upper critical field $\tilde{H}_{c2}$. This kind of supercondutors are classified as type-II superconductors and are characterized by a Ginzburg-Landau parameter of greater than $1/\sqrt{2}$.

Studies using the Bitter method showed that penetration of the magnetic field results in some small regions changing from superconductive to normal state. These small regions in normal state are of cylindrical shape and regularly arranged in the superconductor, as shown in Fig. 2.1. Each cylindrical region is called a vortex (or magnetic field line). The vortex lines are similar to the vortex structure formed in a turbulent flow of fluid.

It was shown through both theoretical analysis and experimental measurement that the magnetic flux associated with one vortex is exactly equal to one magnetic flux quantum, ϕ_0. When the applied field $\tilde{H} \geq \tilde{H}_{c_1}$, the magnetic field penetrates into the superconductor in the form of vortex lines, increased one by one. For an ideal type-II superconductor, stable vortices are distributed in triagonal pattern, and the superconducting current and magnetie field distributions are also shown in Fig. 2.1. For other, non-ideal type-II superconductors, the triagonal distribution can be observed in local regions, even though its overall distribution is disordered. It is evident that the vortex-line structure is quantized and this has been verified by many experiments. It can be considered a result of quantization of magnetic flux. Furthermore, it is possible to determine the energy of each vortex line and the

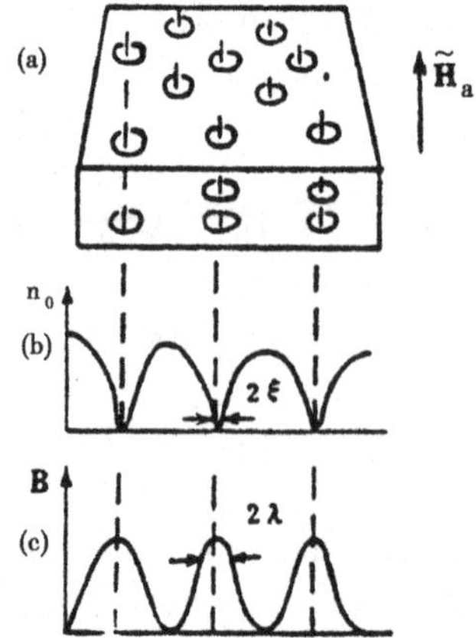

Fig. 2.1 Current and magnetic field distributions in a type-II superconductor.

interaction energy between the vortex lines. Parallel magnetic field lines are found to repel each other while anti-parallel magnetic lines attract each other.

2.1.1.3 *Josephson effect*

As known in the LQM, microscopic particles such as electrons have wave property and they can penetrate through potential barriers. For example, if two pieces of metals are separated by an insulator of width of tens or hundreds of angstroms, an electron can tunnel through the insulator and travel from one metal to the other. If a voltage is applied across the insulator, a tunnel current can be produced. This phenomenon is referred to as a tunnelling effect. If the two pieces of metals in the above experiment are replaced by two superconductors, tunneling current can also occur when the thickness of the dielectric is reduced to about 30 Å. However, this effect is fundamentally different from the tunnelling effect discussed above and is referred to as the Josephson effect.

Evidently, this is due to the long-range coherent effect of the superconductive electron pairs. Experimentally it was demonstrated that such an effect can be produced in many types of junctions involving a superconductor, for example, a superconductor-metal-superconductor junction, superconductor-insulator-superconductor junction and superconductor bridge. These junctions can be considered as superconductors with a weak link. On one hand, they have properties of bulk superconductors, for example, they are capable of carrying certain superconducting current. On the other hand, these junctions possess unique properties which a bulk superconductor does not. Some of these properties are summarized in the following.

(1) When a direct current (dc) passing through a superconductor is smaller than a critical value I_c, the voltage across the junction does not change with current. The critical current I_c can range from a few tens of μA to a few tens of mA. Figure 2.2

shows the characteristics of the dc Josephson current of a Sn-SnO-Sn junction.

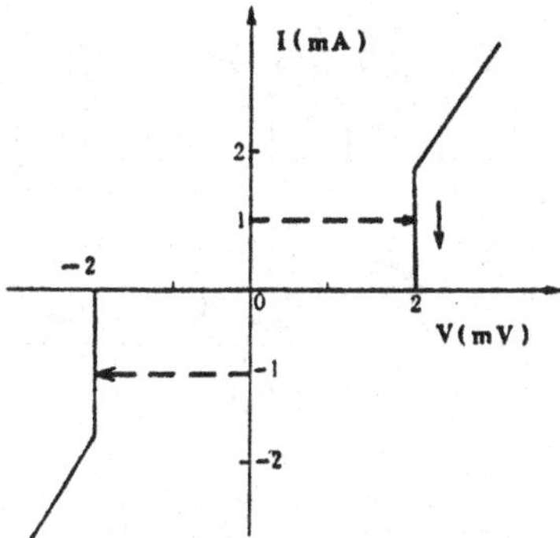

Fig. 2.2 dc Josephson current of a Sn-SnO-Sn junction.

(2) If a constant voltage is applied across the junction and the current passing through the junction is greater than I_c, a high frequency sinusoidal superconducting current occurs in the junction. The frequency is given by $\nu = 2eV/h$, in the microwave and far-infrared regions ($5 - 1000 \times 10^9$ Hz). The junction radiates coherent electromagnetic waves with the same frequency.

This phenomenon can be explained as follows. The constant voltage applied across the junction produces an alternating Josephson current which in turn generates electromagnetic waves of frequency ν. These waves propagate along the planes of the junction. When they reach the surface of the junction (interface between the junction and its surrounding), part of the electromagnetic wave is reflected from the interface and the rest is radiated, resulting in radiation of coherent electromagnetic waves. The power of radiation depends on the compatibility between the junction and its surrounding.

(3) When an external magnetic field is applied over the junction, the maximum dc current I_c is reduced due to the effect of the magnetic field. Furthermore, I_c changes periodically as the magnetic field increases. The $I_c - \tilde{H}$ curve resembles the distribution of light intensity in the Fraunhofer diffraction experiment, and the latter is shown in Fig. 2.3. This phenomena is called quantum diffraction of superconducting junction.

(4) When a junction is exposed to a microwave of frequency ν and if the voltage applied across the junction is varied, it was found that the dc current passing through the junction increases suddenly at certain discrete values of electric potential. Thus, a series of steps appear on the dc $I - V$ curve, and the voltage at a given step is related to the frequency of the microwave radiation by $n\nu = 2eV_n/h$ ($n = 1, 2, 3, \cdots$). More than 500 steps have been observed in experiments.

These phenomena were first derived theorctically by Josephson and each was experimentally verified subsequently. All these phenomena are called Josephson

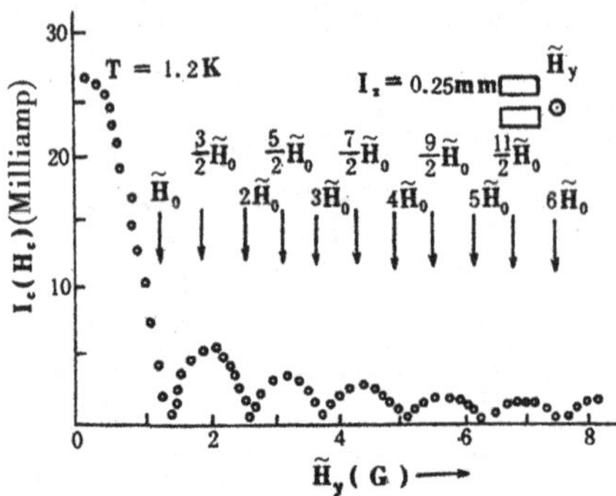

Fig. 2.3 Quantum diffraction effect in superconductor junction

effects. In particular, (1) and (3) are referred to as dc Josephson effects while (2) and (4) are referred to as ac Josephson effects.

Evidently the Josephson effects are macroscopic quantum effects which can be well explained by the macroscopic quantum wave function. If we consider a superconducting juntion as a weakly linked superconductor, then the wave functions of the superconducting electron pairs in the superconductors on both sides of the junction are correlated due to a definite difference in their phase angles. This results in a preferred direction for the drifting of the superconducting electron pairs, and a dc Josephson current is developed in this direction. If a magnetic field is applied in the plane of the junction, the magnetic field produces a gradient of phase difference which makes the maximum current oscillate along with the magnetic field and radiation of electromagnetic wave occurs. If a voltage is applied across the junction, the phase difference will vary with time and results in the Josephson effect. In view of this, the change in the phase difference of the wave functions of superconducting electrons plays an important role in the Josephson effect, which will be discussed in more details in the next section.

The discovery of the Josephson effect opened the door for a wide range of applications of superconductors. Properties of superconductors have been explored to produce superconducting quantum interferometer – magnetometer, sensitive ameter, voltmeter, electromagnetic wave generator, detector and frequency-mixer, and so on.

2.1.2 *Macroscopic quantum effect in liquid helium*

Helium is a common inert gas. It is also the most difficult gas to be liquidified. There are two isotopes of helium, ^{4}He and ^{3}He, with the former being the majority in a normal helium gas. The boiling temperatures of ^{4}He and ^{3}He are 4.2 K and 3.19 K, respectively. Its critical pressure is 1.15 atm for ^{4}He. Because of their light masses, both ^{4}He and ^{3}He have extremely high zero-point energies, and remain in gaseous form from room temperature down to a temperature near the absolute zero. Helium becomes solid due to cohesive force only when the interatomic distance becomes sufficiently small under high pressure. For example, a pressure of 25 – 34 atm is required in order to solidify ^{3}He. For ^{4}He, when it is cystallized at a temperature below 4 K, it neither absorbs nor releases heat. *i.e.* the entropies of the crystalline and liquid phases are the same and only its volume is changed in the crystallization process. However, ^{3}He absorbs heat when it is crystallized at a temperature $T < 0.319$ K under pressure. In other words, the temperature of ^{3}He rises during crystallization under pressure. Such an endothermic crystallization process is called the Pomeranchuk effect. This indicates that the entropy of liquid ^{3}He is smaller than that of its crystalline phase. In other words, the liquid phase represents a more ordered state. These peculiar characteristics are due to the unique internal structures of ^{4}He and ^{3}He.

Both ^{4}He and ^{3}He can crystallize in the body-centered cubic or the hexagonal close-stacked structures. A phase transition occurs at a pressure of 1 atm and a temperature of 2.17 K for ^{4}He. Above this temperature, ^{4}He is no different from a normal liquid and this liquid phase is referred to as He I. However, when the temperature is below 2.17 K, the liquid phase, referred to as He II, is completely different from He I and it becomes a superfluid. This superfluid can pass through, without experiencing any resistance, capillaries of diameters less than 10^{-6} cm. The superfluid has a low viscosity ($< 10^{-11}$ P) and its velocity is independent of the pressure difference over the capillary and its length. If a test tube is inserted into liquid He II contained in a container, the level of liquid He II inside the test tube is the same as that in the container. If the test tube is pulled up, the He II inside the test tube would rise along the inner wall of the tube, climb over the mouth of the tube and then flow back to the container along the outer wall of the tube, until the liquid level inside the test tube reaches the same level as that in the container. On the other hand, if the test tube is lifted up above the container, the liquid in test tube would drip directly into the container until the tube becomes empty. Such a property is called the superfluidity of ^{4}He.

Osheroff and others discovered two phase transitions of ^{3}He, occuring at 2.6 mK and 2 mK, respectively, when cooling a mixture of solid and liquid ^{3}He in 1972. Further experiments showed that ^{3}He condenses into liquid at 3.19 K and becomes a superfluid at temperatures below 3×10^{-3} K. In the absence of any external field, ^{3}He can exist in two superfluid phases, ^{3}He A and ^{3}He B. Under a strong magnetic field of a few thousand Gausses, there can be three superfluid phases: ^{3}He A_1, ^{3}He A and ^{3}He B. The T_c is splited into T_{c_1} and T_{c_2}, with normal ^{3}He liquid above T_{c_1}, the ^{3}He A_1 phase between T_{c_1} and T_{c_2}, the ^{3}He A phase between T_{c_2} and T_{AB}, and the ^{3}He B phase below T_{AB}. ^{3}He A is anisotropic and could be a superfluid with ferromagnetic characteristics. The magnetism results from ordered arrangement of magnetic dipoles. ^{3}He B was believed to be isotropic. However, it was known through experiments that ^{3}He B can also be anisotropic below certain temperature. Thus, ^{3}He not only shows characteristics of superconductivity and liquid crystal, it can also be a superfluid, like ^{4}He. This makes ^{3}He is a very special liquid system.

Experiments have shown that quantization of current circulation and vortex structure, similar to that of magnetic flux in a superconductor, can exist in the ^{4}He II, ^{3}He A and ^{3}He B superfluid phases of liquid helium. In terms of the phase of the macroscopic wave function, θ, the velocity $\vec{v}_s$ of the superfluid is given by $\vec{v}_s = \hbar\nabla\theta/M$, where M is the mass of the helium atom. $\vec{v}_s$ satisfies the following quantization condition:

$$\oint \vec{v}_s \cdot d\vec{r} = n \cdot \frac{h}{M}, \quad (n = 0, 1, 2, \cdots).$$

This suggests that the circulation of the velocity of the superfluid is quantized with a quantum of h/M. In other words, as long as the superfluid is rotating, a new whirl in the superfluid is developed whenever the circulation of the current is increased by h/M, *i.e.*, the circulation of the whirl (or energy of the vortex lines) is quantized. Experiment was done in 1963 to measure the energy of the vortex lines. The results obtained were consistent with the theoretical prediction. The quantization of circulation was thus proved.

If the superfluid helium flows, without rotation, through a tube with a varying diameter, then $\nabla \times \vec{v}_s = 0$, and it can be shown, based on the above quantization condition, that the pressure is the same everywhere inside the tube, even though the fluid flows faster at a point where the diameter is smaller and slower where the diameter of the tube is larger. This is completely different from a normal fluid, but it has been proved by experiment.

The macroscopic quantum effect of ^{4}He II was observed experimentally by Firbake and Maston in the U.S.A. once again. When the superfluid ^{4}He II was set into rotational motion in a cup, a whirl would be formed when the temperature of the liquid is reduced to below the critical temperature. In this case, an effective viscosity develops between the fluid and the cup which is very similar to normal fluid in a cup being stired. The surface of the superfluid becomes inclined at a certain angle and the cross-section of the liquid surface is in a shape of a parabola, due to the combined effects of gravitation and centrifugal forces. Fluid away from the center has a tendency to converge towards its center which is balanced by the centrifugal force and a dynamic equilibrium is reached. The angular momentmn of such a whirl is very small and consists of only a small number of discrete quantum packets. The angular momentum of the quantum packets can be obtained by the quantum theory. In other words, the whirl can exist only in discrete form over a certain range in certain materials such as superfluid helium. Firbake and Maston managed to obtain sufficiently large angular momentum in their experiment and were able to observe the whirl's surface shape using visible light. They used a thin layer of rotating superfluid helium in their experiment. While the rotating superfluid was illuminated from both top and bottom by laser beams of wavelength of 6328 Å from a He-Ne laser, the whirl formed were observed. Alternate bright and dark interference fringes were formed when the reflected beams were focused on an observing screen. Analysis of the interference pattern showed that the surface was indeed inclined at an angle. Based on this, the shape of the surface can be correctly constructed. The observed interference pattern was found in excellent agreement with those predicted by the theory. This experiment further confirmed the existence of quantized vortex rings in the superfluid ^{4}He.

How to theoretically explain the superfluidity and the macroscopic quantum effect of ^{4}He is still a subject of current research. In the 1940s, Bogoliubov calculated the critical temperature of Bose-Einstein condensation in ^{4}He based on an ideal Boson gas model. The value he obtained, 3.3 K was quite close to the experimental value of 2.17 K. At a temperature below T_c, some ^{4}He atoms condense to the state with zero momentum. At absolute zero, all the ^{4}He atoms condense to such a state. According to the relation $\lambda = \hbar/p$, the wave length of each ^{4}He atom would be infinite in this case and an ordered state over the entire space can be formed which leads to the macroscopic quantum effects. Pang believed that this phenomenon can be attributed to Bose-Einstein condensation of the ^{4}He atoms. When the temperature of ^{4}He is below T_c, the symmetry of the system breaks down due to nonlinear interaction in the system. The ^{4}He atoms spontaneously condense which results in a highly ordered and long-range coherent state.

^{3}He is different from ^{4}He. ^{3}He atoms are fermions and obey the Fermi statistical law, rather than Bose statistical law. The mechanism of condensation and superfluidity of ^{3}He is similar to that of superconductivity. At a temperature below T_c, two ^{3}He atoms form a loosely bound atomic-pair due to nonlinear interaction within the system. The two atoms with parallel spins revolve around each other and form a pair with a total angular momentum of $J = 1$. This gives rise to a highly ordered and long-range coherent state and condensation of ^{3}He atomic pairs. The macroscopic quantum effect is thus observed. The mechanisms of superfluidity and vortex structure in ^{4}He and ^{3}He have been extensively studied recently and the readers are referred to reference [**??**] for a review of recent work.

2.1.3 *Other macroscopic quantum effects*

Macroscopic quantum effects were also observed in other materials. A few of which are relevant to the topic of this book are briefly introduced in the following.

2.1.3.1 *Quantum Hall effect*

When a longitudinal electric field and a transverse magnetic field are applied to a metal, electric charges accumulate on surfaces parallel to the plane defined by the external fields, producing an electric field and electric current in the direction perpendicular to the external fields. This phenomenon is referred to as the normal Hall effect which everyone is familiar with. The Hall effect to be discussed here has special characteristics and is referred to as a quantum Hall effect. At an extremely low temperature and in the presence of a strong magnetic field, the measured Hall electric potential and the Hall resistance show a series of steps for certain materials in the quantum regime. It appears that they can only have values which are integer multiples of a basic unit, *i.e.* they are quantized. Obviously, this is a macroscopic quantum effect. The concept of quantum Hall effect was first proposed by Tsuneyu Anda of Tokyo University and was experimentally verified first by Klitzing and coworkers.

The approach by Klitzing *et al.* was based on the fact that the degenerate electron gas in the inversion layer of a metal-oxide-semiconductor field-effect transistor is fully quantized when the transistor is operated at the helium temperature and in the presence of a strong magnetic field ($\sim$ 15 T). The electric field applied perpendicular to the oxide-semiconductor interface (gate field) produces a potential well and electrons are confined within the potential and their motion in the direction perpendicular to the interface (z-direction) is limited. If a magnetic field is applied in the z-direction and an electric field is applied in a direction (x) perpendicular to the magnetic field, then the electron will depart from the x direction and drift in the y-direction due to the Lorenz force, resulting in the normal Hall effect. However, when the magnetic field is increased to above 150 KG, the degenerate electron ground state splits into Landau levels. The energy of an electron occupying the nth Landau energy level is $(n + l/2)/\hbar\omega_c$, where ω_c is the angular frequency which is directly proportional to the magnetic field $\tilde{H}_z$. These Landau energy levels can be viewed as semiclassical electron orbits of approximately 70Å in radius. When the "gate voltage" is modulated so that the Fermi energy of this system lies between two Landau energy levels, all the Landau energy levels below the Fermi level are occupied while all energy levels above the Fermi level are vacant. However, due to the combined effect of the electric field $\tilde{E}_x$ and magnetic field $\tilde{H}_z$, the electrons occupying the Landau energy levels move in the y-direction, but do not produce current in the direction of the electric field.

When a current flows through the sample in a direction perpendicular to the direction of the electric field, the diagonal terms of the Hall conductivity $\sigma_{xx} = \sigma_{yy} = 0$ and the off-diagonal terms σ_{xy} is given by $Ne/\tilde{H}_z$, where N is the density of the 2D electron gas. If n Landau energy levels are filled up, then $N = n\tilde{H}_z/h$, and the Hall resistance $R_{\tilde{H}} = h/ne^2$ which is meterial independent. This suggests that the Hall resistance is quantized and its basic unit is h/e^2. Klitzing *et al.* conducted their experiment using a sample of 400 μm $\times$ 50 μm in size and a distance of 130 μm between the potential probes. The measured Hall resistance corresponding to the nth plateau was exactly equal to h/ne^2.

More recently, same phenomena was also observed in GaAs/Al_xGa_{1-x}As heterojunction by scientists in Bell Laboratary under a slightly low magnetic field and temperature. Abnormal Hall effect with n being a fractional number were also observed subsequently. These phenomena allow us to experimentally determine the value of e^2/h and the fine structural constant $a = e^2/h(\mu_0 c/2)$ (μ_0 is the magnetic permeability in vacuum) with high accuracy.

Laughlin was the first to try to deduce the result of the quantum Hall effect from the gauge invariance principles. He believes that this effect depends on the gauge invariance of electromagnetic interaction and the existence of a "drifting" energy gap. Based on his theory, Laughlin successfully derived $R_{\tilde{H}} = h/ne^2$.

2.1.3.2 *Spin polarized atomic hydrogen system*

Hydrogen atoms are Bosons and obey the Bose-Einstein statistics. Because of its extremely light mass, hydrgon atoms have extremely high zero-point vibrational energy. Thus, under ambient pressure, hydrgens remain in gaseous phase with weak interatomic interaction until the temperature approaches to 0 K. Under high pressure, its state undergoes a change when its density reaches an extremely high value. Above this critical density, excess atoms are transfered to a state which corresponds to an energy minimum, and condensation occurs. In other words, when the temperature is below the critical value T_c and the de Broglie wave length of atomic hydrogen is comparable to the interatomic spacing, a considerable amount of atomic hydrogens in the system occupy the same quantum states through the attraction and coherence among them. Bose-Einstein condensation of these atoms thus occurs which is acompanied by sudden changes in specific heat and susceptilility, *etc.* Theoretical calculations show that such a condensation occurs when the density of atomic hydrogen $\rho > 10^{16}$ cm^{-3}. The critical temperature T_c increases with increase of ρ. At $\rho = 10^{17}$ cm^{-3}, $T_c = 8$ mK; and at $\rho = 10^{19}$ cm^{-3}, $T_c = 100$ mK.

It was also shown theoretically that the Bose-Einstein condensation can only occur when the atomic hydrogens are in a triplet states with parallel down spins, which is a stable state with an extremely low energy. Therefore, in order for the condensation to occur, a magnetic field gradient must be applied to the system to select the polarized hydrogen atoms with parallel spins and to hinder the probability of recombination of atomic hydrogens into hydrogen molecules. A magnetic field higher than $9 - 11$ T is normally required. At present, atomic hydrogens cannot be compressed to the density of 10^{19} cm^{-3}. The methods for avoiding recombination of atomic hydrogens are thus limited. It has not been possible to directly observe the condensation of spin polarized hydrogen atoms and the associated macroscopic quantmn effects.

2.1.3.3 *Bose-Einstein condensation of excitons*

It is well known that an electron and a hole can form a bound state, or an exciton, due to the Coulomb interaction between them in many materials such as silicon, germanium, cadmium sulphide, arsenical bromine, silicon carbide, *etc.*. An exciton has its own mass, energy and momentum. It not only can rotate around its own center of mass but also move freely in the crystal lattice. At high exciton density, it is possible for two excitons with opposite spins to form an exciton molecule with the characteristics of a Boson, due to attractive force betweeen them.

When the temperature is below a certain critical value, a considerable amount of exciton molecules condense to the ground state with zero momentum, resulting in the Bose-Einstein condensation, and the superfluidity as in the liquid helium. Because the ratio of the average mass of an exciton or an exciton molecule to the mass of an electron is $2-3$ orders of magnitude smaller than that in ^{4}He, the quantum effects in an exciton system is more obvious than in the liquid helium.

When the same quantum state is occupied by many excition molecules, relatively stronger lines can be observed in the emission spectrum. In 1974, scientists from Japan, Switzerland and other countries detected a very strong line in the low energy region of free exciton recombination radiation of a high purity AgBr sample. Also observed was a stair pattern similar to that observed in the above superconductive diffraction experiment. Very sharp emission lines were later observed at $E_{ex} \sim 1.8195$ eV using a narrow ranged excitation power ($50-100 \times 10^6$ W/cm^2) in the same material. This is also a macroscopic quantum effect. Hang *et al.* predicated that Bose condensation in exciton systems can result in energy superfluid, similar to the above superfluid liquid helium and superconductors.

The Bose-Einstein condensation phenomenon was a subject of extensive studies, both experimentally and theoretically, in the last few decades. Numerious results have been published in Physical Review Letters, Physical Review, Physics Letters, *etc.*. For a review, the readers are referred to a recent book by C. F. Barenghi *et al.*

Macroscopic quantum effects are also expected to occur in ferromagnets, white dwarf and neutron stars. Recently, V. R. Khalilov investigated theoretically macroscopic quantum effects in a degenerate strongly magnetized neutron star. However, these have not been vertified experimentally.

2.2 Analysis on the Nature of Macroscopic Quantum Effect

From the above discussion, we know that the macroscopic quantum effect is a quantum effect that occurs at the macroscopic scale. Such quantum effects have been observed for physical quantities such as resistance, magnetic flux, vortex line and voltage, *etc.* The macroscopic quantum effect is obviously different from the microscopic quantum effect in which physical quantities depicting microscopic particles, such as energy, momentum and angular momentum, are quantized. Thus it is reasonable to believe that the fundamental nature and the rules governing these effects are different. We know that the microscopic quantum effect is described by the LQM theory. But what are the mechanisms for the macroscopic quantum effects? How can these effects be properly described? These questions apparently need to be addressed.

We all know that materials are composed of a great number of microscopic particles such as atoms, electrons, nuclei and so on, which exhibit quantum features. We can then infer or assume that the macroscopic quantum effect results from the collective motion and excitation of these particles under certain conditions, such as extremely low temperature, high pressure or high density. Under such conditions, a huge number of microscopic particles pair with each other and condense, resulting in a highly ordered and long-range coherent low-energy state. In such a highly ordered state, the collective motion of a large number of particles is the same as the motion of a single particle, and since the latter is quantized, the collective motion of the many-particle system gives rise to the macroscopic quantum effect. Thus, the condensation of the particles and the coherent state play an essential role in the macroscopic quantum effect.

What is condensation? At the macroscopic scale, the process of a gas transforming into a liquid, such as that of changing vapor into water, is condensation. This, however, represents a change in the state of molecular positions, and is referred to as a condensation of positions. The phase transition from a gaseous state to a liquid state is a first order transition in which the volume of the system changes and latent heat is produced, but thermodynamic quantities of the systems are continuous and have no singularities. The word condensation in the context of macroscopic quantum effect has its special meaning. The condensation being discussed here is similar to the phase transition from a gas to a liquid, in the sense that the pressure depends only on temperature, but not on volume during the process. Therefore, it is fundamentally different from the first-order phase transition such as that from vapor to water. It is not the condensation of particles into a high density material in normal space. On the contrary, it is the condensation of particles to a single energy state or to a low energy state with a constant or zero momentum. It is thus also called a condensation of momentum. This differs from a first-order phase transition and theoretically it should be classified as a third order phase transition, even though it is really a second order phase transition, because it is related to the discontinuity of the third derivative of a thermodynamnic function. Discontinuities can be clearly observed in measured specific heat, magnetic susceptibility of certain systems when the condensation occurs. The phenomenon results from a spontaneous breaking of symmetries of the system due to nonlinear interaction within the system under some special conditions, such as extremely low temperature and high pressure. Different systems have different critical temperatures for the condensation. For example, the condensation temperature of a superconductor is its critical temperature T_c, and from previous discussions, the condensation temperatures of superfluids ^{4}He and ^{3}He are 2.17 K and 10^{-3} K respectively under the ambient pressure (1 atm.).

From the above discussions on properties of superconductors, superfluids ^{4}He and ^{3}He, spin polarized hydrogen atoms and excitons, we know that even though the elementary particles involved can be either Bosons or Fermions, those being actually condensed, are either Bosons or quasi-Bosons, since Fermions are bound into pairs. For this reason, the condensation is referred to as Bose-Einstein condensation, since Bosons obey the Bose-Einstein statistics. Properties of Bosons are different from those of Fermions, they do not follow the Pauli exclusion principle, and there is no limit to the number of particles occupying the same energy levels. At finite temperatures, Bosons can be distributed among many energy states and each state can be occupied by one or more particles, and some states may not be occupied at all. Due to the statistical attraction between Bosons in the phase space (consisting of generalized coordinate and momentum), groups of Bosons tend to occupy one quantum state under certain conditions. Then when the temperature of the system becomes below a critical value, the majority or all Bosons condense to the same energy level (e.g. the ground state), resulting in Bose condensation and a series of interesting macroscopic quantum effects. Different macroscopic quantum phenomena are observed bacause of differences in the fundamental properties of the constituting particles and their interactions in different systems.

In the highly ordered state the behavior of each condensed particle is closely related to the properties of the systems. In this case, the wave function $\phi = fe^{i\theta}$ or $\phi = \sqrt{\rho}e^{i\theta}$ of the macroscopic state is also the wave function of an individual condensed particle. The macroscopic wave function is also called the order parameter of the condensed state. It was used to describe the superconductive and superfluid states in the study of these macroscopic quantum effects. The essential features and fundamental properties of the macroscopic quantum effect are given by the macroscopic wavefunction ϕ and it can be further shown that the macroscopic quantum states such as the superconductive and superfluid states, are coherent and Bose condensed states formed through second-order phase transitions after the symmetry of the system is broken due to nonlinear interaction in the system.

In the absence of any external field, the Hamiltonian of a given physical system can be written as

$$H = \int dx\mathcal{H} = \int dx\left[-\frac{1}{2}|\nabla\phi|^2 - \alpha|\phi|^2 + \lambda|\phi|^4\right]. \tag{2.1}$$

The unit system in which $m = \hbar = c = 1$ will be used for convenience. If an external field does exist, then the Hamiltonian given above should be replaced by

$$H = \int dx\mathcal{H} = \int dx\left[-\frac{1}{2}|\nabla - ie^*A\phi|^2 - \alpha|\phi|^2 + \lambda|\phi|^4 + \frac{\tilde{H}^2}{2}\right]$$

or equivalently

$$H = \int dx\mathcal{H} = \int dx\left[-\frac{1}{2}|(\partial_j - ie^*A_j)\phi|^2 - \alpha|\phi|^2 + \lambda|\phi|^4 + \frac{1}{4}F_{jl}\cdot F^{jl}\right] \tag{2.2}$$

where $F_{jl} = \partial_j A_l - \partial_l A_j$ is the covariant field intensity, $\vec{\tilde{H}} = \nabla\times\vec{A}$ is the magnetic field intensity, e is the charge of an electron, and $e^* = 2e$, $\vec{A}$ is the vector potential of the field, α and λ are some interaction constants.

The Hamiltonians given in (2.1) and (2.2) have been extensively used by many scientists in their studies of superconductivity, for example, Jacobs and Rebbi, de Gennes, Saint-James, Laplae *et al.*, Kivshar and Bullough, Nemikovskii *et al.*, Roberts *et al.*, Berloff *et al.*, Huepe *et al.*, Sonin and Svistelnov, to name only a few. The readers are referred to reference [??] for further details of their work. The Hamiltonians given in (2.1) and (2.2) resemble that in the Davydov's superconductivity theory and can be derived from the free energy expression of a superconductive system given by Landau *et al.*. As a matter of fact, the Lagrangian of a superconducting system can be obtained from the well known Ginzberg-Landau (Ginzburg-Landau) equation using the Lagrangian method. The Hamiltonian of the sytem can then be derived from the Lagrangian. The results of course are the same as (2.1) and (2.2). Therefore, the Hamiltonians given in (2.1) and (2.2) are expected to describe the macroscopic quantum states including superconducting and superfluid states. These problems are treated in more details in the following.

Obviously, Hamiltonians (2.1) and (2.2) possess the $U(1)$ symmetry. That is, $\mathcal{H}$ remains unchanged under the following transformation

$$\phi_j(x,t) \rightarrow \phi_j'(x,t) = e^{-i\theta Q_j}\phi_j(x,t) \tag{2.3}$$

where Q_j is the charge of the particle. In the case of one dimension, each term in the Hamitonian (2.1) or (2.1) contains the product of the $\phi_j(x,t)$s,

$$\phi_1(x,t)\cdots\phi_n(x,t) \rightarrow e^{i(Q_1+Q_2+\cdots+Q_n)\theta}\phi_1(x,t)\cdots\phi_n(x,t).$$

Since charge is invariant under the transformation and neutrality is required for $\mathcal{H}$. There must be $(Q_1+Q_2+\cdots+Q_n)=0$ in such a case. Furthermore, since θ is independent of x, it is necessary that $\nabla\phi_j \rightarrow e^{-i\theta Q_j}\nabla\phi_j$. Thus, each term in the Hamiltonian is invarient under the transformation of (2.3), or they possess the $U(1)$ symmetry.

If we rewrite (2.1) as the following

$$\mathcal{H} = -\frac{1}{2}(\nabla\phi)^2 + U_{\text{eff}}(\phi), \quad U_{\text{eff}}(\phi) = -\alpha\phi^2 + \lambda\phi^4, \tag{2.4}$$

we can see that the effective petential energy has two sets of extrema, $\phi_0 = 0$ and $\phi_0 = \pm\sqrt{\alpha/2\lambda}$, and the minimum is located at

$$\phi_0 = \pm\sqrt{\frac{\alpha}{2\lambda}} = \langle 0|\phi|0\rangle, \tag{2.5}$$

rather than at $\phi_0 = 0$. This means that the energy at $\phi_0 = \pm\sqrt{\alpha/2\lambda}$ is lower than that at $\phi_0 = 0$. Therefore, $\phi_0 = 0$ corresponds to the normal ground state while $\phi_0 = \pm\sqrt{\alpha/2\lambda}$ is the ground state of the macroscopic quantum systems (MQS). In this case, the macroscopic quantum state is the stable state of the system. Therefore, the Hamiltonian of a normal state differs from that of the superconducting state.

Meanwhile, the two ground states satisfy $\langle 0|\phi|0\rangle \neq -\langle 0|\phi|0\rangle$ under the transformation $\phi \to -\phi$. That is, they no longer have the $U(1)$ symmetry. In other words, the symmetry of the ground states has been destroyed. The reason for this is obviously due to the nonlinear term $\lambda|\phi|^4$ in the Hamiltonian of the system. Therefore, this phenomenon is referred to as a spontaneous breakdown of symmetry. According to Landau's theory of phase transition, the system undergoes a second-order phase transition in such a case, and the normal ground state $\phi_0 = 0$ is changed to the superconducting ground state $\phi_0 = \pm\sqrt{\alpha/2\lambda}$. We will give a proof in the following.

In order to make the expectation value in a new ground state zero, we make the following transformation,

$$\phi' = \phi + \phi_0 \tag{2.6}$$

so that

$$\langle 0|\phi'|0\rangle = 0. \tag{2.7}$$

After this transformation, the Hamiltonian of the system becomes

$$\begin{aligned}\mathcal{H}(\phi+\phi_0) = &-\frac{1}{2}|\nabla\phi|^2 + (6\lambda\phi_0^2 - \alpha)\phi^2 + 4\lambda\phi_0\phi^3 \\ &+(4\lambda\phi_0^3 - 2\alpha\phi_0)\phi + \lambda\phi^4 - \alpha\phi_0^2 + \lambda\phi_0^4.\end{aligned} \tag{2.8}$$

Substituting (2.5) into (2.8), we have

$$\langle \phi_0 \left| 4\lambda\phi_0^2 - 2\alpha \right| \phi_0\rangle = 0.$$

Consider now the expectation value of the variation $\delta\mathcal{H}/\delta\phi$ in the ground state, *i.e.*

$$\left\langle 0 \left| \frac{\delta\mathcal{H}}{\delta\phi} \right| 0 \right\rangle = 0,$$

then from (2.1), we get

$$\left\langle 0 \left| \frac{\delta\mathcal{H}}{\delta\phi} \right| 0 \right\rangle = \langle 0\left| -\nabla^2\phi + 2\alpha\phi - 4\lambda\phi^3 \right| 0\rangle = 0. \tag{2.9}$$

After the transformation (2.6), this becomes

$$\begin{aligned}\nabla^2\phi_0 + (4\lambda\phi_0^2 - 2\alpha)\phi_0 + 12\lambda\phi_0\langle 0|\phi^2|0\rangle + 4\lambda\langle 0|\phi^3|0\rangle - \\ 2(\alpha - 6\lambda\phi_0^2)\langle 0|\phi|0\rangle = 0\end{aligned} \tag{2.10}$$

where the terms $\langle 0|\phi^3|0\rangle$ and $\langle 0|\phi|0\rangle$ are both zero, but the fluctuation $12\lambda\phi_0\langle 0|\phi^2|0\rangle$ of the ground state is not zero. For a homogeneous system at $T = 0$ K, the term $\langle 0|\phi^2|0\rangle$ is very small and can be neglected. Then (2.10) can be written as

$$-\nabla^2\phi_0 - \phi_0(4\lambda\phi_0^2 - 2\alpha) = 0. \tag{2.11}$$

Two sets of solutions, $\phi_0 = 0$ and $\phi_0 = \pm\sqrt{\alpha/2\lambda}$ can then be obtained from the above equation.

If the displacement is very small, *i.e.* $\phi_0 \rightarrow \phi_0 + \delta\phi_0 = \phi_0'$, then the equation satisfied by the fluctuation $\delta\phi_0$ relative to the normal ground state $\phi_0 = 0$ is

$$\nabla^2 \delta\phi_0 - 2\alpha\delta\phi_0 = 0. \tag{2.12}$$

Its solution attenuates exponentially, indicating that the ground state $\phi_0 = 0$ is unstable.

On the other hand, the equation satisfied by the fluctuation $\delta\phi_0$ relative to the ground state $\phi_0 = \pm\sqrt{\alpha/2\lambda}$ is

$$\nabla^2 \delta\phi_0 + 2\alpha\delta\phi_0 = 0.$$

Its solution $\delta\phi_0$ is an oscillatory function, and thus, the MQS ground states $\phi_0 = \pm\sqrt{\alpha/2\lambda}$ are stable. Further calculation shows that the energy of the MQS ground state is lower than that of the normal state by $\varepsilon_0 = -\alpha^2/4\lambda < 0$. Therefore, the ground state of the normal phase and that of the condensed phase are separated by an energy gap of $\alpha^2/(4\lambda)$, and at $T = 0$ K, all particles condense to the ground state of the condensed phase, rather than filling the ground state of the normal phase. Based on this energy gap, we can conclude that the specific heat of the systems has exponential dependence on the temperature, and the critical temperature is given by $T_c = 1.14\omega_p \exp[-1/(3\lambda/\alpha)N(0)]$. This is a feature of the second-order phase transition. The results are in agreement with that of the BCS theory of superconductivity. Therefore, the transition from the state $\phi_0 = 0$ to the state $\phi_0 = \pm\sqrt{\alpha/2\lambda}$ and the corresponding condensation of particles are second-order phase transition. It is obviously the results of spontaneous breakdown of symmetry due to nonlinear interaction.

In the presence of an electromagnetic field with a vector potential $\vec{A}$, the Hamiltonian of the systems is given by (2.2). It still possesses the $U(1)$ symmetry. That is, under the following gauge transformation

$$\begin{cases} \phi(x) \longrightarrow \phi'(x) = e^{-i\theta(x)}\phi(x) \\ \vec{A}(x) \longrightarrow \vec{A}'(x) = \vec{A}(x) - \dfrac{1}{e^*}\nabla_j\theta(x), \quad \text{or} \quad \delta A_j(x) = \dfrac{1}{e^*}\partial_j\theta(x) \end{cases} \tag{2.13}$$

$\mathcal{H}$ is invariant, since $F_{jl}F^{jl}/4$ is invariant under this transformation. In other words, the Hamiltonian (2.2) has the $U(1)$ invariance.

Because of the existence of the nonlinear terms in (2.2), a spontaneous breakdown of symmetry can be expected. Now consider the following transformation,

$$\phi(x) = \frac{1}{\sqrt{2}}[\phi_1(x) + i\phi_2(x)] \longrightarrow \frac{1}{\sqrt{2}}[\phi_1(x) + \phi_0 + i\phi_2(x)]. \tag{2.14}$$

Since $\langle 0|\phi_i|0\rangle = 0$ under this transformation, equation (2.2) becomes

$$\mathcal{H} = \frac{1}{4}(\partial_i A_j - \partial_j A_i)^2 - \frac{1}{2}(\nabla\phi_2)^2 - \frac{1}{2}(\nabla\phi_1)^2 + \frac{(e^*)^2}{2}[(\phi_1+\phi_0)^2 + \phi_2^2]A_i^2$$
$$-e^*\phi_0 A_i \nabla\phi_2 + e^*(\phi_2\nabla\phi_1 - \phi_1\nabla\phi_2)A_i$$
$$-\frac{1}{2}(-12\lambda\phi_0^2 + 2\alpha)\phi_1^2 - \frac{1}{2}(12\lambda\phi_0^2 + 2\alpha)\phi_2^2 + 4\lambda\phi_0\phi_1(\phi_1^2 + \phi_2^2)$$
$$+4\lambda(\phi_1^2 + \phi_2^2)^2 - \phi_0(4\lambda\phi_0^2 + 2\alpha)\phi_1 - \alpha\phi_0^2 + \lambda\phi_0^4. \tag{2.15}$$

It may be seen that the classical interaction energy of ϕ_0 is still given by

$$U_{\text{eff}}(\phi_0) = -\alpha\phi_0^2 + \lambda\phi_0^4, \tag{2.16}$$

in agreement with that given in (2.4). Therefore, using the same argument, we can conclude that spontaneous symmetry breakdown and the second-order phase transition occur in the system. The system is changed from the ground state of the normal phase, $\phi_0 = 0$, to the ground state $\phi_0 = \pm\sqrt{\alpha/2\lambda}$ of the condensed phase. The above results can also be used to explain the Maissner effect and to determine its critical temperature. Thus, we can conclude that regardless the existence of any external field, macroscopic quantum states, such as the superconducting state of a superconductor, are formed through a second-order phase transition following a spontaneous symmetry breakdown due to nonlinear interaction in the systems.

We can prove that this is a condensed state using either the second quantization theory or the solid state quantum field theory. As discussed above, when $\partial\mathcal{H}/\partial\phi = 0$, we have

$$\nabla^2\phi - 2\alpha\phi + 4\lambda|\phi|^2\phi = 0. \tag{2.17}$$

This is the , and it is a time-independent nonlinear Schrödinger equation. Expanding ϕ in terms of the creation and annihilation operators, b_p^+ and b_p, we have

$$\phi = \frac{1}{\sqrt{\overline{V}}}\sum_p \frac{1}{\sqrt{2\varepsilon_p}}\left(b_p e^{-ip.x} + b_p^+ e^{ip.x}\right) \tag{2.18}$$

where $\overline{V}$ is the volume of the system. After a spontaneous breakdown of symmetry, ϕ_0, the ground state of ϕ, is no longer zero, but $\phi_0 = \pm\sqrt{\alpha/2\lambda}$. The operation of the annihilation operator on $|\phi_0\rangle$ no longer gives zero, *i.e.*

$$b_p|\phi_0\rangle \neq 0. \tag{2.19}$$

A new field ϕ' can then be defined according to the transformation (2.6), where ϕ_0 is a scalar field and satisfies (2.11) in such a case. Evidently, ϕ_0 can also be expanded into

$$\phi_0 = -\frac{1}{\sqrt{\overline{V}}}\sum_p \frac{1}{\sqrt{\varepsilon_p}}(\xi_p e^{-ip.x} + \xi_p^+ e^{ip.x}). \tag{2.20}$$

The transformation between the fields ϕ' and ϕ is obviously a unitary transformation, that is,

$$\phi' = U\phi U^{-1} = e^{-S}\phi e^{S} = \phi + \phi_0 \tag{2.21}$$

where

$$S = i\int dx' \left[\phi(x',t)\phi_0(x',t) - \phi_0(x',t)\phi(x',t)\right]. \tag{2.22}$$

ϕ and ϕ' satisfy the following commutation relation

$$[\phi'(x,t), \phi(x',t)] = i\delta(x - x'). \tag{2.23}$$

From (2.7) we have $\langle 0|\phi'|0\rangle = \phi_0' = 0$. The ground state $|\phi_0'\rangle$ of the field ϕ' thus satisfies

$$b_p|\phi_0'\rangle = 0. \tag{2.24}$$

From (2.21), we can obtain the following relation between the annihilation operator a_p of the new field ϕ' and the annihilation operator b_p of the ϕ field,

$$a_p = e^{-S}b_p e^{S} = b_p + \xi_p \tag{2.25}$$

where

$$\xi_p = \frac{1}{(2\pi)^{3/2}}\int \frac{dx}{\sqrt{2\varepsilon_p}}\left[\phi_0(x,t)e^{ip.x} + i\phi_0^*(x,t)e^{-ip.x}\right]. \tag{2.26}$$

Therefore, the new ground state $|\phi_0'\rangle$ and the old ground state $|\phi_0\rangle$ are related through $|\phi_0'\rangle = e^{S}|\phi_0\rangle$. Thus we have

$$a_p|\phi_0'\rangle = (b_p + \xi_p)|\phi_0'\rangle = \xi_p|\phi_0'\rangle. \tag{2.27}$$

According to the definition of coherent state, equation (2.27) shows that the new ground state $|\phi_0'\rangle$ is a coherent state. Because such a coherent state is formed after the spontaneous breakdown of symmetry of the systems, it is referred to as a spontaneous coherent state. But when $\phi_0 = 0$, the new ground state is the same as the old state which is not a coherent state.

The same conclusion can be directly derived from the BCS theory. In the BCS theory, the wave function of the ground state of a superconductor is written as

$$\begin{aligned}\Phi_0 &= \prod_k \left(u_k + v_k\hat{a}_k^+\hat{a}_{-k}^+\right)\phi_0 = \prod_k \left(u_k + v_k\hat{b}_{k-k}^+\right)\phi_0 \\ &\sim \eta\exp\left(\sum_k \frac{v_k}{u_k}\hat{b}_{k-k}^+\right)\phi_0\end{aligned} \tag{2.28}$$

where $\hat{b}_{k-k}^+ = \hat{a}_k^+\hat{a}_{-k}^+$. This equation shows that the superconducting ground state is a coherent state. Hence, we can conclude that the spontaneous coherent state in superconductors is formed after the spontaneous breakdown of symmetry.

Recently, by reconstructing a quasiparticle-operator-free new formulation of the Bogoliubov-Valatin transformation parameter dependence, Lin *et al.* demonstrated that the BCS state is not only a coherent state of single-Cooper-pairs, but also the squeezed state of double-Cooper-pairs, and reconfirmed the coherent feature of BCS superconductive states.

In the following, we employ the method used by Bogoliubov in the study of superfluid liquid helium ^{4}He to prove that the above state is indeed a Bose condensed state. To do that, we rewrite (2.18) in the following form

$$\phi(x) = \frac{1}{\sqrt{V}} \sum_p q_p e^{-ip.x}, \quad q_p = \frac{1}{\sqrt{2\varepsilon_p}} (b_p + b_{-p}^+). \tag{2.29}$$

Since the field ϕ describes a Boson, such as a Cooper electron-pair in a superconductor, Bose condensation can occurr in the system. Following traditional method in quantum field theory, we consider the following transformation

$$b_p = \sqrt{N_0}\delta(p) + \gamma_p, \quad b_p^+ = \sqrt{N_0}\delta(p) + \beta_p, \tag{2.30}$$

where N_0 is the number of Bosons in the system and

$$\delta(p) = \begin{cases} 0, & \text{if } p \neq 0 \\ 1, & \text{if } p = 0. \end{cases}$$

Substituting (2.29) and (2.30) into (2.1), we can get

$$\begin{aligned} \mathcal{H} = & \frac{4\lambda N_0^2}{\varepsilon_0^2 \overline{V}} - \frac{2N_0\alpha}{\varepsilon_0} + \sum_p \frac{4\lambda}{\varepsilon_0 \varepsilon_p} \frac{N_0}{\overline{V}} \\ & + \left(\frac{4\lambda N_0^2}{\varepsilon_0^2 \overline{V}} - \frac{\alpha}{\varepsilon_0} \right) \sqrt{N_0} \left(\gamma_0 + \gamma_0^+ + \beta_0 + \beta_0^+ \right) \\ & + \sum_p \left(\frac{4\lambda N_0}{\varepsilon_0 \varepsilon_p \overline{V}} - \varepsilon_p \right) \left(\gamma_p^+ \beta_{-p}^+ + \gamma_p \beta_{-p} \right) \\ & + \frac{\lambda N_0}{\varepsilon_0 \overline{V}} \sum_p \frac{1}{\varepsilon_p} \left(\beta_p^+ \beta_{-p}^+ + \beta_p \beta_{-p} + \gamma_p^+ \gamma_{-p}^+ + \gamma_p \gamma_{-p} + 2\gamma_p^+ \beta_p + 2\beta_p^+ \gamma_p \right) \\ & + \sum_p \left(\varepsilon_p - \frac{\alpha}{2\varepsilon_0} + \frac{4\lambda N_0}{\varepsilon_0 \varepsilon_p \overline{V}} \right) \left(\gamma_p^+ \gamma_p + \beta_p^+ \beta_p \right) \\ & + O\left(\frac{\sqrt{N_0}}{\overline{V}} \right) + O\left(\frac{N_0}{\overline{V}^2} \right). \end{aligned} \tag{2.31}$$

Because the condensed density $N_0/\overline{V}$ must be finite, the higher order terms $O(\sqrt{N_0}/\overline{V})$ and $O(N_0/\overline{V}^2)$ may be neglected. Next we perform the following canonical transformation

$$\begin{aligned} \gamma_p &= u_p^* c_p + v_p c_{-p}^+, \\ \beta_p &= u_p^* d_p + v_p d_{-p}^+, \end{aligned} \tag{2.32}$$

where v_p and u_p are real and satisfy

$$\left(u_p^2 - v_p^2\right) = 1.$$

Introducing another transformation

$$\zeta_p = \frac{1}{\sqrt{2}}(u_p\gamma_p^+ - v_p\gamma_{-p} + u_p\beta_p^+ - v_p\beta_{-p}), \tag{2.33}$$
$$\eta_p^+ = \frac{1}{\sqrt{2}}(u_p\gamma_p^+ - v_p\gamma_{-p} - u_p\beta_p^+ + v_p\beta_{-p}),$$

the following relations can be obtained

$$[\zeta_p, \mathcal{H}] = g_p\zeta_p + M_p\zeta_{-p}^+, \quad [\eta_p, \mathcal{H}] = g_p'\eta_p + M_p'\eta_{-p}^+, \tag{2.34}$$

where

$$\begin{cases} g_p = G_p\left(u_p^2 + v_p^2\right) + F_p 2u_p v_p, & M_p = F_p\left(u_p^2 + v_p^2\right) + G_p 2u_p v_p \\ g_p' = G_p'\left(u_p^2 + v_p^2\right) + F_p' 2u_p v_p, & M_p' = F_p'\left(u_p^2 + v_p^2\right) + G_p' 2u_p v_p \end{cases} \tag{2.35}$$

while

$$\begin{cases} G_p = \varepsilon_p - \dfrac{\alpha}{2\varepsilon_p} + 6\xi_p', & F_p = -\dfrac{\alpha}{2\varepsilon_p} + 6\xi_p', \\ G_p' = \varepsilon_p - \dfrac{\alpha}{2\varepsilon_p} + 2\xi_p', & F_p' = \dfrac{\alpha}{2\varepsilon_p} - 2\xi_p', \end{cases} \tag{2.36}$$

where

$$\xi_p' = \frac{\lambda N_0}{\varepsilon_0\varepsilon_p\bar{V}}.$$

We now discuss two cases.

(A) Let $M_p' = 0$, then it can be seen from (2.34) that η_p^+ is the creation operator of elementary excitation and its energy is given by

$$g_p' = \sqrt{\varepsilon_p^2 + 4\varepsilon_p\xi_p' - 2\alpha}. \tag{2.37}$$

Using this, we can obtain the following from (2.35) and (2.36)

$$(u_p')^2 = \frac{1}{2}\left(1 + \frac{G_p'}{g_p'}\right) \quad \text{and} \quad (v_p')^2 = \frac{1}{2}\left(-1 + \frac{G_p'}{g_p'}\right). \tag{2.38}$$

From (2.34), we know that ζ_p^+ is not a creation operator of the elementary excitation. Thus, another transformation must be made

$$B_p = \chi_p\zeta_p' + \mu_p\zeta_p'^+, \quad |\chi_p|^2 - |\mu_p|^2 = 1. \tag{2.39}$$

We can then prove that

$$[B_p, \mathcal{H}] = E_p B_p, \tag{2.40}$$

where

$$E_p = \sqrt{12\varepsilon_p\xi'_p + \varepsilon_p^2 - 2\alpha}.$$

Now, inserting (2.32), (2.39), (2.40) and $M'_p = 0$ into (2.31), and after some reorganizations, we have

$$\mathcal{H} = U + E_0 + \sum_{p>0} \left[E_p(B_p^+ B_p + B_{-p}^+ B_{-p}) + g'_p(\eta_p^+ \eta_p + \eta_{-p}^+ \eta_{-p})\right] \tag{2.41}$$

where

$$U = \frac{\lambda N_0^2}{\varepsilon_0^2 V} - \frac{2\alpha N_0}{\varepsilon_0} + \sum_p 4\xi'_p + \sum_p \left(\varepsilon_p + \frac{\alpha}{2\varepsilon_p} + 4\xi'_p\right) 4v_p'^2 + \sum_{p>0} 4\eta \cdot 2u'_p v'_p,$$

$$E_0 = -2\sum_{p>0} E_p|\mu_p|^2 = -\sum_{p>0}(g'_p - E_p). \tag{2.42}$$

Both U and E_0 are now independent of the creation and annihilation operators of the Bosons. $U + E_0$ gives the energy of the ground state. N_0 can be determined from the condition

$$\frac{\delta(U + E_0)}{\delta N_0} = 0$$

which gives

$$\frac{N_0}{V} = \frac{\alpha\varepsilon_0}{4\lambda} = \frac{1}{2}\varepsilon_0\phi_0^2. \tag{2.43}$$

This is the condensed density of the ground state ϕ_0. From (2.36), (2.37) and (2.40), we can get

$$g'_p = \sqrt{\varepsilon_p^2 - \alpha}, \quad E_p = \sqrt{\varepsilon_p^2 - \alpha}. \tag{2.44}$$

These correspond to the energy spectra of η_p^+ and B_p^+, respectively, and they are similar to the energy spectra of the Cooper pair and phonon in the BCS theory. Substituting (2.44) into (2.38), we have

$$u_p'^2 = \frac{1}{2}\left(1 + \frac{2\varepsilon_p^2 - \alpha}{2\sqrt{(\varepsilon_p^2 - \alpha\ \varepsilon_p}}\right), \quad v_p'^2 = \frac{1}{2}\left(-1 + \frac{2\varepsilon_p^2 - \alpha}{2\sqrt{(\varepsilon_p^2 - \alpha\ \varepsilon_p}}\right). \tag{2.45}$$

(B) In the case of $M_p = 0$, a similar approach can be used to get the energy spectrum corresponding to ζ_p^+ as

$$E_p = \sqrt{\varepsilon_p^2 + \alpha}$$

while that corresponding to $A_p^+ = \chi_p\eta_p^+ + \mu_p\eta_{-p}$ is

$$g'_p = \sqrt{\varepsilon_p^2 + \alpha},$$

where

$$u_p^2 = \frac{1}{2}\left(1 + \frac{2\varepsilon_p^2 + \alpha}{2\varepsilon_p\sqrt{(\varepsilon_p^2 + \alpha)}}\right), \quad v_p^2 = \frac{1}{2}\left(-1 + \frac{2\varepsilon_p^2 + \alpha}{2\varepsilon_p\sqrt{(\varepsilon_p^2 + \alpha)}}\right). \tag{2.46}$$

From quantum statistical physics we know that the occupation number of the level with the energy ε_p for a system in thermal equilibrium at temperature T ($\neq 0$) is given by

$$N_p = \langle b_p^+ b_p \rangle = \frac{1}{e^{\varepsilon_p/K_BT} - 1} \tag{2.47}$$

where $\langle \cdots \rangle$ denotes Gibbs average, defined as

$$\langle \cdots \rangle = \frac{\mathrm{SP}[e^{-\mathcal{H}/K_BT} \ldots]}{\mathrm{SP}[e^{-\mathcal{H}/K_BT})]}.$$

Here SP denotes the trace in Gibbs statistics. At low temperature, or $T \to 0$ K, the majority of the Bosons or Cooper pairs in a superconductor condense to the ground state with $p = 0$. Therefore,

$$\langle b_0^+ b_0 \rangle \approx N_0,$$

where N_0 is the total number of Bosons or Cooper pairs in the system and $N_0 \gg 1$, *i.e.*

$$\langle b^+ b \rangle = 1 \ll \langle b_0^+ b_0 \rangle.$$

As can be seen from (2.29) and (2.30), the number of particles is extremely large when they lie in the condensed state, that is

$$\phi_0 = \phi_{p=0} = \frac{1}{\sqrt{2\varepsilon_0 V}}\,(b_0 + b_0^+). \tag{2.48}$$

Because $\gamma_0|\phi_0\rangle = 0$ and $\beta_0|\phi_0\rangle = 0$, b_0 and b_0^+ can be taken to be $\sqrt{N_0}$. The average value of $\phi^*\phi$ in the ground state then becomes

$$\langle \phi_0|\phi^*\phi|\phi_0 \rangle = \langle \phi^*\phi \rangle_0 = \frac{1}{2\varepsilon_0 V} \cdot 4N_0 = \frac{2N_0}{\varepsilon_0 V}. \tag{2.49}$$

Substituting (2.43) into (2.49), we can get

$$\langle \phi^*\phi \rangle_0 = \frac{\alpha}{2\lambda},$$

or

$$\langle \phi^* \rangle_0 = \pm\sqrt{\frac{\alpha}{2\lambda}}$$

which is the ground state of the condensed phase or the superconducting phase that we have known. Thus, the density of states, N_0/V, of the condensed phase or the superconducting phase formed after the Bose condensation coincides with the average value of the Boson's (or Copper pair's) field in the ground state. We can then conclude that the macroscopic quantum state or the superconducting ground state formed after the spontaneous symmetry breakdown is indeed a Bose-Einstein condensed state.

In the last few decades, Bose-Einstein condensation was observed in a series of remarkable experiments for weakly interacting atomic gases, such as vapors of rabidium, sodium lithium, or hydrogen. Its formation and properties have been extensively studied. These studies show that the Bose-Einstein condensation is a nonlinear phenomenon, analogous to nonlinear optics, the state is coherent, and can be described by the following nonlinear Schrödinger equation or the Gross-Pitaerskii equation

$$i\frac{\partial \phi}{\partial t'} = -\frac{\partial^2 \phi}{\partial x'^2} - \lambda|\phi|^3 + V(x')\phi, \tag{2.50}$$

where $t' = t/\hbar$, $x' = x\sqrt{2m}/\hbar$.

This equation was used to discuss realization of Bose-Einstein condensation in the $d+1$ dimensions ($d = 1, 2, 3$) by Bullough *et al.*. The corresponding Hamillonian of a condensate system was given by Elyutin *et al.* as follows

$$\mathcal{H} = \left|\frac{\partial \phi}{\partial x'}\right|^2 + V(x')|\phi|^2 - \frac{1}{2}\lambda|\phi|^4 \tag{2.51}$$

where the nonlinearity parameter λ is defined as $\lambda = -2Naa_l/a_0^2$, with N being the number of particles trapped in the condensed state, a the ground state scattering length, a_0 and a_l the transverse (y, z) and the longitudinal (x) condensate sizes (without self-interaction), respectively. (Integrations over y and z have been carried out in obtaining the above equation). λ is positive for condensation with self-attraction (negative scattering length). The coherent regime was observed in Bose-Einstein condensation in lithium. The specific form of the trapping potential $V(x')$ trapping potential depends on the details of the experimental setup. Work on Bose-Einstein condensation based on the above model Hamiltonian were carried out and are reported by Barenghi *et al.* [??].

It is not surprising to see that (2.50) is exactly the same as (2.17), and (2.51) is the same as (2.1). This further confirms the correctness of the above theory for Bose-Einstein condensation. As a matter of fact, immediately after the first experimental observation of this condensation phenomenon it was realized that coherent dynamics of the condensated macroscopic wave function can lead to the formation of nonlinear solitary waves. For example, self-localized bright, dark and vortex solitons, formed by increased (bright) or decreased (dark or vortex) probability density, respectively, were experimentally observed, particularly for the vortex soliton which has the same form as the vortex lines found in type II-superconductors and superfluids. These experimental results were in concordance with the results of the above theory. In the following sections, we will study the soliton motions of quasiparticles in macroscopic quantum systems. We will see that the dynamic equations in MQS have such soliton solutions.

From the above discussion we clearly understand the nature and characteristics of an MQS such as a superconducting system. It would be interesting if a comparison is made between the macroscopic quantum effect and the microscopic quantum effect which is well known. Here we give a summary of the main differences between them.

(1) Concerning the origins of these quantum effects, the microscopic quantum effect is produced when microscopic particles which have the particle-wave duality are confined in a finite space, while the macroscopic quantum effect is due to the collective motion of the microscopic particles in systems with nonlinear interaction. It occurs through second-order phase transition following the spontaneous breakdown of symmetry of the systems.

(2) From the point of view of their characteristics, the microscopic quantum effect is characterized by quantization of physical quantities such as energy, momentum, angular momentum, *etc.* of the microscopic particles. On the other hand, the macroscopic quantum effect is represented by discontinuities in macroscopic quantities such as resistance, magnetic flux, vortex lines, voltage, *etc.*. The macroscopic quantum effects can be directly observed in experiments, while the microscopic quantum effects can only be inferred from other effects related to them.

(3) The macroscopic quantum state is a condensed and coherent state, but the microscopic quantum effect typically occurs in bound states. Certain quantization condition is satisfied for a microscopic quantum effect and the particles involved can be either Bosons or Fermions. But so far only Bosons or combination of Fermions are found in macroscopic quantum effects.

(4) The microscopic quantum effect is a linear effect and motions of particles are described by linear differential equations, such as the Schrödinger equation, the Dirac equation, and the Klein-Gordon equations. On the other hand, the macroscopic quantum effect is a nonlinear effect and the motions of the particles are described by non-linear partial differential equations such as the nonlinear Schrödinger equation (2.17).

Thus, the fundamental nature and characteristics of the macroscopic effect are different from those of the microscopic quantum effects. For these reasons, a different approach must be used to describe the macroscopic quatum effect. Also based on these discussions, it is clear that a nonlinear quantum theory would be necessary to describe these macroscopic quantum effects.

2.3 Motion of Superconducting Electrons

In this section, we discuss the motions of quasiparticles, such as superconductive electrons or Cooper pairs in superconductors and superfluid helium, in the MQS. The properties and motion of quasiparticles are not only important for understanding the relevant quantum effects, but also provides the basis for establishing the new nonlinear quantum mechanical theory.

From earlier discussions, we know that the superconductive state is a highly ordered coherent state or a Bose-Einstein condensed state formed after a spontaneous breakdown of the symmetry of the system due to the nonlinear interaction in the system. A macroscopic quantum wave-function, ϕ, can be used to describe the superconducting electrons (Cooper pairs)

$$\phi(\vec{r},t) = f(\vec{r},t)\phi_0 e^{i\theta(\vec{r})}, \tag{2.52}$$

where

$$\phi_0^2 = \frac{\alpha}{2\lambda}.$$

According to the Ginzberg-Landau theory of superconductivity, the free energy density function of a superconducting system is given by

$$f_s = f_n - \frac{\hbar^2}{2m}|\nabla\phi|^2 - \alpha|\phi|^2 + \lambda|\phi|^4 \tag{2.53}$$

in the absence of any external field, where f_n is the free energy of the normal state, α and λ are interaction constants among the particles. On the other hand, if the system is subjected to an electromagnetic field specified by a vector potential $\vec{A}$, the free energy of the system is

$$f_s = f_n - \frac{\hbar^2}{2m}\left|\left(\nabla - \frac{ie^*}{c\hbar}\vec{A}\right)\phi\right|^2 - \alpha|\phi|^2 + \lambda|\phi|^4 + \frac{1}{8\pi}\tilde{H}^2, \tag{2.54}$$

where $e^* = 2e$, $\tilde{H} = \nabla \times \vec{A}$. The free energy of the system is

$$F_s = \int f_s d^3x.$$

In terms of the convariant field, $F_{jl} = \partial_j A_l - \partial_l A^j$, $(j,l = 1,2,3)$, the term $\tilde{H}^2/8\pi$ can be written as $F_{jl}F^{jl}/4$. Then from $\delta F_s = 0$, we get

$$\frac{\hbar^2}{2m}\nabla^2\phi - \alpha\phi + 2\lambda\phi^3 = 0 \tag{2.55}$$

and

$$\frac{\hbar^2}{2m}\left(\nabla - \frac{ie^*}{c\hbar}\vec{A}\right)^2\phi - \alpha\phi + 2\lambda\phi^3 = 0 \tag{2.56}$$

in the absence and presence of an external field, respectively, and

$$\vec{\mathbf{J}} = +\frac{e^*\hbar}{2mi}(\phi^*\nabla\phi - \phi\nabla\phi^*) - \frac{e^*}{mc}|\phi|^2\vec{A}. \tag{2.57}$$

Equations (2.55) – (2.57) are the well-known Ginzburg-Landau equations in a steady state. Here (2.55) is the Ginzburg-Landau equation in the absence of external fields.

It is the same as (2.15) which was obtained from (2.1). Equation (2.56) can also be obtained from (2.2). Therefore, equations (2.1) – (2.2) are the Hamiltonians corresponding to the free energy (2.53) – (2.54).

From (2.53) – (2.57), we see clearly that superconductors are nonlinear systems. The Ginzburg-Landau equations are the fundamental equations of the superconductors and describe the motion of the superconductive electrons. However, the equations contain two unknown functions ϕ and $\vec{A}$ which make them extremely difficult to be solved. We first consider the case of no external field.

2.3.1 *Motion of electrons in the absence of external fields*

We consider only an one-dimensional pure superconductor. Let

$$\phi = f\phi_0, \quad \frac{\hbar^2}{2m|\alpha|} = \zeta^2(T), \quad x' = \frac{x}{\zeta(T)}, \tag{2.58}$$

where $\zeta(T)$ is the coherent length of the superconductor which depends on temperature. For a uniform pure superconductor,

$$\zeta(T) = 0.94\zeta_0\sqrt{\frac{T_c}{T_c - T}},$$

where T_c is the critical temperature and ζ_0 is the coherent length at $T = 0$. Then, equation (2.55) can be written as

$$-\frac{d^2 f}{dx'^2} + f - f^3 = 0 \tag{2.59}$$

and its boundary conditions are

$$f(x' = 0) = 1, \quad \text{and} \quad f(x' \to \pm\infty) = 0. \tag{2.60}$$

After integration, we get the following solutions:

$$f = \pm\sqrt{2}\,\text{sech}\left[\frac{x - x_0}{\sqrt{2}\zeta(T)}\right], \tag{2.61}$$

$$\phi = \pm\sqrt{\frac{\alpha}{\lambda}}\,\text{sech}\left[\frac{x - x_0}{\zeta(T)}\right] = \pm\sqrt{\frac{\alpha}{\lambda}}\,\text{sech}\left[\frac{\sqrt{2m\alpha}}{\hbar}(x - x_0)\right].$$

This is the well-known wave packet-type soliton solution. It represents a bright soliton compatible to that in the Bose-Einstein condensate found by Perez-Garcia, *et al.*. If the signs of α and λ in (2.55) are reversed, we then get a kink soliton solution under the boundary conditions of $f(x' = 0) = 0$ and $f(x' \to \pm\infty) = \pm 1$,

$$\phi = \pm\sqrt{\frac{\alpha}{2\lambda}}\tanh\left[\sqrt{\frac{m\alpha}{\hbar^2}}(x - x_0)\right]. \tag{2.62}$$

The energy of the soliton (2.61) is given by

$$E_{\rm sol} = \int_{-\infty}^{\infty} \mathcal{H} dx = \int_{-\infty}^{\infty} \left[\frac{\hbar^2}{2m} \left(\frac{d\phi}{dx} \right)^2 - \alpha\phi^2 - \lambda\phi^4 \right] dx = \frac{4\hbar\alpha^{3/2}}{3\sqrt{2m}\lambda}. \tag{2.63}$$

We have assumed that the lattice constant $r_0 = 1$. The above soliton energy can be compared with the ground state energy of the superconducting state, $E_{\rm ground} = -\alpha^2/4\lambda$. The difference is

$$E_{\rm sol} - E_{\rm ground} = \frac{\alpha^{3/2}}{4\lambda} \left(\sqrt{\alpha} + \frac{16\hbar}{3\sqrt{2m}} \right) > 0. \tag{2.64}$$

This indicates clearly that the soliton is not at the ground state, but at an excited state of the system. Therefore, the soliton is a quasiparticle.

From the above discussion, we can see that in the absence of external fields, the superconductive electrons move in the form of solitons in a uniform system. These solitons are formed by nonlinear interaction among the supercondivite electrons which suppresses the dispersive behavior of the electrons. A soliton can carry a certain amount of energy while moving in superconductors. It can be demonstrated that these soliton states are very stable.

2.3.2 *Motion of electrons in the presence of an electromagnetic field*

We now consider the motion of superconductive electrons in the presence of an electromagnetic field. Assuming the London gauge $\nabla \cdot \vec{A} = 0$, substitution of (2.52) into (2.56) and (2.57) yields

$$J = \frac{e^*\phi_0^2}{m} \left(\hbar\nabla\theta - \frac{e^*}{c}\vec{A} \right) f^2 \tag{2.65}$$

and

$$\nabla^2 f - \left[\left(\nabla\theta - \frac{e^*}{\hbar c}\vec{A} \right)^2 f \right] - \frac{2m}{\hbar^2} \left(\alpha - 2\lambda\phi_0^2 f^2 \right) f = 0 \tag{2.66}$$

For bulk superconductors, J is a constant (permanent current) for a certain value of $\vec{A}$, and it thus can be taken as a parameter. Letting

$$B^2 = \frac{m^2 J^2}{\hbar^2 (e^*)^2 \phi_0^4}, \quad b = \frac{2m\alpha}{\hbar^2} = \frac{1}{\zeta^2},$$

equations (2.65) and (2.66) can be written as

$$\left(\hbar\nabla\theta - \frac{e^*}{c}\vec{A} \right) = \frac{Jm}{e^*\phi_0^2 f^2}, \tag{2.67}$$

$$\nabla^2 f = -bf^3 + bf + \frac{B^2}{f^3}. \tag{2.68}$$

Equation (2.68) evidently has nonlinear characteristics. When $J = 0$, equation (2.68) is reduced to

$$\nabla^2 f = -bf^3 + bf,$$

which is the same as (2.55).

In one-dimensional case, equation (2.68) can be written as

$$\frac{d^2 f}{dx^2} = -\frac{d}{df} U_{\text{eff}}(f), \tag{2.69}$$

where

$$U_{\text{eff}} = \frac{B^2}{2f^2} - \frac{1}{2}bf^2 + \frac{1}{4}bf^4$$

is the effective potential which is schematically shown in Fig. 2.4. Comparing this

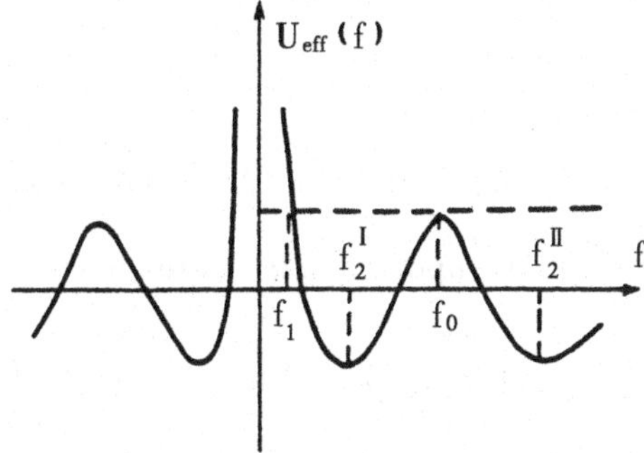

Fig. 2.4 The effective potential energy defined in Eq. (2.69).

case with that in the absence of external fields, we found that the equations have the same form and the electromagnetic field $\vec{A}$ changes only the effective potential of the superconductive electron. When $\vec{A} = 0$, the effective potential well is characterized by double wells. In the presence of an electromagnetic field, there are still two minima in the effective potential, corresponding to the two ground states of the superconductor. This shows that the spontaneous breakdown of symmetry still occurs in the superconductor, and the superconductive electrons move also in the form of solitons.

To obtain the soliton solution, we integrate (2.69) which gives

$$x = \int_{f_1}^{f} \frac{df}{\sqrt{2[E - U_{\text{eff}}(f)]}} \tag{2.70}$$

where E is a constant of integration which is equivalent to energy, the lower limit of the integral, f_1, is determined by the value of f at $x = 0$, *i.e.*

$$E = U_{\text{eff}}(f_0) = U_{\text{eff}}(f_1).$$

Introducing the following dimensionless quantities

$$f^2 = u, \quad E = \frac{b}{2}\varepsilon, \quad 2\tilde{d} = \frac{4J^2 m\lambda}{(e^*)^2\alpha^2},$$

equation (2.70) can be written as the following upon performing the transformation $u \to -u$,

$$-\sqrt{2b}x = \int_{u_1}^{u} \frac{du}{\sqrt{u^3 - 2u^2 - 3\varepsilon u - 2\tilde{d}^2}}. \tag{2.71}$$

It can be seen from Fig. 2.4 that the denominater of the integrand in (2.71) appoaches to zero linearly when $u = u_1 = f_1^2$, but appoaches to zero quadratically when $u = u_2 = f_0^2$. Therefore, it can be rewritten as

$$\sqrt{u^2 - 2u^2 - 2\varepsilon u - 2\tilde{d}^2} = \sqrt{(u - u_1)(u - u_0)^2}. \tag{2.72}$$

Thus, we have

$$2u_0 + u_1 = 2, \quad u_0^2 + 2u_0 u_1 = -2\varepsilon, \quad \text{and} \quad u_1 u_0^2 = 2\tilde{d}^2. \tag{2.73}$$

Substituting (2.72a) into (2.71), upon completing the integral in (2.71), we can get

$$u(x) = f^2(x) = u_0 - g\,\text{sech}^2\left(\sqrt{\frac{1}{2}gb}\; x\right) = u_1 + g\tanh^2\left(\sqrt{\frac{1}{2}gb}\; x\right) \tag{2.74}$$

where $g = u_0 - u_1$ and satisfies

$$(2+g)^2(1-g) = 27\tilde{d}^2. \tag{2.75}$$

It can be seen in (2.74) that for a large piece of sample, u_1 is very small and may be neglected, the solution u is very close to u_0. We then get from (2.74)

$$f(x) = f_0 \tanh\left(\sqrt{\frac{1}{2}gb}\; x\right). \tag{2.76}$$

Substituting the above into (2.67), the electromagnetic field $\vec{A}$ in the superconductors can be obtained

$$A = -\frac{Jmc}{(e^*)^2\phi_0^2}\cdot\frac{1}{f^2} - \frac{\hbar c}{e^*}\nabla\theta = \frac{Jmc}{(e^*)^2\phi_0^2 f_0^2}\coth^2\left(\sqrt{\frac{1}{2}gb}\; x\right) - \frac{\hbar c}{e^*}\nabla\theta.$$

For a large piece of superconductor, the phase change is very small. Using $\vec{\tilde{H}} = \nabla \times \vec{A}$, the magnetic field can be determined and is given by

$$\vec{\tilde{H}} = \nabla \times \vec{A} = \frac{Jmc\sqrt{2gb}}{(e^*)^2\phi_0^2 f_0^2}\left[\coth^3\left(\sqrt{\frac{1}{2}gb}\; x\right) + \coth\left(\sqrt{\frac{1}{2}gb}\; x\right)\right]. \tag{2.77}$$

Equations (2.76) and (2.77) are analytical solutions of the Ginzburg-Landau equations (2.57) and (2.58) in the one-dimensional case. The solutions are shown in Fig. 2.5. Equations (2.76), or (2.74) shows that the superconductive electron in the presence of an electromagnetic field is still a soliton. However, its amplitude, phase and shape are changed compared with those in a uniform superconductor and in the absence of external fields, (2.61). The soliton here is obviously influenced by the electromagnetic field, as reflected by the change in the solitary wave form.

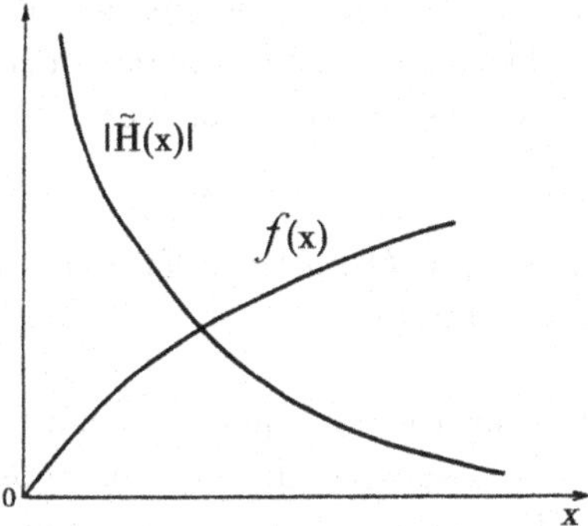

Fig. 2.5 The functions $f(x)$ and $\tilde{H}$ defined in (2.76) and (2.77), respectively.

This is why a permanent superconducting current can be established by the motion of electrons along certain direction in the a superconductor, because solitons have the ability to maintain their shape and velocity while in motion.

It is also clear from Fig. 2.5 that $\tilde{H}(x)$ is larger where $\phi(x)$ is small, and vice versa. When $x \to 0$, $\tilde{H}(x)$ reaches a maximum while ϕ approaches to zero. On the other hand, when $x \to \infty$, ϕ becomes very large while $\tilde{H}(x)$ approaches to zero. These are exactly the well-known behaviors of vortex lines – magnetic flux lines in type-II superconductors.

Recently, vertex solitons in the Bose-Einstein condensates were observed by Caradoc-Daries *et al.*, Matthews *et al.* and Madison *et al.*. Tonomure observed expermentally magnetic vorties in superconductors. These vortex lines in the type-II-superconductors are quantized. The macroscopic quantum effects are well described by the nonlinear theory discussed above, demonstrating the correctness of the theory.

We now proceed to determine the energy of the soliton given by (2.76) and

(2.68). From earlier discussion, the energy of the soliton is given by

$$\begin{aligned} E &= \int_{-\infty}^{\infty} \left[\frac{1}{2} \left(\frac{df}{dx} \right)^2 + \frac{b}{2} f^2 - \frac{1}{4} b f^4 - \frac{B^2}{2f^2} \right] dx \\ &\approx f_0^2 \left[\frac{2bf_0^2}{3} - 1 + \frac{b}{2} \left(1 - \frac{f_0^2}{2} \right) \right] - \frac{B^2}{2f_0^2}. \end{aligned} \tag{2.78}$$

which depends on the interaction between superconductive electrons and external electromagnetic field.

Motion of superconductive electrons in the three-dimensional case can be studied similarly.

From the above discussion, we understand that for bulk superconductor, the superconductive electrons behave as solitons, regardless of the presence of external fields. Therefore, the superconductive electrons are a special type of solitons. They move in the form of solitary wave in the superconducting state. In the presence of external electromagnetic fields, a permanent superconductive current is established, vortex lines or magnetic flux lines also occur in type-II superconductors.

2.4 Analysis of Macroscopic Quantum Effects in Inhomogeneous Superconductive Systems

In this section, we discuss macroscopic quantum effects in the superconductor-normal conductor junction and superconductive junctions, based on the nonlinear theory, the corresponding Hamiltonian, and the Ginzburg-Landau dynamic equations.

2.4.1 *Proximity effect*

Proximity effect refers to the phenomenon that a normal conductor (N) in contact with a superconductor (S) can be superconductive. This is obviously the result of long-range coherent property of superconductive electrons. It can be regarded as the penetration of electron pairs from the superconductor into the normal conductor, or a result of diffraction and transmission of superconductive electron wave.

In this phenomenon, superconductive electrons exist in the normal conductor, but their amplitudes are much small compared to that in the superconductive region, and the nonlinear term $\lambda|\phi|^2\phi$ in the Ginzburg-Landau equations, (2.56)-(2.57), can thus be neglected. Because of these, the Ginzburg-Landau equations in the normal and superconductive regions have different forms. On the S side of the S-N junction, the Ginzburg-Landau equation is

$$\frac{\hbar^2}{2m} \left(\nabla - \frac{ie^*}{ch} \vec{A} \right) \phi - \alpha\phi + 2\lambda\phi^3 = 0, \tag{2.79}$$

while that on the N side of the junction is

$$\frac{\hbar^2}{2m}\left(\nabla - \frac{ie^*}{ch}\vec{A}\right)\phi - \alpha\phi = 0. \tag{2.80}$$

The expression for $\vec{J}$ remains the same on both sides.

$$\vec{J} = \frac{e^*\hbar}{2mi}\left(\phi^*\nabla\phi - \phi\nabla\phi^*\right) - \frac{(e^*)^2}{mc}|\phi|^2\vec{A}. \tag{2.81}$$

In the S region, we have obtained the solutions of (2.79) in the previous section, and it is given by (2.74) or (2.76) and (2.77). In the N region, we assume that the solution is of the form of (2.52). Using the same analysis, equations (2.80) and (2.81) can be written as

$$\vec{J} = \frac{e^*\phi_0^2}{m}\left(\hbar\nabla\theta - \frac{e^*}{c}\vec{A}\right)f^2, \tag{2.82}$$

and

$$\nabla^2 f - b'f - \frac{B^2}{f^3} = 0, \tag{2.83}$$

where

$$b' = \frac{2m\alpha'}{\hbar^2} = \frac{1}{\xi'^2}, \quad B^2 = \frac{m^2\vec{J}^2}{\hbar^2(e^*)^2\phi_0^4}.$$

Here $\vec{J}$ is still treated as a parameter. If $\vec{J} = 0$ (or $B = 0$), equation (2.83) has a plane wave solution. However, $B \neq 0$ in general. In this case, the following solution can be obtained by integrating (2.83)

$$u = \frac{1}{2}\sqrt{(\varepsilon')^2 - 4\tilde{d}^2}\sin(2\sqrt{b'}x) + \frac{\varepsilon'}{2} \tag{2.84}$$

where

$$f^2 = u, \quad E' = \frac{b'}{2}\varepsilon', \quad 2\tilde{d}^2 = \frac{4\vec{J}^2 m\lambda}{(e^*)^2\alpha^2}.$$

The constant of integration has been set to zero. The solution is then given by

$$\begin{cases} f^2 = \frac{1}{2}\sqrt{(\varepsilon')^2 - 4\tilde{d}^2}\ \sin(2\sqrt{b'}x) + \frac{\varepsilon'}{2}, \\ \phi_N^2 = f^2\phi_0^2 e^{-2i\theta} = \frac{1}{2}\sqrt{\varepsilon'^2 - 4\tilde{d}^2\phi_0^2}\sin(2\sqrt{b'}x)e^{-i2\theta} + \frac{\varepsilon'}{2}\phi_0^2 e^{-i2\theta} \end{cases}. \tag{2.85}$$

Graph of ϕ vs. x in both the S and the N regions, as shown in Fig. 2.6, coincide with that obtained by Blackbunu. The solution given in (2.85) is the analytical solution in this case. On the other hand, Blackbunu's result was obtained by expressing the solution in terms of elliptic integrals and then integrating numerically. From this, we see that the proximity effect is caused by diffraction or transmission of the superconductive electrons.

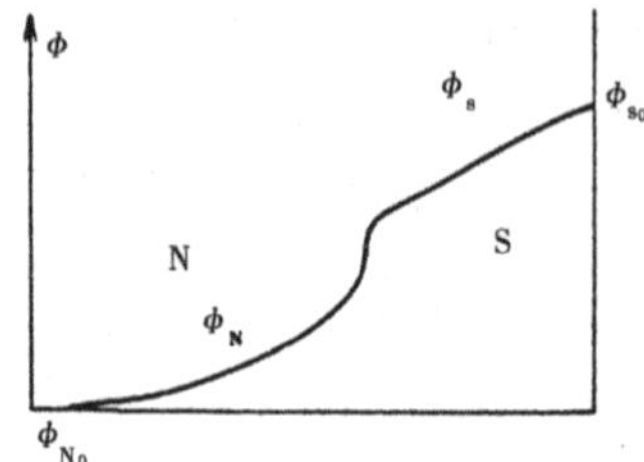

Fig. 2.6 Proximity effect in S-N junctions.

2.4.2 *Josephson current in S-I-S and S-N-S junctions*

A superconductor – normal conductor – superconductor junction (S-N-S) or a superconductor – insulator – superconductor junction (S-I-S) consists of a normal conductor or an insulator sandwhiched between two superconductors and is schematically shown in Fig. 2.7(a). The thickness of the normal conductor or the insulator layer is assumed to be L and we choose the x coordinate such that the the normal conductor or the insulator layer is located at $-L/2 \le x \le L/2$. The S-I-S junctions was studied by Jacobson *et al.* based on the Schrödinger equation. We will treat this problem using a different approach here.

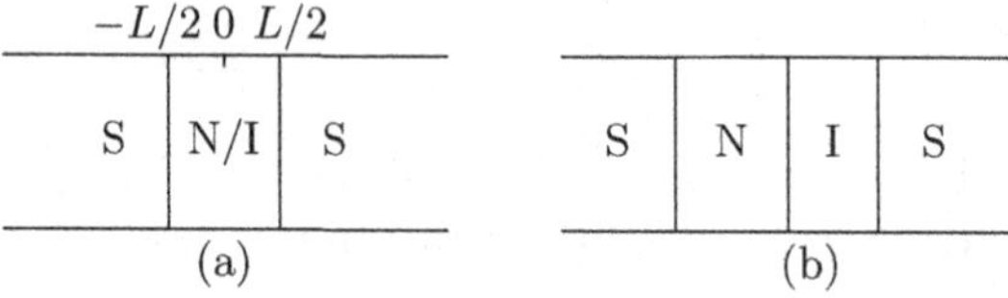

Fig. 2.7 Supereonductive Junctions for S-N-S and S-I-S.

The electrons in the superconducting regions ($|x| \ge L/2$) are described by the Ginzburg-Landau equation, (2.79). Its solution was given earlier in (2.76). After eliminating u_1 from (2.73), we have

$$J = \frac{1}{2} e^* \alpha u_0 \sqrt{\frac{\alpha}{m\lambda}} (1 - u_0).$$

Setting $dJ/du_0 = 0$, we get the maximum current

$$J_c = \frac{e^* \alpha}{3} \sqrt{\frac{\alpha}{3m\lambda}}.$$

This is the critical current of a perfect superconductor, corresponding to the three-fold degenerate solution of (2.72a), *i.e.*, $u_1 = u_0$.

From (2.82), we have:

$$\vec{A} = -\frac{m\vec{J}c}{(e^*)^2 \phi_0^2 f^2} + \frac{hc}{e^*} \nabla\theta.$$

Using the London guage, $\nabla \cdot \vec{A} = 0$, we can get

$$\frac{d^2\theta}{dx^2} = \frac{mJ}{e^*\phi_0^2\hbar}\frac{d}{dx}\left(\frac{1}{f^2}\right).$$

Integrating the above equation twice, we get

$$\triangle\theta = \frac{mJ}{e^*\phi_0^2\hbar}\int\left(\frac{1}{f^2} - \frac{1}{f_\infty^2}\right)dx, \tag{2.86}$$

where $f^2 = u$, and $f_\infty^2 = u_0$. We have used the following de Gennes boundary conditions in obtaining Eq. (2.86).

$$\left.\frac{d\phi}{dx}\right|_{|x|\to\infty} = 0, \quad \left.\frac{d\theta}{dx}\right|_{|x|\to\infty} = 0, \quad \phi(|x| \to \infty) = \phi_\infty. \tag{2.87}$$

If we substitute (2.71) – (2.74) into (2.86), the phase shift of the wave function from an arbitrary point x to infinity can be obtained directly from the above integral.

$$\triangle\theta_L(x \to \infty) = -\tan^{-1}\sqrt{\frac{u_1}{u_0 - u_1}} + \tan^{-1}\sqrt{\frac{u_1}{u - u_1}}. \tag{2.88}$$

For the S-N-S or S-I-S junction, the superconducting regions are located at $|x| \geq L/2$ and the phase shift in the S region is thus

$$\triangle\theta_s = 2\triangle\theta_L\left(\frac{L}{2} \to \infty\right) \approx 2\tan^{-1}\sqrt{\frac{u_1}{u_s - u_1}}, \tag{2.89}$$

where the factor of 2 is due to the contributions from superconductors at both sides of the N or I layer. Therefore,

$$\sqrt{u_s - u_1} - \frac{\sqrt{u_1}}{\tan(\triangle\theta_s/2)} = \sqrt{u_1}\ \mathrm{ctg}(\triangle\theta_s/2). \tag{2.90}$$

According to the results in (2.80) – (2.85), the solution in the I or N region of the S-N-S or S-I-S junction may be expressed as

$$u = g'\cos^2(\sqrt{b'}x) + h' \tag{2.91}$$

where

$$g' = u_b' - u_1', \quad h' = u_1', \quad u_1'u_0' = (g' + h')h' = \frac{2\dot{J}m\lambda}{(e^*)^2\alpha^2}. \tag{2.92}$$

Substituting (2.91) into (2.87), the phase shift over the I or N region of the junction can be obtained and is given by

$$\triangle\theta_N = -2\tan^{-1}\left[\frac{2e^*h'}{\dot{J}}\sqrt{\frac{\alpha^2}{8m\lambda}}\tan\left(\frac{\sqrt{b'}L}{2}\right)\right] + \frac{m\dot{J}L}{2e^*h'u_0} \tag{2.93}$$

where $m\dot{J}L/2e^*h'u_0$ is an additional term introduced to satisfy the boundary conditions (2.87), and may be neglected in the case being studied. Then

$$h' = \sqrt{\frac{8m\lambda}{\alpha^2}\frac{\dot{J}}{2e^*}\frac{\tan(\triangle\theta_N/2)}{\tan(\sqrt{b'}L/2)}}. \tag{2.94}$$

Near the critical temperature ($T < T_c$), the current passing through a weakly linked superconductive junction is very small ($\dot{J} \ll 1$), we then have

$$u_1' = \frac{4\dot{J}^2 m\lambda}{(e^*)^2\alpha^2} = 2\bar{A}^2, \quad \text{and} \quad g' = 1. \tag{2.95}$$

Since ηf^2 and df^2/dx are continuous at the boundary $x = L/2$, we have

$$\eta_s u_s|_{x=L/2} = \eta_N u_N|_{x=L/2}, \quad \left.\frac{du_s}{dx}\right|_{x=L/2} = \left.\frac{du_N}{dx}\right|_{x=L/2}. \tag{2.96}$$

These give

$$2\sqrt{b'}\bar{A}\sin(2\triangle\theta_N) = \varepsilon_1[1-\cos(2\triangle\theta_s)]\sin(\sqrt{b'}L), \tag{2.97}$$

$$\cos(\sqrt{b'}L)\sin(2\triangle\theta_s) = \varepsilon\sin(2\triangle\theta_N) + \sin(2\triangle\theta_s + \triangle\theta_N), \tag{2.98}$$

where $\varepsilon_1 = \eta_N/\eta_s$. From (2.97) and (2.98), we can get

$$\sin(\triangle\theta_s + \triangle\theta_N) = \frac{2\sqrt{2m\lambda}\dot{J}}{e^*\alpha}\sqrt{b'}\sin(\sqrt{b'}L).$$

Thus

$$\dot{J} = \dot{J}_{\max}\sin(\triangle\theta_s + \triangle\theta_N) = \dot{J}_{\max}\sin(\triangle\theta), \tag{2.99}$$

where

$$\dot{J}_{\max} = \frac{e^*\alpha_s}{2\sqrt{2m\lambda b'}}\cdot\frac{1}{\sin(\sqrt{b'}L)}, \quad \triangle\theta = \triangle\theta_s + \triangle\theta_N. \tag{2.100}$$

Equation (2.99) is the well-known Josephson current. From Section 2.1 we know that the Josephson effect is a macroscopic quantum effect. We have seen now that this effect can be explained based on the nonlinear theory.

From (2.100), we can see that the Josephson critical current is inversely proportional to $\sin(\sqrt{b'}L)$ which means that the current increases suddenly whenever $\sqrt{b'}L$ approaches to $n\pi$, suggesting some resonant phenomena in the system. This has not been observed before. Moreover, $\dot{J}_{\max}$ is proportional to $e^*\alpha_s/2\sqrt{2m\lambda b'} = (e^*\hbar/4m\sqrt{\lambda})(\alpha_s/\alpha_N)$ which is related to $(T - T_c)^2$. Finally, it is worthwhile to mention that no explicit assumption was made in the above on whether the junction is a potential well ($\alpha < 0$) or a potential barrier ($\alpha > 0$). The results are thus valid and the Josephson effect (2.99) occurs for both potential well and potential barrier.

2.4.3 *Josephson effect in SNIS junction*

A superconductor-normal conductor-insulator-superconductor junction (SNIS) is shown schematically in Fig. 2.7(b). It can be regarded as a multilayer junction consists of the S-N-S and S-I-S junctions. If appropriate thicknesses for the N and I layers are used (approximately 20 Å– 30 Å), the Josephson effect similar to that discussed above can occur in the SNIS junction.

Since the derivations are similar to that in the previous sections, we will skip much of the details and give the results in the following. The Josephson current in the SNIS junction is given by

$$\dot{J} = \dot{J}_{\max} \sin(\triangle\theta) \tag{2.101}$$

where

$$\triangle\theta = \triangle\theta_{s1} + \triangle\theta_N + \triangle\theta_I + \triangle\theta_{s2},$$

and

$$\begin{aligned}\dot{J}_{\max} = & \frac{1}{\sqrt{b'_N}}\left\{\frac{\varepsilon_1 \sinh(\sqrt{b'_N}L)}{2[\cosh(\sqrt{b'_N}L) - \cos(2\triangle\theta_N)]}\right\} \times \\ & \frac{1}{\sqrt{[1+\cos(2\triangle\theta_N)][1+\cos(2\triangle\theta_I)]} - \sqrt{[1-\cos(2\triangle\theta_N)][1-\cos(2\triangle\theta_I)]}} - \\ & \frac{1}{\sqrt{b'_N}}\left\{\frac{\varepsilon_1\sqrt{1-\cos^2(2\triangle\theta_N)}\ \sinh(\sqrt{b'_N}L)}{[\cosh(\sqrt{b'_N}L) - \cos(2\triangle\theta_N)]^2 - 1 + \cos^2(2\triangle\theta_N)}\right\} \times \\ & \frac{1}{\sqrt{[1-\cos(2\triangle\theta_N)][1-\cos(2\triangle\theta_I)]} + \sqrt{[1-\cos(2\triangle\theta_N)][1+\cos(2\triangle\theta_I)]}}.\end{aligned}$$

The phase shifts $\triangle\theta_{s1}, \triangle\theta_N, \triangle\theta_I$ and $\triangle\theta_{s2}$ satisfy the following relations

$$\begin{aligned}4\sqrt{b'_N}\bar{A}\sin(2\triangle\theta_N) &= \varepsilon_1\left[1 - \cos(2\triangle\theta_{s1})\sinh\left(\sqrt{b'_N}L\right)\right],\\ 4\sqrt{b'_I}\bar{A}\sin(2\triangle\theta_I) &= \varepsilon_1\left[1 - \cos(2\triangle\theta_{s2})\sin\left(\sqrt{b'_I}L\right)\right],\\ \tan(\triangle\theta_I) &= \varepsilon_2\frac{\tan(\sqrt{b_N}L/2)}{\tanh(\sqrt{b_N}L/2)}\tan(\triangle\theta_N)\sqrt{\frac{b'_N}{b'_I}},\\ \cosh\left(\sqrt{b'_N}L\right)\sin(2\triangle\theta_{s1}) &= \sin(2\triangle\theta_N) + \sin(2\triangle\theta_{s1} + 2\triangle\theta_N),\\ \cos\left(\sqrt{b'_I}L\right)\sin(2\triangle\theta_{s2}) &= \sin(2\triangle\theta_I) + \sin(2\triangle\theta_{s2} + 2\triangle\theta_I).\end{aligned}$$

It can be shown that the temperature dependence of $\dot{J}_{\max}$ is $\dot{J}_{\max} \propto (T_c - T_0)^2$, which is quite similar to the results obtained by Blackburm *et al.* for the SNIS junction and those by Romagnan *et al.* using the Pb-PbO-Sn-Pb junction. Here, we obtained the same results using a complete different approach.

2.5 Josephson Effect and Transmission of Vortex Lines Along the Superconductive Junctions

We have learned that in a homogeneous bulk superconductor, the phase $\theta(\vec{r},t)$ of the electron wave function $\phi = f(\vec{r},t)e^{i\theta(\vec{r},t)}$ is constant, independent of position and time. However, in an inhomogeneous superconductor such as a superconductive junction discussed above, θ becomes dependent of $\vec{r}$ and t. In the previous section, we discussed the Josephson effects in the S-N-S or S-I-S, and SNIS junctions starting from the Hamiltonican and the Ginzburg-Landau equations satisfied by $\phi(\vec{r},t)$, and showed that the Josephson current, whether dc or ac, is a function of the phase change, $\varphi = \triangle\theta = \theta_1 - \theta_2$. The dependence of the Josephson current on φ is clearly seen in (2.99) or (2.101). This clearly indicates that the Josephson current is caused by the phase change of the superconductive electrons. Josephson himself derived the equations satisfied by the phase difference φ, known as the Josephson relations, through his studies on both the dc and ac Josephson effects.

The Josephson relations for the Josephson effects in superconductor junctions can be summarized as the following,

$$\begin{cases} J_s = J_m \sin\varphi, & \hbar\dfrac{\partial\varphi}{\partial t} = 2eV, \\ \hbar\dfrac{\partial\varphi}{\partial x} = \dfrac{2ed'}{c}\tilde{H}_y, & \hbar\dfrac{\partial\varphi}{\partial y} = \dfrac{2ed'}{c}\tilde{H}_x, \end{cases} \tag{2.102}$$

where d' is the thickness of the junction. Because the voltage V and magnetic field $\vec{\tilde{H}}$ are not determined, equation (2.102) is not a set of complete equations. Generally, these equations are solved simultaneously with the Maxwell equation $\nabla \times \vec{\tilde{H}} = (4\pi/c)\vec{J}$. Assuming that the magnetic field is applied in the xy plane, *i.e.*, $\vec{\tilde{H}} = (\tilde{H}_x, \tilde{H}_y, 0)$, the above Maxwell equation becomes

$$\frac{\partial}{\partial x}\tilde{H}_y(x,y,t) - \frac{\partial}{\partial y}\tilde{H}_x(x,y,t) = \frac{4\pi}{c}J(x,y,t). \tag{2.103}$$

In this case, the total current in the junction is given by

$$J = J_s(x,y,t) + J_n(x,y,t) + J_i(x,y,t) + J_0.$$

In the above equation, J_n is the normal current density in the junction ($J_n = V/R(V)$ if the resistance in the junction is $R(V)$ and a voltage V is applied at two ends of the junction), J_d is called a displacement current and it is given by $J_d = CdV(t)/dt$, where C is the capacity of the junction, and J_0 is a constant current density. Solving the equations in (2.102) and (2.103) simultaneously, we can get

$$\nabla^2\varphi - \frac{1}{v_0^2}\left(\frac{\partial^2\varphi}{\partial t^2} + \gamma_0\frac{\partial\varphi}{\partial t}\right) = \frac{1}{\lambda_J^2}\sin\varphi + I_0 \tag{2.104}$$

where

$$v_0 = \sqrt{\frac{c^2}{4\pi C d'}}, \quad \gamma_0 = \frac{1}{RC}, \quad \lambda_J = \sqrt{\frac{c^2\hbar}{4\pi d e^*}}, \quad I_0 = \frac{4J_0\pi e^*}{\hbar c^2}.$$

Equation (2.104) is the equation satisfied by the phase difference. It is a Sine-Gordon equation (SGE) with a dissipative term. From (2.102), we see that the phase difference φ depends on the external magnetic field $\tilde{H}$, thus the magnetic flux in the junction

$$\Phi' = \int \tilde{H} ds = \oint \vec{A}\cdot d\vec{l} = \frac{c\hbar}{e^*}\oint \varphi dl$$

can be specified in terms of φ. Equation (2.104) represents transmission of superconductive vortex lines. It is a nonlinear equation. Therefore, the Josephson effect and the related transmission of the vortex line, or magnetic flux, along the junctions are nonlinear problems. The Sine-Gordon equation given above has been extensively studied by many scientists including Kivshar and Malomed. We will solve it here using different approaches.

Assuming that the resistance R in the junction is very high, so that $J_n \to 0$, or equivivalently $\gamma_0 \to 0$, setting also $I_0 = 0$, equation (2.104) reduces to

$$\nabla^2\varphi - \frac{1}{v_0^2}\frac{\partial^2\varphi}{\partial t^2} = \frac{1}{\lambda_J^2}\sin\varphi. \tag{2.105}$$

Define

$$X = \frac{x}{\lambda_J}, \quad T = \frac{v_0 t}{\lambda_J}.$$

Then in one-dimension, the above equation becomes

$$\frac{\partial^2\varphi}{\partial X^2} - \frac{\partial^2\varphi}{\partial T^2} = \sin\varphi$$

which is the 1D Sine-Gordon equation. If we further assume that

$$\varphi = \varphi(X,T) = \varphi(\theta'),$$

with

$$\theta' = X' - X_0' - vT', \quad X' = \frac{x}{\sqrt{\hbar c/2LI_0 e}}, \quad T' = T\sqrt{\frac{2eI_0}{\hbar c^2}},$$

it becomes

$$(1-v^2)\varphi_{\theta'}^2(\theta') = 2(A - \cos\varphi),$$

where A is a constant of integration. Thus

$$\int_{\varphi_0}^{\varphi(\theta')} \frac{d\varphi}{\sqrt{A-\cos\varphi}} = \sqrt{2}\delta\gamma\theta'$$

where

$$\gamma = \frac{1}{\sqrt{1-v^2}}, \qquad \delta = \pm 1.$$

Choosing $A = 1$, we have

$$\int_{\pi}^{\varphi(\theta')} \frac{d\varphi}{\sin(\varphi/2)} = 2\gamma\theta'.$$

A kink soliton solution can be obtained as follows

$$\pm\gamma\theta' = \ln\left[\tan\left(\frac{\varphi}{2}\right)\right],$$
$$\varphi(\theta') = 4\tan^{-1}\left(e^{\pm\gamma\theta'}\right),$$

or

$$\varphi(X', T') = 4\tan^{-1}\left[e^{\delta\gamma(X'-X_0'-vT')}\right]. \tag{2.106}$$

From the Josephson relations, the electric potential difference across the junction can be written as

$$V = \frac{h}{2e}\frac{d\varphi}{dT'} = \frac{\varphi_0}{2\pi c}\frac{d\varphi}{dT'} = 2\delta v\gamma\sqrt{\frac{2I_0 e}{\hbar c^2}}\,\frac{\varphi_0}{2\pi c}\,\mathrm{sech}\left[\gamma(X'-X_0'-vT')\right],$$

where $\varphi_0 = \pi\hbar c/e = 2\times 10^{-7}$ Gauss/cm^{-2} is a quantum fluxon, c is the speed of light. A similar expression can be derived for the magnetic field

$$\tilde{H}_x = \frac{h}{2e}\frac{d\varphi}{dX'} = \frac{\varphi_0}{2\pi c}\frac{d\varphi}{dX'} = \pm 2\delta\gamma\sqrt{\frac{2LI_0 e}{\hbar c^2}}\,\frac{\varphi_0}{2\pi c}\,\mathrm{sech}\left[\gamma(X'-X_0'-vT')\right].$$

We can then determine the magnetic flux through a junction with a length of L and a cross section of 1 cm^2. The result is

$$\Phi' = \int_{-\infty}^{\infty} \tilde{H}_x(x,t)dx = B_0'\int_{-\infty}^{\infty} \tilde{H}_x(X',T')dX' = \delta\varphi_0.$$

Therefore, the kink ($\delta = +1$) carries a single quantum of magnetic flux in the extended Josephson junction. Such an excitation is often called a fluxon, and the Sine-Gordon equation or (2.104) is often referred to as transmission equation of quantum flux or fluxon. The excitation corresponding to $\delta = -1$ is called an antifluxon. Fluxon is an extremely stable formation. However, it can be easily controlled with the help of external effects. It may be used as a basic unit of information.

This result shows clearly that magnetic flux in superconductors is quantized and this is a macroscopic quantum effect as mentioned in Section 2.1. The transmission of the quantum magnetic flux through the superconductive junctions is described by the above nonlinear dynamic equation (2.104) or (2.105).

The energy of the soliton can be determined and it is given by

$$E = \frac{8m^2}{\beta},$$

where

$$\frac{m^4}{\beta} = \frac{1}{\lambda_J^2}.$$

The boundary conditions must be considered for real superconductors. Various boundary conditions have been considered and studied. For example, we can assume the following boundary conditions for a 1D superconductor

$$\varphi_x(0,t) = \varphi_x(L,t) = 0.$$

Lamb obtained the following soliton solution for the SG equation (2.105)

$$\varphi(x,t) = 4\tan^{-1}[h(x)g(x)] \tag{2.107}$$

where h and g are the general Jacobian elliptical functions and satisfy the following equations

$$[h_x(x)]^2 = a'h^4 + (1+b'')h^2 - c',$$

$$[g_x(x)]^2 = c'g^4 + b''g^2 - a',$$

with a', b'', and c' being arbitrary constants. Coustabile *et al.* also gave the plasma oscillation, breathing oscillation and vortex line oscillation solutions for the SG eqution under certain boundary conditions. All of these can be regarded as the soliton solution under the given conditions.

Solutions of (2.105) in two and three-dimensional cases can also be found. In the two-dimensional case, the solution is given by

$$\varphi = \pm 4\tan^{-1}\left[\frac{g(X,Y,T)}{f(X,Y,T)}\right], \tag{2.108}$$

where

$$X = \frac{x}{\lambda_J}, \quad Y = \frac{y}{\lambda_J}, \quad T = \frac{v_0 t}{\lambda_J},$$

$$f = 1 + a(1,2)e^{y_1+y_2} + a(2,3)e^{y_2+y_3} + a(3,1)e^{y_3+y_1},$$

$$g = e^{y_1} + e^{y_2} + a(1,2)a(2,3)a(3,1)e^{y_1+y_2+y_3},$$

and

$$y_i = p_i X + q_i Y - \Omega_i \tau - y_i^0, \quad p_i^2 + q_i^2 - \Omega_i^2 = 1, \quad (i = 1,2,3),$$

$$a(i,j) = \frac{(p_i - p_j)^2 + (q_i - q_j)^2 - (\Omega_i - \Omega_j)^2}{(p_i + p_j)^2 + (q_i + q_j)^2 - (\Omega_i + \Omega_j)^2}, \quad (1 \le i \le j \le 3).$$

In addition, p_i, q_i and Ω_i satisfy

$$\det \begin{vmatrix} p_1 & q_1 & \Omega_1 \\ p_2 & q_2 & \Omega_2 \\ p_3 & q_3 & \Omega_3 \end{vmatrix} = 0.$$

In the three-dimensional case, the solution is given by

$$\varphi(X,Y,Z,T) = 4\tan^{-1}\left[\frac{g(X,Y,Z,T)}{f(X,Y,Z,T)}\right], \tag{2.109}$$

where X, Y, and T are similarly defined as in the 2D case given above, and $Z = z/\lambda_J$. The functions f and g are defined as

$$g = e^{y_1} + e^{y_2} + e^{y_3} + dX_2 dY_3 dZ_3 e^{y_1+y_2+y_3}$$

$$f = dX_2 e^{y_1+y_2} + dY_3 e^{y_2+y_3} + dZ_3 e^{y_1+y_3} + 1$$

with

$$y_i = a_{i1}X + a_{i2}Y + a_{i3}Z + b_i T + C_i, \quad (i = X, Y, Z),$$

$$a_{il}^2 + a_{i2}^2 + a_{i3}^3 - b_i^2 = 1,$$

$$d_{ij} = \frac{\sum_{k=1}^{3}(a_{ik} - a_{jk})^2 - (b_i - b_j)^2}{\sum_{k=1}^{3}(a_{ik} + a_{jk})^2 - (b_i + b_j)^2}, \quad (1 \le j \le 3).$$

Here y_3 is a linear combination of y_1 and y_2, *i.e.*

$$y_3 = \alpha y_1 + \beta y_2.$$

We now discuss the SG equation with a dissipative term $\gamma_0 \partial\varphi/\partial t$. First we make the following substitutions to simplify the equation

$$X = \frac{x}{\lambda_J}, \quad T = \frac{v_0 t}{\lambda_J} = \frac{t}{\omega_J}, \quad a = \frac{\gamma_0 \lambda_J^2}{v_0}, \quad B' = I_0 \lambda_J^2.$$

In terms of these new parameters, the 1D SG equation (2.104) can be rewritten as

$$\frac{\partial^2\varphi}{\partial X^2} - \frac{\partial^2\varphi}{\partial T^2} - a\frac{\partial\varphi}{\partial T} = \sin\varphi + B'. \tag{2.110}$$

The analytical solution of (2.110) is not easily found. Now let

$$\alpha = \frac{1 - v_0^2}{a^2 v_0^2}, \quad \eta = \frac{1}{\sqrt{\alpha}}\,\frac{X - v_0 T}{a v_0}, \quad q' = \frac{a v_0}{\sqrt{1 - v_0^2}}, \quad \varphi = \pi + \varphi'. \tag{2.111}$$

Equation (2.110) then becomes

$$\frac{d^2\varphi'}{d\eta^2} + q'\frac{d\varphi'}{d\eta} + \sin\varphi' - B' = 0. \tag{2.112}$$

This equation is the same as that of a pendulum being driven by a constant external moment and a frictional force which is proportional to the angular displacement. The solution of the latter is well known, generally there exists an stable soliton solution. Let $Y = d\varphi'/d\eta$, equation (2.112) can be written as

$$\frac{dY}{d\eta} + q'Y + \sin\varphi' - B' = 0. \tag{2.113}$$

For $0 < B' < 1$, we can let $B' = \sin\varphi_0$, $(0 < \varphi_0 < \pi/2)$ and $\varphi' = -\pi - \varphi_0 + \varphi_1$, then, equation (2.113) becomes

$$Y\frac{dY}{d\varphi_1} = -q'Y + \sin\varphi_0 + \sin(\varphi_1 - \varphi_0). \tag{2.114}$$

Expand Y as a power series of φ_1,

$$Y = \sum_n c_n \varphi_1^n.$$

Inserting the above into (2.114), and comparing coefficients of terms of the same power of φ_1 on both sides, we get

$$\begin{aligned} c_1 &= -\frac{q'}{2} \pm \sqrt{\frac{q'^2}{4} + \cos\varphi_0}, \\ c_2 &= \frac{1}{q' + 3c_1}\frac{\sin\varphi_0}{2}, \\ c_3 &= \frac{1}{q' + 4c_1}\left(-2c_2^2 - \frac{\cos\varphi_0}{6}\right), \\ c_4 &= \frac{1}{q' + 5c_1}\left(-5c_2c_3 - \frac{\sin\varphi_0}{24}\right), \end{aligned} \tag{2.115}$$

and so on. Substituting these c_n's into $Y = d\varphi'/d\eta = \sum_n c_n\varphi_1^n$, the solution of φ_1 may be found by integrating $\eta = \int d\varphi_1 / \sum c_n \varphi_1^n$. In general, this equation has soliton solution or elliptical wave solution. For example, when

$$\frac{d\varphi_1}{d\eta} = c_1\varphi_1 + c_2\varphi_1^2 + c_3\varphi_1^3$$

it can be found that

$$\eta = \frac{2}{\sqrt{A-C}} F\left(\sqrt{\frac{A-B}{A-C}},\ \sin^{-1}\sqrt{\frac{A-\varphi_1}{A-B}}\right),$$

where $F(k, \varphi_1)$ is the first Legendre elliptical integral, and A, B and C are constants. The inverse function φ_1 of $F(k, \varphi_1)$ is the Jacobian amptitude $\varphi_1 = \mathrm{am}F$. Thus,

$$\sin^{-1}\sqrt{\frac{A-\varphi_1}{A-B}} = \mathrm{am}\sqrt{\frac{A-C}{A-B}}\eta,$$

or

$$\sqrt{\frac{A-\varphi_1}{A-B}} = \mathrm{sn}\sqrt{\frac{A-C}{A-B}}\eta$$

where $\mathrm{sn}F$ is the Jacobian sine function. Introducing the symbol $\mathrm{csc}F = 1/\mathrm{sn}F$, the solution can be written as

$$\varphi_1 = A - (A-B)\left(\mathrm{csc}\sqrt{\frac{A-C}{A-B}}\eta\right)^2. \tag{2.116}$$

This is a elliptic function. It can be shown that the corresponding solution at $|\eta| \to \infty$ is a solitary wave.

It can be seen from the above discussion that the quantum magnetic flux lines (vortex lines) move along a superconductive junction in the form of solitons. The transmission velocity v_0 can be obtained from $h = \alpha v_0/\sqrt{1-v_0^2}$ and c_n in (2.115) and it is given by

$$v_0 = \frac{1}{\sqrt{1+[\alpha/h(\varphi_0)]^2}}.$$

That is, the transmission velocity of the vortex lines depdends on the current I_0 injected and the characteristic decaying constant α of the Josephson junction. When α is finite, the greater the injection current I_0 is, the faster the transmission velocity will be; and when I_0 is finite, the greater the α is, the smaller the v_0 will be, which are realistic.

2.6 Motion of Electrons in Non-Equilibrium Superconductive Systems

The motions of superconductive electrons in equilibrium states, such as those discussed in Sections 2.3 – 2.5, are described by the time-independent Ginzburg-Landau equation. In such a case, the superconductive electrons move as solitons. However, what are the motions of superconductive electrons in non-equilibrium states? Naturally, the superconductive electrons in non-equilibrium states should be described by the time-dependent Ginzburg-Landau equation. Unfortunately, there are many different forms of the time-dependent Ginzburg-Landau equation, under different conditions. The one given in the following is commonly used when an electromagnetic field $\vec{A}$ is involved,

$$\gamma'\left[\hbar\frac{\partial}{\partial t}-2ie\mu(r)\right]\phi=-\frac{1}{2m}\left(\hbar\nabla-\frac{2ie}{c}\vec{A}\right)^2\phi+\alpha\phi-\lambda|\phi|^2\phi, \tag{2.117}$$

$$\vec{j}=\sigma\left[-\frac{1}{c}\frac{\partial\vec{A}}{\partial t}-\nabla\mu(r)\right]+\frac{ie\hbar}{m}(\phi^*\nabla\phi-\phi\nabla\phi^*)-\frac{4e^2}{mc}\vec{A}|\phi|^2, \tag{2.118}$$

where

$$\nabla\times\nabla\times\vec{A}=\frac{1}{c}\frac{\partial}{\partial t}\left(-\frac{1}{c}\frac{\partial\vec{A}}{\partial t}-\nabla\mu\right)+\frac{4\pi\vec{J}}{c},$$

and σ is the conductivity in the normal state, γ' is an arbitrary constant, and μ is the chemical potential.

In certain situations, the following forms of the time-dependent Ginzburg-Landau equation are also used

$$i\hbar\frac{\partial\phi}{\partial t}=-\frac{\hbar^2}{2m}\left(\nabla-\frac{2ie}{\hbar c}\vec{A}\right)^2\phi+\alpha\phi-\lambda|\phi|^2\phi, \tag{2.119}$$

or

$$i\left(\frac{\partial}{\partial t}-\frac{i2e\mu}{\hbar}\right)\phi=\frac{1}{\Gamma}\left(\alpha-\lambda|\phi|^2\right)\phi+\frac{\xi^2}{\Gamma}\left(\nabla-\frac{2ie}{\hbar c}\vec{A}\right)^2\phi. \tag{2.120}$$

Equations (2.117)-(2.120) are nonlinear partial differential equations having soliton solutions. However, these solutions are very difficult to find, and no analytic solutions have been obtained before. An approximate solution was obtained by Kusayanage *et al.* by neglecting the ϕ^3 term in (2.117) and (2.118), in the case of $\vec{A}=(0,\tilde{H}_x,0)$, $\mu=-K\tilde{E}x$, $\vec{\tilde{H}}=(0,0,\tilde{H})$ and $\vec{\tilde{E}}=(\tilde{E},0,0)$, where $\vec{\tilde{H}}$ is the magnetic field, while $\vec{\tilde{E}}$ is the electric field. We will solve the time-dependent Ginzburg-Landau equation in the case of weak fields in the following.

The time-dependent Ginzburg-Landau equations can be written in the following form when $\vec{A}$ is very small,

$$i\hbar\frac{\partial\phi}{\partial t}+\frac{\hbar^2}{2m\Gamma}\nabla^2\phi+\frac{\lambda}{\Gamma}|\phi|^2\phi=\left(\frac{\alpha}{\Gamma}-2e\mu\right)\phi. \tag{2.121}$$

where α and Γ are material dependent parameters, λ is the nonlinear coefficient, m is the mass of the superconductive electron. Equation (2.121) is a nonlinear Schrödinger equation in a potential field $\alpha/\Gamma-2e\mu$. It was used by Cai, Bhattacharjee *et al.*, and Davydov in their studies of superconductivity. However, this equation is also difficult to solve. Here we only present its solution in the one-dimensional case.

For convenience, let

$$t' = \frac{t}{\hbar}, \quad x' = \sqrt{2m\Gamma}\,\frac{x}{\hbar}.$$

Then (2.121) becomes

$$i\frac{\partial\phi}{\partial t'} + \frac{\partial^2\phi}{\partial x'^2} + \frac{\lambda}{\Gamma}|\phi|^2\phi = \left[\frac{\alpha}{\Gamma} - 2e\mu(x')\right]\phi. \tag{2.122}$$

If we let $\alpha/\Gamma - 2e\mu = 0$, then (2.122) is the usual nonlinear Schrödinger equation whose solution is of the form

$$\phi_s^0 = f_0(x', t')e^{i\theta_0(x',t')}, \tag{2.123}$$

$$f_0(x', t') = \sqrt{\frac{\Gamma(v_e^2 - 2v_c v_e)}{2\lambda}}\,\mathrm{sech}\left[\sqrt{\frac{\lambda(v_e^2 - 2v_c v_e)}{4\Gamma}}\,(x' - v_e t')\right],$$

where

$$\theta_0(x', t') = \frac{1}{2}v_e(x' - v_c t').$$

In the case of $\alpha/\Gamma - 2e\mu \neq 0$, we let $\mu = -K\tilde{E}x'$, where K is a constant, and assume that the solution is of the form

$$\phi = f'(x', t')e^{i\theta(x',t')}. \tag{2.124}$$

Substituting (2.124) into (2.122), we get

$$-f'\frac{\partial\theta}{\partial t'} - f'\left(\frac{\partial\theta}{\partial x'}\right)^2 + \frac{\partial^2 f'}{\partial(x')^2} + \frac{\lambda}{\Gamma}(f')^3 = \left(2Ke\tilde{E}x' + \frac{\alpha}{\Gamma}\right)f', \tag{2.125}$$

$$\frac{\partial f'}{\partial t'} + 2\frac{\partial f'}{\partial x'}\frac{\partial\theta}{\partial x'} + f'\frac{\partial^2\theta}{\partial(x')^2} = 0. \tag{2.126}$$

Now let

$$f'(x', t') = f(\xi), \quad \xi = x' - u(t'), \quad u(t') = -2\tilde{E}Ke(t')^2 + vt' + d \tag{2.127}$$

where $u(t')$ describes the accelerated motion of $f'(x', t')$. The boundary condition at $\xi \to \infty$ requires $f(\xi)$ to approach zero rapidly. Equation (2.126) can be written as

$$-\dot{u}\frac{\partial f}{\partial\xi} + 2\frac{\partial f}{\partial\xi}\frac{\partial\theta}{\partial\xi} + f\frac{\partial^2\theta}{\partial\xi^2} = 0, \tag{2.128}$$

where

$$\dot{u} = \frac{du}{dt'}.$$

If $2\partial\theta/\partial\xi - \dot{u} \neq 0$, equation (2.128) may be written as

$$f^2 = \frac{g(t')}{(\partial\theta/\partial\xi) - \dot{u}/2} \quad \text{or} \quad \frac{\partial\theta}{\partial x'} = \frac{g(t')}{f^2} + \frac{\dot{u}}{2}. \tag{2.129}$$

Integration of (2.129) yields

$$\theta(x', t') = g(t') \int_0^{x'} \frac{dx'}{f^2} + \frac{\dot{u}}{2}x' + h(t') \tag{2.130}$$

where $h(t')$ is an undetermined constant of integration. From (2.130) we can get

$$\frac{\partial\theta}{\partial t'} = \dot{g}(t') \int_0^{x'} \frac{dx'}{f^2} - \frac{g\dot{u}}{f^2} + \left.\frac{g\dot{u}}{f^2}\right|_{x'=0} + \frac{\ddot{u}}{2}x' + \dot{h}(t'). \tag{2.131}$$

Substituting (2.130) and (2.131) into (2.125), we have

$$\frac{\partial^2 f}{\partial(x')^2} = \left[\left(2K\tilde{E}ex' + \frac{\alpha}{\Gamma}\right) + \frac{\ddot{u}}{2}x' + \dot{h}(t') + \frac{\dot{u}^2}{4} + \right.$$
$$\left. + \dot{g}\int_0^{x'} \frac{dx'}{f^2} + \left.\frac{g\dot{u}}{f^2}\right|_{x'=0}\right] f - \frac{\lambda}{\Gamma}f^3 + \frac{g^2}{f^3}. \tag{2.132}$$

Since $\partial^2 f/\partial x'^2 = d^2 f/d\xi^2$, it is a function of ξ only. In order for the right-hand side of (2.132) to be also a functin of ξ only, it is necessary that

$$g(t') = g_0 = \text{const.}$$

$$(2K\tilde{E}ex' + \frac{\alpha}{\Gamma}) + \frac{\ddot{u}}{2}x' + \dot{h}(t') + \frac{\dot{u}^2}{4} + \frac{g_0\dot{u}}{f^2}|_{x'=0} = \overline{V}(\xi). \tag{2.133}$$

Next, we assume that $V_0(\xi) = \overline{V}(\xi) - \beta$ where β is real and arbitrary. Then

$$2K\tilde{E}ex' + \frac{\alpha}{\Gamma} = V_0(\xi) - \frac{\ddot{u}}{2}x' + \left[\beta - \left.\frac{g_0\dot{u}}{f^2}\right|_{x'=0} - \dot{h}(t') - \frac{\dot{u}^2}{4}\right]. \tag{2.134}$$

Clearly in the case being discussed, $V_0(\xi) = 0$, and the function in the brackets in (2.134) is a function of t'. Substituting (2.133) and (2.134) into (2.132), we can get

$$\frac{\partial^2 \tilde{f}}{\partial\xi^2} = \beta\tilde{f} - \frac{\lambda}{\Gamma}\tilde{f}^3 + \frac{g_0^2}{\tilde{f}^3}. \tag{2.135}$$

This shows that $\tilde{f} = f(\xi)$ is the solution of (2.135) when β and g are constants. For large $|\xi|$, we may assume that $|\tilde{f}| \leq \beta'/|\xi|^{1+\triangle}$, where $\triangle$ is a small constant. To ensure that $\tilde{f}$ and $d^2\tilde{f}/d\xi^2$ approach zero when $|\xi| \to \infty$, only the solution corresponding to $g_0 = 0$ in (2.135) can be kept, and it can be shown that this soliton solution is stable in such a case. Therefore, we choose $g_0 = 0$ and obtain the following from (2.129)

$$\frac{\partial\theta}{\partial x'} = \frac{\dot{u}}{2}. \tag{2.136}$$

Thus, we obtain from (2.134)

$$2K\tilde{E}ex' + \frac{\alpha}{\Gamma} = -\frac{1}{2}\ddot{u}x' + \beta - \dot{h}(t') - \frac{1}{4}\dot{u}^2, \tag{2.137}$$

$$h(t') = \left(\beta - \frac{\alpha}{\Gamma} - \frac{1}{4}v^2\right)t' - \frac{4}{3}(\tilde{E}Ke)^2(t')^3 + evK\tilde{E}(t')^2. \tag{2.138}$$

Substituting (2.138) into (2.130) and (2.131), we obtain

$$\theta = \left(-2eK\tilde{E}t' + \frac{1}{2}v\right)x' + \left(\beta - \frac{\alpha}{\Gamma} - \frac{1}{4}v^2\right)t' - \frac{4}{3}(\tilde{E}Ke)^2(t')^3 + evK\tilde{E}(t')^2. \tag{2.139}$$

Finally, substituting the above into (2.135), we can get

$$\frac{\partial^2 \tilde{f}}{d\xi^2} - \beta\tilde{f} + \frac{\lambda}{\Gamma}\tilde{f}^3 = 0. \tag{2.140}$$

When $\beta > 0$, the solution of (2.140) is of the form

$$\tilde{f} = \sqrt{\frac{2\beta\Gamma}{\lambda}}\,\mathrm{sech}(\sqrt{\beta}\xi). \tag{2.141}$$

Thus

$$\begin{aligned}\phi = \sqrt{\frac{2\beta\Gamma}{\lambda}}\,\mathrm{sech}\left[\sqrt{\beta}\left(\sqrt{\frac{2m\Gamma}{\hbar^2}}x + \frac{2eKEt^2 - vt - d}{\hbar}\right)\right] \\ \times \exp\left\{i\left[\left(\frac{-2eK\tilde{E}t}{\hbar} + \frac{v}{2}\right)\sqrt{\frac{2m\Gamma}{\hbar^2}}\,x\right.\right. \\ \left.\left. + \left(\beta - \frac{\alpha}{\Gamma} - \frac{v^2}{4}\right)\frac{t}{\hbar} - \frac{4(eK\tilde{E})^2t^3}{3\hbar^3} + \frac{vKe\tilde{E}t^2}{\hbar}\right]\right\}.\end{aligned} \tag{2.142}$$

This is also a soliton solution, but its shape, amplitude, and velocity have changed relatively to those of (2.123). It can be shown that (2.142) indeed satisfies (2.122). Thus, (2.122) has a soliton solution. It can also be shown that this solition solution is stable.

For the solution (2.142), we may define a generalized time-dependent wave number,

$$k = \frac{\partial\theta}{\partial x'} = \frac{v}{2} - 2eK\tilde{E}t', \tag{2.143}$$

and a frequency

$$\begin{aligned}\omega = -\frac{\partial\theta}{\partial t'} &= 2eK\tilde{E}x' - \left(\beta - \frac{\alpha}{\Gamma} - \frac{1}{4}v^2\right) + 4(\tilde{E}Ke)^2(t')^2 - 2eK\tilde{E}vt' \\ &= 2eK\tilde{E}x' - \beta - \frac{\alpha}{\Gamma} + k^2.\end{aligned} \tag{2.144}$$

The usual Hamilton equations for the supercondactive electron (soliton) in the macroscopic quantum systems (MQS) is still valid here and can be written as

$$\frac{dk}{dt'} = -\left.\frac{\partial \omega}{\partial x'}\right|_k = -2eK\tilde{E}, \tag{2.145}$$

$$v_g = \frac{dx'}{dt'} = \left.\frac{\partial \omega}{\partial k}\right|_{x'} = 2\left(\frac{v}{2} - 2eK\tilde{E}t'\right) = v - 4eK\tilde{E}t'. \tag{2.146}$$

This means that the frequency ω still has the meaning of the Hamiltonian in the case of nonlinear waves. Therefore,

$$\frac{d\omega}{dt'} = \left.\frac{d\omega}{\partial k}\right|_{x'}\frac{dk}{dt'} + \left.\frac{\partial \omega}{\partial x'}\right|_k \frac{dx'}{dt'} = 0 \tag{2.147}$$

which is the same as that in the usual stationary linear medium.

These relations show that the superconductive electrons move as if they were classical particles moving with a constant acceleration in the invariant electric field and the acceleration is given by $-4e\tilde{E}K$. If $v > 0$ the soliton initially travels toward the overdensed region, it then suffers a deceleration and its velocity changes sign. The soliton is then reflected and accelerated toward the underdensed region. The penetration distance into the overdense region depends on the initial velocity v.

From the above discussion, we see that the superconductive electrons behave like solitons in non-equilibrium state of superconductor. Therefore, we can conclude that the superconductive electron is enssentially a soliton in both equibilibrium and nonequilibrium systems.

Before ending this section, it should be pointed out that there are many forms of time-dependcnt Ginzburg-Landau equations which have been used and can be found in literatures. For example,

$$i\phi_t + \frac{1}{2}a\phi_{tt} + |\phi|^2\phi + \gamma|\phi|^4\phi = i\delta\phi + i\varepsilon|\phi|^2\phi + i\beta\phi_{xx} + i\mu|\phi|^4\phi,$$

$$\frac{\partial \phi}{\partial t} = \nabla^2\phi + \phi - |\phi|^2\phi,$$

$$\frac{\partial \phi}{\partial t} = (\alpha + \beta|\phi|^2)\phi + r\frac{\partial^2 \phi}{\partial x^2},$$

$$\frac{\partial \phi}{\partial t} = \left[1 - (1 + i\beta)|\phi|^2\right]\phi + (1 + i\vec{A})\frac{\partial^2 \phi}{\partial x^2},$$

to name only a few. These equations are essentially generalization of (2.121) in the case of dissipation. Various studies showed that these equations have soliton solutions. Thus superconductive electron also behaves like a soliton in media with friction which damps the soliton motion. Even though the form, amplitude, and velocity of the soliton are altered, superconductive electrons being solitons ramains a fact.

2.7 Motion of Helium Atoms in Quantum Superfluid

As mentioned in Section 2.1, liquid helium exhibits macroscopic quantum effect at temperature below 2.17 K. The helium atoms form a quantum liquid without viscosity, or superfluid. Studies show that this is a result of Bose-Einstein condensation of He atoms below the critical temperature. It is very similar to the superconducting state, and thus also described by a macroscopic wave function ϕ similar to (2.52). Here ϕ is also called an order parameter of the superfluid liquid helium or an effective wave function of helium atoms.

It is known that the effective wave function of helium atoms, ϕ, satisfies the Gross-Pitaerskii (GP) equation which was derived by Gross and Pitaerskii in 1950,

$$i\hbar\frac{\partial\phi}{\partial t} = -\frac{\hbar^2}{2m}\nabla^2\phi + \lambda|\phi|^2\phi - \mu'\phi \tag{2.148}$$

where λ and μ' are constants. This equation is similar to the Ginzburg-Landau equation (2.122) for superconducting electrons. This is understandable because the superfluid, similar to a superconducting system, is also a nonlinear system. Equation (2.148) was extensively used by Ventura *et al.* in their studies of superfluidity. A similar equation was derived by Dewitt in 1966. According to Dewitt, λ in (2.148) should be a negative value.

The corresponding Lagrangian function density of the system can be obtained and is given by

$$\mathcal{L} = \frac{i}{2}\left(\phi\frac{\partial\phi^*}{\partial t} - \phi^*\frac{\partial\phi}{\partial t}\right) - |\nabla\phi|^2 - \frac{\lambda}{2}|\phi|^4 + \mu'|\phi|^2. \tag{2.149}$$

In one dimension, if we let $t' = t/\hbar$ and $x' = \sqrt{(2m/\hbar^2)}\,x$, then (2.148) is similar to (2.122). Thus its solution can be obtained following the same procedure as that used in Section 2.6 and the result is

$$\phi = \sqrt{\frac{2\beta}{|\lambda|}}\,\mathrm{sech}\left\{\sqrt{\beta}\left[x' - v_e(t' - t_0')\right]\right\}\exp\left\{i\left[v_e x' - \left(\beta + v_e^2 - \mu'\right)t'\right]\right\} \tag{2.150}$$

where β is an arbitary constant.

Therefore, when the atomic helium system undergoes a second-order phase transition and changes from the normal He-I state to the superfluid He-II state at 2.17 K, the helium atoms behave as a soliton due to the spontaneous Bose condensation as a result of nonlinear interaction in the system. The nonlinear interaction suppresses the dispersion effect of the helium atoms. Because solitons can preserve their energy, momentum, wave form and other properties of quasi-particles throughout their motion, the superfluidity occurs naturally when the liquid helium moves as solitons.

As a matter of fact, we can prove that the superfluid helium atom (soliton) moves with an uniform speed and can determine its speed from the solution (2.150) and the Hamilton equations (2.143) – (2.147). First, from (2.150) we can find that the wave number of the soliton is $k = \partial\theta/\partial x = v_e$ and its frequency is $\omega = -\partial\theta/\partial t = \beta + v_e^2 - \mu' = \beta - \mu' + k^2$. Then the acceleration of the helium-soliton is

$$\frac{dk}{dt} = -\left.\frac{\partial\omega}{\partial x}\right|_k = 0.$$

That is, the speed of the helium-soliton is a constant, v_e, and the helium atoms move in the form of soliton with constant speed in the superfluid state. This is a basic property of superfluidity and the discussion above gives the phenomenon a clear physical interpretation.

The mass of the soliton can be determined from (2.150),

$$M = \int_{-\infty}^{\infty} |\phi|^2 dx = \frac{4}{\lambda h}\sqrt{m\beta} = \text{const.}$$

The energy of the soliton is

$$\begin{aligned} E &= \int_{-\infty}^{\infty} \left[\left|\frac{\partial\phi}{\partial x}\right|^2 + \frac{1}{2}\lambda|\phi|^4 - \mu'|\phi|^2 \right] dx \\ &= \left(\frac{1}{2}v_e^2 - \mu'\right) M + \frac{4\sqrt{2m\beta}}{3\lambda h}(4+\beta) \end{aligned} \tag{2.151}$$

Here the first term is the kinetic energy of the soliton, the second and the third terms are binding energies and the fourth term is the energy of interaction.

We now discuss properties of circulation (vortex line) produced by the superfluid helium atoms. The circulation is defined by the velocity of superfluid helium atom v_s,

$$Q = \oint_r v_s dr.$$

In terms of the phase, $\theta(x,t)$, of the macroscopic wave function, the velocity of the superfluid can be written as

$$v_s = \frac{\hbar}{\sqrt{2m}}\nabla\theta.$$

Earlier we concluded that the velocity of the superfluid is equal to the group velocity of the soliton *i.e.*, $v_s = v_e$. This indicates that the motion of soliton is the motion of superfluid, and the vortex lines in superfluid is a result of soliton motion in the liquid helium atoms. The phase difference along a closed path is given by the line integral corresponding to the circulation of the velocity v_s

$$\triangle\theta = \oint_r \nabla\theta dr = \frac{\sqrt{2m}}{\hbar}\oint_r v_s dr.$$

Thus the circulation is related to the phase difference $\triangle\theta$ of the superfluid helium atoms. If the path of integration lies in a multiply connected domain or it encloses a vortex line, then $\triangle\theta(r) \neq 0$. Furthermore, if $\phi(r) \neq 0$ and it is single valued, we have $\triangle\theta = 2\pi n$, and

$$\oint_r v_s dr = n\frac{h}{\sqrt{2m}} \tag{2.152}$$

where n is an integer.

Equation (2.152) implies that whenever the velocity of the rotatng superfluid helium exceeds a critical velocity, vortex is produced in the liquid. The circulation (the vortex) is quantized and is given by an integer multiple of $\hbar/\sqrt{2m}$. Therefore, the nonliear Gross-Pitaevskii equation indeed gives an adequate description of superfluidity in helium II.

Superfluid can be viewed as a Bose condensate with local interactions (Pitaevskii 1961, Gross 1963). The concept of "quantum vortex" was proposed by Ginzburg and Pitaevskii (1960). Quantization of vortices in superfluids was suggested by Onsager (1949) on the basis of classical vortex flow and turbulence. The superfluidity of ^{4}He and its superfluid vortex lines and loops, weak turbulence and dissipative vortex dynamics in superfluids were studied by Brachet, Barenghi *et al.*, Roberts *et al.*, Pismen and Rica, and many others, using the nonlinear Schrödinger equation or the Gross-Pitaerskii equation given above (see reference [90] and [97]–[99]). Experimental observation of the vortices were reported by Yamchuk, Gordon, and Packard in 1979 and Zieve *et al.* Numerical simulations were presented by Frish, Pomeau and Rica (1992) and Schwarz. The energy of vortex lines was also measured.

If the superfluid liquid does not rotate, then $\nabla \times v_s = (h/m)\nabla \times \nabla\theta = 0$. The superfluid velocity field is a conservative field. This suggests that when the superfluid liquid helium flows through a tube with a gradual decreasing diameter, the pressure inside the tube is the same everywhere and it does not depends on the diameter of the tube. This is completely different from that of a normal fluid, but it has demonstrated experimentally. Moreover, the macroscopic wave function $\phi(x,t)$ given in (2.150) approaches 0 when x approches ∞. That is, $\phi(x,t)$ vanishes at the boundary. This implies that the superfluid density $\rho_s(\propto |\phi|^2)$ should also approach 0 at the boundary. The value of ρ_s was measured in 1970 and it was found that its value dropped from the value in the bulk to zero over a few atomic layers. This gave a direct verification of the theoretical results.

The Gross-Pitaerskii equation (2.148) is neither relativistic nor taking the gravitational field into consideration. Anandan and others extended the theory to include the relativistic effect. The generalized relativistic equation of motion for the quantum superfluid liquid helium is given by

$$\frac{\partial^2\phi}{\partial t^2} - \nabla^2\phi + \alpha^2\phi = -\lambda'|\phi|^2\phi, \tag{2.153}$$

where

$$\alpha^2 = \frac{m^2c^2}{\hbar^2}, \quad \lambda' = \frac{2m\lambda}{\hbar^2}.$$

Equation (2.153) is called the Gross-Pitaevskii-Anandan (GPA) equation. It is a type of the ϕ^4-equation. Anandan did not find its solutions. Instead, he gave an order of magnitude estimate for a $\phi = Ae^{i\theta}$ type of solution using the Einstein-Planck Law. As will be shown below, an exact solution of (2.153) is actually possible.

Let's assume the following trial solution,

$$\phi(x, y, z, t) = f(Z)e^{i\theta} \tag{2.154}$$

where

$$Z = \vec{p}\cdot\vec{r} - \Omega t, \quad \theta = \vec{k}\cdot\vec{r} - \omega t = k_1x + k_2y + k_3z - \omega t.$$

Substituting (2.154) into (2.153), the latter can be written as

$$(\Omega^2 - p^2)\frac{d^2f}{dZ^2} + (\alpha^2 + k^2 - \omega^2)f + \lambda' f^3 = 0 \tag{2.155}$$

in terms of $\vec{k} = (k_1, k_2, k_3)$ and $\vec{p} = (p_1, p_2, p_3)$, and $\omega\Omega = \vec{k}\cdot\vec{p}$. Integrating (2.155), we obtain the solution

$$f(Z) = \sqrt{\frac{w}{R}}\,\mathrm{sech}(\sqrt{w}Z),$$

where

$$w = \frac{\omega^2 - \alpha^2 - k^2}{\Omega^2 - p^2}, \quad R = \frac{\lambda'}{2(\Omega^2 - p^2)},$$

and

$$\phi(x, y, z, t) = f(Z)e^{i\theta} = \sqrt{\frac{w}{R}}\,\mathrm{sech}[\sqrt{w}(\vec{p}\cdot\vec{r} - \Omega t)]e^{i(\vec{k}\cdot\vec{r} - \omega t)}. \tag{2.156}$$

This is a soliton solution of the wave packet type, and its group velocity is v.

In obtaining the above solution, we have set the constant of integral C' to 0. If $C' \neq 0$ but is real, let $D' = 2C'/(\Omega^2 - \beta^2) > 0$, we then have:

$$f = \beta\mathrm{sc}\left[\sqrt{R(\alpha^2 - \beta^2)}Z\right],$$

and

$$\phi = f(Z)e^{i\theta} = \beta\mathrm{sc}\left[\sqrt{R(\alpha^2 + \beta^2)}Z\right]e^{i(\vec{k}\cdot\vec{r} - \omega t)} \tag{2.157}$$

where

$$\alpha^2\beta^2 = \frac{D'}{R}, \quad \alpha^2 - \beta^2 = \frac{w}{R}.$$

In (2.157), $\mathrm{sc}\psi = 1/\mathrm{cn}\psi$ and $\mathrm{cn}\psi$ is Jacobian elliptic cosine function. Equation (2.157) is another solution of (2.153), and it is an elliptic wave. Thus, the relativistic superfluid liquid helium can move as solitons or as elliptic wave. It is worth to point out that both $+f$ and $-f$ are solutions of (2.153). Thus, the solutions obtained above, (2.156) and (2.157), are both double solutions. They are actually the solutions of GPA equation in a flat space.

Generally, in a four-dimensional space, equation (2.153) may be written as

$$\eta_{\mu\nu}\nabla_\mu\nabla_\nu\phi + \alpha^2\phi = -\lambda'|\phi|^2\phi, \quad (\mu, \nu = 1, 2, 3, 4). \tag{2.158}$$

In this case, corresponding to (2.156) and (2.157), we have

$$\phi(X) = \sqrt{\frac{w}{R}}\,\mathrm{sech}\left[\sqrt{w}P^\mu X^\nu \eta_{\mu\nu}\right] e^{i\eta_{\mu\nu}K^\mu X^\upsilon}, \tag{2.159}$$

and

$$\phi(X) = \beta \mathrm{sc}\left[\sqrt{R(\alpha^2+\beta^2)}P^\mu X^\nu \eta_{\mu\nu}\right] e^{i\eta_{\mu\nu}K^\mu X^\nu} \tag{2.160}$$

where

$$X = (x, y, z, ict), \quad K = (\vec{K}, i\omega/c), \quad P = (\vec{P}, iv/c).$$

When gravitational force is considered, there is a transition from a flat space to a curved space which corresponds to a transformation from an orthogonal coordinate ($\eta_{\mu\nu}$) system to an obique coordinate ($g_{\mu\nu}$) system. This is because the space is curved in the presence of the gravitational field. Here $g_{\mu\nu}$ is the gauge tensor and ∇_μ is a covariant derivative. Equation (2.153) can be written as

$$g^{\mu\nu}\nabla_\mu\nabla_\nu\phi + \alpha^2\phi = -\lambda'|\phi|^2\phi. \tag{2.161}$$

This is the GPA equation in the space of gravitational force. $g_{\mu\nu}$ is the contravariant component of $g^{\mu\nu}$. Notice that the scalar product $g_{\mu\nu}A^\mu B^\nu$ is invariant under such a transformation. Thus, there exists a relation $\eta_{\mu\nu}K^\mu Z^\nu = g_{\mu\nu}K^\mu X^\nu$ between the $\eta_{\mu\nu}K^\mu Z^\nu$ of the flat space and the $g_{\mu\nu}K^\mu X^\nu$ of the curved space. Hence, the solutions (2.159) and (2.160) in curved space become

$$\phi(X) = \sqrt{\frac{w}{R}}\,\mathrm{sech}\left[\sqrt{w}g_{\mu\nu}P^\mu X^\nu\right] e^{ig_{\mu\nu}K^\mu X^\upsilon}, \tag{2.162}$$

$$\phi(X) = \beta\ \mathrm{sc}\left[\sqrt{R(\alpha^2+\beta^2)}g_{\mu\nu}P^\mu X^\nu\right] e^{ig_{\mu\nu}K^\mu X^\nu}. \tag{2.163}$$

Equations (2.162) and (2.163) describe the relativistic motion of the superfluid liquid helium. We have shown that superfluid liquid helium can move in the form of either a soliton or an elliptic wave, regardless the space is flat or curved. When there is no gravitational force, there could exist dual solitary wave and dual elliptic wave in the surperfluid. In a gravitational space, although the form of solution remains the same, the envelope $\sqrt{w/R}\mathrm{sech}[\sqrt{w}(\vec{p}\cdot\vec{r}-vt)]$, and the carrier wave, $\exp[i(\vec{k}\cdot\vec{r}-\omega t)]$, are all influenced by the gravitational field $g_{\mu\nu}$. If we choose $g_{\mu\nu}\neq 0$ (when $\mu\neq v$), the effect of the gravitational field $g_{\mu\nu}$ is mainly to change the four-dimensional wave vectors of the enevelope wave and the carrier, *i.e.*,

$$K^{\mu}\to g_{\mu\nu}K^{\mu}=(K_G)_{\mu},\quad P^{\mu}\to g_{\mu\nu}P^{\mu}=(P_G)_{\mu}.$$

Then, the solutions of solitary wave and elliptic wave found in the curved space become the same as those in the flat space.

Therefore, there always exist solutions of dual solitary wave and dual elliptic wave in a superfluid regardless of the existence of gravitational force. They correspond to the He-I and He-II phases of the superfluid. Moreover, when the solitary and elliptic waves in the superfluid are subject to gravitational force, their four-dimensional wave vectors are changed, $K\to K_G$ and $P\to P_G$. This property can be used to detect the existence of the gravitational field and gravitational wave. Superfluid gravitational antenna have been built in many research laboratories and an important use of it is to detect the gravitational wave by making use of the quantum interference effect of superfluid liquid helium.

Bibliography

Abdullaev, F. and Kraenkel, R. A. (2000). Phys. Rev. A **62** (023613) .

Abrikosov, A. A. (1957). Zh. Eksp. Theor. Fiz. **32** (1442) .

Abrikosov, A. A. and Gorkov, L. P. (1960). Zh. Eksp. Theor. Phys. **39** (781) .

Afanasjev, V. V., Akhmediev, N. and Soto-Crespo, J. M. (1996). Phys. Rev. E **53** (1931) and 1190.

Anandan, J. (1981). Phys. Rev. Lett. **47** (463) .

Andenson, M. H.,*et al.* (1995). Science **269** (198) .

Avenel, O.,*et al.* (1994). Proc. 20th Inter. Conf. on Low Temp. Phys., Physica B **194-196** (491) .

Balckburn, J. A.,*et al.* (1975). Phys. Rev. B **11** (1053) .

Bardeen, L. N., Cooper, L. N. and Schrieffer, J. R. (1957). Phys. Rev. **108** (1175) .

Barenghi, C. F., Donnerlly, R. J. and Vinen, W. F. (2001). Quantized vortex dynamics and superfluid turbulence, Springer, Berlin.

Bogoliubov, N. N. (1949). Quantum statistics, Nauka, Moscow.

Bogoliubov, N. N. (1958). J. Exp. Theor. Phys. USSR **34** (58) .

Bogoliubov, N. N., Toimachev, V. V. and Shirkov, D. V. (1958). A new method in the theory of superconductivity, AN SSSR, Moscow.

Bradley, C. C., Sackett, C. A. and Hulet, R. G. (1997). Phys. Rev. Lett. **78** (985) .

Bradley, C. C., Sackett, C. A., Tollett, J. J. and Hulet, R. G. (1995). Phys. Rev. Lett. **75** (1687) .

Bullough, R. K., Bogolyubov, N. M., Kapitonov, V. S., Malyshev, C., Timonen, J., Rybin, A. V., Vazugin, G. G. and Lindberg, M. (2003). Theor. Math. Phys. **134** (47) .
Bullough, R. K. and Caudrey, P. J. (1980). Solitons, Springer-Verlag, Berlin.
Burger, S.,*et al.* (1999). Phys. Rev. Lett. **83** (5198) .
Burstein, E. and Lulquist, S. (1969). Tunneling phenomena in solids, Plenum Press, New York.
Cai, S. Y. and Bhattacharjee, A. (1991). Phys. Rev. A **43** (6934) .
Caradoc-Davies, B. M., Ballagh, R. J. and Bumett, K. (1999). Phys. Rev. Lett. **83** (895) .
Cashar, A.,*et al.* (1969). J. Math. Phys. **9** (1312) .
Cooper, L. N. (1956). Phys. Rev. **104** (11-89) .
Cornish, S. L.,*et al.* (2000). Phys. Rev. Lett. **85** (1795) .
Crasovan, L. C., Malomed, B. A. and Mihalache, D. (2000). Phys. Rev. E **63** (016605) .
Creswick, R. J. and Morriso, H. L. (1980). Phys. Lett. A **76** (267) .
Dalfovo, F., Giorgini, S., Pitaevskii, L. P. and Stringali, S. (1999). Rev. Mod. Phys. **71** (463) .
Davis, K. B.,*et al.* (1995). Phys. Rev. Lett. **75** (3969) .
Davydov, A. S. (1985). Solitons in molecular systems, D. Reidel Publishing, Dordrecht.
de Gennes, P. G. (1966). Superconductivity of metals adn alloys, W. A. Benjamin, New York.
Dewitt, B. S. (1966). Phys. Rev. Lett. **16** (1092) .
Dirac, P. M. (1958). General theory of relativity 2, Oxford, London.
Donnely, R. J. (1991). Quantum vortices in heliem II, Cambridge Univ. Press, Cambridge.
Elyutin, P. V. and Rogoverko, A. N. (2001). Phys. Rev. E **63** (026610) .
Elyutin, P. V.,*et al.* (2001). Phys. Rev. E **64** (016607) .
Fried, D. G.,*et al.* (1999). Phys. Rev. Lett. **81** (3811) .
Frisch, T., Pomeau, Y. and Rica, S. (1992). Phys. Rev. Lett. **69** (1644) .
Ginzberg, V. I. and Landau, L. D. (1956). Zh. Eksp. Theor. Fiz. **20** (1064) .
Ginzberg, V. L. and Kirahnits, D. A. (1977). Problems in high-temperature superconductivity, Nauka, Moscow.
Ginzburg, V. L. (1952). Usp Fiz. Nauk. **48** (25) .
Ginzburg, V. L. and Pitayevsky, L. P. (1968). Sov. Phys. JETP **7** (858) .
Ginzburg, V. L. and Sobianin, A. A. (1976). Usp Theor. Fiz. Nauk. **120** (153) .
Gorkov, L. P. (1958). Sov. Phys. JETP **9** (189) .
Gorkov, L. P. (1959). Sov. Phys. JETP **9** (1364) .
Gross, E. P. (1961). Nuovo Cimento **20** (454) .
Guo, Bai-lin and Pang Xiao-feng, (1987). Solitons, Chin. Science Press, Beijing.
Hirota, R. (1972). J. Phys. Soc. Japan **33** (551) ; Phys. Rev. Lett. **27** (1192) .
Hoch, M. J. R. and Lemmer, R. H. (1991). Low temperature physics, Springer, Berlin.
Holm, J.*et al.* (1993). Physica **68** (35) .
Huebenor, R. P. (1974). Phys. Rep. **24** (127) .
Huepe, C. and Brachet, M. E. (2000). Physica D **140** (126) .
Jacobs, L. and Rebbi, L. (1979). Phys. Rev. B **19** (1486) .
Jacobson, D. A. (1965). Phys. Rev. B **8** (1066) ; Adv. Phys. **14** (419) .
Josephson, B. (1962). Phys. Lett. **1** (251) .
Josephson, B. (1964). Thesis, unpublised, Cambridge University.
Josephson, B. (1965). Adv. Phys. **14** (419) .
Kaper, H. G. and Takac, P. (1998). Nonlinearity **11** (291) .
Kayfield, G. W. and Reif, F. (1963). Phys. Rev. Lett. **11** (305) .
Khalilov, V. R. (2002). Theor. Math. Phys. **133** (1406) .
Kictchenside, P. W.,*et al.* (1981). Soliton and condensed matter physics, eds. A. R.

Bishop,*et al.*, Plenum, New York, p. 297.
Kivshar, Yu. S., Alexander, T. J. and Turitsy, S. K. (2001). Phys. Lett. A **278** (225) .
Kivshar, Yu. S. and Malomed, B. A. (1989). Rev. Mod. Phys. **61** (763) .
Klitzing, K. V.,*et al.*, (1980). Phys. Rev. Lett. **45** (494) .
Kusayanage,*et al.* (1972). J. Phys. Soc. Japan **33** (551) .
Lacaze, R., Lallemand, P., Pomeau, Y. and Rica, S. (2001). Physica D **152-153** (779) .
Lamb, G. L. (1971). Rev. Mod. Phys. **43** (99) .
Laplae, L.,*et al.* (1974). Phys. Rev. C **10** (151) .
Leggett, A. J. (1980). Prog. Theor. Phys. (Supp.) **69** (80) .
Leggett, A. J. (1984). in Percolation, localization and superconductivity, eds. by A. M. Goldlinan, S. A. Bvilf, Plenum Press, New York, pp. 1-41.
Leggett, A. J. (1991). in Low temperature physics, Springer, Berlin, pp. 1-93.
Lindensmith, C. A.,*et al.*, (1996). Proc. 21st Inter. Conf. on Low Temp. Phys., Czech J. Phys. **46** (131) .
Liu, W. S. and Li, X. P. (1998). European Phys. J. **D2** (1) .
London, F. (1950). Superfluids, Vol. 1, Weley, New York.
London, F. and London, H. (1935). Proc. Roy. Soc. (London) A **149** (71) ; Physica **2** (341) .
Louesey, S. W. (1980). Phys. Lett. A **78** (429) .
Madison, K. W.,*et al.* (2000). Phys. Rev. Lett. **84** (806) .
Maki, K. and Lin-Liu, Y. R. (1978). Phys. Rev. B **17** (3535) .
Malomed, B. A. (1989). Phys. Rev. B **39** (8018) .
Marcus, P. M. (1964). Rev. Mod. Phys. **36** (294) .
Matthews, M. R.,*et al.* (1999). Phys. Rev. Lett. **83** (2498) .
Niek, I. R. and Fraser, J. C. (1970). J. Low Temp. Phys. **3** (225) .
Ohberg P. and Santos, L. (2001). Phys. Rev. Lett. **86** (2918) .
Onsager, L. (1949). Nuovo Cimento Suppl. **6** (249) .
Ostrovskaya, E. A.,*et al.* (2000). Phys. Rev. A **61** (R3160) .
Packard, R. E., (1998). Rev. Mod. Phys. **70** (641) .
Pang, Xiao-feng (1982). Chin. J. Low Temp. Supercond. **4** (33) ; ibid **3** (62) .
Pang, Xiao-feng (1982). Chin. J. Nature **5** (254) .
Pang, Xiao-feng (1983). Phys. Bulletin Sin. **3** (17) .
Pang, Xiao-feng (1985). Chin. J. Potential Science **5** (16) .
Pang, Xiao-feng (1985). J. Low Temp. Phys. **58** (334) .
Pang, Xiao-feng (1985). Problems of nonlinear quantum theory, Sichuan Normal Univ. Press, Chengdu.
Pang, Xiao-feng (1986). Chin. Acta Biochem. BioPhys. **18** (1) .
Pang, Xiao-feng (1986). J. Res. Met. Mat. Sin. **12** (31) .
Pang, Xiao-feng (1986). J. Science Exploration Sin. **4** (70) .
Pang, Xiao-feng (1986). Phys. Bulletin Sin. **1** (2) .
Pang, Xiao-feng (1987). The properties of soliton motion in superfluid He^4, Proc. ICNP, Shanghai, p. 181.
Pang, Xiao-feng (1988). J. Xinjian Univ. Sin. **3** (33) .
Pang, Xiao-feng (1989). J. Chinghai Normal Univ. Sin. **1** (37) .
Pang, Xiao-feng (1989). J. Kunming Tech. Sci. Univ. **2** (118) ; ibid **6** (54) .
Pang, Xiao-feng (1989). J. Southwest Inst. for Nationalities Sin. **15** (97) .
Pang, Xiao-feng (1989). J. Xinjian Univ. Sin. **3** (72) .
Pang, Xiao-feng (1990). J. Xuntan Univ. Sin. **12** (43) and 65.
Pang, Xiao-feng (1991). J. Kunming Tech. Sci. Univ. **3** (57) .
Pang, Xiao-feng (1991). J. Kunming Tech. Sci. Univ. **4** (13) .

Pang, Xiao-feng (1991). J. Southwest Inst. for Nationalities Sin. **17** (1) and 18.
Pang, Xiao-feng (1991). J. Xuntan Univ. Sin. **13** (63) .
Pang, Xiao-feng (1994). Theory of nonlinear quantum mechanics, Chongqing Press, Chongqing, 1994.
Pang, Xiao-feng,*et al.* (1989). The properties of soliton motion for superconductivity electrons. Proc. ICNP, Shanghai, p. 139.
Papaconstantopoulos, D. A. and Klein, B. M. (1975). Phys. Rev. Lett. **35** (110) .
Pardy, M. (1989). Phys. Lett. A **140** (51) .
Parks, R. D. (1969). Superconductivity, Marcel. Dekker.
Pederser, N. F. (1993). Physica D **68** (27) .
Perez-Garcia, V. M., Michinel, M. and Herrero, H. (1998). Phys. Rev. A **57** (3837) .
Perring, J. K.,*et al.* (1963). Nucl. Phys. **61** (1443) .
Pitaerskii, L. P. (1961). Sov. Phys. JETP **13** (451) .
Pitaevskii, L. P. (1960). Sov. Phys. JETP **10** (553) .
Pomeau, Y. (1992). Nonlinearity **5** (707) .
Pomeau, Y. (1996). Phys. Scr. **67** (141) .
Pomeau, Y. and Rica, S. (1993). Comptes Rendus Acad. Sci. (Paris) **316 (Serie II)** (1523) .
Putterman, P. (1973). Low Temp. Phys. **LT13** (39) .
Putterman, P. (1972). Low Temp. Phys. **LTB1** (39) .
Rieger, T. J.,*et al.* (1976) Phys. Rev. Lett. **27** (1787) .
Rogovin, D. (1974). Ann. Phys. **90** (18) .
Rogovin, D. (1975). Phys. Rev. B **11** (1906) .
Rogovin, D. (1975). Phys. Rev. B **12** (130) .
Rogovin, D. (1976). Phys. Rep. **251** (1786) .
Rogovin, D. and Scully, M. (1969). Ann. Phys. **88** (371) .
Rogovin, D. and Scully, M. (1976). Phys. Rep. **25** (176) .
Ronognan, J. P. (1974). Solid State Commun. **14** (83) .
Schrieffer, J. R. (1964). Theory of superconductivity, Benjamin, New York.
Schrieffer, J. R. (1969). Superconductivity, Benjamin, New York.
Schwarz, K. W. (1990). Phys. Rev. Lett. **64** (1130) .
Schwarz, K. W. (1993). ibid **71** (259) .
Schwarz, K. W. (1993). Phys. Rev. B **47** (12030) .
Scott, A. C.,*et al.* (1973). Proc. IEEE **61** (1443) .
Sonin, E. B. (1994). J. Low Temp. Phys. **97** (145) .
Sorensen, M. P., Malomed, B. A., Vstinov, A. V. and Pedersen, N. F. (1993). Physica **68** (38) .
Suint-James, D.,*et al.* (1966). Type-II superconductivity, Pergamon, Oxford.
Svozil, K. (1990). Phys. Rev. Lett. **65** (3341) .
Tonomara, A. (1998). Foundation Phys. **28** (59) .
Valatin, D. G. (1959). Nuovo Cimento **7** (843) .
Ventura, I. (1990). Solitons in liquid ^{4}He, in Solitons and applications, eds. Makhankov, V. G., Fedyanin, V. K. and Pashaev, O. K, World Scientific, Singapore
Wheatley, J. C. (1975). Rev. Mod. Phys. **47** (415) .
Wiegel, F. W. (1977). Phys. Rev. B **16** (57) .
Yarmchuk, E. J., Gordon, M. J. V. and Packard, R. E. (1979). Phys. Rev. Lett. **43** (214) .
Zieve, R. J., Close, J. D., Davis, J. C. and Packard, R. E. (1993). J. Low Temp. Phys. **90** (243) .
Zieve, R. J., Mukhjarsky, Yu., Close, J. D., Davis, J. C. and Packard, R. E. (1992). Phys. Rev. Lett. **68** (1327) .

Chapter 3

The Fundamental Principles and Theories of Nonlinear Quantum Mechanics

3.1 Lessons learnt from the Macroscopic Quantum Effects

In the previous chapter, we discussed some macroscopic quantum effects observed in experiments involving superconductors, superfluid and so on, and analyzed these effects using quantum solid theory and modern nonlinear theory. Macroscopic quantum effects are results of motions of quasiparticles. Thus, we discussed the dynamics of such quasiparticles as superconducting electrons and superfluid helium atoms on the basis of soliton theory. The observed macroscopic quantum effects in certain systems can then be easily understood based on these theories. This demonstrated that the nonlinear theory is the proper theory for the macroscopic quantum effects and the relevant physical quantities such as the Hamiltonian, Lagrangian as well as the dynamic equations are in the correct forms. In this chapter, we will further explore the physical insights of these concepts and give an in-depth treatment of the theory.

But first, let's summarize what we have learnt from the discussion on the macroscopic quantum effects and the theoretical concepts used to understand these effects.

(1) It has become clear that the macroscopic quantum effects are fundamentally different from the microscopic quantum effects. The latter can be described by the linear quantum theory which, however, failed to describe the macroscopic quantum systems. It is thus very necessary to establish a quantum theory that describes nonlinear systems.

(2) However, why can't the present linear quantum theory describe the macroscopic quantum effects? On what foundation a new theory should be based? Answers to such questions and the key for solving the existing problems of the linear quantum mechanics can only come from a clear understanding of the intrinsic problems of the linear quantum mechanics and the fundamental aspects of the macroscopic quantum effects. The behaviors of quasiparticles in such systems must play an essential role.

It is well know that the macroscopic quantum effects are nonlinear phenomena. The BCS theory of superconductivity and the modern theory of superfluidity, both are nonlinear theories, have been well established. A basic feature of the nonlinear theories is that the Hamiltonian, free energy or Lagrangian functions of the systems are nonlinear functions of the wave function of the microscopic particles, and the dynamic equation becomes a nonlinear equation, due to the nonlinearity. The quasiparticles behave differently in such systems which results in ordered coherent states, such as the Bose-Einstein condensed state. These states occur following a second-order phase transition and spontaneous breaking of symmetry of the systems under the nonlinear interactions. Therefore, the nonlinear interactions play a very important role in the behaviors of the microscopic particles. It also suggests that one should pay particular attention to the nonlinear interactions in order to establish a correct new quantum theory, and the right direction for solving problems encountered by the linear quantum mechanics is to establish a nonlinear quantum theory.

(3) How can a nonlinear quantum theory be established? Again we take a look at what we learned from the macroscopic quantum effects, and try to understand how the superconductive and superfluid theories differ from the linear quantum mechanics? Detailed examination reveals the following.

(a) The Hamiltonian, or Lagrangian function and free energy of these systems, given in (2.1) and (2.2), or (2.53) and (2.54), or (2.149) and (2.151), are all dependent on, and are nonlinear functions of, the wave function $\phi(x,t)$ of the microscopic particles, *i.e.* the superconductive electron or superfluid helium, respectively. These go directly against the fundamental hypothesis of the linear quantum mechanics that the Hamiltonian of the system is independent of the wave functions of the microscopic particle.

It was exactly because of this nonlinear feature in the theories of superconductivity and superfluidity that they were able to correctly describe the nonlinear behaviors of superconductivity and superfluidity and successfully explain these macroscopic quantum effects. On the contrary, lacking of such a nonlinearity was also the reason that other theories failed. For example, although Frohlish's superconducting theory gave the correct superconducting mechanism, *i.e.*, electron-phonon coupling, in 1951, he failed to establish a complete superconducting theory because his theory was based on the linear perturbation theory of the linear quantum mechanics. Therefore, the new nonlinear theory should abandon this hypothesis.

(b) The fundamental dynamic equations in the linear quantum mechanics is the Schrödinger or the Klein-Gordon equation which are wave equations, and are *linear* equations of the wave function of the particles. As a result, solutions of these linear equations cannot describe the wave-corpuscle duality of microscopic particles as discussed in Chapter 1. On the other hand, the Ginzburg-Landau equations, (2.17), (2.55)-(2.57), (2.119), and the GP equation, (2.121), or the GPA equation, (2.148), satisfied by the quasiparticles (e.g. the superconductive electron and superfluid helium atom), as well as the ϕ^4-equation (2.153) and the Sine-Gordon equation (2.104) in superconductors and superfluid are all *nonlinear* equations of the wave function of the quasiparticles. With these nonlinear equations, superconductivity and superfluidity as well as other macroscopic quantum effects observed in experiments can be explained. This suggests that in establishing a new theory, the linear dynamic equation must be replaced with a nonlinear equation, and the superposition principle of wave functions must be abandoned. Fortunately, all the nonlinear equations mentioned above are natural generalizations of the Schrödinger equation or Klein-Gordon equation which are the dynamic equations in linear quantum mechanics. Therefore, the new nonlinear quantum theory should be developed on the basis of the linear quantum mechanics, rather than anything else.

Certainly, it is still necessary to further examine whether these dynamic equations or Hamiltonians given in Chapter 2 have the correct space-time symmetries and what physical invariance they possess. Only nonlinear dynamic equations or Hamiltonians which satisfy the required symmetries and invariances can be adopted in the new theory.

It is known that the nonlinear dynamic equations describing superconductivity and superfluidity states admit stable soliton solutions. This shows that the normal microscopic particles evolve into solitons in nonlinear systems due to nonlinear interactions. It is therefore natural to use the concept of soliton in the description of microscopic particles in nonlinear systems. A soliton is a new form of physical entity which cannot be described by a linear theory. According to modern soliton theory, a soliton, which differs completely from a normal microscopic particle, possesses the wave-particle duality. Its wave property appears in the form of a traveling solitary wave which has all the essential features of wave motion, including frequency, period, amplitude, group and phase velocities, diffraction, transmission, and reflection. Its corpuscle feature is reflected by a stable shape analogous to a classical particle, even after going through a collision with another particle, a definite energy, momentum and mass, and its uniform motion in free space and its motion with a constant acceleration in the presence of a constant external field, and so on. This suggests that modern soliton theory should be an integral part of any new nonlinear quantum-mechanical theory.

To summarize, we see clearly that the direction for developing a new quantum theory is nonlinear quantum mechanics. As a matter of fact, the concept of nonlinear quantum mechanics was proposed by many scientists such as Mielnik, Jordan, Gisin, Weinberg, Doebner and so on. However, the work presented in this book represents a complete different approach. Both this approach and its mathematical treatments are different from those in other publications in many respects. The theories and principles presented here are firmly based on physical foundations Its mathematical basis is the nonlinear partial differential equations and the soliton theory. Its physical basis is the macroscopic quantum effects, and the well established modern theories of superconductivity and superfluidity. With the macroscopic quantum effects as the foundation, and by incorporating nonlinear interactions and soliton motion into a generalized theoretical framework, fundamental principles and theories of nonlinear quantum mechanics can be established, without the hypotheses of the linearity of linear quantum mechanical theory and the independence of the Hamiltonian operator on the wave function of the particles.

3.2 Fundamental Principles of Nonlinear Quantum Mechanics

Based on the earlier discussion, the fundamental principles of nonlinear quantum mechanics (NLQM) may be summarized as the following.

(1) Microscopic particles in a nonlinear quantum system are described by the following wave function,

$$\phi(\vec{r},t) = \varphi(\vec{r},t)e^{i\theta(\vec{r},t)} \tag{3.1}$$

where both the amplitude $\varphi(\vec{r},t)$ and phase $\theta(\vec{r},t)$ of the wave function are functions of space and time.

(2) In the nonrelativistic case, the wave function $\phi(\vec{r},t)$ satisfies the generalized nonlinear Schrödinger equation (NLSE), *i.e.*

$$i\hbar\frac{\partial\phi}{\partial t} = -\frac{\hbar^2}{2m}\nabla^2\phi \pm b|\phi|^2\phi + V(\vec{r},t)\phi + A(\phi), \tag{3.2}$$

or

$$\mu\frac{\partial\phi}{\partial t} = -\frac{\hbar^2}{2m}\nabla^2\phi \pm b|\phi|^2\phi + V(\vec{r},t)\phi + A(\phi), \tag{3.3}$$

where μ is a complex number, V is an external potential field, A is a function of ϕ, and b is a coefficient indicating the strength of the nonlinear interaction.

In the relativistic case, the wave function $\phi(\vec{r},t)$ satisfies the nonlinear Klein-Gordon equation (NLKGE), including the generalized Sine-Gordon equation (SGE) and the ϕ^4-field equation, *i.e.*

$$\frac{\partial^2\phi}{\partial t^2} - \frac{\partial^2\phi}{\partial x_j^2} = \beta\sin\phi + \gamma\frac{\partial\phi}{\partial t} + A(\phi), \quad (j = 1,2,3) \tag{3.4}$$

$$\frac{\partial^2\phi}{\partial t^2} - \frac{\partial^2\phi}{\partial x_j^2} \mp \alpha\phi \pm \beta|\phi|^2\phi = A(\phi), \quad (j = 1,2,3) \tag{3.5}$$

where γ represents dissipative or frictional effects, β is a coefficient indicating the strength of nonlinear interaction and A is a function of ϕ.

These are the only two fundamental hypotheses of the nonlinear quantum mechanics. This is quite different from the linear quantum mechanics which are based on several hypotheses, as discussed in Chapter 1. However, the dynamic equations are generalizations of the linear Schrödinger and linear Klein-Gordon equations of the linear quantum mechanics to nonlinear quantum systems. These equations were used to study the motion of superconducting electrons and helium atoms in the superfluid state. It has been shown that (3.2) – (3.5) indeed describe the law of motion and properties of microscopic particles in nonlinear quantum systems.

This is the basis for having the two hypotheses as the principles of the nonlinear quantum mechanics. Obviously, the nonlinear quantum mechanics is an integration of superconductivity, superfluidity and modern soliton theories, and its experimental foundation is the macroscopic quantum effects. From the two hypotheses, the following can be deduced.

(l) The absolute square of the wave function $\phi(\vec{r},t)$ given in (3.1), $|\phi(\vec{r},t)|^2 = |\varphi(\vec{r},t)|^2 = \rho(\vec{r},t)$, is no longer the probability of finding the microscopic particle at a given point in the space-time, but gives the mass density of the microscopic particles at that point. Thus, the concept of probability or the statistical interpretation of wave function is no longer relevant in nonlinear quantum mechanics. The wave function (3.1) has the similar form as that in the linear quantum mechanics, but their meaning are completely different. Here $\varphi(\vec{r},t)$ is the envelope of the microscopic particle, and it represents the particle feature, or more precisely, the soliton feature of the microscopic particle. Different from that in the linear quantum mechanics, $\varphi(\vec{r},t)$ has its physical meaning and satisfies certain nonlinear equation. $e^{i\theta(\vec{r},t)}$ is a carrier wave of $\phi(\vec{r},t)$. The interpretation of (3.1) will be discussed in more details in the following Chapters.

(2) The wave function $\phi(\vec{r},t)$ represents a soliton or a solitary wave. It is no longer a linear or dispersive wave. Equations (3.2) – (3.5) are nonlinear dynamic equations and have soliton solutions. Therefore, a microscopic particle is a soliton, or is described by a soliton, in nonlinear quantum mechanics. Thus the fundamental nature of microscopic particles in the nonlinear quantum mechanics is different from that in the linear quantum mechanics.

(3) The concept of operator in the linear quantum mechanics is still used in the nonlinear quantum mechanics. However, they are all no longer linear operators, and thus certain properties of linear operators, such as conjugate Hermitian of the momentum and coordinate operators, are no longer required. Instead, nonlinear operators are constructed and used in the nonlinear quantum mechanics. For example, Equation (3.2) may be written as

$$i\hbar\frac{\partial\phi}{\partial t} = \hat{H}(\phi)\phi. \tag{3.6}$$

The Hamiltonian operator $\hat{H}(\phi)$ has a nonlinear dependence on ϕ and is given by

$$\hat{H}(\phi) = -\frac{\hbar^2}{2m}\nabla^2 - b|\phi|^2 + V(\vec{r},t) \tag{3.7}$$

for $A = 0$.

In general, equations (3.2) – (3.5) can be expressed as

$$\phi = \phi(\vec{r},t), \quad \phi_t = \frac{d\phi}{dt} = K(\phi) \tag{3.8}$$

according to Lax, where $K(\phi)$ is a nonlinear operator or hereditary operator. In one dimensional case, a new operator $Q(\phi)$ which is obtained from the generator of the translation group can be used to produce the vectorial field $K(\phi)$, *i.e.*

$$K(\phi) = Q(\phi)\phi_x, \tag{3.9}$$

and the operator $Q(\phi)$ is called a nonlinear recursion operator. The nonlinear Schrödinger equation (3.2) can be generalized to

$$Q(\phi) = -iD + 4i\phi D^{-1}\Re(\bar{\phi}) \tag{3.10}$$

where the operator D denotes the derivative with respect to x, and

$$D^{-1}f(x) = \int_{-\infty}^{x} f(y)dy.$$

From the hereditary property of $K(\phi)$, we can obtain the following vector field

$$K_n(\phi) = Q(\phi)^n\phi_x, \quad n = 0, 1. \tag{3.11}$$

The equation of motion of the eigenvalue λ' of the recursion operator $Q(\phi)$ may be expressed as

$$\lambda'_t = \frac{\partial\lambda'}{\partial t} = K'(\phi)[\lambda'] \tag{3.12}$$

where $K'(\phi)$ is the variational derivative with respect to ϕ and is given by

$$K'(\phi)[\lambda'] = \left.\frac{\partial K'(\phi + \varepsilon\lambda')}{\partial\varepsilon}\right|_{\varepsilon=0}. \tag{3.13}$$

The equation describing the time variation of the recursion operator can be obtained,

$$\frac{\partial}{\partial t}Q(\phi) = K'(\phi)Q(\phi) - Q(\phi)K'(\phi) = [K'(\phi), Q(\phi)]. \tag{3.14}$$

This equation is very similar to the Heisenberg matrix equation in the linear quantum mechanics. However, both $K'(\phi)$ and $Q(\phi)$ here are nonlinear operators.

(4) Because the operators in nonlinear quantum mechanics are nonlinear, it is no longer necessary for their eigenvectors or states $\phi(r, t)$ to satisfy the linear superposition principle, *i.e.* $\phi = \sum_n C_n\phi_n$. This implies that superposition of any two states is not necessarily a state of this system as shown in (1.4). As a matter of fact, superposition of two solitary waves may result in two, three or any number of solitary waves. Therefore, the superposition principle of waves in the linear quantum mechanics must be modified for the nonlinear quantum mechanics. This will be discussed in the next section.

(5) There are also time-independent states and eigenvalue problems in nonlinear quantum mechanics. How the eigenvalue of the nonlinear Schrödinger equation is defined and determined is interesting. Since nonlinear quantum mechanics differs from linear quantum mechanics, these concepts and the method to determine the

eigenvalues are different. The time-independent solution of (3.2) is assumed to have the following form

$$\phi(\vec{r},t) = \varphi(\vec{r})e^{iEt/\hbar}. \tag{3.15}$$

Substituting (3.15) into (3.2) and choose $A(\phi) = 0$, we can get

$$E\varphi(\vec{r}) = -\frac{\hbar^2}{2m}\nabla^2\varphi(\vec{r}) + V(\vec{r})\varphi(\vec{r}) - b|\varphi(\vec{r})|^2\varphi(\vec{r}), \tag{3.16}$$

That is,

$$\hat{H}(\varphi)\varphi(r) = E\varphi \tag{3.17}$$

where

$$\hat{H}(\varphi) = -\frac{\hbar^2}{2m}\nabla^2 + V(\vec{r}) - b|\varphi|^2 = -\frac{\hbar^2}{2m}\nabla^2 + V(\vec{r}) - b\rho(\vec{r}). \tag{3.18}$$

The energy E in (3.18) is the eigenvalue of the Hamiltonian operator $\hat{H}(\varphi)$. Equation (3.17) is very similar to the linear Schrödinger equation in the linear quantum mechanics, but the Hamiltonian operator $\hat{H}(\varphi)$ is a nonlinear operator of wave function φ, *i.e.*,

$$\hat{H}(\varphi) = \hat{H}_0 + b\hat{\rho}(\vec{r}), \tag{3.19}$$

where

$$\hat{H}_0 = -\frac{\hbar^2}{2m}\nabla^2 + V(\vec{r}).$$

This shows that if this Hamiltonian operator acts on a wave vector of a quantum state of a microscopic particle, an eigenvalue independent of time and position cannot always be obtained. It is thus impossible to determine the eigenvalues of the Hamiltonian using the traditional method. Thus the eigenvalue of the nonlinear Schrödinger equation must be redefined.

In general, the eigenequation and eigenvalues of a nonlinear system can be defined and determined using the approach proposed by Lax. For a general nonlinear equation given in (3.8), we know that $K(\varphi)$ is a nonlinear operator. If two linear operators $\hat{L}$ and $\hat{B}$, which depend on ϕ, satisfy the following Lax operator equation

$$iL_{t'} = \hat{B}\hat{L} - \hat{L}\hat{B} = [\hat{B}, \hat{L}], \tag{3.20}$$

where $t' = t/\hbar$ and $\hat{B}$ is a self-adjoint operator, then the eigenvalue E and eigenfunction ψ of the operator $\hat{L}$ may be derived from (3.20), *i.e.*

$$\hat{L}\psi = \lambda\psi; \quad i\psi_{t'} = \hat{B}\psi. \tag{3.21}$$

Thus, the eigenvector and eigenvalue of a nonlinear system are determined by the eigenvectors and eigenvalues of the above two linear operators. It can be shown that λ is an eigenvalue that does not depend on time. In fact, if we differentiate (3.21) and multiply the resulting equation by i, we can get

$$i\left(\psi\frac{d\lambda}{dt'}+\lambda\frac{d\psi}{dt'}\right)=i\left(\hat{L}\psi_{t'}+\frac{\partial\hat{L}}{\partial t'}\psi\right)=i\hat{L}\psi_{t'}+[\hat{B}\hat{L}-\hat{L}\hat{B}]\psi$$
$$=\hat{L}(i\psi_{t'}-\hat{B}\psi)+\lambda\hat{B}\psi.$$

From the above, we get $i\psi(d\lambda/dt')=0$. Thus, λ is a time-independent eigenvalue, and (3.21) is the linear eigenequation corresponding to the nonlinear equation (3.8). Therefore, for any nonlinear equation, we can always find the corresponding linear eigenequation and time-independent eigenvalue. In fact, Lax, Zakharov, Shabat and Ablowitz *et al.* successfully reduced a nonlinear problem into a linear one and then obtained soliton solution by the inverse scattering method.

If both $V(\vec{r},t)$ and $A(\phi)$ in the nonlinear Schrödinger equation (3.2) are zeros, the above operators $\hat{L}$ and $\hat{B}$ are

$$\hat{L}=i\begin{pmatrix}1+s & 0\\ 0 & 1-s\end{pmatrix}\frac{\partial}{\partial x'}+\begin{pmatrix}0 & \phi^*\\ \phi & 0\end{pmatrix},$$
$$\hat{B}=-s\begin{pmatrix}1 & 0\\ 0 & 1\end{pmatrix}\frac{\partial^2}{\partial x'^2}+\begin{bmatrix}|\phi|^2/(1+s) & i\phi_{x'}\\ -i\phi_{x'} & -|\phi|^2/(1-s)\end{bmatrix}, \tag{3.22}$$

and $s^2=1-2/b$, $x'=x\sqrt{2m/\hbar^2}$. The eigenvalues of the nonlinear Schrödinger equation (soliton eigenvalues) are thus determined by

$$L\psi=\lambda\psi,\quad \psi=\begin{pmatrix}\psi_1\\ \psi_2\end{pmatrix}.$$

Its solution can be obtained using the inverse scattering method which will be discussed in detail in chapter 6.

However, it should be pointed out that if $\phi(\vec{r},t)$ or $\psi(\vec{r})$ is further quantized by the creation and annihilation operators of the microscopic particles, then the Hamiltonian operator in (3.18) would be given in terms of the creation and annihilation operators in the second quantization representation, and the eigenvalues of the Hamiltonian operator can also be determined in the same representation. This method was used by Pang *et al.* to obtain the eigenvalues of the Hamiltonian operator in molecular systems, which will be discussed in Chapters 6 and 9, respectively.

(6) Compared to the linear quantum theory, two major breakthroughs were made in the nonlinear qunatum theory. the linearity of the dynamic equation and independence of the Hamiltonian operator on the wave function of the microscopic particles. In the nonlinear quantum mechanics, the dynamic equations are nonlinear in the wave function ϕ, *i.e.*, they are nonlinear partial differential equations. The Hamiltonian operators depend on the wave function. For example,

the Hamiltonian operator (3.7) corresponding to equation (3.2) is no longer simply $H = -(\hbar^2/2m)\nabla^2 + \bar{V}(\vec{r},t)$, but depends on the wave function ϕ. In this respect, the nonlinear quantum mechanics is truly a break-through in the development of modern quantum theory.

3.3 The Fundamental Theory of Nonlinear Quantum Mechanics

Similar to the linear quantum mechanics, the fundamental theory of nonlinear quantum mechanics consists of the following,

(1) principle of nonlinear superposition;
(2) theory of nonlinear Fourier transformation;
(3) method of quantization;
(4) perturbation theory.

A nonlinear quantum mechanical system or problem can be studied based on these fundamental theories. Compare to the linear quantum mechanics, these fundamental theories of the nonlinear quantum mechanics are much complicated. Even though they have been studied in soliton physics, much of it requires further improvement. In view of this, only the fundamental concepts and processes are introduced here and the details will be discussed in subsequent chapters.

3.3.1 *Principle of nonlinear superposition and Bäcklund transformation*

From earlier discussion, we know that the principle of superposition for wave functions of microscopic particles (soliton) in the nonlinear quantum mechanics is not a linear superposition as in the linear quantum mechanics. In the linear quantum theory, a complicated motion can always be taken as a superposition of some basic modes. If $\psi_1(\vec{r},t)$ and $\psi_2(\vec{r},t)$ are solutions of the equation of motion, (1.4), then their superposition $\psi = a\psi_1(\vec{r},t) + b\psi_2(\vec{r},t)$ is also a solution. However, microscopic particles in the nonlinear quantum mechanics do not satisfy this relation. The nonlinear interaction not only complicates the process of superposition, but also gives rise to many different forms of superposition.

(A) Konopelchenko was the first to derive completely the nonlinear superposition principle of integrable nonlinear equations. The following is the differential equations studied by Konopelchenko

$$\frac{\partial}{\partial x}\begin{pmatrix}\psi_1\\ \psi_2\end{pmatrix} = i\lambda'\begin{pmatrix}1 & 0\\ 0 & 1\end{pmatrix}\begin{pmatrix}\psi_1\\ \psi_2\end{pmatrix} + \begin{pmatrix}0 & q(x,t)\\ r(x,t) & 0\end{pmatrix}\begin{pmatrix}\psi_1\\ \psi_2\end{pmatrix} \tag{3.23}$$

which can be written as

$$\frac{\partial P(x,t)}{\partial t} = 2i\Omega(L^+)AP \tag{3.24}$$

where

$$P = \begin{pmatrix} 0 & q \\ r & 0 \end{pmatrix}, \quad A = \begin{pmatrix} 1 & 0 \\ 0 & -1 \end{pmatrix},$$

$$L^+ = -\frac{1}{2}iA\frac{\partial}{\partial x} - \frac{1}{2}i\left\{P(x), \int_{-\infty}^{0} dy[P(y), A^*]\right\},$$

and $\Omega(L^+)$ is an arbitrary mermorphic functions. In the above,

$$\psi(x,t) = \begin{pmatrix} \psi_1 \\ \psi_2 \end{pmatrix}$$

is a two component Jost function, $g(x,t)$ and $r(x,t)$ are wave functions related to the states of microscopic particle with soliton feature, λ' is a constant. Equation (3.24) includes the nonlinear Schrödinger equation, Sine-Gordon equation, *etc.*

The Bäcklund transformation (BT) corresponding to (3.24) from the infinite-dimensional group P to P' for the systems consists of the following

$$\sum_{i=1}^{2} B_i(\Lambda^+)(H_i'P' - PH_i') = 0 \tag{3.25}$$

where

$$H_1' = \begin{pmatrix} 1 & 0 \\ 0 & 0 \end{pmatrix}, \quad H_2' = \begin{pmatrix} 0 & 0 \\ 0 & 1 \end{pmatrix},$$

and $B_1(\Lambda^+)$ and $B_2(\Lambda^+)$ are arbitrary entire functions. The operator Λ^+ has the following property,

$$\Lambda^+\phi = -\frac{i}{2}A\frac{\partial\phi}{\partial x} - \frac{i}{2}\int_{-\infty}^{x} dy[A\phi(y)P'(y) - P(y)A\phi(y)]P'(x)$$
$$+\frac{i}{2}P(x)\int_{-\infty}^{x} dy[A\phi(y)P'(y) - P(y)A\phi(y)].$$

We now consider an arbitrary discrete BT (3.25), *i.e.* let

$$B_1(\Lambda^+) = \prod_{i=1}^{n_1}(\Lambda^+ - \lambda_i), \quad B_2(\Lambda^+) = \prod_{k=1}^{n_2}(\Lambda^+ - \mu_k),$$

where λ_i and μ_k are arbitrary constants and n_1 and n_2 are arbitrary integers. Then, any arbitrary discrete BT may be expressed as

$$B = \prod_{k=1}^{n_2} B_{\mu_k}^{(2)} \prod_{i=1}^{n_1} B_{\lambda_i}^{(1)} \tag{3.26}$$

where $B_{\lambda'}^{(1)}$ and $B_{\mu'}^{(2)}$ are arbitrary elementary Bäcklund transformations (EBT). EBT $B_{\lambda'}^{(1)}$ is the BT (3.25) at $B_1 = \Lambda^+ - \lambda'$, and $B_2 = 1$ (λ' is an arbitrary constant). EBT $B_{\lambda'}^{(2)}$ is the BT (3.25) at $B_1 = 1$, and $B_2 = \Lambda^+ - \mu'$ (μ' is a constant). The explicit expressions of the EBTs are

$$B_{\lambda}^{(1)}(P \to P') : i\frac{\partial q'}{\partial x} - \frac{1}{2}q'^2 r + 2\lambda' q' + 2q = 0, \qquad (3.27)$$
$$i\frac{\partial r}{\partial x} + \frac{1}{2}r^2 q' - 2\lambda' r - 2r' = 0$$

$$B_{\mu}^{(2)}(P \to P') : i\frac{\partial q}{\partial x} - \frac{1}{2}q^2 r' + 2\mu' q + 2q' = 0, \qquad (3.28)$$
$$i\frac{\partial r'}{\partial x} + \frac{1}{2}r'^2 q - 2\mu' r' - 2r = 0.$$

The EBTs $B_{\lambda'}^{(1)}$ and $B_{\mu'}^{(2)}$ commute with each other

$$B_{\lambda'}^{(1)} B_{\mu'}^{(2)} = B_{\mu'}^{(2)} B_{\lambda'}^{(1)} \quad \text{and} \quad B_{\mu'}^{(2)} B_{\lambda'}^{(1)} = B_{\lambda'}^{(1)} B_{\mu'}^{(2)} = 1. \qquad (3.29)$$

Let us consider four solutions of (3.24), (q_0, r_0), (q_1, r_1), (q_2, r_2), and (q_3, r_3), which are related through the commutativity of $B_{\lambda}^{(1)}$ and $B_{\mu}^{(2)}$, respectively. Using (3.27) and (3.28), the following relations can be obtained

$$q_3 = q_0 + \frac{2(\lambda' - \mu')}{r_2/2 + 2/q_1}, \quad r_3 = r_0 + \frac{2(\mu' - \lambda')}{q_1/2 + 2/r_2}. \qquad (3.30)$$

Thus, if three solutions (q_0, r_0), (q_1, r_1) and (q_2, r_2) of (3.24) are known, the fourth solution (q_3, r_3) may be found from the relation given in (3.30). Equation (3.30) can then be considered as the nonlinear superposition formula corresponding to the nonlinear equation (3.24). As a matter of fact, equation (3.30) is applicable to all integrable nonlinear equations.

It can be shown that an infinite number of solutions of (3.24) can be obtained algebraically from the relation (3.30). Let us start from the trivial solution

$$P_{(00)} = \begin{pmatrix} 0 & 0 \\ 0 & 0 \end{pmatrix},$$

and apply all possible discrete BTs (3.26) on $P_{(00)}$, solutions of the entire system may be obtained and can be written as

$$P_{(a_1 a_2)} = \prod_{k=1}^{n_2} B_{\mu_k}^{(2)} \prod_{i=1}^{n_1} B_{\lambda_i}^{(1)} P_{(00)} \qquad (3.31)$$

where (n_1, n_2) is the vertex of $P_{(00)}$. The solutions $P_{(0n)}$ and $P_{(n0)}$ can be found from (3.27) and (3.28) and are given by

$$q_{(0n)} = 0, \quad r_{(0n)} = \sum_{l=1}^{n} \exp[-2i\Omega(\mu_l)t - 2i\mu_l(x - \bar{x}_{01})],$$
$$q_{(n0)} = \sum_{l=1}^{n} \exp[2i\Omega(\lambda_l)t + 2i\lambda_l(x - x_{01})], \quad r_{(n0)} = 0 \tag{3.32}$$

where x_{01} and $\bar{x}_{01}$ are arbitrary constants. Equations in (3.32) are also solutions of the linear equation

$$\frac{\partial P(x,t)}{\partial t} = 2i\Omega\left(-\frac{1}{2}iA\frac{\partial}{\partial x}\right)AP \tag{3.33}$$

which is the linear part of (3.24). The nonlinear superposition formula (3.30) corresponding to (3.31) can now be written as

$$q_{(n_1+1,n_2+1)} = q_{(n_1,n_2)} + \frac{2(\lambda' - \mu')}{r_{(n_1,n_2+1)}/2 + 2/q_{(n_1+1,n_2)}},$$
$$r_{(n_1+1,n_2+1)} = r_{(n_1,\cdot n_2)} + \frac{2(\mu' - \lambda')}{q_{(n_1+1,n_2)}/2 + 2/r_{(n_1,n_2+1)}}. \tag{3.34}$$

Using (3.34), any arbitrary solution $P_{(n_1,n_2)}$ may be found from (3.31). For example, if $P_{(00)}$, $P_{(10)}$ and $P_{(01)}$ are known, then $P_{(11)}$ can be obtained,

$$q_{(11)} = \frac{2(\lambda_1 - \mu_1)}{(1/2)e^{-2i\Omega\mu_1 t - 2i\mu_1(x-\bar{x}_{01})} + 2e^{-2i\Omega\lambda_1 t + 2i\lambda_1(x-x_{01})}}$$
$$r_{(11)} = \frac{2(\mu_1 - \lambda_1)}{(1/2)e^{2i\Omega\lambda_1 t - 2i\lambda_1(x-\bar{x}_{01})} + 2e^{2i\Omega\mu_1 t + 2i\mu_1(x-x_{01})}} \tag{3.35}$$

Similarly, $P_{(12)}$ may be obtained from $P_{(01)}$, $P_{(11)}$ and $P_{(02)}$ using (3.34),

$$q_{(12)} = \frac{2(\lambda_1 - \mu_2)}{r_{(02)}/2 + 2/q_{(11)}},$$
$$r_{(12)} = r_{(01)} + \frac{2(\mu_2 - \lambda_2)}{q_{(11)}/2 + 2/r_{(02)}}, \tag{3.36}$$

where $q_{(11)}$ is given in (3.35). From $P_{(20)}$, $P_{(10)}$ and $P_{(11)}$ we can get $P_{(21)}$

$$q_{(21)} = q_{(10)} + \frac{2(\lambda_2 - \mu_1)}{r_{(11)}/2 + 2/q_{(20)}},$$
$$r_{(21)} = \frac{2(\mu_1 - \lambda_2)}{q_{(20)}/2 + 2/r_{(11)}}, \tag{3.37}$$

where $r_{(11)}$ is given in (3.35). From $P_{(11)}, P_{(12)}$ and $P_{(21)}$ one can get $P_{(22)}$

$$q_{(22)} = q_{(11)} + \frac{2(\lambda_2 - \mu_2)}{r_{(12)}/2 + 2/q_{(21)}},$$
$$r_{(22)} = r_{(11)} + \frac{2(\mu_2 - \lambda_2)}{q_{(21)}/2 + 2/r_{(12)}}. \tag{3.38}$$

Other solutions, such as $P_{(31)}$, $P_{(32)}$, etc., can also be obtained from (3.34). We can see that an arbitrary solution $P_{(n_1n_2)}$ is a simple function of $q_{(10)}, q_{(20)}, \cdots, q_{(n_10)}$ and $r_{(10)}, r_{(02)}, \cdots, r_{(0n_2)}$. Thus, the nonlinear superposition principle (3.34) gives a simple algebraic structure for the family of infinite number of solutions of (3.31).

Solutions $P_{(11)}$, $P_{(22)}$ and $P_{(nn)}$, are soliton-type solutions. They are related through the Bäcklund transformation (3.32). If one of them is known, another can be obtained using (3.34). Let $r = q$ and $r = q^*$, then the solution $P_{(nn)}$ reduces to the n soliton solutions of the Sine-Gordon equation ($\Omega = L^{*-1}$, $r = q$, $\lambda' = -\mu'$) and the nonlinear Schrödinger equation ($\Omega = -2L^{*2}$, $r = q^*$, $\mu' = \lambda'^*$), respectively.

(B) The general nonlinear equation for the microscopic particle depicted by $\phi(x, Y, t)$ may be written as

$$\begin{aligned}\phi_t(x,Y,t) &= 2\beta_0(L,t)\phi_x(x,Y,t) + a_n(L)[\sigma_n, \phi(x,Y,t)] \\ &\quad + \beta_n(L)G\sigma_n + Y(L,t)\frac{\partial}{\partial y}\phi(x,Y,t).\end{aligned} \tag{3.39}$$

Assume that

$$\Phi(x,t) = \int_x^\infty dY \phi(x,Y,t). \tag{3.40}$$

Evidently, $\Phi(x, t)$ satisfies the following boundary conditions

$$\Phi(\pm\infty, t) = \Phi_t(\pm\infty, t) = 0.$$

In this case, the BT may be written as

$$\Phi'_x(x,t) + \Phi_x(x,t) = -\frac{1}{2}[\Phi'(x,t) - \Phi(x,t)][4P + \Phi'(x,t) - \Phi(x,t)] \tag{3.41}$$

where $k = iP$ is the eigenvalue of the following linear Schrödinger equation

$$\psi_{xx} = [\phi - k^2]\psi. \tag{3.42}$$

The nonlinear superposition principle corresponding to (3.41) is

$$(P_1 - P_2)[\Phi_{12} - \Phi_0] + \frac{1}{4}\{[\Phi_{12} - \Phi_0], [\Phi_1 - \Phi_2]\} = -(P_1 + P_2)[\Phi_1 - \Phi_2] \tag{3.43}$$

where P_1 and P_2 satisfy (3.41), $\Phi_0(x, t)$ is the solution of (3.39), $\Phi_1(x, t)$ and $\Phi_2(x, t)$ are also solutions of (3.39) corresponding to P_1 and P_2 respectively, and are related to Φ_0 through (3.41). $\Phi_{12}(x, t)$ is another solution of (3.39) which is related to the solution Φ_1 of (3.41) corresponding to $P = P_2$, or to the solution Φ_2 of (3.41) corresponding to $P = P_1$. The readers are referred to the book by R. H. Bullugh and P. J. Caudrey for more details of this superposition principle.

(C) In regard to the nonlinear superposition principle of $\phi_{\xi\tau} = \sin\phi$ of the Sine-Gordon equation, the following is sometimes used

$$\tan\left(\frac{\phi_3 - \phi_4}{4}\right) = \frac{\alpha_1 - \alpha_2}{\alpha_2 + \alpha_1}\tan\left(\frac{\phi_1 - \phi_2}{4}\right) \tag{3.44}$$

where α_1 and α_2 satisfy the following BT

$$\begin{aligned}\frac{\partial}{\partial\xi}\left(\frac{\phi_1-\phi_2}{2}\right)&=\alpha\sin\left(\frac{\phi_1+\phi_0}{2}\right),\\ \frac{\partial}{\partial\eta}\left(\frac{\phi_1+\phi_0}{2}\right)&=\frac{1}{\alpha}\sin\left(\frac{\phi_1-\phi_0}{2}\right).\end{aligned}\tag{3.45}$$

The derivation for the above will be given in Chapter 7.

3.3.2 *Nonlinear Fourier transformation*

The linear Fourier transformation

$$\psi''(\vec{p},t)=\frac{1}{(2\pi h)^{3/2}}\int\psi''(\vec{r},t)e^{-i\vec{p}\cdot\vec{r}/\hbar}d\tau$$

often used in the linear quantum mechanics no longer holds in the nonlinear quantum mechanics, due to nonlinear interaction. It must be replaced by a nonlinear Fourier transformation which was derived by Zakharov *et al.* An outline of the derivation is given in the following.

According to the Zakharov-Shabat equation, which can be obtained from the nonlinear Schrödinger equation (3.2) at $V(x,t)=A(\phi)=0$, we have

$$\begin{aligned}\psi_{1x}+i\xi'\psi_1&=\frac{\sqrt{2mb}}{\hbar}\phi\psi_2,\\ \psi_{2x}-i\xi'\psi_2&=\frac{\sqrt{2mb}}{\hbar}\phi\psi_1.\end{aligned}\tag{3.46}$$

The corresponding time-dependent forms of the above equations are

$$\begin{aligned}i\psi_{1t}&=-\left[\frac{\hbar}{m}\xi'^2+\left(\frac{b}{2\hbar}\right)\phi^*\phi\right]\psi_1+\sqrt{2mb}\left[i\xi'\phi-\frac{1}{2}\phi_x\right]\psi_2,\\ i\psi_{2t}&=-\left[\frac{\hbar}{m}\xi'^2-\left(\frac{b}{2\hbar}\right)\phi^*\phi\right]\psi_2-\sqrt{2mb}\left[i\xi'\phi^*+\frac{1}{2}\phi_x^*\right]\psi_1.\end{aligned}\tag{3.47}$$

It can be shown that the condition for (3.46) and (3.47) to be solvable is that the following nonlinear Schrödinger equation must be satisfied by ϕ,

$$i\hbar\phi_t=-\left(\frac{\hbar^2}{2m}\right)\phi_{xx}-b(\phi^*\phi)\phi.\tag{3.48}$$

This equation may be solved by using the inverse-scattering method in which the scattering data are

$$S_t=\{\xi'_j,p_j|_{j=1}^J;F(\xi')\ (\xi'=\text{constant})\}\tag{3.49}$$

Assume that ϕ satisfy the following

$$\int_{-\infty}^{\infty}|\phi(x,t)|dx<\infty.\tag{3.50}$$

Then when $x \to \pm\infty$, if

$$\Psi = \begin{pmatrix} \psi_1 \\ \psi_2 \end{pmatrix} = \begin{pmatrix} a(\xi')e^{-i\xi' x} \\ -b'(\xi')e^{i\xi' x} \end{pmatrix}, \tag{3.51}$$

where a and b' satisfy the following equations

$$\begin{aligned} \bar{a}(\xi')\alpha(\xi') + \bar{b}'(\xi')b'(\xi') &= 1, \\ \bar{a}(\xi') &= [a(\xi')]^*, \\ g\bar{b}'(\xi') &= [g\bar{b}'(\xi'^*)]^*, \end{aligned} \tag{3.52}$$

and the continuous frequency spectrum of the scattering data is given by

$$F(\xi') = \frac{b'(\xi')}{a(\xi')}, \tag{3.53}$$

where ξ' is real, then from (3.46), (3.47),(3.52) and (3.53), the nonlinear Fourier transformation for $\phi(x,t)$ determined by (3.47) can be obtained and is given by

$$F(\xi', t) = -\frac{b\sqrt{m}}{\hbar} \int_{-\infty}^{\infty} \phi^*(x,t)e^{-2i\xi' x}dx + O(b^3). \tag{3.54}$$

The lowest order in b of (3.54) gives the normal linear Fourier transformation.

3.3.3 *Method of quantization*

Similar to the linear quantum mechanics, wave function in the nonlinear quantum mechanics can be quantized. The commonly used methods are the canonical quantization method and the path integration method. However, because of the nonlinear effect, the actual procedure becomes much more complicated. The main steps for the canonical quantization are outlined below.

Let $V(\vec{r},t) = 0, A(\phi) = 0$ and using the natural unit system in which $\hbar = m = c = 1$, and replacing b by $2b$, the nonlinear Schrödinger equation (3.2) becomes

$$i\frac{\partial}{\partial t}\phi(t,x) = -\frac{\partial^2}{\partial x^2}\phi(t,x) + 2b\phi^*(t,x)\phi(t,x)\phi(t,x). \tag{3.55}$$

The quantization is best perceived in the Fourier transform space. If one defines approximately the Fourier transform

$$\begin{aligned} \frac{1}{\sqrt{2\pi}} \int_{-\infty}^{\infty} \phi(t,x)e^{i\beta x}dx &\equiv a(t,\beta), \\ \frac{1}{\sqrt{2\pi}} \int_{-\infty}^{\infty} a(t,\beta)e^{-i\beta x}d\beta &\equiv \phi(t,x), \end{aligned} \tag{3.56}$$

one obtains the equation of motion for the amplitude in Fourier transform space

$$i\frac{\partial}{\partial t}a(t,\beta) = \beta^2 a(t,\beta) + 2b\int d\beta_1 d\beta_2 a^*(t,\beta_1)a(t,\beta_2)a(t,\beta+\beta_1-\beta_2). \tag{3.57}$$

The field envelope can be normalized so that it represents the microscopic particle at "time" t. This enables us to identify $a(t,\beta)$ with the microscopic particle annihilation operator at time t, $\hat{a}(t,\beta)$, and $a^*(t,\beta)$ with the creation operator $\hat{a}^+(t,\beta)$. On the right-hand side of (3.57), the first term represents the dispersion effect and the second term represents the third-order nonlinearity. Note that the second term is in the form of a convolution. This is because for a broad-band field one has to integrate over the Fourier-transform space. The quantization is accomplished by assignment of the commutation relations

$$\begin{gathered}[\hat{a}(t,\beta'),\hat{a}^+(t,\beta)] = \delta(\beta-\beta'),\\ [\hat{a}(t,\beta'),\hat{a}(t,\beta)] = [\hat{a}^+(t,\beta'),\hat{a}^+(t,\beta)] = 0\end{gathered} \tag{3.58}$$

The quantized equation is

$$i\frac{\partial}{\partial t}\hat{a}(t,\beta) = \beta^2\hat{a}(t,\beta) + 2b\int d\beta_1 d\beta_2 \hat{a}^+(t,\beta_1)\hat{a}(t,\beta_2)\hat{a}(t,\beta+\beta_1-\beta_2). \tag{3.59}$$

Equation (3.59) can be derived from a well-defined Hamiltonian. That is, one can write (3.59) as

$$i\hbar\frac{d}{dt}\hat{a}(t,\beta) = [\hat{a}(t,\beta),\hat{H}] \tag{3.60}$$

with

$$\begin{aligned}\hat{H} = \hbar\Bigg[&\int \beta^2\hat{a}^+(t,\beta)\hat{a}(t,\beta)d\beta \\ &+ b\int \hat{a}^+(t,\beta)\hat{a}^+(t,\beta_1)\hat{a}(t,\beta_2)\hat{a}(t,\beta+\beta_1-\beta_2)d\beta d\beta_1 d\beta_2\Bigg].\end{aligned} \tag{3.61}$$

By defining new field operators as the inverse Fourier transforms of the annihilation and creation operators and applying the inverse Fourier transform to (3.59), one obtains the quantum nonlinear Schrödinger equation

$$i\frac{\partial}{\partial t}\hat{\phi}(t,x) = -\frac{\partial^2}{\partial x^2}\hat{\phi}(t,x) + 2b\hat{\phi}^+(t,x)\hat{\phi}(t,x)\hat{\phi}(t,x). \tag{3.62}$$

The operators $\hat{\phi}(t,x)$ and $\hat{\phi}^+(t,x)$ are the annihilation and creation operators of a quantum at a "point" x and "time" t.

From the definition of the Fourier transform (3.56) and the commutation relations (3.57), it is easy to prove that the field operators satisfy the following commutation relations

$$\begin{gathered}[\hat{\phi}(t,x''),\hat{\phi}^+(t,x)] = \delta(x-x''),\\ [\hat{\phi}(t,x''),\hat{\phi}(t,x)] = [\hat{\phi}^+(t,x''),\hat{\phi}^+(t,x)] = 0.\end{gathered} \tag{3.63}$$

With the help of (3.63), equation (3.62) can be written as

$$i\hbar\frac{d}{dt}\hat{\phi}(t,x) = \left[\hat{\phi}(t,x),\hat{H}\right] \tag{3.64}$$

with

$$\hat{H} = \hbar\left[\int \hat{\phi}_x^+(t,x)\hat{\phi}_x(t,x)dx + b\int \hat{\phi}^+(t,x)\hat{\phi}^+(t,x)\hat{\phi}(t,x)\hat{\phi}(t,x)dx\right]. \tag{3.65}$$

The quantum nonlinear Schrödinger equation (3.62) is the operator evolution equation of a nonlinear quantum system with the Hamiltonian (3.65). Since the quantum nonlinear Schrödinger equation can be derived from a Hamiltonian, it is a well-defined operator equation.

Equations (3.62) – (3.64) form the representation of nonlinear quantum mechanics in the Heisenberg picture and corresponds to the non-relativistic case. In the Schrödinger picture, the problem is stated in terms of the time evolution of a state $|\Phi\rangle$ of the system

$$i\hbar\frac{d}{dt}|\Phi\rangle = \hat{H}_s|\Phi\rangle, \tag{3.66}$$

with

$$\hat{H}_s = \hbar\left[\int \hat{\phi}_x^+(x)\hat{\phi}_x(x)dx + b\int \hat{\phi}^+(x)\hat{\phi}^+(x)\hat{\phi}(x)\hat{\phi}(x)dx\right] \tag{3.67}$$

where $\hat{\phi}(x)$ and $\hat{\phi}^+(x)$ are the field operators in the Schrödinger picture and satisfy the following commutation relations

$$\begin{gathered}\left[\hat{\phi}(x''),\hat{\phi}^+(x)\right] = \delta(x-x''),\\ \left[\hat{\phi}(x''),\hat{\phi}(x)\right] = \left[\hat{\phi}^+(x''),\hat{\phi}^+(x)\right] = 0.\end{gathered} \tag{3.68}$$

In the Heisenberg picture the quantum nonlinear Schrödinger equation (3.62) can be solved by quantum inverse-scattering method. In the Schrödinger picture, equation (3.66) could be solved from Bether's ansatz. Lai *et al.* expanded the quantum state in the Fock space and substituted it into (3.66). The result is a wave-function equation that has many degrees of freedom (like the equations in many-particle systems). However, for the quantum nonlinear Schrödinger equation, this wave-function equation is in a simple form and can be solved analytically.

Any quantum state of a system can be expanded in Fock space as follows

$$|\Phi\rangle = \sum_n a_n \int \frac{1}{\sqrt{n!}} f_n(x_1,\cdots,x_n,t)\hat{\phi}^+(x_1)\cdots\hat{\phi}^+(x_n)dx_1\cdots dx_n|0\rangle. \tag{3.69}$$

The state $|\Phi\rangle$ is a superposition of states produced from the vacuum state by creating particle at the points $x_1, x_2, \cdots, x_n$ with the weighting functions f_n. Since the particles are Bosons, f_n should be a symmetric function of x_j. We require a_n and f_n to satisfy the following normalization conditions

$$\sum_n |a_n|^2 = 1, \tag{3.70}$$

$$\int |f_n(x_1, \cdots, x_n, t)|^2 dx_1 \cdots dx_n = 1. \tag{3.71}$$

Substituting (3.69) and (3.67) into (3.66) and using (3.68), Lai *et al.* obtained an equation for $f_n(x_1, \cdots, x_n, t)$

$$i\frac{d}{dt}f_n(x_1, \cdots, x_n, t) = \left[-\sum_{j=1}^{n}\frac{\partial^2}{\partial x_j^2} + 2b\sum_{1\le i\le j\le n}\delta(x_j - x_i)\right] f_n(x_1, \cdots, x_n, t). \tag{3.72}$$

This is the Schrödinger equation for a one dimensional Boson system with a δ-function interaction. The t-dependence in (3.72) can be factored out by assuming a solution of the form

$$f_n(x_1, \cdots, x_n, t) = f_n(x_1, \cdots, x_n, t)e^{-iE_n t}. \tag{3.73}$$

The equation for $f_n(x_1, \cdots, x_n)$ is

$$\left[-\sum_{j=1}^{n}\frac{\partial^2}{\partial x_j^2} + 2b\sum_{1\le i\le j\le n}\delta(x_j - x_i)\right] f_n(x_1, \cdots, x_n) = E_n f_n(x_1, \cdots, x_n). \tag{3.74}$$

This shows that the quantum nonlinear Schrödinger equation is equivalent to the evolution equation of an one-dimensional Boson system with a δ-function interaction. It is surprising that the quantum nonlinear Schrödinger equation can be solved exactly. It was first solved by Betpher's ansatz and then by quantum inverse-scattering method.

We now introduce the quantization approach used by Lee *et al.* Consider real quantum field ϕ^i with N-components. Its corresponding Lagrangian function density can be written as

$$\mathcal{L} = -\frac{1}{2}\sum_i\left(\frac{\partial\phi^i}{\partial x_i}\right)^2 - \frac{1}{g^2}V(g\phi^i) \tag{3.75}$$

where the parameter g plays a role of a coupling constant. Now assume that

$$\phi^i_{cl}(r,t) = \frac{1}{g}\varphi^i(r, t, z_1, \cdots, z_k) \tag{3.76}$$

is a classical solution of a single microscopic particle (soliton) and the function $\varphi^i(r, t, z_1, \cdots, z_k)$ satisfies the following equation

$$\frac{\partial^2\varphi^i}{\partial x_\mu^2} - \frac{\partial V(\varphi^i)}{\partial\varphi^i} = 0, \quad (\mu = 1, 2, 3, 4). \tag{3.77}$$

For the Sine-Gordon equation, φ is given by

$$\varphi(x, z) = 4\tan^{-1}\left[e^{r_\mu(x-z)}\right].$$

The method of canonical quantization is to expand the soliton solution (3.77) in terms of the classical soliton solution φ/g, *i.e.*,

$$\phi^i(r,t) = \frac{1}{g}\varphi^i(r, z_1, \cdots, z_k) + \sum_{n=k+1}^{\infty} q_n(t)\varphi_n^i(r, z_1, \cdots, z_k) \tag{3.78}$$

where $q_n(t)$ satisfies the canonical commutative relation

$$\left[q_i(x,t), q_j^+(y,t)\right] = \delta_{ij}\delta(x-y). \tag{3.79}$$

The N-component function φ_n^i satisfies the following constraint,

$$\sum_{i=1}^{N}\int \varphi_n^i \frac{\partial \varphi^i}{\partial z_k} d\tau = 0. \tag{3.80}$$

Under orthogonal condition, we have

$$\sum_{i=1}^{N}\int \varphi_n^i \varphi_{n'}^i d\tau = \delta_{nn'}. \tag{3.81}$$

After some derivations, we can get

$$i\delta_k P_k \mu T + \int_0^{\mu_T} P_k dz_k - \sum_l \left(N_l' + \frac{1}{2}\right)\omega_n = 2\pi n,$$

or

$$\int_0^{\mu_T} \left\{P_k(z_k) + \frac{1}{\mu}\left[\varepsilon_a - \frac{1}{2}\sum_n \omega_n(z_n)\right]\right\} dz_k = 2\pi n, \tag{3.82}$$

where N_l' is an occupation number and $N_l' = 0, 1, 2, \cdots$, ω_n is a single vibrational frequency of static classical soliton solution $\varphi(r, z_1, \cdots, z_k)$, under the approximation of small oscillation. If we further assume that $|p\rangle$ is the eigenstate of both energy and momentum, that is

$$\hat{p}|p\rangle = p|p\rangle, \quad \hat{H}|p\rangle = E|p\rangle,$$

where $E(p) = \sqrt{p^2 + M^2}$. Then the nonlinear relation of the matrix element of the canonical commutation relation for the quantum field ϕ may be written as

$$\langle p|[\phi(x,t), \phi(y,t)]|p'\rangle = i\delta(x-y)2\pi\delta(p-p'). \tag{3.83}$$

The above is the collective coordinate method of canonically quantizing the classical soliton solution proposed by Lee, *et al.* The key in this approach is to find the classical soliton solution. However, this method is only suitable to the case of a nonlinear scalar field ϕ with its internal symmetry G being the Abel group.

From the above discussion we see that the even though the two quantization methods are somewhat different, they all quantize essentially the wave function of the particle or the solution of the dynamic equation.

3.3.4 *Nonlinear perturbation theory*

Most dynamic equations in nonlinear quantum mechanics cannot be solved analytically. Therefore perturbative approach is often used to solve these equations, when certain terms are small relatively to the nonlinear energy or the dispersive energy in these equations. In contrast to the linear quantum mechanics, a systematic perturbative approach and a universal formula are impossible to obtain in the nonlinear quantum mechanics due to nonlinear interactions. As a result, different perturbation methods exist in the nonlinear quantum mechanics which will be discussed in Chapter 7 in details. In this section, one of such methods is briefly described.

Assume that $\phi_0(\vec{r},t)$ is the soliton solution of the nonlinear dynamic equation of a unperturbed system. The wave function of the perturbed system can be written as

$$\phi = \phi_0 + \varepsilon\phi_1 + \varepsilon^2\phi_2 + \cdots . \tag{3.84}$$

where ε is a small quantity. Substituting (3.84) into the original nonlinear equation, equations of $\phi_1, \phi_2, \cdots$ corresponding to different orders of ε can be obtained, respectively. Thus $\phi_1, \phi_2, \cdots$ can be determined by solving these equations, and ϕ is obtained from (3.84).

For example, the Sine-Gordon equation:

$$\phi_{xx} - \phi_{tt} + \sin\phi + \Gamma\phi_t = A_1, \tag{3.85}$$

may be solved using the perturbation method when Γ and A_1 are small. Its solution is given by

$$\phi(x,t) = \phi_0(x - vt) + \phi_1(x,t) \tag{3.86}$$

where $\phi_1(x,t)$ is a small quantity. Equation for ϕ_1 can be obtained by substituting (3.86) into (3.85),

$$\phi_{1,tt} - \phi_{1,xx} - \left(1 - 2\mathrm{sech}^2 x\right)\phi_1 + \gamma\Gamma\phi_{1t} - \beta\gamma\phi_{1x} = A_1 + 2\beta\gamma\Gamma\mathrm{sech}x. \tag{3.87}$$

To solve the above equation, we expand $\phi_1(x,t)$ using the complete set $\{f_b(x), f_k(x)\}$

$$\phi_1(x,t) = \phi_{1b}(t)f_b(x) + \int_{-\infty}^{\infty} dk\phi_1(k,t)f_k(x) \tag{3.88}$$

where

$$\begin{aligned} f_b(x) &= \frac{1}{\sqrt{2}}\mathrm{sech}x, \\ f_k(x) &= \frac{1}{\sqrt{2\pi}}\frac{e^{ikx}}{\Omega_k}(k + i\tanh x). \end{aligned} \tag{3.89}$$

The solutions can be found and are given by

$$\begin{cases} \phi_1(k,t) = \dfrac{\pi k \delta(k) A_1}{\sqrt{2\pi}\Omega_k^3 \sinh(\pi k/2)}, \\ \\ \phi_{1b}(t) = 2\sqrt{2}\left(\beta + \dfrac{\pi A_1}{4\gamma\Gamma}\right)\left(1 - e^{-\gamma\Gamma t}\right). \end{cases} \tag{3.90}$$

Thus, the complete solution is

$$\phi_1(x,t) = 2R\sin[2R(x-\bar{x})]e^{-2iqx-2i\sigma'} \tag{3.91}$$

where

$$R_t = 2\Gamma R - \frac{\pi d}{2}\operatorname{sech}\left(\frac{q\pi}{2R}\right)\sin x,$$

$$q_t = \frac{\pi dq}{2R}\operatorname{sech}\left(\frac{q\pi}{2R}\right)\sin x.$$

We have given a very brief account of the four basic components of the nonlinear quantum mechanics. In the following chapters, these basic theories and their applications will be discussed in more details. Using these principles and theories, nonlinear quantum mechanical problems can be studied. But applications to real systems require detailed information of the nonlinear systems which will be discussed in the next section.

3.4 Properties of Nonlinear Quantum-Mechanical Systems

In the last section we described the fundamental principles of the nonlinear quantum mechanics which are used to describe nonlinear quantum systems. However, what are nonlinearly quantum-mechanical systems? What properties do these systems have? We will look into these in this section.

(1) The nonlinear quantum mechanics describes Hamiltonian systems. That is, behaviors of the systems can be determined by a set of canonical conjugate variables. Using these variables one can determine the Poisson bracket and write the equations in the form of Hamilton's equations. For equation (3.2) with $V(\vec{r},t) = A(\phi) = 0$, the variables ϕ and ϕ^* satisfy the Poisson bracket

$$\left\{\phi^{(a)}(x), \phi^{*(b)}(y)\right\} = i\delta^{ab}\delta(x-y) \tag{3.92}$$

where

$$\{A,B\} = i\int_{-\infty}^{\infty}\left(\frac{\delta A}{\delta\phi}\frac{\delta B}{\delta\phi^*} - \frac{\delta B}{\delta\phi}\frac{\delta A}{\delta\phi^*}\right).$$

The corresponding Lagrangian density $\mathcal{L}$ associated with (3.2) with $A(\phi) = 0$ can be written in terms of $\phi(x,t)$ and its conjugate ϕ^* viewed as independent variables

$$\mathcal{L} = \frac{i\hbar}{2}(\phi^*\phi_t - \phi\phi_t^*) - \frac{\hbar^2}{2m}(\nabla\phi \cdot \nabla\phi^*) - V(x)\phi^*\phi + b(\phi^* \cdot \phi)\phi. \tag{3.93}$$

The action of the system can be written as

$$S(\phi, \phi^*) = \int_{t_b}^{t_1} \int_D \mathcal{L} dx dt. \tag{3.94}$$

and its variation for infinitesimal $\delta\phi$ and $\delta\phi^*$

$$\delta S = S\{\phi + \delta\phi, \phi^* + \delta\phi^*\} - S\{\phi, \phi^*\} \tag{3.95}$$

can be written as

$$\delta S = \int_{t_0}^{t} \int_D \left[\frac{\partial \mathcal{L}}{\partial \phi}\delta\phi + \frac{\partial \mathcal{L}}{\partial \nabla\phi}\nabla\delta\phi + \frac{\partial \mathcal{L}}{\partial \phi_t}\delta\phi_t\right] dx dt + c.c. \tag{3.96}$$

where $\partial\mathcal{L}/\partial(\nabla\phi)$ denotes the vector with components $\partial\mathcal{L}/\partial(\partial_i\phi)$ $(i = 1, 2, 3)$. After integrating by parts, we get

$$\delta S = \int_{t_0}^{t_1} \int_D \left[\frac{\partial \mathcal{L}}{\partial \phi} - \nabla \cdot \left(\frac{\partial \mathcal{L}}{\partial \nabla\phi}\right) - \partial_t\left(\frac{\partial \mathcal{L}}{\partial \phi_t}\right)\right] \delta\phi dx dt + \left[\frac{\partial \mathcal{L}}{\partial \phi_t}\delta\phi\right]_{t_0}^{t_1} + c.c. \tag{3.97}$$

A necessary and sufficient condition for a function $\phi(x,t)$ with known values $\phi(x,t_0)$ and $\phi(x,t_1)$ to yield an extremum of the action S is that it must satisfy the Euler-Lagrange equation

$$\frac{\partial \mathcal{L}}{\partial \phi} = \nabla \cdot \left(\frac{\partial \mathcal{L}}{\partial \nabla\phi}\right) + \partial_t\left(\frac{\partial \mathcal{L}}{\partial \phi_t}\right). \tag{3.98}$$

Equation (3.98) gives the nonlinear Schrödinger equation (3.2) if the Lagrangian (3.93) is used. Therefore, the dynamic equation, or the nonlinear Schrödinger equation, in the nonlinear quantum mechanics can be derived from the Euler-Lagrange equation, if the Lagrangian function of the system is known. This is different from the linear quantum mechanics, in which a dynamic equation, or the linear Schrödinger equation cannot be obtained from the Euler-Lagrange equation. This is a unique property of the nonlinear quantum mechanics.

The above derivation for the nonlinear Schrödinger equation based on the variational principle is a foundation for other methods such as the "collective coordinates", the "variational approach", and the "Rayleigh-Ritz optimization principle", where a solution is assumed to maintain a prescribed approximate profile (often bell-type). Such methods greatly simplify the problem, reducing it to a system of ordinary differential equations for the evolution of a few characteristics of the systems.

The Hamiltonian density corresponding to (3.2) or (3.93) can be written as

$$\mathcal{H} = \frac{i\hbar}{2}(\phi^*\partial_t\phi - \phi\partial_t\phi^*) - \mathcal{L} = \frac{\hbar^2}{2m}(\nabla\phi \cdot \nabla\phi^*) + V(x)\phi^*\phi - b(\phi^*\phi)^2. \quad (3.99)$$

Introducing the canonical variables

$$\begin{aligned} q_1 &= \frac{1}{2}(\phi + \phi^*), \qquad p_1 = \frac{\partial \mathcal{L}}{\partial(\partial_t q_1)}; \\ q_2 &= \frac{1}{2i}(\phi - \phi^*), \qquad p_2 = \frac{\partial \mathcal{L}}{\partial(\partial_t q_2)}, \end{aligned} \quad (3.100)$$

the Hamiltonian density takes the form

$$\mathcal{H} = \sum_i p_i\partial_t q_i - \mathcal{L} \quad (3.101)$$

and the corresponding variation of the Lagrangian density can be written as

$$\delta\mathcal{L} = \sum_i \frac{\delta\mathcal{L}}{\delta q_i}\delta q_i + \frac{\delta\mathcal{L}}{\delta(\nabla q_i)}\delta(\nabla q_i) + \frac{\delta\mathcal{L}}{\delta(\partial_t q_i)}\delta(\partial_t q_i). \quad (3.102)$$

From (3.102), the definition of p_i, and the Euler-Lagrange equation

$$\frac{\partial\mathcal{L}}{\partial q_i} = \nabla \cdot \frac{\partial\mathcal{L}}{\partial(\nabla q_i)} + \partial_t p_i$$

one obtains the variation of the Hamiltonian in the form

$$\delta H = \sum_i \int (\partial_t q_i \delta p_i - \partial_t p_i \delta q_i) dx. \quad (3.103)$$

Thus the Hamilton equations can be derived

$$\frac{\partial q_i}{\partial t} = \frac{\delta H}{\delta p_i}, \quad \frac{\partial p_i}{\partial t} = -\frac{\delta H}{\delta q_i} \quad (3.104)$$

or in complex form

$$i\partial_t\phi = \frac{\delta H}{\delta\phi^*}. \quad (3.105)$$

This is also interesting. It shows that dynamic equations, such as the nonlinear Schrödinger equation, can also be obtained from the Hamilton equation, if the Hamiltonian of the system is known. Obviously, such method of finding dynamic equations is impossible in the linear quantum mechanics.

The Euler-Lagrange equation and the Hamilton equations are important equations in classical theoretical (analytic) mechanics, they were used to describe motions of classical particles. Now these equations are used to depict motions of microscopic particles in the nonlinear quantum mechanics. This shows clearly the classical nature of microscopic particles in the nonlinear quantum mechanics. From these, we obtain new ways of finding the equation of motion of microscopic particles in the nonlinear quantum mechanics. If the Hamiltonian or the Lagrangian of a system is known in the coordinate representation, then we can obtain the equation of motion from the Euler-Lagrange or Hamilton equations.

(2) The nonlinear dynamic equations such as the nonlinear Schrödinger equation can be written either in the Lax form, given in (3.20) and (3.21), or in the Hamilton form of a compatibility condition for overdetermined linear spectral problem. General nonlinear equations with soliton solutions are often expressed as (3.8), where $K(\phi)$ is defined as a nonlinear operator in some suitable function space. If linear operators $\hat{L}$ and $\hat{B}$ which depend on solution ϕ satisfy the Lax operator equation (3.20), where $\hat{B}$ is a self-adjoint operator, the eigenvalue λ and eigenfunction ψ of operator $\hat{L}$ may be derived from (3.21). Thus, eigenvectors and eigenvalues of nonlinear dynamic equations can be determined by the eigenvectors and eigenvalues of linear operators. Then, λ is a time-independent eigenvalue. Equation (3.21) is the corresponding linear eigenequation of the nonlinear equation (3.8). For the nonlinear Schrödinger equation (3.2) with $V(\vec{r}) = A(\phi) = 0$, the operators $\hat{L}$ and $\hat{B}$ are of the forms of (3.22).

We now assume that the eigenfunction of the operator $\hat{L}$ satisfies the following equation

$$i\psi_{t'} = \hat{B}\psi + f(\hat{L})\psi \tag{3.106}$$

where the function $f(\hat{L})$ may be chosen according to convenience. In this case we can write the overdetermined system of linear matrix equations as

$$\begin{aligned} \psi_{x'} &= U(x', t', \lambda)\psi, \\ \psi_{t'} &= V'(x', t', \lambda)\psi, \end{aligned} \tag{3.107}$$

where U and V' are 2×2 matrices, $t' = t/\hbar$, $x' = (\sqrt{2m}/\hbar)x$. The compatibility condition of this system is obtained by differentiating the first equation of (3.107) with respect to t' and the second one with respect to x' and then subtracting one from the other

$$U_{t'} - V'_{x'} - [U, V'] = 0. \tag{3.108}$$

We emphasize that the operators U and V' depend not only on t' and x' but also on some parameter, λ, which is called a spectral parameter. Condition (3.108) must be satisfied by λ. In this case the operators U and V' have the form

$$U = -i\lambda \begin{bmatrix} (1+s)^{-1} & 0 \\ 0 & (1-s)^{-1} \end{bmatrix} + \begin{bmatrix} 0 & i(1+s)^{-1}\phi^* \\ i(1+s)^{-1}\phi & 0 \end{bmatrix},$$

$$V' = -i\lambda^2 \begin{bmatrix} s(1+s)^{-2} & 0 \\ 0 & s(1-s)^{-2} \end{bmatrix} \tag{3.109}$$

$$+2i\lambda \begin{bmatrix} 0 & \dfrac{s\phi^*}{(1+s)(1-s^2)} \\ \dfrac{s\phi}{(1-s)(1-s^2)} & 0 \end{bmatrix} + \begin{pmatrix} -\dfrac{i(\phi\phi^*)}{1-s^2} & \dfrac{\phi_{x'}}{1+s} \\ -\dfrac{\phi_{x'}}{1-s} & \dfrac{i(\phi\phi^*)}{1-s^2} \end{pmatrix}.$$

The presence of a continuous time-independent parameter is a reflection of the fact that the nonlinear Schrödinger equation (3.2) with $V(\vec{r},t) = A(\phi) = 0$ describes a Hamiltonian system with a set of infinite number of conservation laws, which will be discussed in chapters 4. In the case of a system with a finite number (N') of degrees of freedom one can succeed sometimes in finding $2N'$ first integrals of motion between which the Poisson brackets are zero (they are said to be in involution). Such a system is called completely integrable. In such a case, equation (3.108) with (3.109) is referred to as the compatibility condition, and its consequence is the nonlinear Schrödinger equation (3.2) with $V(x,t) = A(\phi) = 0$. This means that for each solution, $\phi(x',t')$, of (3.2), there is always a set of basis function ψ, parameterized by λ, which can be obtained through solving the set of linear equations (3.107) – (3.109). Therefore, the nonlinear quantum systems described by the nonlinear quantum mechanics is completely integrable. This concept was generalized by Zakharov and Faddeev in quantum field theory.

(3) The systems described by the above equations have infinite countable (or in the case of the internal symmetry, sets of) conservation laws (local or non-local) and integrals of motion.

We now write the wave function in terms of the phase and amplitude, $\phi = \sqrt{\rho}e^{i\theta}$. This transformation is usually called Madelung's transformation. After substitution in the nonlinear Schrödinger equation (3.2) with $V(x,t) = 0$ and $A(\phi) = 0$ and separation of the real and imaginary parts of the equation, one obtains (after rescaling the time by a factor of 2)

$$\rho_t + \nabla \cdot (\rho\nabla\theta) = 0, \quad \text{or} \quad \rho_t = -\nabla \cdot \vec{j}, \quad \vec{j} = \rho\nabla\theta, \tag{3.110}$$

and

$$\theta_t + \frac{1}{2}|\nabla\theta|^2 - \frac{b}{2}\rho = \frac{1}{2\sqrt{\rho}}\Delta\sqrt{\rho}. \tag{3.111}$$

Taking $\rho = |\phi|^2 = |\varphi|^2$ as a density and θ as an hydrodynamic potential, equations (3.110) – (3.111) with $b < 0$ can be identified with the equations for an irrotational barotropic gas. Hence, (3.110) is a continuum equation that occurs in macrofluid hydrodynamics. Thus, equation (3.110) is also called the fluid-dynamical form of the nonlinear Schrödinger equation. This shows the features of classical particles of the microscopic particle described by the nonlinear Schrödinger equation in the nonlinear quantum mechanics. From this interpretation of $\rho = |\phi|^2$ we know clearly that the meaning of $|\phi|^2 = |\varphi|^2$ is in truth far from the concept of probability in the linear quantum mechanics. It essentially represents the mass density of the microscopic particles in the nonlinear quantum mechanics. If (3.110) is integrated over x one finds the integral of motion $N = \int_{-\infty}^{\infty} \rho(x,t)dx$ with $j \to 0$ when $x \to \pm\infty$. N is then the mass of the microscopic particle, and (3.110) is nothing but the mass conservation for the microscopic particle. This discussion enhances our understanding on the meaning of wave function of the microscopic particle, $\phi(\vec{r},t)$, in nonlinear quantum mechanics.

(4) In some cases, the inverse scattering problem for the operator $\hat{L}$, in (3.20), can be solved, and a potential can be obtained from the scattering data. The role of the potential is played through the function ϕ. This means that for the equations discussed one can solve the Cauchy problem and the behavior of the integrable system is strictly determinate. The localized solutions of the integrable equations which correspond to the discrete spectrum of the operator $\hat{L}$ are usually related to the solitons. In the case of the simplest integrable systems the soliton dynamics is trivial.

(5) For integrable systems, an exact quantum approach can be developed that allows one to determine the ground state and excitation spectra of the system. This was studied by Faddeev *et al.* and Thacker *et al.*. It thus connects classical description and quantum objects with sufficient rigor. For integrable equations, one may construct a perturbation theory and investigate the structural and "initial" stability problems. In the first case the equation itself is weakly perturbed and the perturbation can be non-Hamiltonian. In the second case the initial state studied (in particular it may be a one-soliton solution) is weakly perturbed. For Hamiltonian systems with a finite number of degrees of freedom there is a rigorous theory of structural stability. These problems were studied in details in the literature.

Bibliography

Ablowitz, M. J., Kaup, D. J. Newell, A. C. and Segur, H. (1973). Phys. Rev. Lett. **30** (1462) ; ibid **31** (125) .

Ablowitz, M. J., Kaup, D. J. Newell, A. C. and Segur, H. (1974). Stud. Appl. Math. **53** (249) .

Alfinito, E., *et al.* (1996). Nonlinear physics, theory and experiment, World Scientific, Singapore.

Bullough, R. K. and Caudrey, P. J. (1980). Solitons, Springer-Verlag, Berlin.

Burt, P. B. (1981). Quantum mechanics and nonlinear waves, Harwood Academic Publishers, New York.

Calogero, F. (1978). Nonlinear evolution equations solvable by the spectral transform, London, Plenum Press.

Calogero, F. and Degasperis, A. (1976). Nuovo Cimento **32B** (201) .
Chen, X. R., Gou, Q. Q. and Pang, X. F. (1996). Chin. Phys. Lett. **13** (660) .
Chen, X. R., Gou, Q. Q. and Pang, X. F. (1997). Chin. J. Atom. Mol. Phys. **14** (393) ; Chin. J. Chem. Phys. **10** (145) .
Chen, X. R., Gou, Q. Q. and Pang, X. F. (1998). Acta Phys. Sin. **7** (329) ; J. Sichuan University (nature) Sin. **35** (362) .
Chen, X. R., Gou, Q. Q. and Pang, X. F. (1999). Acta Phys. Sin. **8** (1313) ; Chin. J. Chem. Phys. **11** (240) ; Chin. J. Comput. Phys. **16** (346) ; Commun. Theor. Phys. **31** (169) .
Christ, N. H. (1976). Phys. Rep. C **23** (294) .
Doebner, H. D., *et al.* (1995). Nonlinear deformed and irreversible quantum systems, World Scientific, Singapore.
Doebner, H. D., Manko, V. I. and Scherer, W. (2000). Phys. Lett. A **268** (17) .
Faddeev, L. (1979). JINR D2-12462, Dubna.
Faddeev, L. D. and Takhtajan, L. A. (1987). Hamiltonian methods in the theory of solitons, Springer, Berlin.
Friedberg, R., Lee, T. D. and Sirlin, A. (1976). Phys. Rev. D **13** (2739) .
Gisin, N. (1990). Phys. Lett. A **143** (1) .
Gisin, N. and Gisin, B. (1999). Phys. Lett. A **260** (323) .
Gisin, N. and Rigo, M. (1995). J. Phys. A **8** (7375) .
Jordan, T. F. and Sariyianni, Z. E. (1999). Phys. Lett. A **263** (263) .
Karpman, K. I. (1979). Phys. Lett. A **71** (163) .
Konopelchenko, B. G., (1979). Phys. Lett. A **74** (189) .
Konopelchenko, B. G., (1981). Phys. Lett. B **100** (254) .
Konopelchenko, B. G., (1982). Phys. Lett. A **87** (445) .
Konopelchenko, B. G. (1987). Nonlinear integrable equations, Springer, Berlin.
Konopelchenko, B. G. (1995). Solitons in multidimensions, World Scientific, Singapore.
Lai, Y. and Haus, H. A. (1989). Phys. Rev. A **40** (844) and 854.
Lee, T. D., *et al.* (1975). Phys. Rev. D **12** (1606) .
Makhankov, V. G. and Fedyanin, V. K. (1984). Phys. Rep. **104** (1) .
Mielnik, B. (2001). Phys. Lett. A **289** (1) .
Miura, R. M. (1976). Bäcklund transformations and the inverse scattering method, solitons and their applications, Springer, Berlin.
Pang, X. F. and Chen, X. R. (2000). Chin. Phys. **9** (108) .
Pang, X. F. and Chen, X. R. (2001). Commun. Theor. Phys. **35** (323) .
Pang, X. F. and Chen, X. R. (2001). Inter. J. Infr. Mill. Waves **22** (291) .
Pang, X. F. and Chen, X. R. (2001). J. Phys. Chem. Solids **62** (793) .
Pang, X. F. and Chen, X. R. (2002). Commun. Theor. Phys. **37** (715) .
Pang, X. F. and Chen, X. R. (2002). Inter. J. Infr. Mill. Waves **23** (375) .
Pang, X. F. and Chen, X. R. (2002). Phys. Stat. Sol. (b) **229** (1397) .
Pang, Xiao-feng (1983). Physica Bulletin Sin. **5** (6) .
Pang, Xiao-feng (1984). Principle and theory of nonlinear quantum mechanics, Proc. 3rd National Conf. on Quantum Mechanics, p. 27.
Pang, Xiao-feng (1985). Chin. J. Potential Science **5** (16) .
Pang, Xiao-feng (1985). J. Low Temp. Phys. **58** (334) .
Pang, Xiao-feng (1985). Problems of nonlinear quantum theory, Sichuan Normal Univ. Press, Chengdu.
Pang, Xiao-feng (1990). The elementary principle and theory for nonlinear quantum mechanics, Proc. 4th APPC (Soul Korea) p. 213.
Pang, Xiao-feng (1991). The theory of nonlinear quantum mechanics, Proc. ICIPES, Beijing, p. 123.

Pang, Xiao-feng (1992). Proc. IWPM-6 (Shenyang), p. 83.
Pang, Xiao-feng (1993). The theory of nonlinear quantum mechanics, in research of new sciences, eds. Lui Hong, Chin. Science and Tech. Press, Beijing, p. 16.
Pang, Xiao-feng (1994). Acta Phys. Sin. **43** (1987) .
Pang, Xiao-feng (1994). The theory for nonlinear quantum mechanics, Chongqing Press, Chongqing.
Pang, Xiao-feng (1995). Chin. J. Phys. Chem. **12** (1062) .
Pang, Xiao-feng (1995). J. Huanghuai Sin. **11** (21) .
Pang, Xiao-feng (1997). J. Southwest Inst. Nationalities, Sin. **23** (418) .
Pang, Xiao-feng (1998). J. Southwest Inst. Nationalities, Sin. **24** (310) .
Pang, Xiao-feng (2003). Soliton physics, Sichuan Sci. and Tech. Press, Chengdu.
Sabatier, P. C. (1990). Inverse method in action, Springer, Berlin.
Sulem, C and Sulem, P. L. (1999). The nonlinear Schrödinger equation: self-focusing and wave collapse, Springer-Verlag, Berlin.
Toda, M. (1989). Nonlinear waves and solitons, Kluwer Academic Publishers, Dordrecht.
Weinberg, S. (1989). Ann. Phys. (N.Y.) **194** (336) .
Weinberg, S. (1989). Phys. Rev. Lett. **62** (485) .
Wright, E. M. (1991). Phys. Rev. A **43** (3836) .
Zakharov, B. E. and Shabat, A. B. (1971). Zh. Eksp. Teor. Fiz. **61** (118) .
Zakharov, B. E. and Shabat, A. B. (1972). Sov. Phys. JETP **34** (62) .
Zakharov, V. and Fadeleev, I. (1971). Funct. Analiz **5** (18) .

Chapter 4

Wave-Corpuscle Duality of Microscopic Particles in Nonlinear Quantum Mechanics

In this chapter, we will look into physical properties of microscopic particles which are described by the nonlinear quantum mechanics. In particular, we focus on the wave-corpuscle duality of microscopic particles and its description in nonlinear quantum mechanics. Whether microscopic particles have wave-corpuscle duality is a key and central issue of quantum mechanics. Unless the nonlinear quantum mechanics can give a correct description for the wave-corpuscle duality of microscopic particles, it cannot be accepted as a proper theory and a different theory from the linear quantum mechanics which only describes the wave feature of microscopic particles properly. Therefore, in this chapter we present an extensive and in-depth study on this feature of microscopic particles, from their static state to dynamic properties, based on the fundamental principles of the nonlinear quantum mechanics. Hence, this chapter is the center of this book.

In accordance with the significance of the corpuscle feature of microscopic particles, we focus our discussion on the following issues: (1) the mass, momentum and energy of microscopic particles and the corresponding conservation laws; (2) the size, position and velocity of microscopic particles and the laws of motion of microscopic particles in free space, in the presence of an external field, and in a viscous environment; (3) interaction and collision between microscopic particles, or between microscopic particles and other particles or objects; (4) the stability of the microscopic particles under external perturbations; and so on. From these discussions, we will see clearly that microscopic particles in the nonlinear quantum mechanics become solitons, which have properties of a classical particle. It can move over a macroscopic distance, retaining its shape, energy, momentum, and remains stable after collisions and interactions with other particles or external fields.

We will also examine the wave features of a microscopic particle, including its wavevector, velocity, amplitude, the superposition principle of waves, scattering, diffraction, reflection, transmission and tunneling phenomena. We will see that each microscopic particle has a determinate wave vector, amplitude and width, and obeys the physical laws of scattering, diffraction, reflection, transmission, tunneling, *etc.* Therefore, the microscopic particles has not only corpuscle but also wave features, *i.e.*, it has evident wave-corpuscle duality in the nonlinear quantum mechanics.

4.1 Invariance and Conservation Laws, Mass, Momentum and Energy of microscopic particles in the nonlinear quantum mechanics

We have learned in earlier chapters some conservation laws in a system of microscopic particles which are described by the nonlinear Schrödinger equation in the nonlinear quantum mechanics. In practice, these conservation laws are related to the invariance of the action with respect to groups of transformations through the Noether theorem, which was studied by Gelfand and Fomin (1963) and Bluman and Kumei (1989) (See Sulem and Sulem, 1999 and references therein). Therefore, before these conservation laws are given, we first discuss the Noether theorem for the nonlinear Schrödinger equation which was given by Sulem.

To simplify the equation, we introduce the following notations: $\bar{\xi} = (t, x) = (\tilde{\xi}_0, \tilde{\xi}_1, \cdots, \tilde{\xi}_a)$, $\partial_0 = \partial_t$, $\partial = (\partial_0, \partial_1, \cdots, \partial_d)$ and $\Phi = (\Phi_1, \Phi_2) = (\phi, \phi^*)$. According to the Lagrangian (3.93) corresponding to the nonlinear Schrödinger equation, the action of the system

$$S\{\phi\} = \int_{t_0}^{t_1} \int \mathcal{L}(\phi, \nabla\phi, \phi_t, \phi^*, \nabla\phi^*, \phi_t^*) dx dt$$

now becomes

$$S\{\phi\} = \int_D \int_{x'}^{\infty} \mathcal{L}(\Phi, \partial\Phi) d\bar{\xi}. \tag{4.1}$$

Under the transformation T^ε which depends on the parameter ε, we have $\bar{\xi} \to \tilde{\xi}'(\bar{\xi}, \Phi, \varepsilon)$, $\Phi \to \tilde{\Phi}(\bar{\xi}, \Phi, \varepsilon)$, where $\tilde{\xi}$ and $\tilde{\Phi}$ are assumed to be differentiable with respect to ε. When $\varepsilon = 0$, the transformation reduces to the identity. For infinitesimally small ε, we have $\tilde{\xi}' = \bar{\xi} + \delta\tilde{\xi}$, $\tilde{\Phi} = \Phi + \delta\Phi$. At the same time, $\Phi(\bar{\xi}) \to \tilde{\Phi}(\tilde{\xi}')$ by the transformation T^ε, and the domain of integration D is transformed into $\tilde{D}$,

$$S\{\phi\} \to \tilde{S}\{\tilde{\phi}\} = \int_{\tilde{D}} \int_{x'}^{\infty} \mathcal{L}(\tilde{\Phi}, \tilde{\partial}\tilde{\Phi}) d\tilde{\xi}',$$

where $\tilde{\partial}$ denotes differentiation with respect to $\tilde{\xi}'$. The change $\delta S = \tilde{S}\{\tilde{\phi}\} - S\{\phi\}$ in the limit of ε under the above transformation can be expressed as

$$\delta S = \int_D \int_{x'}^{\infty} \left[\mathcal{L}(\tilde{\Phi}, \tilde{\partial}\tilde{\Phi}) - \mathcal{L}(\Phi, \partial\Phi)\right] d\bar{\xi} + \int_D \int_{x'}^{\infty} \mathcal{L}(\Phi, \partial\Phi) \sum_{v=0}^{d} \frac{\partial\delta\tilde{\xi}_v}{\partial\tilde{\xi}_v} d\bar{\xi}$$

where we used the Jacobian expansion

$$\frac{\partial(\tilde{\xi}'_0, \cdots, \tilde{\xi}'_d)}{\partial(\tilde{\xi}_0, \cdots, \tilde{\xi}_d)} = 1 + \sum_{v=0}^{d} \frac{\partial\delta\tilde{\xi}_v}{\partial\tilde{\xi}_v},$$

and $\mathcal{L}(\tilde{\Phi}, \tilde{\partial}\tilde{\Phi})$ in the second term on the right-hand side has been replaced by the leading term $\mathcal{L}(\Phi, \partial\Phi)$ in the expansion.

Now define

$$\delta\tilde{\Phi}_i = \tilde{\Phi}_i(\tilde{\xi}') - \Phi_i(\bar{\xi}) = \partial_v\Phi_i\delta\tilde{\xi}_v + \delta\Phi_i(\bar{\xi}),$$
$$\tilde{\partial}_v\tilde{\Phi}_i(\tilde{\xi}) - \partial_v\Phi_i(\bar{\xi}) = (\tilde{\partial}_v - \partial_v)\tilde{\Phi}_i(\tilde{\xi}) + \partial_v[\tilde{\Phi}_i(\tilde{\xi}) - \Phi(\bar{\xi})], \tag{4.2}$$

with

$$\partial_v = \frac{\partial\tilde{\xi}'_\mu}{\partial\tilde{\xi}_v}\tilde{\partial}_\mu = \left(\delta_{v\mu} + \frac{\partial\delta\tilde{\xi}_\mu}{\partial\tilde{\xi}_v}\right)\tilde{\partial}_\mu = \tilde{\partial}_v + \frac{\partial\delta\tilde{\xi}_\mu}{\partial\tilde{\xi}_v}\tilde{\partial}_\mu.$$

We then have

$$\mathcal{L}(\tilde{\Phi}, \tilde{\partial}\tilde{\Phi}) - \mathcal{L}(\Phi, \partial\Phi) = \frac{\partial\mathcal{L}}{\partial\Phi_i}[\tilde{\Phi}_i(\tilde{\xi}') - \Phi_i(\bar{\xi})] + \frac{\partial\mathcal{L}}{\partial(\partial_v\Phi_i)}[\tilde{\partial}_v\tilde{\Phi}_i(\tilde{\xi}') - \partial_\mu\Phi_i(\bar{\xi})]$$
$$= \frac{\partial\mathcal{L}}{\partial\Phi_i}\partial\Phi_i + \partial_\mu(\mathcal{L}\delta\tilde{\xi}_v) - \mathcal{L}\frac{\partial\delta\tilde{\xi}_v}{\partial\tilde{\xi}_v} + \partial_v\left[\frac{\partial\mathcal{L}}{\partial(\partial_v\Phi_i)}\right]\delta\Phi_i - \partial_\mu\left[\frac{\partial\mathcal{L}}{\partial(\partial_v\Phi_i)}\right]\delta\Phi_i.$$

Equation (3.96) can now be replaced by

$$\delta S = \int_D\int_{x'}^{\infty}\left\{\frac{\partial\mathcal{L}}{\partial\Phi_i} - \frac{\partial}{\partial\tilde{\xi}_v}\left[\frac{\partial\mathcal{L}}{\partial(\partial_v\Phi_i)}\right]\right\}\delta\Phi_i d\bar{\xi} + \int_D\int_{x'}^{\infty}\frac{\partial}{\partial\xi_v}\left[\mathcal{L}\delta\tilde{\xi}_v + \frac{\partial\mathcal{L}}{\partial(\partial_v\Phi_i)}\delta\Phi_i\right]d\bar{\xi}$$

where we have used

$$\frac{\partial}{\partial\tilde{\xi}_v}(\mathcal{L}\delta\tilde{\xi}_v) = \mathcal{L}\frac{\partial\delta\tilde{\xi}_v}{\partial\tilde{\xi}_v} + \frac{\partial\mathcal{L}}{\partial\Phi_i}\partial_v\Phi_i\delta\tilde{\xi}_v + \frac{\partial^2\mathcal{L}}{\partial(\partial_\mu\Phi_i)}\partial^2_{v\mu}\Phi_i\delta\tilde{\xi}_v,$$

$$\frac{\partial\mathcal{L}}{\partial(\partial_v\Phi_i)}\partial_v\int_{x'}^{\infty}\delta\Phi_i = \frac{\partial}{\partial\tilde{\xi}_v}\left[\frac{\partial\mathcal{L}}{\partial(\partial_v\Phi_i)}\delta\Phi_i\right] - \frac{\partial}{\partial\tilde{\xi}_v}\left[\frac{\partial\mathcal{L}}{\partial(\partial_v\Phi_i)}\right]\delta\Phi_i.$$

Using the Euler-Lagrange equation, the first term on the right-hand side in the equation of δS vanishes. We can get the Noether theorem.

(I) if the action (4.1) is invariant under the infinitesimal transformation of the dependent and independent variables $\phi \to \phi+\delta\phi$, $\bar{\xi} \to \bar{\xi}+\delta\bar{\xi}$ where $\bar{\xi} = (t, x_1, \cdots, x_d)$, the following conservation law holds

$$\frac{\partial}{\partial\xi_v}\left[\mathcal{L}\delta\xi_v - \frac{\partial\mathcal{L}}{\partial(\partial_v\Phi_i)}\delta\Phi_i\right] = 0,$$

or

$$\frac{\partial}{\partial\xi_v}[\mathcal{L}\delta\xi_v + \frac{\partial\mathcal{L}}{\partial(\partial_v\Phi_i)}(\delta\Phi_i - \frac{\partial\Phi_i}{\partial\tilde{\xi}_\mu}\delta\tilde{\xi}_\mu)] = 0 \tag{4.3}$$

in terms of $\delta\hat{\Phi}_i$ defined above.

(II) If the action is invariant under the infinitesimal transformation

$$\begin{aligned} t \to \tilde{t} &= t + \delta t(x,t,\phi), \\ x \to \tilde{x} &= x + \delta x(x,t,\phi), \\ \phi(x,t) \to \tilde{\phi}(\tilde{t},\tilde{x}) &= \phi(t,x) + \delta\phi(t,x), \end{aligned}$$

then

$$\int \left[\frac{\partial \mathcal{L}}{\partial \phi_t}(\partial_t \phi \partial t + \nabla\phi \cdot \delta\vec{x} - \delta\phi) + \frac{\partial \mathcal{L}}{\partial \phi_t^*}(\partial_t \phi^* \partial t + \nabla\phi^* \cdot \delta\vec{x} - \delta\phi^*) - \mathcal{L}\delta t \right] dx$$

is a conserved quantity.

For the nonlinear Schrödinger equation in (3.2) with $A(\phi) = 0$, we have

$$\frac{\partial \mathcal{L}}{\partial \phi_t} = \frac{i}{2}\phi^*, \quad \text{and} \quad \frac{\partial \mathcal{L}}{\partial \phi_t^*} = -\frac{i}{2}\phi,$$

where $\mathcal{L}$ is given in (3.93). Several conservation laws and invariances can be obtained from the Noether theorem.

(1) *Invariance under time translation and energy conservation law.*

The action (4.1) is invariant under the infinitesimal time translation $t \to t + \delta t$ with $\delta x = \delta\phi = \delta\phi^* = 0$, then equation (4.3) becomes

$$\partial_t \left[\nabla\phi \cdot \nabla\phi^* - \frac{b}{2}(\phi^*\phi)^2 + V(x,t)\phi^*\phi \right] - \nabla \cdot (\phi_t \nabla\phi^* + \phi_t^* \nabla\phi) = 0.$$

This results in the conservation of energy

$$E = \int \left[\nabla\phi \cdot \nabla\phi^* - \frac{b}{2}(\phi^*\phi)^2 + V(x,t)\phi^*\phi \right] dx = \text{constant}. \tag{4.4}$$

(2) *Invariance of phase shift or gauge invariance and mass conservation law.*

It is very clear that the action related to the nonlinear Schrödinger equation is invariant under the phase shift $\tilde{\phi} = e^{i\theta}\phi$, which for infinitesimal θ gives $\delta\phi = i\theta\phi$, with $\delta t = \delta x = 0$. In this case, equation (4.3) becomes

$$\partial_t |\phi|^2 + \nabla \cdot \{i(\phi\nabla\phi^* - \phi^*\nabla\phi)\} = 0. \tag{4.5}$$

This results in the conservation of mass or number of particles.

$$N = \int |\phi|^2 dx = \text{constant}$$

and the continuum equation

$$\frac{\partial N}{\partial t} = \nabla \cdot \vec{j},$$

where $\vec{j}$ is the mass current density

$$\vec{j} = -i(\phi\nabla\phi^* - \phi^*\nabla\phi).$$

Equation (4.5) is the same as (3.110).

(3) *Invariance of space translation and momentum conservation law.*

If the action is invariant under an infinitesimal space translation $x \to x + \delta x$ with $\delta t = \delta\phi = \delta\phi^* = 0$, then (4.3) becomes

$$\partial_t[i(\phi\nabla\phi^* - \phi^*\nabla\phi)] + \nabla \cdot \{2(\nabla\phi^* \times \nabla\phi + \nabla\phi \times \nabla\phi^* + \mathcal{L})] = 0.$$

This leads to the conservation of momentum

$$\vec{P} = i\int(\phi\nabla\phi^* - \phi^*\nabla\phi)dx = \text{constant}. \tag{4.6}$$

Note that the center of mass of the microscopic particles is defined by

$$\langle x\rangle = \frac{1}{N}\int x|\phi|^2 dx,$$

we then have

$$\begin{aligned} N\frac{d\langle x\rangle}{dt} &= \int x\partial_t|\phi|^2 dx = -\int x\nabla \cdot [i(\phi\nabla\phi^* - \phi^*\nabla\phi)]dx \\ &= \int i(\phi\nabla\phi^* - \phi^*\nabla\phi)dx = \vec{P} = -\vec{J} = -\int \vec{j}dx. \end{aligned} \tag{4.7}$$

This is the definition of momentum in classical mechanics. It shows clearly that the microscopic particles described by the nonlinear Schrödinger equation has the feature of classical particles.

(4) *Invariance under space rotation and angular momentum conservation law.*

If the action (4.1) is invariant under a rotation of angle $\delta\theta$ around an axis $\vec{I}$ such that $\delta t = \delta\phi = \delta\phi^* = 0$ and $\delta\vec{x} = \delta\theta\vec{I} \times \vec{x}$, this leads to the conservation of the angular momentum

$$\vec{M} = i\int \vec{x} \times (\phi^*\nabla\phi - \phi\nabla\phi^*)dx.$$

Besides the above, Sulem also derived other invariance of the nonlinear Schrödinger equation from the Noether theorem.

(5) *Galilean Invariance*

If the action is invariant under the Galilean transformation

$$x \to x'' = x - vt,$$
$$t \to t'' = t,$$
$$\phi(x,t) \to \phi''(x'',t'') = -i\left[\frac{1}{2}vx'' + \frac{1}{2}\vec{v}\cdot\vec{v}t''\right]\phi(x,t),$$

then the nonlinear Schrödinger equation can also retain its invariance. For an infinitesimal velocity v, $\delta\vec{x} = -vt$, $\delta t = 0$ and $\delta\hat{\phi} = \phi''(x'',t'') - \phi(x,t) = -(i/2)vx\phi(x,t)$. After integration over the space variables, equation (4.3) leads to the conservation law (4.7) which implies that the velocity of the center of mass of the microscopic particle is a constant.

(6) *Scale invariance of power law nonlinearities in the nonlinear Schrödinger equation with $V(x,t) = A(\phi) = 0$*

If $|\phi|^2$ in the nonlinear term, $b|\phi|^2\phi$, in the nonlinear Schrödinger equation (3.2) is replaced by $b|\phi|^{2\sigma}$, where σ is a constant, then the nonlinear Schrödinger equation is invariant under the scale transformation

$$x \to x'' = \frac{x}{\lambda'},$$
$$t \to t'' = \frac{t}{\lambda^2},$$
$$\phi(x,t) \to \phi''(x'',t'') = \lambda'^{1/\sigma}\phi(\lambda' x'', \lambda'^2 t'').$$

But the action scales as $\lambda'^{2/\sigma - 1/d}$ under the same transformation, where d is the size of the system. In the critical case, $\sigma = 2d$, the action is invariant under the scale transformation while the nonlinear Schrödinger equation remains invariant. $\sigma = 2d$ also happens to be the critical condition for a singular solution and is usually referred to as the critical dimension. In such a case, the invariance under the scale transformation leads to an additional conservation law. For an infinitesimal transformation, $\lambda' = 1 - \varepsilon$, with $|\varepsilon| \ll 1$, $\delta x = \varepsilon x$, $\delta t = 2\varepsilon t$ and $\delta\hat{\phi} = -(\varepsilon/\sigma)\phi(x,t)$. Then (4.3) results in

$$\int\left\{2\varepsilon t\mathcal{L} - \frac{i}{2}\phi^*\left(\frac{\varepsilon}{\sigma}\phi + \varepsilon\vec{x}\cdot\nabla\phi + 2\varepsilon t\phi_t\right) + \frac{i}{2}\phi\left(\frac{\varepsilon}{\sigma}\phi^* + \varepsilon\vec{x}\cdot\nabla\phi^* + 2\varepsilon t\phi_t^*\right)\right\}dx = \varepsilon C_1$$

or

$$i\int\vec{x}\cdot(\phi\nabla\phi^* - \phi^*\nabla\phi)dx - 4Et = 2C_1$$

where C_1 is a constant, E is the energy of the system.

(7) *Pseudo-conformal invariance at critical dimension*

Talanov (1970) proved that when the nonlinearity is a power law $b|\phi|^2\phi$ at the critical dimension $\sigma = 2d$, the nonlinear Schrödinger equation has other additional symmetry. This invariance is given here.

Obviously, the nonlinear Schrödinger equation (3.2) with $V(x,t) = A(\phi) = 0$ at the critical dimension is invariant by the pseudo-conformal transformation:

$$\vec{x} \to \vec{x}'' = \frac{\vec{x}}{l(t)}, \quad t \to t'' = \int_0^t \frac{1}{l^2(s)} ds = \frac{t_0^2 t}{t^*(t^*-t)},$$
$$\phi(x,t) \to \phi''(x'',t'') = l^{d/2}\phi(x,t)\cdot e^{-il_t|x|^2/4l} = l^{d/2}\phi(x,t)e^{ia|x|^2/4l^2}$$

with

$$a = -l\frac{dl}{dt} = -\frac{dl}{ldt'} = \frac{t^*-t}{t_0^2},$$

where t_0 appears as an arbitrary time unit. This shows that there exists certain symmetry in the explicit solution of the nonlinear Schrödinger equation at the critical dimension which is singular at some finite time t^*. It is referred to as pseudo-conformal invariance of the nonlinear Schrödinger equation at the critical dimension.

Expressing ϕ'' in terms of ϕ, the transformed Hamiltonian

$$H' = \int \left[(\nabla''\phi'')^2 - \frac{q}{1+\sigma}|\phi''|^{2\sigma+2} \right] dx,$$

where ∇'' corresponds to x'', becomes

$$H' = \int \left(\left| l\nabla\phi + \frac{i}{2}\vec{x}\phi \right|^2 - \frac{l^2 q}{\sigma+1}|\phi|^{2\sigma+2} \right) dx.$$

We can show that the energy corresponding to the Hamiltonian is conserved. In fact, choosing $t^* = 0$, it yields the conservation of

$$C_2 = \int \left(|\vec{x}\phi + 2it\nabla\phi|^2 - \frac{4t^2 q}{\sigma+1}|\phi|^{2\sigma+2} \right) dx.$$

Kuznetsov and Turitsyn (1985) (see Sulem and Sulem 1999) pointed out that the above transformation preserves the action, and the invariance given by the above equation is a result of Noether theorem. This can be seen by considering a transformation close to identity by choosing $t_0 = t^*$, and by assuming that $\delta\lambda' = 1/t^*$ is infinitesimally small. In such a case $\delta\vec{x} = xt\delta\lambda'$, $\delta t = t^2\delta\lambda'$, and $\delta\hat{\phi} = (-dt/2 + i|\vec{x}|^2/4)\phi\delta\lambda'$. Upon evaluation of the space integration, equation (4.3) leads to the conservation of

$$\int \left[t^2 \left(-|\nabla\phi|^2 + \frac{q}{1+\sigma}|\phi|^{2\sigma+2} \right) + \frac{i}{2} t\vec{x}\cdot(\phi\nabla\phi^* - \phi^*\nabla\phi) - \frac{1}{4}|\vec{x}|^2|\phi|^2 \right] dx = C_2.$$

Kuznetsov and Turitsyn (see Sulem and Sulem 1999) further proved that the pseudo-conformal transformation also holds for the two dimensional nonlinear

Schrödinger equation (in the natural unit system)

$$i\phi_{t'} = -\frac{1}{2}\left(\frac{\partial^2\phi}{\partial x'^2} + \frac{\partial^2\phi}{\partial y'^2}\right) - b|\phi|^2\phi,$$

where $t' = t/\hbar$, $x' = (\sqrt{2m}/\hbar)x$, $y' = (\sqrt{2m}/\hbar)y$.

It is well known from classical physics that the invariance and conservation laws of mass, energy and momentum and angular momentum are some fundamental and universal laws of matters in nature, including classical particles. In this section we demonstrated that the microscopic particles described by the nonlinear Schrödinger equation in the nonlinear quantum mechanics also have such properties. This shows that the microscopic particles in the nonlinear quantum mechanics also have corpuscle feature. Therefore the proposed nonlinear quantum mechanical theory reflects the common rules of motions of matters in nature.

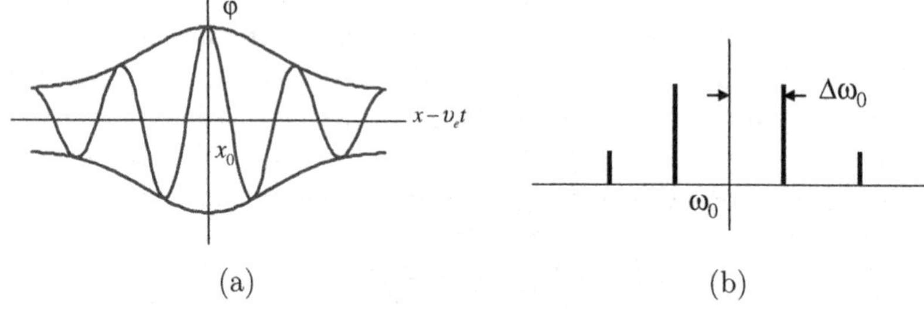

Fig. 4.1 Non-topological solitary wave of nonlinear Schrödinger equation and its frequency spectrum

In the one-dimensional case with $V(x,t) = A(\phi) = 0$, Zakharov *et al.* found the solution of (3.2) using the inverse scattering method. The solution is a bell-type non-topological soliton

$$\phi(x,t) = A_0 \operatorname{sech}\left\{\frac{A_0\sqrt{bm}}{\hbar}\left[(x-x_0) - v_e t\right]\right\} e^{iv_e m[(x-x_0)-v_c t]/2\hbar} \tag{4.8}$$

where

$$A_0 = \sqrt{\frac{(v_e^2 - 2v_c v_e)m}{2b}}.$$

Therefore the microscopic particle in nonlinear quantum mechanics is a soliton. Here $\varphi(x,t) = A_0 \operatorname{sech}\{A_0\sqrt{bm}[(x-x_0) - v_e t]/\hbar\}$ is the envelope of the soliton, and $\exp\{iv_e m(x-x_0) - v_e t)]/2\hbar\}$ is its carrier wave. The form of the soliton is shown in Fig. 4.1. The envelope $\varphi(x,t)$ is a slow varying function and A_0 is its amplitude. The width of the soliton is given by $W = 2\pi\hbar/(\sqrt{mb}A_0)$. Thus the size of the soliton, $A_0W = 2\pi\hbar/\sqrt{mb}$, is a constant. It depends only on the nature of the microscopic particles, but not on its velocity. This shows the corpuscle feature of the microscopic particles. In (4.8), v_e is the group velocity of the soliton, and

v_c is the phase speed of the carrier wave. From Fig. 4.1 and (4.8), we see that the carrier wave carries the envelope to propagate in space-time. The envelope is typical of a solitary wave, its changes in the space-time are slow and its form and amplitude are invariant in the transport. Thus the frequency spectrum of the envelope has a localized structure around the carrier frequency ω_0, as shown in Fig. 4.1(b). This figure shows the width of the frequency spectrum of the envelope $\varphi(x,t)$. For a certain system, v_e and v_c and the size of the soliton are determinate and do not change with time. This shows that the soliton (4.8) or the microscopic particle moves freely with a uniform velocity in space-time, and it has certain mass, momentum and energy, which are given by

$$N_s = \int_{-\infty}^{\infty} |\phi|^2 dx' = \frac{2\sqrt{2}A_0}{\sqrt{b}},$$

$$p = -i\int_{-\infty}^{\infty} (\phi^*\phi_{x'} - \phi\phi^*_{x'})\, dx' = 2\sqrt{2}A_0 v_e = \frac{\sqrt{m}}{2\sqrt{2}} N_s v_e = \text{const.} \cdot v_e,$$

$$E = \int_{-\infty}^{\infty} \left[|\phi_{x'}|^2 - \frac{b}{2}|\phi|^4 \right] dx' = E_0 + \frac{1}{2} M_{\text{sol}} v_e^2,$$

where $M_{\text{sol}} = N_s = 2\sqrt{2}A_0$ is the effective mass of the microscopic particles and it is a constant. These results clearly show that the energy, mass and momentum of a microscopic particle are invariant in nonlinear quantum mechanics and they are not dispersed in its free motion. Thus the microscopic particles can move over a macroscopic distance with the group velocity v_e and retain its energy and momentum. It shows that the microscopic particle has corpuscle characteristics like a classical particle.

4.2 Position of Microscopic Particles and Law of Motion

Besides mass, momentum and energy of microscopic particles mentioned above, position of a microscopic particle can also be precisely determined in nonlinear quantum mechanics. We know that the solution (4.8) shown in Fig. 4.1, of the nonlinear Schrödinger equation (3.2) with $V(x',t') = A(\phi) = 0$, has the behavior

$$\left.\frac{\partial \phi}{\partial x'}\right|_{x'=x'_0} = 0.$$

Thus we can infer that the position of the soliton is $x' = x'_0$ at $t' = 0$. Next, at $|x'| \to \infty$, $\phi(x',t') \to 0$, this means that

$$\frac{\partial}{\partial t'}\int_{-\infty}^{\infty} \phi^*\phi dx' = 0, \quad \text{or} \quad \int \rho(x')dx' = \text{constant}.$$

which is the mass conservation of the microscopic particle. Therefore, $\phi^*\phi dx' = \rho(x')dx'$ can be regarded as the mass in the interval of x' to $x' + dx$. In the light of definition in (4.7), the center of mass of the microscopic particle at $x_0' + v_e t'$ is given by

$$\langle x' \rangle = x_g' = \frac{\displaystyle\int_{-\infty}^{\infty} \phi^* x' \phi dx'}{\displaystyle\int_{-\infty}^{\infty} \phi^* \phi dx'} \tag{4.9}$$

where $x' = x/\sqrt{\hbar^2/2m}$, $t' = t/\hbar$. We can thus obtain the law of motion for the center of mass of the microscopic particles described by (3.2) with $A(\phi) = 0$ and its conjugate equation. Under such a condition, we have

$$\begin{aligned} \frac{d}{dt'} \int_{-\infty}^{\infty} \phi^* \phi_{x'} dx' &= \int_{-\infty}^{\infty} \phi_{t'}^* \phi_{x'} dx' + \int_{-\infty}^{\infty} \phi^* \frac{\partial}{\partial x'} \phi_{t'} dx' \\ &= i \left\{ \int_{-\infty}^{\infty} (\phi^* \phi_{x'x'x'} - \phi_{x'} \phi_{x'x'}) dx' + \right. \\ \int_{-\infty}^{\infty} &[b(\phi^*)^2 \phi \phi_{x'} + b\phi^* \phi^2 \phi_{x'}^*] dx' - \left. \int_{-\infty}^{\infty} \phi^* \frac{\partial V}{\partial x'} \phi dx' \right\} \end{aligned} \tag{4.10}$$

with

$$\begin{aligned} \int_{-\infty}^{\infty} (\phi^* \phi_{x'x'x'} - \phi_{x'x'}^* \phi_{x'}) dx' &= \phi^* \phi_{x'x'}|_{-\infty}^{\infty} - \int_{-\infty}^{\infty} \phi_{x'}^* \phi_{x'x'} dx' \\ &- \phi_{x'}^* \phi_{x'}|_{-\infty}^{\infty} + \int_{-\infty}^{\infty} \phi_{x'}^* \phi_{x'x'} dx' = 0, \end{aligned} \tag{4.11}$$

$$\int_{-\infty}^{\infty} \left[b\phi^* \phi^2 \phi_{x'}^* + b(\phi^*)^2 \phi_{x'} \right] dx = -\frac{1}{2} \int_{-\infty}^{\infty} (\phi^*)^2 \left(\frac{\partial b}{\partial x'} \right) \phi^2 dx' = 0, \tag{4.12}$$

where b is constant. Using (4.11) and (4.12), equation (4.10) becomes

$$\frac{d}{dt'} \int_{-\infty}^{\infty} \phi^* \phi_{x'} dx' = i \int_{-\infty}^{\infty} \phi^* \frac{\partial V}{\partial x'} \phi dx'. \tag{4.13}$$

From (4.7) and (4.9), the acceleration of the center of mass of the microscopic particle can be defined as

$$\frac{d^2 \langle x' \rangle}{d(t')^2} = \frac{\partial^2}{\partial t^2} \frac{\displaystyle\int_{-\infty}^{\infty} \phi^* x \phi dx'}{\displaystyle\int_{-\infty}^{\infty} \phi^* \phi dx'} = -2i \frac{d}{dt} \frac{\displaystyle\int_{-\infty}^{\infty} \phi^* \phi_{x'} dx'}{\displaystyle\int_{-\infty}^{\infty} \phi^* \phi dx'}. \tag{4.14}$$

Since

$$\int_{-\infty}^{\infty} \phi^* \phi dx' = \text{const},$$

$$\frac{d}{dt'}\int_{-\infty}^{\infty}\phi^* x'\phi dx' = \int_{-\infty}^{\infty}(\phi_{t'}^* x'\phi + \phi^* x'\phi_{t'})\,dx' = -2i\int_{-\infty}^{\infty}\phi^*\phi_{x'}dx'$$

we can obtain the following from (4.13) and (4.14)

$$\frac{d^2\langle x'\rangle}{d(t')^2} = -\frac{2\int_{-\infty}^{\infty}\phi^*(\frac{\partial V}{\partial x'})\phi dx'}{\int_{-\infty}^{\infty}\phi^*\phi dx'} = -2\left\langle\frac{\partial V}{\partial x'}\right\rangle, \tag{4.15}$$

where $V = V(x')$ is the external potential field experienced by the microscopic particle.

We now expand $\partial V/\partial x'$ in (4.15) around the center of mass $x' = \langle x'\rangle$ as

$$\frac{\partial V(x')}{\partial x'} = \frac{\partial V(\langle x'\rangle)}{\partial\langle x'\rangle} + \sum_{j=1}^{3}(x'_j - \langle x'_j\rangle)\frac{\partial^2 V(\langle x'\rangle)}{\partial\langle x'_j\rangle\partial\langle x'_j\rangle}$$
$$+\frac{1}{2}\sum_{k,j=1}^{3}(x'_k - \langle x'_k\rangle)(x'_j - \langle x'_j\rangle)\frac{\partial^3 V(\langle x'\rangle)}{\partial\langle x'_j\rangle\partial\langle x'_j\rangle\partial\langle x'_k\rangle} + \cdots.$$

Taking the expectation values for both sides of the above equation, we can get

$$\left\langle\frac{\partial V}{\partial x'}\right\rangle = \frac{\partial V(\langle x'\rangle)}{\partial\langle x'\rangle} - \frac{1}{2}\sum_{k,j=1}^{3}\Delta_{j,k}\frac{\partial^3 V(\langle x'\rangle)}{\partial\langle x'_j\rangle\partial\langle x'_j\rangle\partial\langle x'_k\rangle}$$

where

$$\Delta_{j,k} = \langle[(x'_k - \langle x'_k\rangle)(x'_j - \langle x'_j\rangle)]\rangle = \langle x'_j x'_k\rangle - \langle x'_j\rangle\langle x'_k\rangle.$$

For the non-topological soliton solution, (4.8), the position of its center of mass is determined as mentioned above, thus $\Delta_{j,k} = 0$. Therefore the above equation becomes

$$\left\langle\frac{\partial V(x')}{\partial x'}\right\rangle = \frac{\partial V(\langle x'\rangle)}{\partial\langle x'\rangle}.$$

Thus the acceleration of the center of mass of the microscopic particles in nonlinear quantum mechanics, (4.15), can be expressed as

$$\frac{d^2\langle x'\rangle}{d(t')^2} = -2\frac{\partial V(\langle x'\rangle)}{\partial\langle x'\rangle}. \tag{4.16}$$

Returning to the original variables, equation (4.16) becomes

$$m\frac{d^2\langle x\rangle}{dt^2} = -\frac{\partial V(\langle x\rangle)}{\partial\langle x\rangle}. \tag{4.17}$$

This is the equation of motion of the center of mass of the microscopic particle in the nonlinear quantum mechanics. It is analogous to the Newton's equation for a classical object. It shows that the acceleration of the center of mass of a microscopic particle depends directly on the external potential field. It also states that motion of a microscopic particle in nonlinear quantum mechanics obeys the classical laws. For a free particle, $V = 0$, thus $d^2\langle x\rangle/dt^2 = 0$. Hence, the center of mass of the microscopic particle moves with a uniform speed, which is consistent with results of earlier discussion. We can find from the definition (4.9) that the speed is the group velocity of the microscopic particle (soliton), v_e in (4.8). From (4.9) and (4.13), the velocity of the center of mass of the microscopic particle can be expressed as

$$v_g = \frac{d\langle x'\rangle}{dt'} = \frac{d}{dt'}\left\{\frac{\int \phi^* x'\phi dx'}{\int \phi^*\phi dx'}\right\} = -2i\frac{\int_{-\infty}^{\infty}\phi^*\frac{\partial\phi'}{\partial x'}dx'}{\int \phi^*\phi dx'}. \tag{4.18}$$

We have used the following,

$$\begin{aligned}
&\frac{d}{dt}\int_{-\infty}^{\infty}\phi^* x'\phi dx' \\
&\quad = \int_{-\infty}^{\infty}(\phi_{t'}^* x'\phi + \phi^* x'\phi_{t'})dx' \\
&\quad = i\int_{-\infty}^{\infty}\left\{\phi^* x'\left[\phi_{x'x'} + b\phi^*\phi^2 - V\phi\right] - \phi x'\left[\phi_{x'x'}^* + b\phi(\phi^*)^2 - V\phi^*\right]\right\}dx' \\
&\quad = -2i\int_{-\infty}^{\infty}\phi^*\phi_{x'}dx',
\end{aligned}$$

$$\lim_{|x'|\to\infty}\phi^*\phi(x',t') = \lim_{|x'|\to\infty}\phi_{x'}(x',t') = \lim_{|x'|\to\infty}\phi^*\phi_{x'} = \lim_{|x'|\to\infty}\phi_{x'}^* x'\phi = 0,$$

$$\int_{-\infty}^{\infty}\phi^*\phi dx' = \text{const.}$$

and the boundary condition $\phi(x',t') = 0$ at $|x'| \to \infty$. Substituting (4.8) into (4.18) we can obtain

$$v_g = \frac{d\langle x'\rangle}{dt'} = v_e = \text{const.} \tag{4.19}$$

which is the same as the results obtained from (4.16) above. It shows clearly that for a microscopic particle described by the nonlinear Schrödinger equation with $V = 0$, not only its velocity of the center of mass is the group velocity of the soliton, but it also moves with a uniform speed v_e in the space-time.

The above equation of motion of microscopic particles can also be derived from the nonlinear Schrödinger equation (3.2) with $A(\phi) = 0$ by means of another method. As it is known from Chapter 3, the energy E and momentum P of a microscopic particle described by the nonlinear Schrödinger equation (3.2) with

$A(\phi) = 0$ takes the form of

$$E = \int_{-\infty}^{\infty} \left[\phi_{x'}^* \phi_{x'} - \frac{b}{2}(\phi^*\phi)^2 + 2V(x')\phi^*\phi \right] dx',$$
$$P = -i \int_{-\infty}^{\infty} (\phi^* \phi_{x'} - \phi_{x'}^* \phi) dx'. \tag{4.20}$$

For this system the energy E and quantum number $N_s = \int_{-\infty}^{\infty} |\phi|^2 dx'$ are integral invariant. However, the momentum P is not conserved and has the following property

$$\frac{dP}{dt'} = \int_{-\infty}^{\infty} 2V(x') \frac{\partial}{\partial x'} |\phi|^2 dx' = -2 \int_{-\infty}^{\infty} \frac{\partial V}{\partial x'} |\phi|^2 dx' \tag{4.21}$$

where the boundary condition is $\phi(x') \to 0$ as $|x'| \to \infty$. For slowly varying inhomogeneities (in comparison with soliton scale), *i.e.*, $W_s \gg L$, where L is the inhomogeneity scale, W_s the soliton width, expanding (4.21) into a power series in W_s/L and keeping only the leading term, we can get

$$\frac{dP}{dt'} = -2 \frac{\partial V(x_0')}{\partial x_0'} N_s \tag{4.22}$$

where x_0' is the position of the center of mass of the microscopic particle. Equation (4.21) or (4.22) is essentially consistent with (4.17) which is in the form of the equation of motion for a classical particle. Indeed, if we write the one-soliton solution as

$$\phi(x', t') = \varphi(x' - x_0', t') e^{ip(x' - x_0') + i\theta}. \tag{4.23}$$

It is easy to verify that (4.23) is a solution of (3.2) with $A(\phi) = 0$. Inserting (4.23) into (4.20), we get

$$P = pN_s. \tag{4.24}$$

Let $p = dx_0'/dt'$ for the center of mass of the particle. Then we can obtain from (4.22) and (4.24) that

$$\frac{d^2 x_0'}{dt'^2} = -2 \frac{\partial V}{\partial x_0'}, \quad \text{or} \quad m \frac{d^2 x_0}{dt^2} = -\frac{\partial V}{\partial x_0}. \tag{4.25}$$

This is the same as (4.16) or (4.17) and it is the Newton's equation for a classical particle. Equations (4.22) and (4.25) indicate that the center of mass of the microscopic particle moves like a classical particle in a weakly inhomogeneous potential field $V(x_0')$. Let the function φ in (4.23) has the form

$$\varphi(x', t') = \sqrt{2\omega}\, \mathrm{sech}(\sqrt{2\omega} x') e^{i\omega t'},$$

where $\omega = (N/2)^2$. Equation (4.25) coincides with the equation obtained from the adiabatic approximation of inverse scattering transformation (IST) based perturbation theory for solitons (see Chapter 7).

We now consider a particular one-dimensional case. Let $V(x') = \alpha x'$ in (3.2), where α is constant, and make the following transformation

$$\phi(x',t') = \phi'(\tilde{x}',\tilde{t}')e^{-i\alpha\tilde{x}'\tilde{t}'-i\alpha^2(\tilde{t}')^3/3},$$
$$x' = \tilde{x}' - \alpha\tilde{t}'^2, \quad t' = \tilde{t}',$$

then (3.2) becomes

$$i\phi'_{\tilde{t}'} + \phi'_{\tilde{x}'\tilde{x}'} + 2|\phi'|^2\phi' = 0, \tag{4.26}$$

where $b = 2$. The exact and complete solution of the above equation is well known. We can thus obtain the complete solution of the nonlinear Schrödinger equation with $V(x') = \alpha x'$. Its single soliton solution is given by

$$\begin{aligned}\phi(x',t') = 2\eta \operatorname{sech}\left[2\eta(x' - 4\xi t' + 2\alpha t'^2 - x'_0)\right] \times \\ \exp\left\{-i\left[2(\xi - \alpha t')x' + \frac{4\alpha^2 t'^3}{3} - 4\alpha\xi t'^2 + 4(\xi^2 - \eta^2)t' + \theta_0\right]\right\}.\end{aligned} \tag{4.27}$$

This is the same as the solution (2.142) of (2.122) obtained by Pang *et al.*. However, Chen and Liu first obtained the above soliton solution using the inverse scattering method. The same authors also gave the soliton solution of the nonlinear Schrödinger equation with $V_0(x') = \alpha^2 x'^2$, which is

$$\begin{aligned}\phi = 2\eta \operatorname{sech}\left\{2\eta x' - \frac{4\xi\eta}{\alpha}\sin[2\alpha(t' - t'_0)]\right\} \times \\ \exp\left\{i\left[2\xi x' \cos 2\alpha(t' - t'_0) - \frac{\xi^2}{\alpha}\sin[4\alpha(t' - t'_0)] + 4\eta^2(t' - t'_0) + \theta'_0\right]\right\}.\end{aligned} \tag{4.28}$$

Detailed derivation of these solutions will be given in Chapter 8.

In each of the above two cases, with two different external potential fields, the characteristics of motion of the microscopic particle can be determined according to (4.17) or (4.22). The acceleration of the center of mass of the microscopic particle is given by

$$\frac{d^2x'_0}{dt'^2} = -2\frac{\partial V(\langle x'\rangle)}{\partial\langle x'\rangle} = -2\alpha = \text{constant} \tag{4.29}$$

for $V(x') = \alpha x'$, and

$$\frac{d^2x'_0}{dt'^2} = -4\alpha^2 x'_0 \tag{4.30}$$

for $V(x') = \alpha^2 x'^2$, respectively.

These results can also be obtained using the following method. From de Broglie relations $E = hv = \hbar\omega$ and $\vec{P} = \hbar\vec{k}$ for microscopic particles which have the wave-corpuscle duality in quantum theory, the frequency ω retains its role as the Hamiltonian of the system even in this complicated, nonlinear systems and

$$\frac{d\omega}{dt'} = \left.\frac{\partial\omega}{\partial k}\right|_{x'} \frac{dk}{dt} + \left.\frac{\partial\omega}{\partial x'}\right|_{k} \frac{\partial x'}{\partial t'} = 0$$

as in the usual stationary media. From Chapter 3, we also knew that the usual Hamiltonian equations for nonlinear quantum mechanical systems remain valid for microscopic particles (solitons). At present, the Hamilton equations are

$$\frac{dk}{dt'} = -\left.\frac{\partial\omega}{\partial x'}\right|_{k}, \quad \frac{dx'}{dt'} = \left.\frac{\partial\omega}{\partial k}\right|_{x'}, \tag{4.31}$$

where $k = \partial\theta/\partial x'$ is the time-dependent wave number of the microscopic particle, $\omega = -\partial\theta/\partial t'$ is its frequency, θ is the phase of the wave function of the microscopic particle. Equations (4.31) are essentially the same as the Hamilton equation (3.104). From (4.27) and (4.28), we know that

$$\theta = 2(\xi - \alpha t')x' + \frac{4\alpha^2 t'^3}{3} - 4\alpha\xi t'^2 + 4(\xi^2 - \eta^2)t' + \theta_0,$$

for $V(x') = \alpha x'$, and

$$\theta = 2\xi x' \cos[2\alpha(t' - t_0')] + \left(\frac{\xi^2}{\alpha}\right) \sin[4\alpha(t' - t_0')] + 4\eta^2(t' - t_0') + \theta_0',$$

for $V(x') = \alpha^2 x'^2$, respectively. From (4.31) we can find that for $V(x') = \alpha x'$,

$$k = 2(\xi - \alpha t'), \quad \omega = 2\alpha x' - 4(\xi - \alpha t')^2 + (2\eta)^2 = 2\alpha x' - k^2 + (2\eta)^2.$$

Thus, the group velocity of the microscopic particle is

$$v_g = \frac{d\tilde{x}'}{dt'} = \left.\frac{\partial\omega}{\partial k}\right|_{x'} = 4(\xi - \alpha t'),$$

and the acceleration of its center of mass is given by

$$\frac{d^2\bar{x}'}{dt'^2} = \frac{dk}{dt'} = -2\alpha = \text{constant}. \tag{4.32}$$

Here $\bar{x}' = x_0'$.

For $V(x') = \alpha^2 x'^2$, we have

$$\begin{aligned} k &= 2\xi \cos 2\alpha(t' - t_0'), \\ \omega &= 4\alpha\xi x' \sin 2\alpha(t' - t_0') - 4\xi^2 \cos 4\alpha(t' - t_0') - 4\eta^2 \\ &= 2\alpha x'(4\xi^2 - k^2)^{1/2} - 2k^2 + 4(\xi^2 - \eta^2), \end{aligned}$$

Thus the group velocity of the microscopic particle is

$$v_g = \left.\frac{\partial\omega}{\partial k}\right|_{x'} = \frac{\alpha x'}{\xi} \frac{k}{\sqrt{1 - k^2/4\xi^2}} - 2k = 2\alpha x' \text{ctg}[2\alpha(t' - t_0')] - 4\xi \cos[2\alpha(t' - t_0')],$$

while its acceleration is

$$\frac{dk}{dt'} = -\left.\frac{\partial\omega}{\partial x'}\right|_k = -2\alpha\sqrt{4\xi^2 - k^2} = -4\xi\alpha\sin[2\alpha(t' - t'_0)].$$

Since

$$\frac{d^2\bar{x}'}{dt'^2} = \frac{dk}{dt'},$$

we have

$$\frac{dk}{dt'} = \frac{d^2\bar{x}'}{dt'^2} = -4\xi\alpha\sin[2\alpha(t' - t'_0)],$$

and

$$\bar{x}' = \frac{\xi}{\alpha}\sin[2\alpha(t' - t'_0)]. \tag{4.33}$$

Finally the acceleration of the microscopic particle is

$$\frac{dk}{dt'} = \frac{d^2\bar{x}'}{dt'^2} = -4\alpha^2\bar{x}'. \tag{4.34}$$

Equations (4.32) and (4.34) are exactly the same as (4.29) and (4.30). It shows that (4.17) or (4.22) and (4.25) have same effects and function as (4.31) and (3.104) in nonlinear quantum mechanics. On the other hand, it is well known that a macroscopic object moves with a uniform acceleration, when $V(x') = \alpha x'$ which corresponds to the motion of a charged particle in an uniform electric field, and when $V(x') = \alpha^2 x'^2$ which is a harmonic potential, the macroscopic object performs localized vibration with a frequency $\omega = 2\alpha$ and an amplitude x'_0, and the corresponding classical vibrational equation is $x' = x'_0 \sin\omega t'$. The equations of motion of the macroscopic object are consistent with (4.17) and (4.29) – (4.30) or (4.32) – (4.34) for the center of mass of microscopic particles in the nonlinear quantum mechanics. These correspondence between a microscopic particle and a macroscopic object shows that microscopic particles in the nonlinear quantum mechanics have exactly the same properties as classical particles, and their motion satisfy the classical laws of motion. We have thus demonstrated clearly from the dynamic equations (nonlinear Schrödinger equation), the Hamiltonian or the Lagrangian of the systems, and the solutions of equations of motion, in both uniform and inhomogeneous systems, that microscopic particles in the nonlinear quantum mechanics really have the corpuscle property.

We now look into the motion of microscopic particles described by the Sine-Gordon equation (3.4). In the absence of external fields and damping (*i.e.*, $\gamma = 0$, $A = 0$), the Sine-Gordon equation can be written as

$$\phi_{tt} - v_0^2\phi_{xx} + \omega_0^2\sin\phi = 0. \tag{4.35}$$

Under certain perturbation, the state of the particle is changed and can be denoted by

$$\phi(x,t) = \phi^0(x,t) + \varphi(x,t) \tag{4.36}$$

where $\phi^0(x,t) = 4\tan^{-1}[e^{(x-vt)/(1-v^2)}]$ represents the state of the microscopic particle in the absence of the perturbation, $\varphi(x,t)$ is its change resulting from the perturbation which is assumed to be small in the case of weak perturbation. Inserting (4.36) into (4.35), we get

$$\varphi_{tt} - v_0^2\varphi_{xx} + \omega_0^2\left[1 - 2\mathrm{sech}^2\left(\frac{\omega_o Z}{2}\right)\right]\varphi = 0, \tag{4.37}$$

where $Z = (x - vt)/(1 - v^2)$. Now assume that $\varphi(x,t) = f(x)e^{-i\omega t}$, and insert it into (4.37), we can get

$$-v_0^2 f_{xx} - \omega_0^2\left[1 - 2\mathrm{sech}^2\left(\frac{\omega_o Z}{2}\right)\right] f = \omega^2 f. \tag{4.38}$$

Since the Lagrangian function L and the Hamiltonian function H of the systems are invariant under translation, the frequency spectrum of a single microscopic particle (soliton) must contain the translation mode of $\omega = 0$ (Goldstone mode), *i.e.*, $\omega = \omega_{b,1} = 0$. Thus the solution of (4.38) can be represented by

$$f_{b,1} = \sqrt{\frac{\omega_0}{2v_0}}\,\mathrm{sech}\left(\frac{\omega_0 Z}{2}\right),$$

which satisfies

$$f_{b,1} = \sqrt{\frac{A}{M_k}}\frac{d\phi^0(x)}{dx}, \quad \int_{-\infty}^{\infty} f_{b,1}^2(x)dx = 1.$$

Thus the complete wave function of the system is

$$\begin{aligned}\phi(x,t) &= \phi^0(x) + \varepsilon f_{b,1}(x) = \phi^0 + \varepsilon\sqrt{\frac{A}{M_K}}\frac{d\phi^0(x)}{dx} \\ &\approx \phi^0\left[x + \varepsilon\sqrt{\frac{A}{M_k}}\right] = \phi(x,X),\end{aligned} \tag{4.39}$$

where ε is small quantity and $X = -\varepsilon\sqrt{A/M_k}$. Therefore, the microscopic particle moves a distance $X = -\varepsilon\sqrt{A/M_k}$ when it experiences a small perturbation. This shows that the motion of the center of mass of a microscopic particle described by the Sine-Gordon equation follows the same law of motion of classical particles, similar to that of microscopic particles described by the nonlinear Schrödinger equation.

4.3 Collision between Microscopic Particles

The corpuscle feature of particles can be clearly manifested in the collision process. In the following, we consider collision between microscopic particles described by the nonlinear Schrödinger equation (3.2) with $V(x',t') = A(\phi) = 0$, *i.e.*,

$$i\phi_{t'} + \phi_{x'x'} + b|\phi|^2\phi = 0 \tag{4.40}$$

where $x' = x/\sqrt{\hbar^2/2m}$, $t' = t/\hbar$. This problem has been investigated by many scientists including Gorss, Pitavski, Tsuxuki, Zakharov *et al.*, and Pang *et al.* The discussion below follows closely the approach of Zakharov *et al.*

4.3.1 *Attractive interaction (b > 0)*

We first consider the case of $b > 0$. Equation (4.40) can be solved exactly using the inverse scattering method. As discussed in Chapter 3, the nonlinear Schrödinger equation can be written in the form of Lax equation (3.20), where $\hat{L}$ and $\hat{B}$ are linear differential operators containing the unknown function $\phi(x',t')$ in the form of a coefficient. Their explicit forms are given in (3.22), where $b = 2/(1-s^2) > 2$ and $s^2 > 0$. By virtue of $\lim_{|x'|\to\infty}\phi(x',t') \to 0$, we can examine the scattering problem for operator $\hat{L}$. To this end, we consider (3.21), where

$$\psi = \begin{pmatrix}\psi_1\\ \psi_2\end{pmatrix},$$

and λ is an eigenvalue of $\hat{L}$, and make the following change of variables

$$\begin{cases}\psi_1 = \sqrt{1-s}e^{-i\lambda x'/(1-s^2)}\Psi_2',\\[2ex] \psi_2 = \sqrt{1+s}e^{-i\lambda x'/(1-s^2)}\Psi_1'\end{cases} \tag{4.41}$$

Equation (3.21) can be rewritten in the form of the Zakharov-Shabat equation (3.46) which may be written as

$$\begin{cases}\Psi'_{1x'} + i\zeta\Psi_1' = q\Psi_2',\\[2ex] \Psi'_{2x'} - i\zeta\Psi_2' = -q^*\Psi_1',\end{cases} \tag{4.42}$$

where

$$q = \frac{i\phi}{\sqrt{1-s^2}}, \quad \zeta = \frac{\hbar}{(2m)^{1/2}}\xi' = \frac{\lambda s}{1-s^2}, \quad \zeta = \xi + i\eta.$$

For real $\zeta = \xi$, the solutions of (4.42) are the Jost functions Ψ' and ψ with the following asymptotic forms

$$\Psi' \to \begin{pmatrix} 1 \\ 0 \end{pmatrix} e^{-i\xi x'}, \quad \text{when } x' \to -\infty,$$

$$\psi \to \begin{pmatrix} 0 \\ 1 \end{pmatrix} e^{i\xi x'}, \quad \text{when } x' \to +\infty.$$

The pair of solutions ψ and $\bar{\psi}$ form a complete system and therefore

$$\Psi' = a(\xi)\bar{\psi} + b(\xi)\psi,$$

$$\Psi'(x', \xi) = c_j\psi(x', \xi), \quad j = 1, \cdots, N, \tag{4.43}$$

$$|a^2(\xi)|^2 + |b^2(\xi)|^2 = 1,$$

where $a(\xi)$, $b(\xi)$ and c_j are referred to as scattering data of the nonlinear Schrödinger equation (4.40). From (4.42) and $i\partial\psi/\partial t' = \hat{B}\psi$ in (3.21), we can find that

$$\begin{aligned} b(\xi, t') &= b(\xi, 0)e^{-i4\xi^2 t'}, \\ a(\xi, t') &= a(\xi, 0), \\ a(\xi) &= a^*(-\xi), \\ \zeta_i &= \xi_i + io, \\ c_j(t') &= c_j(0)e^{i4\xi^2 t'}. \end{aligned} \tag{4.44}$$

Let us now find $\phi(x', t')$ from these scattering data $a(\xi)$, $b(\xi, t)$, $(-\infty < \xi < \infty)$ and $c_j(t')$, $j = 1, \cdots, N$. The values of these quantities at $t' = 0$ are determined from the initial conditions. In the inverse scattering problem, the time t' plays the role of a parameter. Therefore, it suffices to consider the reconstruction of the coefficient $q(x')$ in (4.42) from $a(\xi)$, $b(\xi)$ and c_j. We introduce the function

$$\Phi(\zeta) = \Phi(\zeta, x') = \begin{cases} \dfrac{1}{a(\zeta)}\Psi'(x', \zeta)e^{i\zeta x'}, & \text{Im}\zeta > 0 \\ \begin{pmatrix} \psi_2^*(x', \zeta^*) \\ -\psi_1^*(x', \zeta^*) \end{pmatrix} e^{i\zeta x'}, & \text{Im}\zeta < 0 \end{cases}$$

and use $\Psi(\xi)$ to denote the discontinuity of this function across the real axis

$$\Psi(\xi) = \Phi(\xi + i0) - \Phi(\xi - i0).$$

Assuming that the zeroes ζ_1, ζ_2, $\cdots$, ζ_N of $a(\zeta)$ are simple, we can get the formula that reconstructs the piecewise-analytic function $\Phi(\zeta)$ from the discontinuity $\Psi(\xi)$ and the residues at the poles ζ_j

$$\Phi(\zeta) = \begin{pmatrix} 1 \\ 0 \end{pmatrix} + \sum_{k=1}^{N} \frac{\Psi'(x', \zeta_k)e^{i\zeta_k x'}}{(\zeta - \zeta_k)a'(\zeta_k)} + \frac{1}{2\pi i}\int_{-\infty}^{+\infty} \frac{\Psi(\xi)}{\xi - \zeta}d\xi,$$

or

$$\Phi = \Phi^{(1)} + \Phi^{(2)}, \tag{4.45}$$

where

$$\Phi^{(1)}(\zeta) = \begin{pmatrix} 1 \\ 0 \end{pmatrix} + \sum_{k=1}^{N} \frac{e^{i\zeta_k x'}}{\zeta - \zeta_k} \tilde{c}_k \psi(x', \zeta_k),$$

$$\Phi^{(2)}(\zeta) = \frac{1}{2\pi i} \int_{-\infty}^{+\infty} \frac{\Psi(\xi)}{\xi - \zeta} d\xi,$$

and

$$\tilde{c}_k = \frac{c_k}{a'(\zeta_k)}, \quad a'(\xi_k) = \frac{da(\xi)}{d\xi}|_{\xi=\xi_k}.$$

The tilde over c_k will henceforth be omitted. From (4.43) we get

$$\Psi(\xi) = \frac{b(\xi)}{a(\xi)} e^{i\xi x'} \psi(x', \xi). \tag{4.46}$$

According to the inverse scattering theory, the equations for the function $\Psi(\xi)$, $\xi < +\infty$ and for the parameters $\psi(x', \xi_j)$ (x' is fixed), $(j = 1, \cdots, N)$, can be obtained by putting $\zeta = \zeta_j^*$, $(j = 1, \cdots, N)$ in (4.44). For $\zeta = \xi - i0$, equation (4.44) yields

$$\begin{pmatrix} \psi_2^*(x', \xi) \\ -\psi_1^*(x', \xi) \end{pmatrix} e^{i\xi x'} = -\frac{1}{2}(1 - J)\Psi(\xi) + \Phi^{(1)}(\xi).$$

Here J is the Hilbert transformation

$$(J\Psi)(\xi) = \frac{1}{\pi i} \int_{-\infty}^{+\infty} \frac{\Psi(\xi'')}{\xi'' - \xi} d\xi'',$$

$$(J\Psi)^* = -J\Psi^*.$$

Zakharov *et al.* then obtained the following

$$\psi_2^*(x', \xi) e^{i\xi x'} + \frac{1}{2}(1 - J)\Psi_1 = \Phi_1^{(1)}(\xi),$$

$$-\psi_1(x', \xi) e^{-i\xi x'} + \frac{1}{2}(1 + J)\Psi_2^* = \Phi_2^{(1)*}(\xi).$$

Multiplying the first of the above equations by $c^*(x', \xi)$ and the second by $c(x', \xi) = [b(\xi)/a(\xi)]e^{2i\xi x'}$, and using again (4.46), we get

$$\Psi_1 - c(x', \xi) \frac{1 + J}{2} \Psi_2^* = -c(x', \xi) \sum_{k=1}^{N} \frac{e^{-i\zeta_k^* x'}}{\xi - \zeta^*} c_k^* \psi_2^*(x', \zeta_k), \tag{4.47}$$

$$c^*(x', \xi) \frac{1 - J}{2} \Psi_1 + \Psi_2^* = c^*(x', \xi) + c^*(x', \xi) \sum_{k=1} \frac{\exp(-i\zeta_k x')}{\xi - \zeta_k} c_k \psi_1(x', \zeta_k).$$

In addition to (4.47), we obtain $2N$ equations for $\psi_1(x',\zeta_j)$ and $\psi_2^*(x',\zeta_j)$, by putting $\zeta=\zeta_j^*$, $(j=1,\cdots,N)$, in (4.45)

$$\psi_1(x',\zeta_j)e^{-i\zeta_j^* x'}+\sum_{k=1}^{N}\frac{e^{-i\zeta_k^* x'}}{\zeta_j-\zeta_k^*}c_k^*\psi_2^*(x',\zeta_k)=\frac{1}{2\pi i}\int_{-\infty}^{+\infty}\frac{\Psi_2^*(\xi)}{\xi-\zeta_j}d\xi, \tag{4.48}$$

$$-\sum_{k=1}^{N}\frac{e^{i\zeta_k x'}}{\zeta_j-\zeta_k}c_k\psi_1(x',\zeta_k)+\psi_2^*(x',\zeta_j)e^{i\zeta_j^* x'}=1+\frac{1}{2\pi i}\int_{-\infty}^{+\infty}\frac{\Psi_1(\xi)}{\xi-\zeta_j^*}d\xi.$$

Equations (4.47) and (4.48) make it possible to obtain $\Psi(\xi)$ or $\Psi(x',\xi)$ and $\psi(x',\zeta_j)$ from the scattering data. From (4.45) and (4.42), we have

$$\begin{pmatrix}\psi_2(x',\zeta)\\ -\psi_1(x',\zeta)\end{pmatrix}e^{-i\zeta x'}=\begin{pmatrix}1\\0\end{pmatrix}+\frac{1}{\zeta}\left[\sum_{k=1}^{N}c_k^* e^{-i\zeta_k^* x'}\psi^*(x',\zeta_k)+\frac{1}{2\pi i}\int\Psi^*(\xi)d\xi\right]+O\left(\frac{1}{\zeta^2}\right),$$

$$\psi(x',\zeta)e^{-i\zeta x'}=\begin{pmatrix}0\\1\end{pmatrix}+\frac{1}{2i\zeta}\begin{pmatrix}q(x')\\ \int_{x'}^{\infty}|q|^2(p)dp\end{pmatrix}+O\left(\frac{1}{\zeta^2}\right).$$

From these two equations we get

$$q(x')=-2i\sum c_k^* e^{-i\zeta_k^* x'}\psi_2^*(x',\zeta_k)-\frac{1}{\pi}\int\Psi_2^*(\xi)d\xi, \tag{4.49}$$

$$\int_{x'}^{\infty}|q(p)|^2dp=-2i\sum c_k\exp(i\zeta_k x')\psi_1(x',\zeta_k)+\frac{1}{\pi}\int\Psi_1(\xi)d\xi.$$

Equation of the Marchenko type can then be obtained from (4.47) – (4.48) by a Fourier transformation with respect to ξ. Let

$$F(x')=\frac{1}{2n}\int_{x'}^{\infty}\frac{b(\xi)}{a(\xi)}e^{i\xi x'}d\xi+\sum_{k=1}^{N}c_k e^{i\zeta_k x'}, \tag{4.50}$$

then

$$K_1(x',y)=F^*(x'+y)+\int_{x'}^{\infty}K_2^*(x',p)F^*(p+y)dp, \tag{4.51}$$

$$K_2^*(x',y)=-\int_{x'}^{\infty}K_1(x',p)F(p+y)dp, \tag{4.52}$$

where the kernel $K(x',y)$ is connected to $\psi(x',\xi)$ by

$$\psi(x',\zeta)=e^{i\zeta x'}+\int_{x'}^{\infty}K(x,p)e^{i\zeta x'}dp,\quad \mathrm{Im}\zeta\geq 0.$$

Then (4.49) may take the form

$$q(x') = -K_1(x', x'),$$
$$\int_{x'}^{\infty} |q(p)|^2 dp = -2K_2(x', x').$$

We consider the inverse scattering problem in the case of $b(\xi, t) = 0$. The solution of the inverse problem then reduces to the solution of a finite system (4.48) of linear algebraic equations. We rewrite this system in a more symmetrical form

$$\psi_{1j} + \sum_{k=1}^{N} \frac{\lambda_j \lambda_k^*}{\zeta_j - \zeta_k^*} \psi_{2k}^* = 0, \tag{4.53}$$
$$-\sum_{k=1}^{N} \frac{\lambda_k \lambda_j^*}{\zeta_j^* - \zeta_k} \psi_{1k} + \psi_{2j}^* = \lambda_j^*.$$

Here

$$\psi_j = \begin{pmatrix} \psi_{1j} \\ \psi_{2j} \end{pmatrix} = \sqrt{c_j}\psi(x', \zeta_j), \quad \lambda_j = \sqrt{c_j} e^{i\zeta_j x'}.$$

Equation (4.49) also assumes a simpler form in this case

$$q(x') = -2i \sum_{k=1}^{N} \lambda_k^* \psi_{2k}^*, \quad \int_{x'}^{\infty} |q(p)|^2 dp = -2i \sum_{k=1}^{N} \lambda_k \int_{x'}^{\infty} \psi_{1k}. \tag{4.54}$$

If $N = 1$ and $a(\zeta)$ has only one zero in the upper half-plane, then (4.53) becomes

$$\psi_1 + \frac{|\lambda|^2}{2i\eta} \psi_2^* = 0,$$
$$\frac{|\lambda|^2}{2i\eta} \psi_1 + \psi_2^* = \lambda^*. \tag{4.55}$$

It can be easily verified that (4.55) describes a soliton of the following form

$$\phi(x', t') = 2\eta\sqrt{2/b}\, \text{sech}[2\eta(x' - x_0') + 8\eta\xi t'] e^{-4i(\xi^2 - \eta^2)t' - 2i\xi x' + i\theta}, \tag{4.56}$$

where

$$x_0' = \frac{1}{2\eta} \ln \frac{|\lambda(0)|^2}{2\eta}, \quad \theta = -2 \arg \lambda(0).$$

In general (4.53) describes an N-soliton solution. This system is non-degenerate and

$$\int_{x'}^{\infty} |q(p)|^2 dp = i\left(\sum \lambda_k \psi_{1k} - \sum \lambda_k^* \psi_{1k}^*\right) = \frac{d}{dx'} \ln(\det \|A\|) \tag{4.57}$$

where $\|A\|$ is the matrix of (4.53). For (4.57), we have

$$\frac{d}{dx'}\ln(\det\|A\|)=\frac{1}{\det\|A\|}\sum_{k=1}^{N}\det\|A_k\|$$

where the matrix $\|A_k\|$ differs from $\|A\|$ in their kth column, the kth column of $\|A_k\|$ is the derivative of the corresponding column of $\|A\|$. For $l\le k\le N$, the column of the aforementioned matrix $\|A_k\|$ is

$$i\lambda_k\begin{Vmatrix}0\\ \vdots\\ \lambda_1^*\\ \vdots\\ \lambda_N^*\end{Vmatrix}.$$

Using Cramer' rule, Zakharov *et al.* obtained

$$i\lambda_k\psi_{1k}=\frac{\det\|A_k\|}{\det\|A\|}.$$

Thus,

$$i\sum_{k=1}^{N}\lambda_k\psi_{1k}=\frac{1}{\det\|A\|}\sum_{k=1}^{N}\det\|A_k\|.$$

To prove (4.56), it remains to verify that

$$-i\sum\lambda_k^*\psi_k^*=\frac{1}{\det\|A\|}\sum_{k=N+1}^{2N}\det\|A_k\|. \tag{4.58}$$

To this end, Zakharov *et al.* rewrote (4.53) as

$$\psi_{2j}+\sum_{k=1}^{N}\frac{\lambda_j\lambda_k^*}{\zeta_j-\zeta_k^*}(-\psi_{1k}^*)=\lambda_j,$$
$$-\sum_{k=1}^{N}\frac{\lambda_j^*\lambda_k}{\zeta_j^*-\zeta_k}\psi_{2k}+(-\psi_{1j}^*)=0.$$

The matrix of this system relative to $\{\psi_{21},\cdots,\psi_{2N},-\psi_{11}^*,\cdots,-\psi_{1N}^*\}$ coincides with $\|A\|$, from which (4.57) – (4.58) follow. For the N-soliton solutions they finally obtained explicitly

$$|\phi(x',t')|^2=\sqrt{2b}\frac{d^2}{dx'^2}\ln(\det\|A\|)=\sqrt{2b}\frac{d^2}{dx'^2}\ln(\det\|B'B'^*+1\|)$$

where

$$B'_{jk} = \frac{\sqrt{c_j c_k^*}}{\zeta_j - \zeta_k^*} e^{i(\zeta_j - \zeta_k^*)x'}.$$

The time-dependence of c_j has been given in (4.45).

We now study the behavior of the N-soliton solution at large $|t'|$ to determine the rules for collision of microscopic particles (solitons) described by the nonlinear Schrödinger equation (4.40). We confine ourselves to the case where all the ξ_j are different, *i.e.*, there are no two solitons having the same velocity. In this case the N-soliton solution breaks up into diverging solitons as $t' \to \pm\infty$. To verify this, Zakharov *et al.* arranged the ξ_j in decreasing order, $\xi_1 > \xi_2 > \cdots > \xi_N$. From (4.44), one can get

$$\lambda_j(t', x') = \lambda_j(0) e^{-\eta_j(x' + 4\xi_j t') + i[\xi_j x' + 2(\xi_j^2 - \eta_j^2)t']},$$

and

$$|\lambda_j(x', t')| = |\lambda_j(0)| e^{-\eta_j y_j},$$

where $y_j = x' + 4\xi_j t'$.

Let us consider the asymptotic form of the N-soliton solution on the straight line $y_m = \text{const.}$ as $t' \to \infty$. Then

$$\begin{aligned} y_j &\to +\infty,\ |\lambda_j| \to 0, \quad &&\text{for } j < m; \\ y_j &\to -\infty,\ |\lambda_j| \to \infty, \quad &&\text{for } j > m. \end{aligned}$$

It follows from (4.53) that $\psi_{1j}, \psi_{2j} \to 0$, when $j < m$. In this limit, we have a reduced system of equations for the $2(N - m - 1)$ functions of y_m: $\psi_{1m}, \psi_{2m}, \Psi_{1k} = \psi_{1k}\lambda_k, \Psi_{2k}^* = \psi_{2k}\lambda_{x'}^*$, $k > m$, namely

$$\begin{aligned} \psi_{1m} + \frac{|\lambda_m|^2}{2i\eta_m}\psi_{2m}^* &= -\lambda_m \sum_{k=m+1}^{N} \frac{1}{\zeta_m - \zeta_k^*}\Psi_{2k}^*, \\ \frac{|\lambda_m|^2}{2i\eta_m}\psi_{1m} + \psi_{2m}^* &= \lambda_m^* + \lambda_m^* \sum_{k=m+1}^{N} \frac{1}{\zeta_m^* - \zeta_k}\Psi_{1k}, \end{aligned} \tag{4.59}$$

and

$$\begin{aligned} \sum_{k=m+1}^{N} \frac{1}{\zeta_m - \zeta_k^*}\Psi_{2k}^* &= -\frac{\lambda_m^*}{\zeta_m - \zeta_k^*}\psi_{2m}^*, \\ \sum_{k=m+1}^{N} \frac{1}{\zeta_m^* - \zeta_k}\Psi_{1m} &= -1 - \frac{\lambda_m}{\zeta_m^* - \zeta_k}\psi_{1m}. \end{aligned} \tag{4.60}$$

Solving the last system with respect to Ψ_{1k} and Ψ^*_{2k}, we obtain

$$\Psi_{1k} = a_k + \frac{2i\eta_m}{a_m}\frac{a_k}{\zeta_k - \zeta_m}\psi_{1m}\lambda_m,$$

$$\Psi^*_{2k} = -\frac{2i\eta_m}{a^*_m}\frac{a^*_k}{\zeta^*_k - \zeta^*_m}\psi^*_{2m}\lambda^*_m.$$

Here

$$a_k = \frac{\prod\limits_{p=m+1}^{N}(\zeta_k - \zeta^*_p)}{\prod\limits_{m<p\neq k}^{N}(\zeta_k - \zeta_p)}, \qquad a_m = 2i\eta_m \prod_{p=m+1}^{N}\frac{\zeta_m - \zeta^*_p}{\zeta_m - \zeta_p}.$$

Substituting the expressions for Ψ_{1k} and Ψ^*_{2k} in (4.59), Zakharov *et al.* obtained

$$\psi_{1m} + \frac{|\lambda^+_m|^2}{2i\eta_m}\psi^*_{2m} = 0, \qquad \frac{|\lambda^+_m|^2}{2i\eta_m}\psi_{1m} + \psi^*_{2m} = (\lambda^+_m)^*, \tag{4.61}$$

with

$$\lambda^+_m = \lambda_m \prod_{p=m+1}^{N}\frac{\zeta_m - \zeta_p}{\zeta_m - \zeta^*_p}.$$

Equation (4.61) coincides with (4.55) and describes a soliton with a displaced center x'^+_0 and phase θ^+, given by

$$\begin{aligned} x'^+_{0m} - x'_{0m} &= \frac{1}{\eta_m}\prod_{p=m+1}^{N}\left|\frac{\zeta_m - \zeta_p}{\zeta_m - \zeta^*_p}\right| < 0, \\ \theta^+_m - \theta_m &= -2\prod_{p=m+1}^{N} arg\left(\frac{\zeta_m - \zeta_p}{\zeta_m - \zeta^*_p}\right). \end{aligned} \tag{4.62}$$

The case of $t' \to -\infty$ can be analyzed similarly and we can get from (4.55)

$$\lambda = \lambda^-_m = \lambda_m \prod_{p=1}^{m-1}\frac{\zeta_m - \zeta_p}{\zeta_m - \zeta^*_p}. \tag{4.63}$$

On the straight lines $y = x' + \xi t'$, where ξ does not coincide with any of the ξ_ms as $t' \to \pm\infty$, the reduced system becomes asymptotically homogeneous, and the solution approaches to zero at an asymptotic rate, thus proving the asymptotic breakdown of the N-soliton solution into individual solitons.

The above results make it possible to describe the soliton scattering process. As $t' \to \infty$, the N-soliton solution breaks up and the resulting solitons move in such a way that the fastest soliton is in the front while the slowest at the rear. However, this is reversed as $t' \to -\infty$. As time t varies from $-\infty$ to $+\infty$, the quantity

λ_m changes by a factor of λ_m^+/λ_m^-, corresponding to a total change, $\triangle x'_{0m}$, in the coordinate of the soliton center, given by

$$\triangle x'_{0m} = x'^{+}_{0m} - x'^{-}_{0m} = \frac{1}{\eta_m}\left(\sum_{k=m+1}^{N} \ln\left|\frac{\zeta_m - \zeta_k}{\zeta_m - \zeta_k^*}\right| - \sum_{k=1}^{m-1} \ln\left|\frac{\zeta_m - \zeta_k}{\zeta_m - \zeta_k^*}\right|\right), \tag{4.64}$$

and a total change in its phase

$$\triangle\theta_m = \theta_m^+ - \theta_m^- = 2\sum_{k=1}^{m-1} \arg\left(\frac{\zeta_m - \zeta_k}{\zeta_m - \zeta_k^*}\right) - 2\sum_{k=m+1}^{N} \arg\left(\frac{\zeta_m - \zeta_k}{\zeta_m - \zeta_k^*}\right). \tag{4.65}$$

Equation (4.64) can be understood by assuming that the microscopic particles (solitons) collide pairwise and every microscopic particles collides with all others. In each paired collision, the faster microscopic particle moves forward by an amount of $\eta_m^{-1}\ln|(\zeta_m - \zeta_k^*)/(\zeta_m - \zeta_k)|$, $\zeta_m > \zeta_k$, and the slower one shifts backwards by an amount of $\eta_k^{-1}\ln|(\zeta_m - \zeta_k^*)/(\zeta_m - \zeta_k)|$. The total shift of the soliton is equal to the algebraic sum of those of the pair during the paired collisions, so that the effect of multi-particle collisions is insignificant. The situation is the same with the phases. This rule of collision of the microscopic particles in nonlinear quantum mechanics is the same as that of classical particles. This demonstrates the feature of classical motion of the microscopic particles, or, corpuscle feature of the microscopic particles in nonlinear quantum mechanics.

Using the above properties of two-particle collision and following the approach of Zakharov and Shabat, Desem and Chu obtained a solution for the interacting two-particle system from (4.50), it is given in terms of the solution of the nonlinear Schrödinger equation corresponding to two discrete eigenvalues $\zeta_{1,2}$,

$$\phi(x',t') = \frac{|\alpha_1|\cosh(a_1 + i\theta_1)e^{i\theta'_2} + |\alpha_2|\cosh(a_2 + i\theta_2)e^{i\theta'_1}}{\alpha_3\cosh(a_1)\cosh(a_2) - \alpha_4[\cosh(a_1 + a_2) - \cos(A')]}, \tag{4.66}$$

where

$$\theta'_{1,2} = 2\left[2(\eta_{1,2}^2 - \xi_{1,2}^2)t' - x'\xi_{1,2}\right] + (\theta'_0)_{1,2},$$

$$A' = \theta'_2 - \theta'_1 + (\theta_2 - \theta_1),$$

$$a_{1,2} = 2\eta_{1,2}(x' + 4t'\xi_{1,2}) + (a_0)_{1,2},$$

$$|\alpha_{1,2}|e^{i\theta_{1,2}} = \pm\left\{\left[\frac{1}{2\eta_{1,2}} - \frac{\eta}{(\triangle\xi^2 + \eta^2)}\right] \pm i\frac{\triangle\xi}{(\triangle\xi^2 + \eta^2)}\right\},$$

$$\alpha_3 = \frac{1}{4\eta_1\eta_2}, \quad \alpha_4 = \frac{1}{2(\eta^2 + \triangle\xi^2)},$$

$$\zeta_{1,2} = \xi_{1,2} + i\eta_{1,2}, \quad \triangle\xi = \xi_2 - \xi_1, \quad \eta = \eta_1 + \eta_2,$$

η and ξ are the same as those in (4.56), and represent the velocities and amplitudes of the microscopic particle (soliton), $(a_0)_{1,2}$ the position, and $(\theta'_0)_{1,2}$ the phase. They are all determined by the initial conditions.

Of particular interest here is an initial pulse waveform,

$$\phi(0, x') = \text{sech}(x' - x'_0) + \text{sech}(x' + x'_0)e^{i\theta} \tag{4.67}$$

which represents the motion of two soliton-like microscopic particles into the system. Equation (4.67) will evolve into two solitons described by (4.66) and a much smaller non-soliton part which decays like a dispersive tail. The interaction between the two microscopic particles given in (4.67) can therefore be analyzed through the two-soliton function, equation (4.66). Given the initial separation x'_0, phase difference θ between the two microscopic particles, the eigenvalues $\zeta_{1,2}$, a_0 and θ'_0 can be evaluated by solving the Zakbarov and Shabat eigenvalue equation (4.42), using (4.67) as the initial condition. Substituting the eigenvalues obtained into (4.66), we then obtain the description of the interaction between the two microscopic particles.

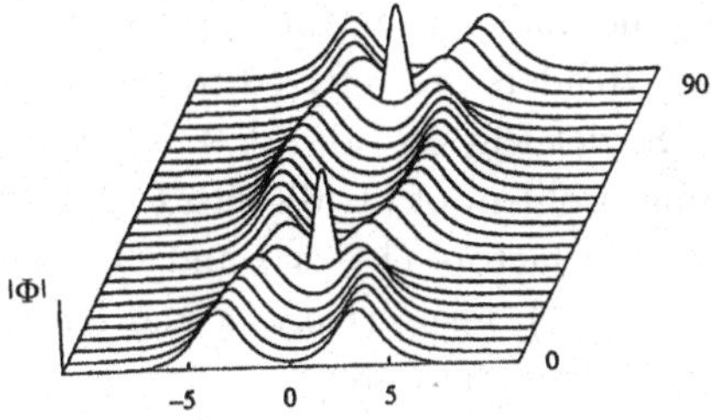

Fig. 4.2 Interaction with two equal amplitude microscopic particles. Initial microscopic particles separation = 3.5 pulse width (pw).

The two microscopic particles described by (4.67) interact through a periodic potential in t', through the $\cos A'$ term. The period is given by $\pi/(\eta_2^2 - \eta_1^2)$. The propagation of two microscopic particles with initial conditions $\theta = 0$, $\xi_1 = \xi_2 = 0$, $(\theta'_0)_1 = (\theta'_0)_2 = 0$, obtained by Desem and Chu is shown in Fig. 4.2. The two microscopic particles, initially separated by x'_0, coalesce into one microscopic particle at $A' = \pi$. Then they separate and revert to the initial state with separation x'_0 at $A' = 2\pi$, and so on. An approximate expression for the spacial variation of the separation between the microscopic particles can be obtained provided the two microscopic particles are well resolved in such a case (Gordon 1983; Karpman and Solovev 1981). Assuming that the separation between the solitons is sufficiently large, one can obtain the separation $\triangle x$ as

$$\triangle x' = \ln\left[\frac{2}{a}|\cos(at')|\right], \quad a = 2e^{-x'_0}.$$

Thus the period of oscillation is approximately given by $t'_p = (\pi/2)e^{x'_0}$ (Blow and Doran 1983; Gordon 1983).

4.3.2 *Repulsive interaction ($b < 0$)*

We now discuss the case of $b < 0$. In this case, equation (4.40) also has soliton solution, when $\lim_{|x'|\to\infty} |\phi(x',t')|^2 \to$ constant, and $\lim_{|x'|\to\infty} \phi_{x'} = 0$. The solitons are dark (hole) solitons, in contrast to the bright soliton when $b > 0$. The bright soliton was observed experimentally in focusing fibers with negative dispersion, and the hole soliton was observed in defocusing fibers with normal dispersion effect by Emplit *et al.* and Krökel. In practice, it is an empty state without matter in light or a hole in microscopic world. Therefore $b > 0$ corresponds to attractive interaction between Bose particles, and $b < 0$ corresponds to repulsive interaction between them. Thus, reversing the sign of b not only leads to changes in the physical picture of the phenomenon described by the nonlinear Schrödinger equation (4.40), but also requires considerable restructuring of the mathematical formalism for its solution. Solution of the nonlinear Schrödinger equation must be analyzed and collision rules of the microscopic particles must be obtained separately for the case of $b < 0$, using the inverse scattering method. Again, the approach of Zakharov and Shabat is outlined in the following. Readers are referred to the original paper by Zakharov and Shabat published in Soviet Physics, JETP, **37** (5) (1973) 823 for details.

In the direct scattering problem of (4.40) with $b < 0$, equations (4.41) – (4.44) are still valid, as long as the factors $(1 - s^2)$ and $(1 - s)$ in these equations are replaced by $(s^2 - 1)$ and $(s - 1)$, respectively. a and b now satisfy $|a|^2 - |b|^2 = 1$. The corresponding the scattering data are

$$\begin{aligned} a(\lambda, t') &= a(\lambda, 0), \\ b(\lambda, t') &= b(\lambda, 0)e^{-4i\lambda\zeta t'}, \\ c_j(t') &= c_j(0)e^{4\lambda_j v_j t'}, \\ \zeta_j &= i\sqrt{1-\lambda_j^2} = iv_j. \end{aligned}$$

The above equations enable us to calculate the scattering matrix at an arbitrary instant of time from the initial data. This yields $\phi(x',t')$ at an arbitrary instant of time. The integral equations (Marchenko equations) can then be reconstructed to determine the potential $q(x') = \phi(x',t')/(s^2-1)$ from the scattering data. Zakharov *et al.* obtained the following

$$\phi(x') = 1 - 2i\Phi_{1,2}(x',x') = \frac{(\lambda + i\nu)^2 + (\nu/\mu)e^{2\nu x'}}{1 + (\nu/\mu)e^{2\nu x'}}, \tag{4.68}$$

where

$$\Phi_{1,2}(x',y) = \frac{\nu(\lambda + i\nu)}{1 + (\nu/\mu)e^{2\nu x'}}e^{\nu(x'-y)}, \quad \nu = \sqrt{1-\lambda^2}.$$

Here y is an arbitrary argument.

Of particular physical interest is the state $|\phi(x',t')|^2 \to$ const., $\phi_{x'} \to 0$ as $x' \to \pm\infty$, which corresponds to the propagation of a wave through a condensate of constant density. In this case the soliton

$$\sqrt{\frac{b}{2}}\phi(x',t') = \frac{(\lambda - i\nu)^2 + e^{2\nu(x'-x'_0)-2\lambda t'}}{1 + e^{2\nu(x'-x'_0)-2\lambda t'}} \tag{4.69}$$

can move with a constant velocity. Thus

$$\frac{b}{2}|\phi(x',t')|^2 = 1 - \frac{\nu^2}{\cosh^2[\nu(x' - x'_0 - 2\lambda t)]}. \tag{4.70}$$

The parameter λ characterizes the amplitude and velocity of the microscopic particles, and x'_0 is the position of its center at $t' = 0$, where $d(\ln\mu)/dt' = 4\lambda\nu$ or $\ln\mu = 2\nu(x'_0 + 2\lambda t')$.

We can verify that the eigenvalue λ corresponds to a soliton moving with velocity 2λ and that the collision between these microscopic particles satisfies the rule for $b > 0$.

Let us consider the interaction of two microscopic particles (solitons) with velocities $2\lambda_2$ and $2\lambda_1$. To this end it is necessary to obtain the corresponding two-soliton solution. Owing to the complexity of the explicit formula for this solution, we limit our discussion to the asymptotic behavior as $t' \to \pm\infty$.

Following the same procedure as before, we can show that as $t' \to \pm\infty$, the two-soliton solution breaks up into individual solitons

$$\begin{aligned}
\phi(x',t') &\to \phi_0(x' - 2\lambda_1 t', \lambda_1, x_1'^{+}) + \phi_0(x' - 2\lambda_2 t', \lambda_2, x_2'^{+}), \quad t' \to +\infty,\\
\phi(x',t') &\to \phi_0(x' - 2\lambda_1 t', \lambda_1, x_1'^{-}) + \phi_0(x' - 2\lambda_2 t', \lambda_2, x_2'^{-}), \quad t' \to -\infty.
\end{aligned}$$

The scattering of the solitons by each other gives rise to the following displacements of their centers

$$\begin{aligned}
\delta x'_1 &= x_1'^{+} - x'^{-},\\
\delta x'_2 &= x_2'^{+} - x_2'^{-}.
\end{aligned}$$

To determine these displacements, we note first that the following discontinuity in the phase of the wave function of the condensate occurs for a microscopic particle (soliton) with velocity $2\lambda_i$

$$\alpha'_i = \arg\phi(-\infty) - \arg\phi(+\infty) = 2\tan^{-1}\left(\frac{\nu_i}{\lambda_i}\right).$$

The total phase discontinuity for N eigenvalues is

$$\alpha' = \sum_{k=1}^{N} \alpha'_i,$$

which, however, does not depend on the relative positions of the solitons. In the case of two solitons, the Jost function $\psi_1'(x',\lambda)$ takes the following form as $x' \to -\infty$,

$$\psi_1'(x',\lambda) \to e^{\nu x'}\begin{bmatrix} 1 \\ (\nu-\lambda)e^{i(\alpha_1'+\alpha_2')} \end{bmatrix}, \quad (-1<\lambda<1).$$

Thus, the asymptotic form of $\psi_1'(x',\lambda)$ as $x' \to +\infty$ is

$$\psi_1'(x',\lambda) \to \begin{cases} a'(\lambda,i\nu)e^{\nu x'}\begin{bmatrix} 1 \\ i\nu-\lambda \end{bmatrix}, & \lambda \neq \lambda_1,\lambda_2, \\ b_{1,2}(t)e^{-\nu x'}\begin{bmatrix} i\nu_{1,2}-\lambda_{1,2} \\ 1 \end{bmatrix}, & \lambda = \lambda_1,\lambda_2. \end{cases} \tag{4.71}$$

Zakharov *et al.* represented the coefficient a in the form of $a' = a_1'a_2'$, where

$$a_{1,2}'(\lambda,\nu) = \frac{i\nu+\lambda-i\nu_{1,2}-\lambda_{1,2}}{i\nu+\lambda+i\nu_{1,2}+\lambda_{1,2}}$$

is the component of "single-soliton" scattering matrix. Let $\lambda_1 > \lambda_2$. As $t' \to -\infty$, the microscopic particles move apart and separated by a large distance and $|\phi| \to 1$. Assume that particle 2, with velocity $2\lambda_2$, is located to the right, then $\phi \to \exp(i\alpha_2')$ in the region between the microscopic particles. As $t' \to \infty$, particle 1, with velocity $2\lambda_1$, is located to the right, then $\phi \to \exp(i\alpha_1')$ in the region between the microscopic particles. We consider in this region the asymptotic form of the Jost functions $\psi_1'(x',\lambda)$ as $t' \to -\infty$. We can get

$$\psi_1' \approx \begin{cases} a_1'(\lambda,i\nu)e^{\nu x'}\begin{bmatrix} 1 \\ e^{i\alpha_2'(i\nu-\lambda)} \end{bmatrix}, & (\lambda \neq \lambda_1), \\ b_1^-(t')e^{-\nu_1 x'}\begin{bmatrix} e^{i\alpha_2'(i\nu_1-\lambda_1)} \\ 1 \end{bmatrix}, & (\lambda = \lambda_1), \end{cases}$$

here $b_1^-(t')$ is an unknown function. To determine $b_1^-(t')$, we use $a(\lambda,-i\nu) = 1/a^*(\lambda,i\nu)$. Through microscopic particle 2, the variation of the function ψ_1' in accordance with this rule, and comparison with (4.71), we finally get

$$b_1^-(t') = a_2'^*(\lambda_1,i\nu_1)b_1(t').$$

Using the same procedure, $b_2^-(t') = b_2(t')/a_1(\lambda_2,i\nu_2)$ can be determined. $b_1^-(t')$ and $b_2^-(t')$ determine the positions of the microscopic particles as $t' \to -\infty$. Similar quantities, $b_1^+(t')$ and $b_2^+(t')$, can be introduced to describe the positions of the solitons as $t' \to \infty$. A similar analysis as above results in the following

$$b_1^+(t') = \frac{b_1(t')}{a_2'(\lambda_1,i\nu_1)}, \quad b_2^+(t') = b_2(t')a_1'(\lambda_2,i\nu_2).$$

Thus, the displacements of the microscopic particles (solitons) after the collision were found to be

$$\begin{aligned}
\delta x_1' &= \frac{1}{2\nu_1}\ln\left(\frac{b_1^+}{b_1^-}\right) = \frac{1}{2\nu_1}\ln\left(\frac{1}{|a_2(\lambda_1, i\nu_1)|^2}\right) = \frac{Y}{2\nu_1}, \\
\delta x_2' &= \frac{1}{2\nu_2}\ln\left(\frac{b_2^+}{b_2^-}\right) = \frac{1}{2\nu_2}\ln\left(|a_1(\lambda_2, i\nu_2)|^2\right) = -\frac{Y}{2\nu_2},
\end{aligned} \tag{4.72}$$

where

$$Y = \ln\left[\frac{(\lambda_1-\lambda_2)^2+(\nu_1+\nu_2)^2}{(\lambda_1-\lambda_2)^2+(\nu_1-\nu_2)^2}\right].$$

The microscopic particle which has the greater velocity acquires a positive shift, and the other a negative shift. The microscopic particles repel each other, like classical particles. From (4.72) we get

$$\nu_1\delta x_1' + \nu_2\delta x_2' = 0.$$

This relation was obtained by Tsuzuki directly from (4.40) for $b < 0$ by analyzing the motion of the center of mass of a Bose gas. It can be interpreted as the conservation of the center of mass of the microscopic particles during the collision. This shows sufficiently the classical feature of microscopic particles described by (4.40).

The case of N-solitons can be studied similarly and it can be shown that as $t' \to \pm\infty$, an arbitrary N-soliton solution breaks up into individual solitons. The arrangement of the solitons according to their velocity is reversed as the changes from $-\infty$ to $+\infty$, so that each soliton collides with each other soliton. Going through the same analysis given above, we can verify that the total displacement of a soliton, regardless of the details of the collisions, is equal to the sum of the displacements in individual collisions

$$\delta_j = x_j'^{+} - x_j'^{-} = \sum_{i=1}\delta_{ij},$$

where

$$\delta_{ij} = \mathrm{sign}(\lambda_i-\lambda_j)\frac{1}{2\nu_i}\ln\left[\frac{(\lambda_i-\lambda_j)^2+(\nu_i+\nu_j)^2}{(\lambda_i-\lambda_j)^2+(\nu_i-\nu_j)^2}\right]. \tag{4.73}$$

From the above studies we see that collisions of many microscopic particles described by the nonlinear Schrödinger equation (3.2), with both $b > 0$ or $b < 0$, satisfy rules of classical physics. This shows sufficiently the corpuscle feature of microscopic particles in nonlinear quantum mechanics.

4.3.3 *Numerical simulation*

The discussion on collision of microscopic particles presented above are based on analytical analysis using the inverse scattering method, and as a result, only the asymptotic behaviors of the particles can be obtained. The same process can be studied by numerically solving (3.2). Numerical simulation can reveal more detailed features of collision between two microscopic particles in nonlinear quantum mechanics. We begin by dividing (3.2) with V = constant and $A(\phi) = 0$ into the following two-equations

$$\begin{gathered} i\frac{\partial \phi'}{\partial t'} + \frac{\partial^2 \phi'}{\partial x'^2} = \phi' u, \\ \frac{\partial^2 u}{\partial t'^2} - \frac{\partial^2 u}{\partial x'^2} = \frac{\partial}{\partial x'}(|\phi'|^2). \end{gathered} \tag{4.74}$$

Obviously, if we assume $\bar{\xi} = x' - vt'$ in (4.74), we can get the nonlinear constant $b = 1/(1-v^2)$ in (3.2) at $\lim_{|x'|\to\infty} \phi' = \lim_{|x'|\to\infty} \phi'_{x'} = 0$. Equation (4.74) is similar to that in Davydov's or Pang's model for bio-energy transport in the α-helix protein and molecular crystalline-acetanilide (see Chapters 5 and 9), in which the soliton is formed by self-trapping of excitons interacting with lattice phonons. In these models, ϕ' represents the wave function of the exciton, u is the longitudinal displacement of the lattice molecule, b is the nonlinear coefficient of the system resulting from the exciton-phonon interaction. Evidently b is related to the exciton velocity v. The soliton solution of (4.74) can now be written as

$$\begin{gathered} \phi' = \sqrt{2(1-v^2)}\ \eta\ \mathrm{sech}[\eta(x' - x'_0 - vt')] \exp\left[\frac{i}{2}vx' - i(\frac{v^2}{4} - \eta^2)t' + i\theta\right], \\ u = -2\eta^2 \mathrm{sech}^2[\eta(x' - x'_0 - vt')]. \end{gathered} \tag{4.75}$$

The soliton in (4.75) is a self-trapped state of exciton plus deformation of the lattice in such a case. The properties of the soliton depend on three parameters: η, v and θ, η determines the amplitude and width of the soliton, θ is the phase of the sinusoidal factor of ϕ' at $t'=0$. Tan *et al.* carried out numerical simulation of the collision process between solitons using the Fourier pseudo-spectral method with 256 basis functions for the spatial discretization, together with the fourth-order Runge-Kutta method for time-evolution. The system given in (4.74) has two exact integrals of motion

$$N_s = \int_{-\infty}^{\infty} |\phi'|^2 dx', \quad \text{and} \quad E_1 = \int_{-\infty}^{\infty} u dx',$$

which can be used to check the accuracy of the numerical solutions.

For the collision experiments, the initial state consists of two solitary waves separated by distance x_0',

$$\begin{aligned}\phi' = {} & \sqrt{2(1-v_1^2)}\ \eta_1\ \mathrm{sech}(\eta_1 x')\exp\left(\frac{iv_1}{2}x' + i\theta_1\right) + \\ & \sqrt{2(1-v_2^2)}\ \eta_2\ \mathrm{sech}[\eta_2(x'+x_0')]\exp\left[-\frac{v^2}{2}(x_2'+x_0') + i\theta_2\right], \\ u = {} & -2\eta_1^2\mathrm{sech}^2(\eta_1 x') - 2\eta_2^2\mathrm{sech}^2[\eta_2(x'+x_0')],\end{aligned}$$

where the first term in each expression represents one soliton (1) while the second term represents the other soliton (2). It can be shown that the post-collision state of the solitons is strongly dependent on both the initial phases and the initial velocities of the solitons. Since ϕ' can be multiplied by an arbitrary phase factor, $\exp(i\tilde{\theta})$ where $\tilde{\theta}$ is an arbitrary constant, and still remains a solution (with u unchanged), one of the phases is arbitrary, and only the difference of the two initial phases is significant. Therefore, we can set $\theta_1 = 0$ for the convenience of discussion.

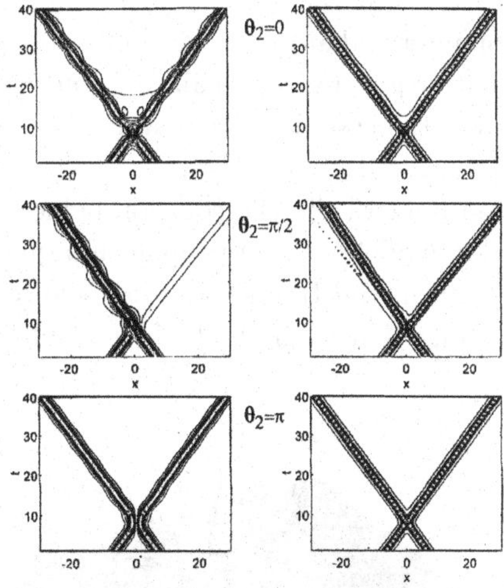

Fig. 4.3 Fast collisions of solitons. The initial ratio of velocities of the fast and slow solitons is 1.8.

Figure 4.3 shows the fast collisions obtained by Tan *et al.*, in which the initial ratio of velocities of the fast and slow solitons is fixed to be 1.8. The absolute value of the ϕ' are shown using contours on the left in each pair of plots, with x' being the horizontal coordinate and t' increasing upward. The right panel shows the absolute value of u. The relative phase increases from 0 (top) to $\pi/2$ (middle) to π (bottom). All cases are identical except that θ_2 is increased by $\pi/2$ in each case, beginning with $\theta_2 = 0$ at the top. Before the collision, each of the initial soliton contributes 0.7600 to N_s in all three cases ($\theta_2 = 0$, $\pi/2$ and π). When the relative phase is zero (top), the solitons penetrate each other freely and then emerge with their shapes and velocity unchanged. When $\theta_2 = \pi/2$ (middle graphs), ϕ' emerges from the collision asymetrically, and a large soliton, which contributes 1.4272, moving to the left, at the same velocity as the initial speed of soliton 2. Another small pulse, contributing 0.0928, travels to the right at the speed which is the same as the initial speed of soliton 1. The post-collision energies are the same as those of pre-collision for ϕ' when $\theta_1 = 0$ and $\theta_2 = \pi$. For all values of θ_2, there is little change in the contributions of the solitons in their u-field to energy E_1, and they are not shown here. When $\theta_2 = \pi$, as shown in the bottom panel of Fig. 4.3, the u-components penetrate freely, but the ϕ'-components bounce off each other and change their directions, without interpenetration. The fourth case, $\theta_2 = 3\pi/2$, is not shown here because it is just the mirror image of the middle figure. That is

$$\phi'\left(x', t', \theta_2 = \frac{3\pi}{2}\right) = \phi'\left(-x', t', \theta_2 = \frac{\pi}{2}\right),$$

$$u\left(x', t', \theta_2 = \frac{3\pi}{2}\right) = u\left(-x', t', \theta_2 = \frac{\pi}{2}\right).$$

The same is true for intermediate and slow collisions processes. However, the reflection principle cannot be generalized to all solutions which are different in their initial phases by π because the cases of $\theta_2 = 0$ and $\theta_2 = \pi$ are quite different in the general case.

Pang *et al.* simulated numerically the collision behaviors of two solitons based on the improved Davydov model with a two-quantum quasi-coherent state and an additional interaction term in the Hamiltonian for the α-helix protein molecules, using the fourth-order Runge-Kutta method. This result is shown in Fig. 4.4 (see Chapter 9 for details).

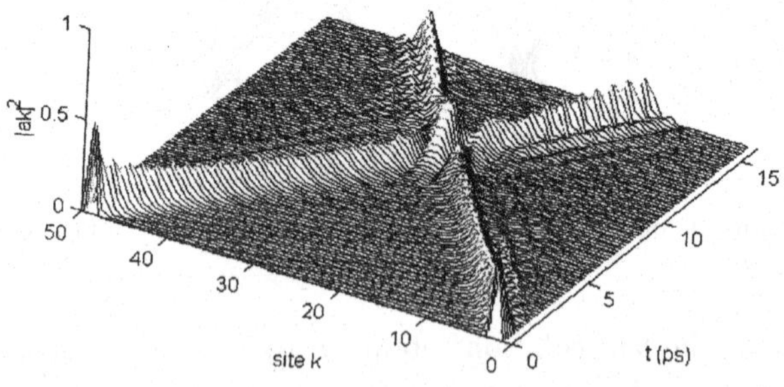

Fig. 4.4 The collision of two microscopic particles described by nonlinear Schrödinger equation in the improved Davydov model.

From Figs. 4.2 – 4.4, we see clearly that the collision between microscopic particles described by the nonlinear Schrödinger equation show features of classical particles. Thus we confirm again that microscopic particles in nonlinear quantum mechanics has the corpuscle property.

4.4 Properties of Elastic Interaction between Microscopic Particles

We now consider further the properties of elastic interaction of microscopic particles (solitons) described by the nonlinear Schrödinger equation (3.2) with $A(\phi) = 0$, following the approach of Gorshkov and Ostrovsky (1981) and Abdullaev *et al.*. We assume that the interaction among the microscopic particles is weak.

For the Euler-Lagrange equation (3.101) corresponding to the nonlinear Schrödinger equation, we further assume that it has a soliton solution $\phi^0(\bar{\xi} = x' - vt', \nu)$ satisfying

$$\frac{d}{d\xi}\left(-v\frac{\partial \mathcal{L}}{\partial \phi_{t'}} + \frac{\partial \mathcal{L}}{\partial \phi_{x'}}\right) - \frac{\partial \mathcal{L}}{\partial \phi} = 0. \tag{4.76}$$

The interaction between the microscopic particles (solitons) in the system is weak, provided they are separated by a sufficiently large distance, and their relative velocity is small. The motion of each soliton can then be analysed by assuming that it is either in the weak field or on the "tail" of other solitons, and expanding the concerned quantities in terms of a small parameter ε. Based on this approach, the solution of (4.76) or (3.2) with $A(\phi) = 0$ in such a case can be expressed as a series

$$\phi(x',t') = \phi^{(0)}[\bar{\xi} - u_n(t'), v] + \sum_{j\neq n}\phi^{(0)}[\lambda - u_j(t'), v] + \sum_i \varepsilon^i\phi^{(i)}[\bar{\xi} - u_i(t'), X, \tau] \tag{4.77}$$

where $X = \varepsilon x'$, $\tau = \varepsilon t'$, n is the number of microscopic particles, $du_n/dt' = O(\varepsilon v)$, and $\varepsilon \approx (v_n - v_j)/(v_n + v_j) \ll 1$. This perturbation method is self-consistent if we assume that the field produced by the tail of the jth microscopic particle at the site of the nth microscopic particle is of the order ε^2. Inserting (4.77) into (4.76) and equating the terms of the same orders in ε^2, we obtain a set of linear equations for $\phi^{(i)}$. In order for the solutions to remain finite at $\bar{\xi} \to \pm\infty$, it is necessary to have a suitable orthogonality condition which, to the first order in ε, has the following form

$$\int_{-\infty}^{\infty} d\bar{\xi}\phi_{\bar{\xi}}^{(0)} R^{(1)} = 0$$

where

$$R^{(1)} = \frac{du_n}{dt'}\left[\frac{d}{d\bar{\xi}}\frac{\partial \mathcal{L}}{\partial \phi_{t'}} + \frac{d}{d\bar{\xi}}\left(-v\frac{\partial^2 \mathcal{L}}{\partial \phi_{t'}^2} + \frac{\partial^2 \mathcal{L}}{\partial \phi_{x'}\partial \phi_{t'}}\right)\phi_{\bar{\xi}}^{(0)} - \frac{\partial^2 \mathcal{L}}{\partial \phi \partial \phi_{t'}}\phi_{\bar{\xi}}^{(0)}\right].$$

In a similar way, Abdullaev *et al.* obtained the equation of motion for the center of mass coordinates of the microscopic particles by setting the determinant of the secular equation up to the second order in ε to zero

$$m\frac{d^2u_n}{dt'^2} = \sum_j f(u_n - u_j). \tag{4.78}$$

These equations are similar to (4.25) and those of classical particles with an effective mass m, and an interaction potential $U(u_n) \approx \int f(u_n)du_n$, given by

$$m = \frac{\partial}{\partial v}\int_{-\infty}^{\infty} d\bar{\xi}\frac{\partial \mathcal{L}}{\partial \phi_{t'}}\phi_{\bar{\xi}}^{(0)} = \frac{\partial P}{\partial v}, \tag{4.79}$$

$$U(u_n - u_j) = \int_{-\infty}^{\infty} d\bar{\xi}\left[\frac{d}{d\bar{\xi}}\left(-v\frac{\partial \mathcal{L}}{\partial \phi_{t'}} + \frac{\partial \mathcal{L}}{\partial \phi_{x'}}\right) - \frac{\partial \mathcal{L}}{\partial \phi}\right]_{\phi=\phi^{(0)}(\bar{\xi}-u_n)} \phi^{(0)}(\bar{\xi} - u_n), \tag{4.80}$$

where P is the momentum of the field.

The above method and results can be easily extended to the multi-dimensional case, where the solution depends on several variables, ξ_1, ξ_2, $\cdots$, ξ_i. For the nonlinear Schrödinger equation (3.2) at $A(\phi) = 0$, ϕ is a decreasing function of ξ_1 as $\xi_1 \to \infty$, and a periodic function in ξ_2. Details of derivation of the corresponding equations for the generalized P_q and U are given by Gorshkov and Ostrovsky (1981). Corresponding to (4.78) – (4.80), we have

$$\left(\frac{d^2u_n}{dt'^2}\right)\frac{\partial P_q}{\partial v_g} = \sum_{j=n} \nabla_{u_n} U(u_{1n} - u_{1j}, \cdots, u_{mn} - u_{mj}),$$

$$P_q = \int_{-\infty}^{\infty} d\bar{\xi}\frac{\partial L}{\partial \phi_{t'}}\phi_{\xi_q}^{(0)},$$

$$U(u_{1n} - u_{1j}, \cdots, u_{mn} - u_{mj}) = \int_{-\infty}^{\infty} d\bar{\xi}\phi^{(0)}(\xi_1 - u_1, \cdots, \xi_m - u_m) \times \left[\frac{d}{d\xi_q}\left(-v_q\frac{\partial \mathcal{L}}{\partial \phi_{t'}} + \frac{\partial \mathcal{L}}{\partial \phi_{x'_q}}\right) - \frac{\partial \mathcal{L}}{\partial \phi}\right]_{\phi=\phi^{(0)}(\xi_1 - u_{1n}, \cdots, \xi_m - u_{mn})}$$

For the ϕ^4-field equation

$$\phi_{tt} - \phi_{xx} - \phi + \phi^3 = 0$$

in the 1+1 dimensional case, according to (4.79), the effective mass of a microscopic particle (kink) is given by

$$m = \frac{\partial}{\partial v}\int_{-\infty}^{\infty} d\bar{\xi}\phi^{(0)}\phi_{\bar{\xi}}^{(0)} = \frac{\partial}{\partial v}\left(-\frac{v}{\sqrt{2(1-v^2)}}\right)\int_{-\infty}^{\infty}\frac{d\bar{\xi}}{\cosh^4\bar{\xi}} = \frac{2\sqrt{2}}{3}\frac{1}{(1-v^2)^{3/2}},$$

in the natural unit system, where

$$\phi^{(0)} = \tanh\left[\pm\frac{x - vt}{\sqrt{2(1-v^2)}}\right].$$

Next, we study the kink-kink and the kink-antikink interactions, by substituting the expression for $\mathcal{L}$ of the ϕ^4-field equation into (4.80)

$$U(u_n - u_j) = \int_{-\infty}^{\infty} d\bar{\xi}\phi^{(0)3}(\bar{\xi} - u_n)\phi^{(0)}(\bar{\xi} - u_j)$$

where the asymptotics $\phi^{(0)}(\xi \pm \infty) = \phi_{\pm}^{(0)}$ have been subtracted from the kink (antikink) fields. We thus obtain an equation of motion for the equivalent classical particle with an effective mass m and moving in the effective potential U

$$m\frac{d^2u_n}{dt'^2} = \pm 8\sqrt{2(1-v)^2}e^{-\sqrt{2(1-v^2)}u_n}.$$

Here the $+$ sign corresponds to the kink-kink or the antikink-antikink interaction, and the $-$ sign corresponds to that of the kink-antikink pair. In other words, the kink and antikink interact through an exponential attractive potential. Consequently, a bound state could be formed between a kink and an antikink. Note, however, that this is valid only for particles moving with low speed ($v < 0.2$). It was shown by Campbell *et al.* (1986), that when the particles move with high speeds, there may be a resonant interaction between the kink and antikink, as well as other inelastic processes (Makhankov, 1979).

For the nonlinear Schrödinger equation (4.40) with $b = 2$, the soliton solution can be written as

$$\phi^{(0)} = \sqrt{2}\lambda''\mathrm{sech}(\lambda\bar{\xi})e^{iv\bar{\xi}/2+\theta},$$

where $\bar{\xi} = x' - vt'$, $\theta = \omega t'$, $(\lambda'')^2 = \omega^2 - v^2/4$. Inserting the above into (4.80), we can obtain the components for the field function and the interaction potential as the following

$$P_{\xi} = 2\pi v\lambda'', \quad P_{\theta} = 4\pi\lambda'',$$
$$U(u_{x'}, u_{\theta}) = 4\pi\lambda'^3\exp(-\lambda''u_{x'})\cos\left[\frac{v}{2}(u_{x'}) - u_{\theta}\right].$$

Making use of these components, the equations of motion can be written as

$$\frac{d^2R}{dt'^2} = \beta e^R,$$

where

$$R = z + i\tilde{\theta}, \quad \beta = 32(\lambda'')^2.$$

In the above,

$$z = -\lambda'' u_{x'}, \quad \tilde{\theta} = -\frac{v}{2} u_{x'} + u_{\theta}.$$

Thus we also got an equation of motion for the equivalent classical particle with an effective mass 1 and moving in the effective potential U, which is consistent with the results of (4.16), (4.17) and (4.25). The solutions depend on four parameters, t'_J, t'_I, R_∞, and R_0, which satisfy the following

$$e^{(R-R_0)/2} = -\cosh\left[R_\infty(t' + t'_J + it'_I)\right].$$

We now consider the interaction between microscopic particles based on the inverse scattering transformation, as it was done by Karpman and Solovev (1981); Anderson and Lisak (1986). Consider again the interaction between microscopic particles (solitons) described by (4.40). We seek a solution with well-separated solitons

$$\phi(x', t') = \phi_1(x', t') + \phi_2(x', t').$$

Then, the interaction between the microscopic particles is described by the following

$$\varepsilon_m R_m(\phi_j) = i(\phi_m \phi_j^{*2} + 2\phi_m \phi_j \phi_j^*), \quad (m, j = 1, 2, m \neq j).$$

Treating $\varepsilon_m R_m$ as perturbation and applying the perturbation theory, Desem and Chu obtained a set of equations for the parameters of the j-th soliton

$$\begin{aligned}
\frac{d\mu_j}{dt'} &= (-1)^j 16\eta^3 e^{-2\eta d}\cos(2\mu d + \tilde{\theta}),\\
\frac{d\eta_j}{dt'} &= (-1)^j 16\eta^3 e^{-2\eta d}\sin(2\mu d + \tilde{\theta}),\\
\frac{d\xi_j}{dt'} &= 2\mu_j + 4\eta e^{-2\eta d}\sin(2\mu d + \tilde{\theta}),\\
\frac{d\delta_j}{dt'} &= 2(\eta_j^2 + \mu_j^2) + 8\mu\eta e^{-2\eta d}\sin(2\mu d + \tilde{\theta}) + 24\eta^2 e^{-2\eta d}\cos(2\mu d + \tilde{\theta})
\end{aligned} \tag{4.81}$$

where

$$d = \xi_1 - \xi_2 > 0, \quad \tilde{\theta} = \delta_1 - \delta_2,$$
$$\eta = \frac{1}{2}(\eta_1 + \eta_2), \quad \mu = \frac{\mu_1 + \mu_2}{2},$$

and $\Delta\mu \ll \mu$, $\Delta\eta = \eta_2 - \eta_1 \ll \eta_1$, $\eta d \gg 1$, $\Delta\eta d \ll 1$, $\Delta\mu = \mu_1 - \mu_2$.

Equations (4.81) have three constants of motion:

$$\begin{aligned}
\mu &= \text{constant},\\
\eta &= \text{constant},\\
Y'^2 &= 16v^2 e^{-2\eta d + i(2\mu d + \tilde{\theta})} = \Lambda^2,
\end{aligned} \tag{4.82}$$

where

$$Y' = \triangle\mu + i\triangle\eta.$$

From this, a soliton solution is readily found

$$Y' = -\Lambda \tanh(2\eta\Lambda t' - \alpha_1 - i\alpha_2).$$

Setting $\Lambda = m + ij$, we obtain the following solution of (4.82)

$$\begin{aligned} d(t') - d(0) &= \frac{1}{2\eta} \log\left[\frac{\cosh(4\eta m t' - 2\alpha_1) + \cos(4\eta j t' - 2\alpha_2)}{\cosh(2\alpha_1) + \cos(2\alpha_2)}\right], \\ \tilde{\theta}(t) - \tilde{\theta}(0) &= -2\tan^{-1}[\tanh(2\eta m t' - \alpha_1)\tanh(2\eta j t' - \alpha_2)] \\ &\quad + 2\tan^{-1}(\tanh\alpha_2) - 2\mu[d - d(0)]. \end{aligned} \tag{4.83}$$

The interaction between microscopic particles in the nearly integrable perturbed nonlinear Schrödinger equation differs from that of the Sine-Gordon equation, for example, in that the binding energy of the two-soliton state of the former is zero. The two-soliton state is unstable with respect to small perturbations. However, there might be situations when perturbation stabilizes a two-soliton state, which enables us to consider multi-particle effects for solitons in nonlinear quantum mechanics, such as interaction between a two-soliton state and another soliton.

One of the features of solitons is their stability against a fairly broad class of perturbations and their elastic interaction upon colliding. The inverse scattering method can be used to study the interaction of solitons in elastic collsion of microscopic particles (solitons). Parameters of the microscopic particles (solitons) do not change, but simply acquire a phase shift δ_m (fast soliton) or $-\delta_p$ (slow soliton),

$$\delta_{m,p} = \frac{1}{2\eta_{m,p}} \ln\left|\frac{\lambda_m - \lambda_p^*}{\lambda_m - \lambda_p}\right|^2, \quad \Re\{\lambda_j\} < \Re\{\lambda_p\}$$

as mentioned in the previous section. For nearly equal velocities the solitons form a bound state.

Using results of inverse scattering method for the two-soliton solution, Gordon (1983) derived expressions for interaction between microscopic particles (solitons). Assuming large separation between solitons, he found that the interaction between the particles decreases exponentially with the distance between the microscopic particles, and depends only on the relative phase. This is consistent with results of the analysis based on the direct perturbation theory (Gorshkov and Ostrovsky, 1981). Gordon showed that with an initial state in the form of

$$\phi_0 = \text{sech}(x' - x_0') + \text{sech}(x' + x_0'), \quad (x_0' \gg 1),$$

the distance d between the microscopic particles is given by

$$d = d_0 + 2\ln|\cos(2t'e^{-x_0'})|. \tag{4.84}$$

It follows that the distance d has an oscillating character and that a bound state of the microscopic particles exists. Numerical simulation (Blow and Doran, 1983; Hermansson and Yervick, 1983) showed that a nonlinear interaction between microscopic particles leads to an attraction between pulses for any value of x'_0, with a relative phase of zero.

We now derive the interaction between two microscopic particles from the system of equations for the soliton parameters (4.81). A general solution is given by (4.83). If the microscopic particles (solitons) have the same amplitude and they are initially at rest, this solution simplifies to

$$d(t') = d(0) + \frac{1}{2\eta} \ln \left[\frac{1}{2} \cosh(4\eta m t') + \cos(4\eta j t') \right],$$

where

$$m = -4\eta e^{-\eta d(0)} \sin \frac{\tilde{\theta}(0)}{2}, \quad \text{and} \quad j = 4\eta e^{-\eta d(0)} \cos \frac{\tilde{\theta}(0)}{2}.$$

If the initial phase of the microscopic particles is zero, $\tilde{\theta}(0) = 0$, then $m = 0$ and $j = 4\eta e^{-\eta d(0)}$. The distance between the microscopic particles is

$$d(t) = d(0) + \frac{1}{\eta} \ln |\cos(2\eta j t')|.$$

Thus the microscopic particles undergo oscillatory motion. This expression coincides with (4.84) if $2\eta = 1$. The separation between the particles $d(t')$ is zero at time

$$t'_c = \frac{1}{2\eta |j|} \cos^{-1}[e^{-\eta d(0)}].$$

Since $\eta d(0) \gg 1$, the period of the oscillation is

$$t'_c = \frac{\pi}{16\eta^2} e^{\eta d(0)}.$$

On the other hand, if the initial phase difference is $\tilde{\theta}(0) = \pi$, then $j = 0$, $m = -4\eta e^{-\eta d(0)}$, and

$$d(t) - d(0) = \frac{1}{\eta} \ln[\cosh(2\eta m t')].$$

In this case, the distance between the microscopic particles increases monotonically. The separation is doubled after a time of

$$t'_d = \frac{1}{2\eta m} \sinh^{-1}[e^{\eta d(0)}].$$

For $\eta d(0) \gg 1$, we have

$$t'_d = \frac{\eta d(0)}{8\eta^2} e^{\eta d(0)}.$$

Desem and Chu gave the changes of $d(t')/d(0)$ with increasing $4\eta mt'$, as shown in Fig. 4.5, where three different situations are shown: a monotonic increasing separation between the microscopic particles (solitons), a quasi-periodic oscillation which gradually reduced to a monotonic increasing function, and an oscillatory mode. Hence, binding of microscopic particles (solitons) can be suppressed by appropriate choice of initial phase difference.

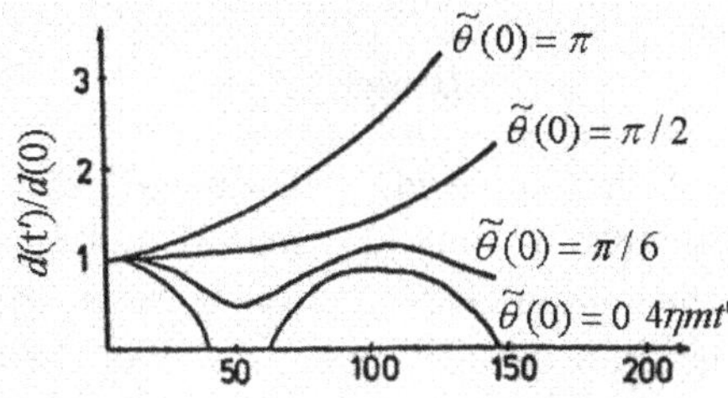

Fig. 4.5 The dependence of distance d between the miroscopic particles (solitons) on the initial phase difference.

4.5 Mechanism and Rules of Collision between Microscopic Particles

The fact that two microscopic particles can survive a collision completely unscathed demonstrates clearly the corpuscle feature of the microscopic particles. This property is frequently used in investigations to separate nonlinear quantum mechanical microscopic particles (solitons) from particles in the linear quantum mechanical regime. During the collision, the microscopic particles interact and exchange positions in the space-time trajectory as if they had passed through each other. After the collision the two microscopic particles may appear to be instantly translated in space and/or time but otherwise unaffected by their interaction. This translation is called a phase shift as mentioned in section 4.3. In one dimension, this process results from two microscopic particles colliding head on from opposite directions, or in one direction between two particles with different amplitudes. This is possible because the velocity of a particle depends on the amplitude.

In the following, we describe a series of laboratory and numerical experiments dedicated to investigation on the detailed structure, mechanism and rules of collision between microscopic particles described by the nonlinear Schrödinger equation in the nonlinear quantum mechanics. The properties and rules of such collision between two microscopic particles have been studied by Aossey *et al.* Both the phase shift of the microscopic particles after their interaction and the range of the interaction are functions of relative amplitude of the two colliding microscopic particles. The microscopic particles preserve their shape after the collision.

For the microscopic particles described by the nonlinear Schrödinger equation

(4.40), we will limit our discussion to the hole (dark) spatial solitons with $b < 0$ which, at present, is given by

$$\phi(x',t') = \phi_0\sqrt{1 - B^2\mathrm{sech}^2(\xi')}\; e^{\pm i\Theta(\xi')},$$

where

$$\Theta(\xi') = \sin^{-1}\left[\frac{B\tanh(\xi')}{\sqrt{1 - B^2\mathrm{sech}^2(\xi')}}\right], \quad \xi' = \mu(x - v_t t').$$

Here, B is a measure of the amplitude ("blackness") of the solitary wave (hole or dark soliton) and can take a value between -1 and 1, v_t is the dimensionless transverse velocity of the soliton center, and μ is the shape factor of the soliton. The intensity (I_d) of the solitary wave (or the depth of the irradiance minimum of the dark soliton) is given by $B^2\phi_0^2$. Zakharov *et al.* and Hasegawa *et al.* showed that the shape factor μ and the transverse velocity v_t are related to the amplitude of the soliton, which can be obtained from the nonlinear Schrödinger equation in optical fibre (see Chapter 5 for its description).

$$\mu^2 = n_0|n_2|\mu_0^2 B^2\phi_0^2, \quad v_t \approx \pm\sqrt{(1 - B^2)\frac{|n_2|\phi_0^2}{n_0}}$$

where n_0 and n_2 are the linear and nonlinear indices of refraction of the optical fibre material. We have assumed $|n_2|\phi_0^2 \ll n_0$. When two microscopic particles (solitons) collide, Aossey *et al.* expressed their individual phase shifts as

$$\delta x_j = \sqrt{\frac{n_0}{|n_2|\phi_0^2}}\frac{1}{2\mu_0 n_0 B_j}\ln\left[\frac{(\sqrt{1-B_1^2}+\sqrt{1-B_2^2})^2 + (B_1+B_2)^2}{(\sqrt{1-B_1^2}+\sqrt{1-B_2^2})^2 + (B_1-B_2)^2}\right]. \quad (4.85)$$

The hole spatial soliton interaction can be easily investigated numerically by using a split-step propagation algorithm which was found, by Skinner *et al.* and Thurston *et al.*, to closely predict experimental results. The results of a simulated collision between two equi-amplitude microscopic particles (solitons) are shown in Fig. 4.6(a), which are similar to that of general (bright) solitons as shown in Figs. 4.3 and 4.4. We note that the two solitons interpenetrate each other, retain their shape, energy and momentum, but experience a phase shift at the point of collision. In addition, there is also a well-defined interaction length in z along the axis of time t that depends on the relative amplitude of the two colliding microscopic particles (solitons).

This case occurs also in the collision of two KdV solitons. Cooney *et al.* studied the overtaking collision, to verify the KdV soliton nature of an observed signal in the plasma experiment. In the following, we discuss a fairly simple model which was used to simulate and to interpret the experimental results on the microscopic particles (solitons) of nonlinear Schrödinger and KdV solitons.

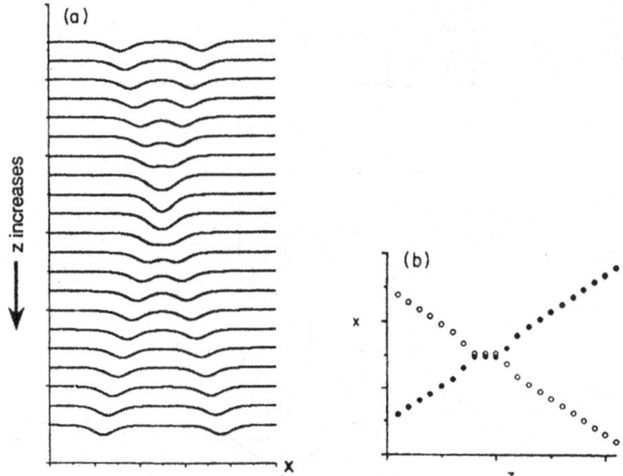

Fig. 4.6 Numerical simulation of an overtaking collision of equiamplitude dark solitons. (a) Sequence of the waves at equal intervals in the longitudinal position z. (b) Time-of-flight diagrams of the signals.

The model is based on the fundamental property of soliton that two solitons can interact and collide, but survive the collision and remain unchanged. Rather than using the exact functional form of sechξ for the microscopic particles (solitons) of the nonlinear Schrödinger equation, the microscopic particles are represented by rectangular pulses with an amplitude A_j and a width W_j where the subscript j denotes the jth microscopic particles.

An evolution of the collision of two microscopic particles is shown in Fig. 4.7(a). In this case, Aossey *et al.* considered two microscopic particles with different amplitudes. The details of what occurs during the collision need not concern us here other than to note that the microscopic particle with the larger-amplitude has completely passed through the one with a smaller amplitude. In regions which can be considered external to the collision, the microscopic particles do not overlap as there is no longer interaction between them. The microscopic particles are separated by a distance, $D = D_1 + D_2$, after the interaction. This manifests itself in a phase shift in the trajectories depicted in Fig. 4.7(b). This was noted in the experimental and numerical results. The minimum distance is given by the half-widths of the two microscopic particles, $D \geq W_1/2 + W_2/2$. Therefore,

$$D_1 \geq \frac{W_1}{2} \quad \text{and} \quad D_2 \geq \frac{W_2}{2}. \tag{4.86}$$

Another property of the microscopic particles (solitons) is that their amplitude and width are related. For the microscopic particles described by the nonlinear Schrödinger equation with $b < 0$ in (4.40) ($W \approx 1/\mu$), we have

$$B_j W_j = \text{constant} = K_1. \tag{4.87}$$

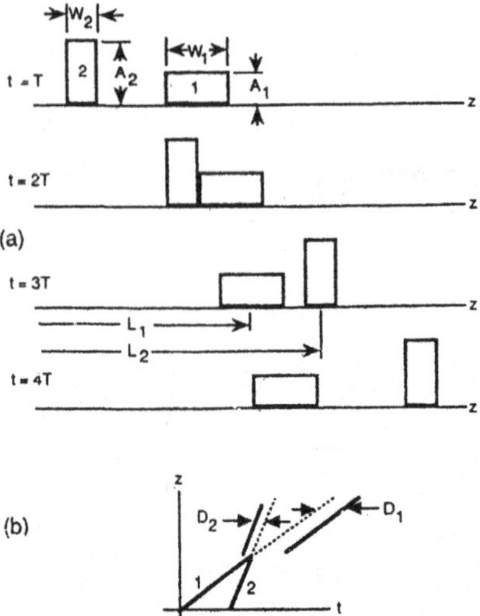

Fig. 4.7 Overtaking collision of two microscopic particles. (a) Model of the interaction just prior to the collision and just after the collision. After the collision, the two microscopic particles are phase shifted. (b) Time-of-flight diagram of the signals. The phase shifts are indicated.

Using the minimum values in (4.86), we find that the ratio of the repulsive shifts for the microscopic particles described by the nonlinear Schrödinger equation is given by

$$\frac{D_1}{D_2} = \frac{B_2}{B_1}. \tag{4.88}$$

Results obtained from simulation of this kind of microscopic particles (solitons) are presented in Fig. 4.8(a). The solid line in the figure corresponds to (4.88).

In addition to predicting the phase shift that results from the collision of two microscopic particles, the model also allows us to estimate the size of the collision region or the duration of the collision. Each soliton depicted in Fig. 4.7 travels with its own amplitude-dependent velocity v_j. For the two microscopic particles to interchange their positions during a time $\triangle T$, they must travel a distance L_1 and L_2, respectively, where

$$L_1 = v_1 \triangle T \quad \text{and} \quad L_2 = v_2 \triangle T. \tag{4.89}$$

The interaction length must then satisfy the relation

$$L = L_2 - L_1 = (v_2 - v_1)\triangle T \geq W_1 + W_2. \tag{4.90}$$

Equation (4.89) can be written in terms of the amplitudes of the two microscopic particles (solitons). For the nonlinear Schrödinger solitons, by combining (4.86) and

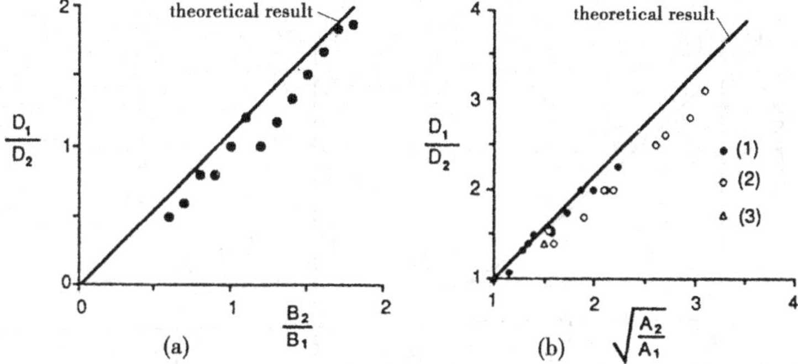

Fig. 4.8 Summary of the ratio of the measured phase shifts as a function of the ratio of the ampliudes. (a) Microscopic particles (solitons) described by the nonlinear Schrödinger equation. The solid line corresponds to (4.88) and the symbols are results of Skinner *et al.* (b) KdV solitons. The symbols represent experimental results of (1) Ikezi *et al.*, (2) Zabusky *et al.*, and (3) Lamb. The solid line corresponds to (4.92).

(4.90), Aossey *et al.* obtained

$$L \geq K_1 \left[\frac{1}{B_1} + \frac{1}{B_2} \right]. \tag{4.91}$$

In Fig. 4.9(a), the results for the microscopic particles described by the nonlinear Schrödinger equation are presented. The dashed line corresponds to (4.91) with $B_2 = 1$ and $K_1 = 6$. The interaction length (solid line) is the sum of the widths of the two microscopic particles, minus their repulsive phase shifts, and multiplied by the transverse velocity of microscopic particle 1. Since the longitudinal velocity is a constant, this scales as the interaction length. From the figure we see that the theoretical result obtained using the simple collision model is in good agreement with that of numerical simulation.

The discussion presented above and the corresponding formulae reveal the mechanism and rule for collision between microscopic particles depicted by the nonlinear Schrödinger equation in the nonlinear quantum mechanics.

In order to verify the validity of this simple collision model, Aossey *et al.* studied collision of solitons using the exact form of $\mathrm{sech}^2\xi$ for the KdV equation, $u_t + uu_x + d'u_{xxx} = 0$, and the collision model shown in Fig. 4.7. For the KdV soliton they found that

$$A_j(W_j)^2 = \text{constant} = K_2 \quad \text{and} \quad \frac{D_1}{D_2} = \frac{W_1/2}{W_2/2} = \sqrt{\frac{A_2}{A_1}}, \tag{4.92}$$

where A_j and W_j are the amplitude and width of the jth KdV soliton, respectively. Corresponding to the above, Aossey *et al.* obtained

$$L \geq K_2 \left(\frac{1}{\sqrt{A_1}} + \frac{1}{\sqrt{A_2}} \right) = \frac{K_2}{\sqrt{A_1}} \left(1 + \sqrt{\frac{A_1}{A_2}} \right) \tag{4.93}$$

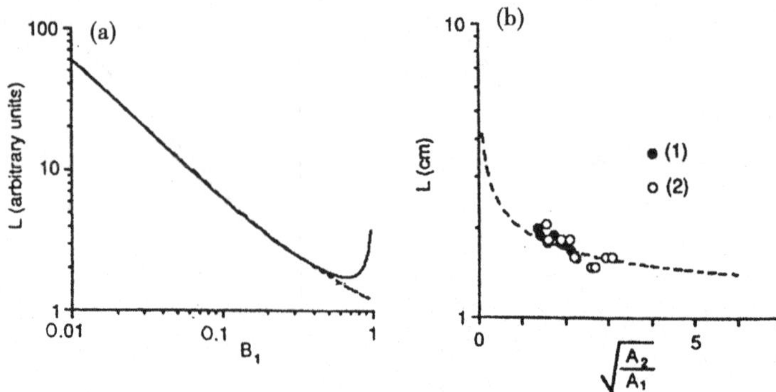

Fig. 4.9 Summary of the measured interaction length as a function of the amplitudes. (a) Nonlinear Schrödinger solitons, the dashed line corresponds to (4.91) with $B_2 = 1$ and $K_1 = 6$. (b) KdV solitons, the symbols represent (1) experimental results of Ikezi *et al.*, and (2) numerical results of Zabusky and Kruskal. The dashed line corresponds to (4.93) with $K_2 = A_1 = 1$.

for the interaction length.

Aossey *et al.* compared their results for the ratio of the phase shifts as a function of the ratio of the amplitudes for the KdV solitons, with those obtained in the experiments of Ikezi *et al.*, and those obtained from numerical work of Zabusky and Kruskal and Lamb, as shown in Fig. 4.8(b). The solid line in Fig. 4.8(b) corresponds to (4.92). Results obtained by Aossey *et al.* for the interaction length are shown in Fig. 4.9(b) as a function of the amplitudes of the colliding KdV solitons. Numerical results (which were scaled) from Zabusky and Kruskal are also shown for comparison. The dashed line in Fig. 4.9(b) corresponds to (4.93), with $A_1 = l$ and $K_2 = l$.

Since the theoretical results obtained by the collision model based on macroscopic bodies in Fig. 4.7 are consistent with experimental data for the KdV soliton, shown in Figs. 4.8(b) and 4.9(b), it is reasonable to believe the validity of the model of collsion presented above, and results shown in Figs. 4.8(a) and 4.9(a) for the microscopic particles described in the nonlinear Schrödinger equation which were obtained using the same model as that shown in Fig. 4.7. Therefore, the above colliding mechanism for the microscopic particles shows clearly the classical corpuscle feature of the microscopic particles in the nonlinear quantum mechanics.

4.6 Collisions of Quantum Microscopic Particles

In previous sections we considered collision of microscopic particles described by the classical nonlinear Schrödinger equation (3.2). However, a microscopic particle is itself quantized. Therefore, it is necessary to consider also collision between such quantum microscopic particles which are described by the quantum nonlinear

Schrödinger equation (3.62).

We have learned that soliton solutions and its properties for the classical nonlinear Schrödinger equation (3.2) can be obtained analytically using the inverse scattering method. From the correspondence principle, it is natural to expect the quantum nonlinear Schrödinger equation (3.62) to serve as an elementary dynamic equation for the quantum microscopic particles in nonlinear quantum mechanics. The quantum nonlinear Schrödinger equation can be solved analytically. In statistical physics, the quantum nonlinear Schrödinger equation is the equation describing the evolution of a one-dimensional Boson system with a δ-function interaction in the second quantized form, as mentioned in (3.72) – (3.74). It was first solved using a method based on the Bethe's ansatz in the 1960s. Since the work of Bethe on the isotropic Heisenberg spin chain in the 1930s, this method has been successfully applied to many models in statistical physics and quantum-field theory. The inverse scattering method has been also applied to the solution of quantum nonlinear Schrödinger equation. Both methods can be used to construct the eigenstates of the Hamiltonian. The quantmn inverse scattering method constructs the creation operators of these eigenstates and derives their commutation relations, while the method based on Bethe's ansatz achieves this by solving the wave equation.

When the coefficient of the nonlinear term in the quantum nonlinear Schrödinger equation is negative, there are bound-state solutions that are the eigenstates of the Hamiltonian with bound wave functions. Surprisingly, many treatments of this problem ended at this stage, leaving important problems unsolved such as how these bound states are related to the soliton motions. Nohl was the first one to give an answer to this question, and later Wadati and Sakagami improved Nohl's results. They introduced a wave packet which is a time-dependent superposition of the fundamental bound states and showed that the main element of the field operator for this wave packet approaches the classical fundamental soliton with zero velocity when the photon number is large. They then generalized their results to moving solitons by a Galilean transformation. Although their results provided a good basis for quantization of nonlinear Schrödinger equation, their approach still leaves some questions open. (1) A soliton state should be a time-independent superposition of the bound states so that it is a solution of the governing equation. (2) It is the expectation value of the field operator that corresponds to the classical soliton field, not the matrix element of the field operator. (3) The construction should be generalized to higher order soliton states to provide infromation about soliton collisions.

Lai and Haus, and Wright constructed soliton states that meet the above crieria. This construction enables us to study the quantun effects of soliton propagation and soliton collisions. An approximate solution was obtained by these authors using the time-dependent Hartree approximation. In this approach, the quantum nonlinear Schrödinger equation is equivalent to the equation of evolution of a one-dimensional Boson system with a δ-function interaction. Using this approach, they constructed approximate fundamental and higher-order soliton states. It was found that a soliton experiences phase-spreading when it propagates. The soliton collisions were also studied by these authors using the same approach.

The collision between quantum microscopic particles described by the nonlinear Schrödinger equation (3.2) is treated in the following using the approach of Lai and Haus. Obviously, in the second quantization representation, the quantum nonlinear Schrödinger equation corresponding to (4.40) can be (3.62), where $\hat{\phi}(x',t')$ and $\hat{\phi}^+(x',t')$ are the annihilation and creation operators, respectively, at a point $x' = x\sqrt{2m}/\hbar$ and time $t' = t/\hbar$, which satisfy the commutation relations (3.63), and (3.64). Any quantum state of this system can be expanded in the Fock space by (3.69) – (3.72). Equation (3.72) can be solved approximately by the time-dependent Hartree approximation. This approximation is valid when the number of particles is large. The basis of the Hartree approximation is the assumption that every particle "sees" the same potential which is due to its interaction with other particles. Therefore we can use a single-particle wave function to describe a system of particles.

Lai and Haus defined explicitly a Hartree wave function based on the following ansutz

$$\int f_n^{(H)}(x_1',\cdots,x_n',t') = \prod_{j=1}^{n}\Phi_n(x_j',t'). \tag{4.94}$$

The function Φ_ns are determined by minimizining the following function

$$\begin{aligned} I = \int f_n^{*(H)}(x_1',\cdots,x_n',t')\left[i\frac{\partial}{\partial t'} + \sum_{j=1}^{n}\frac{\partial^2}{\partial x_j'^2} - 2b\sum_{1\le i\le j\le n}\delta(x_j'-x_i')\right]\times \\ f_n^{(H)}(x_1',\cdots,x_n',t')dx_1',\cdots,dx_n' \\ = n\int \Phi_n^*\left[i\frac{\partial\Phi_n}{\partial t'} + \frac{\partial^2\Phi_n}{\partial x'^2} - (n-1)b\Phi_n^*\Phi_n\Phi_n\right]dx'. \end{aligned} \tag{4.95}$$

It turns out that the above function reaches its minimum value if Φ_n obeys the classical nonlinear Schrödinger equation with the nonlinearity scaled by $n-1$

$$i\frac{\partial\Phi_n}{\partial t'} = -\frac{\partial^2\Phi_n}{\partial x'^2} + 2(n-1)b\Phi_n^*\Phi_n\Phi_n. \tag{4.96}$$

This is one of the connections between quantum theory and classical theory. Equation (4.96) has the following fundamental soliton solution according to (4.56)

$$\Phi_n(x',t') = \frac{2\eta}{\sqrt{|b(n-1)|}}e^{-4i(\xi^2-\eta^2)t'-2i\xi(x'-x_0')}\mathrm{sech}[2\eta(x'-x_0'+4\xi t')]. \tag{4.97}$$

Contrary to the classical case, η cannot be arbitrary because ϕ_n has to satisfy the normalization condition

$$\int |\Phi_n(x',t')|^2 dx' = 1. \tag{4.98}$$

This leads to the following quantization condition

$$\eta = \frac{n-1}{4}|b| \approx \frac{n}{4}|b|. \tag{4.99}$$

Inserting (4.99) into (4.97) and setting $\xi = -p/2$, where p plays the role of momentum, one has

$$\Phi_{np} = \frac{1}{2}\sqrt{(n-1)|b|}e^{i(n-1)^2|b|^2t'/4-ip^2t'+ip(x'-x'_0)} \times \text{sech}\left[\frac{1}{2}(n-1)|b|(x'-x'_0-2pt')\right]. \tag{4.100}$$

With (4.100), we can construct the Hartree product eigenstates according to (4.94)

$$|n,p,t'\rangle_H = \frac{1}{\sqrt{n!}}\left[\int \Phi_{np}(x',t')\hat{\phi}^+(x')dx'\right]^n |0\rangle. \tag{4.101}$$

A superposition of these states using the Poisson distrbution of n gives the fundamental soliton state

$$|\phi_s\rangle_H = \sum_n \frac{\alpha_0^n}{\sqrt{n!}}e^{-|\alpha_0|^2/2}|n,p,t'\rangle_H = \sum_n \frac{\alpha_0^n}{\sqrt{n!}}e^{-|\alpha_0|^2/2}\left[\int \Phi_{np}(x',t')\hat{\phi}^+(x')dx'\right]^n |0\rangle. \tag{4.102}$$

If the photon number is large, $n_0 = |\alpha_0|^2 \gg 1$, the nonlinearity is not excessive, $|b| \ll 1$, and the time of observation is limited, $n_0\sqrt{n_0}|b|^2t \ll 1$, then the summation in (4.102) can be approximated by an exponential term and $|\phi_s\rangle_H$ can be identified as a coherent state

$$|\phi_s\rangle_H \approx \sum_n \frac{\alpha_0^n}{\sqrt{n!}}e^{-|\alpha_0|^2/2}\left[\int \Phi_{n_0p}(x',t')\hat{\phi}^+(x')dx'\right]^n |0\rangle = e^{-|\alpha_0|^2/2}\exp\left[\int \Phi_{n_0p}(x',t')\hat{\phi}^+(x')dx'\right]|0\rangle. \tag{4.103}$$

Here we have ignored the n dependence of Φ_{np} by replacing the variable n by its average n_0. The mean field is

$$_H\langle\phi_s|\hat{\phi}(x')|\phi_s\rangle_H \approx \alpha_0\Phi_{n_0p}(x',t') \approx \frac{n_0-1}{2}\sqrt{|b|} \times e^{i(n_0-1)^2|b|^2t'/4-ip^2t'+ip(x'-x'_0)}\text{sech}\left[\frac{1}{2}(n_0-1)|b|(x'-x'_0-2pt')\right] \tag{4.104}$$

which is the classical solution.

If the time of observation is long enough, then the n dependence of the phase cannot be ignored. The mean field then becomes

$$
{}_H\langle\phi_s|\hat{\phi}(x')|\phi_s\rangle_H \approx \sum_n e^{-|\alpha_0|^2}\frac{|\alpha_0|^{2n}}{n!}\frac{\alpha_0\sqrt{n}}{2}\sqrt{|b|}\times \\
e^{in^2|b|^2t'/4-ip^2t'+ip(x'-x'_0)}\text{sech}\left[\frac{n}{2}|b|(x'-x'_0-2pt')\right]. \quad (4.105)
$$

Equation (4.105) shows that the expectation value of the field is the average of a set of classical solitons. This is surprising. It shows that a simple superposition of solutions of the dynamic equation as the expectation value of the field is not anticipated in nonlinear quantum mechanics when the field propagates in a nonlinear medium. Since in (4.105) components of different n's have different phase velocities, a microscopic particle (soliton) is expected to experience phase spreading during its motion.

Note that here we used a single value of the "momentum" p, rather than a superposition. However, $|n,p,t'\rangle_H$ is not an eigenstate of the momentum operator p and thus a momentum distribution is associated with the state. We can also find that a momentum distribution is necessary to construct a soliton state. Classically the nonlinearity and the dispersion balance exactly to form a soliton. Quantum mechanically only the mean values of the two effects are in balance. There are still higher-order phase-spreading effects and higher-order dispersion effects. The quasi-probability-density method can be used to visualize these effects.

We now use the Hartree approximation to construct two-microscopic particle (soliton) states and study their collision. The construction is not as straightforward as that of the fundamental soliton states because the two-microscopic particle (soliton) states during the collision and other times have to be treated differently. During the collision, the microscopic particles (solitons) are in the same spatial region and interact strongly. All particles behave in the same way and therefore can be represented by the same wave function. However, before and after the collision, the system consists of two independent groups of particles. The particles belong to different groups behave differently and therefore are represented by different wave functions, although particles in the same group still interact and can be represented by the same wave function. Based on this argument, Lai and Haus constructed a two-soliton state that consists of two groups of n_1 and n_2 particles, respectively. They assumeed the following total wave function

$$
f^{(c)}_{n_1n_2}(x'_1,\cdots,x'_{n_1+n_2},t') = \prod_{j=1}^{n_1+n_2}\Phi_{n_1n_2}(x'_j,t') \quad (4.106)
$$

during the collision and

$$
f^{(0)}_{n_1n_2}(x'_1,\cdots,x'_{n_1+n_2},t') = \sum_{|Q|}\prod_{j=1}^{n_1}\Phi^{(1)}_{n_1}(x'_{Q(j)},t')\prod_{j=n_1+1}^{n_1+n}\Phi^{(2)}_{n_1}(x'_{Q(j)},t') \quad (4.107)
$$

otherwise. In the latter expansion the summation is over Q, all possible permutations of $[1, 2, \cdots, n_1 + n_2]$. The grouping of particles into $[1,2,\cdots, n_1]$ and $[n_1 + 1, n_1 + 2, \cdots, n_1 + n_2]$ does not change the results of the summation because $f^{(0)}_{n_1 n_2}$ is symmetric with respect to the x'_j's. All the wave functions Φ_{n_1,n_2}, $\Phi^{(1)}_{n_1}$, $\Phi^{(2)}_{n_1}$ satisfy the normalization condition (4.98). The connection between Φ_{n_1,n_2} and $\Phi^{(1)}_{n_1}$, $\Phi^{(2)}_{n_1}$ can be established by noting that Φ_{n_1,n_2} is the "mean" wave function of a particle. When the two-microscopic particle (soliton) state is not in collision, since there are n_1 particles with wave function $\Phi^{(1)}_{n_1}$ and n_2 particles with wave function $\Phi^{(2)}_{n_1}$, we can conclude that the asymptotic approximation of Φ_{n_1,n_2} should be

$$\Phi_{n_1,n_2} \rightarrow \sqrt{\frac{n_1}{n_1 + n_2}}\Phi^{(1)}_{n_1} + \sqrt{\frac{n_2}{n_1 + n_2}}\Phi^{(2)}_{n2}. \tag{4.108}$$

We shall use (4.108) to establish the connection between the wave functions before and after collision similarly to the WKB method in the linear quantum mechanics. Inserting (4.106) into (4.93) and minimizing the functional, one gets

$$i\frac{\partial\Phi_{n_1,n_2}}{\partial t'} = -\frac{\partial^2\Phi_{n_1,n_2}}{\partial x'^2} + 2(n_1 + n_2 - 1)b|\Phi_{n_1,n_2}|^2\Phi_{n_1,n_2}. \tag{4.109}$$

Substituting (4.107) into (4.95) and minimizing the functional, we have

$$i\frac{\partial\Phi^{(1)}_{n_1}}{\partial t'} = -\frac{\partial^2\Phi^{(1)}_{n_1}}{\partial x'^2} + 2(n_1 - 1)b|\Phi^{(1)}_{n_1}|^2\Phi^{(1)}_{n_1}, \tag{4.110}$$

$$i\frac{\partial\Phi^{(2)}_{n_2}}{\partial t'} = -\frac{\partial^2\Phi^{(2)}_{n_2}}{\partial x'^2} + 2(n_1 - 1)b|\Phi^{(2)}_{n_2}|^2\Phi^{(2)}_{n_2}. \tag{4.111}$$

If (4.108) is substituted into (4.109) and $\Phi^{(1)}_{n_1}$ and $\Phi^{(2)}_{n_2}$ are separated, equations (4.110) – (4.111) can be obtained as well. This shows that (4.108) is consistent with the criteria of the Hartree approximation. Moreover, equations (4.110) and (4.111) are the same equations as (4.96). This indicates that the non-collision two-microscopic particle (soliton) state is a product of two fundamental soliton states. The solution of (4.110) and (4.111) can been obtained as earlier and they are

$$\Phi^{(j)}_{n_j} = \sqrt{\frac{1}{2}(n_j - 1)|b|}\; e^{i(n_j-1)^2|b|^2t'/4 - ip_j^2t' + ip_j(x'-x'_{j0}) + i\theta_j} \times \\ \operatorname{sech}\left[\frac{n_j - 1}{2}|b|(x' - x'_{j0} - 2p_jt')\right], \quad (j = 1, 2). \tag{4.112}$$

However, the phases and mean positions before and after the collision can be different. The difference can be determined by noting that before and after collision,

$$\sqrt{\frac{n_1}{n_1 + n_2}}\Phi^{(1)}_{n_1} + \sqrt{\frac{n_2}{n_1 + n_2}}\Phi^{(2)}_{n_2} \tag{4.113}$$

is the asymptotic approximation of the same Φ_{n_1,n_2}, *i.e.*, the asymptotic solution of the classical nonlinear Schrödinger equation (4.109). It was shown that the classical nonlinear Schrödinger equation has two-soliton solutions. Before the collision, a two-soliton solution is similar to that of two fundamental solitons. After the collision, the solution is of the same form but their phases and positons have been changed. According to Zakharov and Shabat, the magnitudes of these shifts for one of the microscopic particles (solitons) are given by

$$\begin{aligned}\delta\theta_1(n_1,p_1,n_2,p_2) &= -2\arg\left(\frac{\zeta_1-\zeta_2}{\zeta_1-\zeta_2^*}\right) \\ &\approx -2\left\{\tan^{-1}\left[\frac{|b|(n_1+n_2)}{2(p_2-p_1)}\right]-\tan^{-1}\left[\frac{|b|(n_2-n_1)}{2(p_2-p_1)}\right]\right\},\end{aligned} \tag{4.114}$$

$$\begin{aligned}\delta x_1'(n_1,p_1,n_2,p_2) &= \frac{1}{\eta_1}\ln\left(\frac{|\zeta_1-\zeta_2|}{|\zeta_1-\zeta_2^*|}\right) \\ &\approx \frac{2}{n_1|b|}\left\{\ln\left[(p_2-p_1)^2+\frac{|b|^2}{4}(n_2-n_1)^2\right]\right. \\ &\quad \left. -\ln\left[(p_2-p_1)^2+\frac{|b|^2}{4}(n_2+n_1)^2\right]\right\}.\end{aligned} \tag{4.115}$$

where $\zeta_1=\xi_1+i\eta_1$, $\zeta_2=\xi_2+i\eta_2$. Similar shifts occur for the second microscopic particle (soliton). Therefore, the characteristics of collision between quantum microscopic particles described by the quantum nonlinear Schrödinger equation are the same as those of classical microscopic particles described by the classical nonlinear Schrödinger equation in nonlinear quantum mechanics. This result also contains the quantum fluctuations produced in the collision. The $\delta\theta_i$s and $\delta x_i'$s $(i=1,2)$ are functions of n_j $(j=1,2)$ and are determined probabilistically.

In the above calculation, we used the time-dependent Hartree approximation in the constructino of approximate eigenstates of the particle-number operator, instead of the exact eigenstates of the momentum and Hamiltomian operators. The soliton solution thus obtained experiences phase spreading when it propagates. The Hartree approximation suppresses the effect of the momentum uncertainty. A distribution of momentum must be associated with a soliton with a mean position, just as the uncertainty of particle number causes a phase-spreading of its own. Lai and Haus used Bethe's ansatz to construct the exact eigenstates of the Hamiltonian, and soliton states, by superimposing these eigenstates

$$|\phi\rangle = \sum_n a_n \int g_n(p)|n,p,t'\rangle dp, \tag{4.116}$$

where

$$a_n = \frac{\alpha_0^n}{\sqrt{n_1}}e^{-|\alpha_0|^2/2},$$

$$g_n(p) = \frac{1}{(\triangle p)^{1/2}(\bar{n})^{1/4}}\exp\left[-\frac{1}{2}\frac{|p-p_0|^2}{(\triangle p)^2}-inpx_0'\right] = g(p)e^{-inpx_0'}$$

and the a_n's satisfy the following

$$\sum_n |a_n|^2 = 1.$$

Hence both fundamental and higher order soliton states were constructed and their mean fields were also calculated to justify the construction. Classical results can be recovered by taking the limit of large particle number. It was found that due to the uncertainty of momentum a soliton experiences dispersion when it propagates and such effect is very small when the average particle number of the solitons is much larger than unity. Phase and position shifts due to a collision and the uncertainty of these shifts were also obtained by these authors as follows

$$\delta\theta_1 = \theta(n_{10}+1, p_{10}, n_{20}, p_{20}) - \theta(n_{10}, p_{10}, n_{20}, p_{20}),$$
$$\delta x_1' \approx \frac{1}{n_{10}} \frac{\partial\theta(n_{10}, p_{10}, n_{20}, p_{20})}{\partial p_1}$$

for the phase and position shifts, respectively, of the first microscopic particle (soliton) and

$$\delta\theta_2 = \theta(n_{10}, p_{10}, n_{20}+1, p_{20}) - \theta(n_{10}, p_{10}, n_{20}, p_{20}),$$
$$\delta x_2' \approx \frac{1}{n_{20}} \frac{\partial\theta(n_{10}, p_{10}, n_{20}, p_{20})}{\partial p_2}$$

for the second microscopic particle (soliton), respectively. Therefore property of quantum collision of microscopic particles in quantum nonlinear Schrödinger equation is basically the same as that of the microscopic particles in the classical nonlinear Schrödinger equation.

4.7 Stability of Microscopic Particles in Nonlinear Quantum Mechanics

Stability is another important property of macroscopic particle. In this section we will demonstrate that the microscopic particles in nonlinear quantum mechanics has similar property as macroscopic particles.

Let us first define the stability of microscopic particles in nonlinear quantum mechanics. Usually three types of stability of microscopic particles should be considered, (1) with respect to a perturbation of the initial state, (2) with respect to a perturbation of the dynamic equation governing the system dynamics (structural stability); (3) with respect to minimal energy state under an externally applied field.

In the first case, the problem has been investigated in detail using various techniques and approaches (linear and nonlinear). In the linear approximation the stability problem is usually reduced to an eigenvalue problem of linearized equations. In the nonlinear case it is reduced to the study of Lyapunov inequalities.

Study of structural stability of microscopic particles can be done in the framework of dynamic equations under different types of perturbations. Attempts have been made to construct a general perturbation theory of the microscopic particles based on the dynamic equations by using the Green function method and spectral transformation (*i.e.* a transformation from the configuration space (x,t) to the scattering data space based on the well-known two-time formalism). However, the latter technique is suitable only for a rather restricted class of perturbation functionals. Although some results were obtained in this direction, the studies cannot be considered as complete. In some cases numerical studies are very effective tools. From the computational point of view, both problems can be investigated in the framework of a unified approach. In the first case one studies the dynamics, described by an unperturbed dynamic equation, of a perturbed or unperturbed initial state given in the form of a soliton solution, the stability of which is examined. In the second case the initial state evolution is governed by a perturbed equation. In both cases the solutions depend on some parameters which are slowly varying functions of time.

In the first case, a solution is considered stable if initial perturbations are not magnified as the initial state evolves with time. In accordance with this definition, weakly radiating soliton-like solutions which are not destroyed under initial perturbations can be considered stable. Obviously, structure-stable solutions are the solutions which conserve their shape for sufficiently long time. The notion of "sufficiently long time" is relative to the time scale of physical processes occurring in the systems. In the following, a few examples illustrating investigations in both directions will be given. The examples are selected for the purpose of illustrating the problems. They are by no means complete. More details on these problems can be found in the relevant publications of Makhankov *et al.*.

4.7.1 *"Initial" stability*

It is known that in the case of Lagrangian relativistically invariant equations describing a complex field, besides the conventional energy and momentum conservations, there is an additional "charge" conservation which is connected to the Lagrangian $U(1)$ symmetry, $\dot{Q} = 0$ with $Q = \Im \int \phi^*_{t'} \phi dx'$. For non-relativistic models of the nonlinear Schrödinger equation type, instead of Q, the wavefunction normalization (particle number) is conserved, $\dot{N} = 0$ and $N = \Im \int |\phi|^2 dx'$. By the variational principle $Q =$ const. or $N =$ const., one may prove a theorem (the Q-theorem) that formulates sufficient conditions for the stability of complex soliton-like solution. The stability region of these solutions is determined by the inequality $d \ln Q / d \ln \omega < 0$ for the nonlinear Klein-Gordon equation and $dN/d\omega < 0$ $(\omega < 0)$ for nonlinear Schrödinger equation ($\omega < 0$ is necessary for the existence of soliton-like solutions).

To obtain these formulae, we consider the soliton-like solutions of the generalized nonlinear Klein-Gordon equation, $(\Box + 1 - |\phi|^n)\phi = 0$, where $\Box = \partial^2_{t'} - \partial^2_{x'} - \partial^2_{y'} - \partial^2_{z'}$, and the generalized nonlinear Schrödinger equation, $(i\partial_{t'} + \partial^2_{x'} + \partial^2_{y'} + \partial^2_{z'} + |\phi|^n)\phi = 0$, with a $|\phi|^n\phi$ type nonlinearity, in the rest frame with a time dependence of $\phi(x', t') = \varphi(x')e^{-i\omega t'}$, which minimizes the Hamiltonian of these systems. The common form for both equations in the $1+1$ dimensional case is

$$-\varphi_{x'x'} + k^2\varphi - \varphi^{n+1} = 0 \tag{4.117}$$

where $k^2 = 1 - \omega^2$ for the nonlinear Klein-Gordon equation and $k^2 = -\omega$ for the nonlinear Schrödinger equation. The ω dependence of Q and N is easily obtained by the scale transformation $x' \to k^{-1}\xi'$, $\varphi \to k^{2/n}y$. In terms of these variables, equation (4.117) has the following form (free of k)

$$-y_{\xi'\xi'} + y - y^{n-1} = 0, \tag{4.118}$$

with

$$N = \int \varphi^2 dx' = k^{(4-n)/n} \int y^2 d\xi\prime = k^{(4-n)/n} C'(n), \quad Q = \omega N.$$

Calculating the derivatives $\partial_\omega N$ and $\partial_\omega Q$, we find that the stability regions are

$$\begin{aligned} &(1) \quad 1 > \omega^2 > \frac{n}{4}; \\ &(2) \quad n < 4\,. \end{aligned}$$

where $\omega = -k^2$ is arbitrary. That is, in both cases, stable soliton-like solutions exist only for $n < 4$. This result shows that the dynamic equations (3.2) – (3.5) in nonlinear quantum mechanics have stable soliton solutions.

We can also study the initial stability of microscopic particles following the approach of Zakharov and Shabat. As a matter of fact, in the above discussion, we did not consider the general evolution of the initial conditions, where an important role may be played by the "nonsoliton" part of the solution, which is connected to the scattering quantity $b(\xi, t')$ in (4.44). Here we consider only the case when this quantity is small, *i.e.*,

$$\left|\frac{b(\xi, t')}{a(\xi)}\right| = \left|\frac{b(\xi, 0)}{a(\xi)}\right| \ll 1, \tag{4.119}$$

and the coefficient $a(\xi)$ has only one zero, $\xi = \xi_1$, in the upper half-plane. Such a choice for the initial conditions corresponds to posing the problem of the stability of a microscopic particle (soliton) with parameters $\xi_1 + i\eta_1 = \zeta_1$ when the microscopic particle (soliton) is perturbed by a field with a continuous spectrum.

Let us consider the systems (4.47) and (4.48) with $N = 1$ and express the functions $\psi_2^*(x', \zeta_1)$ and $\psi_1^*(x', \zeta_1)$ in (4.48) in terms of

$$\Phi^{(2)}(\zeta_1^*) = \frac{1}{2\pi i} \int \frac{\Psi(\xi)}{\xi - \zeta_1^*} d\xi. \tag{4.120}$$

After substituting it in (4.47), we obtain a system only for the quantities Ψ_1 and Ψ_2^*. This system contains as a coefficient the small quantity $c(x', \xi, t') = b(\xi, t')\exp(i\xi x')/a(\xi)$. Keeping terms up to quadratic in c, we have

$$\Psi_1 = 0, \quad \text{and} \quad \Psi_2 = c(x_1', \xi, t')\left[1 + \frac{\xi_1^* - \zeta_1}{\xi - \zeta_1}\left(1 - \frac{1}{1 + |\lambda|^4}\right)\right], \tag{4.121}$$

where $\lambda = \bar{c}_1(0)e^{2i\zeta_1 x'}$. From (4.49) we now obtain

$$\phi(x', t') = \phi_0(x', t') - \frac{1}{\pi}\int_{-\infty}^{+\infty} \Psi_2^*(x_1', \xi, t'), \tag{4.122}$$

$$\phi_0(x', t') = (\lambda^*)^2/1 + |\lambda|^4, \tag{4.123}$$

where ϕ_0 is the soliton (4.56) with $\xi + i\eta = \zeta_1$. Recognizing that

$$c(x_1', \xi, t') = \frac{b(\xi, 0)}{a(\xi)}e^{2i\xi x' + 4i\xi^2 t'},$$

we find that the integral in (4.122) decreases as $1/\sqrt{t'}$, as $t' \to \infty$. This means that the soliton is stable under the perturbation by a field with a continuous spectrum. As $t' \to \infty$, the solution develops asymptotically into a soliton.

A general perturbation would shift the position of the zero of $a(\zeta)$ on the complex plane and by the same token perturbs the parameters of the microscopic particle (soliton). As $t' - \infty$ the solution evolves asymptotically into this perturbed microscopic particle (soliton).

4.7.2 *Structural stability*

The first studies of structural stability concerned the effect of weak dissipation in the nonlinear Schrödinger equation. The behavior of microscopic particles (solitons) depends essentially on the form of the perturbation term. In the case of power law damping $\gamma \approx \varepsilon k^n$ (k is the wave number, ε is a small number, n is a constant), Makhankov *et al.* demonstrated that only when $n = 2$ the soliton does not change its shape with time (which is connected to the scaling properties of the nonlinear Schrödinger equation). If $n \neq 2$, evolution of the microscopic particle causes its shape variation proportional to the coefficient of the damping term. For $n = 2$, 3, and 4 the inequalities $\varepsilon \leq 0.2$, $\varepsilon < 0.03$ and $\varepsilon < 0.01$ hold, respectively. Structural stability of microscopic particles for nonlinear Schrödinger equation were studied by Yajima *et al.*. One of the main features of perturbed microscopic particles (solitons) is the appearance of oscillatory tails on their back or front. Structural stability of other microscopic particles (kink solitons) was studied by Fogel *et al.* and Currie *et al.*, who worked on the initial value problem of the following Sine-Gordon equation,

$$\Box\phi - \omega_0^2(x)\sin\phi - AF(x) = 0, \tag{4.124}$$

with

$$\phi(x,0) = f(x).$$

Here the x-dependence of the eigenfrequency ω_0 of the microscopic particles can be used to simulate the presence of impurities (non-uniformities) in the system. The third term in (4.124) describes the influence of external actions (fields, currents, and so on). Analytical and numerical results obtained by Currie *et al.* show that for sufficiently small perturbations the Sine-Gordon kinks behave as Newtonian particles with an internal structure in external fields, they may be accelerated (decelerated), may radiate, and change slightly their structure, which is accompanied by transition radiation. However, they can also be trapped in some spatial region. The radiation in the model described by (4.124) would strongly influence the structure and especially the dynamics of microscopic particles, for example, their formation, lifetimes, breaking up.

We now briefly discuss the structural stability of a perturbed nonlinear Schrödinger equation (3.2) with $A(\phi) = 0$, which has the form

$$i\phi_{t'} + \phi_{x'x'} + |\phi|^2\phi = \beta\phi^*_{x'}. \tag{4.125}$$

It is extremely difficult to obtain the Lax representation and the Lagrangian of this equation. Makhankov *et al.* considered at first the plane-wave solutions and their stability. Multiplying (4.125) by ϕ^*, then by $\phi^*_{x'}$, and combining the resulting equation with its complex-conjugate, we get

$$\rho_{t'} = -j_{x'}, \quad |\rho| = |\phi|^2, \tag{4.126}$$

$$p_{t'} = 4\beta\Re\phi^*_{x'} + 2\partial_{x'}[\mathcal{H} - \Re(\beta\phi\phi_{x'} + \phi^*\phi_{x'x'})], \tag{4.127}$$

$$\mathcal{H}_{t'} = -\beta\rho\partial_{x'}(\Im\phi^2) + \partial_{x'}[\rho p - \Im(\beta\phi^2_{x'} - 2\phi^*_{x'}\phi_{x'x'}), \tag{4.128}$$

where

$$j = \beta\Im\phi^2 - p, \quad p = 2\Im\phi^*\phi_{x'}, \quad \mathcal{H} = |\phi_{x'x'}|^2 - \frac{1}{2}|\phi|^4. \tag{4.129}$$

Integrating (4.126) – (4.128) over x', we found that for the entire set of integrals of motion of the non-perturbed nonlinear Schrödinger equation, only the first, $N_{t'} = 0$, survives, *i.e.*, the particle number in the system (4.125) is conserved.

We assume that the solution of (4.125) is of the form

$$\phi(x',t') = D'e^{-i(\omega t' + \vartheta_0)}, \tag{4.130}$$

with

$$\omega = 1 - D'^2 \quad \text{and} \quad \vartheta_0 = \text{constant}.$$

Separating the real and imaginary parts of the solution, $\phi = F - iF'$, equation (4.125) becomes two equations

$$\begin{aligned}F_{t'} &= F' + \beta F'_{x'} - F'_{x'x'} - (F^2 + F'^2)F, \\ F'_{t'} &= -F + \beta F_{x'} + F_{x'x'} + (F^2 + F'^2)F.\end{aligned} \tag{4.131}$$

In the zeroth-order approximation in D'^2, we get the following solution of the linearized system

$$\begin{aligned}F_0 &= D' \cos\vartheta', \\ \vartheta' &= kx' - \omega_0 t', \\ F'_0 &= D'\left(\frac{1+k^2}{\omega_0}\sin\vartheta' - \frac{\beta k}{\omega_0}\cos\vartheta'\right)\end{aligned} \tag{4.132}$$

and the dispersion relation

$$\omega_0^2 = (1+k^2)^2 + \beta^2 k^2.$$

Corrections proportional to D'^2 can be found by the expansion

$$\begin{aligned}\delta F = F - F_0 &= D'\sum_n \left[a_c^{(2n+1)}\cos(2n+1)\vartheta' + a_s^{(2n+1)}\sin(2n+1)\vartheta'\right], \\ \delta F' = F' - F'_0 &= D'\sum_n \left[b_c^{(2n+1)}\cos(2n+1)\vartheta' + b_s^{(2n+1)}\sin(2n+1)\vartheta'\right],\end{aligned} \tag{4.133}$$

where $a_{c,s}^{(2n+1)} \ll a_{c,s}^{(2n-1)}$ and $b_{c,s}^{(2n+1)} \ll b_{c,s}^{(2n-1)}$. The result is

$$\omega^2 = (1+k^2)^2 + \beta k^2 - 2(1+k^2)D'^2. \tag{4.134}$$

To look at the stability of the plane-wave solutions, we consider the solution (4.130) as $D' \to 1$. Assuming

$$\begin{aligned}F &= D'\cos\vartheta_0\left[1 + \delta_1 e^{ikx' - i\Omega t'}\right], \\ F' &= D'\cos\vartheta_0\left[1 + \delta_2 e^{ikx' - i\Omega t'}\right],\end{aligned}$$

($\delta_i^* = \delta_i$), and linearizing the system in (4.131) in δ_1 and δ_2, we can get

$$\left(\frac{\Omega}{k}\right)^2 = k^2 - \beta^2 - 2 - 2i\frac{\beta}{k}\cos\vartheta_0,$$

or

$$\Omega = \pm\sqrt{\frac{1}{2}}a\left(\sqrt{\sqrt{1+\frac{c^2}{a^2}}+1} - i\sqrt{\sqrt{1+\frac{c^2}{a^2}}-1}\right),$$

where $a = k^2(k^2+\beta^2-2)$, $c = 2\beta k\cos 2\vartheta_0$. From these formulae we see that the solution (4.130) is unstable at large amplitudes ($D' \to 1$). The growth rate of the instability has a peak, $\gamma_{\max} = 1 - \beta^2/2$, for perturbations with a wave number $k = 1 - \beta^2/4$.

The stability of solution of (4.131) can be examined similarly by expanding F and F' in harmonics of frequency ω, which, however, is much more tedious. Since the amplitudes are small, *i.e.*, $D' \ll 1$ and $\omega \to 1$, assuming again that $\Omega \ll \omega$ and $k \ll 1$, we can get the following after some lengthy but straightforward derivations

$$\left(\frac{\Omega}{k}\right)^2 = -2D'^2\left(1+\frac{1}{4}\beta^2k^2\right) + k^2\left(1+\frac{1}{2}\beta^2\right) + \frac{i}{4}\beta k D'^4,$$

where $a = k^2(k^2 - \beta^2k^2/2 - 2D'^2)$ and $c = \beta k^3 D'^4/4$, which means that this solution is still unstable.

Therefore, the plane-wave solutions are unstable for both small and large amplitudes. Such an instability is responsible for the breakup of initial packets into solitons in the framework of the nonlinear Schrödinger equation. We note that $\Omega^2 = k^2(2D'^2 - k^2)$ when $\beta = 0$ in (4.134). Although the instability is slightly modified at non-zero, but small β, it still leads to the breakup of the initial packet into soliton-like objects as in the nonlinear Schrödinger equation case. This result shows that the plane wave (4.130) is not a solution of (4.125).

For all the examples considered the perturbation is small for small β. It is interesting to observe its influence on a soliton-type quasi-stationary solution. We can write (4.125) in terms of the amplitude D and phase ϑ of $\phi = D(x',t')e^{i\vartheta(x',t')}$ in the following

$$\begin{aligned} &D_{t'} + 2\vartheta_{x'}D_{x'} - \beta D_{x'}\sin 2\vartheta - \beta\vartheta_{x'}D\cos 2\vartheta + D\vartheta_{x'x'} = 0, \\ &-D\vartheta_{t'} - D + D_{x'x'} - D\vartheta_{x'}^2 + D^3 + \beta D_{x'}\cos 2\vartheta - \beta\vartheta_{x'}D\sin 2\vartheta = 0. \end{aligned} \tag{4.135}$$

Multiplying the first equation by D, we get

$$\partial_{t'}D^2 + \partial_{x'}[(2\vartheta_{x'} - \beta\sin 2\vartheta)D^2] = 0.$$

It is easy to find a solution for $\vartheta_{x'} \to 0$. In that case,

$$D_{t'} = \beta D_{x'}\sin 2\vartheta;$$

$$-D\vartheta_{t'} - D + D_{x'x'} + D^3 + \beta D_{x'}\cos 2\vartheta = 0.$$

Thus D is a function of

$$z = x' + \beta\int_0^{t'} \sin 2\vartheta(\tau)d\tau$$

and a solution for ϑ may be found in the form of

$$\vartheta_{t'} = -\omega + \beta f(z)\cos 2\vartheta.$$

Here the amplitude D satisfies the nonlinear Schrödinger equation

$$(\omega - 1)D + D_{zz} + D^3 = 0,$$

if the condition $f(z) = \partial_z(\ln D)$ is fulfilled. Makhankov *et al.* finally obtained the solution

$$D(z) = \mu_0 \text{sech}\left(\frac{\mu_0 z}{\sqrt{2}}\right),$$

where

$$\mu_0 = \sqrt{2(1-\omega)},$$

or

$$\begin{aligned}\phi = \mu_0 \text{sech}\left[\frac{\mu_0}{\sqrt{2}}\left(x' + \frac{\beta}{2}\cos 2t'\right)\right] \\ \times \exp\left\{-i\left[\left(1 - \frac{\mu_0^2}{2}\right)t' + \frac{\beta\mu_0}{2\sqrt{2}}\tanh\left(\frac{\mu_0 x'}{\sqrt{2}}\right)\sin 2t'\right]\right\}\end{aligned} \tag{4.136}$$

which holds when $\beta\mu_0 \ll 1$ and $\mu_0^2 \ll 1$. This is a perturbed (oscillating) version of the conventional nonlinear Schrödinger equation soliton. It shows that microscopic particles can still maintain the soliton feature in motion under structureal perturbation.

In addition to the above solution (4.136), equation (4.129) has a family of stationary solutions which essentially differ from a soliton solution. To find the stationary solutions, we assume $\partial_{t'} = 0$ in (4.135) and integrate the first equation over x'

$$D^2\left(\vartheta_{x'} - \frac{1}{2}\beta\sin 2\vartheta\right) = \text{const.}$$

Because of the boundary conditions at infinity the constant must vanish. Therefore, the phase ϑ satifies the stationary Sine-Gordon equation

$$\vartheta_{x'x'} = \frac{\beta^2}{4}\sin 4\vartheta,$$

which has the solution

$$\vartheta = \tan^{-1}\left[e^{\beta(x'-x_0')}\right] + \vartheta_\pm,$$

where

$$\vartheta_+ = \left(n + \frac{1}{2}\right)\pi, \quad \vartheta_- = n\pi, \quad (n = 0, \pm 1, \cdots).$$

In this case the second equation in (4.135) is reduced to

$$D_{x'x'} - D + D^3 - \frac{3}{4}\beta^2 D\sin^2 2\vartheta - \beta D_{x'}\cos 2\vartheta = 0.$$

Here the sign of the last term is determined by the phase behavior at $-\infty$. Fixing ϑ_+ or ϑ_- as an "initial" phase, we can get

$$D_{x'x'} - D + D^3 - \frac{3}{4}\beta^2 D \sin 2\sigma \pm \beta D_{x'} \cos 2\sigma = 0.$$

Hence for $\beta \ll 1$ we get

$$D_{x'x'} - D + D^3 \pm \beta D \tanh \beta x' = 0. \tag{4.137}$$

In the first case, the existence of a solution is proved. Equation (4.137) has a countable set of soliton-like solutions D_n, one of which, D_0, is nodeless, and the other ones have an increasing number of nodes n.

The above results show that the influence of the small term $\beta\phi^*_{x'}$ can lead to interesting results. This means that the structural stability of the nonlinear quantum mechanical systems is a rather delicate problem which must be solved either for a specific system or for a restricted class of systems.

4.8 Demonstration on Stability of Microscopic Particles

In addition to the initial and strutural stabilities discussed above, we can also demonstrate the stability of the microscopic particles in nonlinear quantum mechanics by means of the minimum energy concept. A system of microscopic particles is stable, when the particles are located in a finite range which results in the lowest potential energy. This stability principle is very effective, when the microscopic particles are in an extenally applied field. As a matter of fact, since interaction between the microscopic particles is very complicated, it is not easy to define the behavior of each one individually. We cannot use the same strategies as those used in the discussions of initial and structural stabilities in this case. Instead, we apply the fundamental work-energy theorem of classical physics: a mechanical system in the state of minimal energy is said to be stable, and in order to change this state, external energy must be supplied. Pang applied this fundamental concept to demonstrate the stability of microscopic particles described by the nonlinear quantum mechanics, which is outlined in the following.

Let $\phi(x,t)$ represent the field of the particle, and assume that it has derivatives of all orders, and all integrations of it are convergent and finite. The Lagrangian density function corresponding to the nonlinear Schrödinger equation (3.2) with $A(\phi) = 0$ is given in (3.93). The momentum density of this field is defined as $P = \partial\mathcal{L}/\partial\phi$. The Hamiltonian density of the field is given in (3.99). In the general case, the total energy of the field is a function of t', thus (4.4) is replaced by

$$E(t') = \int_{-\infty}^{\infty} \left[\left|\frac{\partial\phi}{\partial x'}\right|^2 - \frac{b}{2}|\phi|^4 + V(x')|\phi|^2 \right] dx'. \tag{4.138}$$

However, in this case, b and $V(x')$ are not functions of t'. So, the total energy of the system is a conservative quantity, *i.e.*, $E(t') = E = \text{const.}$, as given in (4.4).

We can demonstrate that when $x' \to \pm\infty$, the solutions of (3.2) with $A(\phi) = 0$, and $\phi(x', t') = 0$ approach zero rapidly, *i.e.*

$$\lim_{|x'|\to\infty} \phi(x', t') = \lim_{|x|\to\infty} \frac{\partial \phi}{\partial x'} = 0.$$

Then

$$\int_{-\infty}^{\infty} \phi^* \phi dx' = \text{const. (or a function of } t'),$$

$$\lim_{|x'|\to\infty} \phi^* x' \frac{\partial \phi}{\partial x'} = \lim_{|x'|\to\infty} \frac{\partial \phi^*}{\partial x'} x' \phi = 0.$$

The average position or the center of mass of the field ϕ can be represented as $\langle x' \rangle = x'_g$, as given in (4.9). Making use of (4.13), the average velocity or the velocity of the center of mass of the field can be written as (4.18).

However, for different solutions of the same nonlinear Schrödinger equation (3.2), $\int_{-\infty}^{\infty} \phi^* \phi dx'$, $\langle x' \rangle$ and $d\langle x' \rangle / dt'$ can have different values. Therefore, it is unreasonable to compare the energy of one particular solution to that of another solution. The comparison is only meaningful for many microscopic particle (soliton) systems that have the same values of $\int_{-\infty}^{\infty} \phi^* \phi dx' = k$, $\langle x' \rangle = u$ and $d\langle x' \rangle / dt' = \dot{u}$ at the same time t'_0. Based on these, we can determine the stability of the soliton solution of the nonlinear Schrödinger equation. Thus we assume that different solutions of the nonlinear Schrödinger equation (3.2) with $A(\phi) = 0$ satisfy the following boundary conditions at time t'_0

$$\int_{-\infty}^{\infty} \phi^* \phi dx' = K, \quad \langle x' \rangle |_{t'=t'_0} = u(t'_0), \quad \left. \frac{d\langle x' \rangle}{dt'} \right|_{t'=t'_0} = \dot{u}(t'_0). \tag{4.139}$$

Now we assume that the solution of nonlinear Schrödinger equation (3.2) with $A(\phi) = 0$ is of the form

$$\phi(x', t') = \varphi(x', t') e^{i\theta(x', t')}. \tag{4.140}$$

Substituting (4.140) into (4.138), we obtain the energy of the system

$$E = \int_{-\infty}^{\infty} \left[\left(\frac{\partial \varphi}{\partial x'} \right)^2 + \varphi^2 \left(\frac{\partial \theta}{\partial x'} \right)^2 - b\varphi^4 + V(x')\varphi^2 \right] dx'. \tag{4.141}$$

Equations (4.139) then become

$$\int_{-\infty}^{\infty} \varphi^2 dx' = K, \quad \frac{\int_{-\infty}^{\infty} x' \varphi^2 dx'}{\int_{-\infty}^{\infty} \varphi^2 dx'} = u(t'_0), \quad \frac{2\int_{-\infty}^{\infty} \varphi^2 \frac{\partial \theta}{\partial x'} dx'}{\int_{-\infty}^{\infty} \varphi^2 dx'} = \dot{u}(t'_0). \tag{4.142}$$

Finding the extreme value of the functional (4.141) under the boundary conditions (4.142) by means of the Lagrange multiplier method, we obtain the following Euler-Lagrange equation,

$$\frac{\partial^2\varphi}{\partial(x')^2} = \Big\{ V(x') + C_1(t_0')C_2(t_0')[x' - u(t_0')] + C_3(t_0')\left[2\frac{\partial\theta}{\partial t'} - \dot{u}(t_0')\right] + \left(\frac{\partial\theta}{\partial t'}\right)^2\Big\}\varphi - b\varphi^3 = 0, \quad (4.143)$$

$$\frac{\partial^2\theta}{\partial(x')^2}\varphi^2 + 2\frac{\partial\theta}{\partial t'}\varphi\frac{\partial\varphi}{\partial t'} + 2C_3(t_0')\varphi\frac{\partial\varphi}{\partial t'} = 0, \quad (4.144)$$

where the Lagrange factors C_1, C_2 and C_3 are all function of t'. Let

$$C_3(t_0') = -\frac{1}{2}\dot{u}(t_0').$$

If

$$2\frac{\partial\theta}{\partial x'} - \dot{u}(t_0') \neq 0,$$

we can get from (4.144)

$$\frac{2}{\varphi}\frac{\partial\varphi}{\partial x'} = \frac{-\dfrac{\partial^2\theta}{\partial x'^2}}{-\dfrac{\partial\theta}{\partial x'} - \dfrac{1}{2}\dot{u}(t_0')}.$$

Integration of the above equation yields

$$\varphi^2 = \frac{g(t')}{\dfrac{\partial\theta}{\partial x'} - \dfrac{1}{2}\dot{u}(t_0')},$$

or

$$\left.\frac{\partial\theta}{\partial x'}\right|_{t'=t_0'} = \frac{g(t_0')}{\varphi^2} + \frac{\dot{u}(t_0')}{2}, \quad (4.145)$$

where $g(t_0')$ is an integral constant. Thus,

$$\theta(x', t') = g(t_0')\int_0^x \frac{dx'}{\varphi^2} + \frac{\dot{u}(t_0)}{2}x' + M(t_0'). \quad (4.146)$$

Here $M(t_0')$ is also an integral constant. Let again

$$C_2(t_0') = \frac{1}{2}\ddot{u}(t_0'). \quad (4.147)$$

Substituting (4.145) – (4.147) into (4.143), we obtain

$$\frac{\partial^2\varphi}{\partial(x')^2} = \left\{V(x') + \frac{\ddot{u}(t_0')}{2}x' + \left[C_1(t_0') - \frac{\ddot{u}(t_0')}{2}u(t_0') + \frac{u^2(t_0')}{4}\right]\right\}\varphi$$
$$-b\varphi^3 + \frac{g^2(t_0')}{\varphi^3}. \tag{4.148}$$

Letting

$$C_1(t_0') = \frac{u(t_0')\ddot{u}(t_0')}{2} - \frac{\dot{u}^2(t_0')}{2} + M(t_0') + \beta', \tag{4.149}$$

where β' is an undetermined constant, which is t'-independent, and assuming $Z = x' - u(t_0')$, then

$$\frac{\partial^2\varphi}{\partial(x')^2} = \frac{\partial^2\varphi}{\partial Z^2}$$

is only a function of Z. In order for the right-hand side of Eq. (4.149) to be a function of Z, the coefficients of φ, φ^3 and $1/\varphi^3$ must also be funciton of Z. Thus, $g(t_0') = g_0 = \text{const.}$, and

$$V(x') + \frac{\ddot{u}(t_0')}{2}x' + M(t_0') - \frac{u^2(t)}{4} = \tilde{V}_0(Z).$$

Equation (4.149) then becomes

$$\frac{\partial^2\varphi}{\partial(x')^2} = \{\tilde{V}[x' - u(t_0')] + \beta'\}\varphi - b\varphi^3 + \frac{g^2(t_0')}{\varphi^3}. \tag{4.150}$$

Since $\tilde{V}_0(Z) = \tilde{V}_0[x' - u(t_0')] = 0$ in the present case, (4.150) becomes

$$\frac{\partial^2\varphi}{\partial(x')^2} = \beta'\varphi - b\varphi^3 + \frac{g^2(t_0')}{\varphi^3}. \tag{4.151}$$

Therefore, φ is the solution of (4.151) for $\beta' = \text{constant}$ and $g(t_0') = \text{constant}$. For sufficiently large $|Z|$, we may assume that $|\varphi| \leq \tilde{\beta}/|Z|^{1+\triangle}$, where $\triangle$ is a small constant. However, in (4.151) we can only retain the solution $\varphi(Z)$ corresponding to $g(t_0') = 0$ to essure that $\lim_{|\xi|\to\infty} d^2\varphi/dZ^2 = 0$. Thus, equation (4.151) becomes

$$\frac{\partial^2\varphi}{\partial(x')^2} = \beta'\varphi - b\varphi^3. \tag{4.152}$$

As a matter of fact, if $\partial\theta/\partial t' = \dot{u}/2$, then from (4.149) and (4.150), we can verify that the solution given in (4.8) satisfies (4.152). In such a case, it is not difficult to show that the energy corresponding to the solution (4.8) of (4.152) has a miminal value under the boundary conditions given in (4.142). Thus we can conclude that the soliton solution of nonlinear Schrödinger equation, or the microscopic particle in nonlinear quantum mechanics is stable in such a case.

4.9 Multi-Particle Collision and Stability in Nonlinear Quantum Mechanics

One of the most remarkable properties of microscopic particles (solitons) in nonlinear quantum mechanics is that elastic collision between them results only in a phase shift, and the phase shift due to the collision with several microscopic particles (solitons) is the sum of partial phase shifts resulting from separate collisions between a given particle and each other microscopic particle (soliton), as discussed in the previous sections. Their stabilities have a similar property. This property is commonly referred to as absence of multisoliton (or "many-particle") effects in the integrable models. A natural question arises: what about multi-microscopic particle (soliton) collisions in the non-integrable nonlinear quantum mechanical models? It is commonly believed that the main difference between an integrable model and a non-integrable model is due to radiation emitted by the interacting microscopic particles (solitons). Frauenkron *et al.* studied the properties of multi-microscopic particle collisions and demonstrated the existence of nontrivial effects, which do not involve radiation and exist for any value of the perturbation parameter ε in multi-microscopic particle collisions, by extended numerical simulations based on a simple integration scheme. These effects are due to energy exchange between the colliding microscopic particles (solitons) and excitation of internal soliton modes which distinguishes multi-microscopic particle collisions in the integrable and the nonintegrable models.

To understand the multi-particle effects in multi-microscopic particle collisions in nonlinear quantum mechanics systems, we consider a perturbed nonlinear Schrödinger equation with a small quintic nonlinearity,

$$i\frac{\partial\phi}{\partial t'}+\frac{\partial^2\phi}{\partial x'^2}+2|\phi|^2\phi=\varepsilon|\phi|^4\phi \tag{4.153}$$

where ε is the amplitude of the perturbation which is assumed to be small. Equation (4.153) can be used to describe evolution of the electric field of a light in an optical waveguide with an intensity-dependent refractive index $n_{n1}(I=|\phi|^2)$, which slightly deviates from the Kerr dependence. Its solutions will be given in Chapter 8.

In the absence of the perturbation ($\varepsilon=0$), the nonlinear Schrödinger equation (4.153) is known to be exactly integrable and it supports propagation of an microscopic particle (envelope soliton) with amplitude A_0 and velocity v, as given in (4.8). Three elementary integrals are the norm N, field momentum P and energy E, represented here by

$$N_s=2A_0,\quad P_s=A_0,\quad E_s=\frac{1}{2}A_0v^2-\frac{2}{3}A_0^3 \tag{4.154}$$

respectively. In addition, equation (4.153) possesses an infinite number of integrals of motion. Unlike higher (non-elementary) integrals of motion, the above three basic quantities remain conserved in the case of perturbed nonlinear Schrödinger equation,

(4.153). As a matter of fact, the effect of conservative perturbation in (4.153) on the microscopic particles in nonlinear quantum mechanics is trivial. This is consistent with the fact that a perturbed equation has an exact solitary wave solution which is a slightly modified nonlinear Schrödinger soliton. However, interaction of these solitary waves for (4.153) differs drastically from the interaction of two solitons described by the integrable nonlinear Schrödinger equation.

To study multi-microscopic particle collisions in nonlinear quantum mechanics, Frauenkron *et al.* integrated (4.153) using the fourth-order symplectic integrator. The symplectic numerical method is much better than traditional numerical schemes, because it allows conservation of the norm within a relative accuracy of 10^{-11} and conservation of the energy within 10^{-6} during the integration. They used a grid spacing of $dx' = 0.1$, with the total length $L = [-800, 800]$, and a time step $dt' = 0.005$. They studied collisions of three microscopic particles (solitons) and two microscopic particles (solitons), respectively, in nonlinear quantum mechanics. In the case of three microscopic particles, a "fast" particle, with a relatively large velocity, was taken as the exact solution of (4.153) to avoid initial oscillation of its amplitude. The other two "slow" microscopic particles were modeled by an exact two-soliton solution of the unperturbed nonlinear Schrödinger equation to avoid radiation due to strong initial overlapping. The microscopic particles were put on the grid at positions $x_j'^{(0)} = -650$ and $\pm x_{s1}'^{(0)} = \pm 45$ with initial amplitudes $A_j = 1/\sqrt{2}$, $A_{s1} = 0.35$ and velocities $v_j = 3.0$ and $v_{s1} = \pm 0.2$. These values were selected in such a way that all three microscopic particles collide when the two slow particles are overlapping significantly (see Fig. 4.10).

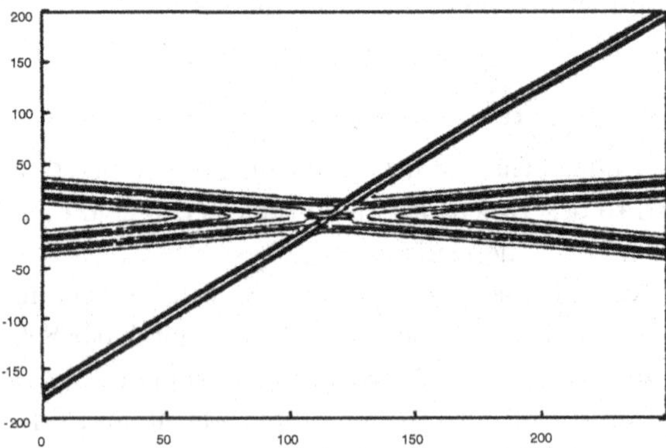

Fig. 4.10 Triple microscopic particle collision in nonlinear quantum mechanics. The fast microscopic particle with amplitude A_j and velocity v_j collides with two slow microscopic particles of equal amplitudes A_{s1} propagating towards each other with velocities $\pm v_{s1}$, so that the three microscopic particles significantly overlap at the moment of collision.

Frauenkron *et al.* first studied collision of two-microscopic particles in nonlinear quantum mechanics. Simulations of the collision were performed using the same initial conditions as mentioned above, but with only one slow microscopic particle (with negative velocity). For $\varepsilon < 0.1$, the changes in velocities of the microscopic particles after the collision were so small that they could not be measured with a sufficient resolution, and they were expected to be on the order of the numerical error. To understand this, we note that for collision of two microscopic particles, inelastic effects may exist due to emission of radiation. The emitted energy $E_{\rm rad}$ has been calculated analytically in the limit of two symmetric microscopic particles with equal amplitudes $A_1 = A_2 = A$ and velocities $v_{1,2} = \pm v$, where $v \gg A$. The result is $E_{\rm rad} = C'\varepsilon^2 A^7[1+F(v/A)]$, where C' is a constant. Similarly, it was found that the radiation-induced change of the norm is $N_{\rm rad} = (4/v^2)E_{\rm rad}$. Because of the symmetry, the total momentum of the two microscopic particles was not affected by the radiation. Due to conservations of the norm and the total energy, we have

$$N_s = N'_s + N_{\rm rad} \quad \text{and} \quad E_s = E'_s + E_{\rm rad},$$

where N'_s and E'_s are given in (4.154). The left-hand sides of the above equations pertain to the microscopic particles before the collisions, whereas the right-hand sides take into account the changes due to the radiation emitted. These two balancing equations for the conserved quantities allow us to determine the changes in the amplitudes $\triangle A$ and velocities $\triangle v$ of the microscopic particles, and they are given by

$$\triangle A_0 = \frac{1}{v^2}E_{\rm rad}, \quad \text{and} \quad \triangle v = \frac{2A_0}{v^3}E_{\rm rad}.$$

This shows that the changes in the microscopic particles' parameters are proportional to ε^2 and inversely proportional to v^2. For collision between two microscopic particles with large velocities v, the interaction time is small and therefore the changes in the microscopic particles' parameters are smaller. For two different microscopic particles these changes will be further reduced due to shorter time of overlapping between the microscopic particles.

Unlike the case of two microscopic particles in nonlinear quantum mechanics, Frauenkron *et al.* obtained nontrivial effects which are in the first order of ε, for the collision of three microscopic particles. Figures 4.11 show the changes in the velocities of the microscopic particles after the collision for different values of the perturbation amplitude ε. It is clear that the change is linear in ε. After the collision small oscillations in the amplitudes of the microscopic particles were observed, which is more prominent for larger ε. It is obvious that these effects are different in collision of two microscopic particles and that of three microscopic particles.

The results also show a nontrivial energy exchange in the first order of the perturbation amplitude ε. To find analytical results, Frauenkron *et al.* considered a collision between a fast microscopic particle with amplitude A_j and velocity v_j, and a symmetric pair of "slow" microscopic particles with equal amplitudes A_{s1} and opposite velocities $\pm v_{s1}$. Assuming $v_j \gg v_{s1} \gg A_{s1}$, they presented a three-soliton

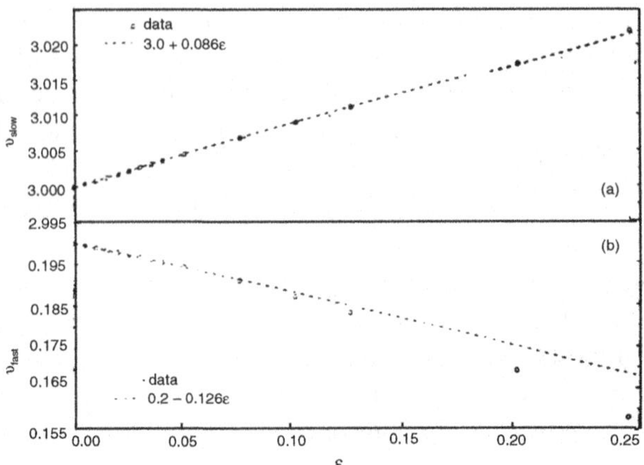

Fig. 4.11 Velocities of the microscopic particles after collision between three microscopic particles, for different values of ε: (a) the fast particle with $v_j = 3.0$; (b) one of the slow particle with $v_{s1} = 0.2$. Open squares and circles are numerical data. Dashed lines are fitted linear functions.

solution in the form of $\phi = \phi_j(v_j) + \phi_{s1}(v_{s1}) + \phi_{s1}(-v_{s1})$, and calculate the changes in the microscopic particles' parameters by means of a perturbation theory based on the inverse scattering method, with the one-soliton Jost functions. If the initial amplitudes of the slow microscopic particles are equal, the microscopic particles parameters after the collision are given by

$$v_j' = v_j + \triangle v_j \quad \text{and} \quad \pm v_{s1}' = \pm v_{s1} \mp \triangle v_{s1},$$

where

$$\triangle v_j = -192\varepsilon \frac{v_{s1} A_{s1}^4}{v_j^2} G(\delta) \quad \text{and} \quad \triangle v_{s1} = 96\varepsilon \frac{A_j A_{s1}^3}{v_j} G(\delta)\,. \tag{4.155}$$

Here $\delta = A_{s1}(x_{s12}'^{(0)} - x_{s11}'^{(0)})$ is the separation between the slow microscopic particles at the moment of collision with the third (fast) microscopic particle. In (4.155),

$$G(\delta) = \frac{1}{\sinh^2\delta}\left[\frac{3(\delta - \tanh\delta)}{\tanh^2\delta} - \delta\right], \tag{4.156}$$

which vanishes as $\delta \to 0$ and $\delta \to \infty$, and is of the same order as the perturbation. There is no change to the amplitudes of the microscopic particles. Because the amplitudes of the microscopic particles do not change after the collision, they further found that $\triangle N = \sum_l \triangle A_l = 0$, and the total momentum is conserved up to the order of v_j^{-2} due to the symmetry of the problem, whereas, the energy of the microscopic particles changes by $\triangle E = (2A_{s1}v_{s1}\triangle v_{s1} + A_j v_j \triangle v_j)$. Using the results of (4.155), they got $\triangle E = 0$. This shows that the results given in (4.155) are consistent with the conservation laws.

To analyze the dependence of the energy exchange between the microscopic particles on the separation of the microscopic particles, Frauenkron *et al.* fixed the perturbation amplitude at $\varepsilon = 0.01$ and varied the initial separation between the two slow microscopic particles. This results in a variation of the average distance between the two slow microscopic particles at the moment of collision, *i.e.*, effectively δ in (4.156). To measure the separation, they considered simultaneously the collision between the two slow microscopic particles under the same conditions but without the third microscopic particles. The initial distance between the slow microscopic particles, x'_D, was measured at the moment when the fast microscopic particle passed the point $x' = 0$. The final velocities were measured by the linear regression analysis. The numerical results are summarized in Fig. 4.15, where the relative changes in the velocities of the microscopic particles, $\triangle v_j$ and $\triangle v_{s1}$, respectively, are shown as functions of x'_D. Indeed, the changes in the velocities of the slow microscopic particles are due to the energy exchange during the collision. This effect strongly depends on the separation between the colliding microscopic particles at the moment of collision and it vanishes for larger separation, which is different from what has been predicted by previous theory. It is also noted that the energy exchange vanishes when the centers of the slow microscopic particles (solitons) approximately coincide.

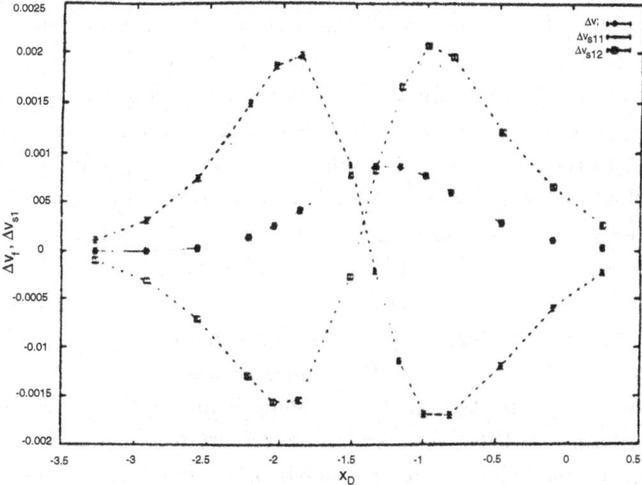

Fig. 4.12 Changes in the velocities of the microscopic particles vs. the extrapolated initial distance between the slow microscopic particles x_D. The dotted line shows the change in the velocity of the fast microscopic particle, $\triangle v_j$, and the dash-dotted line shows that of the slow microscopic particle, $\triangle v_{s1}$. Symbols indicate data obtained from direct numerical simulations at $\varepsilon = 0.01$.

More detailed analysis indicates that the energy exchange is more complicated and it involves the excitation of an internal mode of the fast microscopic particles. In fact, the internal mode appears as a nontrivial localized eigenmode of the linear problem associated with the microscopic particle (soliton) of the perturbed nonlinear Schrödinger equation (4.153). This mode always exists if $\varepsilon > 0$, and describes long-lived oscillations of the amplitude of the microscopic particles. Therefore, internal mode of solitons plays an important role in the energy exchange between microscopic particles during the three particle collisions.

In conclusion, the above discussion shows that, unlike collisions betweeen two microscopic particles, the collision involving three microscopic particles in nonlinear quantum mechanics is accompanied by a radiationless energy exchange among the particles and excitation of internal modes of the colliding microscopic particles. Both effects lead to changes in the soliton velocities which are of first order in the perturbation amplitude ε. This effect depends nontrivially on the relative distance betwween the microscopic particles, and vanishes when the microscopic particles are separated by a large distance.

4.10 Transport Properties and Diffusion of Microscopic Particles in Viscous Environment

We have seen that a remarkable feature of microscopic particles in nonlinear quantum mechanics is their ability to maintain the shape, velocity and energy during propagation in free space. But what about propagation of microscopic particles in viscous systems? Can the microscopic particles retain these properties in such systems at finite temperature? This will be the focus of this section. We will discuss the dynamics of microscopic particles in such viscous systems. The motion of microscopic particles in viscous media can be described using two parameters, the damping and the diffusion coefficients. The former is related to the systematic force applied to the microscopic particles by the environment, while the latter describes the effect of fluctuations in the interaction. We will show that the microscopic particles move in the form of low energy Brownian motion in such a nonlinear system due to the scattering of internal excitation. It also exhibits classical features of microscopic particles in nonlinear quantum mechanics.

As metioned above, in the nonlinear quantum mechanics the position of the center of a microscopic particle (soliton), being a function of time, is sufficient to describe its dynamics in the classical nonlinear Schrödinger equation. In the classical field theory, we may think that the microscopic particles are field configurations which move in space without changing their shape. In quantum field theory the center of mass of the microscopic particles plays also a special role. It can be viewed as a true quantum dynamical variable and the microscopic particle can be treated as a quasi-classical particle. Thus classical and translational invariant theories with the soliton solutions described by dynamic equations in nonlinear quantum mechanics can be quantized using the collective-coordinate method. The problem is that at finite temperature not all the degrees of freedom contribute to the formation of the soliton, and the soliton is never free to move as in the classical theory. In other words, there is always a residual interaction due to quantum degrees of freedom which changes drastically the dynamical properties of the microscopic particles.

This residual interaction results in a damped motion of the microscopic particle (soliton). The microscopic particle thus undergoes a Brownian motion. This problem was studied by Neto *et al.* using a model of dissipation in nonlinear quantum field theory. In the following, we will follow their approach in our discussion on the transport properties of the microscopic particles described by the nonlinear Klein-Gordon equation in a viscous system.

Let's consider, according to (4.1) or (3.75) and (3.94), the following classical static action in the 1+1 dimension for a nonlinear Klein-Gordon scalar field under the rescaling $\phi \to \phi/g$,

$$S = \frac{1}{g^2}\int d^2x \left[\frac{1}{2}\left(\frac{\partial\phi}{\partial x}\right)^2 + V(\phi)\right], \tag{4.157}$$

where g is a coupling constant of the fields. The natural unit system is used in the above equation. The extremum value of the action is given by $\delta S/\delta\phi_s = 0$, namely,

$$\frac{d^2\phi_s}{dx^2} = \frac{dV(\phi_s)}{d\phi_s}. \tag{4.158}$$

Equation (4.158) is similar to (3.77). It is a relativistic dynamic equation of a microscopic particle in nonlinear quantum mechanics. Due to its translational invariance, equation (4.158) must have the soliton solution $\phi_s = \phi_s(x - x_0)$. We can expand (4.157) near the extremum (4.158) in terms of the coupling constant g, according to (3.78), in the first order approximation,

$$\phi_s(x,t) = \phi_s(x - x_0) + g\delta\phi(x,t), \tag{4.159}$$

which gives

$$S[\delta\phi] = S[\phi_s] + \int d^2x\delta\phi(x,t)\left[-\frac{d^2}{dx^2} + \left(\frac{d^2V}{d\phi^2}\right)_{\phi_s}\right]\delta\phi(x,t),$$

plus some higher-order terms in g. Using (4.158), we get

$$m = S[\phi_s] = \frac{1}{g^2}\int d^2x \left(\frac{d\phi_s}{dx}\right)^2, \tag{4.160}$$

which is the classical mass of the microscopic particles. Therefore, up to the leading order in g, the eigenmodes of the system are given by

$$\left[-\frac{d^2}{dx^2} + \left(\frac{d^2V}{d\phi^2}\right)_{\phi_s}\right]\Phi_n(x - x_0) = \omega_n^2\Phi_n(x - x_0). \tag{4.161}$$

Using (4.158) in (4.161), it can be shown that the function $d\phi_s/dx$ is an eigenfunction with an eigenvalue of zero. This zero mode can be understood as follows.

Suppose we displace the center of the microscopic particle (soliton), x_0, by an infinitesimal amount δx_0. Then up to the first order in δx_0, we have

$$\delta_0\phi_s = \phi_s(x_0+\delta x_0) - \phi_s(x_0) = \frac{\partial \phi_s}{\partial x_0}\delta x_0.$$

However, since ϕ_s is a function of the difference $x - x_0$, we have $\delta_0\phi_s = -(d\phi_s/dx)\delta x_0$. Therefore, the eigenmode $d\phi_s/dx$ is related to the movement of the localized solution or the motion of the microscopic particle. Thus,

$$\Phi_0(x-x_0) = \frac{1}{N'}\frac{d\phi_s}{dx},$$

where N' is a normalization constant. We can then get from (4.159)

$$\delta\phi(x,t) = \sum_{n=0}^{\infty} a_n(t)\Phi_n(x-x_0),$$

$$\phi(x,t) = \phi_s(x-x_0) + \frac{g}{N'}a_0(t)\frac{d\phi_s}{dx} + g\sum_{n=0}^{\infty} a_n(t)\Phi_n(x-x_0).$$

It can be seen that the center of the microscopic particle is a true dynamical variable. We rewrite the above expansion as

$$\phi(x,t) = \phi_s[x-x_0(t)] + g\sum_{n=1}^{\infty} a_n(t)\Phi[x-x_0(t)], \tag{4.162}$$

where

$$x_0(t) = x_0 - \frac{g}{N'}a_0(t').$$

Equation (4.162) is known as the collective-coordinate method and is consistent with (4.39) in the first order approximation. It has all the physics of this system. The microscopic particle whose motion is represented by that of its center of mass is a collective excitation (since it is a solution of the nonlinear field equation), but it is also coupled to all other modes by the relative coordinate $x - x_0$. The corresponding classical Hamiltonian can be expressed as

$$H = -m + \frac{1}{2m}\left(P - \sum_{n,i=1}^{\infty} G_{in}p_nq_i\right)^2 + \sum_{n=1}^{\infty}\left(\frac{p_n^2}{2} + \frac{\omega_n^2 q_n^2}{2}\right), \tag{4.163}$$

where P is the momentum canoncally conjugate to x_0, q_n and p_n are also conjugate parts, and

$$G_{in} = -\int dx \frac{d\Phi_i(x)}{dx}\Phi_n(x)$$

couples the microscopic particle to the other modes (phonons) in the system.

This shows that the microscopic particle cannot move freely. Phonons are scattered by the microscopic particle, resulting in the microscopic particle moving like a Brownian particle. Although (4.163) is obtained using a perturbative method, this expression is general and captures the essential physics of this problem.

Making use of the commutation relations, $[\hat{x}_0, \hat{P}] = i\hbar$, $[\hat{q}_n, \hat{p}_i] = i\hbar\delta_{ni}$, and in the context of quantum dissipation, the quantum model for the Hamiltonian (4.163) can be expressed as

$$H = \frac{1}{2m} : \left(\hat{P} - \sum_{n,i=1}^{\infty} \hbar g_{in} \hat{b}_n^+ \hat{b}_n\right)^2 : + \sum_{n=1}^{\infty} \hbar\omega_n \hat{b}_n^+ \hat{b}_n, \tag{4.164}$$

where $: \cdots :$ denotes the normal order, and

$$\hat{b}_n = \left(\frac{\omega_n}{2\hbar}\right)\left(\hat{q}_n + \frac{i\hat{P}_n}{\omega_n}\right),$$

with

$$[\hat{b}_n, \hat{b}_m^+] = \delta_{mn}, \quad [\hat{b}_n^+, \hat{b}_m^+] = 0,$$

and

$$g_{ni} = \frac{1}{2i}\left[\sqrt{\frac{\omega_n}{\omega_i}} + \sqrt{\frac{\omega_i}{\omega_n}}\right] G_{in}$$

is the new coupling constant of the system. Equation (4.164) describes the dynamics of microscopic particles at low energy and can also be used to model other physical systems where microscopic particles are coupled with their environment.

Here we are interested only in the quantum statistical features of the microscopic particles and phonons acting as a source of relaxation in the diffusion processes. Consider the density operator of the system consisting of microscopic particles and phonons, $\hat{\rho}(t)$. This operator evolves in time according to

$$\hat{\rho}(t) = e^{-i\hat{H}t/\hbar} \hat{\rho}(0) e^{i\hat{H}t/\hbar}, \tag{4.165}$$

where $\hat{H}$ is given in (4.164), $\hat{\rho}(0)$ is the density operator at $t = 0$, which is assumed to be decoupled and is given by the product of the density operators of the microscopic particles, $\hat{\rho}_s(0)$, and the phonon, $\hat{\rho}_R(0)$, respectively, *i.e.*, $\hat{\rho}(0) = \hat{\rho}_s(0)\hat{\rho}_R(0)$.

Neto *et al.* studied this problem in which they assumed that the phonons are in thermal equilibrium at $t = 0$, that is,

$$\hat{\rho}_R(0) = \frac{e^{-\beta_T \hat{H}_R}}{Z},$$

where

$$Z = \mathrm{tr}_R(e^{-\beta_T \hat{H}_R}), \quad \beta_T = \frac{1}{K_B T},$$

and $\hat{H}_R$ is the Hamiltonian of the free phonons which is given by the last term in (4.164). We define a reduced density operator $\hat{\rho}_s(t) = tr_R[\hat{\rho}(t)]$, which contains all the information about the system when it is in thermal contact with a reservoir. Projecting $\hat{\rho}(t)$ in the coordinate representation of the microscopic particle system $\hat{x}_0|q\rangle = q|q\rangle$ and in the coherent state representation for the phonons $\hat{b}_n|\alpha_n\rangle = \alpha_n|\alpha_n\rangle$. We have

$$\rho_s(x,y,t) = \int dx'' \int dy' J(x,y,t;x'',y',0)\rho_s(x'',y',0),$$

where J is the superpropagator of the microscopic particle (soliton) which can be written as

$$J = \int_{x''}^{x} dx \int_{y'}^{y} dy e^{i(S_0[x]-S_0[y])/h} F[x,y], \tag{4.166}$$

where

$$S_0[x] = \int_0^t dt'' \left[\frac{m_0 \dot{x}^2(t'')}{2}\right] \tag{4.167}$$

is the classical action for the free microscopic particle. F is the influence functional,

$$F[x,y] = \int \frac{d^2\alpha}{\pi^N} \int \frac{d^2\tilde{\beta}}{\pi^N} \int \frac{d^2\tilde{\beta}'}{\pi^N} \rho_R(\tilde{\beta}^*, \tilde{\beta}') e^{-|\alpha|^2 - |\tilde{\beta}|^2/2 - |\tilde{\beta}'|^2/2} \times$$
$$\int_{\beta}^{\alpha^*} D^2\alpha \int_{\beta^*}^{\alpha} D^2\gamma e^{S_1[x,\alpha]+S_1^*[y,\gamma]}, \tag{4.168}$$

where $\tilde{\beta}'$ denotes the vector $(\beta_1, \beta_2, \cdots\cdots, \beta_N)$ and S_1 is a complex action related to the reservoir and the interaction,

$$S_1[x,\alpha] = \int_0^t dt'' \left[\frac{1}{2}\left(\alpha \cdot \frac{d\alpha^*}{dt''} - \alpha^* \cdot \frac{d\alpha}{dt''}\right) - \frac{i}{\hbar}(H_R - \dot{x}h_1)\right], \tag{4.169}$$

with

$$H_R = \sum_{n=1}^{\infty} \hbar\omega_n \alpha_n^* \alpha_n, \quad h_1 = \sum_{n,m=1}^{\infty} \hbar g_{nm} \alpha_m^* \alpha_n.$$

Now we expand the action (4.169) around the classical solution of its Euler-Lagrange equation. After integrating (4.168), we can get

$$F[x,y] = \prod_{n=1}^{\infty} (1 - \Gamma_{nn}[x,y]\bar{n}_n)^{-1},$$

where

$$\Gamma_{nm} = W_{nm}^*[y] + W_{mn}[x] + \sum_{l=1}^{\infty} W_{lm}^*[y] W_{ln}[x],$$

$$W_{nm}[\tau] = \delta_{mn} + \sum_{n'=1}^{\infty} \int_0^{\tau} dt'' K_{nn'}(t'') W_{n'm}(t''),$$

$$K_{nm}([x], \tau) = i g_{nm} \dot{x}(\tau') e^{i(\omega_n - \omega_m)\tau'},$$

$$\bar{n}_n = (e^{\beta_T h\omega_n} - 1)^{-1}.$$

Here K_{nm} is the kernel of the integral equation, $W_{nm}[\tau]$ is the scattering amplitude from mode k to mode j. The terms that appear in the summation represent the virtual transitions between these modes. If the velocity of the microscopic particle is small, we can use the Born approximation. In matrix form, this can be written as

$$W = (1 - W^0)^{-1} W^0 \approx W^0 + W^0 W^0.$$

In such a case, the terms Γ_{nm} are small. Then we can write approximately

$$F[x, y] \approx \exp\left\{ \sum_{n=1}^{\infty} \Gamma_{nm}[x, y] \bar{n}_n \right\}. \tag{4.170}$$

If the interaction is turned off ($\Gamma \to 0$) at $T = 0$, the functional (4.170) is unity, and, as we would expect, the microscopic particle moves like a free classical particle.

Using the Born approximation for W in (4.170), Neto *et al.* obtained the following.

$$J = \int_{x'}^{x} Dx \int_{y'}^{y} Dy \exp\left\{ \frac{i}{\hbar} \bar{S}[x, y] + \frac{1}{\hbar} G[x, y] \right\}, \tag{4.171}$$

where

$$\begin{aligned}\bar{S} = &\int_0^t dt'' \left\{ \frac{m_0}{2} \left[\dot{x}^2(t'') - \dot{y}^2(t'') \right] + \left[\dot{x}(t'') - \dot{y}(t'') \right] \right. \\ = &\left. \int_0^t dt''' \Gamma_1(t'' - t''') [\dot{x}(t'') + \dot{y}(t''')] \right\},\end{aligned} \tag{4.172}$$

and

$$G = \int_0^t dt'' \int_0^t dt''' \left\{ [\Gamma_R(t'' - t''')][\dot{x}(t'') - \dot{y}(t'')][\dot{x}(t''') - \dot{y}(t''')] \right\}, \tag{4.173}$$

$$\Gamma_R(t) = \hbar \Theta'(t) \sum_{n,n=1}^{\infty} g_{nm}^2 \bar{n}_n \cos(\omega_n - \omega_m)t,$$

$$\Gamma_1(t) = \hbar \Theta'(t) \sum_{n,n=1}^{\infty} g_{nm}^2 \bar{n}_n \sin(\omega_n - \omega_m)t.$$

Here $\Theta'(t)$ is the theta function defined as

$$\Theta'(t) = \begin{cases} 1, & \text{if } t > 0 \\ 0, & \text{if } t < 0 \end{cases}.$$

Introducing the following new variables

$$X = \frac{x+y}{2}, \quad Y = x - y,$$

the equations of motion for the action (4.172) can be written as

$$\begin{aligned} \ddot{X}(\tau) + 2\int_0^t dt''\gamma(\tau - t')\dot{X}(t'') = 0, \\ \ddot{Y}(\tau) - 2\int_0^t dt''\gamma(t'' - \tau)\dot{Y}(t'') = 0, \end{aligned} \tag{4.174}$$

where

$$\gamma(t) = \frac{1}{m_0}\frac{d\Gamma_1}{dt},$$

or

$$\gamma(t) = \frac{\hbar\Theta'(t)}{m_0}\sum_{n,m=1}^{\infty} g_{nm}^2\bar{n}_n(\omega_n - \omega_m)\cos(\omega_n - \omega_m)t \tag{4.175}$$

which is the damping function. The two equations in (4.174) have the same form as that obtained in the case of quantum Brownian motion by Calderia and Leggett. In the limit of the time scale of interest being much greater than the correlation time of the phonon variables, we can write $\gamma(t) = \bar{\gamma}(T)\delta(t)$, where $\bar{\gamma}(T)$ is a damping parameter which is temperature-dependent and $\delta(t)$ is the Dirac delta function. The above form of $\gamma(t)$ is the same as that in the Markovian approximation in these systems. If we use (4.174) and expand the phase of (4.171) around this classical solution, we can get the well-known result for the quantum Brownian motion provided that the damping parameter γ (temperature independent) is replaced by $\bar{\gamma}(T)$ and the diffusive part is replaced by (4.173). Neto *et al.* gave the diffusion parameter in momentum space

$$D(t) = \hbar\frac{d^2\Gamma_R}{dt^2} = -\hbar^2\Theta'(t)\sum_{n,m=1}^{\infty} g_{nm}^2\bar{n}_n(\omega_n - \omega_m)^2\cos(\omega_n - \omega_m)t.$$

As discussed before, in the Markovian limit, we can write $D(t)$ in the Markovian form, $D(t) = \bar{D}(T)\delta(t)$, where $\bar{D}(T)$ and $\bar{\gamma}(T)$ obey the classical fluctuation-dissipation theorem at low temperature. If we define the scattering function as

$$S(\omega - \omega') = S(\omega' - \omega) = \sum_{n,m=1}^{\infty} g_{nm}^2\delta(\omega - \omega_m)^2\cos(\omega' - \omega_m),$$

$D(t)$ can can be written as

$$D(t) = -\frac{\hbar^2}{2}\Theta(t)\int_0^\infty d\omega \int_0^\infty d\omega' S(\omega-\omega')(\omega'-\omega)^2[n(\omega_n)+n(\omega_m)]\cos(\omega-\omega')t.$$

In the above discussion, we established that the Hamiltonian (4.164) leads to Brownian dynamics. That is, a microscopic particle in nonlinear quantum mechanics moves like a classical particle at low energy in a viscous environment where its relaxation and diffusion are due to scattering by the phonons. These phonons are the residual excitations created by the presence of the microscopic particle. We showed that the damping and the diffusion coefficients of the microscopic particle depend on temperature, since phonons must be thermally activated in order to scatter off the microscopic particle. Therefore at absolute zero temperature the microscopic particle moves freely, but its mobility decreases as the temperature increases. This shows again the classical feature of microscopic particles in nonlinear quantum mechanics.

By expanding the perturbation in powers of temperature in the systems, Dziarmaga *et al.* developed a perturbative method to obtain an explicit expression for the diffusion coefficient of microscopic particles (solitons) moving in a dissipation systems with thermal noise. For a microscopic particle described by the ϕ^4-equation in one spatial dimension, we have (in natural units)

$$\Gamma\phi_t = \phi_{xx} + 2(1-\phi^2)\phi + \eta(t,x), \tag{4.176}$$

where Γ is the dissipation coefficient and $\eta(t,x)$ is a Gaussian white noise with correlation $\langle\eta(t,x)\rangle = 0$, and

$$\langle\eta(t_1,x_1)\eta(t_2,x_2)\rangle = 2K_BT\Gamma\delta(t_1-t_2)\delta(x_1-x_2). \tag{4.177}$$

The system is coupled to an ideal heat bath at temperature T. At a finite temperature the field $\phi(x,t)$ performs random walk in its configuration space. If the noise is absent at $T=0$, equation (4.176) admits a static kink solution $\phi(t,x) = \phi_s(x) \equiv \tanh(x)$, and an antikink is given by $\phi_s(-x)$. Small perturbations around the kink take the form $\phi = e^{-\gamma t/\Gamma}\varphi(x)$. Linearization of (4.176) with respect to $\varphi(x)$, for $\eta(t,x)=0$, gives

$$\gamma\varphi(x) = -\frac{d^2\varphi(x)}{dx^2} + \left[4 - \frac{6}{\cosh^2(x)}\right]\varphi(x). \tag{4.178}$$

Its eigenvalues and eigenstates are

$$\gamma = 0, \quad \varphi(x) = \phi_{\rm sx}(x) = -\frac{1}{\cosh^2(x)},$$

$$\gamma = 3, \quad \varphi(x) = \varphi_0(x) \equiv \frac{\sinh(x)}{\cosh^2(x)}, \tag{4.179}$$

$$\gamma = 4 + k^2, \quad \varphi(x) = \varphi_k(x) \equiv e^{ikx}\left[1 + \frac{3ikm\tanh(x) - 3\tanh^2(x)}{1+k^2}\right],$$

where k is the momentum of the microscopic particle. The zero mode ($\gamma = 0$) is separated by a gap from the first excited state (breather mode, $\gamma = 3$). The continuum states are normalized so that

$$\int_{-\infty}^{\infty} dx \varphi_k^*(x)\varphi_{k'}(x) = 2\pi\frac{4+k^2}{1+k^2}\delta(k-k') \equiv N(k)\delta(k-k').$$

The field in the one kink sector for the theory (4.176) can be expanded using the complete set (4.179) as

$$\phi(t,x) = \phi_s[x - \xi(t)] + \sum_{n\neq 0} A_n(t)\varphi_n[x-\xi(t)]. \tag{4.180}$$

Inserting (4.180) into (4.176) and projecting on the orthogonal basis (4.179) gives a set of stochastic nonlinear differential equations,

$$\Gamma N_0\dot{\xi}(t) - \Gamma\dot{\xi}(t)\sum_{n\neq 0} A_n(t)M_{n0} + \sum_{i,j\neq 0} A_i(t)A_j(t)P_{ijo} +$$
$$\sum_{i,j,k\neq 0} A_i(t)A_j(t)A_k(t)R_{ijko} = \eta_0(t),$$

$$\Gamma N_n\dot{A}_n(t) + \gamma_n N_n A_n(t) - \Gamma\dot{\xi}(t)\sum_{i\neq 0} A_i(t)M_{in} + \sum_{i,j\neq 0} A_i(t)A_j(t)P_{ikn} +$$
$$\sum_{i,j,k\neq 0} A_i(t)A_j(t)A_k(t)R_{ijkn} = \eta_n(t), \tag{4.181}$$

where the coefficients are

$$N_n = \int_{-\infty}^{\infty} dx\varphi_n(x)\varphi_n^*(x), \quad M_{ni} = \int_{-\infty}^{\infty} dx\varphi_{nx}(x)\varphi_i^*(x),$$
$$P_{nij} = 6\int_{-\infty}^{\infty} dx\phi_s(x)\varphi_n(x)\varphi_j^*(x), \quad P_{nijk} = 2\int_{-\infty}^{\infty} dx\varphi_n(x)\varphi_i(x)\varphi_j(x)\varphi_k^*(x),$$
$$\eta_n(t) = \int_{-\infty}^{\infty} dx\eta(t,x)\varphi_n^*(x).$$

Applying (4.177) and the orthogonality of the basis (4.179), we get

$$\langle\eta_n(t)\rangle = 0, \quad \langle\eta_n^*(t_1)\eta_i(t)\rangle = 2K_BT\Gamma N_n\delta_{ni}\delta(t_1 - t_2). \tag{4.182}$$

The natural parameter of expansion at low temperature is $\sqrt{K_BT}$. The projected noises, (4.182), are of the same order. Dziarmaga *et al.* introduced the rescaling $\sqrt{K_BT} \to \varepsilon\sqrt{K_BT}$. Terms in (4.181) can then be expanded in powers of ε and a power series solution of the equation can be obtained. ε was set back to 1 at the end of the calculation. With the expansion of the collective coordinates,

$\xi(t) = \varepsilon\xi^{(1)} + \varepsilon^{(2)}\xi^{(2)} + \cdots$ and $A_n = \varepsilon A_n^{(1)} + \varepsilon^{(2)} A_n^{(2)} + \cdots$, equations in (4.181) become, to the leading order in ε,

$$\Gamma N_0 \dot{\xi}^{(1)}(t) = \eta_0(t), \quad \Gamma N_n \dot{A}_n^{(1)}(t) + \gamma_n N_n A_n^{(1)}(t) = \eta_n(t). \tag{4.183}$$

In this approximation the zero mode $\xi^{(1)}(t)$ and the excited modes $A_n^{(1)}(t)$ are uncorrelated and stochastic process driven by their mutually uncorrelated projected noises. $\xi^{(1)}(t)$ represents a Markovian-Wiener process whose only nonvanishing single connected correlation function is $\langle \dot{\xi}^{(1)}(t)\dot{\xi}^{(1)}(t'')\rangle = (2K_BT/\Gamma N_0)\delta(t-t'')$, which is singular at $t \to t''$. According to the first equation in (4.183), the probability, $\rho(t, \xi(t))$, for the kink random walks to $\xi(t)$ at time t satisfies the diffusion equation

$$\frac{\partial \rho}{\partial t} = D\frac{\partial^2 \rho}{\partial \xi^2},$$

where the diffusion coefficient is given by

$$D = \frac{K_BT}{\Gamma N_0} = \frac{3K_BT}{4\Gamma}.$$

Other terms in the expansion of collective coordinates can be recursively worked out and they are

$$\dot{\xi}^{(2)}(t) = \dot{\xi}^{(1)}(t)\sum_{n\neq 0} A_n^{(1)}(t)\frac{M_{no}}{N_0} - \sum_{i,j\neq 0} A_i^{(1)}(t)A_j^{(1)}(t)\frac{P_{ijo}}{\Gamma N_0},$$

$$\begin{aligned} A_n^{(2)}(t) = {} & \sum_{i\neq 0}\frac{M_{in}}{N_n}\int_0^t d\tau e^{-\gamma_n(t-\tau)/\Gamma}\dot{\xi}^{(1)}(\tau)A_i^{(1)}(\tau) - \\ & \sum_{i,j\neq 0}\frac{\rho_{ijn}}{\Gamma N_n}\int_0^t d\tau e^{-\gamma_n(t-\tau)/\Gamma}A_i^{(1)}(\tau)A_j^{(1)}(\tau), \end{aligned}$$

$$\begin{aligned} \dot{\xi}^{(3)}(t) = {} & \sum_{n\neq 0}[\dot{\xi}^{(1)}(t)A_n^{(2)}(t) + (1 \leftrightarrow 2)]\frac{M_{no}}{N_0} - \\ & \sum_{i,j\neq 0}[A_i^{(1)}(t)A_j^{(2)}(t) + (1 \leftrightarrow 2)]\frac{P_{ijo}}{\Gamma N_0} - \\ & \sum_{i,j,k\neq 0} A_i^{(1)}(t)A_j^{(1)}(t)A_k^{(1)}(t)\frac{R_{ijko}}{\Gamma N_0}. \end{aligned}$$

The equilibrium correlations of $\xi(t)$, up to one order higher than the leading order, are

$$\begin{aligned} \langle \dot{\xi}(t')\dot{\xi}(\bar{t})\rangle \approx {} & \varepsilon^2\langle \dot{\xi}^{(1)}(t)\dot{\xi}^{(1)}(\bar{t})\rangle + \varepsilon^4\langle \dot{\xi}^{(2)}(t)\dot{\xi}^{(2)}(\bar{t})\rangle + \varepsilon^4[\langle \dot{\xi}^{(1)}(t)\dot{\xi}^{(3)}(\bar{t})\rangle + (1 \leftrightarrow 3)] \\ = {} & \langle \dot{\xi}^{(1)}(t)\dot{\xi}^{(1)}(\bar{t})\rangle\left[1 + \frac{3K_BT}{N_0^2}\sum_{n\neq 0}\frac{|M_{no}|^2}{\gamma_n N_n} + \frac{2}{N_0}\sum_{n\neq 0} M_{no}\langle A_n^{(2)}(\infty)\rangle\right] + \cdots. \end{aligned}$$

The integrals over the continuous part of the spectrum and the summation over the breather mode yield

$$\langle\dot{\xi}(t)\dot{\xi}(\bar{t})\rangle = \frac{3K_BT}{2\Gamma}\delta(t-\bar{t})[1+1.8164K_BT] + O[(K_BT)^3].$$

Finally, Dziarmaga *et al.* gave the diffusion coefficient in the equilibrium state, which is given by the average over time much longer than the relaxation time of the breather mode. It is given by

$$D(\infty) = \frac{3K_BT}{4\Gamma}(1+1.8164K_BT) + O[(K_BT)^3]. \tag{4.184}$$

Dziarmaga *et al.* performed numerical simulation of (4.181) with $\Gamma = 1$ for a range of temperature values. The numerical results were consistent with the prediction of the perturbative theory, (4.184).

4.11 Microscopic Particles in Nonlinear Quantum Mechanics versus Macroscopic Point Particles

From previous sections of this Chapter, we understand that a microscopic particle in nonlinear quantum mechanics is a soliton which exhibits many features of a macroparticle. However, the microscopic particle cannot be regarded as a macro-point particle. We already know from sections 4.9 and 4.10 that the excitations of the internal mode of soliton – a long-lived oscillation of the soliton amplitude – participates in inelastic and radiationless energy exchange among particles during the three-soliton collisions, and the residual excitations (phonons) created by the presence of the microscopic particles result in damping and diffusion of the particles in viscous systems. These phenomena do not occur in system of macro-point particles. In this section, we will further explore these phenomena and address the point particle limit of microscopic particles in nonlinear quantum mechanics, following the work of Kalbermann. A simple coupling between the center of the microscopic particle (soliton) and a potential will be introduced, which is useful in the discussion of differences and similarities between a microscopic particle in nonlinear quantum mechanics and a point particle in classical physics, as well as the pointlike limit of the microscopic particles.

Kalbermann considered the center of mass of a microscopic particle (soliton) as a candidate for the location of the particle, and coupled it to an external potential. He took a kink mode supplemented by a potential and a Lagrange multiplier to force the interaction with the center of mass. This procedure treats the center of the microscopic particle (soliton) as the collective coordinate, as it is usually done in the case of topological defects such as those by Kivshar *et al.* and Neto *et al.*, given in Section 4.10. In this approach, certain parameters of the microscopic particle, related to the symmetries or zero modes, are identified and coupled to external sources. In this case, translational invariance yields the corresponding symmetry. In other cases, internal symmetries can be identified and the generators of the symmetries can be treated as the collective coordinates of the microscopic particles as a whole. This enables the separation of the bulk characteristics of the microscopic particle from its internal excitation. In this way, the coupling between the center of the microscopic particle and the collective coordinate corresponding to translational invariance breaks the symmetry, which allows a clear distinction between internal excitations and bulk behavior of the microscopic particles as a whole. As our aim is to demonstrate the importance of the extended character of the microscopic particles on its interaction with external sources, the above method seems to be the most straightforward approach.

Kalbermann used the following Lagrangian in his study of the problem

$$\mathcal{L} = \frac{1}{2}\partial_\mu\phi\partial^v\phi + \frac{1}{4}\lambda\left(\phi^2 - \frac{m^2}{\lambda}\right)^2 - U(x_0) + \chi(x,t)\phi\delta(x - x_0), \qquad (4.185)$$

where $\chi(x,t)$ is the Lagrange multiplier that enforces the coupling between the center of the microscopic particle, x_0, and the external potential, $U(x_0)$. For kinks with topological charge $=\pm 1$ for which $\phi(x_0) = 0$, the Lagrange multiplier selects the value of x_0 at $\phi = 0$. To couple the potential to a point at which the microscopic particle vanishes, Kalbermarnn chose

$$U(x_0) = \frac{1}{4}\int dx V(x)\lambda\left(\phi^2 - \frac{m^2}{\lambda}\right)^2. \qquad (4.186)$$

which represents interaction between a microscopic particle and an impurity. The integrand is zero except in the region around the center of the microscopic particle where $\phi \approx 0$. If $V(x)$ is a smooth potential (in contrast to the δ spike impurity), then to the lowest order of the potential, the above equation becomes $U(x_0) = c'V(x_0)$, where c' is a constant depending on the initial parameters of the microscopic particle. The choice for the interaction corresponds to the case in which the microscopic particle propagates in a smooth medium and the impurity is similar to a background metric. If the microscopic particle can be regarded as topological soliton, then the present treatment would correspond to the classical propagation of a microscopic particle in a background potential. In order to see clearly this correspondence, Kalbermarnn chose the interactions which consist of introducing a nontrivial metric for the relevant space time. The metric carries the information of the medium.

The Lagrangian in the scalar field theory for the (1+1)-dimensional system immersed in a background with the metric $g_{\mu v}$ is

$$\mathcal{L} = \sqrt{g}\left[g^{uv}\frac{1}{2}\partial_\mu\phi\partial_v\phi - U(\phi)\right], \qquad (4.187)$$

where g is the determinant of the metric, and U is the self-interaction that makes the existence of the soliton possible. For a weak potential we have

$$g_{00} \approx 1 + V(x), \quad g_{11} = -1, \quad g_{-11} = g_{1-1} = 0, \qquad (4.188)$$

where $V(x)$ is the external space-dependent potential. The equation of motion of the microscopic particle in the background becomes

$$\frac{\partial^2\phi}{\partial t^2} - \frac{1}{\sqrt{g}}\frac{\partial}{\partial x}\left(\sqrt{g}\frac{\partial\phi}{\partial x}\right) + g_{00}\frac{\partial U}{\partial\phi} = 0. \tag{4.189}$$

This equation is identical, for slowly varying potentials, to the equation of motion of a microscopic particle interacting with an impurity $V(x)$. The constraint in (4.185) only insures the proper identification of the center of the microscopic particle in the present case. In the following, we will let $\lambda = m^2 = 1$ in (4.185). Variations of the field ϕ as well as the collective coordinate x_0 in (4.185) yield the equation of motion of the microscopic particle constrained by the Lagrange multiplier,

$$\frac{\partial^2\phi}{\partial t^2} - \frac{\partial^2\phi}{\partial x^2} + \phi(\phi^2-1) = \frac{\partial U}{\partial x_0}B(x,t), \tag{4.190}$$

where

$$B(x,t) = \delta(x-x_0)\left(\frac{\partial\phi}{\partial x}\right)^{-1} \tag{4.191}$$

is the new source term that denotes the coupling between the center of the microscopic particle and the external potential. The limit of a pointlike object is achieved, when the width of the microscopic particle is negligible as compared to the characteristic scale of the potential. This method provides a clean separation between the particulate behavior of the microscopic particle and its wavelike extended character in the nonlinear quantum mechanics. For the potential in (4.187), Kalbermann used a bell-shaped function,

$$U(x_0) = A\,\text{sech}^2\left(\frac{x_0-r}{a}\right), \tag{4.192}$$

where A is the height (depth) of the potential located at r, and a is related to the width, W, of the interaction potential by $a = W/(2\pi)$.

Using a δ-function distribution for $\rho(x)$,

$$\delta(x) = \rho(x) = \lim_{\epsilon\to 0}\rho_\epsilon(x), \quad \rho_\epsilon(x) = \frac{\exp(-x^2/\epsilon)}{\sqrt{\epsilon\pi}}, \tag{4.193}$$

$B(x,t)$ can be expressed as

$$B(x,t) = \frac{\rho(x-x_0)}{\int\rho(x-x_0)(\partial\phi/\partial x)dx}. \tag{4.194}$$

In actual calculations the limit of $\epsilon \to 0$ is achieved when $\epsilon \approx dx$, the spatial step, is small enough compared to the size of the microscopic particle.

In solving the partial differential equation of motion, the spatial boundaries are taken to be $-25 < x < 25$, with a grid of $dx = 0.04$ and a time lapse of $dt = 0.02$ up to a maximal time of $t = 200$ (10000 time steps). The upper time limit of $t = 200$

is sufficient for the resonances to decay, to define clearly the asymptotic behavior of the microscopic particle (soliton) while at the same time to avoid reflection from the boundaries. Kalbermann started with a free microscopic particle impinging from the left, at distance from the potential range, with an initial velocity v,

$$\phi = \pm \tanh \left[\frac{x - x_0 - vt}{\sqrt{2(1 - v^2)}} \right], \tag{4.195}$$

with $x_0(t = 0) = -5$. The total and kinetic energies of the free microscopic particle are

$$E(v) = \frac{3\sqrt{2}}{2\sqrt{1 - v^2}}, \quad E_k(v) = E(v) - E(0), \tag{4.196}$$

respectively.

The value of A in (4.192) was chosen such that there are visible results for the kink with $m = 1$. Kalbermann used $A = 0.1$ and $A = -0.2$. The former allowes transition through the barrier for a pointlike particle with velocity $v \approx 0.43$, and the latter provides enough strength to trap it in a reasonable range of the initial velocity. The location of the potential was fixed at r=3. For the attractive potential, it was found that there were no islands of trapping between transmissions. The microscopic particle either is trapped, when its velocity is below a certain threshold, or passes through the impurity. For a barrier depth of $A = -0.2$ and $W = 1$ which is comparable to the size of the microscopic particle, the critical velocity is $v = 0.583$. At this initial velocity, the microscopic particle drifts extremely slowly through the potential well, and lags behind the free propagation by almost an infinite time, but it eventually goes through. If the velocity is below this critical value, the microscopic particle is always trapped. Contrary to the case of a point particle, there appear to be bound states for velocities below the threshold for positive kinetic energies. These states may be referred to as bound states in the continuum, extraneous to classical particulate behavior. This feature is due to the finite extent of the microscopic particle (soliton) in nonlinear quantum mechanics.

One way to show that such a phenomena is indeed due to space extention of the microscopic particle is by increasing the well width. In order to confirm this assertion, Kalbermann selected an initial velocity of the microscopic particle which was well below the threshold transmission velocity, $v = 0.2$, and a well depth of $A = -0.2$ as before. Starting from $W = 1$ he gradually increased the well width and found that transmission occured when the well width became $W = 5$, which was much wider than the extent of the microscopic particles. When the velocity was above the threshold velocity, the microscopic particle slowly drifted through the well in an effectively infinite time. It remained within the barrier range for almost $t = 200$ and then started to emerge from the opposite side. Figure 4.13 shows the asymptotic velocity of the microscopic particle (soliton), plotted versus the potential width W. In the case of the microscopic particle being trapped, the asymptotic velocity vanishes. The microscopic particle oscillates with zero mean velocity inside the well. It never emerges from it. Its amplitude decreases when the particle moves farther from the center of the well during the oscillatory motion. The energy of the microscopic particle is conserved in all cases. When the microscopic particle is trapped, the initial kinetic energy transforms into deformation and oscillation energies as well as radiation that damps the oscillations.

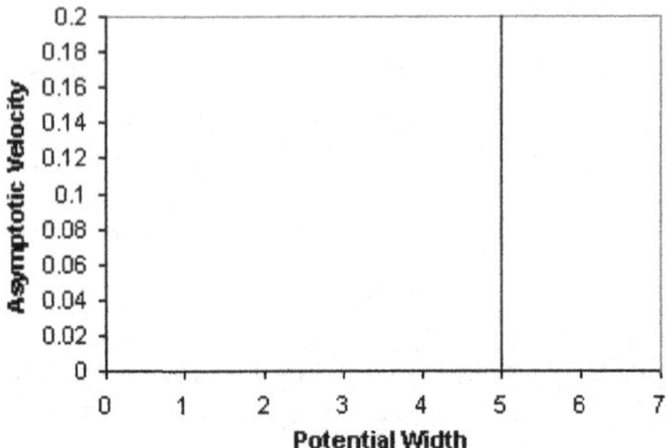

Fig. 4.13 Asymptotic velocity of the microscopic particle (soliton) as a function of the width of the attractive potential.

When the microscopic particles is transmitted, there is no radiation due to the topological conservation of winding number that forces the microscopic particle (soliton) to be exactly the one that impinged onto the potential. This result was borne out in the case of multi-soliton collisions in section 4.8 for which the dominant effect is the inelastic nonradiative exchange of energy between the microscopic particles (solitons). In the present case, there is no possible recipient of such energy. Therefore, the microscopic particle (soliton) will emerge elastically. The asymptotic velocities of the reflected and transmitted microscopic particle were calculated using the actual motion of its center, and so were the theoretical expressions for the kinetic and total energies of the free microscopic particles. This comparison proved energy conservation and served as a measure of numerical accuracy.

For the repulsive potential, the microscopic particle (soliton) is reflected even for initial kinetic energies above the barrier height. When $A = 0.1$, the corresponding initial velocity whose kinetic energy would yield classically to transmission is $v \approx 0.43$. However, this does not occur. For a barrier of width $W = 1$, the microscopic particles goes through, only when the initial velocity is $v = 0.52$, corresponding to an initial kinetic energy of $E_k = 0.16$, well above the barrier height.

The particulate behavior will be restored only when the barrier width is larger than the extent of the microscopic particle (soliton), such that it looks as if almost pointlike. In particular for $v = 0.43$, transmission starts for barrier widths greater than $W = 19$. Fig. 4.14 shows the asymptotic velocity of the microscopic particle as a function of the barrier width. The microscopic particle is repelled by sharp barriers and transmits through wider barriers, while trapping never occurs.

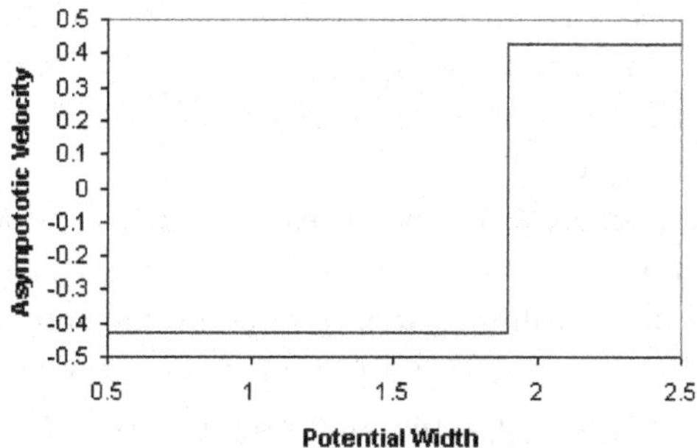

Fig. 4.14 Asymptotic velocity of microscopic particle (soliton) as a function of the repulsive barrier width.

From the above discussion we see that a microscopic particle in nonlinear quantum mechanics behaves as a particle only when the width of the potential is large compared to the size of the particle so that the microscopic particles (soliton) appears pointlike. In such a case the microscopic particle obviously has corpuscle feature of a classical particle. In contrary cases, the coupling between the center of the microscopic particle and a repulsive or an attractive potential can result in bound states or trapped states in the continuum, which does not occur for point-particle objects. Kalbermann simulated numerically the behaviors of the microscopic particles in terms of collective coordinates. The simulation results showed also chaotic behavior depending on the initial position of the center of the microscopic particle. The initial states can transform a trapped state into a reflected or transmitted state. This exhibits clearly the wave feature of the microscopic particle. Therefore, a microscopic particle in nonlinear quantum mechanics has not only corpuscle but also wave features and it is necessary to study its wave feature which will be the topics of discussion in the following sections.

4.12 Reflection and Transmission of Microscopic Particles at Interfaces

As mentioned above, microscopic particles in nonlinear quantum mechanics represented by (3.1) have also wave property, in addition to the corpuscle property. This wave feature can be conjectured from the following reasons.

(1) Equations (3.2) – (3.5) are wave equations and their solutions, (4.8), (4.56) and (4.195) are solitary waves having features of travelling waves. A solitary wave consists of a carrier wave and an envelope wave, has certain amplitude, width, velocity, frequency, wavevector, and so on, and satisfies the principles of superposition of waves, (3.30), (3.43) and (3.44), although the latter are different from classical waves or the de Broglie waves in linear quantum mechanics.

(2) The solitary waves have reflection, transmission, scattering, diffraction and

tunneling effects, just as that of classical waves or the de Broglie waves in linear quantum mechanics. Some of these properties of microscopic particles will be described in the following sections. In this section, we consider first the reflection and transmission of microscopic particles at an interface.

The propagation of microscopic particles (solitons) in a nonlinear and nonuniform medium is different from that in a uniform medium. The nonuniformity here can be due to a physical confining structure or at the interface of two nonlinear materials. One could expect that a portion of a microscopic particles that was incident upon such an interface from one side would be reflected and a portion would be transmitted into the other side. Lonngren *et al.* observed the reflection and transmission of microscopic particles (solitons) in a plasma consisting of a positive ion and negative ion interface, and numerically simulated the phenomena at the interface of two nonlinear materials. In order to illustrate the rules of reflection and transmission of microscopic particles, we introduce here the work of Lonngren *et al.*

Lonngren *et al.* simulated numerically the behaviors of microscopic particle (soliton) described by the nonlinear Schrödinger equation. They found that the signal had the property of soliton. These results are in agreement with numerical investigations of similar problems by Aceves *et al.*, and Kaplan and Tomlinson. A sequence of pictures obtained by Lonngren *et al.* at uniform temporal increments of the spatial evolution of the signal are shown in Fig. 4.15. From this figure, we note that the incident microscopic particles propagating toward the interface between the two nonlinear media splits into a reflected and a transmitted soliton at the interface. From the numerical values used in producing the figure, the relative amplitudes of the incident, the reflected and the transmitted solitons can be deduced.

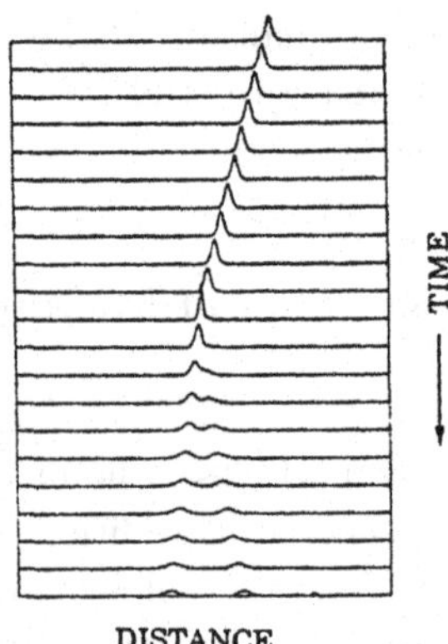

Fig. 4.15 Simulation results showing the collision and scattering of an incident microscopic particle (soliton) described by the nonlinear Schrödinger equation (top) onto an interface. The peak nonlinear refractive index change is 0.67% of the linear refractive index for the incident microscopic particle, and the linear offset between the two regions is also 0.67%.

They assumed that the energy carried by the incident microscopic particle (soliton) is all transferred to either the transmitted or the reflected solitons and none is lost through radiation. Thus

$$E_{\rm inc} = E_{\rm ref} + E_{\rm trans}. \tag{4.197}$$

Lonngren *et al.* gave approximately the energy of a microscopic particle (soliton)

$$E_j = \frac{A_j^2}{Z_c} W_j,$$

where the subscript j refers to the incident, reflected or transmitted solitons. The amplitude of the soliton is A_j and its width is W_j. The characteristic impedance of a material is given by Z_c. Hence, (4.197) can be written as

$$\frac{A_{\rm inc}^2}{Z_{\rm cI}} W_{\rm inc} = \frac{A_{\rm ref}^2}{Z_{\rm cI}} W_{\rm ref} + \frac{A_{\rm trans}^2}{Z_{\rm cII}} W_{\rm trans}. \tag{4.198}$$

Since $A_j W_j$ = constant, for the microscopic particles (solitons) of the nonlinear Schrödinger equation (see (4.8) or (4.87) in which B_j is replaced by A_j), we obtain the following relation between the reflection coefficient $R = A_{\rm ref}/A_{\rm inc}$ and the transmission coefficient $T = A_{\rm trans}/A_{\rm inc}$

$$1 = R + \frac{Z_{\rm cI}}{Z_{\rm cII}} T, \tag{4.199}$$

for microscopic particle (soliton) of the nonlinear Schrödinger equation.

Fig. 4.16 Sequence of signals detected as the probe is moved in 2 mm increments from 30 to 6 mm in front of the reflector. The incident and reflected KdV solitons coalesce at the point of reflection, which is approximately 16 mm in front of the reflector. A transmitted soliton is observed closer to the disc. The signals at 8 and 6 mm are enlarged by a factor of 2.

Lonngren *et al.* conducted an experiment to verify this idea. In the experiment, they found that the detected signal had the characteristics of a KdV soliton. A sequence of pictures as shown in Fig. 4.16 was taken by Lonngren *et al.* using a small probe at equal spatial increments, starting initially in a homogeneous plasma region and then progressing into an inhomogeneous plasma sheath adjacent to a perturbing biased object. In Fig. 4.16, we can clearly see that the probe first detects the incident soliton, and some time later the reflected soliton. These signals are observed, as expected, to coalesce together as the probe passed through the point where the microscopic particle (soliton) was reflected. Beyond this point where the density started to decrease in the steady-state sheath, a transmitted soliton was observed. From Fig. 4.16, the relative amplitudes of the incident, the reflected and the transmitted solitons can be deduced.

For the KdV solitons, there is also $A_j W_j^2 =$ constant (see (4.92)). Thus,

$$1 = R^{3/2} + \frac{Z_{\rm cI}}{Z_{\rm cII}} T^{3/2},$$

for the KdV soliton.

The relations between the reflection and the transmission coefficients for both types of solitons are shown in Fig. 4.17, with the ratio of characteristic impedances set to one. The experimental results on KdV solitons and results of the numerical simulation of microscopic particles depicted by the nonlinear Schrödinger equation are also given in the figure. Good agreement between the analytic results and simulation results can be seen. The oscillatory deviation from the analytic result is due to the presence of radiation modes in addition to the soliton modes. The interference between these two types of modes results in the oscillation in the soliton amplitude. In the asymptotic limit, the radiation will spread, damp the oscillation, and result in the reflection-transmission coefficient curver falling on the analytic curve.

The above rule of propagation of the microscopic particles in nonlinear quantum mechanics is different from that of linear waves in classical physics. Lonngren *et al.* found that a liner wave obeys the following relation

$$1 = R^2 + \frac{Z_{\rm cI}}{Z_{\rm cII}} T^2. \tag{4.200}$$

This can also be derived from (4.198) by assuming linear waves. The widths of the incident, reflected and transmited pulses W_j are the same. For linear waves

$$R = \frac{Z_{\rm cII} - Z_{\rm cI}}{Z_{\rm cII} + Z_{\rm cI}}, \quad \text{and} \quad T = \frac{2Z_{\rm cII}}{Z_{\rm cII} + Z_{\rm cI}},$$

and equation (4.200) is satisfied. Obviously, equation (4.200) is different from (4.199). This shows clearly that the microscopic particles in nonlinear quantum mechanics have wave feature, but it is different from that of linear classical waves and the de Broglie waves in the linear quantum mechanics.

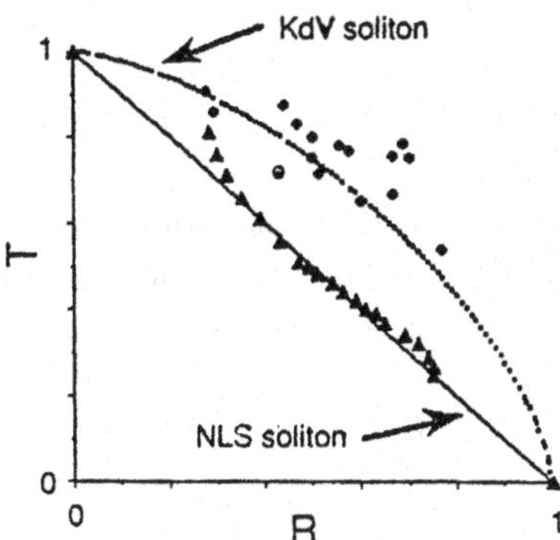

Fig. 4.17 Relationship between the reflection and transmission coefficients of a microscopic particle (soliton), Eq.(4.199). The solid circles are experimental results on KdV solitons, and the open circles represent results of Nishida. The solid triangles indicate numerical results of Lonngren *et al.* for microscopic particles(solitons) described by the nonlinear Schrödinger equation.

However, the above results are based on two assumptions: (1) no radiation modes are excited through the interaction of the incident soliton with the interface; (2) the product of the interaction of the incident soliton with the interface is a single transmitted and a single reflected soliton. If these conditions are not satisfied, the above relation will not hold.

As a matter of fact, the reflection of a microscopic particle (soliton) in nonlinear quantum mechanics is complicated. Alonso discovered a shift in the positions of microscopic particles which occurs in the case of nonzero reflection coefficient. He used the inverse scattering method to obtain the phase shifts of the microscopic particles in (4.40) with $b = 2$. As it is known, the inverse scattering method for solving the nonlinear Schrödinger equation is based on the resolution of the integral equations of the Marchenko type (4.51), which can be written as

$$F_1^*(t',x',y) - \int_0^\infty K^*(t',x'+y+z)F_2^*(t',x',z)dz = K^*(t',x'+y),$$
$$F_2^*(t',x',y) + \int_0^\infty K(t',x'+y+z)F_1(t',x',z)dz = 0, \tag{4.201}$$

where the kernel K is determined from the set of scattering data

$$S(t') = \{\zeta_j = \xi_j + i\eta_j,\ b_j(t') = b_j \exp(4i\zeta_j^2 t'),\ j = 1,\cdots,N;$$
$$c(t',\zeta) = c(\zeta)\exp(4i\zeta^2 t'),\ \zeta \epsilon R\}$$

in the form of

$$K(t',x') = 2\sum_j b_j(t')e^{2i\zeta_j x'} + \frac{1}{\pi}\int_{-\infty}^\infty c(t',\zeta)e^{2i\zeta x'}d\zeta. \tag{4.202}$$

Solution of (4.40) is given by $\phi(t',x') = -F_1(t',x',+0)$. As $t' \to \pm\infty$, $\phi(t',x')$

evolves into a superposition of N freely moving solitons with velocities $v_j = 4\xi_j$, and a radiation component which decays as $|t'|^{-1/2}$. The parameters $q_j^{\pm}$ which characterize the asymptotic trajectories $q_j^{\pm}(t') \approx q_j^{\pm} + v_j t'$ for the solitons as $t' \to \pm\infty$ remain to be determined.

It is known from Section 4.1 that the nonlinear Schrödinger equation is a Galilean-invariant Hamiltonian system and its corresponding generator for the pure Galilean transformation can be written as

$$\begin{aligned} G &= -\frac{1}{2}\int_{-\infty}^{\infty} x'|\phi|^2 dx' \\ &= -\sum_j \ln|b_j \partial_\zeta a(\zeta_j)| - \frac{1}{2\pi}\int_{-\infty}^{\infty} \ln|a(\zeta)|\partial_\zeta \arg F(\zeta) d\zeta, \end{aligned} \tag{4.203}$$

where

$$\begin{cases} a(\zeta) = \prod_j \dfrac{\zeta - \zeta_j}{\zeta - \zeta_j^*} \exp\left[\dfrac{1}{2\pi}\displaystyle\int_{-\infty}^{\infty} \dfrac{\rho(q)}{q - \zeta - i0} dq\right], & \Im\eta > 0 \\ \rho(\zeta) = \ln(1 + |c(\zeta)|^2), \ F(\zeta) = c(\zeta)a(\zeta), & \Im\eta = 0. \end{cases} \tag{4.204}$$

From (4.203) and (4.204), the following alternative expression for G can be obtained

$$\begin{aligned} G = &-\sum_j \left[\ln\frac{|b_j|}{2\eta_j} + \sum_{n\neq j} \ln\left|\frac{\zeta_j - \zeta_n}{\zeta_j - \zeta_n^*}\right| - \frac{\eta_j}{2\pi}\int_{-\infty}^{\infty} \frac{\rho(\zeta)}{|\zeta - \zeta_j|^2} d\zeta\right] \\ &+ \frac{1}{4\pi}\int_{-\infty}^{\infty} \rho(\zeta)\partial_\zeta \left[\arg c(\zeta) + \frac{1}{2\pi} P\int_{-\infty}^{\infty} \frac{\rho(q)}{q - \zeta} dq\right] d\zeta, \end{aligned} \tag{4.205}$$

where P denotes the principle value. We observe that for a pure one-soliton solution ϕ_{sol}, the position of its center is $q(t) = (2\eta)^{-1}\ln[|b(t)|/2\eta]$, and therefore, $G[\phi_{\text{sol}}(t)] = -2\eta q(t)$.

As a consequence of the Galilean invariance of the nonlinear Schrödinger equation, we can define the mass and momentum functionals which have the following forms in terms of scattering data

$$M = \frac{1}{2}\int_{-\infty}^{\infty} |\phi|^2 dx' = 2\sum_j \eta_j - \frac{1}{2\pi}\int_{-\infty}^{\infty} \rho(\zeta) d\zeta,$$

$$P = -i\int_{-\infty}^{\infty} \phi^* \phi_{x'} dx' = -8\sum_j \eta_j \xi_j + \frac{2}{\pi}\int_{-\infty}^{\infty} \zeta\rho(\zeta) d\zeta.$$

According to these expressions, one may think that the field of microscopic particle depicted by the nonlinear Schrödinger equation appears as a Galilean system composed of N particles with masses $2\eta_j$ and velocities $v_j = 4\xi_j$ (solitons), and a has continuous mass distribution with density $-\rho(\zeta)/2\pi$ and velocity $v_g(\zeta) = 4\zeta$

(radiation). This velocity spectrum associated with the scattering data can be understood through the asymptotic analysis of the nonlinear Schrödinger equation field as follows.

From (4.201) we know that the nonlinear Schrödinger equation field ϕ at a given point (t', x'_0) depends only on the restriction of the kernel $K(t', x')$ to the interval $(x'_0, +\infty)$. If the evolution of the scattering data is inserted into (4.202), the modulus of the jth term in the summation propagates with velocity $v_j = 4\xi_j$, while the group velocity of the Fourier modes in the integral term coincides with $v_g(\zeta) = 4\zeta$. Hence, given arbitrary values x'_0 and v_0, as $t' \to \pm\infty$ the restriction of K to intervals of the form $I^{\pm}(t') = [x'_0 + (v_0 \pm \epsilon)t', +\infty]$ where $\in$ is an arbitrary positive number, depends only on those scattering data with velocity v such that $\pm(v - v_0) > 0$. In fact, it can be shown that the contribution to K due to the remaining scattering data has a L^2 norm on $I^{\pm}(t)$ which vanishes asymptotically as $t' \to \pm\infty$.

We now consider the motion of the lth soliton and let us denote by $\phi_{\pm}$ the parts of the nonlinear Schrödinger equation field propagating to the right of the soliton as $t' \to \pm\infty$. Then the relevant kernels for characterizing $\phi_{\pm}$ through the Marchenko equations (4.201) are

$$K_{\pm}(t', x') = 2\sum_{j\neq l} \theta[\pm(v_j - v_l)]b_j(t')e^{2i\zeta_j x'} + \frac{1}{\pi}\int_{-\infty}^{\infty} \theta\{\pm[v_g(\zeta) - v_l]\}c(t', \zeta)e^{2i\zeta x'}d\zeta,$$

where θ is the step function. The sets of scattering data related to $\phi_{\pm}$ are

$$S_{\pm}(t') = \{\zeta_j, b_j(t'), j \text{ such that } \pm(v_j - v_l) > 0; \theta(\pm[v_g(\zeta) - v_l])c(t', \zeta), \zeta\epsilon R\}.$$

The scattering data $S_{\pm}(t')\{\zeta_l, b_l(t')\}$ correspond to the parts $\phi'_{\pm}$ of the nonlinear Schrödinger equation moving with velocity v such that $\pm(v - v_l) \geq 0$ as $t' \to \pm\infty$. In other words, $\phi'_{\pm}$ result from the addition of the lth soliton to $\phi_{\pm}$. Hence, as a result of the dispersive character of the radiation component and the localized form of the soliton, it is clear that as $t' \to \pm\infty$ the difference between the values of the functional Gs for $\phi'_{\pm}$ and $\phi_{\pm}$, respectively, must be $-2\eta_e q_l^{\pm}(t')$ of G at the lth soliton. Using this and the identity (4.205), Alonso obtained

$$q_l^{\pm} = \frac{1}{2\eta_l}\ln\frac{|b_l|}{2\eta_l} + \frac{1}{\eta_l}\sum_{j\neq l}\theta[\pm(v_j - v_l)]\ln\left|\frac{\zeta_l - \zeta_j}{\zeta_l - \zeta_j^*}\right| - \frac{1}{2\pi}\int_{-\infty}^{\infty}\theta\{\pm[v_g(\zeta) - v_l]\}\frac{\rho(\zeta)}{|\zeta - \zeta_l|^2}d\zeta.$$

The phase shift of the lth microscopic particle (soliton) as it interacts with both the other microscopic particles and the radiation component is given by

$$q_l^{\pm} - q_l^{-} = \frac{1}{\eta_l}\sum_{j\neq l}\operatorname{sign}(v_j - v_2)\ln\left|\frac{\zeta_l - \zeta_j}{\zeta_l - \zeta_j^*}\right|$$
$$-\frac{1}{2\pi}\int_{-\infty}^{\infty}\operatorname{sign}[v_g(\zeta) - v_l]\frac{\ln[1+|c(\zeta)|^2]}{|\zeta-\zeta_l|^2}d\zeta.$$

From these results we see that the microscopic particles experience a phase shift when they are reflected due to interaction with other microscopic particles and with the radiation component. It shows again the wave feature of the microscopic particles.

4.13 Scattering of Microscopic Particles by Impurities

A wave can be scattered by an obstruction, and scattering is an essential feature of waves. In this section we discuss scattering of microscopic particles by impurities, and propagation of microscopic particles in nonlinear disordered media in the framework of nonlinear quantum mechanics. It is well known that in linear systems, disorder generally creates Anderson localization, which means that the transmission coefficient of a plane wave decays exponentially with the length of the system. However, the nonlinearity may drastically modify properties of propagation of microscopic particles in disorder systems. We will find that besides reflection and transmission, there are also excitations of internal modes and impurity modes in the systems, which again reflects clearly the wave nature of microscopic particles. These properties of microscopic particles depend on the nature of the impurity and its velocity. As the first step towards a full understanding of these phenomena, we start by discussing scattering of a microscopic particle by a single impurity, which was studied by Kivshar *et al.*, Malomed, Zhang Fei, *et al.*, and others.

In such a case, the dynamics of the perturbed microscopic particle in nonlinear quantum mechanics is described by the following nonlinear Schrödinger equation

$$i\phi_{t'} + \phi_{x'x'} + 2|\phi|^2\phi = \varepsilon f(x')\phi, \tag{4.206}$$

where ε is a real, small parameter, $f(x')$ is a localized potential function arising from the impurity in an otherwise homogeneous medium, which satisfies $f(x' \to \infty) = 0$. If $\varepsilon = 0$, equation (4.206) has a soliton solution which is given by (4.56), with the amplitude and velocity of the soliton given by 2η and $+4\xi$, respectively. Equation (4.56) can be interpreted as a bound state of $N = 4\eta$ quasi-particles with a binding energy $E_b = 4\eta^2$.

We now consider the scattering of the microscopic particle (soliton) by the impurity described by (4.206) using the Born approximation. We assume that the velocity of the microscopic particle does not change during the scattering under the condition $|\varepsilon|\eta \ll \xi^2$. The spectral density of the "quasi-particles", $n(\lambda)$, due to the scattered microscopic particle, can be calculated following the inverse scattering method of Zakharov *et al.*, and the result is given in Section 4.3, $n(\lambda) \approx \pi^{-1}|b(\lambda)|^2$, where $b(\lambda)(|b(\lambda)|^2 \ll 1)$ is the Jost coefficient used in the inverse scattering method, and λ is a real spectral parameter that determines the wave number $k(\lambda) = 2\lambda$ and the frequency $\omega(\lambda) = k^2\lambda = 4\lambda^2$ of the emitted linear waves. The perturbation-induced evolution of the Jost coefficient $b(\lambda)$ was obtained by Karpman as

$$\begin{aligned}\frac{\partial b(\lambda,t')}{\partial t'} = 4i\lambda^2 b(\lambda,t') + \varepsilon\int_{-\infty}^{\infty} dx' f(x')\,[\phi(x',t')\Psi_1'(x',t',\lambda]\psi_2^*(x',t',\lambda)\\ - \phi^*(x',t')\Psi_2'(x',t')\psi_1^*(x',t',\lambda')]\,,\end{aligned} \tag{4.207}$$

where $\Psi_{1,2}(x',t',\lambda)$ and $\psi_{1,2}(x',t',\lambda)$ are the components of the Jost functions. For the one-soliton solution (4.56), these functions were obtained by Karpman.

Kivshar *et al.* assumed that before the scattering, *i.e.* at $t' \to -\infty$, all the "quasi-particles" were in the bound state, or the soliton state (4.56). This means that the initial condition for (4.207) should be in the form of $b(\lambda, t' = -\infty) = 0$. Then, the total density $n_{\text{rad}}(\lambda)$ of the linear waves emitted by the microscopic particles during the scattering is

$$\begin{aligned}n_{\text{rad}}(\lambda) &= \frac{1}{\pi}|b(\lambda,t'=+\infty)|^2\\ &= \varepsilon^2\tilde{n}_{\text{rad}}^{(1)}(\lambda)\int_{-\infty}^{\infty}dx_1\int_{-\infty}^{\infty}dx' f(x_1)f(x')e^{i\beta(\lambda)(x_1-x')},\end{aligned} \tag{4.208}$$

where

$$\begin{aligned}\tilde{n}_{\text{rad}}^{(1)}(\lambda) &= \frac{\pi}{2^8\xi^4}\beta^2(\lambda)\text{sech}^2\left[\frac{\pi(\lambda^2-\xi^2+\eta^2)}{4\eta\xi}\right];\\ \beta(\lambda) &= \frac{(\lambda-\xi^2)+\eta^2}{\xi}.\end{aligned} \tag{4.209}$$

The number of "quasi-particles", N, emitted by the microscopic particle (soliton) in the backward direction (*i.e.* the reflected "quasi-particles") can be calculated from

$$N_r = \int_{-\infty}^{\infty} n_{\text{rad}}(\lambda)d\lambda.$$

The reflection coefficient R of the microscopic particle can be defined as the ratio of N_r to the total number of "quasi-particles", $N = 4\eta$

$$R = \frac{1}{4\eta}\int_0^{\infty} n_{\text{rad}}(-\lambda)d\lambda. \tag{4.210}$$

Kivshar *et al.* studied the scattering of a microscopic particle by an isolated point impurity in which $f(x')$ is a delta-function. From (4.208), we get the emitted spectral density $n_{ren} = \varepsilon^2 \tilde{n}^{(1)}_{\rm rad}(\lambda)$. The reflection coefficient can then be written as

$$R^{(1)} = \frac{\pi\varepsilon^2}{2^{10}\alpha\xi^2}\int_0^\infty \frac{(x'+1)^2+\alpha^2}{\cosh^2[\pi(x'^2+\alpha^2-1)/4\alpha]}dx', \qquad (4.211)$$

where $\alpha = \eta/\xi$. At $\alpha = 0$, $R^{(1)}$ reduces to that of the linear wave packet, *i.e.*, $R_0^{(1)} \approx \varepsilon^2/4k_0^2 = \varepsilon^2/16\xi^2$. For small α, $R^{(1)}(\alpha)/R_0^{(1)}$ first increases slowly to 1.004 (at $\alpha = \alpha_c \approx 0.178$) and then rapidly decreases to zero, so that for $\alpha \gg 1$ the reflection coefficient approaches zero exponentially.

$$R^{(1)} \approx \frac{\pi\alpha^{7/2}}{16\sqrt{2}} R_0^{(1)} e^{-\pi\alpha/2}, \quad (\alpha \gg 1). \qquad (4.212)$$

The fact that $R^{(1)}$ differs from $R_0^{(1)}$ shows that the wave nature of the microscopic particles in nonlinear quantum mechanics differs from linear wave packet.

The number of transmitted "quasi-particles" N_t, *i.e.*, those emitted by the microscopic particle (soliton) in the forward direction, may be defined by

$$N_t = \int_0^\infty n_{\rm rad}(\lambda)d\lambda.$$

Note that for $\alpha \ll 1$ this quantity is essentially less than N_r, since it has the asymptoic expansion $N_t \approx \varepsilon^2\alpha^4/120\xi^2 N$. However, for $\alpha \gg 1$, N_t asymptotically approaches N_r.

We define the density, $\varepsilon_{\rm rad}(\lambda) = 4\lambda^2 n_{\rm rad}(\lambda)$, of the emitted energy

$$E_{\rm rad} = \int_{-\infty}^\infty dx'(|\phi_{x'}|^2 - |\phi|^4)$$

by means of the inverse scattering method. The total energy emitted by the microscopic particles in the forward (E_t) and backward (E_r) directions can be found by

$$E_{t,r} = \int_0^\infty d\lambda \varepsilon_{\rm rad}(\pm\lambda).$$

For the microscopic particles described by the nonlinear Schrödinger equation (4.206), Kivshar *et al.* found that

$$E_r = \varepsilon^2\eta\left(1 - \frac{1}{3}\alpha^2 - \frac{2}{15}\alpha^4\right), \quad E_t = \frac{2}{15}\varepsilon^2\eta\alpha^4, \quad \text{for } \alpha \ll 1,$$

$$E_r \approx E_t = \frac{\varepsilon^2\eta\alpha^{9/2}}{16\sqrt{2}} e^{-\pi\alpha/2}, \quad \text{for } \alpha \gg 1.$$

Kivshar *et al.* also considered scattering of a microscopic particle (soliton) by two point impurities separated by a distance a. The perturbation $\varepsilon f(x')$ take the form of $\varepsilon f(x') = \varepsilon_1\delta(x') + \varepsilon_2\delta(x'-a)$. It can be shown that

$$n_{\rm rad}(\lambda) = n_{\rm rad}^{(1)}\left\{(\varepsilon_1-\varepsilon_2)^2 + 4\varepsilon_1\varepsilon_2\cos^2\left[\frac{1}{2}a\beta(\lambda)\right]\right\},$$

where $\tilde{n}_{\rm rad}^{(1)}$ and $\beta(\lambda)$ have been defined in (4.209). In this case the reflection coefficient of the microscopic particle, $R^{(2)}$, is determined by two factors, corresponding to the soliton scattering by an isolated point impurity with the effective intensity $(\varepsilon_1-\varepsilon_2)$, and to the resonant effect during the scattering, respectively, and it is given by

$$R^{(2)} = \frac{\pi\varepsilon^2}{2^8\alpha\xi^2}\int_0^\infty dx'\frac{[(x'+1)^2+\alpha^2]^2}{\cosh^2[\pi(x'^2+\alpha^2-1)/4\alpha]}\cos^2\left[\frac{1}{4}d(x'+1)^2+\alpha^2\right], \quad (4.213)$$

where $d = 2a\xi$. When $\alpha \ll 1$ and $\alpha^2 d \ll 1$, equation (4.213) reduces to

$$R^{(2)} = 2R_0^{(1)}\left[1 + \frac{2\alpha d}{\sinh(2\alpha d)}\cos(2d)\right]. \quad (4.214)$$

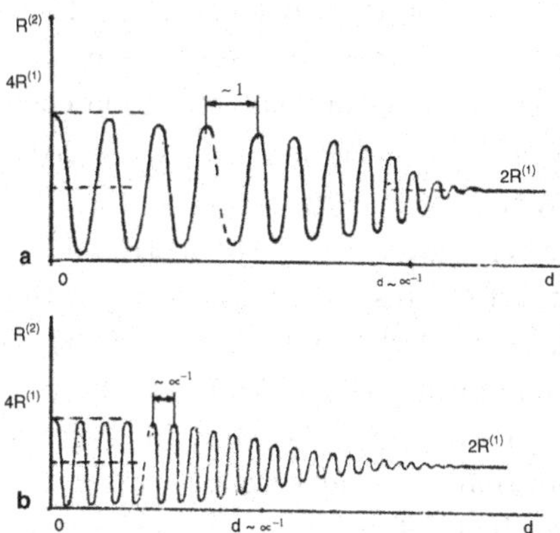

Fig. 4.18 The reflection coefficient of the microscopic particle described by the nonlinear Schrödinger equation scattered by two identical point impurties, as a function of $d = 2a\xi$. (a) $\alpha \ll 1$; (b) $\alpha \gg 1$; $R^{(1)}$ is the reflection coefficient for scattering by a single isolated impurity.

The dependence of $R^{(2)}$ on d for small α is shown in Fig. 4.18(a). If the size of microscopic particle ($\approx \eta^{-1}$) exceeds considerably the distance between the two impurities (*i.e.* $\alpha d \approx a\eta \ll 1$), we can distinguish the interference phenomena which manifest themselves as oscillation of $R^{(2)}$ against the parameter d. Different from the linear approximation, $R^{(2)}$ does not approach zero, because $x' < \sinh x'$ for $x' \neq 0$. When $d = d_{\rm min} = (1-2\alpha^2/3)d_n^0$, where $d_n^0 = \pi/2+n\pi$ $(n = 0, 1, \cdots)$, $R^{(2)}$ has the minimum values: $R_{\rm min}^{(2)} = 2(\alpha d_{\rm min})^2/3 \approx (2\alpha^2/3)(\pi/2+n\pi)^2$. The maximum values, $R_{\rm max}^{(2)} = 4R_0^{(1)}[(1-n\alpha\pi/3)^2]$, which occur at $d = d_{\rm max} = (1-2\alpha^2/3)n\pi$, decrease as d increases. When the size of the microscopic particle becomes comparable to the separation of the impurities, a (*i.e.* $\alpha d \approx 1$), the interference decays and for $\alpha d \gg 1$, $R^{(2)}$ approaches the value $2R_0^{(1)}$ (see Fig. 4.18(a)). That is, the scattering by each impurity is independent of the other. Thus, the microscopic particle shows both wave ($\alpha d \ll 1$) and corpusclar ($\alpha d \gg 1$) properties during the scattering by the impurities.

In the nonlinear case ($\alpha \gg 1$), we can obtain from (4.213)

$$R^{(2)} = 2R^{(1)}(\alpha)\left[1 + (1+c')^{1/4}\cos\left(\frac{1}{2}\alpha^2 d + y'\right)\right], \tag{4.215}$$

where $c' = 2\alpha\eta/\pi$, $y' = \tan^{-1}(c')/2$, and $R^{(1)}(\alpha)$ is given in (4.212). The dependence of $R^{(2)}$ on d for $a \gg 1$ is shown in Fig. 4.18(b). For large values of d (when $\alpha d \approx c' \gg 1$), the reflection coefficient becomes equal to $2R^{(1)}$, which characterizes the non-resonant scattering of the microscopic particle in which the scattering intensities by separate impurities are simply added. For $d\alpha \leq 1$ ($c' < 1$), the reflection coefficient oscillates with increasing d and the period of the oscillation is approximately $\alpha^{-2} \ll 1$. The frequency of the internal oscillation of the microscopic particle for $\alpha \gg 1$ is then $\approx \eta^2$, and the resonant condition is $\alpha^2 d \approx 1$. For $\alpha \ll 1$, the period of the internal oscillations is of the order of ξ^{-2}, and the resonant condition is $\alpha\xi \approx d \approx 1$. However, in the case of $\varepsilon_1 = \varepsilon_2 = \varepsilon$, scattering of linear wave is known to be resonant and the reflection coefficient $R_0^{(2)}$ in the Born approximation oscillates as a function of $d = ak_{\bar{0}} = 2\alpha\xi$ as $R_0^{(2)} = 4R_0^{(1)}\cos^2 d$, where $R_0^{(1)} = \varepsilon^2/16\xi^2$. When $d_n^{(0)} = \pi/2 + n\pi$ ($n = 0, 1, \cdots$), $R_0^{(2)} = 0$, which corresponds to the situation where the energy of the scattered "quasi-particle", $k_{\bar{0}}^2 = 4\xi^2$, coincides with the energy of the resonant state between the two delta-function potentials. The maximum value, $4R_0^{(1)}$ of $R_0^{(2)}$, corresponds to scattering of a linear wave by a single effective impurity with twice the intensity ($\varepsilon \to 2\varepsilon$). Therefore, we see that scattering of microscopic particles described by the nonlinear Schrödinger equation is different from that of linear waves.

The above results can be easily generalized to the case of a periodic system of N point impurities, with separation a between adjacent impurities. Tedious but straightforward calculation leads to the following when $N\alpha^2 d \ll 1$,

$$R^{(N)} = R_0^{(1)}\left(N + 2\sum_{k,l\neq 0}^{(N-1)/2} C_N^{2k+1}C_N^{2l+1}\sum_{m=0}^{k-1}(-1)^m C_{k-1}^m 2^{-2(N-m-1)}\right.$$
$$\left.\times\sum_{p=1}^{N-m-1} C_{2(N-m-1)}^{N-m-1+p}\frac{2\alpha dp}{\sinh(2\alpha dp)}\cos(2pd)\right),$$

where C_N^K are the binomial coefficients and ( ) stands for the integer part. In the case of $\alpha Nd \gg 1$, the microscopic particle is scattered as a corpuscle and it can be shown that $R^{(N)} \approx NR_0^{(1)}$.

However, transmission always occurs in the scattering process of the microscopic particles by the impurities except in the case of complete reflection, the condition for which is given above. The transmission coefficients are defined by $T^{(E)} = E_t/E_i$ for the energy and $T^{(N)} = N_t/N_i$ for the excitation, where E_i and N_i are incident energy and number of microscopic particles, respectively, E_t and N_t are the transmitted energy and number, respectively. Because $E_i = E_t + E_r$ and $N_i = N_t + N_r$, then $T^{(E,N)} = 1 - R^{(E,N)}$. Thus the transmission coefficients can be easily calculated from the reflection coefficient given above

When the impurity concentration ρ is low, the average distance between two impurities is larger than the size of microscopic particles. In this case, the scattering by many impurities can be considered independent, *i.e.*, $T = \prod_j T_j$, where T_j is the transmission coefficient of the jth impurity. Because the transmitted microscopic particle from scattering by the jth impurity is the incident microscopic particle for scattering by the $j+1$th impurity, we have

$$\begin{aligned} E_{j+1} &= E_J T_j^{(E)}(E_j, N_j), \\ N_{j+1} &= N_j T_j^{(N)}(E_j, N_j) \end{aligned}$$

and

$$\begin{aligned} \Delta E_{j+1} &= E_{j+1} - E_j = -E_j R_j^{(E)}(E_j, N_j), \\ \Delta N_{j+1} &= N_{j+1} - N_j = -N_j R_j^{(N)}(E_j N_j). \end{aligned} \tag{4.216}$$

When $\alpha \ll 1$,

$$T^{(N,E)}(x') = \frac{N(x')}{N(0)} = \frac{E(x')}{E(0)} = e^{-x'/\lambda_0},$$

where

$$\lambda_0 = \frac{16\xi^2(0)}{\rho\varepsilon^2} = \frac{1}{\rho R^{(1)}},$$

$R^{(1)}$ is the reflection coefficient due to a single impurity. This shows that the transmission coefficent decays exponentially.

When $\alpha = \eta/\xi \geq 1$, Kivshar *et al.* found that the asymptotic change in $T^{(N,E)}(z = x'/x_0')$, where $x_0' = 64/\pi\rho\varepsilon^2$, depends essentially on the parameter $\alpha(0) = \eta(0)/\xi(0)$ which is related to the nonlinearity of the incoming microscopic particle. The greater α is, the larger the number of excitations in the microscopic particle (soliton) becomes, and the smaller its spatial extension. On the contrary, if α is small, the microscopic particle looks very similar to a linear wave packet. Therefore, for initial conditions $\alpha(0) \ll \alpha_c = 1.28505$, the system evolves to a final state in which N decreases exponentially to zero while the speed of the microscopic particle, v, reaches a constant positive value. If $\alpha_c > \alpha \approx 1$, the decay consists of an initial slow transient which followed by a fast exponential behavior. Finally, the initial condition $\alpha(0) > \alpha_c$ leads to a situation in which both N and v become practically constants.

These results show that strong nonlinearity can completely inhibit the localization effect stimulated by the disorder. This effect appears over a threshold nonlinearity. Below this threshold, the transmission coefficient decreases to zero as the size of the system increases, either exponentially throughout or exponentially after a short transient. Above the threshold value this model shows undistorted motion of the microscopic particle in the disordered systems, *i.e.*, the transmission coefficient does not decay and localization does not occur because of the small width of the microscopic particle compared to the large values of α.

In the above discussion, the only effect of the impurity is to give rise to an effective potential on the microscopic particle. In particular, a microscopic particle (soliton) may be trapped by an attractive potential due to the impurity as a result of loss of its kinetic energy through radiation. However, the impurity is not a "hard" object, and it may support a localized oscillating state, or the so-called impurity mode. As a consequence, a microscopic particle may be able to transfer part of its kinetic energy to the impurity and thus excite the impurity mode during an inelastic interaction. Hence, there is a different type of interaction between a microscopic particle and an impurity when the impurity supports a localized impurity mode. The microscopic particle can be totally reflected in the case of an attractive impurity if its initial velocity lies in certain resonance "windows". We now discuss these problems.

Let us considered the Sine-Gordon model in the 1+1 dimensional case, including a local impurity

$$\phi_{tt} - \phi_{xx} + \sin\phi = \varepsilon\delta(x)\sin\phi, \tag{4.217}$$

where $\delta(x)$ is the Dirac δ-function. The natural unit system is assumed in (4.217). It is known that the Sine-Gordon model supports a topological soliton (kink) at $\varepsilon = 0$, given by

$$\phi_k = 4\tan^{-1}\left\{\exp\left[\frac{x - X(t)}{\sqrt{1-v^2}}\right]\right\}, \tag{4.218}$$

where $X(t) = vt$ is the kink coordinate and $v = \dot{X}$ is its velocity. For $\varepsilon > 0$, the impurity in (4.217) creates an effective attractive potential well for the microscopic particle (kink). Kivshar *et al.* studied the scattering of a microscopic particle (kink) by a pointlike impurity, and integrated (4.217) using a conservative numerical scheme. Simulations were carried out in the spatial interval $(-40, 40)$ which was discretized by a step size of $\Delta x = 2\Delta t = 0.04$. The Dirac δ-function was approximated by $1/\Delta x$ at $x = 0$, and zero elsewhere. The initial conditions were chosen such that the microscopic particle (kink) always centers at $X = -6$, moving toward the impurity with an initial velocity $v_i > 0$. Fixed boundary conditions, $\phi(-40) = 0$ and $\phi(40) = 2\pi$, were used in the simulation. Only the atractive interaction by the impurity, *i.e.*, $\varepsilon > 0$ were considered in the work of Kivshar *et al.*

Depending on whether the microscopic particle passes through the impurity, is captured by the impurity, or is reflected from the impurity, there are three different windows of the initial velocity of the incoming microscopic particle, and a critical velocity v_c. If the initial velocity of the microscopic particle is greater than v_c, it will pass through the impurity inelastically and escape to the positive direction, losing part of its kinetic energy through radiation and excitation of an impurity mode during the process. In this case, there is a linear relationship between the squares of the initial velocity v_i and the final velocity v_e of the microscopic particle, *i.e.*, $v_e = \alpha'(v_i^2 - v_c^2)$, where α' is a constant. For $\varepsilon = 0.7$, the critical velocity is approximately 0.2678, and $\alpha' \approx 0.887$.

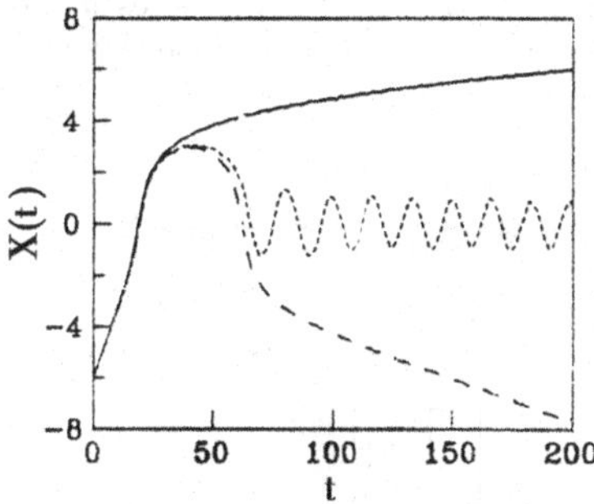

Fig. 4.19 The kink coordinate $X(t)$ versus time for initial velocity, v_i, of the microscopic particle in different windows: passing through (solid line, $v_i = 0.268$), being captured (dotted line, $v_i = 0.257$), and being reflected (dashed line, $v_i = 0.255$).

If the initial velocity of the incoming microscopic particle is smaller than v_c, the microscopic particle cannot escape to infinity from the impurity after the first interaction, but will stop at a certain distance and return, due to the attractive force of the impurity, and to interact with the impurity again. For most of the velocity values, the microscopic particle will lose energy again in the second interaction and eventually get trapped by the impurity (see Fig. 4.19). However, for certain values of the initial velocity, the microscopic particle may escape to the negative infinity after the second interaction, *i.e.*, the microscopic particle may be totally reflected by the impurity (see Figs. 4.19 and 4.20). This effect is very similar to the resonance phenomena in the kink-antikink collision which was explained by the resonant energy exchange mechanism proposed by Campbell *et al.*. The reflection of the particle (kink) is possible only if the initial velocity of the particle is within some resonance windows. By numerical simulation, Campbell *et al.* found eleven such windows. The details are shown in Fig. 4.21.

Similar resonance phenomena for the microscopic particle (kink) – impurity interaction can also be observed in the ϕ^4-model,

$$\phi_{tt} - \phi_{xx}[1 - \varepsilon\delta(x)](\phi - \phi^3) = 0,$$

which is given in the natural unit system. The inelastic interaction of the micro-

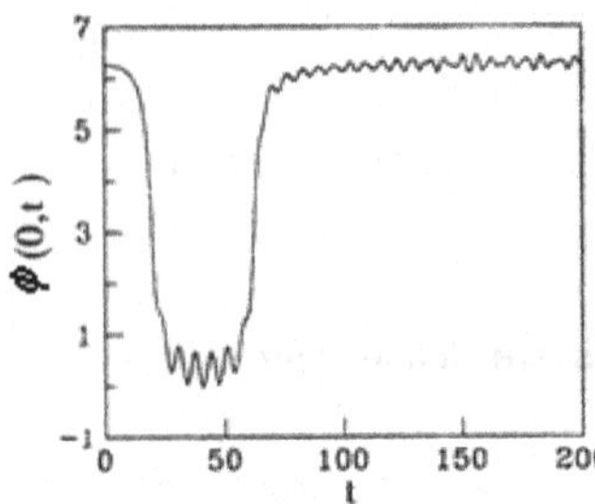

Fig. 4.20 $\phi(0,t)$ versus time in the case of resonance ($v_i = 0.255$). Note that between the two interactions there are four small bumps which show the oscillation of the impurity mode, and after the second interaction the energy of the impurity mode is resonantly transferred back to the microscopic particle (kink).

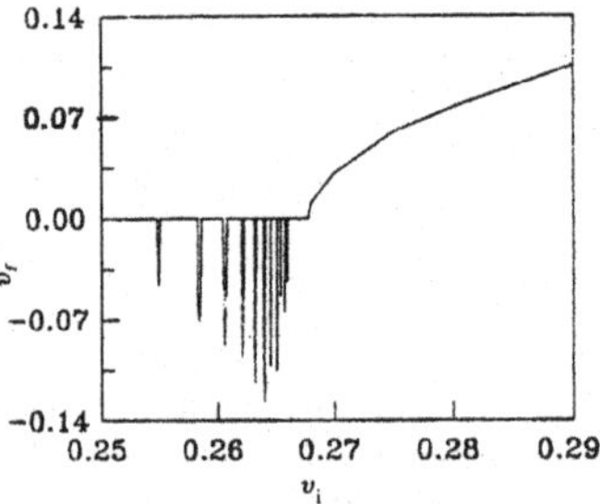

Fig. 4.21 Final velocity of the microscopic particle (kink) as a function of the initial velocity for $\varepsilon = 0.7$. Zero final velocity means that the microscopic particle is captured by the impurity.

scopic particle with an impurity was first studied by Belova *et al.* However, they ignored the impurity mode, and attempted to explain the resonance effects based on energy exchange between the translational mode of the microscopic particle and its internal mode. Later, Kivshar *et al.* studied again the ϕ^4-kink – impurity interaction by intensive numerical simuiation and found that both the internal mode and the impurity mode take part in the resonant interaction. For example, at $\varepsilon = 0.5$, they found six resonance windows below the critical velocity $v_c \approx 0.185$. The resonance structure in the ϕ^4-kink – impurity system has an internal mode which also can be considered as an effective oscillator. These problems can be solved using the collective-coordinate approach taking the three dynamical variables into account.

To summarize, we described here a new type of microscopic particle – impurity interaction when the impurity supports a localized mode. In particular, we demonstrated that a microscopic particle can be totally reflected by an attractive impurity if its initial velocity is in a certain resonance window, which shows that the microscopic particle has similar properties as a classical particle. These resonance phenomena can be explained by the mechanism of resonant energy exchange between the translational mode of the microscopic particle and the impurity mode.

Other phenomena, for example, the transmission, the soliton modes and the impurity modes which occur in the interactions, can be understood by the wave feature of the microscopic particle. Therefore, the scattering of microscopic particles in nonlinear quantum mechanics by impurities also exhibit wave-corpuscle duality.

4.14 Tunneling and Fraunhofer Diffraction

In the last section we discussed the resonant behaviors of microscopic particles through trapping, reflection and excitation of impurity modes, when an microscopic particle is scattered by impurities with an attractive δ-function potential well. The use of the δ-function impurity potential allowed us to investigate the dependence of reflection and capture processes by the impurities on the initial velocity of the incoming microscopic particle. In these discussions, however, we did not consider the effects of finite width of the potential well or barrier. If this is taken into consideration, tunneling of the microscopic particle through the well or barrier would occur due to its wave property in the nonlinear quantum mechanics. Kalbermann studied numerically the interaction of an microscopic particle (kink) in the Sine-Gordon model with an impurity of finite width, and attractive, repulsive or mixed interactions, respectively. In these studies, the Lagrangian function of the system was represented by (in natural units)

$$\mathcal{L} = \partial_\mu \phi \partial^\mu \phi + \frac{1}{4}\Lambda\left(\phi^2 - \frac{m^2}{g}\right)^2, \tag{4.219}$$

where g is a constant, $\Lambda = g + U(x)$, and $U(x)$ is the impurity potential which is given by

$$U(x) = h_1 \cosh^{-2}\left(\frac{x - x_1}{a_1}\right) + h_2 \cosh^{-2}\left(\frac{x - x_2}{a_2}\right). \tag{4.220}$$

Here h_1, h_2 and x_1, x_2 are the strengths and positions of the two impurities, respectively, a_1 and a_2 are some constants. If $h_1 > 0$ and $h_2 < 0$, this potential describes a repulsive impurity and an attractive impurity. The equations of motion were solved by Kalbermann using a finite difference method. A microscopic particle (soliton) was incident from $x = -3$ with an initial velocity v_0, on an impurity located at $x = 3$. The spatial boundaries were taken to be $-40 < x < 40$, with a grid of $dx = 0.04$. A time step of $dt = 0.02$ was used and the simulation was carried out for a total elapsed time of $t = 200$ (or 10000 time steps), which was sufficient to allow for resonant tunneling to decay, permitting a clear definition of the asymptotic behavior of the microscopic particle. The asymptotic velocities for the reflected and transmitted cases were calculated using the actual motion of the center of the microscopic particle.

In his calculation, Kalbermann used $g = m^2$ in (4.219), and 0.7, 1.0, and 1.5 respectively for m, which correspond to the cases where the size of the microscopic particle, $\approx 1/m$, is larger, comparable, and smaller, respectively than the barrier width, $\approx a/6$, where a is the parameter in the argument of $U(x)$ in (4.220). A width of $a_1 = 1$ was assumed for the repulsive barrier while the attractive barrier has a width of $a_2 = 0.3$. These values were chosen to illustrate all effects in a reasonable range of velocity values around $v_0 \approx 0.25$. These considerations, plus trial and error, led to the choices of $h_1 = 1$ and $h_2 = -6$. Fig. 4.22 shows the impinging microscopic particle (soliton) as well as the barriers. Figs. 4.23 – 4.25 show the final velocity v' as a function of the initial velocity v_0 for the repulsive ($h_1 = 1$ and $h_2 = 0$), attractive ($h_1 = 0$ and $h_2 = -6$), and attractive-repulsive cases, respectively.

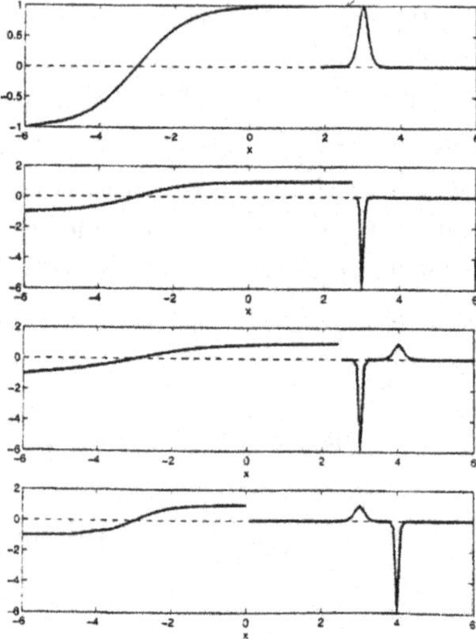

Fig. 4.22 From top to bottom: Kink with $m = 1$ impinging from left onto a repulsive barrier; Kink with $m = 1$ impinging from left onto an attractive impurity; Kink with $m = 0.7$ impinging from left onto an attractive-repulsive system; Kink with $m = 1.5$ impinging from left onto a repulsive-attractive arrangement.

In the repulsive case (Fig. 4.23), the microscopic particle is reflected if the final velocity of the particle $v' < 0$, up to a certain value of the initial speed for which the effective barrier height becomes comparable to the kinetic energy, and then a sudden jump to transmission occurs. In all three cases shown in Fig. 4.23, the transmission starts at the same kinetic energy, with minor differences due to the effective barrier.

In the attractive case there are islands of reflection between trappings and resonant behavior in which the microscopic particle remains inside the impurity and oscillates, exciting the impurity mode as mentioned in the previoius section. Again the higher the mass, the smaller the critical velocity for which transmission starts.

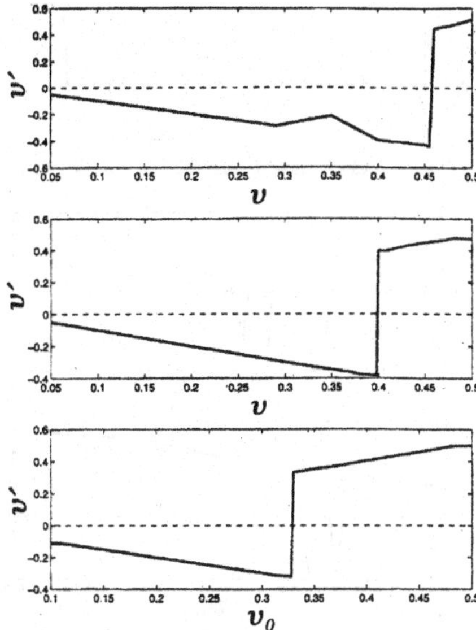

Fig. 4.23 Final velocity v' versus the initial velocity v_0 for $m = 0.7$ (upper panel), $m = 1$ (middle panel) and $m = 1.5$ (lower panel) for the repulsive barrier.

The details of the reflection islands depend strongly on the parameters, but the general trend is similar for all three cases of different masses.

In the case of combination of attractive and repulsive impurities shown in Fig. 4.25, the reflection dominates at low velocities which is induced by the repulsive impurity. Trapping and resonant behavior occur with islands of reflection. At high velocities, the transmission occurs which is essentially dictated by the same impurity. The repulsive-attractive case is similar to the repulsive case for velocity below transimission, and the critical speed here is determined mainly by the attractive impurity that can drag back the microscopic particle (soliton) after it passes through the barrier. It appears that the larger the mass (the thinner the microscopic particle), the more the attractive impurity is capable of trapping, thereby producing a somewhat counterintuitive behavior in which a massive microscopic particle needs a higher initial velocity in order to traverse them. Concerning the duration of the microscopic particle inside the barrier, there is always a time delay in the impurities, in contrast to the quantum-mechanical Hartmann effect. Furthermore, the energy of the microscopic particle is conserved in the scattering.

The above discussion addressed the tunneling effect of microscopic particles in the one-dimensional case. It shows again the wave property of microscopic particles in the nonlinear quantum mechanics. However, in order to relate more closely to actual tunneling of microscopic particles in nonlinear quantum mechanics, one has to consider higher dimensions, such as the $O(3)$ two-dimensional case, and eventually including rotations of the microscopic particle (soliton), and other effects such as

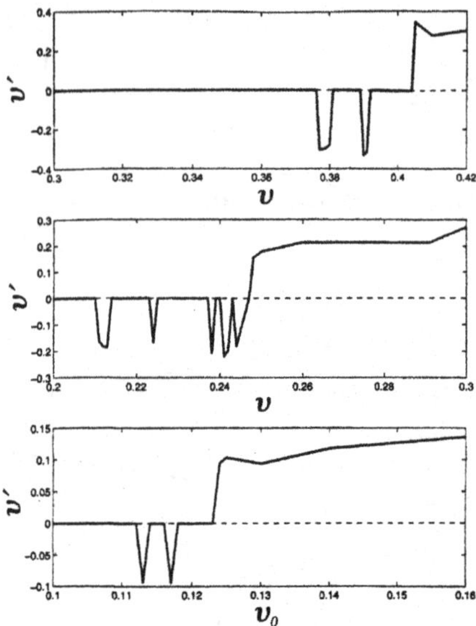

Fig. 4.24 Same as Fig. 4.23 but for the attractive case.

fluctuations. Moreover, actual barriers can be dynamic. There is then a need to allow for more flexibility in the modeling of impurities as well as the possibility of energy dissipation.

In the following, we consider the Fraunhofer diffraction of microscopic particles described by the nonlinear Schrödinger equation (4.40) with a small scale initial phase modulation and a rather large initial intensity. These conditions allow us to use a perturbution approach based on multiscale expansions to investigate such problems, as was done by Konotop.

Following the approach of Lax and the inverse scattering method, the linear spectral problem corresponding to the nonlinear Schrödinger equation satifies the Zakharov-Shabat equation (4.42), where λ is the spectral parameter, $\phi_0(x') = \phi(x', t' = 0)$ being the initial condition for the nonlinear Schrödinger equation (4.40). Konotop assumed the following for for $\phi_0(x')$,

$$\phi_0(x') = b' g(x') f(x'), \tag{4.221}$$

here $g(x')$ is a random and statistically homogeneous function

$$\begin{aligned} \langle g(x') \rangle &= g, \\ \langle g(x') g^*(x'_1) \rangle &= Q_{d_1}(x' - x'_1), \\ \langle g(x') g(x'_1) \rangle &= G_{d_2}(x' - x'_1), \end{aligned} \tag{4.222}$$

where $\langle \cdots \rangle$ denotes the average over all realizations of the random function $g(x')$, d_1 and d_2 are correlation radii (from now on, we assume $d_1 = d_2 = d$ for simplicity),

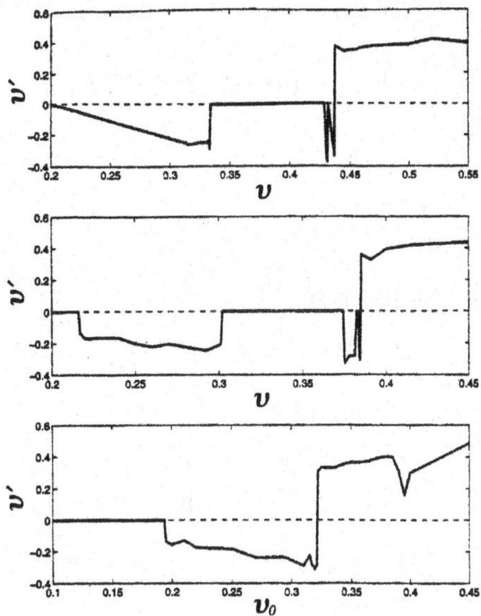

Fig. 4.25 Same as Fig. 4.23 but for the attractive-repulsive case.

$f(x')$ is a regular function varying on a unit scale, $|f(x')| \sim |g(x')| \sim 1$, b' represents the amplitude of the initial pulse and is a nonlinearity parameter.

The solution of the Zakharov-Shabat equation is known to be of the form

$$\Psi = \begin{pmatrix} \Psi_1 \\ \Psi_2 \end{pmatrix} = C(x')\psi_1(x', \lambda) + D(x')\psi_2(x', \lambda), \tag{4.223}$$

with

$$\psi_1(x', \lambda) = \begin{pmatrix} e^{i\lambda x'} \\ 0 \end{pmatrix}, \quad \psi_2(x', \lambda) = \begin{pmatrix} 0 \\ e^{-i\lambda x'} \end{pmatrix}. \tag{4.224}$$

$C(x')$ and $D(x')$ in (4.223) satisfy the "initial" conditions: $C(x' = -\infty) = C_0$, $D(x' = -\infty) = 0$. The reflection coefficient $R(\lambda)$ was determined by Konotop, which is given by

$$R(\lambda) = \lim_{x\to\infty} r(x')e^{-2i\lambda x'}, \quad r(x') = \frac{D(x')}{C(x')}e^{2i\lambda x'}. \tag{4.225}$$

From (4.221), $r(x')$ satisfies the Riccati differential equation

$$\frac{dr}{dx'} = 2i\lambda r + i\phi_0^*(x')b' - i\phi_0(x')r^2. \tag{4.226}$$

Konotop restricted the "potential" $\phi_0(x)$ in the interval $[0, L]$, that is, $\phi_0(x') = 0$ for $x' < 0$ and $x' > L$. In this case, the initial conditions for (4.226) are $r(0) = 0$, and $r(x' = \infty) = r(L)$. Equation (4.226) then becomes

$$\frac{dr}{d\xi'} = i\epsilon\left[Ar + f^*(d\xi')g'^*(\xi') - f(d\xi')g'(\xi')r^2\right], \tag{4.227}$$

where

$$A = \frac{\lambda}{b'}, \quad \epsilon = db', \quad g(\xi') = g(x'), \quad \xi' = \frac{x'}{d}.$$

For small correlation radius $d \ll k^{-1}$ ($\epsilon \ll 1$), $r(\xi')$ may be expressed as a series

$$r(\xi') = r^{(0)}(\xi', \xi_1, \cdots) + \epsilon r^{(1)}(\xi', \xi_1, \cdots,) + \cdots \tag{4.228}$$

where $\xi_i = \epsilon^i \xi'$ and correspondingly

$$\frac{d}{d\xi'} = \frac{\partial}{\partial \xi'} + \epsilon\frac{\partial}{\partial \xi_1} + \cdots. \tag{4.229}$$

In the case of $f(d\xi') = f(x_1)$ (*i.e.* $b' = 1$ and $\epsilon = d$), from (4.227) – (4.229), Konotop obtained

$$\frac{\partial r^{(0)}}{\partial \xi'} = 0, \tag{4.230}$$

and

$$\frac{\partial r^{(1)}}{\partial \xi'} = -\frac{\partial r^{(0)}}{\partial \xi_1} + iAr^{(0)} + if^*(\xi_1)g'^*(\xi') - if(\xi_1)g'(\xi')[r^{(0)}]^2, \tag{4.231}$$

where ξ', ξ_i ($i = 1, 2, \cdots$) are independent variables. Equation (4.230) implies that $r^{(0)} \equiv r^{(0)}(\xi_1)$. An expression for $r^{(1)}$ follows directly from (4.231)

$$r^{(1)} = -\xi'\frac{\partial r^{(0)}}{\partial \xi_1} + iA\xi' r^{(0)} + i\xi' f^*(\xi_1)F^*(\xi') - i\xi' f(\xi_1)F(\xi')[r^{(0)}]^2. \tag{4.232}$$

Here $F(\xi')$ represents the "average"

$$F(\xi') = \frac{1}{\xi'}\int_0^{\xi'} dx' g(x'),$$

introduced by Konotop. $r^{(0)}(\xi_1)$ is then determined by requiring the secular terms on the right hand side of (4.232) to vanish. That is

$$\frac{\partial r^{(0)}}{\partial \xi_1} = iAr^{(0)} + if^*(\xi_1)F^*(\tilde{L}) - if(\xi_1)F(\tilde{L})[r^{(0)}]^2 \tag{4.233}$$

where $\tilde{L} = L/\epsilon$. In the case under consideration the average reflection coefficient is given by $r(L) = r^{(0)}(L) + O(\epsilon^2)$.

Now let

$$G(\xi_1, x') = if^*(\xi_1)F^*(x') - if(\xi_1)F(x')[r^{(0)}]^2,$$

where

$$\begin{cases} \Re G(\xi', x') = G_1, \\ \Im G(\xi', x') = G_2, \end{cases} \quad \text{and} \quad \begin{cases} \Re r^{(1)} = r_1^{(1)}, \\ \Im r^{(1)} = r_2^{(1)}. \end{cases}$$

Then, (4.232) may be rewritten in the following form

$$r_i^{(1)}(\xi') = \xi'[G_i(\xi_1, \xi') - G_i(\xi_1, L)].$$

In the case of stochastic initial conditions, *i.e.*, $g(x')$ is a random function, the reflection coefficient $r_i^{(1)}(\xi')$ is a random function too. Hence, there is

$$\epsilon \left| \left\langle r_i^{(1)}(\xi', \xi_1) \right\rangle \right| \ll 1, \tag{4.234}$$

and

$$\epsilon \sqrt{W[r_i^{(1)}(\xi', \xi_1)]} \ll 1, \tag{4.235}$$

where $W[r] = \langle r^2 \rangle - \langle r \rangle^2$ is the dispersion, and it is given by

$$\begin{aligned} W\left[r_i^{(1)}(\xi')\right] &= W\left[\xi' G_i(\xi_1, \xi')\right] + \frac{\xi'}{\tilde{L}} W\left[\tilde{L} G_i(\xi_1, L)\right] \\ &\quad - \frac{2\xi'}{\tilde{L}} \left[\left\langle \xi' G_i(\xi_1, \xi') \tilde{L} G_i(\xi_1, L) \right\rangle - \langle \xi' G_i(\xi_1, \xi') \rangle \left\langle \tilde{L} G_i(\xi_1, L) \right\rangle \right]. \end{aligned} \tag{4.236}$$

Since our discussion is limited to the statistically homogeneous function $g(x')$ with finite dispersion $|Q(x')| \approx |W(x')| \approx 1$, an estimate for $W[r]$ follows from (4.236), $W[r] \approx \tilde{L}$. Therefore, the inequalities (4.234) and (4.235) may be rewritten in the form of $L \ll \epsilon^{-1}$. If we further assume that $g(x)$ is an ergodic process, the function $F(L)$ reduces to $\langle g(L) \rangle + O(\epsilon L^{-1})$ and $\langle r^{(1)}(L) \rangle = 0$ with an uncertainty of ϵL^{-1}. Konotop finally obtained, from (4.233), the following in the interval $\epsilon \ll L \ll \epsilon^{-1}$,

$$\frac{\partial r^{(0)}}{\partial \xi_1} = iAr^{(0)} + if^*(\xi_1)g^* - if(\xi_1)g(x')[r^{(0)}]^2. \tag{4.237}$$

Konotop also studied Fraunhofer diffraction of an microscopic particle (soliton) which has features of a phase modulated noncoherent wave through a slit. The diffraction pattern detected on the screen is also affected by the nonlinearity of the medium behind the screen which was taken into consideration by Konotop. The corresponding initial condition for the nonlinear Schrödinger equation in this diffraction is taken to be a square potential with a random phase,

$$\phi_0(x') = \begin{cases} \phi_0 e^{i\theta(x') + i\theta_0}, & \text{for } 0 \le x' \le L, \\ 0, & \text{for } x' < 0 \text{ or } x' > L. \end{cases}$$

where $\theta(x')$ is a Gaussian noise with statistical characteristics $\langle \theta(x) \rangle = 0$, $\langle \theta(x)\theta(x') \rangle = Q(x' - x_1')$, $Q(0) = \sigma^2$. For $0 < x' < L$, it is easy to show that

$$\langle \phi_0(x') \rangle = \phi_0 e^{-\sigma^2/2} \quad \text{and} \quad \langle \phi_0(x')\phi_0(x_1') \rangle \approx Q(x' - x_1'). \tag{4.238}$$

If $Q(x') \to 0$ as $x' \to \infty$, the process $\phi_0(x')$ is ergodic, *i.e.*, one can use (4.237) to determine the reflection coefficient

$$r(L) = -\frac{(-i)^{(\delta+1)/2}\phi_0 \tanh\left[i^{(\delta+1)/2}L\Delta_\delta\right] e^{-i\theta_0 - 2i\lambda L}}{(-i)^{(\delta+1)/2}\lambda \tanh\left[i^{(\delta+1)/2}L\Delta_\delta\right] + i\Delta_\delta}, \tag{4.239}$$

where

$$\Delta_\delta = \sqrt{\phi_0^2 e^{-\delta^2} + \delta\lambda^2}, \quad (\delta = \pm 1). \tag{4.240}$$

We can also determine the property of the Fraunhofer diffraction in such a case. In fact, in the diffraction problem, $\delta = 1$ and $\delta = -1$ correspond to focusing and defocusing medium, respectively. The diffraction picture in the Fraunhofer zone is determined by the Manakov's formula

$$|\phi(x', t')|^2 = \frac{1}{4\pi t} \ln\left|1 + \delta |r(L)|^2\right|^\delta. \tag{4.241}$$

From (4.238), (4.240) and (4.241), it follows that the diffraction picture of the phase modulated microscopic particle is similar to that of the regular wave with a smaller amplitude. As far as the phase modulations are concerned, there are no limitations on the phase fluctuation dispersion, *i.e.*, $\langle e^{i[\theta(x')+\theta(x_1')]}\rangle \leq 1$ for any $Q(x' - x_1')$.

Similar results may be obtained for a wave with an amplitude modulation of the type

$$\phi_0(x') = \begin{cases} \phi_0 e^{\theta(x')}, & \text{for } 0 \leq x' \leq L, \\ 0, & \text{for } x' < 0 \text{ or } x' > L. \end{cases}$$

The solution takes the form of (4.238), with $-\sigma^2$ substituted in the place of σ^2. But for the case of a focusing medium one must take the possibility of creation of a microscopic particle (soliton) into account. This leads to an upper bound for σ^2 for applicability of (4.241) in the case of $\delta = 1$.

Equation (4.233) for $r^{(0)}$ does not contain L explicitly. However, if we extend the above results to infinite initial conditions with high decreasing speed, the time interval in which the initial condition $\phi_0(x')$ is essentially nonzero can be used as parameter L.

From the above discussion we see that the Fraunhofer diffraction phenomenon takes place when a microscopic particle described by the nonlinear Schrödinger equation moves in a perturbed systems with random initial condition and a small correlation radius. This phenomenon also occurs for pulse waves with slowly varying initial phase modulations (amplitude and phase modulations) which propagate in a dispersive medium. Once more, this shows the wave nature of microscopic particles described by the nonlinear Schrödinger equation.

Fraunhofer diffraction of microscopic particles, described by (4.40), from a belt in a nonlinear defocusing medium, was also studied by Zakharov and Shabat, under the initial conditions

$$\phi|_{t'=0} = \begin{cases} 0, & \text{for } |x'| < a', \\ 1, & \text{for } |x'| > a'. \end{cases} \tag{4.242}$$

Here t' and x' are regarded as the longitudinal and transverse coordinates, and a' is the half-width of the belt required for the diffraction, the amplitude of the infinite wave is set equal to unity.

In accordance with the results of Section 4.3, it is necessary to solve the eigenvalue problem (3.21) and (4.42) with the function $q = \phi|_{t'=0}$. We have

$$\Psi = \begin{cases} c_1 \begin{bmatrix} 1 \\ 0 \end{bmatrix} e^{-i\lambda x'} + c_2 \begin{bmatrix} 0 \\ 1 \end{bmatrix} e^{i\lambda x'}, & (\text{for } |x'| < a'), \\ c_2 \begin{bmatrix} i(1-\lambda^2)^{1/2} - \lambda \\ 1 \end{bmatrix} e^{-\sqrt{1-\lambda^2}x'}, & (\text{for } x' > a'), \\ c_1 \begin{bmatrix} 1 \\ i(1-\lambda^2)^{1/2} - \lambda \end{bmatrix} e^{-\sqrt{1-\lambda^2}x'}, & (\text{for } x' < -a'), \end{cases} \tag{4.243}$$

where c_1 and c_2 are the Jost coefficients. Requiring the solution (4.243) to be continuous at $x' = \pm a$, we obtain the following eigenvalue equation after some elementary transformations

$$\cos 2\lambda a' = \lambda. \tag{4.244}$$

Equation (4.244) has a set of zeros, $\pm\lambda_n$, which are symmetric and approaching zero, and $|\lambda_n| < 1$, as expected. For sufficiently small a', there is only one pair of zeros. For large a', the number of pairs of zeros, N, can be estimated from $N \approx 2a'/\pi$.

Local darker bands in the diffraction pattern, given by the zeros of (4.244), can be found which correspond microscopic particles (solitons) that propagate along straight lines in the (x', t') plane. The microscopic particle (soliton) with the eigenvalue λ_n moves in a direction that makes an angle $\theta_n = \tan^{-1}(2\lambda_n)$ with the direction of propagation of the wave. The minimum number of such bands is two. Thus, the diffraction of a microscopic particles by a belt in a nonlinear medium differs in principle from the diffraction in a linear medium in that it can be observed at an arbitrarily large distance from the belt, whereas in a linear medium the diffraction picture becomes "smeared out" at large distances.

As a matter of fact, the Fraunhofer diffraction for superconducting electron (soliton) was experimentally observed in the superconducting junctions (See Fig. 2.3). Therefore, it is confirmed beyond doubt that Fraunhofer diffraction can occur for microscopic particles in nonlinear quantum mechanics.

4.15 Squeezing Effects of Microscopic Particles Propagating in Nonlinear Media

Interferometer experiments of Wu *et al.* showed that propagation of a laser field along an optical fiber produces a squeezing effect, which refers to the reduction of narrow-band quantum fluctuation in one of the quadrature phases of the field to a level below the usual limit for a coherent state. The squeezing of a laser field can be induced by four-wave mixing, resulting from quantum noise. Calculation on the quantum fluctuations in a dispersionless medium predicted the squeezing effect in one of the quantum phases of a propagating continuous wave (CW) field that is consistent with experimental results.

In nonlinear quantum mechanics such squeezing phenomena also occur for quantum microscopic particles described by the nonlinear Schrödinger equation (3.62), *i.e.*, propagation of a microscopic particle in a dispersive nonlinear medium can lead to a wide-band squeezing of quantum fluctuations in the vicinity of the microscopic particle (soliton). This problem was studied by Carter *et al.*. In their work, source of quantum noise was added in the original nonlinear Schrödinger equation as a stochastic field, and the nonlinear Schrödinger equation becomes the following stochastic form

$$i\frac{\partial\phi(x',t')}{\partial t'} = \frac{1}{2}\left[1 \pm \frac{\partial^2}{\partial x'^2}\right]\phi(x',t') - |\phi|^2\phi(x',t') + i\sqrt{\frac{i}{\bar{n}}}\eta(x',t')\phi(x',t'), \quad (4.245)$$

where $\bar{n} = k''/bt'_0$ is the average number of excitation of the particle field, or photon number in the CW field, $k'' = \partial^2 k/\partial\omega^2|_{k_0}$ is the dispersion of overall medium group velocity, $t'_0 = \sqrt{v_g|k''|/(2\Delta\omega)}$ is the time scale, b is the effiective nonlinearity, v_g is the group speed of the microscopic particle (soliton). The real and stochastic field η and η^+ are introduced here with

$$\begin{gathered}\langle\eta\left(x'_1,t'_1\right)\eta\left(x'_2,t'_2\right)\rangle = \langle\eta^+\left(x'_1,t'_1\right)\eta^+\left(x'_2,t'_2\right)\rangle = \delta\left(x'_1 - x'_2\right)\delta\left(t'_1 - t'_2\right), \\ \langle\eta^+(x'_1,t'_1)\eta(x'_2,t'_2)\rangle = 0,\end{gathered} \quad (4.246)$$

where the $(+,-)$ signs in (4.245) correpond to the cases of normal ($k'' > 0$) and anormalous ($k'' < 0$) dispersions, respectively. Equation (4.245) can be derived from quantum theory of propagation in dispersive nonlinear media in which quantum fluctuations are handled via the coherent-state positive-P representation of Carter *et al.* In (4.245), the stochastic part provides the quantum fluctuations in the problem, and is the source of inherently squeezed quantum noise. This term is easily modified to include thermal phase-modulation noise.

Carter *et al.* first studied the squeezing effect of a CW field. They first defined the spectrum of quadrature fluctuations in the CW field at location x' with a phase angle θ, following the approach of Gardiner *et al.*,

$$S(\bar{\omega}, x', \theta) \equiv \frac{4\pi\bar{n}t_0'}{T_p}\Re\left[e^{-2i\theta}\langle\Delta\tilde{\phi}(\bar{\omega}, x')\Delta\tilde{\phi}^+(-\bar{\omega}, x')\rangle\right. \\ \left. +\langle\Delta\tilde{\phi}(\bar{\omega}, x')\Delta\tilde{\phi}^+(-\bar{\omega}, x')\rangle\right]. \tag{4.247}$$

A similar equation for $\Delta\tilde{\phi}^+$ can be obtained

$$\Delta\tilde{\phi}(\bar{\omega}, x') \equiv \frac{1}{\sqrt{2\pi}}\int_{-T/2t_0'}^{T/2t_0'} \Delta\phi(t', x')e^{i\bar{\omega}t'}dt',$$
$$\Delta\phi(t', x') = \phi(t', x') - \langle\phi(t', x')\rangle.$$

Squeezing occurs when $S < 0$. In (4.247), T_p represents a pulse duration defined for normalization purpose. In the CW case, $T_p = T$, which is the total observation time. The quantum limit of squeezing is given by $S \geq -1$.

When one deals with relatively small quantum fluctuations, the results obtained from a linearized fluctuation equation are good approximations. Hence we write ϕ as $\phi_0 + \Delta\phi$. Here $\phi_0 = \langle\phi\rangle$ is a first order approximate classical solution to the nonlinear Schrödinger equation. It corresponds to a coherent-state input of $\phi_0(\tau, 0)$ at $x' = 0$. Linearization requires $\bar{n} \gg 1$, although squeezing does not depend on $\bar{n}$ initially. For large propagation distances ($x' \gg 1$), linearization can break down even for $\bar{n} \gg 1$, if there is exponential growth in fluctuations.

Considering now the CW field, the constant solution is $\phi_0^2 = 1/2$. This implies an input power of $h\omega_0\bar{n}/2t_0$, where ω_0 is a renormalized frequency. The constant solution is therefore applicable to any CW of input power W provided that $\Delta\omega = v_g Wb/(\hbar\omega_0)$, with $v_g = (\partial\omega/\partial k)|_{k_0}$ being the group velocity of the moving frame. The resultant squeezing at the phase for the maximum fluctuation reduction is

$$S(\bar{\omega}, x')_{\max} = \frac{1-\cos(\gamma x')}{\gamma^2} - \left|\frac{i\sin(\gamma x)}{\gamma} + \frac{\beta[\cos(\gamma x') - 1]}{\gamma^2}\right|, \tag{4.248}$$

where $\beta = 1 \pm \bar{\omega}^2$; $\gamma = \bar{\omega}\sqrt{\bar{\omega}^2 \pm 2}$. This expression can also be derived using operator techniques, and in the nondispersive limit it gives a rigorous basis. The spectrum, in the anomalous-dispersion case, is shown in Fig. 4.26. It has an exponential fluctuation reduction at finite frequency ($\bar{\omega}^2 < 2$), with perfect squeezing in the limit of $x' \to \infty$. This implies exponential growth in the complementary quadratures, so that linearization will not be valid for sufficiently large x'. Similar modulation instabilities are known to exist for propagation of a classical CW.

In the case of a microscopic particle (soliton) in nonlinear quantum mechanics, an identical computational technique can be used. Carter *et al.* extended the above study to the case of a known classical soliton solution. Here the fluctuations are defined relative to a time-dependent solution of $\phi_0(t') = \text{sech}(t')$. $\Delta\omega$ is chosen to give a characteristic time scale, t_0, corresponding to the size of the microscopic particle. The peak power is $\hbar\omega_0|k''|/bt_0^2$. The linearized equations are similar to

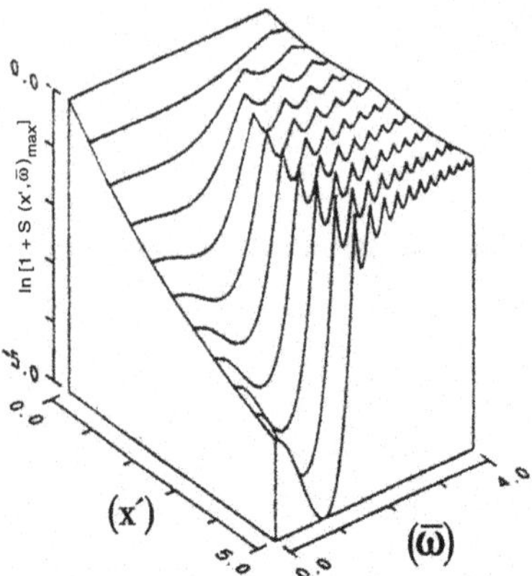

Fig. 4.26 Graph of the logarithmic spectrum $\ln[1 + S(x', \bar{\omega})_{\max}]$ of maximum squeezing in the case of anomalous-dispersion.

those given above in the case of CW, except for the time-dependent four-wave mixing and stochastic terms. These can be treated numerically as a set of coupled linear differential equations in x'. The maximum-squeezing curves are shown in Fig. 4.27. The results clearly demonstrate a wide-band squeezing over the spectral width of the input microscopic particle (soliton). The phase angle of the largest squeezing is approximately $\pi/4$.

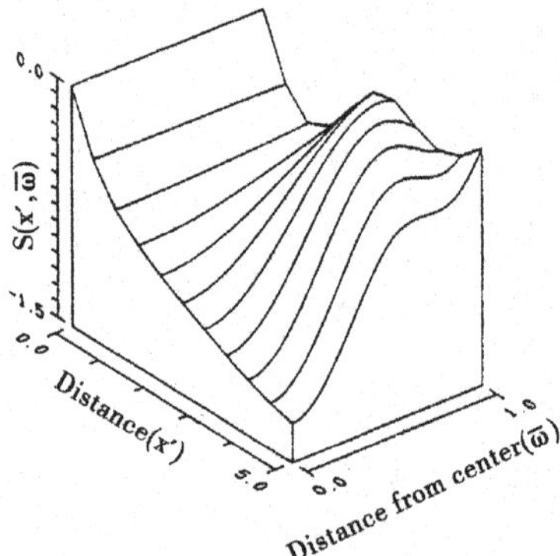

Fig. 4.27 Graph of the spectrum of maximum squeezing in the case of anomalous-dispersion microscopic particle (soliton).

Note that quantum limits in the CW and microscopic particle in nonlinear quantum mechanics are somewhat different. The fluctuations are now localized over a time interval $\approx 2t_0$ near the mass center of the microscopic particle. This corresponds to squeezing in a localized mode of the radiation field copropagating with the microscopic particle. Hence the pulse duration used here to normalize the spectrum is $T_p = 4t_0$, which is approximately the size of a classical microscopic particle (soliton). This is necessary in order to obtain a finite result for a pulsed field. Squeezing can also be defined relative to more general mode functions.

In summary, the quantum noise source is added into the general nonlinear Schrödinger equation as a stochastic field, which reduces the quadrature fluctuations in one quantum at a phase of $\pi/4$ relative to the carrier. This intrinsic squeezing initially increases linearly with distance. At high relative frequencies, the effects of linear dispersion cause a phase rotation of the squeezing generated at different spatial locations, which causes interference in the spectrum, and reduced squeezing. For long propagation distances the quantum noise undergoes further modifications due to four-wave mixing and nonlinear dispersion. In the normal-dispersion regime, this reduces the extent of the squeezing. In the anomalous-dispersion regime, there is a range of frequencies where the linear dispersion counterbalances the nonlinear dispersion. This results in enhanced squeezing and exponential noise reduction with distance.

These general comments of the squeezing effect hold for both CW and the microscopic particle (soliton) inputs. This squeezing effect of the solitons shows that a microscopic particle has wave property in nonlinear quantum mechanics, because only wave is known to have the squeezing property. When the input is CW, the results are well known. In the case of a coherent microscopic particle (soliton) input, the results were not obtained before. These demonstrate the squeezing of quantum fluctuations below the vacuum level in a microscopic particle (soliton). Therefore, this result suggestes that the squeezing is an universal property of propagation of microscopic particles described by the nonlinear Schrödinger equation in quantum limit.

4.16 Wave-corpuscle Duality of Microscopic Particles in a Quasiperiodic Perturbation Potential

The behaviors of microscopic particles subject to arbitrary perturbations are very complex in general. However, for certain types of perturbations, sufficiently simple behavior can be expected even for relatively strong perturbations, which allow an analytic description in terms of a small number of collective variables. In nonlinear systems there always exist inhomogeneities or disorder, and the interactions involved will certainly have an anharmonic component. These can give rise to interconversion of excitations. Thus the microscopic particle (soliton) can emit radiation which can self-focus to form coherently localized excitations. Meanwhile the microscopic particles can be scattered inelastically or be broken up. In this section, we examine the behaviors of microscopic particles in nonlinear quantum mechanics subject to a spatially quasiperiodic perturbation potential in nonlinear systems. This problem was studied by Scharf *et al.*

The perturbed nonlinear Schrödinger equation (3.2) in such a case can be written as

$$i\phi_{t'} - \phi_{x'x'} + 2|\phi|^2\phi = \varepsilon V(x')\phi, \tag{4.249}$$

where $V(x') = \cos(kx')$ and ε is a small parameter. As mentioned above, for a single localization of a microscopic particle moving in the presence of the perturbation, we make a collective variable ansatz

$$\phi(x', t') = 2\eta e^{iqx'/2-i\theta}\mathrm{sech}[2\eta(x'-q)], \tag{4.250}$$

where q is the location of the microscopic particle (soliton). Equation (4.249) possesses two integrals of motion, the norm N and energy E. For $\phi(x,t)$ given in (4.250), they are given by

$$N = \int_{-\infty}^{+\infty} |\phi|^2 dx' = 4\eta, \tag{4.251}$$

and

$$\begin{aligned} E &= \int_{-\infty}^{+\infty} \left[|\phi_{x'}|^2 - |\phi|^4 + \varepsilon|\phi|^2\cos(kx')\right] dx' \\ &= \eta\left[\dot{q}^2 - \frac{16}{3}\eta^2\right] + \frac{\varepsilon k\pi}{\sinh(k\pi/4\eta)}\cos(kq), \end{aligned} \tag{4.252}$$

respectively.

The conservation of the norm of ϕ leads to η = const. + $O(\varepsilon^2)$. The time dependence of the phase θ is decoupled from the time dependence of the position q of the microscopic particle, which is the relevant dynamic variable. The total energy given above depends only on the position $q(t)$ of the microscopic particle. Conservation of the total energy E leads to an equation of motion for q which can be derived from the following effective single-particle Hamiltonian

$$H_1 = \frac{p^2}{2\eta} + \frac{\varepsilon k\pi}{2\sinh(k\pi/4\eta)}\cos(kq). \tag{4.253}$$

The effective potential of the particle is

$$V_{\text{eff}}(q) = \varepsilon k\pi\ \mathrm{csch}\left(\frac{\pi k}{4\eta}\right)\cos(kq), \tag{4.254}$$

which is derived from

$$V_{\text{eff}}(q) = 4\eta^2\varepsilon\int_{-\infty}^{\infty} \mathrm{sech}[2\eta(x-q)]V(x).$$

Scharf *et al.* found the following time-dependent equation for the collective coordinate

$$\ddot{q}(t) = -\frac{1}{2\eta}V'_{\text{eff}}(q), \tag{4.255}$$

where $V'_{\text{eff}} = dV_{\text{eff}}/dt$.

Equation (4.255) resembles (4.17) and (4.25). It shows that the microscopic particle satisfies the classical equation of motion, *i.e.*, the microscopic particle has corpuscle property just as classical particles, and the mass of the particle is 2η in such a case. Equation (4.254) shows also that a periodic potential with short wavelength becomes an effective potential which is exponentially decreasing for large k/η in such a case.

We now neglect the perturbation ($\varepsilon = 0$) and focus on a collision between two microscopic particles (solitons). When the two microscopic particles collide their centers of inertia, q_i, as well as their phases θ_i, suffer a shift as mentioned earlier. As the perturbed nonlinear Schrödinger equation (4.249) is $U(1)$ invariant, we neglect the dynamics of the phases. The shifts of the positions q_i ("space shifts") of the microscopic particles amount to an attractive interaction between the microscopic particles. For example, the microscopic particle that was on the right hand side at $t \to -\infty$ and on the left hand side at $t \to +\infty$ has the form

$$\lim_{t' \to \pm\infty} |\phi| = \operatorname{sech}[2\eta(x' - q) \pm a].$$

For the other microscopic particle, the shift a should be replaced by $-a$. The shift a is given by

$$a = \frac{1}{2} \ln \left[\frac{16(\eta_1 + \eta_2)^2 + (v_1 - v_2)^2}{(v_1 - v_2)^2} \right], \tag{4.256}$$

where v_1 and v_2 are the asymptotic velocities of the two separated microscopic particles and η_1 and η_2 their amplitude parameters [see (4.250)]. A simple calculation shows that the following two-particle Hamiltonian gives rise to the same space shifts as (4.256)

$$H_2 = \frac{p_1^2}{2\eta_1} + \frac{p_2^2}{2\eta_2} - 8\eta_1\eta_2(\eta_1 + \eta_2) \operatorname{sech}^2 \left[\frac{2\eta_1\eta_2(q_1 - q_2)}{\eta_1 + \eta_2} \right]. \tag{4.257}$$

Combining the single-particle Hamiltonian H_1 with the attractive two-microscopic particles interaction in (4.257), we can find the following effective N-particle Hamiltonian:

$$\begin{aligned} H_{\text{eff}} = & \sum_{i=1}^{N} \left[\frac{p_i^2}{2\eta_i} + \frac{\varepsilon k\pi \cos(kq_i)}{2\sinh(k\pi/4\eta_i)} \right] \\ & -8 \sum_{1 \le i < j \le N} \eta_i\eta_j(\eta_i + \eta_j) \operatorname{sech}^2 \left[\frac{2\eta_i\eta_j(q_i - q_j)}{\eta_i + \eta_j} \right]. \end{aligned} \tag{4.258}$$

This Hamiltonian is similar to that describing the motion of N anharmonically coupled nonlinear pendula which becomes nonintegrable for $N = 2$.

The microscopic particle (soliton) depicted by (4.250) has a spatial width and thereby selects a certain length scale. Thus two parameters, the size of the microscopic particle, L_s, and the width of the perturbation potential, L_p, respectively, can be introduced. Two situations, $L_s > L_p$ and $L_s < L_p$, can be distinguished. In the former, the microscopic particle (soliton) covers many wiggles of the potential, while in the latter the microscopic particle experiences only negligibe potential differences over its size. In the intermediate case, $L_s \approx L_p$, the behavior of the microscopic particle is much more complicated and bears no resemblance to the unperturbed dynamics.

Scharf *et al.* carried out numerical simulation of (4.249) – (4.258) which gave rich and fascinating behaviors of the microscopic particle. In summary, if the width of the microscopic particle (soliton) is small compared to the length of perturbation potential, it is natural to neglect all degrees of freedom of the microscopic particle besides its center of mass motion and therefore treat it as a particle. In contrast to the full dynamics this reduced dynamics will still be integrable for the nonlinear systems. Its properties are governed by an effective Hamiltonian (4.258) or (4.253) and (4.257). If the width of a microscopic particle is large compared to the typical lengthscale of the perturbation, then one can expect the effect to be an average over the fine spatial details of the potential. Thus the microscopic particle shows obvious wave property. This was verified by detailed investigations.

Starting from (4.249), with $V(x') = \cos(kx')$ and with a large microscopic particle ($\eta \ll k$) of the form given by (4.250), Scharf calculated the correction to (4.250) due to the perturbation up to the second order in the small parameter εk^{-2}. He found that

$$\phi_\varepsilon(x', t') = \phi_0(x', t')\left[1 + \frac{\varepsilon}{k^2}\chi(kx')\right],$$

with $\phi_0(x', t')$ given by (4.250) and the spatial modulation

$$\chi(kx') = -\frac{\cos(kx') - i(\dot{q}/k)\sin(kx')}{1 - (\dot{q}/k)^2}. \tag{4.259}$$

The bare soliton $\phi_0(x', t')$ of the unperturbed equation acquires a dressing in the presence of the perturbation. It can be shown that the dressed microscopic particle (soliton) fulfills an unperturbed effective nonlinear Schrödinger equation with renormalized parameters. The dressed microscopic particles behave like bare solitons when subject to additional long-wavelength perturbation. Again, their center of mass motion can be described by an effective single particle Hamiltonian. When two dressed microscopic particles collide they reemerge essentially unchanged thereby illustrating the near-integrability of the dynamics. As can be seen from (4.259) there are two types of dressed microscopic particles, namely the slow and the fast particles depending on whether $\dot{q} \ll k$ or $\dot{q} \gg k$. The slow dressed microscopic particles are spatially modulated, with the maxima appearing at minima of the perturbing potential, while the maxima of the fast dressed microscopic particles occur at the maxima of the potential. These two regimes are devided by a "phase resonance" leading to a destruction of the microscopic particles.

This analysis has been extended to more general potentials with a quasiperiodic part containing many short wavelengths and an arbitrary long wavelength part. The resulting solitary solutions are dressed microscopic particles (solitons) of more complicated forms moving like particles in a long wavelength effective potential. Lengthscale competition in general leads to complicated behavior in space and time. In the case of the nonlinear Schrödinger equation this can happen either through the "phase resonance" ($\dot{q} \approx k$) as mentioned above or a "shape resonance" ($\eta \approx k$).

The effective particle approximation for the microscopic particles in the nonlinear quantum mechanics shows that the nonintegrability of the perturbed nonlinear Schrödinger equation can manifest itself through "microscopic particle (soliton) chaos" induced by the long-wavelength part of the perturbation. The effective decoupling of a few degrees of freedom leads to simple dynamics of the microscopic particle through nonintegrable behavior in space and time. Inelastic effects can show up in processes involving only few effective coordinates or many degrees of freedom. Larger curvature of the long-wavelength potential (*i.e.*, $|V''(x')| \approx \eta^2|$) can lead to a decay of the microscopic particles through radiation. The radiative power can be calculated using perturbation theory.

Collisions of two microscopic particles with nearly vanishing relative velocity in a perturbing potential show effects no longer described by the effective two-particle Hamiltonian H_{eff} given in (4.258). The soliton parameters η_i which played the role of masses in the collective variable description are no longer constant, but become dynamic variables themselves. As the amount of radiation generated might still be negligible an extended collective variable description seems to be feasible. Simultaneous collision of three and more microscopic particles in the presence of a perturbation lead to effects which in general can no longer be treated by effective two-particle interactions. The power of radiation generated by "dressed microscopic particles" appears to be of higher order than η/k, as concluded from numerical simulation. Collisions of two "dressed microscopic particles" in the case of the nonlinear Schrödinger equation as well as kinks and antikinks for Sine-Gordon equation lead to a noticible increase in the power of radiation, probably to the order of η/k. These phenomena are due to the wave feature of the microscopic particles. Therefore, the above discussion gave evidence to the wave-corpuscle duality of microscopic particles in the nonlinear quantum mechanics.

In order to justify this statement, Scharf *et al.* carried out further studies on the properties of microscopic particles described by the following perturbed Sine-Gordon equation (using the natural unit system)

$$\phi_{tt} - \phi_{xx} + [1 + \varepsilon \cos(kx)] \sin\phi = 0. \tag{4.260}$$

This equation of motion is generated by the Hamiltonian

$$H = \int_{-\infty}^{+\infty} dx \left\{ \frac{1}{2}\phi_t^2 + \frac{1}{2}\phi_x^2 + [1 + \varepsilon \cos(kx)](1 - \cos\phi) \right\}. \tag{4.261}$$

They used two different kinds of exact solutions of the unperturbed Sine-Gordon equation ($\varepsilon = 0$) as initial conditions, the breathers and the kink-antikink (K-$\bar{K}$) solutions. The breather at rest has the form

$$\phi^{br}(x,t) = 4\tan^{-1}\left\{\tan\mu \frac{\sin[(t-t_0)\cos\mu]}{\cosh[(x-x_0)\sin\mu]}\right\}, \tag{4.262}$$

where μ is the parameter governing the breather shape and size. As $\mu \to 0$, the breather becomes shallower, its frequency, given by $\omega_{br} = \cos\mu$, grows, and it can be effectively described by the nonlinear Schrödinger equation. On the other hand, when $\mu \to \pi$, the breather frequency goes to zero and it is actually very close to the K-$\bar{K}$ pair. As for the K-$\bar{K}$ solution, its expression at rest is given by

$$\phi^{K\text{-}\bar{K}}(x,t) = 4\tan^{-1}\left\{\frac{\sinh[\gamma v(t-t_0)]}{v\cosh[\gamma(x-x_0)]}\right\}, \tag{4.263}$$

with $v^2 < 1$, $\gamma \equiv 1/\sqrt{1-v^2}$. Equation (4.263) can also be obtained from (4.262), by letting $\mu = \pi/2 + i\sigma'$, where $e^{\sigma'} = \gamma(1+v)$. Finally, both (4.262) and (4.263) can be derived using standard methods which will be discussed in Chapter 8.

The energy of an unperturbed breather at rest, as can be obtained from (4.260) and (4.261), is $E_0^{br} = 16\sin\mu$, whereas the same computation for an unperturbed K-$\bar{K}$ solution with its center of mass at rest is $E_0^{K\text{-}\bar{K}} = 16\gamma$. In the absence of any perturbation the energies for the breather and the K-$\bar{K}$ solutions at rest obviously fulfill $E_0^{br} < 16 < E_0^{K\text{-}\bar{K}}$. The perturbation may shift this borderline between the breather and the K-$\bar{K}$ solutions. Of relevance to this analysis will be the potential energy of an excitation in the absence of any perturbation

$$E^0_{\text{pot}} = \int_{-\infty}^{+\infty} dx \left(\frac{1}{2}\phi_x^2 + 1 - \cos\phi\right).$$

The amount by which the total energy is changed in the presence of the perturbation is given by

$$V_{\text{eff}} = \int_{-\infty}^{+\infty} dx'\varepsilon\cos(kx')(1-\cos\phi). \tag{4.264}$$

For a breather at rest at $t_0 = 0$, Sharf *et al.* obtained

$$E^0_{\text{pot}} = 8\sin\mu + \frac{8\tanh^2 z}{\sin\mu} - \frac{8z\sin\mu(1-\cot^2\mu\sinh^2 z)}{\sinh z\cosh^3 z},$$

and

$$V^{br}_{\text{eff}}(x_0, z) = \frac{4\pi\varepsilon\sinh z\cos(kx_0)}{\sin\mu\cosh^2 z\ \sinh(K\pi/2)}\left[\frac{\sin(Kz)}{\cosh z} + K\cos(Kz)\sin z\right],$$

where $K \equiv k/\sin\mu$ and z, defined by $\sinh z \equiv \tan\mu\sin(t\cos\mu)$, is a measure of the distance between the kink and the antikink bound in the breather.

For a K-$\bar{K}$ solution at rest at $t_0 = 0$, they found correspondingly

$$E^0_{\text{pot}} = 8\gamma + \frac{8\gamma z(1 + v^2 \sinh^2 z)}{\sinh z \cosh^3 z},$$

and

$$V^{K\text{-}\bar{K}}_{\text{eff}}(x_0, z) = \frac{4\pi\varepsilon \sinh z \cos(kx_0)}{\gamma \cosh^2 z \, \sinh(K\pi/2)} \left[\frac{\sin(Kz)}{\cosh z} + K\cos(Kz)\sin z\right], \tag{4.265}$$

where $K = k/\gamma$ and z, defined by $\sinh z \equiv v^{-1}\sinh(\gamma vt)$, is a measure of the distance between the kink and the antikink.

The effective potential V_{eff} depends on the distance between the (virtual) kink and antikink, z, and on the center of mass, x_0, of the K-$\bar{K}$ solution or the breather. For nonrelativistic velocities ($\gamma \approx 1$), $V_{\text{eff}}(x_0, z)$ can be used to calculate the influence of the potential on the motion of the center of the excitation as well as on the relative distance between the kink and the antikink in an adiabatic approximation, assuming that the parameter z can be considered as a second collective variable.

A breather is a bound state solution of the Sine-Gordon equation which oscillates around $\phi = 0$ with a period $T = 2\pi/\cos\mu$. For sufficiently strong perturbations the total potential energy $E_{\text{pot}} = E^0_{\text{pot}} + V_{\text{eff}}$ can have other minima besides $\phi = 0$ around which the solution can oscillate.

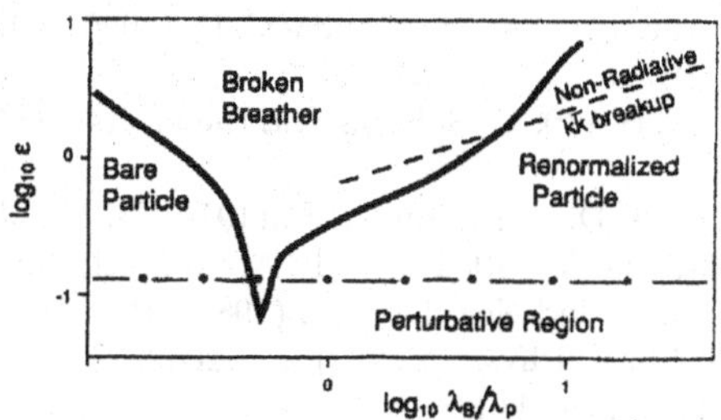

Fig. 4.28 Schematic "phase diagram" for the microscopic particle in the perturbed Sine-Gordon equation.

Scharf *et al.* introduced two characteristic length scales, which are the breather width (λ_B) and the perturbation period (λ_p) and carried out a number of numerical simulations spanning large intervals of the parameters, namely, the length ratio (λ_B/λ_p) and the potential strength (ε). They found that there are basically three possible behaviors for the breather, largely independent of its amplitude. If $\lambda_B < \lambda_p$ (see Fig. 4.28) the breather can indeed be considered as a particle in the external periodic potential, and if $\lambda_B > \lambda_p$, this is still valid except that now the effective potential in which the particle moves is not the original but a renormalized one. On the contrary, if $\lambda_B \approx \lambda_p$ (competing lengths), this particle-like behavior ceases to be true even for small ε values, and the breather rapidly breaks up, either into a kind-antikink (K-$\bar{K}$) pair (if its amplitude is large enough) or into two or more breathers, involving always a great amount of radiation. Interestingly, they also found that breather breakup happens also for noncompeting lengths when ε is above a certain threshold ε_τ, which depends on λ_B/λ_p. They observed nonradiative splitting of large-amplitude breathers into K-$\bar{K}$ pairs for large λ_B/λ_p, which seems to have its origin in energetic considerations and not in the length competition. All these phenomena are smoothed out by dynamic effects, when the breathers are moving, as observed in defail in the nonlinear Schrödinger equation. Therefore, the features shown in Fig. 4.28 indicate that microscopic particles described by the Sine-Gordon equation and the nonlinear Schrödinger equation have wave-corpuscle duality because they have either corpuscle feature in one case, or wave feature in other case. Thus it gives us a satisfactory answer to the question of this chapter - whether microscopic particles have the wave-corpuscle duality in nonlinear quantum mechanics.

Bibliography

Abdullaev, F. (1994). Theory of solitons in inhomogeneous media, Wiley and Sons, New York.

Abdullaev, F., Bishop, A. R. and Pnevmtikos, St. (1991). Nonliearity with disorder, Springer-Verlag, Berlin.

Abdullaev, F., Darmanyan, S. and Khabibullaev, P. (1990). Optical solitons, Springer, Berlin.

Ablowitz, M. J. and Segur, H. (1981). Solitons and the inverse scattering transform, SIAM, Philadelphia.

Ablowitz, M. J., Kruskal, M. D. and Ladik, J. F. (1979). SIAM. J. Appl. Math. **36** (428) .

Abrowitz, M. J. and Kodama, Y. (1982). Study Appl. Math. **66** (159) .

Aceves, A. B., Moloney, J. V. and Newell, A. C. (1989). Phys. Rev. A **39** (1809) and 1839.

Alonso, L. M. (1983). J. Math. Phys. **24** (982) and 2652.

Alonso, L. M. (1984). Phys. Rev. D **30** (2595) .

Alonso, L. M. (1985). Phys. Rev. Lett. **54** (499) .

Anandan, J. (1981). Phys. Rev. Lett. **47** (463) .

Andenson, M. H.,*et al.* (1995). Science **269** (198) .

Andersen, D. (1979). Phys. Scr. **20** (345) .

Andersen, D. R. and Regan, J. J. (1989). J. Opt. Soc. Am. **B6** (1484) .

Andersen, D. R., Cuykendall, R. and Regan, J. J. (1988). Comput. Phys. Commun. **48** (255) .

Anderson, D. and Lisak, M. (1985). Phys. Rev. A **32** (2270) .

Anderson, D. and Lisak, M. (1986). Opt. Lett. **11** (174) .

Aossey, D. W., Skinner, S. R., Cooney, J. T., Williams, J. E., Gavin, M. T., Amdersen, D. R. and Lonngren, K. E. (1992). Phys. Rev. A **45** (2606) .

Bass, F. G., Kirshar, Yu. S. and Konotop, V. V. (1987). Sov. Phys.-JETP **65** (245) .

Berryman, J. (1976). Phys. Fluids **19** (771) .

Bishop, A. R. (1978). Soliton and physical perturbations, in Solitons in actions, eds. by K. Lonngren and A. C. Scott, Academic Press, New York, pp. 61-87.

Boyd, J. P. (1989). Chebyshev and Fouier Spectral methods, Springer, New York, p.792.

Campbell, D. K., Peyrard, M. and Sodano, P. (1986). Physica **D19** (165) .

Campbell, D. K., Schonfied, J. E. and Wingate, C. A. (1983). Physica **D9** (1) and 33.
Carter, S. J., Drummond, P. D., Reid, M. D. and Shelby, R. M. (1987). Phys. Rev. Lett. **58** (1843) .
Castro Neto, A. H. and Caldeira, A. O. (1991). Phys. Rev. Lett. **67** (1960) .
Castro Neto, A. H. and Caldeira, A. O. (1992). Phys. Rev. B **46** (8858) .
Castro Neto, A. H. and Caldeira, A. O. (1993). Phys. Rev. E **48** (4037) .
Chen, H. H. (1978). Phys. Fluids **21** (377) .
Chen, H. H. and Liu, C. S. (1976). Phys. Rev. Lett. **37** (693) .
Cooney, J. L., Gavin, M. T. and Lonngren, K. E. (1991). Phys. Fluids **B3** (2758) .
Cooney, J. L., Gavin, M. T., Williams, J. E., Aossey, D. W. and Lonngren, K. E. (1991). Phys. Fluids **B3** (3277) .
Currie, J., Trullinger, S., Bishop, A. and Krumhansl, J. (1977). Phys. Rev. B **15** (5567) .
Davydov, A. S. and Kislukha, N. I. (1976). Sov. Phys.-JETP **44** (571) .
Desem, C. and Chu, P. L. (1987). Opt. Lett. **12** (349) ; IEE. Proc.-J. Optoelectron. **134** (145) .
Desem, C. and Chu, P. L. (1992). Soliton-soliton interactions in optical solitons, ed. Taylor,J. R., Cambridge University Press, Cambridge, pp. 107-351.
Dziarmaga, J. and Zakrzewski, W. (1999). Phys. Lett. A **251** (193) .
Faddeev, L. (1958). Dokl. Akad. Nauk SSSR **121** (63) .
Fei, Z., Kivshar, Yu. S. and Vazquez, L. (1992). Phys. Rev. A **46** (5214) ; ibid **45** (6019) .
Fogel, M. B., Trullinger, S. E., Bishop, A. R. and Krumhansl, J. A. (1976). Phys. Rev. Lett. **36** (1411) .
Fogel, M. B., Trullinger, S. E., Bishop, A. R. and Krumhansl, J. A. (1977). Phys. Rev. B **15** (1578) .
Frauenkron, H. and Grassberger, P. (1995). J. Phys. A **28** (4987) .
Frauenkron, H. and Grassberger, P. (1996). Phys. Rev. E **53** (2823) .
Frauenkron, H., Kivshar, Yu. S. and Malomed, B. A. (1996). Phys. Rev. E **54** (R2344) .
Friedberg, R., Lee, T. D. and Sirlin, A. (1976). Phys. Rev. D **13** (2739) .
Garbalzewski, P. (1985). Classical and quantum field theory of exactly soluble nonlinear systems, World Scienfic, Singapore.
Gardiner, C. W. and Savage, C. M. (1984). Opt. Commun. **50** (173) .
Gardner, C. S., Greene, J. M., Kruskal, M. D. and Miure, R. M. (1967). Phys. Rev. Lett. **19** (1095) .
Gordon, J. P. (1983). Opt. Lett. **8** (596) .
Gordon, J. P. (1986). Opt. Lett. **11** (662) .
Gorshkov, K. A. and Ostrovsky, L. A. (1981). Physica D **3** (428) .
Gorss, E. P. (1961). Nuovo Cimento **20** (454) .
Hasegawa, H. and Brinkman, W. (1980). IEEE. J. Quantum Electron **16** (694) .
Hasegawa, H. and Tappert, F. (1973). Appl. Phys. Lett. **23** (171) .
Hermansson, Y. B. (1983). Opt. Commun. **47** (101) .
Ichikava, Y. (1979). Phys. Scr. **20** (296) .
Ikezi, H., Taylor, R. J. and Baker, D. R. (1970). Phys. Rev. Lett. **25** (11) .
Infeld, E. and Rowlands, G. (1990). Nonlinear waves, solitons and chaos, Cambridge Univ. Press. Cambridge.
Kalbermann, G. (1997). Phys. Rev. E **55** (R6360) .
Kalbermann, G. (1999). Phys. Lett. A **252** (33) .
Kaplan, A. E. (1976). JETP Lett. **24** (114) .
Karpman, V. I. and Solovev, V. V. (1981). Physica **D3** (487) .
Keener, J. and Mclaughlin, D. (1977). Phys. Rev. A **16** (777) .
Kivshar, Yu. S. and Gredeskul, S. A. (1990). Phys. Rev. Lett. **64** (1693) .
Kivshar, Yu. S. and Malomed, B. A. (1986). Phys. Lett. A **115** (377) .

Kivshar, Yu. S. and Malomed, B. A. (1989). Rev. Mod. Phys. **61** (763) .

Kivshar, Yu. S., Fei, Z. and Vazquez, L. (1991). Phys. Rev. Lett. **67** (1177) .

Kivshar, Yu. S., Gredeskul, S. A., Sanchez, A. and Vazquez, L. (1990). Phys. Rev. Lett. **64** (1693) .

Klyatzkin, V. I. (1980). Stochastic equations and waves in randomly inhomogeneous media, Nauka, Moscow (in Russian).

Konotop, V. V. (1990). Phys. Lett. A **146** (50) .

Lai, Y. and Haus, H. A. (1989). Phys. Rev. A **40** (844) and 854.

Lamb, G. L. (1980). Elements of soliton theory, Wiley, New York, pp. 118-125.

Lawrence, B., *et al.*, (1994). Appl. Phys. Lett. **64** (2773) .

Lax, P. D. (1968). Comm. Pure and Appl. Math. **21** (467) .

Leedke, E. and Spatschek, K. (1978). Phys. Rev. Lett. **41** (1798) .

Lifshits, I. M., Gredeskul, S. A. and Pastur, L. A. (1982). Introduction to theory of disordered systems, Nauka, Moscow.

Lonngren, K. E. (1983). Plasma Phys. **25** (943) .

Lonngren, K. E., Andersen, D. R. and Cooney, J. L. (1991). Phys. Lett. A **156** (441) .

Lonngren, K. E. and Scott, A. C. (1978). Solitons in action, Academic, New York, p. 153.

Louesey, S. W. (1980). Phys. Lett. A **78** (429) .

Makhankov, A. and Fedyamn, V. (1979). Phys. Scr. **20** (543) .

Makhankov, V. (1980). Comp. Phys. Com. **21** (1) .

Makhankov, V. G. (1978). Phys. Rep. **35** (1) .

Makhankov, V. G. and Fedyanin, V. K. (1984). Phys. Rep. **104** (1) .

Marchenko, V. A. (1955). Dokl. Akad. Nauk SSSR **104** (695) .

Miles, J. W. (1984). J. Fluid Mech. **148** (451) .

Miles, J. W. (1988). J. Fluid Mech. **186** (718) .

Miles, J. W. (1988). J. Fluid Mech. **189** (287) .

Nakamura, Y. (1982). IEEE Trans. Plasma Sci. **PS-10** (180) .

Nishida, Y. (1982). Butsuri **37** (396) .

Nishida, Y. (1984). Phys. Fluids **27** (2176) .

Pang, Xiao-feng (1985). J. Low Temp. Phys. **58** (334) .

Pang, Xiao-feng (1989). Chin. J. Low Temp. Supercond. **10** (612) .

Pang, Xiao-feng (1990). J. Phys. Condens. Matter **2** (9541) .

Pang, Xiao-feng (1993). Chin. Phys. Lett. **10** (381) , 417 and 570.

Pang, Xiao-feng (1994). Phys. Rev. E **49** (4747) .

Pang, Xiao-feng (1994). The theory for nonlinear quantum mechanics, Chongqing Press, Chongqing.

Pang, Xiao-feng (1999). European Phys. J. B **10** (415) .

Pang, Xiao-feng (2000). Chin. Phys. **9** (89) .

Pang, Xiao-feng (2000). European Phys. J. E **19** (297) .

Pang, Xiao-feng (2000). J. Phys. Condens. Matter **12** (885) .

Pang, Xiao-feng (2000). Phys. Rev. E **62** (6898) .

Pang, Xiao-feng (2003). Phys. Stat. Sol. (b) **236** (43) .

Pang, Xiao-feng (2003). Soliton physics, Sichuan Sci. and Tech. Press, Chengdu.

Perein, N. (1977). Phys. Fluids **20** (1735) .

Peyrard, M. and Campbell, D. K. (1983). Physica **D9** (33) .

Pitaerskii, L. P. (1961). Sov. Phys. JETP **13** (451) .

Rytov, S. M. (1976). Introduction to statistical radiophysics, Part I. Random Processes, Nauka, Moscow.

Sanchez, A., Scharf, R., Bishop, A. R. and Vazquez, L. (1992). Phys. Rev. A **45** (6031) .

Sanders, B. F., Katopodes, N. and Bord, J. P. (1998). United pseudospectral solution to

water wave equations, J. Engin. Hydraulics of ASCE, p. 294.
Satsuma, J. and Yajima, N. (1974). Prog. Theor. Phys. (Supp) **55** (284) .
Scharf, R. and Bishop, A. R. (1991). Phys. Rev. A **43** (6535) .
Scharf, R. and Bishop, A. R. (1992). Phys. Rev. A **46** (2973) .
Scharf, R. and Bishop, A. R. (1993). Phys. Rev. A **47** (1375) .
Scharf, R., Kivshar, Yu. S., Sanchez, A. and Bishop, A. R. (1992). Phys. Rev. A **45** (5369) .
Shabat, A. B. (1970). Dinamika sploshnoi sredy (Dynamics of a continuous Medium, No. 5, 130.
Shelby, R. M., Leverson, M. D., Perlmutter, S. H., Devoe, P. S. and Walls, D. F. (1986). Phys. Rev. Lett. **57** (691) .
Skinner, S. R., Allan, G. R., Allan, D. R., Andersen, D. R. and Smirl, A. L. (1991). IEEE Quantum Electron. **QE-27** (2211) .
Skinner, S. R., Allan, G. R., Andersen, D. R. and Smirl, A. L. (1991). in Proc. Conf. Lasers and Electro-Optics, Baltimore, (Opt. Soc. Am.).
Snyder, A. W. and Sheppard, A. P. (1993). Opt. Lett. **18** (482) .
Sulem, C and Sulem, P. L. (1999). The nonlinear Schrödinger equation: self-focusing and wave collapse, Springer-Verlag, Berlin.
Takeno, S. (1985). Dynamical problems in soliton systems, Springer Serise in synergetics, Vol. 30, Springer Verlag, Berlin.
Tanaka, S. (1975). KdV equation: asymptotic behavior of solutions, Publ. RIMS, Kyoto Univ.**10** pp. 367-379.
Tan, B. and Bord, J. P. (1998). Phys. Lett. A **240** (282) .
Tappert, F. D. (1974). Lett. Appl. Math. Am. Math. Soc. **15** (101) .
Taylor, J. R. (1992). Optical soliton-theory and experiment, Cambridge Univ. Press, Cambridge.
Thurston, R. N. and Weiner, A. M. (1991). J. Opt. Soc. Am. **B8** (471) .
Tominson, W. J., Gordon, J. P., Smith, P. W. and Kaplan, A. E. (1982). Appl. Opt. **21** (2041) .
Tsuauki, T. (1971). J. Low Temp. Phys. **4** (441) .
Wright, E. M. (1991). Phys. Rev. A **43** (3836) .
Wu, L., Kimble, H. J., Hall, J. L. and Wu, H. (1986). Phys. Rev. Lett. **57** (252) .
Yajima, N., Dikawa, M., Satsuma, J. and Namba, C. (1975). Res. Inst. Appl. Phys. Rep. **XXII70** (89) .
Zabusky, N. J. and Kruskal, M. D. (1965). Phys. Rev. Lett. **15** (240) .
Zakharov, B. E. and Shabat, A. B. (1972). Sov. Phys. JETP **34** (62) .
Zakharov, V. E. (1968). Sov. Phys.-JETP **26** (994) .
Zakharov, V. E. (1971). Sov. Phys.-JETP **33** (538) .
Zakharov, V. E. and Manakov, S. V. (1976). Sov. Phys.-JETP **42** (842) .
Zakharov, V. E. and Manakov, S. V. (1976). Sov. Phys.-JETP **44** (106) .
Zakharov, V. E. and Shabat, A. B. (1973). Sov. Phys.-JETP **37** (823) .
Zhou, X. and Cui, H. R. (1993). Science in China A **36** (816) .

Chapter 5

Nonlinear Interaction and Localization of Particles

As mentioned in Chapter 4, in the nonlinear quantum mechanics, microscopic particles have wave-corpuscle duality. Obviously, this is due to nonlinear interactions in the nonlinear quantum systems. However, what is the relationship between the nonlinear interactions and localizations of microscopic particles? What are the functions and related mechanisms of the nonlinear interactions? These are all important issues, which are worth further exploration. We will discuss these problems in this chapter.

5.1 Dispersion Effect and Nonlinear Interaction

We consider first the effects of the dispersion force and nonlinear interaction, from the evolution of the solutions of the linear Schrödinger equation (1.7) and the nonlinear Schrödinger equation (3.2), respectively.

In the linear quantum mechanics, the dynamic equation is the linear Schrödinger equation (1.7). When the external potential field is zero ($V = 0$), the solution is a plane wave,

$$\Psi\left(\vec{r}, t\right) = |\Psi\rangle = Ae^{i(\vec{p}\cdot\vec{r}-Et)/\hbar}. \tag{5.1}$$

It denotes the state of a freely moving microscopic particle with an eigenenergy of

$$E = \frac{p^2}{2m} = \frac{1}{2m}(P_x^2 + P_y^2 + P_z^2), \quad (-\infty < P_x, P_y, P_z < \infty). \tag{5.2}$$

This is a continuous spectrum. It states that the particle has the same probability to appear at any point in space. This is a direct consequence of the dispersion effect of the microscopic particle in the linear quantum mechanics. We will see that this nature of the microscopic particle cannot be changed with variations of time and external potential V.

In fact, if a free particle is confined in a rectangular box of dimension a, b and c, then the solutions of the linear Schrödinger equation are standing waves

$$\Psi\left(x, y, z, t\right) = A\sin\left(\frac{n_1\pi x}{a}\right)\sin\left(\frac{n_2\pi y}{b}\right)\sin\left(\frac{n_3\pi z}{c}\right)e^{-iEt/\hbar}. \tag{5.3}$$

In such a case, there is still a dispersion effect for the microscopic particle, namely, the microscopic particle still appears with a determinate probability at each point in the box. However, its eigenenergy is quantized, *i.e.*,

$$E = \frac{\pi^2 \hbar^2}{2m}\left(\frac{n_1^2}{a^2} + \frac{n_2^2}{b^2} + \frac{n_3^2}{c^2}\right). \tag{5.4}$$

The corresponding momentum is also quantized. This shows that microscopic particles confined in a box possess evident quantum features. Since all matters are composed of different microscopic particles, they always have certain bound states. Therefore, microscopic particles always have quantum characteristics.

When a microscopic particle is subject to a conservative time-independent field, $V(\vec{r},t) = V(\vec{r}) \neq 0$, the microscopic particle satisfies the time-independent linear Schrödinger equation

$$-\frac{\hbar^2}{2m}\nabla^2\psi + V(\vec{r})\psi = E\psi, \tag{5.5}$$

where

$$\Psi = \psi(\vec{r})e^{-iEt/\hbar}. \tag{5.6}$$

Equation (5.5) is referred to as eigenequation of the energy of the microscopic particle. In many physical systems, a microscopic particle is subject to a linear potential field, $V = \vec{F}\cdot\vec{r}$, for example, the motion of a charged particle in an electrical field, or a particle in the gravitational field, where $\vec{F}$ is a constant field independent of $\vec{r}$. For a one dimensional uniform electric field, $V(x) = -e\varepsilon x$, the solution of (5.5) is

$$\psi = A\sqrt{\xi}H_{1/2}^{(1)}\left(\frac{2}{3}\xi^{3/2}\right), \quad \left(\xi = \frac{x}{l} + \lambda\right) \tag{5.7}$$

where $H^{(1)}(x)$ is the Hankel function of the first kind, A is a normalized constant, l is the characteristic length and λ is a dimensionless quantity. The wave in (5.7) is still a dispersed wave. When $\xi \to \infty$, it approaches

$$\psi(\xi) = A'\xi^{-1/4}e^{-2\xi^{3/2}/3}, \tag{5.8}$$

which is a damped wave.

If $V(x) = Fx^2$ (such as the case of a harmonic oscillator), the eigen wavevector and eigenenergy of (5.5) are

$$\psi(x) = N_n e^{-a^2x^2/2}H_n(\alpha x), \quad E_n = \left(n + \frac{1}{2}\right)\hbar\omega, \quad (n = 0, 1, 2, \cdots) \tag{5.9}$$

respectively, where $H_n(\alpha x)$ is the Hermite polynomial. The solution obviously has a decaying feature.

We can keep changing the form of the external potential field $V(\vec{r})$, but soon we will find out that the dispersion and decaying nature of the microscopic particle persists no matter what form the potential field takes. The external potential field $V(\vec{r})$ can only change the shape of the microscopic particle, *i.e.*, its amplitude and velocity, but not its fundamental property such as dispersion. This is due to the fact that the kinetic energy term, $-(\hbar^2/2m)\nabla^2 = \hat{p}^2/2m$ itself is dispersive.

Because microscopic particles are in motion, the dispersion effect of the kinetic energy term always exists. It cannot always be balanced and suppressed by the external potential field $V(\vec{r}, t)$. Therefore, microscopic particles in linear quantum mechanics always exhibit the wave feature, not the corpuscle feature.

However, nonlinear interactions present in nonlinear quantum mechanics, for example, in the nonlinear Schrödinger equation (3.2), and such nonlinear interactions depend directly on the wave function of the state of the microscopic particle. It becomes possible to change the state and nature of the microscopic particle. If the nonlinear interaction is so strong that it can balance and suppress the dispersion effect of the kinetic term in (3.2), then its wave feature will be suppressed, and the microscopic particle becomes a soliton with the wave-corpuscle duality.

This can be verified using the nonlinear Schrödinger equation (4.40). Both kinetic energy dispersion and nonlinear interaction exist in the nonlinearly dynamic equation and the equation has soliton solutions (4.8) and (4.56) with constant energy, shape and momentum. This demonstrates clearly the suppression of the dispersion effect by the nonlinear interaction, and localization of the microscopic particle. As can be expected, the localization of the microscopic particle cannot be changed by altering the external potential field $V(x)$ in (3.2) at $A(\phi) = 0$. As a matter of fact, when $V(x)$ is a constant, as in (2.148) for the superfluid, its soliton solution is given by (2.150). Here $\text{sech}(x)$ is a stable function. If $V(x') = e\varepsilon x$, as in (2.122) in a superconductor, its soliton solution is given by (2.141) and (4.26), respectively. If $V(x') = \alpha^2 x'^2$, its soliton solution is given in (4.28). These solutions are all of the form $\text{sech}(x)$, which has localized feature and stable.

For a more complicated external potential $V(x)$, for example $V(x) = kx^2 + A(t)x + B(t)$, the soliton solution of (3.2) with $A(\phi) = 0$ can be written as

$$\phi = \varphi\left(x - u(t)\right) e^{i\theta(x,t)}, \tag{5.10}$$

where

$$\varphi\left(x - u(t)\right) = \sqrt{\frac{2B}{b}}\,\text{sech}\left(ax - u(t)\right),$$

$$\theta(x,t) = \left[-2\sqrt{k}\sin(2\sqrt{kt} + \beta) + \frac{u_0}{2}\right]x - \int_0^1 \left\{\left[u_0(t') - k(2\cos(2 - \sqrt{kt'} + \beta)\right]^2 \right.$$
$$\left. + B(t') + \left[\frac{u_0}{2} - 2\sqrt{k}\sin(2\sqrt{kt'} + \beta)\right]\right\} dt' + \lambda_0 t + g_0.$$

and

$$u(t) \equiv 2\cos(2\sqrt{kt} + \beta) + u_0(t).$$

When $A(t) = B(t) = 0$,

$$u(t) = 2\cos(2\sqrt{kt}) + u_0(t),$$

$$\theta(x,t) = -\int_0^1 \left\{\left[-k(2\cos(2\sqrt{kt} + u_0(t)\right]^2 \left(-2\sqrt{k}\sin(2\sqrt{kt'}\right) + \frac{u_0}{2}\right\} dt'$$
$$-2\sqrt{k}\sin\left(2\sqrt{kt} + \frac{u_0}{2}x\right) + g_0.$$

This solution is similar to (4.28) and is still a sech(x)-type soliton, and is stable, although it oscillates with a frequency which is directly proportional to k. It will be discussed in more details in Chapter 8.

The localization feature or wave-corpuscle duality of a microscopic particle cannot be changed by changing the form of the external potential $V(x)$ in nonlinear quantum mechanics. This shows also that the influence of the external potential $V(x)$ on soliton feature of microscopic particles is secondary, the fundamental nature of microscopic particles is mainly determined by the combined effect of dispersion forces and nonlinear interaction in nonlinear quantum mechanics. It is the nonlinear interaction that makes a dispersive microscopic particle a localized soliton. To further clarify these, we consider in the following the effects and functions of the dispersion force and nonlinear interaction.

What exactly is a dispersion effect? When a beam of white light passes through a prism, it is split into beams with different velocities. This phenomenon is called a dispersive effect, and the medium in which the light or wave propagates is referred to as a dispersive medium. The relation between the wave length and frequency of the light (wave) in this phenomenon is called a dispersion relation, which can be expressed as $\omega = \omega(\vec{k})$ or $G(\omega, \vec{k}) = 0$, where

$$\det \frac{\partial^2 \omega}{\partial k_i \partial k_j} \neq 0,$$

or

$$\frac{\partial^2 \omega}{\partial^2 k^2} \neq 0,$$

in one-dimensional case. It specifies how the velocity or frequency of the wave (light) depends on its wavelength or wavevector. The equation determines wave propagation in a dispersive median and is called an dispersion equation. The linear Schrödinger equation (1.7) in the linear quantum mechanics is a dispersion equation. If (5.1) is inserted into (1.7), we can get $\omega = \hbar k^2/2m$ from (5.2), here $E = \hbar\omega$, $\vec{p} = \hbar\vec{k}$. The quantity $v_p = \omega/k$ is called the phase velocity of the microscopic particle (wave). The wave vector $\vec{k}$ is a vector designating the direction of the wave propagation. Thus the phase velocity is given by $\vec{v_e} = (\omega/k^2)\vec{k}$. This is a standard dispersion relation. Therefore the solution of the linear Schrödinger equation (1.7) is a dispersive wave.

But how does the dispersive effect influence the state of a microscopic particle? To answer this question, we consider the dispersive effect of a wave-packet which is often used to explain the corpuscle feature of microscopic particles in linear quantum mechanics. The wave-packet is formed from a linear superposition of several plane waves (5.1) with wavevector k distributed in a range of $2\Delta k$. In the one dimensional case, a wave-packet can be expressed as

$$\Psi(x,t) = \frac{1}{2\pi}\int_{k_0-\Delta k}^{k_0+\Delta k} \psi(k,t)e^{i(kx-\omega t)}dk. \tag{5.11}$$

We now expand the angular frequency $\omega(k)$ at k_0 by

$$\omega = \omega_0 + \left(\frac{d\omega}{dk}\right)_{k_0}\Delta k + \frac{1}{2!}\left(\frac{d^2\omega}{dk^2}\right)_{k_0}(\Delta k)^2 + \cdots. \tag{5.12}$$

If we consider only the first two terms in the dispersive relation, *i. e.*,

$$\omega = \omega_0 + \left(\frac{d\omega}{dk}\right)_{k_0}\xi,$$

where $\xi = \Delta k = k - k_0$, then,

$$\begin{aligned}\Psi(x,t) &= \psi\left(k_0\right)e^{i(k_0x-\omega_0t)}\int_{-\Delta k}^{\Delta k} d\xi e^{i[x-(d\omega/dk)_{k_0}t]\xi} \\ &= 2\psi(k_0)\frac{\sin\{[x-(d\omega/dk)_{k_0}t]\Delta k\}}{x-(d\omega/dk)_{k_0}t}e^{i(k_0x-\omega_0t)},\end{aligned} \tag{5.13}$$

where the coefficient of $e^{i(k_0x-\omega_0t)}$ in (5.13) is the amplitude of the wave-packet. Its maximum is $2\psi(k_0)\Delta k$ which occurs at $x = 0$. It is zero at $x = x_n = n\pi/\Delta k$ $(n = \pm1, \pm2, \cdots)$.

Figure 5.1 shows that the amplitude of the wave-packet decreases with increasing distance of propagation due to the dispersion effect. Hence, the dispersion effect results directly in damping of the microscopic particle (wave). We can demonstrate from (1.11) and (1.13) that a wave-packet could eventually collapse with increasing time. Therefore, a wave-packet is unstable and cannot be used to describe the corpuscle feature of microscopic particles in the linear quantum mechanics.

When nonlinear interactions exist in a system, such nonlinear interaction can balance and suppress the dispersion effect of the microscopic particle. Thus the microscopic particle becomes a localized soliton as mentioned repeatedly, and its corpuscle feature can be observed. Therefore, microscopic particles have wave-corpuscle duality as described in Chapter 4. As the nonlinear interaction increases, the corpuscle feature of microscopic particles becomes more and more obvious. The localization nature of the microscopic particle can always persist no matter what form the externally applied potential field takes. Therefore, the localization and wave-corpuscle duality are a fundamental property of the microscopic particles in nonlinear quantum mechanics.

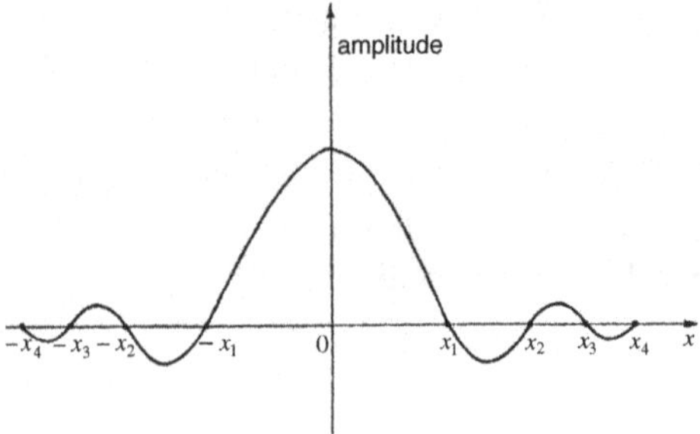

Fig. 5.1 Amplitude of a plane wave propagating along the x-direction.

5.2 Effects of Nonlinear Interactions on Behaviors of Microscopic Particles

What would be the effects of nonlinear interactions on microscopic particles? To answer this question, we first consider carefully the motion of water wave in a sea. When a wave approaches the beach, its shape varies gradually from a sinusoidal cross section to triangular, and eventually to a crest which moves faster than the rest, as shown in Fig. 5.2.

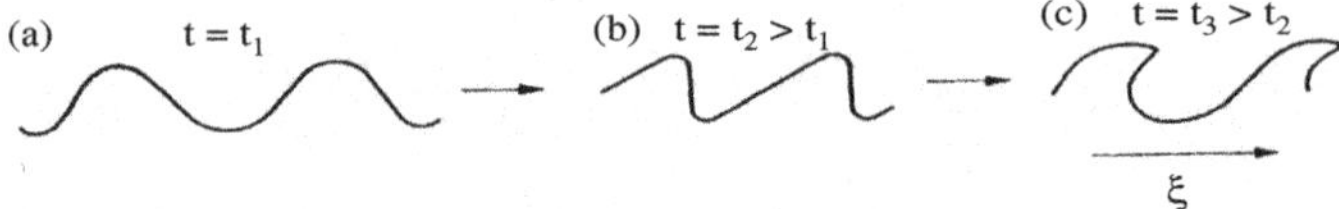

Fig. 5.2 Changes of the shape of a water wave approaching a beach.

This is a result of nonlinear nature of the wave. As the water wave approaches the beach, the wave will be broken up due to the fact that the nonlinear interaction is enhanced. Since the speed of wave propagation depends on the height of the wave, this is a nonlinear phenomenon. If the phase velocity of the wave, v_p, depends weakly on the height of the wave, h, then,

$$v_p = \frac{\omega}{k} = v_{p0} + \delta_1 h, \tag{5.14}$$

where

$$\delta_1 = \left.\frac{\partial v_p}{\partial h}\right|_{h=h_0},$$

h_0 is the average height of the wave surface, v_{p0} is the linear part of the phase velocity of the wave, δ_1 is a coefficient denoting the nonlinear effect. Therefore, the nonlinear interaction results in changes in both the form and the velocity of the waves. This effect is similar to that of dispersion, but their mechanisms and rules are different. When the dispersive effect is weak, the velocity of the wave can be denoted by

$$v_p = \frac{\omega}{k} = v'_{p_0} + \delta_2 k^2, \tag{5.15}$$

where v_{p0} is the dispersionless phase velocity, and

$$\delta_2 = \left.\frac{\partial^2 v_p}{\partial k^2}\right|_{k=k_0}$$

is the coefficient of the dispersion of the wave. Generally speaking, the term proportional to k in the expansion of the phase velocity gives rise to the dissipation effect, and the lowest order dispersion is proportional to k^2.

To further explore the effects of nonlinear interaction on the behaviors of microscopic particles, we consider a simple situation

$$\phi_t + \phi\phi_x = 0, \tag{5.16}$$

where the time evolution is given solely by the dispersion. In (5.16), $\phi\phi_x$ is a nonlinear interaction. There is no dispersive term in this equation. It is easy to verify that an arbitrary function of the variable $x - \phi t$,

$$\phi = \Phi'(x - \phi t) \tag{5.17}$$

satisfies (5.16). Fig. 5.3 shows ϕ as a function of x for various values of t, obtained from (5.16) and (5.17). In this figure, each point ϕ proceeds with the velocity ϕ. Thus, as time elapses, the front side of the wave gets steeper and steeper, until it becomes a triple-valued function of x due to the nonlinear interaction, which does not occur for a normal wave equation. Obviously, this wave deformation is the result of the nonlinear interaction.

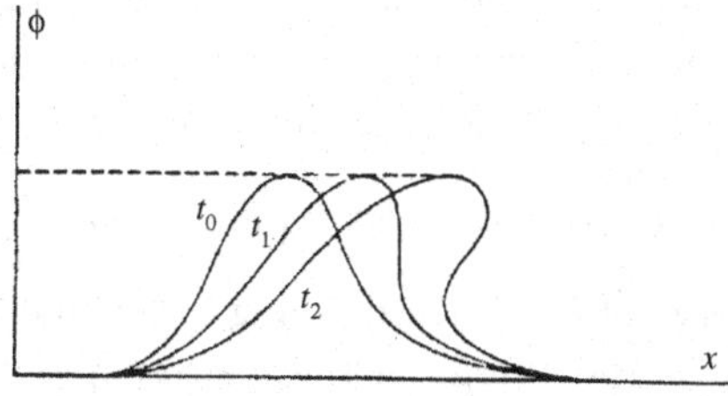

Fig. 5.3 Form of the wave given in (5.16).

If we let $\phi = \Phi' = \cos \pi x$ at $t = 0$, then at $x = 0.5$ and $t = \pi^{-1}$, $\phi = 0$ and $\phi_x = \infty$. The time $t_B = \pi^{-1}$ at which the wave becomes very steep is called destroy period of the wave. The collapsing phenomenon can be suppressed by adding a dispersion term ϕ_{xxx}, as in the KdV equation. Then, the system has a stable soliton, $\text{sech}^2(X)$ in such a case. Therefore, a stable soliton, or a localization of particle can occur only if the nonlinear interaction and dispersive effect exist simultaneously in the system, so that they can be balanced and their effects cancel each other. Otherwise, the particle cannot be localized, and a stable soliton cannot be formed.

However, if ϕ_{xxx} is replaced by ϕ_{xx}, then (5.16) becomes

$$\phi_t + \phi\phi_x = v\phi_{xx}, \quad (v > 0). \tag{5.18}$$

This is the Burgur's equation. In such a case, the term $v\phi_{xx}$ cannot suppress the collapse of the wave, arising from the nonlinear interaction $\phi\phi_{xx}$. Therefore, the wave is damped. In fact, using the Cole-Hopf transformation

$$\phi = -2\gamma\frac{d}{dx}\left(\log\psi'\right),$$

equation (5.18) becomes

$$\frac{\partial\psi'}{\partial t} = v\frac{\partial^2\psi'}{\partial x^2}.$$

This is a linear equation of heat conduction (the diffusion equation), which has a damping solution. Therefore, the Burgur's equation (5.18) is essentially not a equation for solitary wave, but a one-dimensional Navier-Stokes equation in viscous fluids. It only has a damping solution, not soliton solution.

This example tells us that the deformational effect of nonlinearity on the wave can only be suppressed by the dispersive effect. Soliton solution of dynamic equations, or localization of particle can then occur. This analysis and conclusion are also valid for the nonlinear Schrödinger equation and the nonlinear Klein-Gordon equation with dispersive effect and nonlinear interaction. This example also demonstrated that a stable soliton or localization of particle cannot occur in the absence of nonlinear interaction and dispersive effect.

In order to understand the conditions for localization of microscopic particles in the nonlinear quantum mechanics, we examine again the property of nonlinear Schrödinger equation (3.2). From (3.17) – (3.19), we get

$$E(\rho) \approx E_0 - b\rho = E_0 + (-b\rho), \tag{5.19}$$

where E_0 is the eigenenergy of $\hat{H}_0$, corresponding to the linear Schrödinger equation. To be more accurate, it is the energy of the microscopic particle depicted by the linear quantum mechanics. If we consider $E(\rho)$ as the energy of a microscopic particle in nonlinear quantum mechanics, then it should depend on the mass and state wave function of the microscopic particle. From (5.19), we know that the energy is lower than that of the particle in the linear quantum mechanics by $b\rho$. This indicates that in the nonlinear quantum mechanics, the microscopic particle is more stable than that in linear quantum mechanics. Obviously, this is due to the localization of the microscopic particle. Thus a energy gap between the normal and the soliton states exists in the energy-spectrum of the particles which is seen in superconductors in Chapter 2. The term $-b\rho$ (< 0) here is the binding energy or localization energy of the microscopic particle (soliton). Since the energy decrease is related to the state wave function of the particle, $-b|\phi|^2$, this phenomenon is called a self-localization effect of the particle, and this energy is referred to as self-localization energy. If there is no nonlinear interaction, *i.e.*, $b = 0$, then the self-localization energy is zero, and the microscopic particle cannot be localized. Therefore, the nonlinear interaction is a necessary condition for the localization of the microscopic particle in nonlinear quantum mechanics.

Known that nonlinear interaction is essential for localization of microscopic particle, the next question would be how could a microscopic particle be localized? To answer this question, we look further at the potential function of the system. Equation (3.17) can be written as

$$-\frac{\hbar^2}{2m}\nabla^2\phi + [V(\vec{r}) - E]\,\phi - b|\phi|^2\phi = 0. \tag{5.20}$$

For the purpose of showing the properties of this system, we assume that $V(\vec{r})$ and b are independent of $\vec{r}$. Then in one-dimensional case, equation (5.20) may be written as

$$\frac{\hbar^2}{2m}\frac{\partial^2\phi}{\partial x^2} = -\frac{d}{d\phi}V_{\text{eff}}\,(\phi)\,, \tag{5.21}$$

with

$$V_{\text{eff}}(\phi) = \frac{1}{4}b|\phi|^4 - \frac{1}{2}(V - E)|\phi|^2. \tag{5.22}$$

When $V > E$ and $V > 0$, the relationship between $V_{\text{eff}}(\phi)$ and ϕ is shown in Fig. 5.4. From this figure we see that there are two minimum values of the potential, corresponding to the two ground states of the microscopic particle in the system, *i.e.*,

$$\phi_0 = \pm\sqrt{\frac{V-E}{b}}.$$

This is a double-well potential, and the energies of the two ground states are $-(V - E)^2/4b \leq 0$. It shows that the microscopic particle can be localized because it has negative binding energy. The localization is achieved through repeated reflection of the microscopic particle in the double-well potential field. The two ground states limit the energy diffusion. Obviously, this is a result of the nonlinear interaction because the particle is in the normal, expanded state if $b = 0$. In that case, there is only one ground state of the particle which is $\phi_0' = 0$. Therefore, only if $b \neq 0$, the system can have two ground states, and the microscopic particle can be localized. Its localization or binding energy is negative, $-b\rho$. This nonlinear interaction is an attractive interaction. The nonlinear attraction can be produced by the following three mechanisms by means of interaction among particles and interaction between

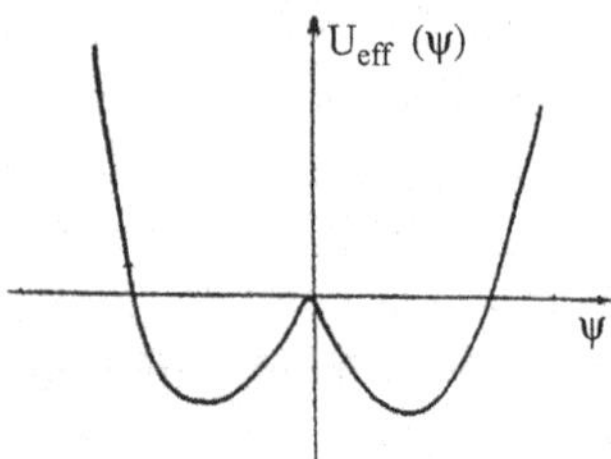

Fig. 5.4 The nonlinear effective potential in (5.22).

the medium and the particles. In the first mechanism, the attractive effect is due to interactions between the microscopic particle and other particles. This is called a self-interaction. A familiar example is the Bose-Einstein condensation mechanism of microscopic particles, which is due to attraction among the Bose particles, and is also referred to as self-condensation. In the second mechanism, the medium has itself anomalous dispersion effect (*i. e.*, $k'' = \partial^2 k/\partial\omega^2|_{\omega_0} < 0$) and nonlinear features resulting from the anisotropy and nonuniformity. Motion of the microscopic particles in the system are modulated by these nonlinear effects. This mechanism is called self-focusing. The third mechanism is referred to as self-trapping. It is produced by interaction between the microscopic particles and the lattice or medium. In the subsequent sections, these mechanisms will be discussed in more details.

From (5.22), we know that when $V > 0$, $E > 0$ and $V < E$, or $|V| > E$, $E > 0$ and $V < 0$ for $b > 0$, the microscopic particle may not be localized by the mechanisms mentioned above. On the other hand, we see from (5.20) – (5.22) that if the nonlinear self-interaction is of a repelling type (*i.e.*, $b < 0$), then, equation (3.2) with $A(\phi) = 0$ becomes

$$i\hbar\frac{\partial}{\partial t}\phi + \frac{\hbar^2}{2m}\frac{\partial^2\phi}{\partial x^2} - |b||\phi|^2\phi = V(x,t)\phi. \tag{5.23}$$

It is impossible to obtain a soliton solution, with full matter features, of this equation. However, if $V(x,t) = V(x)$ or a constant, solution of kink soliton type exists.

Inserting (3.15) into (5.23), we can get

$$\frac{\hbar^2}{2m}\frac{\partial^2\varphi}{\partial x^2} - |b|\varphi^3 + [E - V(x)]\varphi = 0. \tag{5.24}$$

If V is independent of x and $0 < V < E$, equation (5.24) has the following solution

$$\varphi = \frac{\sqrt{2(E-V)}}{|b|}\tanh\left[\sqrt{\frac{2(E-V)}{h^2}}(x - x_0)\right]. \tag{5.25}$$

This is the kink soliton solution when $|V| > E$ and $V < 0$. In the case of $V(x) = 0$, Zakharov and Shabat obtained dark soliton solution (see Section 4.3) which was experimentally observed in optical fiber and was discussed in the Bose-Einstein condensation model by Bargeretal.

Equation (5.24) may be written as

$$E\varphi = \left(H_0 + |b||\varphi|^2\right)\varphi = (E_0 + |b|\rho)\varphi. \tag{5.26}$$

Then, $E = E_0 + |b|\rho$. This corresponds to a upward shifting of energy levels. It is associated with nonlinear localization of holes in nonlinear semiconductors.

5.3 Self-Interaction and Intrinsic Nonlinearity

Self-interaction and intrinsic nonlinearity of the microscopic particles often occur in many body and many particle systems, and are described in quantum field theory and condensed matter physics. For the purpose of discussing self-interaction and intrinsic nonlinearity of microscopic particles, we briefly introduce nonlinear field theory here.

According to quantum field theory, the Hamiltonian operator corresponding to interaction between two fermions can be expressed using the creation and annihilation operators as

$$\mathcal{H}_{\text{int}} = G\psi_p^+\gamma_\mu^+\psi_n\psi_e^+\gamma^\mu\psi_v + c.c., \tag{5.27}$$

where the subscripts refer to proton, neutron, electron, and neutrino, respectively, $\psi_p^+(\psi_p)$ and $\psi_e^+(\psi_e)$ are the creation (annihilation) operators of the proton and electron, respectively, γ_μ is the Dirac matrices, and G is the coupling constant for the interactions of these fields which is the analog of the electric charge. This is a local interaction which occurs at a point in the space-time.

Shortly after Fermi's theory appeared, Yukawa proposed that the interaction between an electron and a proton was mediated by an intermediate Boson field – the meson. The meson was coupled directly and locally to both baryons and leptons with

$$\mathcal{H}_b = g\psi_p^+\psi_n\phi, \quad \text{and} \quad \mathcal{H}_l = g\psi_e^+\psi_\nu\phi, \tag{5.28}$$

where ϕ is the meson field and g is their coupling constant. In the field theory, each particle is represented by a separate field, but an unified theory has been established. One method for the unification is the intrinsically nonlinear and self-interacting theory. An early example of such a theory is the nonlinear spinor theory proposed by Heisenberg *et al.*, in which the basic field describes a spin 1/2 system. Bosons are expected to appear as excitations of Fermion-antifermion pairs. The interaction current density in such a case has the form

$$j = \eta\psi^+\sigma_\nu\psi\sigma^\nu\psi, \tag{5.29}$$

where η is a coupling constant and $\sigma_\nu = (1, \vec{\sigma})$, with $\vec{\sigma}$ being the Pauli matrices. This is an intrinsically nonlinear theory since only one type of field appears in the field equation. Recent unification schemes are to unify interactions, such as weak and electromagnetic in combination with local, intrinsic symmetries in order to reduce the number of distinct fields. These theories also rely on the intrinsic nonlinearities.

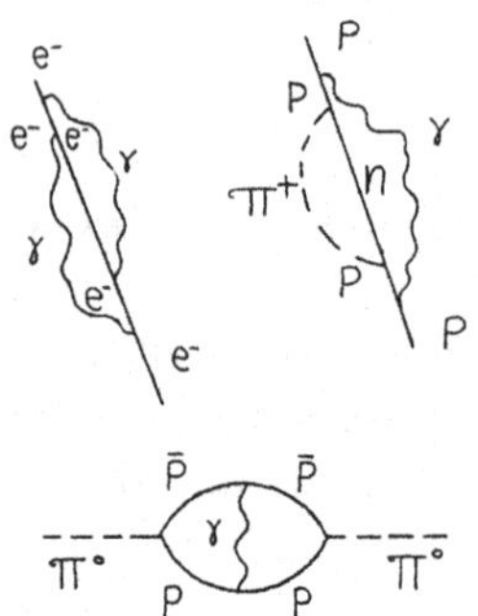

Fig. 5.5 Feynman diagrams of various virtual processes leading to the interaction (see book by Burt).

Self-interactions of mesons are commonly observed in experiments. For example, a pair of π-mesons is observed to form a ρ-meson, which subsequently decays into a pion pair. Self-interactions have been studied in connection with the self-energy problem of an interacting field theory, which exists in any interaction field theory, whether it is quantum or classical. It introduces the persistent interaction, which is not present in the classical scattering of two point particles, into the field theory. In quantum theory, the concept of an isolated system becomes meaningless. The familiar three-stage interaction of particles, *i.e.*, initial noninteracting, interacting, final noninteracting, or "essentially" free motion, interaction, essentially free motion, no longer holds. For example, electrons described by the Daric theory coupled with Maxwell's theory in quantized form, are always interacting. Due to the possibility of virtual transition, a single electron continuously experiences interactions with its own electromagnetic field, showing that the electron is in self-interaction or persistent interaction. The Feynman diagrams of self-interactions of some virtual process are shown in Fig. 5.5. These processes can be described by conventional perturbation theory. That is, "bare" electrons and "free electromagnetic fields" are combined in the perturbation theory to "dress" the electron, to provide its physical mass and charge. The persistent interaction leads to the physical particle.

The field equation including self-interaction of Bosons can generally be expressed as

$$\partial_\mu \partial^\mu \phi + G(m^2, \lambda_i : \phi) = 0, \tag{5.30}$$

where m^2 is the mass and λ_i $(i = 1, \cdots n)$ are coupling constants, ϕ is the Boson field. The common element in these interactions is that they create and annihilate the constituents in some processes. Burt proposed the following classes of intrinsic nonlinearities generated from the self-interactions.

(1) The self-interaction Hamiltonian $\lambda\phi^4$. This interaction can lead to a creation process in which a proton-antiproton pair annihilates into a virtual pion pair with subsequent creation of a real pion pair through the self-interaction mechanism. The Feynman diagram corresponding to the interaction $\lambda\phi^4$ is shown in Fig. 5.6. The baryon-pion interaction can be depicted by, for example, (5.28).

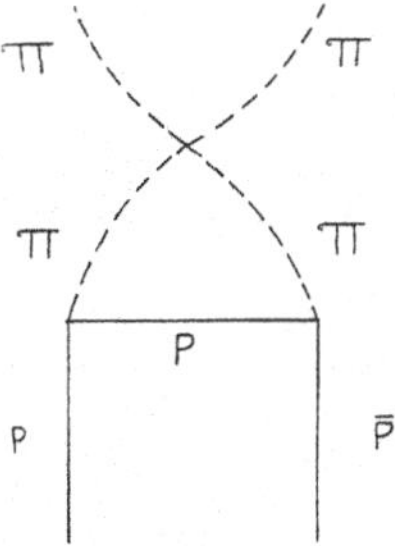

Fig. 5.6 A possible mechanism for conversion from a proton-antiproton pair to a pion pair mediated by the 4π self-interaction. (see book by Burt)

(2) The self-interaction Hamiltonian $\alpha\phi^5$. This interaction can lead to the situation where a 2π system annihilates into a K^0-$\bar{K}^0$ system which decays into a 3π system. The Feynman diagram of the $\alpha\phi^5$ interaction is shown in Fig. 5.7.

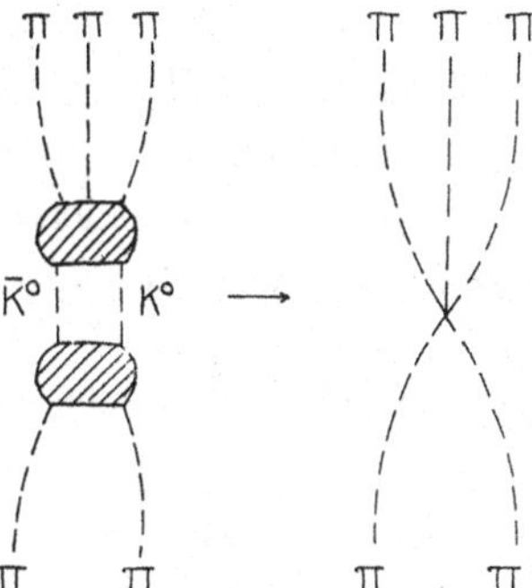

Fig. 5.7 A 2π system can be related to a 3π system through a K^0-$\bar{K}^0$ intermediate state which can describe the self-interaction on the right side (see book by Burt).

(3) Polynomial currents of the form $J_p(\phi) = \sum \alpha_n \phi^n$. This interaction can lead to processes of higher multiplicity. If pions have a Yukawa coupling to nucleons, as low energy nuclear physics seems to indicate, π-π interactions occur through the creation of a virtual nucleon-antinucleon state. This process has a scattering

amplitude which in the lowest order is given by

$$M(\hat{p}_1, \hat{p}_2, \hat{q}_1, \hat{q}_2) = -g^4 \int \frac{d^4k}{(\pi\alpha)^2} \frac{1}{d(\hat{k}-\hat{p}_1)d(\hat{k}-\hat{q}_1)d(\hat{k}-\hat{q}_1-\hat{q}_2)d(\hat{k})} \tag{5.31}$$

$$\times \mathrm{Tr}[\gamma_s(k+\phi_1+m)\gamma_s(k-\phi_1+m)\gamma_s(k-\phi_1-\phi_2+m)\gamma_s(k+m)] = -g^4 \int \frac{d^4k}{|\hat{k}|^4},$$

as shown in Fig. 5.8. In (5.31),

$$d(\hat{k}) = \hat{k}^2 - m^2 + i\varepsilon, \tag{5.32}$$

when the interaction Hamiltonian is

$$H_{\pi\mathrm{n}} = ig\psi^+\gamma_s\psi\phi + c.c..$$

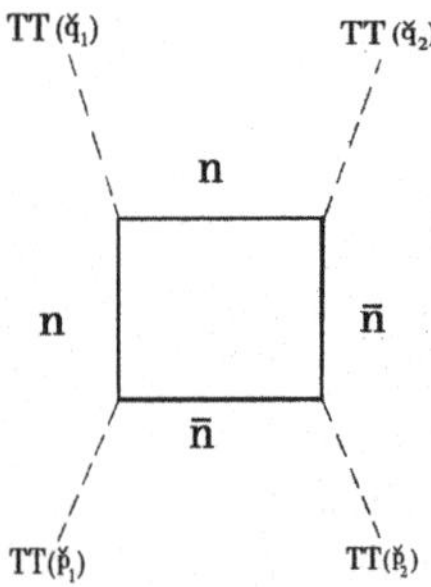

Fig. 5.8 Pions, coupled to necleons, have a self-interaction depicted in lowest order perturbation by annihilation diagrams (see book by Burt).

Since the interaction in (5.31) can extend over all momentum space, the matrix element is infinite. However, the Yukawa coupling is accompanied by a self-interaction, $H_{\pi\pi} = \lambda\phi^4$. Then both interactions lead to a finite result in the perturbation theory, if the value of λ is adjusted to cancel the parts which are infinite in (5.31).

(4) The self-interactions consisting of a transcendental current

$$J_{\mathrm{SG}} = \frac{m^2}{g}\sin(g\phi), \tag{5.33}$$

and a rational current

$$J_r = \sigma\phi^{2p+1} + \beta\phi^{p+1} + \gamma\phi^{-p+1} + \delta\phi^{-2p+1} \tag{5.34}$$

are also interesting. They were the basis for nonperturbative mass generation and nondispersive solitary waves. All of the three currents, polynomial, transcendental and rational correspond to a generalized solitary wave field. Each class, however, has its unique features, and different applications in physics.

For a Boson field, when the self-interaction is taken into account, the field equation with nonlinear self-interaction (5.31) is

$$\partial\mu\partial^{\mu}\phi + m^2\phi + J(\lambda, \phi) = 0, \tag{5.35}$$

where J is one of the self-interaction currents described above. If $J = 0$, the solutions of this equation, containing either creation or annihilation operators, describe quanta of the field. Here we consider the persistent self-interaction by first seeking solution of (5.35) that also contains either creation or annihilation operators, but not both. Obviously, equation (5.35) is a nonlinear field equation, including both the ϕ^4-equation and the Sine-Gordon equation. It can describe microscopic particles. Burt pointed out that the solutions of this equation have the following properties, (i) $\phi = \phi_{\vec{k}}^{(\pm)}(\hat{x}) = f(A_k^{(\pm)}e^{\pm i\hat{k}\cdot\hat{x}})$, (ii) $\phi_{\vec{k}}^{(\pm)}$ contains the coupling constants for all times, (iii) $\phi_{\vec{k}}^{(\pm)}$ is not perturbative. The exact solutions of the field equation are special cases of nonlinear wave solutions. Thus, these new nonlinear fields differ from classical fields by having the integration constants replaced by creation or annihilation operator. In this respect, they are nonlinear generalization of the particle solutions of the free field solutions.

Whitham proved that these particular solutions of the nonlinear, dispersive field equations are similar to classical solitary wave solutions. They propagate with constant phase speed, and are nonperturbative and exact solutions of the field equations for all times. They describe new, intrinsically localized nonlinear modes of the systems. A major difference between these solutions and the solitary waves of classical theory is that the matrix elements of the quantized fields describe unlocalized solution containing oscillating tails.

Letting the current in (5.35) take the form

$$J_p = \alpha\phi^{2p+1} + \lambda\phi^{4p+1}, \tag{5.36}$$

Burt obtained the solution of (5.35) as

$$\phi_{\vec{k}}^{(\pm)}(\hat{x}) = U_{\vec{k}}^{(\pm)}(\hat{x})\left\{\left(1 - \frac{\alpha U_{\vec{k}}^{(\pm)}(\hat{x})^{2p}}{4(1+p)m^2}\right)^2 - \frac{\lambda U_{\vec{k}}^{(\pm)}(\hat{x})^{4p}}{4\,(1+2p)\,m^2}\right\}^{-1/2p}, \tag{5.37}$$

where $p \neq 0, -1/2, -1$, $U_{\vec{k}}^{(\pm)}(\hat{x}) = A_{\vec{k}}^{(\pm)}e^{\mp i\hat{k}\cdot\hat{x}}(D\omega V)^{-1/2}$, $\omega = \sqrt{k^2+m^2}$, D is an arbitrary constant, V is the volume of the system, $A_k^{(\pm)}$ is a coefficient. The solution (5.37) has the first two features mentioned above. The exponentials are purely oscillatory. So the field satisfies the nonlinear field equation at all times. If $\alpha = \lambda = 0$, the field reduces to the solution of the free field equation, $U_{\vec{k}}^{(\pm)}(\hat{x})$. In order for the persistently interacting fields to be nonperturbative, it is necessary to write $\phi_{\vec{k}}^{(\pm)}(\hat{x})$ as a series of positive powers of $U_{\vec{k}}^{(\pm)}(\hat{x})$. To do this Burt used the generating function for the Gegenbauer polynomials

$$\left(1 - 2wZ + Z^2\right)^{-1/(2p)} = \sum_n C_n^{1/(2p)}(w) Z^n, \tag{5.38}$$

where C_n^m is a Genenbauer polynomial. Letting

$$w = \frac{\alpha}{4(p+1)m^2 b'}, \quad Z = b' U_{\vec{k}}^{(\pm)}(\hat{x})^{2p},$$

$$b' = \left\{ \left[\frac{\alpha}{4(p+1)m^2} \right]^2 - \frac{\alpha}{4(2p+1)m^2} \right\}^{1/2},$$

Burt then obtained the following

$$\phi_{\vec{k}}^{(\pm)}(\hat{x}) = \sum_{n=0}^{\infty} C_n^{1/(2p)}(w) b'^n U_{\vec{k}}^{(\pm)}(\hat{x})^{2pn+1}. \tag{5.39}$$

Therefore, the operator solution of the nonlinear field equation is a positive power series expansion of creation or annihilation operators. With this expansion, we can show that this solution is nonperturbative. In practice, if we choose the following new solution

$$\phi_{\vec{k}}^{(\pm)}(\hat{x}) = \frac{1}{b'^{(2p)}} U_{\vec{k}}^{(\pm)}(\hat{x}) \left\{ 1 - \frac{\alpha U_{\vec{k}}^{(\pm)}(\hat{x})^{2p}}{2(p+1)m^2 b'} + U_{\vec{k}}^{(\pm)}(\hat{x})^{4p} \right\}^{-1/(2p)}, \tag{5.40}$$

when $\alpha \to 0$ and $\lambda \to 0$, we can find that the result depends on the order in which the limit is taken. First, let $\alpha = 0$, such that

$$b' = \sqrt{-\frac{\lambda}{4(2p+1)m^2}}.$$

Then

$$\phi_{\vec{k}}^{(\pm)}(\hat{x}) = \frac{U_{\vec{k}}^{(\pm)}(\hat{x})}{\left\{ b' \left[1 + U_{\vec{k}}^{(\pm)}(\hat{x})^{4p} \right] \right\}^{1/(2p)}}. \tag{5.41}$$

If $\alpha \neq 0$ and $\lambda = 0$, the result is

$$\phi_{\vec{k}}^{(\pm)}(\hat{x}) = \frac{1}{b'^{1/(2p)}} \frac{U_{\vec{k}}^{(\pm)}(\hat{x})}{\left[1 - U_{\vec{k}}^{(\pm)}(\hat{x})^{2p} \right]^{1/p}}, \tag{5.42}$$

where

$$b' = \frac{1}{(p+1)m^2}.$$

In the first case, the solutions are singular for $\lambda = 0$ and the corrections are proportional to α^2/λ for small α. In the second case the solutions are singular for

$\alpha = 0$ and the corrections are proportional to λ/α^2 for small λ. Hence, no such solutions could be found from a perturbation series in α and λ. This shows that the system has intrinsic nonlinear modes. Obviously, it results from the self-interaction consisting of the nonlinear current.

These intrinsic nonlinear modes have many classical counterparts. The analogy with classical solitary waves is very useful. In the presence of nonlinear interactions a quantized system can respond by either a perturbative or a persistent interacting mode. The intrinsic nonlinearity has important meanings in nonlinear fields. The intrinsic nonlinear field is not a perturbation of the quantum field. It also cannot be reached from the field by successive infinitesimal motions.

The above solutions, (5.39) – (5.42), show that there are intrinsically nonlinear modes in the self-interaction fields. For a classical system, its feature can be designated by the wave amplitude. This can mean a small (linear) oscillation, a large amplitude solitary wave, in which the amplitude is so large that discontinuities develop in the systems. Quantum systems also have these features, but the matrix elements of the persistently interacting field can become singular, corresponding to overgrowth in the classical systems. Davidson proved that when the density of the system or the interaction coupling constant becomes too large or the mass of the particle in the system is too small, the persistently interaction field has singular matrix elements. This development of a singularity is similar to solitary wave instability and breaking up in the classical systems. If the wave amplitude or nonlinear strength is very large, or the dispersion, denoted by $m^2\phi$ in (5.35), is so small that it cannot balance the nonlinearity, the system cannot support the persistent interaction. Therefore, this balance depends on the form of the interaction current.

The intrinsic nonlinear fields discussed above possess many features in common with classical solitary waves, but there are also considerable differences between them. The most important difference is the relationship between speed of a solitary wave and that of accompanying linear waves. The classical systems have acoustic (linear) modes in which the solitary waves are supersonic. For the quantum systems, the linear and nonlinear fields have identical (superluminal) phase speeds. This feature of the persistent interacting quantized fields enables us to interpret these fields as coherent excitations of the systems. From (5.39) we see that the field creates $2pn+1$ quantum states from the vacuum, each quantum having the same momentum and mass. This is similar to the coherent many-photon states of the electromagnetic field. However, for the persistent nonlinear fields there is no velocity dependence in the amplitude, which is contrary to the classical solitary waves. Burt proved further that the persistent self-interacting fields describe noncanonical excitations of the systems because the field equation cannot be obtained from Hamiltonian of the systems, likely the linear equations. Thus the intrinsically nonlinear fields would form a class of motions canonical with respect to the new Hamiltonian.

5.4 Self-localization of Microscopic Particle by Inertialess Self-interaction

In the previous section we discussed the self-interaction among the elementary particles and the intrinsic nonlinearity in the classical and quantum field theories. In practice, the self-interactions occur also in many condensed matters. The detection of quasi-one-dimensional metal conductivity in metal-organic complexes and some organic salts such as TTF (tetrathiafulvalene), TCNQ (tetralyanoquinodimethane), TCNE (tetracyanoethylene), TTF-TCNQ, Q (quinolinium) - $TCNQ_2$, NMQ (N-methylquinolinium) - TCNQ and HMTTF (hexamethylenetrathiafulvalene) - TCNQ, and so on, stimulated significant interest in the study of nonlinear localization of the concerned microscopic particles, the electrons. This mechanism was first proposed by Fröhlich who discussed the superconductivity of electrons in one-dimensional systems without using the concept of Cooper pairs.

Fröhlich pointed out that charge density wave (CDW) could reduce the energy of electrons in metals, when the electrons are restricted to one-dimensional metal chains in which the Fermi surface consists of two parallel planes. This effect results from the interaction of the electrons with a sinusoidal potential field due to displacements of the ionic charge density in metals ("the jellium model") which lead to an increase of the energy gap in the electronic energy spectrum and also to a periodic change in the electronic density in the form of nondamped waves (CDW) under the cyclic boundary conditions. The undamped wave is a soliton, which arises from the self-localization of the electrons due to self-interactions in the one-dimensional conductors. Davydov and Pestryakov studied the nonlinear localization phenomenon of a complex scalar field of spinless quasiparticles with inertialess self-interaction. The Hamiltonian density of the scalar field with self-interaction in one-dimensional infinite space is given by

$$\mathcal{H} = \frac{\hbar^2}{2m}\left|\frac{\partial \phi}{\partial x},\right|^2 \mp b|\phi|^4 \tag{5.43}$$

where b is a nonlinear parameter which is independent of the velocity, the "$-$" and "$+$" signs correspond to attraction and repulsion self-interactions, respectively. Such a self-interaction can be realized by local interaction between the microscopic particles which correspond to the field or by local interaction of these quasiparticles with the field of inertialess displacements of the density of other particles which are not considered explicitly here. This model resembles that of the $\lambda\phi^4$-type self-interaction which is similar to that discussed in Section 5.3.

If the self-interaction is absent, $b = 0$, the field $\phi(x, t)$ describes the states of noninteracting quasiparticles with mass m. Their motion is described by a plane wave with wave vector k and energy $\hbar^2k^2/2m$. If $b \neq 0$, the corresponding equation of motion is

$$\left(i\hbar\frac{\partial}{\partial t}+\frac{\hbar^2}{2m}\frac{\partial^2}{\partial x^2}\mp 2b|\phi|^2\right)\phi=0. \tag{5.44}$$

This is a standard nonlinear Schrödinger equation. If we assume that $\phi(x,t)$ satisfies the periodic condition $\phi(x,t)=\phi(x+r_0L,t)$ and the normalization condition

$$\int_0^L |\phi(x,t)|^2\,dx=1,$$

equation (5.44) has soliton solution as that given by (4.8). Therefore, electrons in these conductors are self-localized due to self-interaction. In other word, equation (5.44) has stationary solutions in the form of modulated plane waves, with an inhomogeneous spatial density distribution of the microscopic particle moving along the x-axis with a small constant velocity ν. In order to elucidate the nature of the particle density distribution, we rewrite the solution in the following form

$$\phi(x,t)=\frac{1}{r_0}\varphi(\xi)\exp\left\{i\left[kx-\left(\omega+\frac{\hbar^2k^2}{2m}\right)t+\eta\right]\right\}, \tag{5.45}$$

where ξ is a dimensionless coordinate in the reference frame, $\xi=(x-x_0-vt)/r_1$, moving with the speed $v=\hbar k/m$, $\hbar\omega$ is the energy of the particle relative to the reference frame, η is an arbitrary phase, r_0 is length unit in the system being studied, which is given by $r_0=\hbar/\sqrt{4mb}$. The periodic condition given above holds relative to the moving reference frame, if the wave vector k and the velocity v takes discrete values,

$$k=\frac{2\pi n}{r_0L},\quad (n=0,\pm1,\pm2,\pm3\cdots).$$

Therefore, k is quantized.

Substituting (5.45) into (5.44), we can get

$$\left[\frac{d^2}{d\xi^2}+\varepsilon+\varphi^2(\xi)\right]\varphi=0, \tag{5.46}$$

where $\varepsilon=\hbar\omega/\varepsilon_0$, and $\varepsilon_0=2b$. The corresponding energy E and momentum P of the field-excited state are

$$E(v)=\frac{1}{2}mv^2+\varepsilon_0\int_0^L\left[\left(\frac{d\varphi}{d\xi}\right)^2\mp\frac{1}{2}\varphi^4\right]d\xi,$$

$$P(v)=mv\int_0^L\varphi^2(\xi)d\xi=mv.$$

Multiplying (5.46) on the left by $\varphi(\xi)$ and integrating from 0 and L, we get

$$\varepsilon=\int_0^\infty\left[\left(\frac{d\varphi}{d\xi}\right)^2\mp\varphi^4\right]d\xi.$$

Thus the energy of the field is

$$E(v) = \frac{1}{2}mv^2 + \varepsilon_0 \left[\varepsilon \pm \frac{1}{2}\int_0^L \varphi^4(\xi)d\xi\right].$$

The momentum of the field is quantized because v is quantized.

In the case of attractive forces, Davydov *et al.* found that $\varphi_a(\xi, k) = A\text{dn}(a\xi, k)$ and

$$E_a(v,k) = \frac{1}{2}mv^2 - \frac{\xi_0^0}{16E(k)}\left[2 - k^2 - \frac{Q_a(k)}{E(k)}\right], \tag{5.47}$$

if $\varepsilon = a^2(2-k^2)$ and $A = 2a^2$, where A and a are determined by the periodicity and normalization conditions mentioned earlier. In (5.47)

$$Q_a(k) = \frac{1}{3}\left[2(2-k^2)E(k) - (1-k^2)K(k)\right],$$

$\text{dn}(u,k)$, $K(k)$ and $E(k)$ are the Jacobian elliptic function, and the complete elliptic integrals of the first and second kinds, respectively. When $k_1^2 = 1 - k^2 \to 0$ and $k_1^2 \to 0$, $\varphi_a(\xi,k) \approx \text{sech}(\xi/4)/\sqrt{8}$, $E_a \approx mv^2/2 - \varepsilon_0/48$ in an infinite system. When $k_1^2 \to 1$,

$$E_a(v,k) = \frac{1}{2}mv^2 - \frac{2mb}{\pi^2\hbar}\left(1 - \frac{1}{2}k^2\right).$$

In the case of a repulsive force, Davydov *et al.* found that

$$\varphi_r = \frac{1}{\sqrt{2\pi}}\left\{\sin(u) + \frac{1}{4}k^2\cos(u)\left[\sin(u)\cos(u) - u\right]\right\},$$

and

$$E_r(v,k) = \frac{1}{2}mv^2 + \frac{mb^2}{8\pi^2}\left(1 + k^2\right),$$

if $k^2 \ll 1$, where $u = \xi/(2\pi)$, $|\xi| \ll L$.

5.5 Nonlinear Effect of Media and Self-focusing Mechanism

Many physical systems themselves have nonlinear effects resulting from their anisotropy and nonuniformity, etc. Such systems could provide nonlinear interactions to microscopic particles moving in the media. To understand the mechanism of such nonlinear interaction, we consider a familiar dispersive medium, an optical fiber, in which microscopic particles (photons) or light waves propagate. In this case, Hasegawa obtained a nonlinear equation for propagation of the electric field $\vec{E}(x,t)$ of the light due to the nonlinearity of the medium (fiber) – the Kerr effect.

In this section, we derive the equation of motion of microscopic particle in a nonlinear medium following Hasegawa's approach, and give its localization properties due to the self-focusing mechanism arising from this nonlinearity.

The refractive index of the medium, n, is defined by the ratio of the speed of light, c, in the vacuum to the phase velocity of the wave in the medium, $n = c/v_p = ck/\omega$. The refractive index of a dielectric medium is related to the frequency of the a wave in the medium. This phenomenon is called the dispersive effect which arises from molecular structure of the medium as described in Section 5.1. Therefore, in the dispersive medium, the quantities such as $\partial n/\partial\omega$, $\partial^2 n/\partial\omega^2$, *etc.* always have finite values. We can expand the frequency of the wave $\omega(= ck/n)$ around the wavevector k_0 for the carrier wave shown in Fig. 4.1

$$\omega - \omega_0 = \left.\frac{\partial\omega}{\partial k}\right|_{k_0}(k-k_0) + \frac{1}{2!}\left.\frac{\partial^2\omega}{\partial k^2}\right|_{k_0}(k-k_0)^2 + \frac{1}{3!}\left.\frac{\partial^2\omega}{\partial k^2}\right|_{k_0}(k-k_0)^3 + \cdots. \tag{5.48}$$

It gives the change of the frequency of the modulated wave due to the deviation of its wavevector k from the carrier wave vector k_0.

Due to the fact that the envelope function of the wave, $\varphi(x,t)$ (or $\vec{E}(x,t)$, the electric field of the light wave) in Fig. 4.1 is a slowly varying function in both x and t, we can perform a Fourier transform on this function and then use the Fourier space variables, $\Delta k(= k - k_0)$ and $\Delta\omega(= \omega - \omega_0)$, which are defined by

$$\bar{\varphi}(\Delta K, \Delta\omega) = \frac{1}{2\pi}\int_{-\infty}^{\infty}\int_{-\infty}^{\infty}\varphi(x,t)e^{i(\Delta kx - \Delta\omega t)}dtdx. \tag{5.49}$$

The corresponding inverse transform is

$$\varphi_{(x,t)} = \frac{1}{2\pi}\int_{-\infty}^{\infty}\int_{-\infty}^{\infty}\bar{\varphi}(\Delta k, \Delta\omega)e^{-i(\Delta kx - \Delta\omega t)}d(\Delta k)d(\Delta\omega).$$

From (5.48) and (5.49), we see that $\partial\varphi/\partial t$ and $\partial\varphi/\partial x$ can be identified as the Fourier transform of $i\Delta\omega\bar{\varphi}$ and $-i\Delta k\bar{\varphi}$, respectively. Making use of these, we can rewrite $\Delta\omega$ and Δk as $-i\partial/\partial t$ and $i\partial/\partial x$, respectively. In terms of these operators, equation (5.48) can then be written as

$$-i\frac{\partial}{\partial t} = i\omega'\frac{\partial}{\partial x} - \frac{\omega''}{2!}\frac{\partial^2}{\partial x^2} - i\frac{\omega'''}{3!}\frac{\partial^3}{\partial x^3} + \cdots. \tag{5.50}$$

Operating (5.50) on the envelope function $\varphi(x,t)$ (or $\vec{E}(x,t)$) and keep terms up to the order of the second derivative with respect to the carrier frequency, we then have

$$i\left(\frac{\partial}{\partial t} + \omega'\frac{\partial}{\partial x}\right)\varphi - \frac{\omega''}{2!}\frac{\partial^2\varphi}{\partial x^2} = 0, \tag{5.51}$$

where

$$\omega' = \left.\frac{\partial\omega}{\partial k}\right|_{k_0}, \quad \omega'' = \left.\frac{\partial^2\omega}{\partial k^2}\right|_{k_0},$$

and the group velocity v_g of the envelop modulated wave is defined by

$$v_g = \frac{\partial \omega}{\partial k} = \omega', \quad \omega'' = \frac{\partial v_g}{\partial k}.$$

Thus ω'' is given by the wavevector dependence of the group velocity of the wave, and it expresses the dispersion feature of the group velocity of the wave. If $\omega'' = 0$, the solution of (5.51) can be represented by an arbitrary function of $x - v_g t$, *i.e.*, $\varphi(x - v_g t)$. This shows that the envelope wave of the microscopic particle propagates at the group speed. Furthermore, we may introduce a new coordinate system which moves with the group velocity, $\tau = \varepsilon^2 t$, $\xi = \varepsilon(x - \omega' t)$, where $\varepsilon = \Delta k_0/k_0$ is a small quantity which characterizes the width of the spectrum. Then (5.51) becomes

$$i\frac{\partial \varphi}{\partial \tau} - \frac{\omega''}{2}\frac{\partial^2 \varphi}{\partial \xi^2} = 0. \tag{5.52}$$

Equation (5.52) is a dispersive equation which is obtained from the dispersive relation (5.48). It shows that the envelope wave of the microscopic particle deforms in proportion to the distance of propagation due to the group velocity dispersion. If a microscopic particle with a short wavelength is transmitted through a medium, $\partial^2\varphi/\partial x^2$ (or $\partial^2\vec{E}/\partial x^2$ in optical fiber) may become large, and inversely proportional to the square of the width of the microscopic particle (pulse). Thus the form of the microscopic particle (pulse) distorts during the transmission. In the case of a medium (fiber) with a certain cross-sectional size which is comparable to the wavelength of the microscopic particle (or light), the group velocity dispersion ω'' is determined by the property of the medium itself. For a shorter wavelength, ω'' is negative, *i.e.*, the group velocity decreases with increase of the wave number. But at a longer wavelength, ω'' is positive and the group velocity increases with increasing wave number. The domain of the wavelength corresponding to negative ω'' is called a normal dispersion region, while that corresponding to positive ω'' is called an anomalous dispersion region. If one chooses the wavelength that corresponds to the minimum loss of the medium (or fiber), then the group dispersion ω'' may become positive and the shape of the microscopic particle (or pulse) distorts considerably due to the group velocity dispersion.

However, for some nonlinear media, for example, optical fiber, the nonlinear interaction can change the property of propagation of the microscopic particle (or light). Kerr discovered that the refractive index of a fiber increases in proportion to the square of the electric field $\vec{E}(x,t)$ (or $\varphi(x,t)$). This is called effect and it is caused by a deformation of the electron orbits in the glass molecules due to the applied electric field. This effect can be expressed by $n = n_0 + n_2|\varphi|^2$ (or $n = n_0 + n_2|\vec{E}|^2$), where n_2 is called the coefficient in the case of a fiber. Obviously, this is a nonlinear effect because n is a nonlinear function of the state wave function $\varphi(x,t)$ (or the electric field $\vec{E}(x,t)$). For a grass fiber, n_2 is about 1.2×10^{-22} m^2/V^2. For a typical optical fiber with a 69 μm^2 affective cross-sectional area and 10^6 V/m electric field produced by a 100 mW optical pulse, the refractive index increases by a factor of about 10^{-10}. Because of this, the change in the frequency of the microscopic particle (or light wave) in the medium variates by a factor of $-n_2|\varphi|^2ck/(nn_0) = -2\pi n^2|\varphi|^2\omega/n_0$. When the microscopic particle (or the light wave) propagates in the medium, it experiences this effect. Therefore, the nonlinear interaction should be included in the above equation of motion (5.52). With this, equation (5.52) becomes

$$i\frac{\partial\varphi}{\partial t} - \frac{\omega''}{2!}\frac{\partial^2\varphi}{\partial x^2} + b|\varphi|^2\varphi = 0, \tag{5.53}$$

where $b = 2\pi n_2\rho\omega/(n_0\varepsilon^2)$ and ρ denotes the reduction factor due to variation of the microscopic particle (or intensity of light wave) in the cross-section of the medium (or fiber), and is approximately taken as constant in most case (or 1/2 for the fiber). Evidently, equation (5.53) is a nonlinear Schrödinger equation. In this model, the nonlinear interaction results from the interaction of the microscopic particle (or light wave, photons) with the nonlinear medium. The nonlinear medium makes the microscopic particles (or photons) localized, or a light self-focused to a light soliton. This mechanism of localization of microscopic particles is called a self-focus. Hasegawa was the first one to obtain the above nonlinear Schrödinger equation for the electric field $\vec{E}(x,t)$ and its soliton solution in optical fibers.

This self-focusing mechanism is a new type for generating nonlinear interaction in nonlinear quantum mechanics. It is also the mechanism of nonlinear interaction in hydrogen-bonded molecular systems and DNA in living systems. In these systems the neighboring heavy ion groups (or the bases in DNA) provides a nonlinear double-well potential (or the Morse potential in DNA) for the hydrogen atom (or proton). Motion of protons in these systems can self-localize to form a soliton which travels along the molecular chains. These systems will be discussed in Chapter 10.

To further clarify this mechanism, we derive the equation of motion of a microscopic particle in a nonlinear medium following the approach of Sulem. In the linear quantum mechanics, a linear wave equation often involves a linear operator $\hat{L}$ which consists of ∂_t and ∇,

$$\hat{L}(\partial_t, \nabla)\Psi = 0.$$

This equation has an approximate monochromatic wave solution,

$$\Psi = \varepsilon\varphi e^{i(\vec{k}\cdot\vec{x}-\omega t)},$$

with a constant amplitude $\varepsilon\varphi$. The frequency ω and the wave vector are real quantities, related by the dispersion relation

$$L(-i\omega, ik) = 0,$$

which is often written as $\omega = \omega(\vec{k})$.

To understand the accumulative effect or canonical resonant character of nonlinear interaction, which arises from the long time and long distance propagation of microscopic particles (or wave) in the nonlinear quantum systems, it is convenient to rewrite the above linear dispersion relation in the following form

$$[i\partial_t - \omega(-i\partial_{\vec{x}})]\varphi e^{i(\vec{k}\cdot\vec{x}-\omega t)} = 0, \tag{5.54}$$

where $\partial_{\vec{x}}$ is the gradient with respect to $\vec{x}$ and $\omega(-i\partial_{\vec{x}})$ is the pseudo-differential operator obtained by replacing k with $-i\partial_{\vec{x}}$ in $\omega(\vec{k})$.

In a weak nonlinear medium which responds adiabatically (or instantaneously) to a wave of finite amplitude, the nonlinearity is expected to affect the dispersion relation of the carrying wave (in addition to the generation of harmonics of smaller amplitudes). The frequency of the microscopic particle (or wave) becomes intensity dependent. It is then necessary to replace the frequency $\omega(\vec{k})$ by a function $\Omega(\vec{k}, \epsilon^2|\varphi|^2)$ with $\Omega(\vec{k}, 0) = \omega(\vec{k})$. This is called a self-focusing effect. Moreover, a complex wave amplitude φ of a microscopic particle is no longer a constant, but is modulated in space and time, and becomes dependent on the slow variables $X = \epsilon x$ and $T = \varepsilon t$. In (5.54), the derivatives ∂_t and $\partial_{\vec{x}}$ are thus replaced with $\partial_t + \epsilon\partial_T$ and $\partial_{\vec{x}} + \varepsilon\nabla$, where ∇ now denotes the gradient with respect to the slow spatial variable X. As a result, equation (5.54) becomes

$$\left[i\partial_t + i\epsilon\partial_T - \Omega(-i\partial_x - i\epsilon\nabla, \epsilon^2|\varphi|^2)\right] \varphi e^{i\left(\vec{k}\cdot\vec{x}-\omega t\right)} = 0.$$

It is then natural to have the weakly nonlinear dispersion relation

$$\left[\omega + i\epsilon\partial_T - \Omega(\vec{k} - i\epsilon\nabla, \epsilon^2|\varphi|^2)\right] \varphi = 0.$$

Here the parameter ϵ is small. If we expand various quantities in this equation in powers of ϵ and keep terms up to the second order, the linear dispersion relation obeyed by the carrying wave becomes

$$i\left(\partial_T + \vec{v}_g \cdot \nabla\right)\varphi + \varepsilon\left[\nabla\cdot\nabla\varphi + b|\varphi|^2\varphi\right] = 0, \tag{5.55}$$

where $v_g = \nabla_k\omega$ is the group velocity, the coupling coefficient b is related to the expansion coefficient of intensity of the microscopic particle (or wave) and is given by $\partial\Omega/\partial(|\varphi|^2)$ evaluated at $|\varphi|^2 = 0$ and at the carrier wave vector $\vec{k}$. If we consider (5.55) as an initial value problem in time, it can be written in a reference frame moving at the group velocity, by defining $\vec{\xi} = \vec{X} - T\vec{v}_g$. Rescaling the time in the form of $\tau = \varepsilon T$, we get the following nonlinear Schrödinger equation

$$i\frac{\partial\varphi}{\partial\tau} + \nabla^2\varphi + b|\varphi|^2\varphi = 0.$$

where the spatial derivatives are now taken with respect to ξ. This equation is the same as (5.53).

This nonlinear Schrödinger equation was also derived by Zakharov, L'yov, and Falkovich (1992) in their work on plasma physics (see Sulem and Sulem 1999). Their approach was based on the Fourier-mode coupling formalism in which the modulation corresponds to broadenings of frequency and wave vector spectra of the carrier wave, ω and $\vec{k}$. This mechanism also results in the localization of microscopic particles or waves, to form solitary wave, and is referred to as self-focusing as well.

Considering now a wave vector $\vec{K}$ close to k and defining $\vec{\kappa} = \vec{K} - \vec{k}$ such that $|\vec{\kappa}| \ll k = |\vec{k}|$, the corresponding frequency of the wave can be approximated by

$$\omega(\vec{K}) \approx \omega(\vec{k}) + \vec{v}_g \cdot \vec{\kappa} + \frac{1}{2}\omega''_{jl}\vec{\kappa}_j \cdot \vec{\kappa}_l$$

in such a nonlinear medium.

In the linear theory, the related Fourier-mode $\varphi(\vec{K}, t)$ is proportional to $e^{-i\omega(\vec{K})t}$ and satisfies (5.54). At present, it is of the form

$$\partial_t \varphi(\vec{K}, t) + i\omega(\vec{K})\varphi(\vec{K}, t) = 0. \tag{5.56}$$

The nonlinearity contributes mainly through four-wave interactions in this mechanism where approximately equal wave vectors are assumed. This leads to the replacement of (5.56) in the nonlinear regime by

$$\begin{aligned}
&\partial_t \varphi(\vec{K}, t) + i\omega(\vec{K})\varphi(\vec{K}, t) \\
&\qquad = i\int g_{\vec{K},\vec{K}_1,\vec{K}_2,\vec{K}_3}\varphi^*(\vec{K}_1)\varphi(\vec{K}_2)\varphi(\vec{K}_3)\delta(\vec{K} + \vec{K}_1 - \vec{K}_2 - \vec{K}_3)d\vec{K}_{123},
\end{aligned}$$

where $d\vec{K}_{123} = d\vec{K}_1 d\vec{K}_2 d\vec{K}_3$. For a narrow wave packet centered at the wave vector $\vec{k}$, the kernel g is approximately given by

$$g_{\vec{K},\vec{K}_1,\vec{K}_2,\vec{K}_3} \approx g_{\vec{k},\vec{k},\vec{k},\vec{k}} \equiv \frac{b}{(2\pi)^3}.$$

Thus we have

$$\begin{aligned}
&\partial_t \varphi(\vec{K}) + i\left[\omega\vec{k}) + \vec{v}_g \cdot \vec{\kappa} + \frac{1}{2}\omega''_{jl}\vec{\kappa}_j \cdot \vec{\kappa}_l\right]\varphi(\vec{K}) \\
&\qquad - \frac{ib}{(2\pi)^3}\int \varphi^*(\vec{K}_1)\varphi(\vec{K}_2)\varphi(\vec{K}_3)\delta\left(\vec{K} + \vec{K}_1 - \vec{K}_2 - \vec{K}_3\right)d\vec{K}_{123} = 0.
\end{aligned}$$

For a fixed $\vec{k}$, we can define

$$\phi'(\vec{\kappa}, t) = e^{i\omega(\vec{k})t}\varphi(\vec{\kappa} + \vec{k}, t).$$

This is a slowly varying function of t and obeys

$$\begin{aligned}
&\partial_t \phi'(\vec{\kappa}) + i(\vec{v}_g \cdot \vec{\kappa})\phi'(\vec{\kappa}) + \frac{i}{2}\omega''_{jl}\vec{\kappa}_j \cdot \vec{\kappa}_l \phi'(\vec{\kappa}) \\
&\qquad - \frac{ib}{(2\pi)^3}\int \phi'^*(\vec{\kappa}_1)\phi'(\vec{\kappa}_2)\phi'(\vec{\kappa}_3)\delta(\vec{\kappa} + \vec{\kappa}_1 - \vec{\kappa}_2 - \vec{\kappa}_3)d\vec{\kappa}_{123} = 0.
\end{aligned} \tag{5.57}$$

The complex field can then be obtained

$$\begin{aligned}\phi'(\vec{x},t) &= \frac{1}{(2\pi)^{3/2}}\int \phi'(\vec{\kappa},t)e^{i\vec{\kappa}\cdot\vec{x}}d\vec{\kappa}\\ &= \frac{1}{(2\pi)^{3/2}}e^{i\omega(\vec{k})t}\int \varphi(\vec{\kappa}+\vec{k},t)e^{i\vec{\kappa}\cdot\vec{x}}d\vec{\kappa}\\ &= \frac{1}{(2\pi)^{3/2}}e^{-i(\vec{k}\cdot\vec{x}-\omega(\vec{k})t)}\int \varphi(\vec{K},t)e^{i\vec{K}\cdot\vec{x}}d\vec{K},\end{aligned}$$

which appears as the envelope of the wave packet $(2\pi)^{-3/2}\int \varphi(\vec{K},t)e^{i\vec{K}\cdot\vec{x}}d\vec{K}$. From (5.57), we know that $\phi'(\vec{x},t)$ satisfies the nonlinear Schrödinger equation

$$i\partial_t\phi' + i\vec{v}_g\cdot\nabla\phi' - \frac{1}{2}\omega''_{ij}\partial_i\partial_j\phi' - ib|\phi'|^2\phi' = 0.$$

Since the amplitude of the wave is small, the function ϕ' varies slowly in space and time. The simple rescaling $\phi'(\vec{x},t) = \epsilon\phi(\vec{X},T)$ with $\vec{X} = \epsilon\vec{x}$ and $T = \varepsilon t$ then reproduces (5.53) or (5.55).

The above discussion illustrates that the motion of a microscopic particle (or wave) is always described by the nonlinear Schrödinger equation under this self-focusing mechanism which results in the nonlinear interactions in the systems.

5.6 Localization of Exciton and Self-trapping Mechanism

We are interested in the collective excitations in one-dimensional molecular systems in which excitons generated by intramolecular excitation or extra electrons can be self-trapped as a localized exciton-soliton, through its interaction with the molecular lattice. These problems were first studied by Davydov and co-workers. Examples of one-dimensional systems include α-helical protein molecules in living systems and organic molecular crystal-acetanilide (ACN). In these systems, the peptide groups (or amino acid molecules) which consists of four atoms (N, H, O and C) bound by chemical forces, are arranged along the molecular chains in the form of $\cdots$ H-N-C=O $\cdots$ H-N-C=O $\cdots$ H-N-C=O $\cdots$. The neighboring peptide groups in the chain are linked by hydrogen bonds. In acetanilide, two close chains of hydrogen bonded amide-I groups run through the crystal. In α-helix protein molecules there are three amino acid molecular chains, of which $\cdots$H-X is a hydrogen bond, X is N-C=O in which an amide group, C=O is a carbohyl group. The collective excitation in these systems are intramolecular excitations which has a very specific nature and is different from ordinary elementary excitations in three-dimensional crystals. In protein molecules and acetanilide the intramolecular excitations include amide-I vibrational quantum with an energy of 0.205 eV and an electric dipole moment of 0.3 Debye directed along the axis of the molecular chains. This intramolecular excitation or quantum of amide-I vibration was referred to as an exciton by Davydov who studied its property.

Consider an infinite chain of weakly bound molecules (or groups of atoms) of mass M, which are separated at a distance r_0. We assume that the excitons in the chains have energy ε_0 and electric dipole moment $\vec{d}$ directed along the chains. When the molecule is excited the forces of the excitons interacting with neighboring molecules are changed. This results in a change in the equilibrium distance between the molecules. If the molecules are identical, there will be some additional resonance interactions which cause transfer of the excitations from one molecule to another. The states of the excitons with energy ε_0 can be described by the Hamiltonian

$$\mathcal{H}_{\text{ex}} = \sum_n (\varepsilon_0 - D)B_n^+ B_n - J\left(B_n^+ B_{n+1} + B_{n+1}^+ B_n\right), \tag{5.58}$$

where B_n^+ (B_n) is the exciton creation (annihilation) operator at site n with an energy ε_0 (0.205 eV). They satisfy the commutation relation,

$$[B_n, B_m^+] = \delta_{nm}, \qquad [B_n, B_m] = 0.$$

Also in (5.58), $J = 2\vec{d}^2/r_0^3$ is the resonance (or dipole-dipole) interaction that determines the transition of excitation from one molecule to another, D is the deformation excitation energy, and is approximately a constant.

The collective excitations in the systems are represented by the wave function

$$|\Psi\rangle = \sum_n \psi_n(t) B_n^+ |0\rangle_{\text{ex}}, \tag{5.59}$$

where $|0\rangle_{\text{ex}}$ is the ground state of the exciton, the coefficient $\psi_n(t)$ satisfies the normalization condition

$$\sum_n |\psi_n(t)|^2 = 0.$$

We assume that the wave functional $|\Psi\rangle$ satisfies the following Schrodinger equation,

$$i\hbar \frac{\partial}{\partial t}|\Psi\rangle = H|\Psi\rangle. \tag{5.60}$$

Then we can get

$$i\hbar \frac{\partial \psi_n}{\partial t} = (\varepsilon_0 - D)\psi_n - J\left(\psi_{n+1} - \psi_{n-1}\right). \tag{5.61}$$

In the continuum approximation,

$$\psi_n(t) = \psi(x,t),$$
$$\psi_{n\pm 1}(t) = \psi(x,t) \pm r_0 \frac{\partial \psi}{\partial x} \mp \frac{1}{2!} r_0^2 \frac{\partial^2 \psi}{\partial x^2},$$

we can get the equation of motion of the excitons

$$i\hbar \frac{\partial \psi(x,t)}{\partial t} = (\varepsilon_0 - D - 2J)\psi(x,t) - \frac{\hbar^2}{2m}\frac{\partial^2 \psi(x,t)}{\partial x^2}, \tag{5.62}$$

where $m = \hbar^2/2Jr_0^2$ is the effective mass of the exciton. Equation (5.62) shows that the exciton, an elementary excitation, satisfies the linear Schrödinger equation in the linear quantum mechanics. Therefore, the exciton is a standard microscopic particle with wave features. It is described by linear quantum mechanics and has the same properties of the electron.

We now determine the energy of the exciton. Making the transformation

$$B_n = \frac{1}{\sqrt{N}}\sum_k A_k e^{iknr_0}, \quad A_n = \frac{1}{\sqrt{N}}\sum_k B_k e^{-iknr_0}, \tag{5.63}$$

the periodic boundary condition requires that

$$k = \frac{2\pi}{r_0}m, \quad (m = 0, \pm 1, \pm 2 \cdots).$$

Inserting (5.63) into(5.58), we get

$$H_{\text{ex}} = \sum_\kappa E(k) A_k^+ A_k, \tag{5.64}$$

where $E(k) = \varepsilon - D - 2J\cos\kappa r_0$, which is the energy of the exciton. The wave function of the exciton is given by

$$\psi_k(x,t) = \frac{1}{\sqrt{N}}\sum_\kappa e^{i[knr_0 - \omega(k)t]}\psi'(x - nr_0), \tag{5.65}$$

where $\hbar\omega(k) = E(k)$, and $\psi'(x - nr_0)$ is the amplitude of the exciton at site n. Equation (5.65) shows that once the exciton is formed, all components in the chains are excited, *i.e.*, the exciton state with energy $E(k)$ spreads over the entire chain, rather than localizes at one molecule. Hence, it is an expanded and collective excitation states.

Equation (5.65) may be rewritten as

$$\psi_k(x,t) = A(k)e^{i[kx - \omega(k)t]}, \tag{5.66}$$

where

$$A(\kappa) = \frac{1}{\sqrt{N}}\sum_n e^{ik(nr_0 - x)}\psi'(x - nr_0).$$

This shows that the exciton moves as a plane wave with a wavelength $\lambda = 2\pi/k$, but its amplitude is modulated. When its wavelength is larger than the spacing between the molecules, *i.e.*, $kr_0 \ll 1$, the energy of the exciton can be written as

$$E(k) = E_0 + \frac{\hbar^2 k^2}{2m}, \tag{5.67}$$

where $E_0 = \varepsilon - D - 2J$. Therefore, the energy spectrum is continuum. In such a state, the excitation energy is uniformly distributed over the whole chain, and thus cannot propagate along the chain. Excitation encompassing a small part of the chain, ℓ_0, is not stationary.

Quasi-stationary excitation localized over a segment ℓ_0 is described by the wave packet

$$\Psi_{\text{ex}}(x,t) = \int_{k_0-\Delta k}^{k_0+\Delta k} A(k) e^{i[kx-\omega(k)t]} dk, \tag{5.68}$$

where $\Delta k = \pi/2\ell_0$. The group speed of the wave packet is

$$v_g = \left.\frac{\partial\omega(k)}{\partial k}\right|_{k=k_0} = \frac{\hbar k_0}{m}.$$

Here $\hbar k_0$ is the average momentum of the wave packet. When the resonant interaction $|J|$ increases, the effective mass m^* decreases, and the velocity v_g increases for a given k_0. Therefore, the exciton with energy $E(k_0)$ localized in a segment ℓ_0 at time t becomes the wave packet with a length

$$\ell_\tau = \sqrt{\ell_0^2 + \left(\frac{\hbar\tau}{m\ell_0}\right)^2} > \ell_0 = \frac{\pi}{2\Delta k}$$

at time $t+\tau$. This indicates that the wave packet is spread out and dispersed with increasing time. When $\tau > m\ell_0/k$, the wave packet will collapse gradually as described in Chapter 1, and (5.11) – (5.13). The exciton is essentially unstable.

However, since the exciton is produced by vibrations of the amide-I (or the C=O stretching) in the peptide groups, it is a intramolecular excitation. When it moves and decays, the states and positions of the peptide groups will be changed. If we consider this effect and the low-frequency vibration of the peptide group, and assume that the displacement of the peptide groups is u_n, then the low-frequency vibrational Hamiltonian of the peptide groups and the interaction Hamiltonian between the exciton and the vibration of the peptide group can be written as

$$H_{\text{ph}} = \sum_n \left[\frac{P_n^2}{2M} + \frac{1}{2}w(u_n - u_{n-1})\right], \tag{5.69}$$

$$H_{\text{int}} = \sum_n \chi(u_{n+1} - u_{n-1}) B_n^+ B_n, \tag{5.70}$$

respectively, where M is the mass of the peptide group, w is the force constant of the molecular chain, P_n is the conjugate momentum of u_n, $\chi = \partial J/\partial u_n$ is the coupling constant. In such a case, the Hamiltonian of the system can be written as

$$H = H_{\text{ex}} + H_{\text{ph}} + H_{\text{int}}. \tag{5.71}$$

Davydov used the following wave function to describe the collective excitation state

$$|\Phi(t)\rangle = \sum_n \phi_n(t) B_n^+ |0\rangle_{\text{ex}} \exp\left\{-\frac{i}{\hbar}\sum_n [\beta_n(t)P_n - \pi_n(t)u_n]\right\}|0\rangle_{\text{ph}}, \tag{5.72}$$

where $|0\rangle_{\text{ph}}$ is the ground state of the phonon. Using the variational approach and the functional $\langle\Phi(t)|H|\Phi(t)\rangle$, we can get

$$i\hbar\frac{\partial\phi_n}{\partial t} = [\varepsilon_0 - D + W + \chi(\beta_{n+1} - \beta_{n-1})]\phi_n - J(\phi_{n+1} + \phi_{n-1}), \tag{5.73}$$

$$\frac{\partial^2\beta^2}{\partial t^2} + \frac{w}{M}(2\beta_n - \beta_{n-1} - \beta_{n+1}) = \frac{\chi}{M}\left(|\phi_{n+1}|^2 - |\phi_{n-}|^2\right), \tag{5.74}$$

where

$$\beta_n(t) = \langle\Phi(t)|u_n|\Phi(t)\rangle,$$

$$\pi_n = \langle\Phi(t)|p_n|\Phi(t)\rangle = M\frac{\partial\beta_n}{\partial t},$$

$$W = \frac{1}{2}\sum_n\left[M\left(\frac{\partial\beta_n}{\partial t}\right)^2 + w(\beta_n - \beta_{n-1})^2\right].$$

In the continuum approximation, equations (5.73) and (5.74) become

$$\left[i\hbar\frac{\partial}{\partial t} - \Lambda + \frac{\hbar^2}{2m}\frac{\partial^2}{\partial x^2} - 2\chi\frac{\partial\beta(x,t)}{\partial x}\right]\phi(x,t) = 0, \tag{5.75}$$

$$\left[\frac{\partial^2}{\partial t^2} - v_0^2\frac{\partial^2}{\partial x^2}\right]\beta(x,t) - \frac{2\chi r_0}{M}\frac{\partial}{\partial x}|\phi(x,t)|^2 = 0, \tag{5.76}$$

where $\Lambda = \varepsilon_0 - 2J - D + W$, $v_0 = r_0\sqrt{\omega/M}$ is the sound speed of the molecular chain. Now let $\xi = x - x_0 - vt$, $G = 4\chi^2/[w(1-s^2)]$, $s = v/v_0$, $\mu = G/4J$, we can get from (5.75) and (5.76) that

$$i\hbar\frac{\partial\phi}{\partial t} - \Lambda\phi + \frac{\hbar^2}{2m}\frac{\partial^2\phi}{\partial x^2} + G|\phi|^2\phi = 0, \tag{5.77}$$

$$\frac{\partial\beta(x,t)}{\partial x} = -\frac{2r_0\chi\,|\phi(x,t)|^2}{w(1-s^2)}. \tag{5.78}$$

Equation (5.77) is a nonlinear Schrödinger equation. It has a soliton solution as given in (2.150). we now write the solution as

$$\phi(x,t) = \sqrt{\frac{u}{2}}\,\text{sech}\left[\frac{\mu}{r_0}(x - x_0 - vt)\right]\exp\left\{i\left[\frac{\hbar v}{2Jr_0^2}(x - x_0) - \frac{Et}{\hbar}\right]\right\}, \tag{5.79}$$

$$\beta(x,t) = -\frac{\chi r_0^2}{w(1-s^2)}\tanh\left[\frac{\mu}{r_0}(x - x_0 - vt)\right]. \tag{5.80}$$

Equations (5.77) – (5.80) show clearly that the exciton now is localized and becomes a soliton due to the nonlinear interaction $G|\phi|^2$. The energy of the exciton-soliton is

$$E = E_0 + \frac{1}{2} m_{\text{sol}} v^2, \tag{5.81}$$

where v is the soliton velocity, E_0 is the rest energy of the soliton, given by

$$E_0 = \varepsilon_0 - D - 2J - \frac{\chi^2}{3w^2 J} < E_{\text{ex}} = \varepsilon_0 - D - 2J, \tag{5.82}$$

and

$$m_{\text{sol}} = m + \frac{4\chi^4(1 + 3s^2/2 - s^4/2)}{3w^2 J v_0^2 \left(1 - s^2\right)^3} > m.$$

Equation (5.82) shows that the soliton energy is lower than that of the exciton given in (5.64) by $E_B = -\chi^4/3w^2 J$, which is the binding energy of the soliton. Obviously, E_B increases with increasing coupling interaction between the exciton and vibration of the peptide group for a fixed chain. This result reveals not only that the soliton is more stable, but also the mechanism of the soliton formation. This mechanism of exciton localization is called a self-trapping of the exciton due to the nonlinear interaction, $G|\phi|^2$, generated by the interaction between the exciton and the vibration of the peptide groups. The self-trapping mechanism of excitons works as follows. The exciton or vibrational energy of the amide-I (or C=O bond stretching oscillation) acts through a phonon (vibration of the peptide group) coupling effect to distort the structure of the molecular chains. The chain deformation reacts, again through phonon coupling, to trap the exciton and to prevent its dispersion. Thus the exciton is localized on the chain and becomes a soliton. This is the self-trapping mechanism of the exciton. This concept or mechanism of self-trapping was early proposed by Landau on the motion of an electron in a crystal lattice. He suggested that an effect of a localized electron would be to polarize the crystal which, in turn, would lower the energy of the electron. Landau's suggestion was discussed in detail by Pekar, Fröhlich and by Holstein. Other examples of the self-trapping include electromagnetic energy in a plasma and hydrodynamic energy in a water tank which will be discussed in Section 5.8. Besides Davydov, transport of vibrational energy in α-helix proteins and acetanilide have been studied extensively by many scientists. An improved theory was proposed by Pang for energy transport in protein molecules which will be discussed in Chapter 9.

5.7 Initial Condition for Localization of Microscopic Particle

As it is known from earlier discussion, the necessary and sufficient conditions for localization of microscopic particle in nonlinear quantum mechanics are that the dispersion effect and nonlinear interaction occur simultaneously and balance each other. However, an appropriate initial perturbation or excitation is also an important factor for the localization of microscopic particle. Proper initial perturbation will be able to enhance the nonlinear interaction on microscopic particle. Therefore, it is necessary to study the influence of time evolution of the initial excitation on the localization. Brizhik translated the mechanism of soliton generation in a molecular chains into a problem of time evolution of an initial excitation, in the form of an exponential damping in space in accordance with the nonlinear Schrödinger equation (4.40) in which b is now replaced by $2b$, with the following initial condition

$$\phi(x',0) = \begin{cases} \sqrt{2\alpha}e^{(ik-\alpha)x'} & \text{if } x' \geq 0, \\ 0 & \text{if } x' < 0, \end{cases} \tag{5.83}$$

where α is a constant. The nonlinear Schrödinger equation was integrated by the inverse scattering method by Zakharov and Shabat. The same method can also be used to investigate the time evolution of the initial impulses in the form of rectangular steps. According to this method, the nonlinear Schrödinger equation with an initial state corresponding to an arbitrary function $\phi(x',0)$ which decreases rapidly at infinity is consistent with the linear scattering problem for the two-component eigenvector ψ, satisfying (4.42) with eigenvalues $\zeta = \xi + i\eta$ and potential $q(x)$ determined via the initial state (5.83),

$$q(x') = i\sqrt{b}\phi(x',0).$$

In the region of $x' < 0$, equation (4.42) has the solution

$$\psi = \begin{pmatrix} 1 \\ 0 \end{pmatrix} e^{-i\xi x'}, \tag{5.84}$$

while in the region of $x' > 0$, the solution is expressed in terms of the Bessel functions,

$$\psi_1 = \left[AJ_{-m}(y) + BJ_m(y)\right] e^{(ik-\alpha)x'/2},$$
$$\psi_2 = -i\left[AJ_{-m}(y) - BJ_{m-1}(y)\right] e^{-(\alpha+ik)x'/2},$$

where

$$y = je^{-\alpha x'}, \quad j = \sqrt{\frac{2b}{\alpha}}, \quad m = \frac{(\alpha - ik)/2 - i\zeta}{\alpha}. \tag{5.85}$$

The constants of integration are determined by matching the solutions (5.84) – (5.85) at $x' = 0$.

$$A = \frac{J_{m-1}(j)}{D}, \quad B = \frac{J_{-m+1}(j)}{D},$$

where

$$D = J_m(j)J_{-m+1}(j) + J_{-m}(j)J_{m-1}(j).$$

The asymptotic behavior of this solution at $x' \to \infty$ is of the form

$$\psi = a(\zeta)\begin{pmatrix}1\\0\end{pmatrix}e^{-i\zeta x'} + b'(\zeta)\begin{pmatrix}0\\1\end{pmatrix}e^{i\zeta x'},$$

where $a(\zeta)$ and $b(\zeta)$ are the transmission and reflection coefficients, respectively, given by

$$a(\zeta) = \left(\frac{2\alpha}{b}\right)^{m/2}\frac{A}{\Gamma(1-m)}, \quad b'(\zeta) = i\left(\frac{b}{2\alpha}\right)^{(m-1)/2}\frac{B}{\Gamma(m)}, \tag{5.86}$$

where $\Gamma(m)$ is the gamma function.

According to the inverse scattering method, the bound states or localization of the microscopic particle corresponds to solitons and are determined by the zeros of the transmission coefficient. The velocities of the solitons are proportional to the real part of the corresponding eigenvalues, $v_i = -4\xi_i$, $\xi_i = \Re\zeta_i$, and the amplitudes (and widths) are proportional to the imaginary parts, $\alpha = 2\eta_i$, $\eta_i = \Im\zeta_i$. From (5.86) we know that the zeros of the transmission coefficient are determined from the equation

$$J_{m-1}(j) = 0. \tag{5.87}$$

According to (5.85), the soliton velocity is proportional to the wave number of the initial excitation, $\xi = -k/2$. Thus

$$m - 1 = \mu = \frac{2\eta - \alpha}{\alpha}. \tag{5.88}$$

According to the theorems on the absence of multiple zeros (except for $j = 0$), and on the alternation of the roots of linearly independent real Bessel functions, (5.87) yields the amplitude of the microscopic particle (soliton), η, as a function of the width of the initial impulse α. The lower and upper bounds of the amplitude of the microscopic particle are given by the Shafheitlin's theorem,

$$\frac{\alpha}{2} - \sqrt{2b\alpha} < \eta < \frac{\alpha}{2} + \sqrt{2b\alpha}.$$

Because η is related to the index of the Bessel function, μ, which in turn depends on the root j according to (5.87), one can get the α dependence of η as follows

$$\frac{d\eta}{d\alpha} = \frac{\eta}{\alpha} + \frac{bF(\eta,\alpha)}{2\eta - \alpha}, \tag{5.89}$$

where

$$F(\eta,\alpha) = J_{\mu+1}^2(j)\left|\int_0^j J_\mu^2(z)\frac{dz}{z}\right|^{-1} > 0.$$

From (5.89), Brizhik obtained that at very small values of α the amplitude of the microscopic particle increases with α, till the value of α_0 which is determined from the transcendental equation

$$\alpha_0 F(\eta, \alpha_0)\, b = \frac{\alpha_0 - 2\eta}{\eta}.$$

Beyond α_0, the derivative in (5.89) changes sign. Therefore, we can conclude from (5.87) and (5.88) that soliton solutions exist in the region $j > 1.5$, *i.e.*,

$$\alpha < \alpha_{\text{cr}} = 0.889b. \tag{5.90}$$

In the interval $\alpha_{\text{cr}}^{ls} < \alpha < \alpha_{\text{cr}}$, where $\alpha_{\text{cr}}^{ls} = 0.08b$, a single soliton is generated, and its amplitude increases from zero to $\eta = 0.176b$, and eventually reaches the maximum value $\eta_{\max} = 0.180b$ at $\alpha_0 = 0.12b$ (see Fig. 5.9).

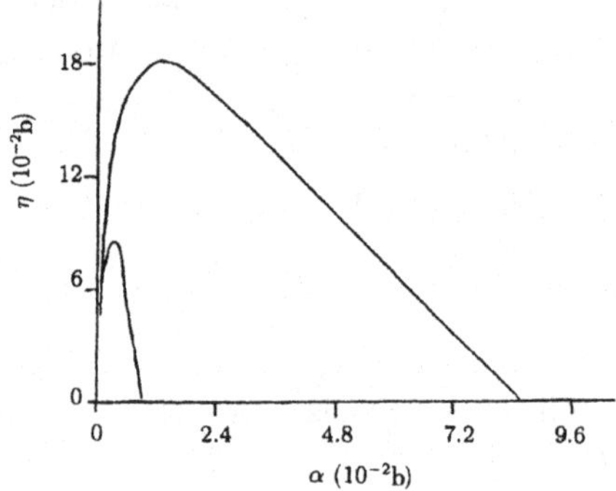

Fig. 5.9 Dependence of soliton amplitude η on the inverse width α of the initial condition (b is the nonlinearity parameter of the system).

When α has values in the interval $\alpha_{\text{cr}}^{2s} < \alpha < \alpha_{\text{cr}}^{ls}$, where $\alpha_{\text{cr}}^{2s} = 3.6 \times 10^{-2}b$, equation (5.87) admits a two-soliton solution with parameters $\mu_{i,\min}^{2s} < \mu_i < \mu_{i,\max}^{2s}$, ($i = 1, 2$). In this case, the amplitude of the soliton with larger initial amplitude decreases with decreasing α, while the amplitude of the second soliton increases with decreasing α, but is always smaller than that of the first soliton, $\eta_2 < \eta_1$. If $\alpha \ll 1$, the initial component decays with time into N solitons with small amplitudes $\eta_i/b \ll 1$, which approach to zero at $\alpha \to 0$.

The propagation of microscopic particle (soliton) is accompanied by an oscillating tail which decreases with time, and is determined by the reflection coefficient (5.86).

The above discussions show clearly that for a given initial condition the exponentially decreasing initial impulse admits soliton solution, or localization of the microscopic particle, if the nonlinearity of the system exceeds the critical value, which, according to (5.77), is inversely proportional to the width α of the initial impulse, $b > b_{\text{cr}} = 1.125\alpha$.

The existence of the critical value of the nonlinearity parameter was shown by numerical calculations for Davydov's solitons in molecular systems by Hyman *et al.* and Scott, and by analytical investigation of the rectangular step evolution (see book by Christiansen *et al.*).

5.8 Experimental Verification of Localization of Microscopic Particle

The theoretical prediction on localization of microscopic particles discussed above is subject to experimental verification. However, it is almost impossible to experimentally observe the localization of microscopic particles in a material using available instruments. How then is it possible to verify these predictions? It is fortunate that the nonlinear Schrödinger equation describing localization of microscopic particle in nonlinear quantum mechanics can also be used to depict the nonlinear behaviors of condensed matter consisting of a large number of molecules, such as water, and transmission of light in optical fibers. The nonlinear Schrödinger equation describing the self-focus mechanism of microscopic particle in a nonlinear medium such as an optical fiber has been given in Section 5.5. The fluid dynamical form of the nonlinear Schrödinger equation was also given in (3.110) – (3.111). As a matter of fact, a nonpropagating solitary wave in water can be described by a time-independent nonlinear Schrödinger equation. Therefore, water-soliton and light-soliton (or photon-soliton) can be used to verify the nonlinear behaviors or localization of microscopic particles.

Nonpropagating solitary wave was first found in water by Wu *et al.* in 1984. Subsequently, it was confirmed experimentally by Cui *et al.*. Larraza *et al.*, Miles, and Pang *et al.* demonstrated theoretically that this phenomenon can be very well described by a nonlinear Schrödinger equation. Larraza *et al.* derived that the surface wave in water troughs satisfies the following nonlinear Schrödinger equation

$$C^2\frac{d^2\phi}{dx^2} + \left(\omega_1^2 - \omega^2\right)\phi - A|\phi|^2\phi = 0, \qquad (5.91)$$

based on the velocity potential of the fluid which satisfies the Laplace's equation and the corresponding boundary conditions.

For $A > 0$ and $k\phi > 1.022$, the soliton solution is of the form

$$\phi = \sqrt{\frac{2\left(\omega_1^2 - \omega^2\right)}{A}}\ \text{sech}\left[\frac{(\omega_1^2 - \omega^2)x}{C^2}\right], \qquad (5.92)$$

where

$$C^2 = \frac{g}{2k}\left[\tilde{T} + kd\left(1 - \tilde{T}^2\right)\right],$$

$$A = \frac{1}{8}k^4\left(6\tilde{T}^4 - 5\tilde{T}^2 + 16 - 9\tilde{T}^{-2}\right),$$

and d is the depth of the water, $k = \pi/\tilde{b}$, where $\tilde{b}$ is the width of the water troughs, $\tilde{T} = \tanh(kd)$, $\omega_1^2 = gk\tilde{T}$, where g is the acceleration due to gravity, and ω is the frequency of the applied field.

A slightly different form was obtained by Miles

$$B\phi_{xx} + \left(\beta + A|\phi|^2\right)\phi + \nu\phi^* = 0, \tag{5.93}$$

which has the following soliton solution

$$\phi = e^{i\theta}\text{sech}\left(\sqrt{\frac{A'}{2B}}X\right), \tag{5.94}$$

at $\nu = 0$, where

$$B = \tilde{T} + kd\left(1 - \tilde{T}^2\right), \quad A' = \frac{1}{8}\left(6\tilde{T}^4 - 5\tilde{T}^2 + 16 - 9\tilde{T}^{-2}\right),$$

$$\beta = \frac{\omega^2 - \omega_1^2}{2\varepsilon\omega_1^2}, \quad \gamma = \frac{\omega^2 a_0}{\varepsilon g}, \quad \omega_1 = \sqrt{gk\tanh(kd)}, \quad X = 2\sqrt{\varepsilon\tilde{T}}\ kx,$$

a_0 and θ are constants, ε is a small, positive scaling parameter.

Pang *et al.* obtained the following nonlinear Schrödinger equation

$$\frac{\partial^2\phi_n}{\partial x^2} - K^2\phi_n(x) + \gamma'|\phi_n|^2\phi_n = 0, \tag{5.95}$$

which has the following soliton solution

$$\phi = B_n\phi_n(x)\cos(k_n y)\frac{\cos[k_n'(z+d)]}{\cos(k_n^1 d)},$$

with

$$\phi_n(x) = B_n\sqrt{\frac{2k^2}{\gamma'}}\ \text{sech}\left[\sqrt{k^2}(x - x_0)\right]\frac{\cos[k_n'(z+d)]}{\cos(k_n' d)}\cos(k_n y) \tag{5.96}$$

where

$$K^2 = k_n^2 + k_n'^2, \quad k_n \to \frac{\pi}{b}n, \quad k_n' = \frac{\pi}{d}n, \quad (n = 1, 2, 3\cdots),$$

$$\gamma' = \frac{\alpha^2 B_n^2}{\rho^2(\omega^2 - \omega_0^2)}.$$

Here $\omega_0^2 = \beta/\rho$, β is surface tension constant, ρ is the density of the water, α is the nonlinear coefficient of the surface water, B_n is a constant. From (5.91) – (5.96), we see that the water-soliton satisfies the time-independent nonlinear Schrödinger equation, and has a bell-shaped form.

Likewise, the discussions in Section 5.5 showed also that the light transmission in optical fibers satisfies the nonlinear Schrödinger equation and light can be self-focused to form a light soliton which was obtained by Hasegawa *et al.* These phenomena can be clearly observed experimentally. In the following, results of these two experiments are briefly presented.

5.8.1 *Observation of nonpropagating surface water soliton in water troughs*

The experimental apparatus used by Cui *et al.* was quite simple. It included a organic glass trough of 38 cm long and 2.0 – 5.0 cm wide, which is filled with various liquids to a depth of $d = 2.0 - 5.0$ cm; a loudspeaker, whose cone is driven by a low frequency $(7-15$ Hz$)$ signal in the vertical direction and another signal $(12-25$ Hz$)$ in the horizontal direction, respectively; a power amplifier and some measurement instruments. The water trough was placed on the vibrational platform which was driven by the power amplifier. During the initial period when the frequency of the driving signal was below a certain threshold, only some small ripples were observed and no significant wave generation took place in the surface of the water trough. In this case, the density of water molecules were uniform. However, above the threshold driving frequency, if a nonlinear initial disturbance was applied to the surface of water trough, a parametrically excited wave was observed at half the driven frequency. A large number of water molecules assembled and got together to form a bell-shaped nonpropagating solitary wave on the surface of the water in the trough, due to nonlinear interaction among the water molecules arising from the surface tension, as shown in Fig. 5.10. The soliton was the same type as that of the nonlinear Schrödinger equation in nonlinear quantum mechanics.

Fig. 5.10 The bell-type nonpropagating surface water soliton occurred in the water trough.

An obvious peculiarity in this phenomenon is that the density of water molecules where the soliton occurs is much larger than other places in the water trough, as shown in Fig. 5.11. Therefore, the mechanism forming the water soliton through gathering of the water molecules can be called self-localization of water molecules due to the nonlinear interaction arising from surface tension.

With different liquids and different initial disturbances, one, two, or even more solitary waves can be established on the water surface. However, the shapes of these solitary waves are different in different liquids. If the driving amplitude is further increased, more solitary waves can be formed.

From these experiments, the following conditions for forming solitary wave were derived.

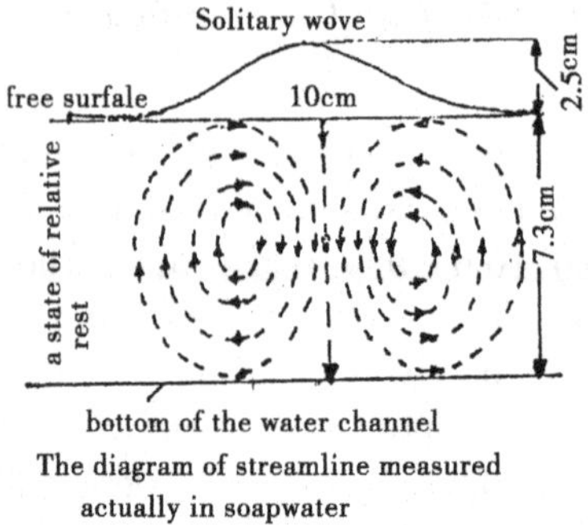

Fig. 5.11 The change of density of water and illustration of direction of motion of water molecules.

(1) The liquid must have a certain depth.
(2) The coefficient of surface tension of the liquid must be smaller than a certain value. This upper bound depends on the characteristics of the liquid.
(3) The vertical driving frequency is about twice the intrinsic frequency, and the horizontal frequency is about the same as the intrinsic frequency. The driving amplitude must be in a certain range.
(4) There must be a nonlinear initial excitation.
(5) The channel should not be too wide.

Another interesting behavior is that the soliton has a fixed position in the x direction and does not propagate, but it oscillates in the y direction, and the frequency of oscillation is about the same as the intrinsic frequency. In the x direction, the profile of the soliton is a bell-shaped curve. In glycrini-water, the soliton curve can be fitted to the expression: $\phi = 1.7\text{sech}(x/1.25)$ cm. The measured profiles of soliton for four other liquids are shown in Fig. 5.12. They are given in the order of increasing surface tension. Curve I is that of an ideal fluid whose surface tension is zero. Curve V is the wave of glycrine-water. It is clear that the amplitude of the soliton increases, while its width decreases as the surface tension increases. If the coefficient of surface tension is greater than the upper bound, no soliton can be formed.

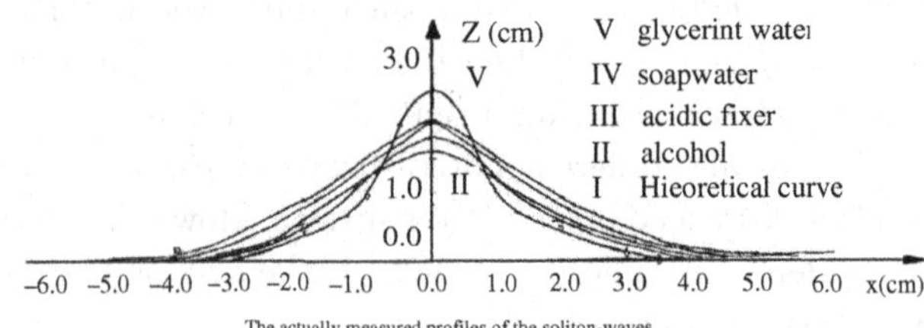

Fig. 5.12 Measured profiles of soliton-waves.

The surface water solitons can move under the effect of external forces, for example, when they are pushed by a little oar, or when they are blown by wind. If saltwater is added into the channel so as to increase the surface tension, the soliton moves in the direction of greater surface tension. If the channel is tilted to a slope of 0.05, the soliton moves toward the shallow side and the height of the solitary wave is reduced. As soon as the soliton reaches the point where the depth of water is the minimum for forming solitons, the motion of the soliton stops. It often occurs that the motion of a soliton can be prevented by another soliton which is out of phase and has a small amplitude. These dynamic properties of water soliton are the same as those of classical particles as well as microscopic particles in nonlinear quantum mechanics, as described in Sections 4.1 and 4.2. Therefore, these results were regarded as an experimental verification of the dynamic properties of the microscopic particle in nonlinear quantum mechanics discussed in Chapter 4.

The following two types of interaction or collision between two solitons should be distinguished. (1) Out-of-phase solitons repel each other and do not merge, as shown in Fig. 5.13. (2) Interactions between two solitons that are in-phase show a cyclic process of attraction $\rightarrow$ merging $\rightarrow$ separation $\rightarrow$ re-attraction, as the driving amplitude is increased, as shown in Fig. 5.14.

These properties of collision between solitons are basically the same as those obtained in Sections 4.3 – 4.7 for microscopic particles. Thus the properties of collision of microscopic particles in nonlinear quantum mechanics, and the fact that the forms of the microscopic particles can be retained after a collision, are experimentally confirmed.

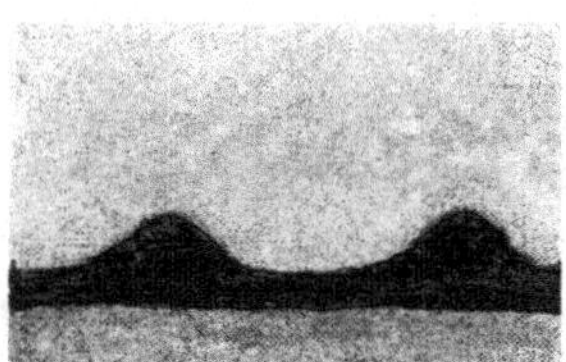

Fig. 5.13 Collision between two solitons which are out-of-phase.

In the case of horizontal driving force, the driving frequency required for soliton formation increases when the depth of the liquid increases. If the channel is rotated horizontally by upto 40 degrees while the driving signal remains the same, the soliton still remains at its original place. Similar phenomena can also be observed in non-rectangular channels, for example, in round, ring or trapezoid channels, and V- or X-shaped channels.

Solitons can be easily generated and are very stable if the liquid is magnetized. The collision process between two solitary waves can also be easily observed in magnetized liquid.

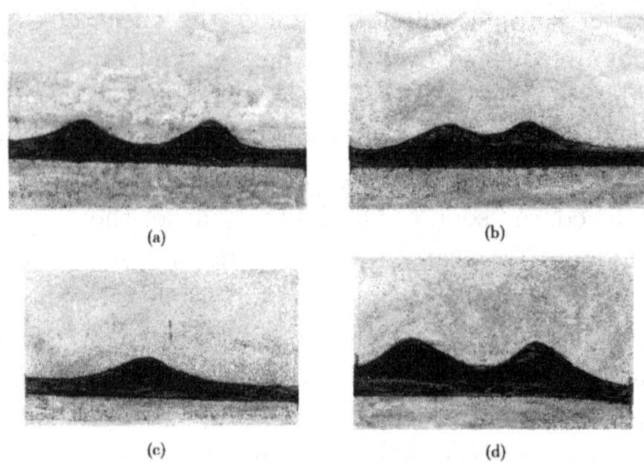

Fig. 5.14 Collision between two solitons which are in-phase.

5.8.2 *Experiment on optical solitons in fibers*

From earlier discussion, we know that optical solitons generated by nonlinear interaction arising from the Kerr effect in optical fibers can be well described by the nonlinear Schrödinger equation. The form and properties of the solitons can be verified experimentally. Thus, we can understand the mechanism of localization of microscopic particles in nonlinear quantum mechanics by examining the optical solitons in details.

In order to experimentally verify the propagation of a soliton in an optical fiber, it is necessary to generate a short optical pulse with sufficiently high power and use a fiber which has a loss rate less than 1 dB/km, and it is required that the spectral width of the laser is narrower than the inverse of the pulse width in time. In other words, it requires the generation of a pulse with a narrow spectrum. In 1980, Mollenauer *et al.* at the Bell Laboratories succeeded for the first time in experimentally verifying soliton transmission in an optical fiber. This was achieved by using an F^{2+} color center laser which is pumped by a Nd:YAG laser. Using a fiber with a relatively large cross-section (10^{-6} cm^2) and a length of 700 meters, they transmitted an optical pulse with a 7 ps width and measured the shape of the output pulse by means of autocorrelation. Their experimental results are shown in Fig. 5.15. The pulse shape is measured by autocorrelation at the output side of the fiber, for different input power levels. It can be seen clearly from Fig. 5.15 that while the output pulse width increases for a power below the threshold of 1.2 W, it continuously decreases for an input power above 1.2 W. The appearance of two peaks in the case of an input power of 11.4 W is a result of phase interference of three solitons, which are produced simultaneously in such a case. This is consistent with results of numerical calculation obtained by Hasegawa *et al.*. The periodic behavior of the higher order solitons was confirmed by Stolen *et al.* in a later experiment. Most experimental observations of light solitons were achieved by using the color center laser because of the need to produce the Fourier transform limited pulse. However, light solitons have also been successfully observed using other types of lasers, such as the dye laser or laser diodes. But it is necessary to control frequency chirping and to have a narrow spectrum for a well-defined soliton.

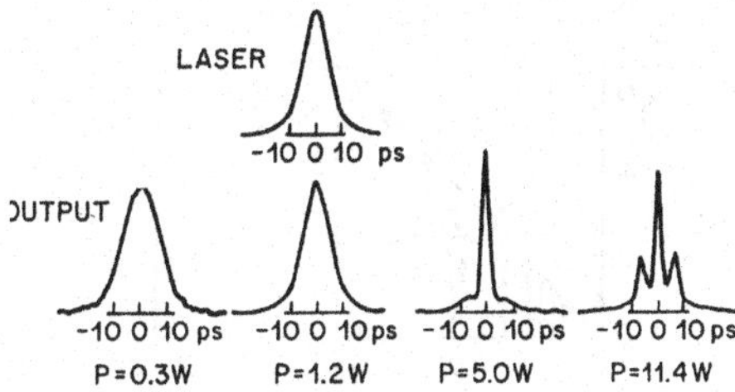

Fig. 5.15 Experimental observation of soliton formation by Mollenauer *et al.*

The interaction between optical solitons has been experimentally observed by Mitschke and Mollenauer in 1987. Both the repulsion and the attraction regimes have been observed, depending on the soliton phase difference. In this experiment, solitons were generated using a soliton laser. The pulse first passed through a Michelson interferometer, giving a pair of pulses. The length of one arm of the interferometer permitted a change in the distance between the pulses from zero up to several picoseconds. The soliton phase difference was also measured. In this experiment, they used a polarization preserving 340 m low-loss (> 0.3 dB/km) fiber with $D = 14.5$ ps/(nm·km) at an operating wavelength of 1.52 μm. The pulse duration was ≈ 1 ps. Therefore the fiber length is 10 soliton periods. Such a long fiber enables the researchers to clearly observe interaction between two well-separated soliton pulses. The results of the measurements are shown in Fig. 5.16 for both the repulsive and attraction regimes (See book by Abdullaev *et al.*). As it can be seen, the observed result is different from that predicted by theory of optical soliton in the case of attractive interaction and in the region where there is a small initial distance between the pulses. According to the theoretical calculation discussed in Chapter 4, two interacting microscopic particles (solitons) pass through each other. This is denoted in Fig. 5.16 by the oscillating structure of the theoretical curve. The experimental data deviate somewhat from such a feature. In the unstable region the attractive force becomes repulsive, as a result of the influence of the Raman self-frequency shift.

Mollenauer and Smith (1989) discovered that between optical solitons separated at long distances (more than 1000 km), there exists a long-range phase-independent interaction. Dianov *et al.* (1990) showed that the electrostrictional mechanism may be responsible for the observed anomalous interaction. Reynaud and Barthelemy (1990) observed interaction of spatial solitons between two soliton beams. These results demonstrated the existence of a mutual interaction between two close propagating soliton beams which depends on the relative phase and the distance between

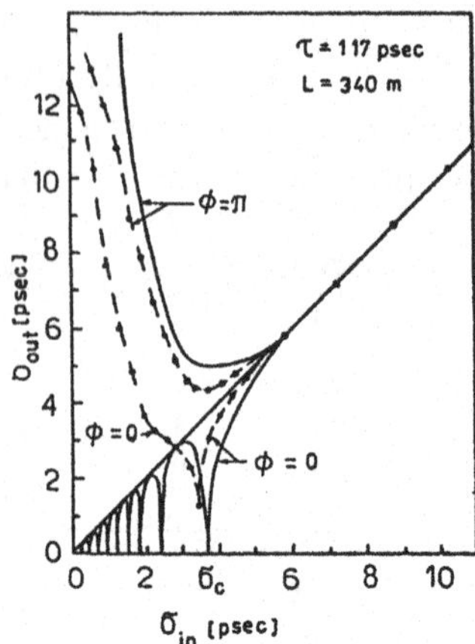

Fig. 5.16 Pulse separation at output (σ_{out}) vs. the pulse separation at input (σ_{in}) in a 340 m fiber. Solid curve: theory; dashed line: experiment. Without interaction, all points would fall on the 45° line (Courtesy of Mitschke and Molenauer, 1987)

the beams. (see book by Abdullabev *et al.*)

In addition, Vysloukh and Cherednik (1986) developed an effective method based on a combination of the inverse scattering theory and numerical methods to describe soliton interaction, and to follow the evolution of an arbitrary pulse. The method is useful for studying dynamics of multi-soliton states, and for studying the interaction of optical solitons in nearly integrable systems. These experimental results of soliton collisions are basically consistent with that of the microscopic particles described by nonlinear Schrödinger equation in Chapter 4.

From Fig. 5.15, we see that a bell-type soliton, which is the same as that in Fig. 4.1, can exist in optical fibers. This exhibits clearly that light wave with dispersive feature can be self-localized or self-focused due to the nonlinear interaction to form a stable soliton which is described by the nonlinear Schrödinger equation. This makes us believe that microscopic particles in nonlinear quantum mechanics can be self-localized or self-trapped as solitons with wave-corpuscle duality under nonlinear interactions and the theories discussed in Chapters 4 and 5 give a appropriate description of the microscopic particles in nonlinear quantum mechanics.

Bibliography

Abdullaev, F., Darmanyan, S. and Khabibullaev, P. (1994). Optical solitons, Springer, Berlin.

Barut, A. D. (1978). Nonlinear equations in physics and mathematics, D. Reidel Publishing, Dordrecht.

Beehgaard, K. and Jerome, D. (1982). Sci. Am. **247** (50) .

Beehgaard, K., Jerome, D. (1981). J. Am. Chem. Soc. **103** (2440) .

Borken, J. D. and Drell, S. D. (1964). Relativistic quantum mechanics, McGraw-Hill, New

York.
Borken, J. D. and Drell, S. D. (1965). Relativistic quantum fields, McGraw-Hill, New York.
Brizhik, L. S. (1993). Phys. Rev. B **48** (3142) .
Brizhik, L. S. and Davydov, A. S. (1983). Phys. Stat. Sol. (b) **115** (615) .
Burger, S.,*et al.* (1999). Phys. Rev. Lett. **83** (5198) .
Burt, P. B. (1974). Phys. Rev. Lett. **32** (1080) .
Burt, P. B. (1977). Lett. Nuovo. Cimento **18** (547) .
Burt, P. B. (1978). Proc. R. Soc. London **A359** (479) .
Burt, P. B. (1979). Phys. Lett. A **71** (19) ; Phys. Lett. B **82** (423) .
Burt, P. B. (1981). Quantum mechanics and nonlinear waves, Harwood Academic Publishers, New York.
Christiansen, P. L. and Scott, A. C. (1990). Self-trapping of vibrational energy, Plenum Press, New York.
Cui, Hong-nong,*et al.* (1988). J. Hydrodynamics **3** (43) .
Cui, Hong-nong, Yang Xue-qun, Pang Xiao-feng and Xiang Longwan, (1991). J. Hydrodynamics **6** (18) .
Davidson, R. C. (1972). Methods in nonlinear plasma theory, Academic Press, New York.
Davydov, A. S. (1979). Phys. Scr. **20** (387) .
Davydov, A. S. (1980). Phys. Stat. Sol. (b) **102** (275) .
Davydov, A. S. (1985). Solitons in molecular systems, D. Reidel Publishing, Dordrecht.
Davydov, A. S. and Kislukha, N. I. (1973). Phys. Stat. Sol. (b) **59** (465) .
Davydov, A. S. and Kislukha, N. I. (1976). Phys. Stat. Sol. (b) **75** (735) .
Davydov, A. S. and Pestryakov, G. M. (1981). Excitation states of the field with inertial self-action, preprint TTP-112R, Inst. Theor. Phys. Kiev (in Russian).
Dianov, E. M., Mamyshev, P. V., Prokhorov, A. M., and Chernikov, S. V. (1990). Opt. Lett. **14** (18) .
Dianov, E. M., Luchnikov, A. V., Pilipetskii, A. N., and Starodumov, A. N. (1990). Opt. Lett. **15** (314) .
Erdely, A. (1953). Bateman manuscript project, McGraw-Hill, New York.
Fermi, E. (1951). Elementary particles, Yale, New Haven.
Feynman, R. P. and Hibbs, A. R. (1965). Quantum mechanics and path integrals, McGraw-Hill, New York.
Fröhlich, H. (1954). Adv. Phys. **3** (325) .
Fröhlich, H. (1954). Proc. Roy. Soc. London A **223** (296) .
Guo, Bai-lin and Pang Xiao-feng, (1987). Solitons, Chin. Science Press, Beijing.
Hasegawa, A. (1983). Opt. Lett. **8** (342) .
Hasegawa, A. (1984). Opt. Lett. **9** (288) ; Appl. Opt. **23** (3302) .
Hasegawa, A. (1989). Optical solitons in fiber, Berlin, Springer.
Heisenberg, W. (1966). Introduction to the unified field theory of elementary particles, Interscience, London.
Holstein, T. (1959). Ann. Phys. **8** (325) and 343.
Hyman, J. M., Mclaughlin, D. W. and Scott, A. C. (1981). Physica **D3** (23) .
Keller, H. J. (1974). Low dimensional conductive phenomena, Plenum Press, New York.
Landau, L. D. (1933). Phys. Z. Sowjetunion **3** (664) .
Landau, L. D. and Lifshitz, E. M. (1987). Quantum mechanics, Pergamon Press, Oxford.
Larraza A. and Putterman, S. (1984). J. Fluid. Mech. **148** (443) ; Phys. Lett. A **103** (15) .
Miles, J. W. (1984). J. Fluid Mech. **148** (451) .
Mitschke, F. M. and Mollenauer, L. F. (1986). Opt. Lett. **11** (659) .
Mitschke, F. M. and Mollenauer, L. F. (1987). Opt. Lett. **12** (355) .

Mittal, R. and Howard, I. A. (1999). Physica D **125** (79) .

Mollenauer, L. F. and K. Smith (1988). Opt. Lett. **13** (675) .

Mollenauer, L. F. and K. Smith (1989). Opt. Lett. **14** (1284) .

Mollenauer, L. F., Gordon, J. P. and Islam, M. N. (1986). IEEE. J. Quantum Electron. **22** (157) .

Mollenauer, L. F., Stolen, R. H. and Islam, M. N. (1986). Opt. Lett. **10** (229) .

Mollenuaer, L. F.,*et al.* (1979). Opt. Lett. **4** (247) .

Mollenuaer, L. F., Stolen, R. H. and Gordon, J. P. (1980). Phys. Rev. Lett. **45** (1095) .

Pang, Xiao-feng (1994). Theory of nonlinear quantum mechanics, Chongqing Press, Chongqing, 1994.

Pang, Xiao-feng (2003). Soliton physics, Sichuan Sci. and Tech. Press, Chengdu.

Pang, Xiao-feng,*et al.* (1984). Proc. ICNP, Shanghai, p. 168.

Pekar, S. (1946). J. Phys. USSR. **10** (341) and 347.

Reynaud, F. and Barthelemy, A. (1990). Europhysics Lett. **12** (401) .

Schuster, H. (1975). One-dimensional conductors, Springer, Berlin.

Scott, A. C. (1984). Phys. Scr. **29** (284) .

Stolen, R. H. and Lin, C. (1978). Phys. Rev. A **17** (1448) .

Sulem, C and Sulem, P. L. (1999). The nonlinear Schrödinger equation: self-focusing and wave collapse, Springer-Verlag, Berlin.

Toda, M. (1989). Nonlinear waves and solitons, KTK Scientific, Tokyo; Kluwer, Dordrecht.

Vysloukh, V. A. and Cherednik, I. V. (1986). Dokl. Akad. Nauk SSSR **289** (336) , (in Russian).

Vysloukh, V. A., Ivanov, A. V. and Cherednik, I. V. (1987). Izv. Vysch. Uchebn. Zaved. Radiofiz **30** (980) .

Whitham, G. B. (1975). Linear and nonlinear waves, Wiley, New York.

Whitham, G. B. (1979). J. Phys. A **12** (L1) .

Wu, R. J., Keolian, R. and Rudnik, I. (1984). Phys. Rev. Lett. **52** (1421) .

Chapter 6

Nonlinear versus Linear Quantum Mechanics

In this chapter we discuss the relations between the nonlinear quantum mechanics and the linear quantum mechanics in relation to the self-consistency of the nonlinear quantum mechanics, the differences of the uncertainty relation and causality in the two quantum mechanical theories, and the method of calculating eigenenergy spectra of the Hamiltonian in the nonlinear quantum mechanics, which are also different from that in the linear quantum mechanics. The calculated eigenenergy spectra can be compared with experimental data. Incorporating results obtained in Chapters 2, 4 and 5, we can conclude that the nonlinear quantum mechanical theory presented in Chapter 3 is a correct, complete and self-consistent theory.

The validity of the theory means that it has solid theoretical and experimental foundation, as mentioned in Chapter 2, that the laws of conservation and invariance in the nonlinear quantum mechanics are consistent with natural laws, that the motion of particles in the nonlinear quantum mechanics can be simulated and verified by experiments, and the energy spectra obtained in the nonlinear quantum mechanics can be compared with experimental data. The completeness means that it can completely describe the properties of the microscopic particle including its wave-corpuscle duality in nonlinear systems. Finally, self-consistency means that it is compatible and correspondent with the linear quantum mechanics and it is itself self-compatible. These are necessary conditions for the nonlinear quantum mechanics to be a correct theory.

6.1 Nonlinear Quantum Mechanics: An Inevitable Result of Development of Quantum Mechanics

As far as the validity of the nonlinear quantum mechanics is concerned, it can be demonstrated from the following four aspects.

(1) The nonlinear quantum mechanics proposed here is a complete and entire theory because it can give the properties and states of microscopic particles in different systems and under various conditions, including their wave-corpuscle duality, which was confirmed by the experiments discussed in earlier chapters, and the linear

quantum mechanics failed to explain.

(2) The nonlinear quantum mechanics can give the universal laws of motion of matters in nature, for example, conservation laws of the mass, energy and momentum, as mentioned in Section 4.1. This shows that the nonlinear quantum mechanics can correctly describe motion of microscopic matters.

(3) The nonlinear quantum mechanics is an inevitable result of the development of the linear quantum mechanics in nonlinear systems. It is consistent with the linear quantum mechanics.

(4) The nonlinear quantum mechanics is self-consistent or self-correspondent.

In this and the following sections we will discuss the last two aspects.

To show that the nonlinear quantum mechanics is a necessary result of development of the linear quantum mechanics in nonlinear systems, we focus our discussion to the evolutions of dynamic equations, the Schrödinger equation, from linear to nonlinear cases in quantum mechanics. We know now that the nonlinear Schrödinger equation and the linear Schrödinger equation, are fundamental equations in the nonlinear quantum mechanics and linear quantum mechanics, respectively. The linear Schrödinger equation describes the motion of microscopic particles in linear systems. In these systems, there are no nonlinear interactions or very small nonlinear interactions. In such a case, microscopic particles move in the form of waves in the whole space-time due to the dispersion effect of the media. According to the theory of nonlinear interactions described in Chapter 4, these linear quantum mechanical systems can only be simple atoms and molecules consisting of a few elementary particles such as the hydrogen atom, or hydrogen molecule, or the helium atom, and so on. In these simple systems there are no mechanisms that generate nonlinear interactions. In other words, they have only very small self-interaction. In fact, we confirmed that the nonlinear two-body and three-body self-interaction energy in these systems is small, compared to its dispersive energy, through calculation of the corresponding Feynmann diagrams using the method given in Section 5.3. Such a small nonlinear interaction cannot balance and cancel the dispersion effect of the microscopic particle. Thus the microscopic particle cannot be localized, and it can only has wave or dispersive feature. The linear quantum mechanics is then sufficient to describe the states of motion of the particles. As a matter of fact, linear quantum mechanics indeed gives very accurate description for these systems, and the results compare well with the experimental data, as mentioned in Chapter 1. Therefore, the linear quantum mechanics is appropriate for describing these systems, and it would be unnecessary to use the nonlinear quantum mechanics in such cases.

However, for systems consisting of many bodies or many particles and other complex systems in which there are significant nonlinear interactions generated by self-interaction, self-focusing, self-trapping, self-condensation and self-localization, *etc.*, these nonlinear interactions cannot be neglected. These nonlinear interaction can be so large that it can balance and suppress the dispersion effect of these media. In such a case, the nature of the microscopic particles changes. It becomes a soliton with the wave-corpuscle duality. Then, the microscopic particle must be described by the nonlinear quantum mechanics. This is a very natural transition for the motion of the particle.

Vice versa. For nonlinear quantum systems depicted by the nonlinear quantum mechanics, if the nonlinear interactions are very small relative to the dispersion effect, we may use the linear quantum mechanics to describe them. But strictly speaking, all physical systems have nonlinear interactions to a certain extent. Therefore, the nonlinear quantum mechanics is an more general theory, and the linear quantum mechanics can be considered as a special case or an approximation of the nonlinear quantum mechanics in the case of weak nonlinear interactions. We can therefore say that the relation between the nonlinear quantum mechanics and the linear quantum mechanics resembles that between Einstein theory of relativity and Newton's classical mechanics. Furthermore, the dynamic equations, including the nonlinear Schrödinger equation and the nonlinear Klein-Gordon equation, can be obtained by adding nonlinear interactions to the corresponding dynamic equations, the linear Schrödinger equation or the linear Klein-Gordon equation, in the linear quantum mechanics. This correspondence relation advances also the consistency between the nonlinear quantum mechanics and the linear quantum mechanics. It is therefore clear that the nonlinear quantum mechanics is consistent with the linear quantum mechanics.

If we chose other approaches or other dynamic equations to establish the nonlinear quantum mechanics, then we could not obtain the consistency and compatibility between the nonlinear and the linear quantum mechanics. Therefore, the present nonlinear quantum mechanics is a correct theory and it is a necessary result of the development of the linear quantum mechanics.

However, we must remember that solutions of dynamic equations in nonlinear quantum mechanics in the limit of weak nonlinear interaction are not exactly the solutions of the dynamic equations in the linear quantum mechanics. To see this clearly, we first examine the velocity of the periphery of the soliton given in (4.8), which is rewritten as

$$\phi\left(x', t'\right) = 2\sqrt{2}k \,\mathrm{sech}\left[\sqrt{2}k\left(x' - v_e t'\right)\right] e^{iv_e(x'-v_c t')/2}, \tag{6.1}$$

for $b = 1$, $V(\vec{r}, t') = A(\phi) = 0$ and $x' = \sqrt{2m/\hbar^2}x$, $t' = t/\hbar$ in the nonlinear Schrödinger equation (3.2). In (6.1), v_e and v_c are the group and the phase velocities, respectively. As it is known, the nonlinear term in (3.2) sharpens the peak, while its dispersion term has the tendency to broaden it. Thus for weak nonlinear interaction and small periphery $\phi(x', t')$, it may be approximated by (for $x > v_e t$).

$$\phi = 2\sqrt{2}ke^{-\sqrt{2}k(x'-v_e t')}e^{iv_e(x'-v_c t')/2}. \tag{6.2}$$

The small term $|\phi|^2\phi$ in (3.2) can be approximated by

$$i\phi_{t'} + \phi_{x'x''} \approx 0. \tag{6.3}$$

Substituting (6.2) into (6.3) we get

$$v_e \approx 2\sqrt{2}k, \tag{6.4}$$

which is the group speed of the soliton. (Near the top of the peak, we must take both the nonlinear and dispersion terms into account because their contributions are of the same order. The result is the group speed.) Here we have only considered the region where $\phi(x,t)$ is small, that is, when a soliton is approximated by (6.2), it satisfies the approximate wave equation (6.3) with $v_e \approx 2\sqrt{2}k$.

However, if (6.3) is treated as an linear equation, its solution is of the form

$$\phi'(x,t) = Ae^{i(kx-\omega t)}. \tag{6.5}$$

We now have $\omega = k^2$, which gives the phase velocity ω/k as $v_c = k$ and the group speed $\partial\omega/\partial k = v_{\text{gr}} = k$. Apparently this is different from $v_e = 2\sqrt{2}k$. This is because the solution (6.2) is essentially different from (6.5). Therefore, the solution (6.5) is not the solution of the nonlinear Schrödinger equation (3.2) with $V(x) = 0$ and $A(\phi) = 0$ in the case of weak nonlinear interactions. Solution (6.2) is a "divergent solution" ($\phi(x,t) \to \infty$ at $x \to -\infty$), which is not an "ordinary plane wave". The concept of group speed does not apply to a divergent wave. Thus we can say that the soliton is made from a divergent solution, which is abandoned in linear waves. The divergence is development by the nonlinear term to yield waves of finite amplitude. When the nonlinear term is very weak, the soliton will be diverge, and suppression of divergency will result in no soliton. These circumstances are clearly seen from the following soliton solution in the case of the nonlinear coefficient $b \neq 1$

$$\phi(x,t) = 2\sqrt{\frac{2}{b}}k \,\text{sech}\left[\sqrt{2}k(x' - v_e t')\right] e^{iv_e(x'-v_c t')/2}. \tag{6.6}$$

If the nonlinear term approaches zero ($b \to 0$), the solitary wave diverges ($\phi(x,t) \to \infty$). If we want to suppress the divergence, then we have to set $k = 0$. In such a case, we get (6.5) from (6.6). This illustrates that the nonlinear Schrödinger equation or nonlinear quantum mechanics can reduce to the linear Schrödinger equation or linear quantum mechanics if and only if the nonlinear interaction and the group speed of the soliton are zero. Therefore we can conclude that the solutions of the nonlinear quantum mechanical dynamic equations in the weak nonlinear interaction limit is not the same as that in the linear quantum mechanics, and only if the nonlinear interaction is zero, the nonlinear quantum mechanics reduces to the linear quantum mechanics.

However, real physical systems or materials are made up of a great number of microscopic particles, and nonlinear interactions always exist extensively in the systems or among the particles. The nonlinear interactions arise from interactions among the microscopic particles or between the microscopic particles and the environment. Therefore, the nonlinear quantum mechanics should be the correct and more appropriate theory for real systems. It should be used often and extensively, even in weak nonlinear interaction cases. The linear quantum mechanics, on the other hand, is an approximation to the more general nonlinear quantum theory and can be used to study motions of microscopic particles in systems in which there exists only very weak and negligible nonlinear interaction.

6.2 Relativistic Theory and Self-consistency of Nonlinear Quantum Mechanics

As we know, there are nonrelativistic and relativistic theories in the linear quantum mechanics. In the relativity theory the dynamic equations are the linear Klein-Gordon equation and Daric equation, which can reduce to the linear Schrödinger equation in nonrelativistic case. Therefore, the linear quantum mechanics is itself self-consistent. In the nonlinear quantum mechanics there are also relativistic and nonrelativistic theories. The nonlinear Klein-Gordon equation in (3.3) and (3.4) is the dynamic equation of microscopic particles in the relativity case in the nonlinear quantum mechanics, which have been well studied by many scientists and have been extensively used to test various rules concerning motions of elementary particles in nonlinear fields and nonlinear systems, in particle physics and condensed matter physics. It is also a fundamental equation in the nonlinear field theory. This equation can reduce to the nonlinear Schrödinger equation in the nonrelativistic case, *i.e.*, there is also a consistency between the nonlinear Schrödinger equation and nonlinear Klein-Gordon equation, which will be the topic of discussion in this section.

The nonlinear Klein-Gordon equation in the relativistic case in the nonlinear quantum mechanics can be, in general, represented as

$$\phi_{tt} - \phi_{xx} = -\frac{\partial U}{\partial \phi} = -U'(\phi), \tag{6.7}$$

and the corresponding Hamiltonian is given by

$$H = \int dx \left[\frac{1}{2}\phi_t^2 - \frac{1}{2}\phi_x^2 + U(\phi),\right] \tag{6.8}$$

where $F(\phi)$ is a field force, $U(\phi)$ is the potential of an applied field which is a nonlinear function of the wave function or the field ϕ. The nonlinear interaction is generated by self-interactions between the microscopic particles which are described in Section 5.3. The natural unit system is used here. The solution of (6.7), as a function of $\xi = x - vt$, can be obtained by integration,

$$\int \frac{d\phi}{\sqrt{2[U(\phi) - C]}} = \pm\gamma(\xi - \xi_0),$$

where

$$\gamma = \frac{1}{\sqrt{1-v^2}}.$$

Under the "zero derivative" boundary condition

$$\left(\frac{d\phi}{d\xi}\right)_{\xi\to\pm\infty} \Rightarrow 0,$$

i.e., $C = 0$, the solution becomes

$$\pm(\xi - \xi_0) = \sqrt{\frac{1}{2\gamma^2}} \int \frac{d\phi}{\sqrt{|U(\phi)|}}.$$

The exact solution can be obtained for a given potential and it depends only on $\xi = x - vt$. Its dependence on v follows from the covariant form of the original equation (6.7). The energy associated with a single soliton is easily obtained,

$$E_k(v) = \sqrt{E_k^2(o) + P^2} = \gamma E_k(o),$$

where $P = \gamma M_k v$ and

$$E_k(o) = M_k = 2\int U[\phi_k(\xi)]\, dx = \sqrt{2}\int_{\phi_1}^{\phi_2} \sqrt{|U|}d\phi$$

is the soliton energy and mass in its rest frame, ϕ_1 and ϕ_2 are minima of the localized potential. The + and the − signs in the above equation correspond to the kink ($d\phi/d\xi > 0$) and antikink ($d\phi/d\xi < 0$) soliton solutions, respectively, and allow one to define the topological charge $N = N_+ - N_-$, *i.e.*, the difference between the kink and antikink numbers.

There are many forms of the nonlinear Klein-Gordon equation. For example, if $U'(\phi) = \sin\phi$, it becomes the Sine-Gordon equation. $U'(\phi) = \sin\phi + \lambda\sin 2\phi$, gives the dual Sine-Gordon equation. If $U'(\phi) = \phi - \phi^3$, it becomes the ϕ_+^4-field equation. If $U'(\phi) = -\phi + \phi^3$, it is the ϕ_-^4-field equation. In practice, the ϕ^4-field is a special case of the Sine-Gordon equation because $\sin\phi \approx \phi - \phi^3/3! + \cdots$. In the relativistic nonlinear quantum mechanics we concentrate mainly on the Sine-Gordon equation and the ϕ^4-field equation. These are fundamental equations in nonlinear field theory in particle physics and condensed physics and they have wide applications. However, when $U'(\phi) = \phi$, (6.7) reduces to $\phi_{tt} - \phi_{xx} = -\phi$, which is an equation of motion of the filed for a free Boson with a unit mass, *i.e.*, it is the linear Klein-Gordon equation in the linear quantum mechanics, and its solution is a plane wave. Therefore, the nonlinear Klein-Gordon equation (6.7) is a natural generalization of the linear Klein-Gordon equation in the nonlinear case. Thus, this relativistic theory in the nonlinear quantum mechanics is a necessary result of the development of that in the linear quantum mechanics.

6.2.1 *Bound state and Lorentz relations*

We can obtain the solutions of the nonlinear Klein-Gordon equation. For example, for the ϕ_-^4-field equation,

$$\phi_{tt} - \phi_{xx} - \phi + \phi^3 = 0, \tag{6.9}$$

its kink soliton solution is given by

$$\phi(x,t) = \pm\tanh\left(\frac{x - x_0 - vt}{2\sqrt{1-v^2}}\right), \tag{6.10}$$

where the + (−) sign is for the positive (negative) kink.

For the ϕ_+^4-field equation,

$$\phi_{tt} - \phi_{xx} + \phi - \phi^3 = 0, \tag{6.11}$$

its solution is of the form

$$\phi(x,t) = \phi_0 \text{sech}\left(\frac{x - x_0 - vt}{2\sqrt{1-v^2}}\right). \tag{6.12}$$

For the Sine-Gordon equation

$$\phi_{tt} - \phi_{xx} + m^2 \sin\phi = 0, \tag{6.13}$$

its solution is given in (2.106), and is written down here for convenience,

$$\phi(x,t) = \pm 4\tan^{-1}\left\{\pm\left[\exp\left(\pm m\frac{x - x_0 - vt}{\sqrt{1-v^2}}\right)\right]\right\}, \tag{6.14}$$

where the + (−) sign is for the positive (negative) kink, $x(t) = x_0 + vt$ is the position of the kink at time t. The width of the running kink is $\sqrt{1-v^2}/m$, *i.e.*, $\sqrt{1-v^2}$ times the static kink width. In (6.14), m is the mass of the microscopic particle. We can show the dependence of $\phi(x,t)$ on m, (6.14), using a graph, see clearly that the region of variation of $\phi(x,t)$ is centered at the point $x(t) = q = x_0 + vt$. Furthermore, the "length" of the variational region decreases as m increases. Thus, if the function $\phi(x,t)$ is used to denote a localized disturbance, then the quantity $q_0 = x_0 + vt$ and m^{-1} can be considered as its "position" and "width", respectively. The function $\phi^{\pm}(x,t)$ in (6.14) is the static solution of the Sine-Gordon equation with mass m at $t = 0$. Because there is a Lorentz factor $\gamma = \sqrt{1-v^2}$ in (6.14) and the position of the microscopic particle $q(t) = x_0 + vt$ is moving with a linear velocity v, we have a "Lorentz" contraction for the length of the microscopic particle when it moves with velocity v. This is exactly analogous to the motion of a relativistic particle.

The analogy between the kink $(+\phi(x,t))$ and a positively charged particle, or between the antikink $(-\phi(x,t))$ and a negatively charged particle is also readily exhibited for the Sine-Gordon equation. In Fig. 6.1, we show results obtained by Dodd *et al.* for the time evolution of an initial state consisting of two static kinks

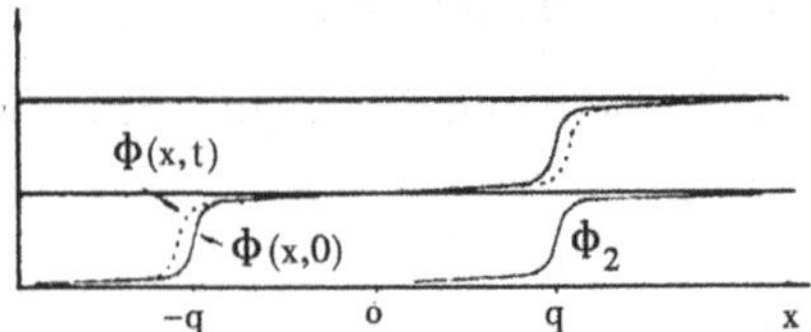

Fig. 6.1 Interaction between kinks: Two $N = 1$ kinks are initially at rest at $x = -q$ and $x = q$ respectively. The dotted line is $\phi(x, t)$ for $t > 0$ but small, showing that the kinks repel each other.

localized at $x = +q$ and $-q$, respectively. The boundary conditions, $\phi(t, -\infty) = \phi(t, +\infty) = 0$, or 2π, were used. We see that the kinks move away from each other. Fig. 6.2 shows the time evolution of an initial state consisting of a kink at $x = +q$ and an antikink at $x = -q$. In this case, the two disturbances move toward each other. This is an attractive effect for the kink-antikink pair. This diagram makes it intuitively clear why this must happen. It arises as a direct consequence of boundary conditions. For two like-kinks the solutions of the Sine-Gordon equation which evolves out of the initial configuration has to change by 2π over a distance. This results in high field gradients and consequently large contributions to the energy. The system moves so as to reduce its energy and as a result the kinks separate. For the solution which develops from the initial condition consisting of two kinks with opposite helicities, no such change is required and the energy is lower if the kinks are nearer. Hence, two dissimilar kinks approach each other. These properties of microscopic particles described by the Sine-Gordon equation show that microscopic particles have corpuscle feature.

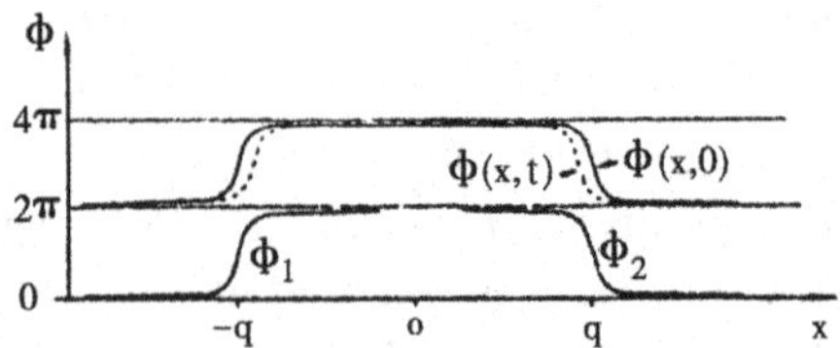

Fig. 6.2 Same as in Fig. 6.1, except that now we have an antikink at $x = q$ ($N_2 = -1$). Here $\phi(x, t)$ shows that a kink and an antikink attract each other.

The energy of the running kink is given by

$$E = \int_{-\infty}^{\infty} \frac{1}{2}\left[\phi_x^2 + \phi_t^2 + 4m^2 \sin^2\left(\frac{\phi}{2}\right)\right] dx = \frac{8m}{\sqrt{1-v^2}}. \tag{6.15}$$

The rest mass of the kink is a constant, given by

$$M = \int_{-\infty}^{\infty} |\phi|^2 dx = 8m.$$

These also show the corpuscle feature of microscopic particles.

Clearly, there is an important factor $\gamma = \sqrt{1-v^2}$ in (6.9) - (6.15). This is the Lorentz factor, or the Lorentz contraction coefficient. It's appearance in the above equations represents the relativistic invariance of the nonlinear quantum mechanical theory which exhibits the right relativistic dependence on the speed v of the microscopic particle (kink). Therefore, the model described above has Lorentz invariance. We can write down its Lagrangian, Hamiltonian and momentum as follows,

$$L = \int dx \mathcal{L} = \int dx \left[\frac{1}{2} \left(\phi_t^2 - \phi_x^2 \right) - m^2(1 - \cos\phi) \right], \tag{6.16}$$

$$H = \int dx \mathcal{H} = \int dx \left[\frac{1}{2} \phi_x^2 + m^2(1 - \cos\phi) \right], \tag{6.17}$$

$$P = \int dx \phi_x \phi_t. \tag{6.18}$$

Applying the Lorentz transformation,

$$\begin{pmatrix} t' \\ x' \end{pmatrix} = \sqrt{1-v^2} \begin{pmatrix} 1 & -v \\ -v & 1 \end{pmatrix} \begin{pmatrix} t \\ x \end{pmatrix}, \tag{6.19}$$

we can get

$$\begin{pmatrix} H' \\ p' \end{pmatrix} = \sqrt{1-v^2} \begin{pmatrix} 1 & -v \\ -v & 1 \end{pmatrix} \begin{pmatrix} H \\ p \end{pmatrix}. \tag{6.20}$$

This shows the invariance of the momentum and Hamiltonian of the microscopic particle under the Lorentz transformation. The ϕ^4-field equation also has Lorentz invariance. These are just the Lorentz or relativistic invariance for the nonlinear field theory. Therefore, the above theory or equations are actually appropriate to describe the motion of the microscopic particle in relativistic case in the nonlinear quantum mechanics.

The Sine-Gordon equation satisfies the following Heisenberg equation,

$$\phi_{tt} = \{\phi_t, H\} = \int \left(\frac{\partial \phi_t}{\partial \phi} \frac{\partial H}{\partial \phi_t} - \frac{\partial \phi_t}{\partial \phi_t} \frac{\partial H}{\partial \phi} \right) dx, \tag{6.21}$$

and the corresponding conservation laws are

$$\frac{\partial}{\partial t} \tilde{C}_n(\phi) + \frac{\partial}{\partial x} \tilde{T}(\phi) = 0, \tag{6.22}$$

where

$$\tilde{C}_1 = H - P = \int \phi_x^2 dx,$$

$$\tilde{C}_3 = \int \left(\phi_x^5 - \phi\phi_{xx}^2 \right) dx,$$

$$\tilde{C}_5 = \int \left(\phi_x^6 - 20\phi_x^2\phi_{xx}^2 + B\phi_{xxx}^2 \right) dx.$$

These problems were studied by Fogel, Olsen, Rubisten and Finkelstein, and others.

6.2.2 *Interaction between microscopic particles in relativistic theory*

Interaction between microscopic particles in relativistic theory is a very interesting problem. As a matter of fact, collision of microscopic particles described by the Sine-Gordon equation or the ϕ_+^4-field equation also has similar properties as those of microscopic particles described by the nonlinear Schrödinger equation as mentioned in Chapter 4. Interaction between two microscopic particles in relativistic theory will be discussed here, following the approach of Rubinstein.

For short periods of time the deformation of several interacting microscopic particles (kinks) will be small. Rubinstein looked for solutions of the form

$$\phi(x,t) = \phi'(x,t) + \sum_{i=1}^{n} \phi_i(x,t), \tag{6.23}$$

with

$$\phi_i(x,t) = 4\tan^{-1}\left\{\exp\left[\frac{m\delta_i(x - v_i t - x_i(0))}{\sqrt{1-v_i^2}}\right]\right\}, \quad \delta_i = \pm 1, \tag{6.24}$$

and $\phi'(x,t) \ll 1$. In fact, with

$$v_i = \partial_t \phi'(x,0) = \phi'(x,0) = 0, \tag{6.25}$$

such a solution can determine the direction and strength of the force acting on each kink at rest at $t = 0$. For simplicity we restrict ourselves to $n = 2$, and let $-x_1 = x_2 = q$. Substituting (6.23) into the Sine-Gordon equation (6.13), Rubinstein obtained, at $t = 0$, that

$$\begin{aligned}\frac{1}{m^2}\partial_t^2\phi'(x,0) &= \sum \sin\phi_i - \sin\left(\sum \phi_i\right)\\ &= -4\frac{\delta_1 \sinh[m(x+q)] + \delta_2 \sinh[m(x-q)]}{\cosh^2[m(x+q)]\cosh^2[m(x-q)]}.\end{aligned} \tag{6.26}$$

In Figs. 6.1 and 6.2, we plotted $\phi(x,0) = \phi_1(x,0) + \phi_2(x,0)$ using solid line and $\phi(x,t) \approx \phi(x,0) + \partial_t^2\phi'(x,0)t^2/2$ (t is small) using dotted line. The figures show that the kinks repel if $\delta_1 = \delta_2$ and attract if $\delta_1 = -\delta_2$ (the cases $\delta_1 = \delta_2 = -1$ and $-\delta_1 = \delta_2 = +1$ are represented in Figs. 6.1 and 6.2 by $\phi \to -\phi$). If $q \gg 1/m$ and $x \simeq q$, Rubinstein obtained, from (6.26), the change in ϕ_2 as

$$\frac{1}{m^2}\partial_t^2\phi'(x,0) \simeq \frac{-4\delta_1 \sinh[m(x+q)]}{\cosh^2[m(x+q)]\cosh^2[m(x-q)]} \simeq \frac{-8\delta_1 e^{-2mq}}{\cosh^2[m(x-q)]}. \tag{6.27}$$

From (6.24), the change in $\phi_2(x,0)$ generated by a small change in q, *i.e.*, a rigid displacement of the kink is

$$\delta\phi_2 = \frac{\partial\phi_2(x,0)}{\partial q}\delta q = \frac{-2\delta_2 m}{\cosh[m(x-q)]}\delta q, \tag{6.28}$$

and, if δq changes in time,

$$\partial_t^2(\delta\phi_2) = \frac{-2\delta_2 m}{\cosh[m(x-q)]}\delta\ddot{q}. \tag{6.29}$$

Comparing (6.29) and (6.27) at $x \simeq q$, we see that due to the interaction the microscopic particles tend to contract and that the force acting on ϕ_2 is

$$(8m)(\delta\ddot{q}) \simeq 32m^2\delta_1\delta_2 e^{-m(2q)}. \tag{6.30}$$

This formula approximates the classical law of motion of the particle. It shows that microscopic particles described by the Sine-Gordon equation obviously have corpuscle feature.

Two exact solutions of (6.13) were obtained by Perring and Skyrme. Letting $m = 1$ and $\phi + 2\pi = \phi''$, the solutions for interaction between the two kinks are

$$\begin{aligned}\phi_a'' &= 4\tan^{-1}\left[-\frac{\cosh(\gamma vt)}{v\sinh(\gamma x)}\right] \\ &\underset{x\to-\infty}{\longrightarrow} 4\tan^{-1}\left[\frac{2}{v}e^{\gamma x}\cosh(\gamma vt)\right] \\ &\underset{t\to-\infty}{\longrightarrow} 4\tan^{-1}\left[\frac{1}{v}e^{\gamma(x-vt)}\right] \\ &\underset{t\to\infty}{\longrightarrow} 4\tan^{-1}\left[\frac{1}{v}e^{\gamma(x+vt)}\right];\end{aligned} \tag{6.31}$$

$$\begin{aligned}\phi_a'' &= 4\tan^{-1}\left[\frac{v\sinh(\gamma x)}{\cosh(\gamma vt)}\right] + 2\pi \\ &\underset{x\to\infty}{\longrightarrow} 4\tan^{-1}\left[\frac{ve^{\gamma x}}{2\cosh(\gamma vt)}\right] + 2\pi \\ &\underset{t\to-\infty}{\longrightarrow} 4\tan^{-1}\left[ve^{\gamma(x+vt)}\right] + 2\pi \\ &\underset{x\to\infty}{\longrightarrow} 4\tan^{-1}\left[ve^{\gamma(x-vt)}\right] + 2\pi;\end{aligned} \tag{6.32}$$

$$\begin{aligned}\phi_b'' &= 4\tan^{-1}\left[\frac{v\cosh(\gamma x)}{\sinh(\gamma vt)}\right] \\ &\underset{t\to-\infty}{\longrightarrow} 4\tan^{-1}\left[2v\cosh(\gamma x)e^{\gamma vt}\right] \\ &\underset{x\to-\infty}{\longrightarrow} 4\tan^{-1}\left[ve^{-\gamma(x-vt)}\right] \\ &\underset{x\to\infty}{\longrightarrow} 4\tan^{-1}\left[ve^{\gamma(x+vt)}\right];\end{aligned} \tag{6.33}$$

$$\begin{aligned}
\phi_b'' &= 4\tan^{-1}\left[\frac{\sinh(\gamma vt)}{\cosh(\gamma x)}\right] + 2\pi \\
&\xrightarrow[t\to\infty]{} 4\tan^{-1}\left[\frac{e^{\gamma vt}}{2v\cosh(\gamma x)}\right] + 2\pi \\
&\xrightarrow[x\to-\infty]{} 4\tan^{-1}\left[\frac{1}{v}e^{\gamma(x+vt)}\right] + 2\pi \\
&\xrightarrow[x\to\infty]{} 4\tan^{-1}\left[\frac{1}{v}e^{-\gamma(x-vt)}\right] + 2\pi.
\end{aligned} \tag{6.34}$$

Except for the unimportant factors $1/v$ and v (which can be absorbed in the phase), (6.31) and (6.32) show that ϕ_a'' represents two $\delta = 1$ kinks approaching the origin with speeds $\pm v$ for $t \to -\infty$ and moving away from it for $t \to +\infty$. If v is small enough (≤ 0.05), ϕ_a'' departs little from a linear combination of two kinks, with the distance of the closest approach larger than twice the kink size, so that ϕ_a'' can be recognized as the bouncing of two kinks. For larger $v's$, ϕ_a'' is strongly distorted and such an interpretation is impossible.

Similarly, ϕ_b'' represents, at $t = -\infty$, a $\delta = -1$ kink approaching the origin with speed v from $x = -\infty$ and a $\delta = 1$ kink coming from $x = \infty$, with speed $-v$. They pass through each other, and, at $t = \infty$, ϕ_{-1} is at $x = \infty$ and ϕ_{+1} at $x = -\infty$, $\phi_b''(x, +\infty)$ has the form of $\phi(x, 0)$ in Fig. 6.2.

The above discussions show that the interaction or collision of microscopic particles described by the Sine-Gordon equation is the same as that described by the nonlinear Schrödinger equation, *i.e.*, the form and energy of the microscopic particles can be retained after the collision. Considering also results of the Sine-Gordon equation discussed in Chapter 4, we can conclude that microscopic particles in the relativistic theory also have the wave-corpuscle duality.

6.2.3 *Relativistic dynamic equations in the nonrelativistic limit*

In the following, we show that the dynamic equations in the relativistic theory can reduce to the nonlinear Schrödinger equation in the nonrelativistic limit. We start from the following ϕ^4-field equation

$$\frac{\partial^2\phi}{\partial t^2} - \frac{\partial^2\phi}{\partial x^2} - \frac{\partial^2\phi}{\partial y^2} - m^2\phi + b\phi^3 = 0, \tag{6.35}$$

with the boundary conditions, $\phi \approx 0$ at $y = 0$ and $y = a$.

Obviously, the solution of (6.35), in the linear case ($b = 0$), is of the form

$$\phi(x, y, t) = \tilde{\phi}(y)e^{i(kx-\omega t)}. \tag{6.36}$$

Inserting (6.36) into (6.35), we get

$$\tilde{\phi}(y) = A\sin(Ky) + B\cos(Ky), \tag{6.37}$$

where $K^2 = \omega^2 - k^2 - m^2$. Applying the boundary conditions, we get

$$B = 0, \quad K^2 = \left(\frac{n\pi}{a}\right)^2, \quad (n = 1, 2, 3 \cdots).$$

Thus there are many modes of motion given by the dispersion relation $\omega(k, n) = 0$ in such a case. For given k and n, ω can be determined. We now consider the nonrelativistic case of slow changes or small velocity in (6.35). Thus we can assume that $\xi = \varepsilon(x - vt)$, $\tau = \varepsilon^3 t$ in such a case, where ε is a small factor. Let

$$\frac{\partial}{\partial x} \to \frac{\partial}{\partial x} + \varepsilon\frac{\partial}{\partial \xi}, \quad \frac{\partial}{\partial t} \to \frac{\partial}{\partial t} + \varepsilon^3\frac{\partial}{\partial \tau} - \varepsilon v\frac{\partial}{\partial \xi}. \tag{6.38}$$

Then the solution of (6.35) becomes

$$\phi = \sum_{n=1}^{\infty} \varepsilon^n \phi^{(n)}, \tag{6.39}$$

where

$$\phi^{(n)} = \sum_{\ell=-\infty}^{\infty} \phi_\ell^{(n)}(\xi, \tau, y)e^{i(kx-\omega t)}.$$

From the dispersion relation we get

$$v = \frac{\partial \omega}{\partial k} = \frac{k}{\omega}. \tag{6.40}$$

Substituting (6.39) into (6.35), the equation in ε is of the form

$$\phi_1^{(1)} = \varphi(\xi, \tau)\sin(Ky), \quad \phi_\ell^{(1)} = 0, \quad (|\ell| \neq 1). \tag{6.41}$$

The equation in ε^2 is

$$2i\ell(k - v\omega)\frac{\partial \phi_\ell^{(1)}}{\partial \xi} + \frac{\partial^2 \phi_\ell^{(2)}}{\partial y^2} + \left(\ell^2\omega^2 - \ell^2 k^2 - m^2\right)\phi_\ell^{(2)} = 0. \tag{6.42}$$

Because $v = k/\omega$, the first term in (6.42) is zero. Thus we get $\phi_\ell^{(2)} = 0$ for $|\ell| \neq 1$. The equation in ε^3 now becomes

$$\frac{\partial^2 \phi_\ell^{(3)}}{\partial y^2} + K^2\phi_\ell^{(3)} + \left[2i\omega\frac{\partial \varphi}{\partial \tau} + (1 - v^2)\frac{\partial^2 \varphi}{\partial \xi^2}\right]\sin(Ky) - 3k|\varphi|^2\varphi\sin(Ky) = 0. \tag{6.43}$$

Multiplying (6.43) from left by $\sin(Ky)$ and integrating again over y from 0 to a, then the first and the second terms become zero in (6.43) according to match conditions of functions. Finally we get

$$i\varphi_\tau - \frac{\omega^2 - k^2}{2\omega^3}\frac{\partial^2 \varphi}{\partial \xi^2} - \frac{3k}{2\omega}|\varphi|^2\varphi = 0. \tag{6.44}$$

This is a nonlinear Schrödinger equation. Therefore, the relativistic equation of motion reduces to the nonlinear Schrödinger equation in the nonrelativistic case.

This again shows that the nonlinear quantum mechanics established in Chapter 3 is self-consistent.

For the Sine-Gordon equation, $\phi_{xx} - \phi_{tt} = \sin\phi$, we can also derive the nonlinear Schrödinger equation in nonrelativistic and small amplitude cases following a similar procedure, and making the same assumptions, (6.38) – (6.40). For the equation in ε^3, we obtain the following nonlinear Schrödinger equation,

$$i\varphi_\tau + \frac{1-(v')^2}{2\omega'}\varphi_{\xi\xi} + \frac{1}{6\omega'}|\varphi|^2\varphi = 0,$$

where $\omega' = \pm\sqrt{1+k^2}$, $v' = \pm k/\sqrt{1+k^2}$.

In addition, the above Sine-Gordon equation has the following breather solution

$$\tanh\left(\frac{\phi}{4}\right) = \frac{v}{\sqrt{(1-v^2)}}\frac{\sin\sqrt{1-v^2}\xi'}{\cosh(v\xi)},$$

with $\xi = dx + t/d$, $\xi' = dx - t/d$, where $d > 0$. If the function is expanded in term of $\tanh^{-1}\Phi$ to the first order, then we get

$$\phi \approx \frac{2iv}{\sqrt{1-v^2}}\mathrm{sech}(v\xi)e^{i(1-v^2/2)\xi'} + \text{c.c.} = \varphi(\xi,\xi')e^{i\xi'} + \text{c.c.},$$

where

$$\varphi(\xi,\xi') = \frac{2iv}{\sqrt{1-v^2}}e^{-iv^2\xi'/2}\mathrm{sech}(v\xi).$$

Inserting $\varphi(\xi,\xi')$ into the Sine-Gordon equation given above, in the nonrelativistic case, keeping only the $O(v^3)$ term, we get the following nonlinear Schrödinger equation

$$i\varphi_{\xi'}(\xi,\xi') - \frac{1}{2}\varphi_{\xi\xi} - \frac{1}{6}|\varphi|^2\varphi = 0.$$

Therefore, both the ϕ^4-field equation and the Sine-Gordon equation can reduce to the nonlinear Schrödinger equation in the nonrelativistic case.

This can also be demonstrated in other ways. In fact, for the ϕ^4_+ equation

$$\phi_{tt} - \phi_{xx} + m^2\phi - b\phi^3 = 0,$$

we can assume that its solution is of the form

$$\phi(x,t) = \varphi(x,t)e^{-imt}. \tag{6.45}$$

Inserting (6.45) into (6.44), and taking into account the relation, $i\partial\varphi/\partial t \ll m\varphi$, in the nonrelativistic case, we then get

$$i\frac{\partial\varphi}{\partial t} = -\frac{1}{2m}\varphi_{xx} - \frac{b}{2m}\varphi^3.$$

This is also the nonlinear Schrödinger equation. Therefore, we can always reduce these relativistic equations of motion to the nonlinear Schrödinger equation in the nonrelativistic limit, and thus conform that the nonlinear quantum mechanics established in Chapter 3 is self-consistent.

6.2.4 *Nonlinear Dirac equation*

In linear quantum mechanics, Fermi particles are described by the following Dirac equation in the relativistic theory

$$i\hbar\frac{\partial}{\partial t}\phi = (c\hat{\alpha}\cdot\hat{p})\phi + mc^2\beta\phi. \tag{6.46}$$

It is natural to extend the above equation by including self-interaction discussed in Section 5.4, to arrive at the following nonlinear Dirac equation

$$i\hbar\frac{\partial}{\partial t}\phi = (c\hat{\alpha}\cdot\hat{p})\,\phi + mc^2\beta\phi + b|\phi^2|\phi. \tag{6.47}$$

The corresponding matrix equation is

$$\begin{pmatrix} i\hbar\phi_1 \\ i\hbar\phi_2 \\ i\hbar\phi_3 \\ i\hbar\phi_4 \end{pmatrix} = \begin{pmatrix} mc^2 & 0 & CP_z & C(P_z - ip_y) \\ 0 & mc^2 & c(p_x+ip_y) & cp_z \\ cp_x & c(p_x-ip_y) & -mc^2 & 0 \\ c(p_z+ip_y) & cp_x & 0 & -mc^2 \end{pmatrix} \begin{pmatrix} \phi_1 \\ \phi_2 \\ \phi_3 \\ \phi_4 \end{pmatrix}$$
$$+\frac{b}{4}\left[(\phi_1^*\phi_2^*\phi_3^*\phi_4^*)\begin{pmatrix} \phi_1 \\ \phi_2 \\ \phi_3 \\ \phi_4 \end{pmatrix}\right]\begin{pmatrix} \phi_1 \\ \phi_2 \\ \phi_3 \\ \phi_4 \end{pmatrix}. \tag{6.48}$$

This is a vector nonlinear equations. Its solutions have not been obtained previously. Here we are interested in whether this equation can have a soliton solution, and if so, what properties it has. These problems cannot be answered yet. But nevertheless, they are interesting and worth further investigation.

For the nonlinear Dirac equation in terms with a cubic nonlinear term in the Dirac matrix γ_μ $(\mu = 1, 2, 3, 4)$

$$\gamma_\mu\frac{\partial\phi}{\partial x_\mu} - b^2\gamma_\mu\gamma_v\phi(\phi^*\gamma_\mu\gamma_v\phi) = 0,$$

Heisenberg obtained its single soliton solution which is given by

$$\phi_\lambda = \frac{\gamma_5\gamma_\lambda}{b}\sqrt{\frac{A}{2\hbar}\mathrm{sech}\left(\frac{A}{\hbar}\gamma_\mu x_\mu\right)}\exp\left(\frac{i}{\hbar}\xi - \frac{A}{2\hbar}\gamma_v x_v\right),$$

where

$$A = i\gamma_\mu\frac{\partial\xi}{\partial x_\mu}, \quad \xi = x - u(x)t.$$

6.3 The Uncertainty Relation in Linear and Nonlinear Quantum Mechanics

6.3.1 *The uncertainty relation in linear quantum mechanics*

The uncertainty relation in quantum mechanics is an important formula and also a problem that troubled many scientists. Whether this is an intrinsic property of microscopic particles or an artifact of the linear quantum mechanics or measuring instruments has been a long-lasting controversy. Obviously, this is related to the wave-corpuscle duality of microscopic particles. Since we have established nonlinear quantum mechanics which differs from the linear quantum mechanics, we could expect that the uncertainty relation in the nonlinear quantum mechanics is different from that in the linear quantum mechanics. The significance of the uncertainty relation can be revealed by comparing the linear and nonlinear quantum theories.

It is well known that the uncertainty relation in the linear quantum mechanics can be obtained from

$$I(\xi) = \int \left| \left(\xi \triangle \hat{A} + i \triangle \hat{B}\right) \psi(\vec{r},t) \right|^2 d\vec{r} \geq 0, \tag{6.49}$$

or

$$\bar{\hat{F}}(\xi) = \int d\vec{r} \psi^*(\vec{r},t) \hat{F}\left[\hat{A}(\vec{r},t), \hat{B}(\vec{r},t)\right] \psi(\vec{r},t). \tag{6.50}$$

In the coordinate representation, $\hat{A}$ and $\hat{B}$ are the operators of two physical quantities, for example, position and momentum, or energy and time, and satisfy the commutation relation $[\hat{A}, \hat{B}] = i\hat{C}$, $\psi(x,t)$ and $\psi^*(x,t)$ are wave functions of the microscopic particle satisfying the linear Schrödinger equation and its conjugate equation, respectively, $\hat{F} = (\triangle A\xi + \triangle B)^2$, $(\triangle \hat{A} = \hat{A} - \bar{A}$, $\triangle \hat{B} = \hat{B} - \bar{B}$, $\bar{A}$ and $\bar{B}$ are the average values of the physical quantities in the state denoted by $\psi(x,t))$, is an operator of physical quantity related to $\hat{A}$ and $\hat{B}$, ξ is a real parameter.

After some simplifications, we can get

$$I = \bar{\hat{F}} = \overline{\triangle \hat{A}^2}\xi^2 + 2\overline{\triangle \hat{A} \triangle \hat{B}}\xi + \overline{\triangle \hat{B}^2} \geq 0,$$

or

$$\overline{\triangle \hat{A}^2}\xi^2 + \bar{\hat{C}}\xi + \overline{\triangle \hat{B}^2} \geq 0. \tag{6.51}$$

Using mathematical identities, this can be written as

$$\overline{\triangle \hat{A}^2 \triangle \hat{B}^2} \geq \frac{\bar{\hat{C}}^2}{4}. \tag{6.52}$$

This is the uncertainty relation in the linear quantum mechanics. From the above derivation we see that the uncertainty relation was obtained based on the fundamental hypothesizes of the linear quantum mechanics, including properties of operators of the mechanical quantities, the state of the particle represented by the wave function, which satisfies the linear Schrödinger equation, the concept of average values of mechanical quantities and the commutation relations and eigenequation of operators. Therefore, we can conclude that the uncertainty relation (6.52) is a necessary result of the linear quantum mechanics. Since the linear quantum mechanics only describes the wave nature of microscopic particles, the uncertainty relation is a result of wave feature of microscopic particles, and it inherits the wave nature of microscopic particles. This is why its coordinate and momentum cannot be determined simultaneously. This is an essential interpretation for the uncertainty relation (6.52) in the linear quantum mechanics. It is not related to measurement, but closely related to the linear quantum mechanics. In other word, if linear quantum mechanics could correctly describe the states of microscopic particles, then the uncertainty relation should also reflect the peculiarities of microscopic particles.

Equation (6.51) can be written in the following form

$$\bar{\hat{F}} = \overline{\triangle \hat{A}^2}\left(\xi + \frac{\overline{\triangle \hat{A} \triangle \hat{B}}}{\overline{\overline{\triangle \hat{A}^2}}}\right)^2 + \overline{\triangle \hat{B}^2} - \frac{(\overline{\triangle \hat{A} \triangle \hat{B}})^2}{\overline{\overline{\triangle \hat{A}^2}}} \geq 0,$$

or

$$\overline{\triangle \hat{A}^2}\left(\xi + \frac{\bar{\hat{C}}}{4\overline{\overline{\triangle \hat{A}^2}}}\right)^2 + \overline{\triangle \hat{B}^2} - \frac{(\bar{\hat{C}})^2}{4\overline{\overline{\triangle \hat{A}^2}}} \geq 0. \tag{6.53}$$

This shows that $\overline{\triangle \hat{A}^2} \neq 0$, if $(\overline{\triangle \hat{A} \triangle \hat{B}})^2$ or $\bar{\hat{C}}^2/4$ is not zero. Or else, we cannot obtain (6.52) and $\overline{\triangle \hat{A}^2 \triangle \hat{B}^2} \geq \overline{(\triangle B \triangle B)}^2$ because when $\overline{\triangle \hat{A}^2} = 0$, (6.53) does not hold. Therefore, $(\overline{\triangle \hat{A}^2}) \neq 0$ is a necessary condition for the uncertainty relation (6.52). $\overline{\triangle \hat{A}^2}$ can approach zero, but cannot be exactly zero. Therefore, in the linear quantum mechanics, the right uncertainty relation should take the form

$$\overline{\triangle \hat{A}^2 \triangle \hat{B}^2} > \frac{(\bar{\hat{C}})^2}{4}. \tag{6.54}$$

6.3.2 *The uncertainty relation in nonlinear quantum mechanics*

We now return to the uncertainty relation in the nonlinear quantum mechanics. Since microscopic particles are solitons and they have wave-corpuscle dualityin the nonlinear quantum mechanics, we can also expect that uncertainty relation in nonlinear quantum mechanics is different from (6.52).

We now derive this relation for position and momentum of a microscopic particle described by the nonlinear Schrödinger equation (3.2) with $V = 0$, and $A = 0$, given in (4.40), with a soliton solution, ϕ_s, as given in (4.56). The function $\phi_s(x', t')$ is a square integrable function localized at $x'_0 = 0$ in the position space. If the microscopic particle (soliton) is localized at $x'_0 \neq 0$, it satisfies the nonlinear Schrödinger equation,

$$i\phi_{t'} + \frac{1}{2}\phi_{x'x'} + |\phi|^2\phi = 0$$

for $b = 1$, where $t' = t/\hbar$, $x' = \sqrt{m}x/\hbar$. The Fourier transform of this function is

$$\phi_s(p,t') = \frac{1}{\sqrt{2\pi}}\int_{-\infty}^{\infty}\phi_s(x',t')e^{-ipx'}. \tag{6.55}$$

It shows that $\phi_s(p,t')$ is localized at p in the momentum space. For (4.56), the Fourier transform is explicitly given by

$$\phi_s(p,t') = \sqrt{\frac{\pi}{2}}\mathrm{sech}\left[\frac{\pi}{2\eta}(p - 2\sqrt{2}\xi)\right] e^{4i(\eta^2+\xi^2-p\xi/2\sqrt{2})t' - i(p-2\sqrt{2}\xi)x'_0}. \tag{6.56}$$

The results in (4.56) and (6.56) show that the microscopic particle is localized not only in position space and moves in the form of soliton, but also in the momentum space, as a soliton. For convenience, we introduce the normalization coefficient A_0 in (4.56) and (6.56), then $A_0^2 = 1/(4\sqrt{2}\eta)$. According to the definition in Section 4.2, the position of the center of mass of the microscopic particle, $\langle x'\rangle$, and its square, $\langle x'^2\rangle$, at $t' = 0$ are given by

$$\langle x'\rangle = \int_{-\infty}^{\infty} dx'|\phi_s(x')|^2, \quad \langle x'^2\rangle = \int_{-\infty}^{\infty} dx' x'^2|\phi_s(x')|^2.$$

We can thus find that

$$\langle x'\rangle = 4\sqrt{2}\eta A_0^2 x'_0, \quad \langle x'^2\rangle = \frac{A_0^2\pi^2}{12\sqrt{2}\eta} + 4\sqrt{2}A_0^2\eta x_0'^2, \tag{6.57}$$

respectively. Similarly, the momentum of the center of mass of the microscopic particle, $\langle p\rangle$, and its square, $\langle p^2\rangle$, are given by

$$\langle p\rangle = \int_{-\infty}^{\infty} p|\hat{\phi}_s(p)|^2 dp, \quad \langle p^2\rangle = \int_{-\infty}^{\infty} p^2|\hat{\phi}_s(p)|^2 dp,$$

which yield

$$\langle p\rangle = 16A_0\eta\xi, \quad \langle p^2\rangle = \frac{32\sqrt{2}}{3}A_0^2\eta^3 + 32\sqrt{2}A_0^2\eta\xi^2. \tag{6.58}$$

The standard deviations of position $\Delta x' = \sqrt{\langle x'^2\rangle - \langle x'\rangle^2}$ and momentum $\Delta p = \sqrt{\langle p^2\rangle - \langle p\rangle^2}$ are given by

$$\begin{aligned}(\Delta x')^2 &= A_0^2\left[\frac{\pi^2}{12\eta} + 4\eta x_0'^2(1 - 4\sqrt{2}\eta A_0^2)\right] = \frac{\pi^2}{96\eta^2},\\ (\Delta p)^2 &= 32\sqrt{2}A_0^2\left[\frac{1}{3}\eta^3 + \eta\xi^2(1 - 4\sqrt{2}\eta A_0^2)\right] = \frac{8}{3}\eta^2,\end{aligned} \tag{6.59}$$

respectively. Thus we obtain the uncertainty relation between position and momentum for the microscopic particle, (4.56), in nonlinear quantum mechanics,

$$\triangle x'\triangle p = \frac{\pi}{6}. \tag{6.60}$$

This result is not only valid for microscopic particle (soliton) described by the nonlinear Schrödinger equation, but holds in general. This is because we did not use any specific property of the nonlinear Schrödinger equation in deriving (6.60) π in (6.60) comes from the integral constant $1/\sqrt{2\pi}$. For a quantized microscopic particle, π in (6.60) should be replaced by $\pi\hbar$, because (6.56) is replaced by

$$\phi_s(p,t') = \frac{1}{\sqrt{2\pi\hbar}}\int_{-\infty}^{\infty} dx'\phi_s(x',t')e^{-ipx'/\hbar}.$$

The corresponding uncertainty relation of the quantum microscopic particle (soliton) in the nonlinear quantum mechanics is given by

$$\triangle x\triangle p = \frac{\pi\hbar}{6} = \frac{h}{12}. \tag{6.61}$$

The relations (6.60) or (6.61) are different from that in linear quantum mechanics (6.54) *i.e.*, $\triangle x\triangle p > h/2$. However, the minimum value $\triangle x\triangle p = h/2$ has not been observed in practical systems in linear quantum mechanics up to now except for the coherent and squeezed states of microscopic particles. The relation (6.61) cannot be obtained from the solutions of the linear Schrödinger equation. Practically, we can only get $\triangle x\triangle p > h/2$ from (6.54), but not $\triangle x\triangle p = h/2$, in linear quantum mechanics.

As a matter of fact, for an one-quantum coherent state,

$$|\alpha\rangle = \exp(\alpha\hat{b}^+ - \alpha^*\hat{b})|0\rangle = e^{-\alpha^2/2}\sum_{n=0}^{\infty}\frac{\alpha^n}{\sqrt{n-1}}\hat{b}^{+n}|0\rangle,$$

which is a coherent superposition of a large mumbler of microscopic particles (quanta). Thus

$$\langle\alpha|\hat{x}|\alpha\rangle = \sqrt{\frac{\hbar}{2\omega m}}(\alpha+\alpha^*), \quad \langle\alpha|\hat{p}|\alpha\rangle = i\sqrt{\frac{\hbar m\omega}{2}}(\alpha^*-\alpha),$$

and

$$\langle\alpha|\hat{x}^2|\alpha\rangle = \frac{\hbar}{2\omega m}(\alpha^{*2}+\alpha^2+2\alpha\alpha^*+1), \quad \langle\alpha|\hat{p}^2|\alpha\rangle = \frac{\hbar\omega m}{2}(\alpha^{*2}+\alpha^2-2\alpha\alpha^*-1),$$

where

$$\hat{x} = \sqrt{\frac{\hbar}{2\omega m}}\left(\hat{b}+\hat{b}^+\right), \quad \hat{p} = i\sqrt{\frac{\hbar\omega m}{2}}\left(\hat{b}^+-\hat{b}\right),$$

and $\hat{b}^+$ $(\hat{b})$ is the creation (annihilation) operator of the microscopic particle (quantum), α and α^* are some unknown functions, ω is the frequency of the particle, m

is its mass. Thus we can get

$$(\triangle x)^2 = \frac{\hbar}{2\omega m}, \quad (\triangle p)^2 = \frac{\hbar\omega m}{2}, \quad \langle\Delta x\rangle^2\langle\Delta p\rangle^2 = \frac{h^2}{4},$$

and

$$\frac{\triangle x}{\triangle p} = \frac{1}{\omega m},$$

or

$$\triangle p = (\omega m)\triangle x. \tag{6.62}$$

For the squeezing state,

$$|\beta\rangle = \exp\left[\beta\left(b^{+2} - b^2\right)\right]|0\rangle,$$

which is a two-microscopic particle (quanta) coherent state, we can obtain

$$\langle\beta|\triangle x^2|\beta\rangle = \frac{\hbar}{2m\omega}e^{4\beta}, \quad \langle\beta|\triangle p^2|\beta\rangle = \frac{\hbar m\omega}{2}e^{-4\beta},$$

using a similar approach as the above. Here β is the squeezing coefficient and $|\beta| < 1$. Thus

$$\triangle x\triangle p = \frac{h}{2}, \quad \frac{\triangle x}{\triangle p} = \frac{1}{m\omega}e^{8\beta},$$

or

$$\triangle p = \triangle x(\omega m)e^{-8\beta}. \tag{6.63}$$

This shows that the momentum of the microscopic particle (quantum) is squeezed in the two-quanta coherent state compared to that in the one-quantum coherent state.

From the above results, we see that both the one-quantum and the two-quanta coherent states satisfy the minimal uncertainty principle. This is the same as that of the nonlinear quantum states in nonlinear quantum mechanics. We can conclude that a coherent state is a nonlinear quantum state, and the coherence of quanta is a nonlinear phenomenon, instead of a linear effect.

As it is known, the coherent state satisfies classical equation of motion, in which the fluctuation in the number of particles approaches zero, *i.e.*, it is a classical steady wave. According to quantum theory, the coherent state of a harmonic oscillator at time t can be represented by

$$\begin{aligned}|\alpha,t\rangle &= e^{-i\hat{H}t/\hbar}|\alpha\rangle = e^{-i\omega(\hat{b}^+\hat{b}+1/2)t}|\alpha\rangle = e^{-i\omega t/2-|\alpha|^2/2}\sum_{n=0}^{\infty}\frac{\alpha^n e^{-in\omega t}}{\sqrt{n!}}|n\rangle \\ &= e^{-i\omega t/2}|\alpha e^{-i\omega t}\rangle, \quad (|n\rangle = (b^+)^n|0\rangle).\end{aligned}$$

This shows that the shape of a coherent state can be retained during its motion, which is the same as that of a microscopic particle (soliton) in the nonlinear quantum mechanics. The mean position of the particle in the time-dependent coherent state is

$$\begin{aligned}\langle\alpha,t|x|\alpha,t\rangle &= \langle\alpha|e^{iHt/\hbar}xe^{-iHt/\hbar}|\alpha\rangle \\ &= \left\langle\alpha\left|x-\frac{it}{\hbar}[x,H]+\frac{(-it)^2}{2!\hbar^2}[[x,H],H]+\cdots\right|\alpha\right\rangle \\ &= \left\langle\alpha\left|x+\frac{pt}{m}-\frac{1}{2!}t^2\omega^2x+\cdots\right|\alpha\right\rangle \\ &= \left\langle\alpha\left|x\cos\omega t+\frac{p}{m\omega}\sin\omega t\right|\alpha\right\rangle \\ &= \sqrt{\frac{2\hbar}{m\omega}}|\alpha|\cos(\omega t+\theta),\end{aligned} \tag{6.64}$$

where

$$\theta=\tan^{-1}\left(\frac{y}{x}\right),\quad x+iy=\alpha,$$

and

$$[x,H]=\frac{i\hbar p}{m},\quad [p,H]=-i\hbar m\omega^2x.$$

Comparing (6.64) with the solution of a classical harmonic oscillator,

$$x=\sqrt{\frac{2E}{m\omega^2}}\cos(\omega t+\theta),\quad E=\frac{p^2}{2m}+\frac{1}{2}m\omega^2x^2,$$

we find that they are similar, with

$$E=\hbar\omega\alpha^2=\langle\alpha|H|\alpha\rangle-\langle 0|H|0\rangle,\quad H=\hbar\omega\left(b^+b+\frac{1}{2}\right).$$

Thus we can say that the center of the wave packet of the coherent state indeed obeys the classical law of motion, which happens to be the same as the law of motion of microscopic particles in nonlinear quantum mechanics discussed in Section 4.2.

We can similarly obtain

$$\begin{aligned}\langle\alpha,t|p|\alpha,t\rangle &= -\sqrt{2m\hbar\omega}|\alpha|\sin(\omega t+\theta), \\ \langle\alpha,t|x^2|\alpha,t\rangle &= \frac{2\hbar}{\omega m}\left[|\alpha|^2\cos^2(\omega t+\theta)+\frac{1}{4}\right], \\ \langle\alpha,t|p^2|\alpha,t\rangle &= 2m\hbar\omega\left[|\alpha|^2\sin^2(\omega t+\theta)+\frac{1}{4}\right], \\ [\Delta x(t)]^2 &= \frac{\hbar}{2\omega m},\quad [\Delta p(t)]^2=\frac{1}{2}m\omega\hbar,\end{aligned}$$

and

$$\triangle x(t)\triangle p(t) = \frac{h}{2}.$$

This is the same as (6.62). It shows that the minimal uncertainty principle for the coherent state can be retained at all time, *i.e.*, the uncertainty relation does not change with time t.

The mean number of quanta in the coherent state is given by

$$\bar{n} = \langle\alpha|\hat{N}|\alpha\rangle = \langle\alpha|\hat{b}^{+}\hat{b}|\alpha\rangle = \alpha^2, \quad \langle\alpha|\hat{N}^2|\alpha\rangle = |\alpha|^4 + |\alpha|^2.$$

Therefore, the fluctuation of the quantum in the coherent state is

$$\triangle n = \sqrt{\langle\alpha|\hat{N}^2|\alpha\rangle - (\langle\alpha|\hat{N}|\alpha\rangle)^2} = |\alpha|,$$

which leads to

$$\frac{\triangle n}{\bar{n}} = \frac{1}{|\alpha|} \ll 1.$$

It is thus obvious that the fluctuation of quantum in the coherent state is very small. The coherent state is quite close to the feature of soliton or solitary wave.

These properties of coherent states of microscopic particles described by the nonlinear Schrödinger equation, the ϕ^4-equation, or the Sine-Gordon equation in nonlinear quantum mechanics are similar. In practice, the state of a microscopic particle in the nonlinear quantum mechanics can always be represented by a coherent state, for example, the Davydov wave functions, both $|D_1\rangle$ and $|D_2\rangle$ (see (5.72)), the wave function of exciton-solitons in protein molecules and acetanilide (see Chapter 9), and the BCS wave function (2.28) in superconductor, etc. Hence, the coherence of particles is a kind of nonlinear phenomenon which occurs only in nonlinear quantum mechanics. It does not belong to systems described by linear quantum mechanics, because the coherent state cannot be obtained by superposition of linear waves, such as plane wave, de Broglie wave, or Bloch wave, which are solutions of the linear Schrödinger equation in linear quantum mechanics. Therefore, the minimal uncertainty relation (6.63) as well as (6.60) and (6.61) are only applicable to microscopic particles in nonlinear quantum mechanics. In other words, only microscopic particles in nonlinear quantum mechanics satisfy the minimal uncertainty principle. It reflects the wave-corpuscle duality of microscopic particles because it holds only if the duality exists. This uncertainty principle also suggests that the position and momentum of the microscopic particle can be simultaneously determined in certain degree and range. A rough estimate for the size of the uncertainty is given in the following.

We choose $\xi = 140$, $\eta = \sqrt{300/0.253}/2\sqrt{2}$ and $x_0 = 0$ in (4.56) or (6.56), so that $\phi_s(x,t)$ or $\phi(p,t)$ satisfies the admissibility condition, *i.e.*, $\phi_s(0) \approx 0$. (In fact, in such a case, $\phi_s(0) \approx 10^{-6}$, thus the admissibility condition can be considered satisfied). We then get $\triangle x \approx 0.02624$ and $\triangle p \approx 19.893$, according to (6.59) and

(6.60). These results show that the position and momentum of a microscopic particle in nonlinear quantum mechanics can be simultaneously determined within a certain approximation.

Finally, we determine the uncertainty relation of the microscopic particle described by the nonlinear Schrödinger equation, arising from the quantum fluctuation effect in the nonlinear quantum field. The quantum theory presented in Section 4.6 was discussed by Lai and Haus based on the nonlinear Schrödinger equation. As discussed in Chapters 3 and 4, a superposition of a subclass of the bound states $|n, P\rangle$ which are characterized by the number of Bosons, such as photons or phonons, n, and the momentum of the center of mass, P, can reproduce the expectation values of the field of the microscopic particle (soliton) in the limit of a large average number of Bosons (phonons). Such a state formed by the superposition of $|n, P\rangle$ is referred to as a fundamental soliton state. In quantum theory, the field equation is given by (3.62), with the commutation relation (3.63). The corresponding quantum Hamiltonian is given by (3.65). In the Schrödinger picture, the time evolution of the system is described by (3.66). The many-particle state $|\psi'\rangle$ can be built up from the n-particle states given by (3.69).

The quantum theory based on (3.69) describes an ensemble of Bosons interacting via a δ-potential. Note that $\hat{H}$ preserves both the particle number.

$$\hat{N} = \int \hat{\phi}^{+}(x)\hat{\phi}(x)dx \tag{6.65}$$

and the total momentum

$$\hat{P} = i\frac{\hbar}{2}\int \left[\frac{\partial}{\partial x}\hat{\phi}^{+}(x)\hat{\phi}(x) - \hat{\phi}^{+}(x)\frac{\partial}{\partial x}\hat{\phi}(x)\right] dx. \tag{6.66}$$

Lai *et al.* proved that the Boson number and momentum operators commute, so that common eigenstates of $\hat{H}$, $\hat{P}$ and $\hat{N}$ can exist in such a case. In the case of a negative b value, the interaction between the Bosons is attractive and the Hamiltonian (3.63) has bound states. A subset of these bound states is characterized solely by the eigenvalues of $\hat{N}$ and $\hat{P}$. The wave functions of these states are

$$f_{n,p} = N_n \exp\left(ip\sum_{j=1}^{\infty} x_j + \frac{b}{2}\sum_{1\le i,j<n}^{\infty} |x_i - x_j|\right), \tag{6.67}$$

where

$$N_n = \sqrt{\frac{(n-1)!|b|^{n-1}}{2\pi}}.$$

Thus

$$f_n(x_1, \cdots x_n, t) = \int dp g_n(p) f_{n,P}(x_1, \cdots x_n, t) e^{-iE(n,p)t}, \tag{6.68}$$

where

$$g_n = \sqrt{g(p)}e^{-inpx_0}, \quad \text{and} \quad g(p) = \frac{e\phi\left\{-(p-p_0)^2/[2(\Delta p)^2]\right\}}{\sqrt{2\pi(\Delta p)^2}}.$$

Using $f_{n,p}$ given in (6.67), we found that $|n, P\rangle$ decays exponentially with separation between any pair of Bosons. It describes an n-particle soliton moving with momentum $P = \hbar np$ and energy $E(n,p) = np^2 - |b|^2(n^2-1)n/12$. By construction, the quantum number p in this wave function is related to the momentum of the center of mass of the n interacting Bosons, which is now defined as

$$\hat{X} = \lim_{\epsilon\to 0}\int x\hat{\phi}^+(x)\hat{\phi}(x)dx(\epsilon+\hat{N})^{-1}, \tag{6.69}$$

with

$$[\hat{X}, \hat{P}] = i\hbar.$$

The limit of $\epsilon \to 0$ is introduced to regularize the position operator for the vacuum state.

We are interested in the fluctuations of (6.65), (6.66) and (6.69) for a state $|\psi'(t)\rangle$ with a large average Boson number and a well-defined mean field. Kartner and Boivin decomposed the field operator into a mean value and a remainder which is responsible for the quantum fluctuations.

$$\begin{gathered}\hat{\phi}(x) = \langle\psi'(0)|\hat{\phi}^+(x)|\psi'(0)\rangle + \hat{\phi}_1(x),\\ \left[\hat{\phi}_1(x), \hat{\phi}_1^+(x')\right] = \delta(x-x'), \quad \left[\hat{\phi}_1(x), \hat{\phi}_1(x')\right] = 0.\end{gathered} \tag{6.70}$$

Since the field operator $\hat{\phi}$ is time independent in the Schrödinger representation, we can then choose $t = 0$ for definiteness. Inserting (6.70) into (6.65), (6.66) and (6.69) and neglecting terms of second and higher order in the noise operator, Kartner *et al.* obtained that

$$\hat{N} = n_0 + \Delta\hat{n}, \quad \hat{P} = \hbar n_0 p_0 + \hbar n_0 \Delta\hat{p}, \quad \hat{X} = x_0\left(1 - \frac{\Delta\hat{n}}{n_0}\right) + \Delta\hat{x},$$

with

$$\begin{aligned} n_0 &= \int dx\langle\hat{\phi}^+(x)\rangle\langle\hat{\phi}(x)\rangle, & \Delta\hat{n} &= \int dx\langle\hat{\phi}^+(x)\rangle\hat{\phi}_1(x) + c.c.,\\ p_0 &= \frac{i}{n_0}\int dx\langle\hat{\phi}_x^+(x)\rangle\langle\hat{\phi}(x)\rangle, & \Delta\hat{p} &= \frac{i}{n_0}\int dx\langle\hat{\phi}_x^+(x)\rangle\hat{\phi}_1(x) + c.c.,\\ x_0 &= \frac{1}{n_0}\int dxx\langle\hat{\phi}^+(x)\rangle\langle\hat{\phi}(x)\rangle, & \Delta\hat{x} &= \frac{1}{n_0}\int dxx\langle\hat{\phi}^+(x)\rangle\hat{\phi}_1(x) + c.c.,\end{aligned}$$

where $\Delta\hat{x}$ is the deviation from the mean value of the position operator, $\Delta\hat{n}$, $\Delta\hat{p}$, and $\Delta\hat{x}$ are linear in the noise operator. Because the third- and fourth-order correlators of $\hat{\phi}_1$ and $\hat{\phi}_1^+$ are very small, they can be neglected in the limit of large n_0.

Note that $\triangle\hat{p}$ and $\triangle\hat{x}$ are conjugate variables. To complete this set we introduce a variable conjugate to $\triangle\hat{n}$,

$$\triangle\hat{\theta} = \frac{1}{n_0}\int dx\left\{i\left[\langle\hat{\phi}^{+}(x)\rangle + x\langle\hat{\phi}_x^{+}(x)\rangle\right] - p_0 x\langle\hat{\phi}^{+}(x)\rangle\right\}\hat{\phi}_1(x) + c.c$$

As it is known, if the propagation distance is not too large, then to the first order, the mean value of the field is given by the classical soliton solution

$$\langle\hat{\phi}(x)\rangle = \phi_{0,n_0}(x,t)\left[1 + O\left(\frac{1}{n_0}\right)\right],$$

with

$$\begin{aligned}\phi_{0,n_0}(x,t) = {} & \frac{n_0\sqrt{|b|}}{2}\exp\left[i\Omega_{nl} - ip_0^2 t + ip_0(x - x_0) + i\theta_0\right] \\ & \times\text{sech}\left[\frac{n_0|b|}{2}(x - x_0 - 2p_0 t)\right],\end{aligned} \tag{6.71}$$

where the nonlinear phase shift $\Omega_{nl} = n_0^2|b|^2 t/4$. If $p_0 = x_0 = \theta_0 = 0$, we obtain the following for the fluctuation operators in the Heisenberg picture,

$$\triangle\hat{n}(t) = \int dx\left[f_{-n}(x)^* F'_{nl} + c.c\right],$$

$$\triangle\hat{\theta}(t) = \int dx\left[f_{-\theta}(x)^* F'_{nl} + c.c\right],$$

$$\triangle\hat{p}(t) = \int dx\left[f_{-p}(x)^* F'_{nl} + c.c\right],$$

$$\triangle\hat{x}(t) = \int dx\left[f_{-x}(x)^* F'_{nl} + c.c\right],$$

with

$$F'_{nl} = e^{i\Omega nl}\hat{\phi}_1(x,t),$$

and the set of adjoint functions

$$f_{-n}(x) = \frac{n_0\sqrt{|b|}}{2}\text{sech}(x_{n_0}),$$

$$f_{-\theta}(x) = \frac{i\sqrt{|b|}}{2}\left[\text{sech}(x_{n_0}) + x_{n_0}\frac{d}{dx_{n_0}}\text{sech}(x_{n_0})\right],$$

$$f_{-p}(x) = \frac{-in_0\sqrt{|b|^3}}{4}\frac{d}{dx_{n_0}}\text{sech}(x_{n_0}),$$

$$f_{-x}(x) = \frac{1}{n_0\sqrt{|b|}}x_{n_0}\text{sech}(x_{n_0}),$$

where

$$x_{n_0} = \frac{1}{2}n_0\,|b|\,x.$$

For a coherent state defined by

$$\hat{\phi}(x)|\Phi_{0,n_0}\rangle = \phi_{0,n_0}(x)|\Phi_{0,n_0}\rangle, \quad \text{or} \quad \hat{\phi}_1(x)|\phi_{0,n_0}\rangle = 0,$$

where

$$|\Phi_{0,n_0}\rangle = \exp\left\{\int dx \left[\phi_{0,n_0}(x)\hat{\phi}^+(x) - \phi^*_{0,n_0}(x)\hat{\phi}^+(x)\right]\right\}|0\rangle,$$

and ϕ_{0,n_0} has been given (6.71), Kartner and Boivin further obtained that

$$\langle\triangle\hat{n}_0^2\rangle = n_0, \quad \langle\triangle\hat{\theta}_0^2\rangle = \frac{0.6075}{n_0}, \quad \langle\triangle\hat{p}_0^2\rangle = \frac{1}{3n_0\tau_0^2}, \quad \langle\triangle\hat{x}_0^2\rangle = \frac{1.645\tau_0^2}{2n_0},$$

where $\tau_0^2 = 2/n_0|b|$ is the width of the microscopic particle (soliton). The uncertainty products of the Boson number and phase, momentum and position are, respectively,

$$\langle\triangle\hat{n}_0^2\rangle\langle\triangle\hat{\theta}_0^2\rangle = 0.6075 \geq 0.25, \quad \text{and} \quad n_0^2\langle\triangle\hat{p}_0^2\rangle\langle\triangle\hat{x}_0^2\rangle = 0.27 \geq 0.25.$$

Here the quantum fluctuation of the coherent state is white, *i.e.*,

$$\langle\hat{\phi}_1(x)\hat{\phi}_1(y)\rangle = \langle\hat{\phi}_1^+(x)\hat{\phi}_1(y)\rangle = 0.$$

However, the quantum fluctuation of the soliton cannot be white because interaction between the particles introduces correlations between them. Thus, Kartner and Boivin assumed a fundamental soliton state with a Poissonian distribution for the Boson number

$$p_n = \frac{n_0^n}{n!}e^{-n_0}$$

and a Gaussian distribution for the momentum (6.68) with a width of $\langle\triangle\hat{p}_0^2\rangle = n_0|b|^2/4\mu$, where μ is a parameter of the order of unity compared to n_0. They finally obtained the minimum uncertainty values

$$\langle\triangle\hat{\theta}_0^2\rangle = \frac{0.25}{\langle\triangle\hat{n}^2\rangle} = \frac{0.25}{n_0}\left[1 + O\left(\frac{1}{n_0}\right)\right],$$

$$\langle\triangle\hat{x}_0^2\rangle = \frac{0.25}{\langle n_0^2\triangle\hat{p}^2\rangle} = \frac{0.25\mu\tau_0^2}{n_0}\left[1 + O\left(\frac{1}{n_0}\right)\right],$$

up to the order of $1/n_0$, for the corresponding initial fluctuations in soliton phase and timing. Thus at $t = 0$ the fundamental soliton with the given Boson number and momentum distributions is a minimum uncertainty state in the four collective variables, the Boson number, phase, momentum, and position, up to terms of

$O(1/n_0)$,

$$\langle\Delta\hat{n}_0^2\rangle\langle\Delta\hat{\theta}_0^2\rangle = 0.25\left[1 + O\left(\frac{1}{n_0}\right)\right],$$
$$n_0^2\langle\Delta\hat{p}_0^2\rangle\langle\Delta\hat{x}_0^2\rangle = 0.25\left[1 + O\left(\frac{1}{n_0}\right)\right]. \tag{6.72}$$

These are the uncertainty relations arising from the quantum fluctuations in nonlinear quantum field of microscopic particles described by the nonlinear Schrödinger equation. They are the same as (6.61) – (6.63). Therefore, we conclude that the uncertainty relation in the nonlinear quantum mechanics takes the minimum value regardless a state is coherent or squeezed, a system is classical or quantum.

Pang *et al.* also calculated the uncertainty relation of quantum fluctuations and studied their properties in nonlinearly coupled electron-phonon systems based on the Holstein model but using a new ansatz which includes correlations among one-phonon coherent and two-phonon squeezing states and polaron state. Many interesting results were obtained. The minimum uncertainty relation takes different forms in different systems which are related to the properties of the microscopic particles. Nevertheless, the minimum uncertainty relation (6.63) holds for both the one-quantum coherent state and two-quanta squeezed state (see Jajernikov and Pang 1997, Pang 1999, 2000, 2001, 2002). These work enhanced our understanding of the significance and nature of the minimum uncertainty relation.

In light of the above discussion, we can distinguish the motions of particles in the linear quantum mechanics, nonlinear quantum mechanics, and classical mechanics based on the uncertainty relation. When the motion of the particles satisfy $\triangle x\triangle p > h/2$, the particles obey laws of linear quantum mechanics, and they have only wave feature. When the motion of the particles satisfy $\triangle x\triangle p = h/12$ or $\pi/6$, the particles obey laws of motion in the nonlinear quantum mechanics, and the particles are solitons, exhibiting wave-corpuscle duality. If the motion of the particles satisfy $\triangle x\triangle p = 0$, then the particles can be treated as classical particles, with only corpuscle feature. The nonlinear quantum mechanics introduced here thus gives a more complete description of physical systems. Therefore, we can say that the nonlinear quantum mechanics bridges the gap between the classical mechanics and the linear quantum mechanics.

6.4 Energy Spectrum of Hamiltonian and Vector Form of the Nonlinear Schrödinger Equation

Like in the linear quantum mechanics, it is useful to obtain the energy spectrum of the Hamiltonian operator for a given system in the nonlinear quantum mechanics. The energy of a microscopic particle satisfying the nonlinear Schrödinger equation can be obtained from

$$E = \int_{-\infty}^{\infty} \mathcal{H} dx,$$

where $\mathcal{H}$ is the classical Hamiltonian density which depends on the wave function $\phi(x,t)$. However, this only gives the energy of the microscopic particle, not the eigenenergy spectrum of the Hamiltonian operator. As discussed in Chapter 3, we can obtain the eigenvalues of the L operator corresponding to the nonlinear Schrödinger equation (3.2) following Lax's approach. But these eigenvalues are not the eigenenergy spectrum of the Hamiltonian operator of the systems. We must, therefore, use alternative method to calculate its eigenenergy spectrum. In linear quantum mechanics, the eigenenergy spectrum was obtained from the eigenequation of the Hamiltonian operator. Because the latter is independent of the state wavefunction of the particle, there is little difficulty in calculating the eigenenergy in linear quantum mechanics. However, this is not the case in nonlinear quantum mechanics in which the Hamiltonian operator depends on the state wave vector of the microscopic particle, as mentioned in Chapter 3. How do we then obtain the eigenenergy spectra in nonlinear quantum mechanics?

6.4.1 *General approach*

As discussed in Chapter 3, the wave function of a microscopic particle can be quantized by the creation and annihilation operators of the particle in nonlinear quantum mechanics. The Hamiltonian of a system described by the wave function $\phi(x,t)$ can be quantized by introducing creation and annihilation operators in the particle number representation or the second quantization representation. We can then calculate the eigenenergy spectrum using the eigenequation of the quantum Hamiltonian and the corresponding wavevector in the particle number representation. This is basically how the eigenenergy spectrum in the nonlinear quantum mechanics can be obtained. For convenience, we express the nonlinear Schrödinger equation (3.2) with $A(\phi) = 0$ in the following discrete form

$$i\hbar\frac{\partial \phi_j}{\partial t} = -\frac{\hbar}{2mr_0^2}(\phi_{j+1} - 2\phi_j + \phi_{j-1}) - b|\phi_j|^2\phi_j + V(j,t)\phi, \quad (j = 1, 2, 3, \cdots, J), \tag{6.73}$$

where r_0 is a spacing between two neighboring lattice points, j labels the discrete lattice points, J is the total number of lattice points in the system.

The vector form of the above equation is

$$\left[i\hbar\frac{\partial}{\partial t} - \frac{\hbar^2}{mr_0^2} - V(j,t)\right]\bar{\phi} = -\varepsilon M\bar{\phi} - b\,\text{diag.}(|\phi_1|^2, |\phi_2|^2 \cdots |\phi_\alpha|^2)\bar{\phi}, \tag{6.74}$$

where $\bar{\phi}$ is a column vector, $\bar{\phi} = \text{Col.}(\phi_1, \phi_2 \cdots \phi_\alpha)$ whose components are complex. Equation (6.74) is a vector nonlinear Schrödinger equation with α modes of motion. In (6.74), b is a nonlinear parameter and α is the number of modes that exist in the systems, $M = [M_{n\ell}]$ is an $\alpha \times \alpha$ real symmetric dispersion matrix, $\varepsilon = \hbar^2/(2mr_0^2)$. Here n and ℓ are integers denoting the modes of motion.

The Hamiltonian and the particle number corresponding to (6.74) are

$$H = \sum_{n=1}^{\alpha} \left(\hbar\omega_0 |\phi_n|^2 - \frac{1}{2} b |\phi_n|^4 \right) - \varepsilon \sum_{n \neq \ell}^{\alpha} M_{n\ell} \phi_n \phi_\ell, \tag{6.75}$$

$$N = \sum_{n=1}^{\alpha} |\phi_n|^2, \tag{6.76}$$

where

$$\hbar\omega_0 = \frac{\hbar^2}{mr_0^2} + V(j,t).$$

We have assumed that $V(j,t)$ are independent of j and t. In the canonical second quantization theory, the complex amplitudes (ϕ_n^* and ϕ_n) become Boson creation and annihilation operators ($\hat{B}_n^+$ and $\hat{B}_n$) in the number representation. If $|m_n\rangle$ is an eigenfunction of a particular mode, then $\hat{B}_n^+|m_n\rangle = \sqrt{m_n+1}|m_n+1\rangle$, $\hat{B}_n|m_n\rangle = \sqrt{m_n}|m_n-1\rangle$ and $\hat{B}_n|0\rangle = 0$.

Since the no particular ordering is specified in (6.76) and (6.76), we use the averages,

$$|\phi_n|^2 \to \frac{1}{2}\left(\hat{B}_n^+\hat{B}_n + \hat{B}_n\hat{B}_n^+\right),$$

and

$$\begin{aligned} |\phi_n|^4 \to \frac{1}{6}\Big(&\hat{B}_n^+\hat{B}_n^+\hat{B}_n\hat{B}_n + \hat{B}_n^+\hat{B}_n\hat{B}_n^+\hat{B}_n + \hat{B}_n^+\hat{B}_n\hat{B}_n\hat{B}_n^+ \\ &+\hat{B}_n\hat{B}_n^+\hat{B}_n\hat{B}_n^+ + \hat{B}_n\hat{B}_n\hat{B}_n^+\hat{B}_n^+ + \hat{B}_n\hat{B}_n^+\hat{B}_n^+\hat{B}_n\Big), \end{aligned}$$

with the Boson commutation rule $\hat{B}_n\hat{B}_n^+ - \hat{B}_n^+\hat{B}_n = 1$. Equations (6.76) and (6.76) then become

$$H = \sum_{n=1}^{\alpha}\left[\left(\hbar\omega_0 - \frac{1}{2}b\right)\left(\hat{B}_n^+\hat{B}_n + \frac{1}{2}\right) - \frac{1}{2}b\hat{B}_n^+\hat{B}_n\hat{B}_n^+\hat{B}_n\right] - \varepsilon\sum_{n\neq\ell}^{\alpha} M_{n\ell}\hat{B}_n^+\hat{B}_\ell, \tag{6.77}$$

$$N = \sum_{n=1}^{\alpha}\left(\hat{B}_n^+\hat{B}_n + \frac{1}{2}\right). \tag{6.78}$$

From now on, we will use the notation $[m_1 m_2 \cdots m_\alpha]$ to denote the product of the number states $|m_1\rangle|m_2\rangle\cdots|m_\alpha\rangle$. The stationary states of the vector nonlinear Schrödinger equation (6.74) must be eigenfunctions of both $\hat{N}$ and $\hat{H}$. Consider an m-quantum state (*i.e.*, the mth excited level, $m = m_1 + m_2 + \cdots m_j$), with $m < \alpha$.

An eigenfunction of $\hat{N}$ can be established as

$$\begin{aligned}|\Phi_m\rangle = {} & C_1[m,0,0,\cdots,0] + \cdots + C_2[0,m,0,0,\cdots,0] + \cdots + \\ & C_i[0,0,0,\cdots,m] + \cdots + C_{i+1}[(m-1),1,0,\cdots,0] + \cdots + \\ & C_p[0,0,\cdots,0,\underbrace{1,1,\cdots,1}_{\text{(m times)}}].\end{aligned} \tag{6.79}$$

The number of terms in (6.79) is equal to the number of ways that m quanta can be placed on the α sites, which is given by

$$p = \frac{(m+\alpha-1)}{m!\,(\alpha-1)!}.$$

The wave function $|\Phi_m\rangle$ in (6.79) is an eigenfunction of $\hat{N}$ for any values of the C_α's. Thus we are free to choose these coefficients such that

$$\hat{H}|\Phi_m\rangle = E|\Phi_m\rangle. \tag{6.80}$$

Equation (6.80) requires that the column vector $\bar{C} = \text{Col.}(C_1, C_2, \cdots Cp)$ satisfies the matrix equation

$$|H - IE|\bar{C} = 0, \tag{6.81}$$

where H is a $p \times p$ symmetric matrix with real elements. I is a $p \times p$ identity matrix, E is the eigenenergy. Equation (6.80) is an eigenvalue equation of the quantum Hamiltonian operator (6.77) of the system. We can obtain the eigenenergy spectrum E_m of the system from (6.81) for given parameters, ε, ω_0, and b. Scott, Bernstein, Eilbeck, Carr and Pang *et al.* used the above method to calculate the energy-spectra of vibrational excitations (quanta) in many nonlinear systems, for example, small molecules or organic molecular crystals and biomolecules. These results can be compared with experimental data, and will be discussed in Chapter 9.

6.4.2 *System with two degrees of freedom*

We now discuss a simple instance and consider a system with two degrees of freedom (or two modes of motion). This was studied by Scott *et al.* and Pang. It takes the form

$$\left(i\hbar\frac{d}{dt} - \hbar\omega_0\right)\begin{pmatrix}\phi_1\\ \phi_2\end{pmatrix} = \begin{pmatrix}b|\phi_1|^2 & \epsilon\\ \epsilon & b|\phi_2|^2\end{pmatrix}\begin{pmatrix}\phi_1\\ \phi_2\end{pmatrix} = 0. \tag{6.82}$$

The corresponding $\hat{N}$ and $\hat{H}$, in such a case, are given by

$$\hat{N} = \hat{B}_1^+\hat{B}_1 + \hat{B}_2^+\hat{B}_2 + 1, \tag{6.83}$$

$$\hat{H} = \left(\hbar\omega_0 - \frac{b}{2}\right)\hat{N} - \frac{b}{2}\left(\hat{B}_1^+\hat{B}_1\hat{B}_1^+\hat{B}_1 + \hat{B}_2^+\hat{B}_2\hat{B}_2^+\hat{B}_2\right) - \varepsilon\left(\hat{B}_1^+\hat{B}_2 + \hat{B}_1\hat{B}_2^+\right),$$

respectively.

We seek eigenfunction $|\Phi\rangle$ of both the operators. For the ground state, $|\Phi_0\rangle = |0\rangle|0\rangle$, which is the product of the ground state wavefunctions of the two degrees of freedom, we have

$$\hat{N}|\Phi_0\rangle = |\Phi_0\rangle, \quad \hat{H}|\Phi_0\rangle = E_0|\Phi_0\rangle,$$

where

$$E_0 = \hbar\omega_0 - \frac{b}{2}.$$

For the first excited state, we have $|\Phi_1\rangle = C_1|1\rangle|0\rangle + C_2|0\rangle|1\rangle$ with $|C_1|^2 + |C_2|^2 = 1$. Thus $\hat{N}|\Phi_1\rangle = 2|\Phi_1\rangle$. From $\hat{H}|\Phi_1\rangle = E_1|\Phi_1\rangle$, we can get

$$\begin{bmatrix} (2\hbar\omega_0 - E_1) - \frac{1}{2}b & \varepsilon \\ \varepsilon & (2\hbar\omega_0 - E_1) - \frac{1}{2}b \end{bmatrix} \begin{pmatrix} C_1 \\ C_2 \end{pmatrix} = 0.$$

The first excited state splits into a symmetric and an antisymmetric states,

$$|\Phi_{1s}\rangle = \frac{1}{\sqrt{2}}(|1\rangle|0\rangle + |0\rangle|1\rangle),$$
$$|\Phi_{1a}\rangle = \frac{1}{\sqrt{2}}(|1\rangle|0\rangle - |0\rangle|1\rangle),$$

with

$$E_{1s} = E_0 + \hbar\omega_0 - b - \varepsilon,$$
$$E_{1a} = E_0 + \hbar\omega_0 - b + \varepsilon.$$

respectively.

For the second excited state, we have $|\Phi_2\rangle = C_1|2\rangle|0\rangle + C_2|1\rangle|1\rangle + C_3|0\rangle|2\rangle$, with

$$\sum_{i=1}^{3} |C_i|^2 = 1,$$
$$\hat{N}|\Phi_2\rangle = 3|\Phi_2\rangle, \quad \hat{H}|\Phi_2\rangle = E_2|\Phi_2\rangle,$$

and

$$\begin{pmatrix} 3\hbar\omega_0 - E_2 + \frac{7}{2}b & \sqrt{2}\varepsilon & 0 \\ \sqrt{2}\varepsilon & 3\hbar\omega_0 - E_2 + \frac{5}{2}b & \sqrt{2}\varepsilon \\ 0 & \sqrt{2}\varepsilon & 3\hbar\omega_0 - E_2 + \frac{7}{2}b \end{pmatrix} \begin{pmatrix} C_1 \\ C_2 \\ C_3 \end{pmatrix} = 0.$$

Diagonalizing this matrix equation, we get the eigenenergies with $E_{2a} = 2\hbar\omega_0 + E_1 + 7b/2$ corresponding to the antisymmetric state $C_a = (1, 0, -1)/\sqrt{2}$, and $E_{2\pm} = 2\hbar\omega_0 + E_1 + 3b \pm 3\sqrt{b^2 + 16\varepsilon^2}$ corresponding to the symmetric state

$C_\pm = (\varepsilon, \sqrt{b^2+16\varepsilon^2}/2, \sqrt{2}\varepsilon)$, respectively. In the limit of $b \gg \varepsilon$, $C_+ = (0,1,0)$, $C_- = (1/\sqrt{2}, 0, 1/\sqrt{2})$, and the state C_a is a localized mode which has all the energy and the particle has equal probability to be in each of the two sites. In this case, $C_\pm$ have the same structure, but a different relative phase. Linear combinations of these states will first localize all the energy on a particular site.

For the mth excited state, we have

$$|\Phi_m\rangle = C_1|m\rangle|0\rangle + C_2|m-1\rangle|1\rangle + \cdots + C_m|1\rangle|m-1\rangle + C_{m+1}|0\rangle|m\rangle \tag{6.84}$$

with

$$\sum_{i=1}^{m} |C_i|^2 = 1,$$

$$\hat{N}|\Phi_m\rangle = (m+1)|\Phi_m\rangle, \quad \hat{H}|\Phi_m\rangle = E_m|\Phi_m\rangle,$$

and

$$[(m+1)\hbar\omega_0 - E_m - \Omega_m]\,\bar{C} = 0,$$

here $\bar{C} = \text{Col.}(C_1, C_2, \cdots, C_{n+1})$, Ω_m is an $(m+1)\times(m+1)$ tridiagonal matrix. For an odd number n, we have

$$\Omega_n = \begin{pmatrix} D(1) & Q(1) & 0 & \cdots & 0 & \cdots & 0 & \cdots & 0 & 0 & 0 \\ Q(1) & D(2) & Q(2) & \cdots & 0 & \cdots & 0 & \cdots & 0 & 0 & 0 \\ 0 & Q(2) & & \cdots & 0 & \cdots & 0 & \cdots & 0 & 0 & 0 \\ \vdots & \vdots & \vdots & & \vdots & & \vdots & & \vdots & \vdots & \vdots \\ 0 & 0 & 0 & \cdots & D[(m+1)/2] & \cdots & Q[(m+1)/2] & \cdots & 0 & 0 & 0 \\ 0 & 0 & 0 & \cdots & Q[(m+1)/2] & \cdots & D[(m+1)/2] & \cdots & 0 & 0 & 0 \\ \vdots & \vdots & \vdots & & \vdots & & \vdots & & \vdots & \vdots & \vdots \\ 0 & 0 & 0 & \cdots & 0 & \cdots & 0 & \cdots & 0 & Q(2) & 0 \\ 0 & 0 & 0 & \cdots & 0 & \cdots & 0 & \cdots & Q(2) & D(2) & Q(1) \\ 0 & 0 & 0 & \cdots & 0 & \cdots & 0 & \cdots & 0 & Q(1) & D(1) \end{pmatrix},$$

where

$$D(i) = \frac{1}{2}b\left[m+1+(m+1-i)^2+(i-1)^2\right], \quad Q(i) = \varepsilon\sqrt{i(m+1-i)}.$$

For an even number m, we have

$$\Omega_n = \begin{pmatrix} D(1) & Q(1) & 0 & \cdots & 0 & 0 & 0 & \cdots & 0 & 0 & 0 \\ Q(1) & D(2) & Q(2) & \cdots & 0 & 0 & 0 & \cdots & 0 & 0 & 0 \\ 0 & Q(2) & 0 & \cdots & 0 & 0 & 0 & \cdots & 0 & 0 & 0 \\ \vdots & \vdots & \vdots & & \vdots & \vdots & \vdots & & \vdots & \vdots & \vdots \\ 0 & 0 & 0 & \cdots & 0 & Q(m/2) & 0 & \cdots & 0 & 0 & 0 \\ 0 & 0 & 0 & \cdots & Q(m/2) & D(m/2+1) & Q(m/2) & \cdots & 0 & 0 & 0 \\ 0 & 0 & 0 & \cdots & 0 & Q(m/2) & 0 & \cdots & 0 & 0 & 0 \\ \vdots & \vdots & \vdots & & \vdots & \vdots & \vdots & & \vdots & \vdots & \vdots \\ 0 & 0 & 0 & \cdots & 0 & 0 & 0 & \cdots & 0 & Q(2) & 0 \\ 0 & 0 & 0 & \cdots & 0 & 0 & 0 & \cdots & Q(2) & D(2) & Q(1) \\ 0 & 0 & 0 & \cdots & 0 & 0 & 0 & \cdots & 0 & Q(1) & D(1) \end{pmatrix},$$

where

$$D\left(\frac{m}{2}+1\right) = \frac{b}{2}\left(m+1+\frac{m^2}{2}\right), \quad Q\left(\frac{m}{2}\right) = \sqrt{\frac{m}{2}\left(\frac{m}{2}+1\right)}.$$

We can see that if $(m+1)\hbar\omega_0 - E_n^i$ is one of the $(m+1)$ real eigenvalues of Ω_n, and C^i is the corresponding eigenvector, the wave function for the mth excited state can be established by (6.84). For example, for a localized mode with its energy concentrated on a single degree of freedom, the classical system (6.82) has harmonic solutions of the form $\phi_i = \phi_i' e^{-i\omega t}$. For $N > 2\varepsilon$, there is a local mode branch with $\hbar\omega = \hbar\omega_0 - bN$. Along this branch, $E = \hbar\omega_0 N - bN^2/2 - \varepsilon^2/b$. In the quantum case here (let $\varepsilon < \gamma$), there are two states corresponding to the localized modes. One is symmetric with $C_1 = C_{m+1} = 0(1)$, $C_2 = C_m = 0(\varepsilon/b)$, $C_3 = C_{m-1} = O(\varepsilon^2/b^2)$, etc. The other is antisymmetric with $C_1 = -C_{m+1} = 0(1)$, $C_2 = -C_m = 0(\varepsilon/b)$, $C_3 = -C_{m-1} = O(\varepsilon^2/b^2)$, etc. The gap between them is about $E_{ms} - E_{ma} = O(\varepsilon^m/b^{m-1})$. To the order of ε^2, the energy of these modes is

$$E_{ma} - E_0 = E_{ms} - E_0 = \left[\left(\hbar\omega_0 - \frac{1}{2}b\right)m - \frac{1}{2}bm^2 - \frac{m\varepsilon^2}{(m-1)b}\right]. \tag{6.85}$$

When $m \geq 4$, there could be overtone spectra at integer values of m. As $m \to \infty$, it approaches the classical limit given above except that the frequency is reduced by a factor of $b/2$. Therefore, the energy of the local mode remains constant on a particular state in the classical case, but oscillates between the two states with a period $\tau = 0\left(b^{m-1}/\varepsilon^m\right)$ in the quantum case. When $m \to \infty$, $\tau \to \infty$, it returns to the classical result.

6.4.3 *Perturbative method*

From (6.73) – (6.76), we know that $\hbar\omega_0$ contains the applied potential $V(x)$, which in general depends on x. Strictly speaking, the method discussed above is only applicable to the case of $V(x) = 0$ or a constant. If $V(x)$ depends on x, we cannot use the above method to obtain the eigenvalue spectrum, but we can use perturbation method to find an approximate eigenenergy spectrum, when $|V(x)| < b$ or $\hbar^2/2mr_0^2$. Scott *et al.* and Pang studied the eigenenergy spectrum of a Hamiltonian operator with $V(x)$ using a perturbation approach. Their work are briefly described in the following.

In accordance with the general perturbation theory, the Hamiltonian operator of the system in (6.74) is written as

$$H = H_0 + \tilde{\varepsilon}V. \tag{6.86}$$

At the same time, we assume that the eigenvectors and eigenvalues of H can be written as

$$\begin{aligned} \bar{C} &= \bar{C}_0 + \tilde{\varepsilon}\bar{C}_1 + \tilde{\varepsilon}^2\bar{C}_2 + \cdots , \\ E &= E_0 + \tilde{\varepsilon}E_1 + \tilde{\varepsilon}^2 E_2 + \cdots . \end{aligned} \tag{6.87}$$

Inserting (6.87) into (6.81) we can get

$$\begin{aligned} L\bar{C}_0 &= 0, \\ L\bar{C}_1 &= (E_1 - V)\bar{C}_0, \\ L\bar{C}_2 &= (E_1 - V)\bar{C}_1 + E_2\bar{C}_0, \\ L\bar{C}_3 &= (E_1 - V)\bar{C}_2 + E_2\bar{C}_1 + E_2\bar{C}_0, \\ &\cdots\cdots \end{aligned} \tag{6.88}$$

with $L = H_0 - E_0$, where the matrix H_0 is diagonal and its first α elements are equal. Since we are mainly interested in the dependence of the lowest α eigenvalues on ε, $\bar{C}_0$ can be considered the eigenvector corresponding to the eigenenergy E_0. Then L has a null space of dimension α which we must take into consideration as we seek solutions of (6.88).

Solution of this problem was given by Scott *et al.*. The first excited state ($m = 1$) has a wave function of the form

$$|\Phi_1\rangle = C_1[1, 0, 0, \cdots, 0] + \cdots + C_\alpha[0, \cdots, 0, 1].$$

Measuring energy with respect to the ground state

$$H_0 = (\hbar\omega_0 - b)\text{diag.}(1, 1, 1, 1, \cdots, 1). \tag{6.89}$$

Thus $E_0 = (\hbar\omega_0 - b)$, $L = [0]$ and the first equation in (6.87) places no constraints on the selection of $\bar{C}_0$.

Next, Scott *et al.* defined an inner product

$$(\bar{v}, \bar{\omega}) = \sum_{i'=1}^{p} v_{i'}^* \omega_{i'}.$$

Since the $C_{i'}$'s may be complex, and L is self-adjoint, from the Fredholm alternative theorem, we know that (6.88) has solutions if and only if the right hand side of the second equation is orthogonal to (α dimensional) null space of L, as pointed out by Strang. Assuming this null space is spanned by the orthogonal set $[(\bar{C}_0^{(i')})]$, $i' = 1, 2, \cdots, \alpha$ and that $\bar{C}_0 = \bar{C}_0^{(i')}$, the second equation in (6.87) has a solution if and only if

$$E_1\left(\bar{C}_0^{(i')}, \bar{C}_0^{(i')}\right) - \left(\bar{C}_0^{(i')}, M\bar{C}_0^{(j^{i'})}\right) = 0, \tag{6.90}$$

and

$$E_1\left(\bar{C}_0^{(k')}, \bar{C}_0^{(i')}\right) - \left(\bar{C}_0^{(k')}, M\bar{C}_0^{(i')}\right) = 0, \quad \text{for } k' \neq i'. \tag{6.91}$$

Condition (6.90) is satisfied if we choose

$$E_1 = \frac{\left(\bar{C}_0^{(i')}, M\bar{C}_0^{(i')}\right)}{\left(\bar{C}_0^{(i')}, \bar{C}_0^{(i')}\right)}. \tag{6.92}$$

For this value of E_1, condition (6.91) is satisfied if $\bar{C}_0^{(i)}$ is an orthogonal eigenvector of M. Then $\bar{C}_1 = \bar{O}$ leads to $\bar{C}_2 = 0$, etc. and the exact result of the perturbation theory is $\bar{C}_1 = \bar{C}_0$ and $E = E_0 + \varepsilon E_1$. This is essentially a direct determination of the eigenvectors and eigenvalues of $E = (\hbar\omega_0 - b)I + \varepsilon M$.

Next, we consider the second excited level ($m = 2$). With the wave function $|\Phi_n\rangle$ constructed as in (6.79),

$$H_0 = \text{diag.}(\underbrace{E_0, E_0, \cdots, E_0}_{\alpha \text{ times}}, \underbrace{h_{\alpha+1}, h_{\alpha+2}, \cdots, h_p}_{(p-\alpha) \text{ times}}) \tag{6.93}$$

and

$$V = \begin{bmatrix} O & G \\ G^T & R \end{bmatrix} \begin{matrix} \}\alpha \\ \}(p-\alpha). \end{matrix} \quad \underbrace{}_{\alpha}\underbrace{}_{p-\alpha} \tag{6.94}$$

Similarly

$$L = \begin{bmatrix} O & O \\ O & \Delta \end{bmatrix} \begin{matrix} \}\alpha \\ \}(p-\alpha). \end{matrix} \quad \underbrace{}_{\alpha}\underbrace{}_{p-\alpha} \tag{6.95}$$

where

$$\triangle = \text{diag.}(d_1, d_2 \cdots, d_{p-\alpha}).$$

All elements of Δ are nonzero. To be consistent with the partitioning in (6.94) and (6.95), Scott *et al.* defined

$$\bar{C}_i = \begin{bmatrix} W_i \\ Z_i \end{bmatrix} \begin{matrix} \}m \\ \}(p-m) \end{matrix}. \tag{6.96}$$

From the first equation in (6.88) and the structure of L, we have $Z_0 = \bar{O}$. For solvability of the second equation in (6.88) and the structure of V, the Fredholm alternative theorem requires that $E_1 = 0$. Thus (6.88) can be expressed as the following two sets

$$\begin{aligned} Z_1 &= -\Delta^{-1}G_0^T W_0, \\ Z_2 &= -\Delta^{-1}(G^T W_1 + RZ_1), \\ Z_3 &= -\Delta^{-1}(G^T W_3 + RZ_2 - E_2 Z_1), \\ Z_4 &= -\Delta^{-1}(G^T W_3 + RZ_3 - E_2 Z_2 - E_3 Z_1) \\ &\cdots\cdots \end{aligned} \tag{6.97}$$

and

$$\begin{aligned} TW_0 &= 0, \\ TW_1 &= -G\Delta^{-1}RZ_1 - E_3 W_0, \\ TW_2 &= -G\Delta^{-1}(E_2 Z_1 - RZ_2) - E_3 W_1 - E_4 W_0, \\ TW_3 &= G\Delta^{-1}(E_3 Z_1 + E_2 Z_2 - RZ_3) - E_3 W_2 - E_4 W_1 - E_5 W_0, \\ &\cdots\cdots \end{aligned} \tag{6.98}$$

with

$$T = G\Delta^{-1}G^T + E_2.$$

Scott *et al.* gave the following procedure for solving (6.97) – (6.98).

(1) Choose W_0 as an eigenvector of $G\Delta^{-1}G^T$ and $-E_2$ the corresponding eigenvalue, to satisfy the first equation in (6.98), assuming that E_2 is not a multiple eigenvalue.
(2) Obtain Z_1 from the first equation in (6.97).
(3) Since E_0 is not a multiple eigenvalue, the null space of T is simple W_0. From the Fredholm alternative theorem, a necessary and sufficient condition for the second equation in (6.98) to have a solution is

$$E_3 = -\frac{(W_0, G\Delta^{-1}RZ_1)}{(W_0, W_0)}.$$

(4) Solve the second equation in (6.98) for W_1 with the requirement of $(W_1, W_0) = 0$.
(5) Obtain Z_2 from the second equation in (6.97).

(6) The solvability of the third equation in (6.98) requires

$$E_4 = \frac{[W_0, G\Delta^{-1}(E_2 Z_1 - R Z_2)]}{(W_0, W_0)}.$$

Following this procedure, we can find the solutions and the corresponding energy spectrum of the system under the perturbation $\tilde{\varepsilon}V$ in (6.86).

6.4.4 *Vector nonlinear Schrödinger equation*

Since we are dealing with colum vectors in (6.74), it would be useful to have a clear understanding on properties of the vector nonlinear Schrödinger equation. This equation can often be written as

$$i\hbar\phi_t = -\frac{\hbar^2}{2m}\phi_{xx} - b\left(\bar{\phi}\phi\right)\phi, \tag{6.99}$$

where

$$\phi = \begin{pmatrix} \phi^{(1)} \\ \vdots \\ \phi^{(n)} \end{pmatrix}$$

is a column vector of n-components, describing the "isospace" state. The corresponding Lagrangian and Hamiltonian densities of the system are given by

$$\mathcal{L} = \frac{i\hbar}{2}\left(\bar{\phi}\phi_t - \bar{\phi}_t\phi\right) - \frac{\hbar^2}{2m}\left(\bar{\phi}_x\phi_x\right) + \frac{b}{2}\left(\bar{\phi}\phi\right)^2, \tag{6.100}$$

$$\mathcal{H} = \frac{\hbar^2}{2m}\left(\bar{\phi}_x\phi_x\right) - \frac{b}{2}\left(\bar{\phi}\phi\right)^2. \tag{6.101}$$

Here $\bar{\phi} = \phi^+\gamma_0$ and γ_0 is a diagonal matrix, and the internal product $(\bar{\phi}\phi) = \phi^+\gamma_0\phi$ is conserved. Such transformations also conserve $\mathcal{L}$ and $\mathcal{H}$. That is, they are symmetry transformations of the systems which do not change (6.99). When γ_0 is also an n-component unit matrix, the transformations form a compact group $U'(n)$. On the other hand, if

$$\gamma_0 = \text{diag.}(\underbrace{1, \cdots, 1}_{i}, \underbrace{-1, \cdots, -1}_{j}),$$

which plays the role of isotropic space metric and the dagger sign (†) denotes Hermitian conjugate, the iso-transformation belongs to the non-compact group $U(i, j)$. Zakharov and Shabat showed that (6.99) is completely integrable, in the case of $U(1)$ group for both positive ($b > 0$) (the $U(1, 0)$ model) and negative ($b < 0$) (the $U(0, 1)$ model) coupling constants. Furthermore, Manakov integrated (6.99) for the $U(2)$ group (the $U(2, 0)$ model, $b > 0$), and the integrability for the case of $U(1, 1)$ group was shown by Makhankov. These can be considered as special cases of the general system.

In the $U(1,0)$ model, Makhankov *et al.* found that the integrals of the particle number, momentum and energy can be written in the field and the action-angle variables as follows

$$\begin{aligned} N &= \int |\phi|^2 dx = \int n(p)dp + \sum_s N_s, \\ P &= -\Im \int (\phi_x^* \phi) dx = \int pn(p)dp + \sum_{s=1} \frac{1}{2} v_s N_s, \\ E &= \int \left(\frac{\hbar^2}{2m} |\phi_x|^2 - \frac{b}{2} |\phi|^4 \right) dx = \int p^2 n(p)dp + \sum_{s=1} \frac{N_2}{12} \left(3v_s^2 - \frac{1}{4} b^2 N_s^2 \right). \end{aligned} \tag{6.102}$$

The continuous action $n(p)$ and angular variable $\theta(p)$ are related to the scattering matrix element. From (6.102), we see that the angular variables θ and v_s are cyclic and the action functionals corresponding to them are integrals of motion. In quantum language the continuous variables (wave background) correspond to linear "microparticle" or elementary excitations (phonons, magnons, etc.), with dispersion $E = P^2$ and mass $1/2$ (in natural unit system). The nonlinear microscopic particle (solitons) (*i.e.*, the discrete variable) can be interpreted as a special form of bound states with N_s constituent "particles" of mass $1/2$ in the nonlinear quantum field. The energy, momentum and mass of a microscopic particle (soliton) are

$$E_s = \frac{1}{N_s} P_s^2 - \frac{1}{48} b^2 N_s^3, \quad P_s = \frac{1}{2} v_s N_s, \quad M_s = \frac{1}{2} N_s, \quad (N_s \gg 1), \tag{6.103}$$

The first term of E_s is the kinetic energy of the microscopic particle (soliton) with mass $N_s/2$ and the second term is its binding energy.

In the $U(0,1)$ model, the dynamic equation is of the form

$$i\hbar\phi_t + \frac{\hbar^2}{2m}\phi_{xx} - b\left(|\phi|^2 - \rho\right)\phi = 0. \tag{6.104}$$

A term, $b\rho\phi$, is introduced into the equation and a corresponding term, $-b\rho|\phi|^2 = -\mu|\phi|^2$ is introduced into the Hamiltonian. Here the quantity μ plays the role of chemical potential. The integrals of particle number, momentum and energy are renormalized as follows

$$\begin{aligned} \bar{N} &= \int_{-\infty}^{\infty} (|\phi|^2 - \rho) dx, \\ P &= \int_{-\infty}^{\infty} \Im(\phi^* \phi_x) dx + \rho\theta, \\ E &= \int_{-\infty}^{\infty} \left[\frac{\hbar^2}{2m} |\phi_x|^2 + \frac{1}{2} b (|\phi|^2 - \rho)^2 \right] dx, \end{aligned} \tag{6.105}$$

where θ is the phase shift in the solution when the coordinate varies from $-\infty$ to $+\infty$. The discrete part of the spectrum corresponding to the hole excitation mode in quantum case was studied by Lieb. In the classical limit, this mode is denoted by a complex kink solution

$$\phi_k = a_k \tanh[a_k(x - v_k t - x_0)] + i\frac{v_k}{2},$$

where

$$a_k = \sqrt{\frac{b\rho}{2} - \frac{v_k^2}{4}}.$$

Thus we get

$$\bar{N}_k = -\frac{4a_k}{b}, \quad P_k = \frac{2}{b}\left[b\rho\cos^{-1}\left(\frac{v_k}{\sqrt{2b\rho}}\right) + \frac{1}{4}bv_k\bar{N}_k\right], \quad E_k = \frac{4}{3}b\bar{N}_k^3.$$

The negative sign in the expression of $\bar{N}$ corresponds to a particle "deficiency" in the condensate, *i.e.*, the presence of holes in the system. The hole number is given by $-\bar{N}_k = N_h > 0$. Therefore, this model can be used to describe the nonlinear excitation of holes or hole-solitons in the nonlinear quantum mechanical systems.

6.5 Eigenvalue Problem of the Nonlinear Schrödinger Equation

As mentioned in Chapter 3, the eigenvalue problem of the nonlinear Schrödinger equation (3.2) with the Galilei invariance is determined by the eigenequation of the linear operator $\hat{L}$ in the Lax system in (3.20) – (3.22). The eigenequation corresponding to the nonlinear Schrödinger equation is found by the linear Zakharov-Shabat equation (4.42) from Lax equation (3.20), or

$$i\psi_{x'} + \Phi\psi = \lambda\sigma_3\psi. \tag{6.106}$$

This is an eigenequation for an eigenfunction ψ with a corresponding eigenvalue λ and a potential Φ, where,

$$\psi = \begin{pmatrix}\psi_1\\ \psi_2\end{pmatrix}, \quad \sigma_3 = \begin{pmatrix}1 & 0\\ 0 & -1\end{pmatrix}, \quad \Phi = \begin{pmatrix}0 & \phi\\ \phi^* & 0\end{pmatrix}. \tag{6.107}$$

Here ϕ satisfies (4.40). It evolves with time according to (3.21). However, what are the properties of the eigenvalue problems determined by these relations? This deserves further consideration.

As it is known, the eigenequation is invariant under the Galilei transformation. As a matter of fact, if we substitute the following Galilei transformation

$$\begin{cases} \tilde{x} = x' - vt', \\ \tilde{t} = t', \\ \phi'(\tilde{x}, \tilde{t}) = e^{ivx' - iv^2t'/2}\phi(x', t') \end{cases} \tag{6.108}$$

into (6.107), Φ is transformed into

$$\Phi'(\tilde{x}) = \begin{pmatrix} e^{i\theta/2} & 0 \\ 0 & e^{-i\theta/2} \end{pmatrix} \Phi(x') \begin{pmatrix} e^{-i\theta/2} & 0 \\ 0 & e^{i\theta/2} \end{pmatrix}, \tag{6.109}$$

where

$$\theta = vx' - \frac{1}{2}v^2 t' + \theta_0,$$

and θ_0 is an arbitrary constant. If the eigenfunction $\psi(x')$ is transformed as

$$\psi'(\tilde{x}) = \begin{pmatrix} e^{i\theta/2} & 0 \\ 0 & e^{-i\theta/2} \end{pmatrix} \psi(x'), \tag{6.110}$$

then (6.106) becomes

$$i\psi'_{x'} + \Phi'\psi' = \left(\lambda - \frac{v}{2}\right)\sigma_3\psi'. \tag{6.111}$$

It is clear that in the reference frame that is moving with velocity v, the eigenvalue is reduced to $v/2$ compared to that in the rest frame. It shows that the velocity of the microscopic particle (soliton) is given by $2\Re(\lambda_\kappa)$. When θ is constant, *i.e.*, $\theta = \theta_0$, the eigenvalue is unchanged because $v = 0$. This implies that the nonlinear Schrödinger equation is invariant under the gauge transformation, $\phi' = e^{i\theta_0}\phi(x')$.

Satsuma and Yajima studied the eigenfunction of (6.106) that satisfies the boundary condition $\psi = 0$ at $|x| \to \infty$. The eigenvalues and the corresponding eigenfunctions were denoted by $\lambda_1, \lambda_2, \cdots, \lambda_N$ and $\psi_1, \psi_2, \cdots, \psi_N$. For a given eigenfunction, $\psi_n(x')$, equation (6.106) reads

$$i\frac{d\psi_n(x')}{dx'} + \Phi(x')\psi_n(x') = \lambda_n\sigma_3\psi_n(x'), \quad n = 1, 2, \cdots, N. \tag{6.112}$$

$\Phi(x')$ was expressed in terms of the Pauli's spin matrices σ_1 and σ_2,

$$\Phi(x') = \Re[\phi(x')]\sigma_1 - \Im[\phi(x')]\sigma_2. \tag{6.113}$$

Multiplying (6.112) by σ_2 from left and taking the transpose of the resulting equation, we get

$$-i\frac{d\psi_m^T}{dx'}\sigma_2 - \psi_m^T\Phi^*\sigma_2 = i\lambda_m\psi_m^T\sigma_1,$$

where the superscript T denotes transpose. Multiplying the above equation by ψ_n from right and (6.111) by $\psi_m^T\sigma_2$ from left and subtracting one from the other, Satsuma and Yajima obtained the following equation

$$(\lambda_n - \lambda_m)\int_{-\infty}^{\infty} \psi_m^T\sigma_1\psi_n dx' = 0.$$

The boundary conditions, $\psi_n, \psi_m \to 0$ as $|x'| \to \infty$, were used in obtaining the above equation. The following orthonormal condition was then derived.

$$\int_{-\infty}^{\infty} \psi_m^T \sigma_1 \psi_n dx' = \delta_{nm}. \tag{6.114}$$

Satsuma and Yajima further demonstrated that (6.112) has the following symmetry properties.

(I) If $\phi(x')$ satisfies $\phi(-x') = \phi^*(x')$, then replacing x' by $-x'$ in (6.112) and multiplying it by σ_2 from left, we can get

$$i\frac{d}{dx'}\left[\sigma_2\psi_n(-x')\right] + \Phi(x')\left[\sigma_2\psi_n(-x')\right] = \lambda_n\sigma_3\left[\sigma_2\psi_n(-x')\right].$$

Since $\sigma_2\psi_n(-x')$ is also an eigenfunction associated with λ_n, its behavior resembles that of $\psi_n(x')$ in the asymptotic region, *i.e.*, $\sigma_2\psi_n(-x') \to 0$ as $|x'| \to \infty$, Thus ψ_n has the following symmetry

$$\sigma_2\psi_n(-x') = \delta\psi_n(x'), \quad \text{or} \quad \psi_n(-x') = \delta\sigma_2\psi_n(x'), \quad (\delta = \pm 1).$$

Therefore, if $\phi(-x') = -\phi^*(x')$, then $\psi_n(x')$ satisfies the symmetry property $\psi_n(-x') = \sigma_1\psi_n(x')$ with $\delta = \pm 1$. This can be easily verified by replacing σ_1 with σ_2 in the above derivations.

(II) If $\phi(x')$ is a symmetric (or antisymmetric) function of x', *i.e.*, $\phi(-x') = \pm\phi(x')$, then $\psi_n'^{(s)}(x') = \sigma_1\psi_n^*(-x')$ is the eigenfunction belonging to the eigenvalue $-\lambda_n^*$, and $\psi_n'^{(a)}(x') = \sigma_2\psi_n^*(-x')$ is the eigenfunction belonging to the eigenvalue λ_n. The suffix s (or a) to the eigenfunction ψ_n' indicates that ϕ is symmetric (or antisymmetric). Since $\phi(-x') = \phi(x')$, replacing x' with $-x'$ in (6.112) and taking complex conjugate, we get

$$i\frac{d}{dx'}\left[\sigma_1\psi_n^*(-x')\right] + \Phi(x')\left[\sigma_1\psi_n^*(-x')\right] = -\lambda_n^*\sigma_3\left[\sigma_1\psi_n^*(-x')\right].$$

Compared with (6.112), the above equation implies that $-\lambda_n^*$ is also an eigenvalue and the associated eigenfunction $\psi_n'^{(s)}(x')$ is $\sigma_1\psi_n^*(-x')$, with an arbitrary constant. For $\phi(-x') = -\phi(x')$, the same conclusion is obtained by replacing σ_1 with σ_2 in the above derivations.

These symmetry properties are useful in providing a general view of the solution of (4.40). As it is known, the real part of the eigenvalue, ξ_n, corresponds to the velocity of a soliton and the imaginary part, η_n, the amplitude. Then, if $\phi(x', t')$, whose initial value has the symmetry $\phi(x', t' = 0) = \pm\phi(-x', t' = 0)$, breaks into a series of solutions, the decay is bisymmetric, corresponding to the eigenvalues λ_n and $-\lambda_n^*$.

If $\phi(x')$ is real, the above symmetry property yields

$$\psi_n'^{(s)}(-x') = \sigma_1\left[-\delta\sigma_2\psi_n^*(-x')\right] = \delta\sigma_2\psi_n'^{(s)}(x'),$$
$$\psi_n'^{(a)}(-x') = \sigma_2\left[-\delta\sigma_1\psi_n^*(-x')\right] = -\delta\sigma_1\psi_n'^{(a)}(x'),$$

i.e., $\psi_n'^{(s)}(x')$ has the same parity as $\psi_n(x')$, while $\psi_n'^{(a)}(x')$ has the opposite one. When $\phi(-x') = -\phi(x')$, and λ_n is pure imaginary ($\lambda_n = -\lambda_n^*$), the eigenvalues corresponding to the positive and negative parity eigenfunctions degenerate.

(III) If $\phi(x')$ is real but not antisymmetric, then the eigenvalue λ_n is pure imaginary, *i.e.*, $\Re(\lambda_n) = 0$. From (6.112) and its Hermitian conjugate, Satsuma and Yajima found that

$$\Re(\lambda_n)\langle n|\sigma_2|n\rangle = \langle n|\Im[\phi(x')]\sigma_3|n\rangle, \tag{6.115}$$

with

$$\langle m|\sigma_2|n\rangle = \int_{-\infty}^{\infty} \psi_m^+ \sigma_2 \psi_n dx', \tag{6.116}$$

where $[\Phi, \sigma_1] = 2i\Im(\phi)\sigma_3$ was used. From (6.115), we see that $\Re(\lambda_n)$ vanishes if ϕ is real and $\langle n|\sigma_2|n\rangle \neq 0$. When ϕ is a real and an antisymmetric function of x', symmetry property (I) gives

$$\langle n|\sigma_2|n\rangle = \delta^2 \int_{-\infty}^{\infty} \psi_n^+(-x')\sigma_1\sigma_2\sigma_1\psi_n(-x')dx' = -\langle n|\sigma_2|n\rangle.$$

Thus $\langle n|\sigma_2|n\rangle = 0$.

(IV) If the initial value takes the form of $\phi = e^{ivx'}R(x')$, where $R(x')$ is a real but not an antisymmetric function of x', all the eigenvalues have the common real part, $-v/2$. This can be easily shown by the Galilei transformation. In fact, when $\phi(x', t' = 0) = e^{ivx'}R(x')$, the solution does not decay into a series of solitons moving with different velocities, but forms a bound state. In this case, the real parts are common to all the eigenvalues, *i.e.*, the relative velocities of the solitons vanish.

(V) If ϕ is a real non-antisymmetric function of x', it can be shown that

$$\psi_n^*(x') = i\delta\sigma_3\psi_n(x'), \tag{6.117}$$

where $\delta = \pm 1$. Because $\Re(\lambda_n) = 0$, from the complex conjugate of (6.112), one can get $\psi_n^*(x) \propto \sigma_3\psi_n(x)$. Substituting (6.117) into the normalization condition (6.114), one then has $\delta = \pm 1$. If the eigenvalue of (6.106) is real, *i.e.*, $\lambda = \xi$ is real, then

$$i\frac{d\psi}{dx'} + \Phi\psi = \xi\sigma_3\psi, \tag{6.118}$$

and the adjoint function of ψ, $\bar{\psi} = i\sigma_2\psi^*$, is also a solution of (6.118) *i.e.*,

$$i\frac{d\bar{\psi}}{dx'} + \Phi\bar{\psi} = \xi\sigma_3\bar{\psi}.$$

From this and (6.118), Satsuma and Yajima obtained the following

$$\frac{d}{dx'}(\psi^+\psi) = \frac{d}{dx'}(\bar{\psi}^+\psi) = \frac{d}{dx'}(\psi^+\bar{\psi}) = \frac{d}{dx'}(\bar{\psi}^+\bar{\psi}) = 0. \tag{6.119}$$

Using the above boundary conditions, they found that the solutions of (6.106), $\psi_1(x',\xi)$, $\psi_2(x',\xi)$, and $\bar{\psi}_2(x',\xi)$ satisfy the following relations

$$\psi_1^+\psi_1 = \psi_2^+\psi_2 = \bar{\psi}_2^+\bar{\psi}_2 = 1, \quad \bar{\psi}_2^+\psi_2 = \psi_2^+\bar{\psi}_2 = 0.$$

From $\psi_1 = a(\xi)\bar{\psi}_2 + b(\xi)\psi_2$, we get $a = \bar{\psi}_2^+\psi_1$ and $b = \bar{\psi}_2^+\psi_1$, where

$$\psi_1(x',\xi) = \begin{pmatrix} 1 \\ 0 \end{pmatrix} e^{-i\xi x'},$$

as $x' = -\infty$ and

$$\psi_2(x',\xi) = \begin{pmatrix} 0 \\ 1 \end{pmatrix} e^{+i\xi x'},$$

$$\bar{\psi}_2(x',\xi) = \begin{pmatrix} 1 \\ 0 \end{pmatrix} e^{-i\xi x'},$$

as $x' = \infty$.

As it is pointed out earlier, if real (not antisymmetric) initial value is considered, the microscopic particle does not decay into moving solitons, but forms a bound state of solitons pulsating with the proper frequency. Satsuma and Yajima developed a perturbation approach to investigate the conditions for the solutions to evolve and decay into moving solitons.

If the wave function ϕ in (6.106) undergoes a small change, *i.e.*, $\phi \to \phi' = \phi + \triangle\phi$, the corresponding change in Φ is given by

$$\triangle\Phi = \begin{pmatrix} 0 & \triangle\phi \\ \triangle\phi^* & 0 \end{pmatrix}.$$

λ_n and ψ_n change as $\lambda_n + \triangle\lambda_n$ and $\psi_n + \triangle\psi_n$, respectively. To the first order in the variation, equation (6.112) becomes

$$\left[i\frac{d}{dx'} + (\Phi - \lambda_n\sigma_3)\right]\triangle\psi_n + (\triangle\Phi - \triangle\lambda_n\sigma_3)\psi_n = 0.$$

Multiplying the above equation by $\psi_n^T\sigma_2$ from left and integrating with respect to x' over $(-\infty,\infty)$, we get

$$\begin{aligned}\triangle\lambda_n &= -i\int_{-\infty}^{\infty}\psi_n^T\sigma_2\triangle\Phi\psi_n dx' \\ &= -\int_{-\infty}^{\infty}\psi_n^T\Re(\triangle\phi)\sigma_3\psi_n dx' + i\int_{-\infty}^{\infty}\psi_n^T\Im(\triangle\phi)\psi_n dx'.\end{aligned}$$

If ϕ is a real and non-antisymmetric function of x', equation (6.117) holds and

$$\triangle\lambda_n = \delta\langle n|\Im(\triangle\phi)\sigma_3|n\rangle + i\delta\langle n|\Re(\triangle\phi)|n\rangle. \tag{6.120}$$

Equation (6.120) indicates that if $\langle n|\Im(\triangle\phi)\sigma_3|n\rangle \neq 0$, the perturbation $\triangle\phi$ makes the real part of the eigenvalue finite. That is, for the initial value, $\phi(x') + \triangle\phi(x')$, the solution of the nonlinear Schrödinger equation (4.40) breaks up into moving solitons with velocity $2\Re(\Delta\lambda_n)$. If ϕ is a real and is either a symmetric or an antisymmetric function of x', the above symmetry properties of eigenvalues of the nonlinear Schrödinger equation lead to

$$\langle n|\Im(\Delta\phi(x'))\sigma_3|n\rangle = -\langle n|\Im(\Delta\phi(-x))\sigma_3|n\rangle.$$

Therefore, if $\Im(\Delta\phi)$ is a symmetric function, $\langle n|\Im(\Delta\phi)\sigma_3|n\rangle$ vanishes, *i.e.*, $\Re(\Delta\lambda_n) = 0$, and the soliton bound state does not resolve into moving solitons even in the presence of the perturbation $\Delta\phi$.

Satsuma and Yajima also obtained the shifts of the eigenvalues of (6.106) under the double-humped initial values, $\phi(x', t' = 0) = \phi_0(x' - x_0') + e^{i\theta_0}\phi_0(x' + x_0')$, where ϕ_0 is a real and symmetric function of x', x_0' and ϕ_0 are real. The shifts of the eigenvalues were finally written as

$$\begin{aligned}\Delta\lambda_n^{\pm} = \delta &\left[\sin\theta_0\langle n|\sigma_3\phi_0(x' + 2x_0')|n\rangle \mp \sin\left(\frac{\theta_0}{2}\right)\langle n|\sigma_3\phi_0(x')e^{2x_0'(d/dx')}|n\rangle\right] + \\ i\delta &\left[\cos\theta_0\langle n|\phi_0(x' + 2x_0')|n\rangle \pm \cos\left(\frac{\theta_0}{2}\right)\langle n|\phi_0(x')e^{2x_0'(d/dx')}|n\rangle\right]. \end{aligned} \quad (6.121)$$

where

$$\begin{aligned} -\delta\cos\left(\frac{\theta_0}{2}\right)&\langle n|\phi_0(x')e^{2x_0'(d/dx)}|n\rangle - i\delta\sin\left(\frac{\theta_0}{2}\right)\langle n|\sigma_3\phi_0(x')e^{2x_0'(d/dx)}|n\rangle \\ &= \int_{-\infty}^{\infty}\psi_2''^{(n)T}\sigma_2\Phi_2\psi_1''^{(n)}dx' = \int_{-\infty}^{\infty}\psi_1''^{(n)T}\sigma_2\Phi_1\psi_2''^{(n)}dx', \end{aligned}$$

$$\begin{aligned} -\delta\cos(\theta_0)&\langle n|\phi_0(x' + 2x_0')|n\rangle + i\delta\sin\theta_0\langle n|\sigma_3\phi_0(x' + 2x_0')|n\rangle \\ &= \int_{-\infty}^{\infty}\psi_1''^{(n)T}\sigma_2\Phi_2\psi_1''^{(n)}dx' = \int_{-\infty}^{\infty}\psi_2''^{(n)T}\sigma_2\Phi_1\psi_2''^{(n)}dx'. \end{aligned}$$

Here

$$\Phi(x') = \Phi_1(x') + \Phi_2(x'),$$

and

$$\begin{aligned}\Phi_1(x') &= \sigma_1\phi_0(x' - x_0'), \\ \Phi_2(x') &= [\cos(\theta_0)\sigma_1 - \sin(\theta_0)\sigma_2]\phi_0(x' + x_0').\end{aligned}$$

The corresponding eigenvalue equation is given by

$$i\frac{d}{dx'}\psi_n''(x') + \Phi(x')\psi_n''(x') = \lambda_n\sigma_3\psi_n''(x').$$

The eigenfunction $\phi''_n(x')$ satisfies the following symmetry and orthogonality requirements

$$\psi''_{n\pm}(-x') = \pm\delta\left[\cos\left(\frac{\theta_0}{2}\right)\sigma_2 + \sin\left(\frac{\theta_0}{2}\right)\sigma_1\right]\psi''_{n\pm}(x'), \quad \delta = \pm 1,$$

$$\int_{-\infty}^{\infty} \psi''^{T}_{n+}(x')\sigma_1\psi''_{n-}(x')dx' = 0,$$

when $\theta_0 = 0$, $\phi(x')$ is real and symmetric, $\triangle\lambda_n^{(\pm)}$ is pure imaginary. When $\theta_0 = \pi$, $\phi(x')$ is real and antisymmetric, $\triangle\lambda_n^{(\pm)}$ is real,

$$\begin{aligned} \Re\left[\triangle\lambda_n^{\pm}(\theta_0 = \pi)\right] &= \mp\delta\langle n|\sigma_3\phi_0(x')e^{2x'_0(d/dx')}|n\rangle, \\ \Im\left[\triangle\lambda_n^{\pm}(\theta_0 = \pi)\right] &= -\delta\langle n\,|\phi_0(x' + 2x'_0)|\,n\rangle. \end{aligned} \tag{6.122}$$

Thus the solution of the nonlinear Schrödinger equation (4.40) decays into paired solitons and each pair consists of solitons with equal amplitude and moving in the opposite directions with the same speed. For arbitrary θ'_0, we can see from (6.122) that the solution of (4.40) breaks up into an even number of moving solitons with different speeds and amplitudes.

6.6 Microscopic Causality in Linear and Nonlinear Quantum Mechanics

The microscopic causality is an interesting problem in quantum mechanics, and understanding the difference in microscopic causalities in the linear quantum mechanics and the nonlinear quantum mechanics is important for a better understanding of the essential features of the nonlinear quantum mechanics. The microscopic causality will be discussed in this section.

In classical physics, the causal relations, which crystallizes the intuitive notion that a cause must precedes its effects, are often expressed in the Green's function of a theory or by the Kramers-Kronig type of dispersion relations. These forms can be easily generalized to the linear quantum mechanics. In fact, the commutator of two operators or its matrix form contains causality information, and for this reason, it is referred to as an interference function, and the relations are called microcausal. Using solutions of the dynamic equation represented by the creation and annihilation operators, Burt expressed the microscopic causality in terms of probability amplitudes.

The microcauslity concerns two processes or events that must be classified into two independent data sets if the events are separated by a space-like interval. As shown in Fig. 6.3, two such events may appear to have quite different temporal orientations in various coordinate systems. Thus, in the absence of superlumimal signals, the description of the events must be independent. In the canonical quantum theory, the information concerning the interference of the two events is described by

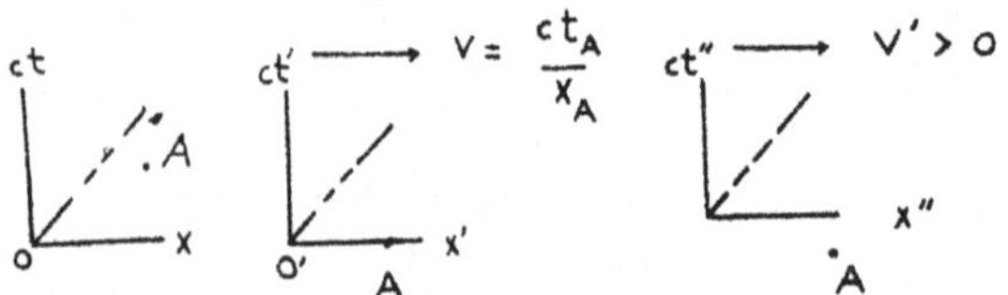

Fig. 6.3 Minkowski diagram illustrating relativity of temporal order of two events separated by a space-like interval in three coordinate systems related by Lorentz transformations (see book by Burt).

the commutator of the operators denoting the two events. This commutator gives the interference when the sequence of events is reversed and the difference taken between the two orderings. The representation in terms of probability amplitudes results in an interference function which can be built by the creation and annihilation operators. The sequence of events to be considered consists of the creation and annihilation of microscopic particles from the vacuum at two points in space-time. In the reversed sequence the points in space-time are exchanged. The difference between the amplitudes for the two sequences defines the interference function. Using the knowledge of quantum field theory, the amplitude for the creation of a particle with positive energy from the vacuum at $\check{x}$ and its subsequent annihilation into the vacuum is represented by

$$\begin{aligned} A^{(+)}_{\rm amp}(\check{y},\check{x}) &= \sum_{\check{k},\check{p},\epsilon} \langle 0|A^{(+)}_{\check{k}}|\epsilon\rangle\langle\epsilon|A^{(+)\dagger}_{\check{p}}|0\rangle e^{i\check{p}\cdot\check{x}-i\check{k}\cdot\check{y}}(D^2\omega_p\omega_k V^2)^{-1/2} \\ &= \sum_{\check{k},\check{p}} \langle 0|A^{(+)}_{\check{k}}A^{(+)\dagger}_{\check{p}}|0\rangle (D^2\omega_p\omega_k V^2)^{-1/2} e^{i\check{p}\cdot\check{x}-i\check{k}\cdot\check{y}}, \end{aligned} \tag{6.123}$$

where V is the volume of the system. Similarly, the sequence for a particle with an negative energy is

$$A^{(-)}_{\rm amp}(\check{y},\check{x}) = \sum_{\check{k},\check{p}} \langle 0|A^{(-)}_{\check{k}}A^{(-)\dagger}_{\check{p}}|0\rangle (D^2\omega_p\omega_k V^2)^{-1/2/} e^{i\check{k}\cdot\check{y}-i\check{p}\cdot\check{x}}. \tag{6.124}$$

Here $A^{\dagger}_{\check{k}}$ and $A_{\check{k}}$ are the creation and annihilation operators of the particle with wave vector $\check{k}$, respectively, $(+)$ and $(-)$ indicate the positive and negative energy states, respectively, the field operators ϕ of the particles is represented by

$$\phi = \sum \frac{1}{\sqrt{D\omega_k V}} \left[A_{\check{k}} e^{i\check{k}\cdot\check{x}} + A^{\dagger}_{\check{k}} e^{-i\check{k}\cdot\check{x}}\right],$$

where ω_k is the frequency of the particle, $\check{k} = (\omega_k, \vec{k})$, $\omega_k^2 = \vec{k}^2 + m^2$, $\check{x} = (\vec{x}, -ict)$, $[A_{\vec{k}}, A^{\dagger}_{\vec{k}'}] = \delta_{\vec{k}\vec{k}'}$, D is a normalization constant, $A^{(+)\dagger}_{\vec{k}} e^{i\check{k}\cdot\check{x}}$ is a creation operator. Thus, the probability amplitude that a positive energy particle with momentum $\check{k}$, created from the vacuum at position $\breve{x}$, is found in the intermediate state $|\varepsilon\rangle$ is

$$\rho_{k\varepsilon}^{(+)} = \frac{1}{\sqrt{D\omega_k V}} \langle \varepsilon \left| A_{\vec{k}}^{(+)\dagger} e^{i\breve{k}\cdot\breve{x}} \right| 0 \rangle .$$

Similarly, the probability amplitude that a negative energy particle annihilated from the vacuum at position $\breve{x}$ with momentum $\breve{k}$ is found in the state $|\varepsilon\rangle$ is

$$\rho_{\vec{k}\varepsilon}^{(-)} = \frac{1}{\sqrt{D\omega_k V}} \langle \varepsilon \left| A_{\vec{k}}^{(-)} e^{i\breve{k}\cdot\breve{x}} \right| 0 \rangle .$$

Then, the amplitude for annihilation of a negative energy particle from the vacuum at $\breve{y}$, followed by the creation (to fill the negative energy state) of a negative energy particle at $\breve{x}$ (the time sequency is reversed) is $\rho_{\vec{k}\varepsilon}^{(-)}(\breve{x})\rho_{\vec{k}\varepsilon}^{(-)}(\breve{y})$. Thus we get the representations (6.123) and (6.124). Since the two amplitudes have equal weight, total amplitude is given by the arithmetic average,

$$A_{\mathrm{amp}}(\breve{y},\breve{x}) = \frac{1}{2}\left[A_{\mathrm{amp}}^{(+)}(\breve{y},\breve{x}) + A_{\mathrm{amp}}^{(-)}(\breve{y},\breve{x})\right]. \tag{6.125}$$

However, from the definition of the vacuum state, it is required that

$$A_{\breve{p}}^{(-)\dagger}|0\rangle = 0. \tag{6.126}$$

Therefore,

$$A_{\mathrm{amp}}(\breve{y},\breve{x}) = \frac{1}{2}A_{\mathrm{amp}}^{(+)}(\breve{y},\breve{x}). \tag{6.127}$$

The amplitude of the reversing sequence can be calculated following the same steps. The total interference amplitude is then given by

$$\begin{aligned} \triangle A_{\mathrm{amp}}(\breve{y},\breve{x}) &= A_{\mathrm{amp}}(\breve{y},\breve{x}) - A_{\mathrm{amp}}(\breve{x},\breve{y}) \qquad (6.128)\\ &= \frac{1}{2}\sum_{\breve{k},\breve{p}} \langle 0|A_{\breve{k}}^{(+)}A_{\breve{p}}^{(+)\dagger}|0\rangle \left(e^{i\breve{p}\cdot\breve{x}-i\breve{k}\cdot\breve{y}} - e^{i\breve{p}\cdot\breve{y}-i\breve{k}\cdot\breve{x}}\right)\left(D^2\omega_p\omega_k V^2\right)^{-1/2}. \end{aligned}$$

For the neutral, spin zero system of particle, using the commutator $[A_{\breve{k}}, A_{\breve{p}}^{\dagger}] = \delta_{\breve{n}_k\breve{n}_p}$ and converting the sum into an integral using

$$\frac{1}{V}\sum_{k} \to \int d^3k \left(\frac{1}{2\pi}\right)^3,$$

the total interference amplitude can be obtained

$$\triangle A_{\mathrm{amp}}(\breve{y},\breve{x}) = \frac{1}{2(2\pi)^3}\int \frac{d^3k}{\omega D}\left[e^{i\breve{k}\cdot(\breve{x}-\breve{y})} - e^{i\breve{k}\cdot(\breve{y}-\breve{x})}\right]. \tag{6.129}$$

If the constant D is set to unity, this becomes the interference function defined in the canonical formalism by Bjorken *et al.* and Bogoliubov *et al.* (see book by

Burt)

$$\triangle A_{\text{amp}}(\check{y},\check{x}) = \triangle(\check{y}-\check{x}) = \frac{1}{2(2\pi)^3}\int \frac{d^3k}{\omega}\left[e^{i\check{k}\cdot(\check{y}-\check{x})} - e^{i\check{k}\cdot(\check{x}-\check{y})}\right]. \tag{6.130}$$

This Lorentz invariant function clearly vanishes for space-like intervals. If $(\check{y}-\check{x})^2 < 0$, a Lorentz transformation can always be found which takes us to a coordinate system in which $x_0 = y_0$. Replacing $\check{k}$ by $-\check{k}$ in the second term does not change the integrand. Hence

$$\triangle(\check{y}-\check{x}) = 0, \quad (\check{y}-\check{x})^2 < 0. \tag{6.131}$$

Properties of this function were studied by Bjorken *et al.* and Bogoliubov *et al.*. The statement of microscopic causality in terms of probability amplitudes is equivalent to the usual canonical relation for free fields in the linear quantum mechanical theory. This idea was generalized to the self-interacting fields in nonlinear quantum mechanics by Burt.

As in the establishment of propagator including persistently interacting fields in the nonlinear quantum mechanics, the interference function describing microcausality with interactions is obtained by employing the superposition principle of quantum theory. The algorithm developed in the linear quantum mechanics is utilized for the persistently interacting field operators since the latter can be interpreted as creation or annihilation operators. For the representation of the interference function in (6.128), Burt generalized it to the following form

$$\begin{aligned}\triangle A_{\text{amp}}(\check{y},\check{x})^{\text{pers}} &= A_{\text{amp}}(\check{y},\check{x})^{\text{pers}} - A_{\text{amp}}(\check{x},\check{y})^{\text{pers}}\\ &= \frac{1}{2}\sum_{\check{k},\check{q}}\langle 0|\phi_{\check{k}}^{(+)}(\check{y})\phi_{\check{q}}^{(+)}(\check{x})^{\dagger} - \phi_{\check{k}}^{(+)}(\check{x})\phi_{\check{q}}^{(+)}(\check{y})^{\dagger}|0\rangle,\end{aligned} \tag{6.132}$$

where $\phi_{\check{k}}^{(\pm)}(\check{x})$ are persistently interacting fields and

$$A_{\text{amp}}(\check{y},\check{x})^{\text{pers}} = \frac{1}{2}A_{\text{amp}}^{(+)}(\check{y},\check{x})^{\text{pers}} = \frac{1}{2}\sum_{\check{k},\check{q}}\langle 0|\phi_{\check{k}}^{(+)}(\check{y})\phi_{q}^{(+)}(\check{x})^{\dagger}|0\rangle. \tag{6.133}$$

Using the general expressions for persistently interacting fields given in section 5.3, the current $\vec{J}$ is denoted by a series containing only positive powers of the creation or annihilation operators,

$$\phi_{\check{k}}^{(\pm)}(\check{x}) = \sum_{n=0}^{\infty}\phi_n(\lambda_i;m^2;p)U_{\check{k}}^{(\pm)}(\check{x})^{(pn+p_0)},$$

where $\phi_n(\lambda_i;m^2;p)$ contains all the coupling constant dependence and depends on the parameters m^2, p and p_0, the p and p_0 are indeces of interaction which are related to the form of the self-interaction current. The interference function can be written as

$$\Delta A_{\rm amp}(\check{y},\check{x})^{\rm pers} = \frac{1}{2}\sum_{\check{k},\check{q},n,s} \phi_n(\lambda_i;m^2;p)\phi_s(\lambda_i;m^2;p)\times \tag{6.134}$$

$$\left\{\langle 0|\left[U_{\check{k}}^{(\pm)}(\check{y})\right]^{pn+p_0}\left[U_{\check{q}}^{(\pm)\dagger}(\check{x})\right]^{ps+p_0} - \left[U_{\check{k}}^{(\pm)}(\check{x})\right]^{pn+p_0}\left[U_{\check{q}}^{(\pm)\dagger}(\check{y})\right]^{ps+p_0}|0\rangle\right\}.$$

Using

$$U_{\check{k}}^{(\pm)}(\check{x}) = A_{\check{k}}^{(\pm)}e^{\mp i\check{k}\cdot\check{x}}(D\omega V)^{-1/2},$$

where D is a constant, V is the volume of the system, $\omega = \sqrt{k^2+m^2}$, and

$$\phi_{\check{k}}^{(-)}(\check{x}) = \phi_{\check{k}}^{(+)}(\check{x}), \quad A_{\check{k}}^{(+)} = a_{\check{k}}, \quad A_{\check{k}}^{(-)} = a_{\check{k}}^{+},$$

with

$$[a_{\check{k}}, a_{\check{k}}^{\dagger}] = \delta_{n_{\check{k}}n_{\check{q}}},$$

equation (6.134) becomes

$$\begin{aligned}\triangle A_{\rm amp}(\check{y},\check{x})^{\rm pers} &= \frac{1}{2}\sum_{\check{k},\check{q},n,s}\phi_n(\lambda_i;m^2;p)\phi_s(\lambda_i;m^2;p)(D\omega V)^{-p(n+s)/2-p_0}\times\\ &\quad \left\{\langle 0|a_{\check{k}}^{pn+p_0}a_{\check{q}}^{\dagger ps+p_0}|0\rangle\left[e^{-i(pn+p_0)\check{k}\cdot\check{y}+i(ps+p_0)\check{q}\cdot\check{x}}\right.\right.\\ &\qquad \left.\left. -e^{-i(pn+p_0)\check{k}\cdot\check{x}+i(ps+p_0)\check{q}\cdot\check{y}}\right]\right\}\\ &= \frac{1}{2}\sum_{\check{k},n}\{\phi_n(\lambda_i;m^2;p)^2(D\omega V)^{-pn-p_0}(pn+p_0)!\times\\ &\quad \left[e^{-i(pn+p_0)\check{k}\cdot(\check{y}-\check{x})} - e^{-i(pn+p_0)\check{k}\cdot(\check{x}-\check{y})}\right]\right\}.\end{aligned} \tag{6.135}$$

If we consider events for which the interval $(\check{y}-\check{x})^2$ is space-like, a Lorentz transformation can be performed to the system in which the events are simultaneous. From (6.135), Burt found that

$$\begin{aligned}\triangle A_{\rm amp}(\check{y}\cdot\check{x})^{\rm pers} &= \frac{1}{2}\sum_{\check{k},n}\phi_n(\lambda_i;m^2;p)^2(D\omega V)^{-pn-p0}(pn+p_0)!\times\\ &\quad \left\{e^{-i(pn+p_0)\check{k}\cdot(\check{y}-\check{x})} - e^{-i(pn+p_0)\check{k}\cdot(\check{x}-\check{y})}\right\}.\end{aligned} \tag{6.136}$$

Replacing $\check{k}$ by $-\check{k}$ in the second term, the total expression vanishes. Since the interference function is Lorentz invariant, the result is independent of the coordinate systems. Thus, Burt obtained that

$$\triangle A_{\rm amp}(\check{y},\check{x})^{\rm pers} = 0, \quad (\check{x}-\check{y})^2 < 0. \tag{6.137}$$

Therefore, the interference between the creation and annihilation operations of the persistently interacting fields vanishes when the processes are separated by space-like intervals. The principle of microscopic causality, stated in terms of probability amplitudes, is satisfied by the persistently interacting fields.

As is the case for a propagator, the causal function $\triangle A_{\text{amp}}(\breve{y}, \breve{x})^{\text{pers}}$ typically reduces to a sum of terms of the form

$$\triangle A_{\text{amp}}(\breve{y}, \breve{x})^{\text{pers}} = \triangle(\breve{y} - \breve{x}) + \triangle_c^{\text{pers}}(\breve{y} - \breve{x}; \lambda_i), \tag{6.138}$$

where $\triangle(\breve{y} - \breve{x})$ is the usual free particle interference function obtained by Bjorken and Bogoliubov, and $\triangle_c^{\text{pers}}(\breve{y} - \breve{x}; \lambda_i)$ contains the interaction effects. Thus, for the $\lambda\phi^3$ current or ϕ^4-field equation, Burt found that

$$\triangle_c^{\text{pers}}(\breve{z}; \lambda_i) = \frac{1}{2} \sum_{\breve{k}, n=1} (2n+1)!(D\omega V)^{-2n-1} \left(\frac{\lambda}{8m^2}\right)^{2n} \left\{e^{-i(2n+1)\breve{k}\cdot\breve{z}} - e^{i(2n+1)\breve{k}\cdot\breve{z}}\right\}.$$

Due to the presence of the factor $(D\omega V)^{-2n-1}$, this function is less singular than $\triangle(\breve{z})$. The effect of the interaction is to smear the singularity into a smooth function with a tail characterized by the Compton wavelength of the higher mass contributions and by strength proportional to the coupling constant. Therefore, there are only somewhat differences between the representations of the causalities in the nonlinear quantum mechanics and linear quantum mechanics, but their results are basically the same.

Bibliography

Asano, N., Taniuti, T. and Yajima, N. (1969). J. Math. Phys. **10** (2020) .

Bjoken, J. D. and Drell, S. D. (1964). Relativistic quantum mechanics, McGraw-Hill, New York.

Bjoken, J. D. and Drell, S. D. (1980). Relativistic quantum fields, McGraw-Hill, New York.

Bogoliubov, N. N. and Shirkor, H. (1959). Introduction to the theory of quantized fields, Interscience, New York.

Bogoliubov, N. N. (1962). Problems of a dynamical theory in statistical physics, North-Holland, Amsterdam.

Burt, P. B. (1981). Quantum mechanics and nonlinear waves, Harwood Academic Publishers, New York.

Carr, J. and Eilbeck, J. C. (1985). Phys. Lett. A **109** (201) .

Chen, X. R., Gou, Q. Q. and Pang, X. F. (1996). Chin. Phys. Lett. **13** (660) .

Chen, X. R., Gou, Q. Q. and Pang, X. F. (1997). Chin. J. Atom. Mol. Phys. **14** (393) ; Chin. J. Chem. Phys. **10** (145) .

Chen, X. R., Gou, Q. Q. and Pang, X. F. (1998). Acta Phys. Sin. **7** (329) ; J. Sichuan University (nature) Sin. **35** (362) .

Chen, X. R., Gou, Q. Q. and Pang, X. F. (1999). Acta Phys. Sin. **8** (1313) ; Chin. J. Chem. Phys. **11** (240) ; Chin. J. Comput. Phys. **16** (346) ; Commun. Theor. Phys. **31** (169) .

Dodd, R. K., Eiloeck, J. C., Gibbon, J. D. and Morris, H. C. (1984). Solitons and nonlinear wave equations, Academic Press, London.
Drazin, P. G. and Johnson, R. S. (1989). Solitons, an introduction, Cambridge Univ. Press, Cambridge.
Eilbeck, J. C., Lomdahl, P. S. and Scott, A. C. (1984). Phys. Rev. B **30** (4703) .
Eilbeck, J. C., Lomdahl, P. S. and Scott, A. C. (1985). Physica **D16** (318) .
Finkelstein, D. (1966). J. Math. Phys. **7** (280) .
Goldstone, J. and Jackiw, R. (1975). Phys. Rev. D **11** (1486) .
Guo, Bai-lin and Pang Xiao-feng, (1987). Solitons, Chin. Science Press, Beijing.
Heisenberg, W. (1974). Across the frontiers, Harper and Row.
Kartner, F. X. and Boivin, L. (1996). Phys. Rev. A **53** (454) .
Kleinert, H., Schulte-Frohlinde, V. (2001). Critical properties of ϕ^4-theories, World Scientific, Singapore.
Kolk, W. R. (1992). Nonlinear system dynamics, Reinhold.
Korepin, V. E., Kulish, P. P. and Faddeev, L. D. (1975). JETP Lett. **21** (138) .
Lai, Y. and Haus, H. A. (1989). Phys. Rev. A **40** (844) and 854.
Lieb, E. (1963). Phys. Rev. **130** (1616) .
Lieb, E. and Liniger, W. (1963). Phys. Rev. **130** (2605) .
London, R. (1986). The quantum theory of light, 2nd ed., Oxford University Press, Oxford.
Majernikov, E. and Pang Xiao-feng (1997). Phys. Lett. A **230** (89) .
Makhankov, V. G. and Fedyanin, V. K. (1984). Phys. Rep. **104** (1) .
Makhankov, V. G., Makhaidiani, N. V. and Pashaev, O. K. (1981). Phys. Lett. A **81** (161) .
Manakov, S. (1974). Sov. Phys. JETP **38** (248) .
Pang, X. F. (1994). Acta Phys. Sin. **43** (1987) .
Pang, X. F. (1995). Chin. J. Phys. Chem. **12** (1062) .
Pang, X. F. and Chen, X. R. (2000). Chin. Phys. **9** (108) .
Pang, X. F. and Chen, X. R. (2001). Commun. Theor. Phys. **35** (323) ; J. Phys. Chem. Solids **62** (793) .
Pang, X. F. and Chen, X. R. (2002). Chin. Phys. Lett. **19** (1096) ; Commun. Theor. Phys. **37** (715) ; Phys. Stat. Sol. (b) **229** (1397) .
Pang, Xiao-feng (1994). Theory of nonlinear quantum mechanics, Chongqing Press, Chongqing.
Pang, Xiao-feng (1999). Acta Physica Sin. **8** (598) ; Chin. Phys. Lett. **16** (129) ; Phys. Lett. A **259** (466) .
Pang, Xiao-feng (2000). J. Phys. Chem. Solid **61** (701) ; Phys. Stat. Sol. (b) **217** (887) .
Pang, Xiao-feng (2001). J. Phys. Chem. Solid **62** (491) .
Pang, Xiao-feng (2002). Inter. J. Mod. Phys. **B16** (4783) .
Perring, J. K. and Skyrme, T. H. (1962). Nucl. Phys. **31** (550) .
Potter, J. (1970). Quantum mechamics, North-Holland, Amsterdam.
Rubinstein, J. (1970). J. Math. Phys. **11** (258) .
Saji, R. and Konno, H. (1998). J. Phys. Soc. Japan **67** (361) .
Satsuma, J. and Yajima, N. (1974). Prog. Theor. Phys. (Supp) **55** (284) .
Scott, A. C. (1982). Phys. Rep. **217** (1) .
Scott, A. C. (1985). Phil. Trans. Roy. Soc. Lond. **A315** (423) .
Scott, A. C. and Eilbeck, J. C. (1986). Chem. Phys. Lett. **132** (23) ; Phys. Lett. A **119** (60) .
Scott, A. C., Bernstein, L. and Eilbeck, J. C. (1989). J. Biol. Phys. **17** (1) .
Scott, A. C., Lomdahl, D. S. and Eilbeck, J. C. (1985). Chem. Phys. Lett. **113** (29) .
Strang, G. (1975). Linear algebra and its applications, Academic Press, New York.

Tomboulis, E. and Woo, G. (1976). Nucl. Phys. B **107** (221) .
Zakharov, V. E. and Shabat, A. B. (1972). Sov. Phys.-JETP **34** (62) .
Zakharov, V. E. and Shabat, A. B. (1973). Sov. Phys.-JETP **37** (823) .

Chapter 7

Problem Solving in Nonlinear Quantum Mechanics

In Chapters 3 – 6, we have established the fundamental principles and theory of the nonlinear quantum mechanics. The next important step is to discuss methods for solving the dynamic equations. In this Chapter, some common approaches for solving these equations will be discussed.

7.1 Overview of Methods for Solving Nonlinear Quantum Mechanics Problems

From earlier discussions, we know clearly that problems in nonlinear quantum phenomena are very complicated. This complexity is due to the different mechanisms of nonlinear interactions. Therefore, understanding the mechanism and property of nonlinear interaction is essential for establishing correct dynamic equations and for solving these equations. The dynamic equations, (3.2) – (3.5), cannot be solved unless the exact form of the nonlinear interaction, b, is known. Therefore, the first step one should take in solving a nonlinear quantum mechanics problem is to carefully study the nonlinear quantum phenomenon and its properties, look into the mechanism of the nonlinear interaction, and come up with an appropriate model for the nonlinear interaction. Based on these understandings, the Lagrangian, Hamiltonian function, and operators of the system can be established. Finally, the dynamical equations of the microscopic particles can be obtained from the corresponding classical quantities, or Lagrangian, using the Euler-Lagrange equation or the Hamilton equation, or from the quantum Hamiltonian using the Schrödinger equation or the Heisenberg equation in the second quantization representation in which the Hamiltonian is given in terms of creation and annihilation operators of the microscopic particle. Once the dynamical equation is known, we can solve the equation to obtain its the solution and study the nonlinear-phenomena. Related systems can also be studied based on the fundamental dynamic equation. For example, external fields or forces can be considered and included in the dynamical equations. This is a general approach for solving nonlinear quantum mechanics problems. Of course, certain problems can be easily solved using some special techniques.

The commonly used methods for obtaining soliton solutions of nonlinear dynamical equations in nonlinear quantum mechanics are summarized in the following.

7.1.1 *Inverse scattering method*

The inverse scattering method was first proposed by Gardner, Green, Kruskal and Miura (GGKM) in 1967, for solving the KdV equation which connects a nonlinear dynamic equation to the linear Schrödinger equation

$$\frac{\partial^2 \psi}{\partial x^2} + (\lambda - \phi)\psi = 0,$$

through the GGKM transformation

$$\phi = \frac{1}{\psi}\frac{\partial^2 \psi}{\partial x^2} + \lambda(t),$$

where ϕ satisfies the nonlinear dynamic equation. In 1968, Lax gave a more general formulation of the inverse scattering method. The nonlinear Schrödinger equation was solved by Zakharov and Shabat in 1971 using the inverse scattering method which was discussed in Section 4.3. In 1973, the Sine-Gordon equation was solved by Ablowitz *et al.*.

According to Lax, given a nonlinear equation

$$\phi_t = K(\phi), \quad \phi = \phi(x,t),$$

the inverse scattering method for solving the equation consists of the following three steps.

(1) Find operators L and B, which depend on the solution ϕ, such that $iL_t = BL - LB$, as given in (3.20) and (3.21).
(2) The scattering operator, L, satisfies its eigenvalue equation, $L\varphi = \lambda\varphi$.
(3) The time dependence of the scattered wave is determined by $i\varphi_t = B\varphi$.

Instead of directly solving the nonlinear equation to obtain $\phi(x,t)$ for the given initial condition $\phi(x,0)$, the inverse scattering method proceeds in the above three steps and obtains solution of the linear integral equations. The major difficulty of the inverse scattering method is the lack of a systematic approach for finding the operators L and B, even if they exist.

7.1.2 *Bäcklund transformation*

Given $d\phi = Pdx + Qdt$, then the integrability of the corresponding first-order differential equations

$$\phi_x = P, \quad \phi_t = Q \tag{7.1}$$

requires $P_t = Q_x$. The Bäcklund transformation consists of the following three steps.

(1) Find P and Q, as functions of ϕ.
(2) Choose a trial function, ϕ_0.
(3) Use $\phi_{1x} = P(\phi_1, \phi_0)$ and $\phi_{1t} = Q(\phi_1, \phi_0)$ to find a new solution ϕ_1.

The Bäcklund transformation method enables one to find a new solution from a given one. When applied repeatedly, the Bäcklund transformation can be used to obtain the breather and the N-soliton solutions of the Sine-Gordon equation. It can also be applied to the nonlinear Schrödinger equation. The corresponding Bäcklund transformation can be (3.30) or (3.43). The difficulty for the Bäcklund transformation method is to find the functions P and Q.

7.1.3 *Hirota method*

The Hirota method changes the independent variables and rewrites the original dynamic equation into the form of

$$F(D_x^m D_t^n) = 0,$$

where D is an operator. Analytic solutions are then obtained using perturbation method. Here,

$$D_x^m D_t^n(f,g) = \left[\left(\frac{\partial}{\partial x} - \frac{\partial}{\partial \tilde{x}}\right)^m \left(\frac{\partial}{\partial t} - \frac{\partial}{\partial \tilde{t}}\right)^n f(x,t)g(\tilde{x},\tilde{t})\right]_{x=\tilde{x}',t=\tilde{t}'}. \tag{7.2}$$

The Hirota method is widely used.

7.1.4 *Function and variable transformations*

Quite often a complicated equation can be transformed into a simpler or a standard equation which is readily solved. The forms of transformation vary and depend on the problem and equation. The following are some commonly used transformations.

7.1.4.1 *Function transformation*

As an example, we consider the transformation

$$\phi = \sqrt{\rho} e^{i\theta(\vec{r},t)}. \tag{7.3}$$

Using this transformation, the nonlinear Schrödinger equation (3.2) with $V(\vec{r},t) = 0 = A(\phi)$ can be written as two equations (3.110) and (3.111), which can be readily solved.

7.1.4.2 *Variable transformation and characteristic line*

The following variable transformation is often used.

$$\xi = x - vt, \quad \eta = x + vt. \tag{7.4}$$

For example, the one-dimensional linear fluid equation,

$$\frac{\partial \phi}{\partial t} + v\frac{\partial \phi}{\partial x} = 0$$

can be written as

$$\frac{\partial \phi}{\partial \eta} = 0, \quad \phi = f(\xi) = f(x - vt)$$

under this transformation. The equation $dx/dt = v$ or $dx/v = dt/1 = d\phi/0$ is called the characteristic equation of the fluid equation, dx/dt is the characteristic direction, $\xi = x - vt =$ constant is referred to as the characteristic line of the equation. The value of ϕ on the characteristic lines is constant, and ϕ is called a Riemann invariance of the fluid equation.

For the one-dimensional nonlinear fluid equation,

$$\frac{\partial \phi}{\partial t} + \phi\frac{\partial \phi}{\partial x} = 0,$$

its characteristic line is $dx/dt = u(x,t)$. The variable transformation in this case is

$$\xi = x - u(x,t)t, \quad \eta = x + u(x,t)t. \tag{7.5}$$

Using (7.5), the above equation becomes

$$\frac{d\phi}{d\eta} = 0.$$

Here $u(x,t) = u =$ constant.

7.1.4.3 *Other variable transformations*

Let

$$\xi = x - y, \quad \eta = x + y,$$

or

$$\xi = \frac{x - t}{2}, \quad \eta = \frac{x + t}{2}.$$

Then the Liouville equation

$$\frac{\partial^2 \phi}{\partial x^2} - \frac{\partial^2 \phi}{\partial y^2} = \lambda e^u,$$

where λ is a constant, can be written as

$$\frac{\partial^2\phi}{\partial\eta\partial\xi} = \frac{\lambda}{4}e^u.$$

Under a similar transformation,

$$\xi = \frac{x-t}{2}, \quad \eta = \frac{x+t}{2},$$

the Sine-Gordon equation (in natural unit system)

$$\frac{\partial^2\phi}{\partial x^2} - \frac{\partial^2\phi}{\partial t^2} = \sin\phi, \tag{7.6}$$

becomes

$$\frac{\partial^2\phi}{\partial\eta\partial\xi} = \sin\phi.$$

Many other function and variable transformations are used to solve the dynamic equations but we cannot list all of them here.

7.1.4.4 *Self-similarity transformation*

Assume that $\phi(x,t)$ is a solution of the nonlinear equation, $P\phi = 0$. If x, t and $\phi(x,t)$ are replaced by

$$\begin{aligned} \tilde{x}' &= x + \varepsilon X(x,t,\phi) + O(\varepsilon^2), \\ \tilde{t}' &= t + \varepsilon T(x,t,\phi) + O(\varepsilon^2), \\ \phi' &= \phi + \varepsilon\Phi(x,t,\phi) + O(\varepsilon^2), \end{aligned} \tag{7.7}$$

respectively, where ε is a small parameter. The transformation (7.7) is called an infinitesimal transformation, where X, T and Φ are coefficients of the first order terms of $\tilde{x}'$, $\tilde{t}'$ and ϕ', respectively. If we require $\phi'(\tilde{x}',\tilde{t}')$ to satisfy the same equation, it can be shown that

$$X\frac{\partial\phi}{\partial x} + T\frac{\partial\phi}{\partial t} = \Phi, \quad \text{and} \quad \frac{dx}{X} = \frac{dt}{T} = \frac{d\phi}{\Phi}.$$

Because these equations are invariant under (7.7), and X, T and Φ are independent, we can assume $T = 1$. We then get $f(x,t) = \text{const}$ and $g(x,t,\phi) = \text{const}$ from $dx/dt = X$ and $d\phi/dt = \Phi$, respectively. If the similarity variables are chosen as ξ and $\tilde{\phi}$, then $\xi = f(x,t)$ and $\tilde{\phi} = g(x,t,\phi)$ are called a self-similarity transformation, the solution $\phi(x,t)$ obtained from $\tilde{\phi} = g(x,t,\phi)$ is called a self-similarity solution. The self-similarity transformation $\xi = f(x,t)$ and $\tilde{\phi} = g(x,t,\phi)$ are often expressed as $\xi = a(t)x$, $\tilde{\phi}(\xi) = \phi(x,t)/\beta(t)$, and the self-similarity solution is then given by $\phi(x,t) = \beta(t)\tilde{\phi}(a(t)x)$.

For the linear heat conduction equation,

$$\frac{\partial \phi}{\partial t} = \gamma \frac{\partial^2 \phi}{\partial x^2},$$

where γ is a constant, its self-similarity transformation is,

$$\xi = \frac{x}{\sqrt{t}}, \quad \tilde{\phi} = \frac{\phi}{t},$$

and its solution is

$$\phi = t\tilde{\phi}\left(\frac{x}{\sqrt{t}}\right).$$

Under this transformation, the equation becomes

$$\gamma \frac{d\tilde{\phi}}{d\xi^2} + \frac{\xi}{2}\frac{d\tilde{\phi}}{d\xi} - \gamma = 0.$$

For the Sine-Gordon equation

$$\frac{\partial^2 \phi}{\partial \xi \partial \eta} = \sin \phi,$$

its self-resemblant transformation is

$$\zeta = \xi\eta, \quad \tilde{\phi} = \phi,$$

and the equation becomes

$$\xi \frac{d^2\phi}{d\zeta^2} + \frac{d\phi}{d\zeta} = \sin \phi,$$

under this transformation. The solution of this equation can be easily obtained.

7.1.4.5 *Galilei transformation*

Function transformation is a general approach for solving dynamic equations. As an example, we consider the nonlinear Schrödinger equation (3.2) with $V(x) = A(\phi) = 0$, and $b = 1$. If we let

$$x' = x\sqrt{\frac{2m}{\hbar^2}}, \quad \text{and} \quad t' = \frac{t}{\hbar},$$

its solution is of the form

$$\phi(x', t') = A\,\text{sech}(Ax')\exp\left(-i\frac{A^2 t'}{2}\right) \tag{7.8}$$

Since this nonlinear Schrödinger equation is invariant under the Galilei transformation (6.108), its solution can be written as

$$\phi(x', t') = A\,\text{sech}[A(x' - vt')]\exp\left[-ivx' + i\frac{(v^2 - A^2)t'}{2}\right]. \tag{7.9}$$

Let $v = 2\xi$, and $A = 2\eta$, then it reduces to the soliton solution (4.56) which was obtained by the inverse scattering method.

7.1.4.6 *Traveling-wave method*

. If we can write the solution of an equation of motion in the form of $\phi(x,t) = \phi(\xi)$, with $\xi = x - vt$, it is called a traveling-wave solution. This method is widely used in wave mechanics and nonlinear quantum mechanics.

7.1.4.7 *Perturbation method*

In reality, due to the existence of boundaries, defects, impurities, dissipation, external fields, *etc.*, a system usually experiences some perturbation. As a matter of fact, when the soliton behavior of microscopic particles in the system is probed, the measurement itself is bound to alter the state of the system to some extent. Thermal excitation at finite temperature is another form of perturbation that no system can avoid. In fact, the appearance of microscopic particles (solitons) itself implies some form of internal excitation has occured.

When such perturbations are relatively weak, they may be treated by perturbation methods. The main spirit of soliton perturbation was illustrated in Chapter 3.

7.1.4.8 *Variational method*

The variational method will be discussed later in this Chapter.

7.1.4.9 *Numerical method*

When analytic solutions cannot be found and perturbation methods are inapplicable, one can always solve the nonlinear equation by numerical methods. Sometimes the numerical solutions may lead to a correct guess of their analytic forms. There are many numerical approaches for solving differential equation such as Monte-Carlo, Runge-Kutta, finite elements *etc.* With the rapid increase in computing power in the last decades, numerical simulations plays a more and more important role in practical calculations.

7.1.4.10 *Experimental simulation*

Some dynamic equations may be simulated by mechanical models or electric circuits. For example, the Sine-Gordon equation was simulated by a chain of torsional pendula. The overdamped Sine-Gordon equation $\phi_{xx} = +\phi_{tt} + \sin\phi$ was studied by submerging the chain of pendula in water and the driven Sine-Gordon equation was simulated using long Josephson junctions and using mechanical models.

7.2 Traveling-Wave Methods

7.2.1 *Nonlinear Schrödinger equation*

For the nonlinear Schrödinger equation given in (3.2) with $V(x,t) = A(\phi) = 0$, a traveling-wave solution can be assumed

$$\phi(x,t) = \phi(\xi)e^{i(kx'-\omega t')},$$

where

$$\xi = x' - vt', \quad x' = \sqrt{\frac{2m}{\hbar^2}}x, \quad t' = \frac{t}{h}.$$

In terms of the new variables, equation (3.2) can be written as

$$\frac{d^2\phi(\xi)}{d\xi} + (\omega - k^2)\phi(\xi) - b\phi^3 = 0.$$

The soliton solution of this equation can be easily obtained when $b > 0$ and $\gamma = \omega - k^2 > 0$,

$$\phi(x,t) = \sqrt{\frac{2\gamma}{b}}\,\mathrm{sech}[\sqrt{\gamma}(\xi - \xi_0)]e^{i(kx'-\omega t')}. \tag{7.10}$$

On the other hand, when $b < 0$ and $\gamma < 0$, the solution is

$$\phi(x,t) = \sqrt{\left|\frac{\gamma}{b}\right|}\tanh\left(\frac{\sqrt{|\gamma|}}{2}(\xi - \xi_0)\right)e^{i(kx'-\omega t')}. \tag{7.11}$$

If we assume the traveling-wave solution is of the form

$$\phi(x',t') = \tilde{\phi}(\xi)e^{i\Omega t'+iQx'}, \tag{7.12}$$

then (3.2) with $b = 2$, $V(x) = A(\phi) = 0$, becomes

$$\tilde{\phi}_{\xi\xi}(\xi) + i(2Q - v)\tilde{\phi}_\xi - \tilde{\phi}(\xi)(\Omega + Q^2) + 2\tilde{\phi}^3(\xi) = 0. \tag{7.13}$$

If the complex coefficient of $\tilde{\phi}_\xi(\xi)$ vanishes, then $Q = v/2$. Furthermore, from $A = Q^2 + \Omega$, we get that $\Omega = -v^2/4 + A$. Finally from (7.13), we obtain

$$\phi_{\xi\xi} - A\phi + 2\phi^3 = 0.$$

This equation can be integrated, which results in

$$(\phi_\xi)^2 = D + A\phi^2 - \phi^4,$$

where D is a integral constant. The solution ϕ is then obtained by inverting an elliptic integral,

$$\int_{\phi_0}^{\phi}\frac{d\phi}{\sqrt{D + A\phi^2 - \phi^4}} = \pm\xi.$$

Let

$$P(\phi) = (\beta_1 - \phi^2)(\phi^2 - \beta_2).$$

we have,

$$\frac{1}{\sqrt{\beta_1}}\{K(k) - F(\phi, k)\} = \pm\xi,$$

where $F(\phi, k)$ is the incomplete elliptic integral, and

$$k = \sqrt{\frac{\beta_1 - \beta_2}{\beta_1}}, \quad \beta_{1,2} = \frac{A}{2} \pm \sqrt{\frac{A^2}{4} + D}.$$

Using these and $\beta_{1,2} = \phi_{1,2}^2$, then we have

$$\phi(\xi) = \phi_1\left\{1 - \left[\left(1 - \frac{\phi_1^2}{\phi_2^2}\right)\mathrm{sn}^2(\xi, k)\right]\right\}^{1/2}.$$

When $D \to 0$, $\phi_1 \to \phi_0$, $k \to 1$, $\phi \to \phi_0\,\mathrm{sech}\phi_0\xi$, where $\phi_0 = \sqrt{A}$, the soliton solution of this nonlinear Schrödinger equation becomes

$$\phi_s(x,t) = \sqrt{A}\,\mathrm{sech}\left[\sqrt{A}(x' - vt')\right]\exp\left[i\left(-\frac{1}{4}v^2t' + \frac{1}{2}vx' + At'\right)\right]. \qquad (7.14)$$

The corresponding canonical form of the one-soliton solution has been given in (7.9) and (4.56).

7.2.2 *Sine-Gordon equation*

Using the traveling-wave method in natural unit system, $\tilde{\xi} = x - vt$, the Sine-Gordon equation,

$$\phi_{tt} - v_0^2\phi_{xx} + h_0^2\sin\phi = 0, \qquad (7.15)$$

can be written as

$$(v^2 - v_0^2)\frac{d^2\phi}{d\tilde{\xi}^2} + h_0^2\sin\phi = 0. \qquad (7.16)$$

If $v^2 > v_0^2$, (7.16) can be written as

$$\frac{d^2\phi}{d\tilde{\xi}^2} + \beta^2\sin\phi = 0, \qquad (7.17)$$

with

$$\beta^2 = \frac{h_0^2}{v^2 - v_0^2}.$$

Equation (7.17) resembles the equation of motion of a undamped pendulum, and can be separated into the following two equations

$$\frac{d\phi}{d\tilde{\xi}} = \Phi, \quad \frac{d\Phi}{d\tilde{\xi}} = -\beta^2 \sin\phi. \tag{7.18}$$

This gives

$$\frac{d\Phi}{d\phi} = -\frac{\beta^2 \sin\phi}{\Phi}.$$

Thus we can get

$$\frac{1}{2}\Phi^2 + \beta^2(1 - \cos\phi) = A,$$

where A is a constant of integration. Using

$$1 - \cos\phi = 2\sin^2\frac{\phi}{2}$$

and (7.18) and letting $A = 2\beta^2 k^2$, the above equation can be written as

$$\left(\frac{d\phi}{d\tilde{\xi}}\right)^2 = 4\beta^2\left[k^2 - \sin^2\left(\frac{\phi}{2}\right)\right]. \tag{7.19}$$

If $0 < k^2 = A/2\beta^2 < 1$, the solution of (7.19) is given by

$$\sin\left(\frac{\phi}{2}\right) = \pm k\mathrm{sn}[\beta(\tilde{\xi} - \tilde{\xi}_0), k]. \tag{7.20}$$

If $k \to 1$, equation (7.20) is replaced by

$$\sin\left(\frac{\phi}{2}\right) = \tanh[\pm\beta(\tilde{\xi} - \tilde{\xi}_0)],$$

or

$$e^{\pm\beta(\tilde{\xi}-\tilde{\xi}_0)} = \left[\frac{1+\sin(\phi/2)}{1-\sin(\phi/2)}\right]^{1/2} = \tan\left(\frac{\phi}{4} + \frac{\pi}{4}\right).$$

Thus the kink solution of the Sine-Gordon equation (7.15) is finally given by

$$\phi_\pm = -\pi + 4\tan^{-1} e^{\pm\beta(\tilde{\xi}-\tilde{\xi}_0)}. \tag{7.21}$$

When $v^2 < v_0^2$, equation (7.19) is replaced by

$$\left(\frac{d\phi}{d\tilde{\xi}}\right)^2 = 4\beta'^2\left[\sin^2\left(\frac{\phi}{2}\right) - k^2\right],$$

where

$$\beta'^2 = \frac{h_0^2}{v_0^2 - v^2}.$$

The kink solution of the Sine-Gordon equation (7.15) in the case of $k' \to 1$ and $k'^2 = 1 - k^2$ becomes

$$\phi = 4\tan^{-1}\left\{e^{\mp\beta'(\tilde{\xi}-\tilde{\xi}_0)}\right\}. \tag{7.22}$$

Using the following dimensionless variables, $t' = h_0 t$, $x' = \delta x$, $\delta = h_0/v_0$, equation (7.15) can be written as

$$\phi_{x'x'} - \phi_{t't'} = \sin\phi. \tag{7.23}$$

According to the properties of (7.21) and (7.22), the solution of (7.23) is of the form

$$\phi = 4\tan^{-1}\left[\frac{Y(x')}{T(t')}\right]. \tag{7.24}$$

In terms of Y and T, equation (7.23) becomes

$$(Y^2+T^2)\left(\frac{Y_{x'x'}}{Y}+\frac{T_{t't'}}{T}\right) - 2[(Y_{x'})^2 + (T_{t'})^2] = T^2 - Y^2. \tag{7.25}$$

Differentiating (7.25) with respect to t' and x', and then dividing the two resulting equations by $YY_{x'}$ and $TT_{t'}$, we can get

$$\frac{1}{YY_{x'}}\left(\frac{Y_{x'x'}}{Y}\right)_{x'} = -\frac{1}{TT_{t'}}\left(\frac{T_{t't'}}{T}\right)t' = 4q, \tag{7.26}$$

where $4q$ is a constant. Integrating (7.26), the following equation can be obtained

$$\frac{Y_{x'x'}}{Y} = 2qY^2 + a_1, \quad \frac{T_{t't'}}{T} = -2qT^2 + a_2, \tag{7.27}$$

and

$$(Y_{x'})^2 = qY^4 + a_1Y^2 + d_1, \quad (T_{t'})^2 = -qT^4 + a_2T^2 + d_2, \tag{7.28}$$

where a_1, a_2 and d_1, d_2 are some integral constants. Inserting (7.27) and (7.28) into (7.25), we can get $a_1 - a_2 = 1$, and $d_1 + d_2 = 0$.

If we choose $a_2 = -a$, $d_1 = d$, then $a_1 = 1 - a$ and $d_2 = d$. When $q = -1$, and $d = 0$, equation (7.28) becomes

$$Y_{x'}^2 = -Y^4 + (1-a)Y^2, \quad T_{t'}^2 = T^4 - aT^2. \tag{7.29}$$

If $0 < a < 1$, integrating (7.29) and taking the integral constant to be zero, we can get

$$-\frac{1}{\sqrt{1-a}}\,\mathrm{sech}^{-1}\left(\frac{Y}{\sqrt{1-a}}\right) = \pm x' = \pm\delta x,$$

$$\frac{1}{\sqrt{a}}\sin^{-1}\left(\frac{\sqrt{a}}{T}\right) = \pm t' = \pm h_0 t.$$

Thus

$$Y = \sqrt{1-a}\,\mathrm{sech}(\sqrt{1-a}\;\delta x), \quad \frac{1}{T} = \pm\frac{1}{\sqrt{a}}\sin(\sqrt{a}h_0 t). \tag{7.30}$$

Substituting (7.30) into (7.24), we get the breather solution of the Sine-Gordon equation (7.15)

$$\tan\left(\frac{\phi}{4}\right) = \pm\sqrt{\frac{1-a}{a}}\frac{\sin(\sqrt{a}h_0 t)}{\cosh(\sqrt{1-a}\delta x)}, \quad (o < a < 1). \tag{7.31}$$

Following the same procedure, we can also obtain the traveling-wave soliton-solution of the ϕ^4-field equation in natural unit system

$$\phi_{tt} - v_0^2\phi_{xx} + \alpha'\phi - \beta'\phi^3 = 0.$$

Since the procedure is the same, we only give the results here,

$$\phi = \pm\sqrt{\frac{\alpha'}{\beta'}}\tanh\left[\sqrt{\frac{\alpha'}{2(v^2-v_0^2)}}(\tilde{\xi}-\tilde{\xi}_0)\right],$$

when $v^2 > v_0^2$, $\alpha' > 0$, $\beta' > 0$ or $v^2 < v_0^2$, $\alpha < 0$, $\beta' < 0$, and

$$\phi = \pm\sqrt{\frac{2\alpha'}{\beta'}}\,\mathrm{sech}\left[\sqrt{\frac{-\alpha'}{v^2-v_0^2}}(\tilde{\xi}-\tilde{\xi}_0)\right],$$

when $v^2 > v_0^2$, $\alpha' < 0$, $\beta' < 0$ or $v^2 < v_0^2$, $\alpha' > 0$, $\beta' > 0$.

7.3 Inverse Scattering Method

We now determine the soliton solution of the nonlinear Schrödinger equation (4.40) with an initial pulse using the inverse scattering method mentioned earlier, following the approach of Zakharov and Shabat, in accordance with the Lax operator equation (3.20), where $\hat{L}$ and $\hat{B}$ are some linear operators with coefficients depending on the function $\phi(x,t)$ and its derivatives, as discussed in Section 4.3. In other words, the equation for $\phi(x,t)$ (nonlinear Schrödinger equation) is a condition of compatibility of auxiliary linear equations (3.20) and (3.21), where λ is a spectral parameter (generally speaking, complex), and ψ in (3.21) is often called a Jost function and is represented by

$$\psi = \begin{pmatrix}\psi_1\\ \psi_2\end{pmatrix}$$

where $\hat{L}$ is given in (3.22) for the nonlinear Schrödinger equation (3.2) with $V = A = 0$. Therefore, equation (3.21) can be rewritten in the form of Zakharov-Shabat equation (4.42), where

$$q = \frac{i\phi}{\sqrt{1-s^2}}, \quad \xi' = \frac{\lambda s}{\sqrt{1-s^2}}, \quad b = \frac{2}{1-s^2},$$

b is the nonlinear coefficient in the nonlinear Schrödinger equation (4.40). Here the characteristic value, $\zeta = \xi' + i\eta$. Equation (4.42) can be further written in the form

$$\tilde{\Psi}_{1x'x'} - \frac{q_{x'}}{q}\tilde{\Psi}_{1x'} + \tilde{\Psi}_1\left(\xi'^2 + |q|^2 - i\xi'\frac{q_{x'}}{q}\right) = 0, \tag{7.32}$$

$$\tilde{\Psi}_2 = \frac{1}{q}[\tilde{\Psi}_{1x'} + i\xi'\tilde{\Psi}_1]. \tag{7.33}$$

According to the inverse scattering method, it is necessary to find the solutions of the system (7.32) which have the following asymptotic behavior for a given function $q(x')$ and real $\zeta = \xi'$

$$\lim \tilde{\Psi}(\xi', x') = \begin{cases} \begin{pmatrix} 1 \\ 0 \end{pmatrix} e^{-i\xi' x'}, & \text{if } x' \to -\infty, \\ a(\xi')\begin{pmatrix} 1 \\ 0 \end{pmatrix} e^{-i\xi' x'} - b(\xi')\begin{pmatrix} 0 \\ 1 \end{pmatrix} e^{i\xi' x'}, & \text{if } x' \to \infty. \end{cases} \tag{7.34}$$

The coefficients $a(\xi')$ and $b(\xi')$ are complex and satisfy

$$|a(\xi')|^2 + |b(\xi')|^2 = 1.$$

The quantity, $1/a(\xi')$, determines the transmission coefficient at x' of a plane wave incident on the potential $q(x')$ except at $x' \approx \infty$, while the ratio, $R(\xi') = b(\xi')/a(\xi')$ determines its reflection coefficient. The functions in (7.34) continue to be analytic into the upper half ($\eta > 0$) of the complex plane of ζ. With the inverse scattering method, we can obtain the values of $a(\xi')$ and $b(\xi')$ at time $t' \neq 0$ from their values at $t' = 0$ using the relations

$$a(\zeta, t') = a(\zeta), \quad b(\zeta, t') = b(\zeta)e^{4i\zeta^2 t'}. \tag{7.35}$$

If for certain values ζ_j, $a(\zeta_j)$ vanishes in the upper half of the complex ζ plane, then the asymptotic behavior of $\tilde{\Psi}(\zeta_j, x')$ is given by

$$\tilde{\Psi}(\zeta_j, x') = \begin{cases} \begin{pmatrix} 1 \\ 0 \end{pmatrix} e^{(\eta_j - i\xi_j')x'}, & x' \to -\infty, \\ b(\zeta_j)\begin{pmatrix} 0 \\ 1 \end{pmatrix} e^{-(\eta_j - i\xi_j')x'}, & x \to \infty \end{cases} \quad (j = 1, \cdots, N). \tag{7.36}$$

For $\eta_j \neq 0$, $\tilde{\Psi}(\zeta_j, x')$ decreases asymptotically when $|x'| \to \infty$ and therefore describes bound states corresponding to complex ζ_j given in (4.42) or in (7.32) – (7.33). Values of ζ_j and $c_j(t')$, $(j = 1, 2, \cdots, N)$, where

$$c_j(t') = b(\zeta_j, x')\left[\frac{\partial a(\zeta)}{\partial \zeta}\right]^{-1}_{\zeta=\xi'} = \frac{ib}{2}e^{4i\xi'^2 t'}, \tag{7.37}$$

and

$$R(\xi', t') = \frac{b(\xi', t')}{a(\xi)}$$

form a set of "scattering data". From these data, Davydov obtained the following auxiliary function

$$F(x', t') = \frac{1}{2\pi}\int_{-\infty}^{\infty} d\xi' R(\xi', t)e^{i\xi' x'} + \sum_{j=1}^{N} c_j(t')e^{i\zeta_j, x'}. \tag{7.38}$$

Inserting (7.38) into the Gelfand-Levitan-Marchenko (GLM) equation

$$K(x', y') = F^*(x'+y', t') - \int_{-\infty}^{\infty} dp \int_{-x'}^{\infty} dz F^*(p+y', t')F(p+z, t')K(x', t'). \tag{7.39}$$

the undetermined function

$$q(x', t') = -2K(x', x') \tag{7.40}$$

and the wave function satisfying the nonlinear Schrödinger equation (4.40), with the initial value $\phi(x', 0)$, can be expressed in terms of the solution of (7.39), with the help of the following

$$\phi(x', t') = -\frac{i\sqrt{2}q(x', t')}{\sqrt{b}}. \tag{7.41}$$

Using the inverse scattering method, Davydov obtained a soliton solution of the nonlinear Schrödinger equation (4.40) with the following initial data

$$\phi(x', t' = 0) = \frac{1}{2\sqrt{2}}\sqrt{b}e^{2ikx'} \operatorname{sech}\left(\frac{bx'}{4}\right), \tag{7.42}$$

where $2k = m_{ex}r_0 v/\hbar$. Here r_0 is the lattice constant, m_{ex} and v are effective mass and velocity of the particle, respectively.

To obtain the scattering data, it is sufficient to find the asymptotic solutions of (7.32) and (7.33) with

$$q(x', 0) = i\sqrt{\frac{b}{2}}\phi(x', 0) = \frac{ib}{4}e^{2ikx'} \operatorname{sech}\left(\frac{bx'}{4}\right). \tag{7.43}$$

If $|x'| \to \infty$, equation (7.32) becomes

$$\tilde{\Psi}_1'' - \left(2ik \pm \frac{b}{4}\right)\tilde{\Psi}_1' + \zeta\left(\zeta - k \mp \frac{ib}{4}\right)\tilde{\Psi}_1 = 0, \tag{7.44}$$

where

$$\tilde{\Psi}' = \frac{d\tilde{\Psi}}{dx'}.$$

The upper and lower signs in (7.44) correspond to $x' \to -\infty$ and $x' \to \infty$, respectively. It can be shown that (7.44) has the following solution

$$\tilde{\Psi}_1 \to \begin{cases} e^{[i(k+m_2)+b/4]x'}, & x' \to -\infty, \\ e^{[i(k-m_2)-b/4]x'}, & x' \to \infty, \end{cases} \tag{7.45}$$

with

$$m_1 = \zeta + k - i\frac{b}{8}, \quad m_2 = \zeta + k + i\frac{b}{8}.$$

If $b \neq 0$, (7.45) coincides with the asymptotic function (7.34) with

$$\zeta_1 = \xi_1' + i\eta_1, \quad \xi_1' = -k, \quad \eta_1 = \frac{b}{8}.$$

From (7.34), we found that as $x' \to \infty$, $\tilde{\Psi}_2(\xi') = b(\xi')e^{i\xi' t'}$. However, (7.45) and (7.33) imply that $\tilde{\Psi}_2(\infty) = 0$. Therefore, for the potential (7.43), $b(\xi')$ and $R(\xi')$ must equal to zero. Such potentials are called non-reflective potentials and cannot be observed by scattering of plane waves coming from infinity. In such a case, the spectral data (7.37) include

$$R(\xi', t') = 0, \quad \zeta_1 = -\kappa + i\frac{b}{8}, \quad c_1(t') = c_0 e^{4i\zeta_1^2 t'},$$

where $c_0 = ib/4$. Thus,

$$F(x', t') = c_1(t')e^{i\zeta_1^2 x'}.$$

Inserting this expression into (7.39), we get

$$K(x', y') = f(x')e^{-i\zeta_1^* y'}.$$

We can then obtain

$$f(x') = c_1^*(t')e^{-i\zeta_1^* x'} \left[1 + \frac{16|c_1(t')|^2}{b^2} e^{-bx'/2}\right]^{-1}.$$

The square of the modulus of the wave function, including $q(x', t')$ in (7.40) is given by

$$|\phi(x', t')|^2 = \frac{2}{b}|q(x', t')|^2 = \frac{b}{8\cosh^2[b(x' - x_0' - 4\kappa t')/4]}, \tag{7.46}$$

where

$$x_0' = \frac{4}{b}\ln\left(\frac{4|c_0|}{b}\right) = 0.$$

This indicates that given an arbitrary nonzero value of b and the initial pulse (7.42), the microscopic particle propagates with velocity v in the form of a single soliton.

Davydov further obtained solution of the nonlinear Schrödinger equation (4.40) with a rectangular step initial pulse

$$\phi(x',0) = \begin{cases} \dfrac{1}{\sqrt{l}} e^{2ikx'}, & \text{if } 0 \le x' \le l \\ 0, & \text{if } x' < 0 \text{ and } x' > l. \end{cases} \tag{7.47}$$

In this case the solution of (4.42) or (7.32) – (7.33) can be written as

$$\tilde{\Psi} = \begin{pmatrix} 1 \\ 0 \end{pmatrix} e^{-i\zeta x'}, \qquad \text{for } x' < 0,$$

$$\tilde{\Psi} = a(\zeta) \begin{pmatrix} 1 \\ 0 \end{pmatrix} e^{-i\zeta x'} + b(\zeta) e^{i\zeta x'}, \qquad \text{for } x' > l,$$

while in the region of $0 \le x \le \ell$,

$$\tilde{\Psi}_1 = [A_1 \sin nx' + A_2 \cos nx'] e^{ikx'},$$

$$\tilde{\Psi}_2 = \frac{1}{Q_0} \{[(\zeta + k)A_1 + inA_2] \sin nx' + [(\zeta - \kappa)A_2 - inA_1] \sin nx'\} e^{-ikx'},$$

with

$$Q_0 \equiv \sqrt{\frac{bl}{2}}, \quad n^2 \equiv Q_0^2 - (\zeta - \kappa)^2. \tag{7.48}$$

It follows from the continuity of the solutions that

$$A_1(\zeta) = -i\frac{k - \zeta}{n}, \qquad A_2 = 1, \tag{7.49}$$

$$a(\zeta) = \frac{1}{n} S(\zeta, k) e^{i(\kappa+\zeta)t'}, \qquad b(\zeta) = \frac{i}{n} Q_0 e^{-i(k+\zeta)t'},$$

where

$$S(\zeta, k) = n\cos(nl) - i(k + \zeta)\sin(nl).$$

Using (7.49), we can find the reflection coefficient at time $t' = 0$

$$R(\xi') = iQ_0 S^{-1}(\xi', k) e^{-2i(k+\xi')\ell}.$$

According to (7.36), the reflection coefficient at time t' is given by

$$R(\xi', t') = R(\xi') e^{4i\xi'^2 t'}. \tag{7.50}$$

The parameters of the bound states are obtained from the condition $S(\zeta, k) = 0$. This equation has the following solution, $\zeta_0 = -k + i\eta_0$, where η_0 is determined from

$$l\eta_0 = -\sqrt{\frac{bl}{2} - l^2\eta_0^2} \cot \sqrt{\frac{bl}{2} - \eta_0^2 l^2}.$$

If $\eta_0 \neq 0$, according to (7.35), (7.37) and (7.48), Davydov obtained that

$$c_0(t') = \tilde{\alpha} e^{4i\zeta_0^2 t'}, \tag{7.51}$$

where

$$\tilde{\alpha} \equiv \frac{n_1^2 e^{2l\eta_0}}{Q_0[n_1 l \cos(n_1 l) - \sin(n_1 l)]}, \quad n_1^2 \equiv Q_0^2 - \eta_0^2.$$

Thus, the auxiliary function $F(x', t')$ defined in (7.38) can be written as

$$F(x', t') = c_0(t')e^{i\zeta_0 t'} - \frac{1}{2\pi}\int_{-\infty}^{\infty} d\xi' R(\xi', t')e^{i\xi' x'}, \tag{7.52}$$

where $c_0(t')$ and $R(\xi', t')$ have been given in (7.51) and (7.50), respectively. The first term in (7.52) represents the soliton solution, while the second term determines the "tail" accompanying the soliton.

For a sufficiently long time t', the integral in (7.52) can be approximated by

$$\frac{1}{2\pi}\int R(\xi', t')e^{i\xi' x'} d\xi' \approx \frac{i\sqrt{b}}{4\sqrt{2t'\pi l}S(0,k)} \exp\left\{-i\left[\frac{(x'-2l)^2}{16t'} + 2\left(kl - \frac{\pi}{8}\right)\right]\right\}.$$

Then, the "tail" decreases with time according to $1/\sqrt{t'}$. Keeping only the first term in (7.52) in the long time limit, obtained $K(x', y')$ and the wave function, by solving (7.39),

$$\phi_s(x', t') = \frac{i2\sqrt{2}\eta_0 \exp\left\{2i\left[kx' - 2(k^2 - \eta_0^2)t'\right]\right\}}{\sqrt{b}\cosh\left[2\eta_0(x' - x_0' - 4kt')\right]}, \tag{7.53}$$

where

$$x_0 = \frac{1}{2\eta_0} \ln \frac{|\tilde{\alpha}|}{2\eta_0}.$$

Thus, given the initial excitation in the form of a pulse (7.47), soliton solutions exist if the nonlinearity parameter b is greater than the critical value $b_{cr} = \pi^2/4\ell$. The amplitude of the soliton increases and its width decreases with increasing b.

7.4 Perturbation Theory Based on the Inverse Scattering Transformation for the Nonlinear Schrödinger Equation

In this section, we establish the perturbation theory for the nonlinear Schrödinger equation based on the inverse scattering method. According to the basic idea of the inverse scattering method described in (3.21), we seek a pair of Lax operators $\hat{L}$ and $\hat{B}$ and the corresponding scattering data. We will first build the general theoretical framework.

The first step in using the inverse scattering method is to solve directly the scattering problem, *i.e.*, to find the eigenfunctions of the spectral equation (3.21) (or scattering data). Since the coefficients of $\hat{L}$ depend on $\phi(x,t)$, we can map this function into the scattering data. In general, the scattering data consist of two components, $S(\lambda)$ and S_n, representing the continuous and discrete spectra. Here n is the number of discrete eigenvalues. The evolution of $\phi(x,t)$ gives evolution of the scattering data $\psi_t = \hat{B}\psi$. The evolution equations for $S(\lambda)$ and S_n have the following form

$$\frac{\partial S(\lambda,t)}{\partial t} = i\Omega(\lambda)S(\lambda,t), \quad \frac{dS_n(t)}{dt} = \Omega_n S_n(t). \tag{7.54}$$

To solve the Cauchy problem for the nonlinear Schrödinger equation, we first determine the initial scattering data, $S(\lambda, t=0)$ and $S_n(t=0)$, corresponding to an initial condition $\phi(x, t=0)$ (direct scattering problem). Next we find $S(\lambda,t)$ and $S_n(t)$ from (7.54). And finally reconstruct $\phi(x,t)$ on the basis of $S(\lambda,t)$ and $S_n(t)$, *i.e.*, solve the inverse scattering problem. The discrete-spectrum scattering data correspond to solitons, and the continuous-spectrum scattering data correspond to radiation. In this way, the variational derivatives $\delta S/\delta\phi(x)$ can be written in terms of the Jost function and the scattering data. Making use of these quantities, we can obtain a generalization of the evolution equations (7.54) for the perturbed nonlinear Schrödinger equation,

$$i\phi_{t'} + \phi_{x'x'} + 2|\phi|^2\phi = \varepsilon P[\phi], \tag{7.55}$$

where $P[\phi]$ represents the perturbation, $x' = \sqrt{2m/\hbar^2}x$ and $t' = t/\hbar$. The following was obtained by Kivshar *et al.*,

$$\begin{aligned}\frac{\partial S(\lambda,t')}{\partial t'} &= \int_{-\infty}^{+\infty} dx' \frac{\delta S(\lambda,t')}{\delta\phi(x',t')} F[\phi] + \epsilon \int_{-\infty}^{+\infty} \frac{\delta S(\lambda,t')}{\delta\phi(x',t')} P[\phi] \\ &= i\Omega(\lambda)S(\lambda,t') + \epsilon \int_{-\infty}^{+\infty} \frac{\delta S(\lambda,t')}{\delta\phi(x',t)} P[\phi].\end{aligned} \tag{7.56}$$

$\partial S_n(t')/\partial t'$ can be obtained similarly,

$$\frac{\partial S_n(\lambda,t')}{\partial t} = \Omega_n S_n(t') + \epsilon \int_{-\infty}^{+\infty} dx \frac{\delta S_n(t')}{\delta\phi(x',t')} P[\phi], \tag{7.57}$$

where

$$F[\phi] = -\phi_{x'x'} - 2|\phi|^2\phi,$$

is the unperturbed part in the nonlinear Schrödinger equation. If ϵ is small, we can substitute the unperturbed $\phi(x',t')$ and the Jost functions into the right-hand side of (7.56) and obtain the lowest order approximation for the perturbed evolution equations for the scattering data. Obviously, this procedure can be iterated to obtain higher orders of approximation. These ideas were independently proposed by Kaup, and by Karpman and Maslov. Their work were further generalized by Kivshar and Malomed.

Following the approach of Zakharov and Shabat discussed in Chapter 4 and (3.22), the operators $\hat{L}$ and $\hat{B}$ corresponding to the unperturbed nonlinear Schrödinger equation (7.55) can now be written as

$$\hat{L} = \begin{bmatrix} -i\dfrac{\partial}{\partial x'} & \phi^*(x') \\ -\phi(x') & -i\dfrac{\partial}{\partial x'} \end{bmatrix}, \tag{7.58}$$

$$\hat{B} = \begin{bmatrix} -4i\lambda^3 + 2\lambda|\phi|^2 + \phi^*\phi_{x'} - \phi\phi^*_{x'} & -4i\lambda^2\phi^* - 2i\lambda\phi^*_{x'} + i\phi^*_{x'x'} + 2i\phi^{*2}\phi \\ -4i\lambda^2\phi + 2i\lambda\phi_{x'} + i\phi_{x'x'} + 2i\phi^2\phi^* & 4i\lambda^3 - 2\lambda\left|\phi\right|^2 - \phi_{x'}\phi^* + \phi\phi^*_{x'} \end{bmatrix}. \tag{7.59}$$

The Jost functions for real spectral parameter λ are defined by the boundary conditions

$$\Psi_\pm(x', \lambda) = e^{i\lambda\sigma_3 x'} + O(1), \quad \text{when } x' \to \pm\infty, \tag{7.60}$$

where σ_3 is the Pauli matrix. This matrix form of the Jost functions can be expressed as $\Psi_+ = (\psi', \bar{\psi})$, $\Psi_- = (-\bar{\varphi}, \varphi')$. Here ψ' and φ' are independent vector columns. The linear involution operation acting on

$$\psi' = \begin{bmatrix} \psi_1 \\ \psi_2 \end{bmatrix}$$

transforms it into

$$\tilde{\psi} = \begin{bmatrix} -\psi_2^* \\ \psi_1 \end{bmatrix}.$$

The monodromy matrix T relates the two fundamental solutions Ψ_+ and Ψ_-,

$$\Psi_-(x', \lambda) = \Psi_+(x', \lambda)T(\lambda).$$

Kivshar, *et al.* gave this matrix in the following form

$$T(\lambda) = \begin{bmatrix} a^*(\lambda) & b(\lambda) \\ -b^*(\lambda) & a(\lambda) \end{bmatrix}, \tag{7.61}$$

where the two Jost coefficients $a(\lambda)$ and $b(\lambda)$ satisfy

$$|a(\lambda)|^2 + |b(\lambda)|^2 = 1. \tag{7.62}$$

The vector Jost functions $\psi'(x', \lambda)$, $\varphi'(x', \lambda)$ and the Jost coefficient $a(\lambda)$ are analytic and continuous in the upper half of the complex λ-plane. The zeros, $\lambda_n = \xi_n + i\eta_n$ $(n = 1, 2, \cdots)$, of the functions $a(\lambda)$ in the upper half-plane give the discrete spectrum of the corresponding linear problem (4.41). The Jost functions $\psi'(x', \lambda_n)$ and $\varphi'(x', \lambda_n)$ are linearly dependent, $\varphi'(x', \lambda_n) = b_n\psi'(x', \lambda_n)$. They decay exponentially as $|x'| \to \infty$. $a(\lambda)$ and $b(\lambda)$ with real λ constitute the continuous-spectrum scattering data, and the set of complex numbers λ_n and b_n

constitutes the discrete-spectrum scattering data of the corresponding scattering problem. Their time evolution gives the linear equation (4.41) of the operator $\hat{B}$ in (7.59),

$$\begin{aligned} a(\lambda, t') &= a(\lambda, 0), \quad b(\lambda, t') = b(\lambda, 0)e^{4i\lambda^2 t'}, \\ \lambda_n(t') &= \lambda_n(0), \quad b_n(t') = b_n(0)e^{4i\lambda_n^2 t'}. \end{aligned} \tag{7.63}$$

The inverse scattering problem for the operator $\hat{L}$ in (7.58) now reduces to a system of singular integral equations

$$\begin{aligned} \tilde{\psi}(x', \lambda)e^{i\lambda x'} &= \begin{pmatrix} 0 \\ 1 \end{pmatrix} + \sum_{n=1}^{N} \frac{b_n \psi(x', \lambda_n)e^{i\lambda_n x'}}{(\lambda - \lambda_n)a'(\lambda_n)}, \\ &+ \frac{1}{2\pi i} \int_{-\infty}^{\infty} \frac{R(\lambda')\psi(x', \lambda')e^{i\lambda' x'}}{\lambda' - \lambda + i0} d\lambda' \end{aligned} \tag{7.64}$$

$$\begin{aligned} \tilde{\psi}(x', \lambda_n)e^{i\lambda_n^* x'} &= \begin{pmatrix} 0 \\ 1 \end{pmatrix} + \sum_{m=1}^{N} \frac{b_m \psi(x', \lambda_m)e^{i\lambda_m x'}}{(\lambda_n^* - \lambda_m)a'(\lambda_m)} \\ &+ \frac{1}{2\pi i} \int_{-\infty}^{\infty} \frac{R(\lambda)\psi(x', \lambda)e^{i\lambda x'}}{\lambda - \lambda_n^*} d\lambda, \end{aligned} \tag{7.65}$$

where

$$a'(\lambda_n) \equiv \left. \frac{\partial a(\lambda)}{\partial \lambda} \right|_{\lambda = \lambda_n}, \quad R(\lambda) \equiv \frac{b(\lambda)}{a(\lambda)}. \tag{7.66}$$

Finally, $\phi(x', t)$ is given in terms of the scattering data and the solutions (7.64) and (7.65),

$$\phi^*(x', t') = -2 \sum_{n=1}^{N} \frac{b_n}{a'(\lambda_n)} \psi_1(x', \lambda_n)e^{i\lambda_n x'} + \frac{1}{\pi i} \int_{-\infty}^{\infty} R(\lambda)\psi_1(x', \lambda)e^{i\lambda x'} dx'. \tag{7.67}$$

The reflectionless potentials $\phi(x', t')$, for which $b(\lambda) \equiv 0$, are soliton solutions of the unperturbed nonlinear Schrödinger equation. The refectionless scattering data with the single zero $\lambda_1 = \xi' + i\eta$ of the function $a(\lambda)$ corresponds to the case of one soliton (4.56) or (7.53).

When a perturbation is present as in (7.55), in the lowest order approximation, Kivshar *et al.* obtained the following from the evolution equations (7.56) and (7.57)

$$\begin{aligned} \frac{\partial a(\lambda, t')}{\partial t'} &= \epsilon \int_{-\infty}^{+\infty} dx' P[\phi]\psi_1(x', \lambda)\varphi_2(x', \lambda) \\ &+ \epsilon^* \int_{-\infty}^{+\infty} dx' P^*[\phi]\psi_2(x', \lambda)\varphi_1(x', \lambda), \end{aligned} \tag{7.68}$$

$$\frac{\partial b(\lambda,t')}{\partial t'} = 4i\lambda^2 b + \epsilon \int_{-\infty}^{+\infty} dx' P[\phi]\psi_1(x',\lambda)\varphi_2^*(x',\lambda) - \epsilon^* \int_{-\infty}^{+\infty} dx' P^*[\phi]\psi_2(x',\lambda)\varphi_1^*(x',\lambda), \tag{7.69}$$

$$\frac{d\lambda_n}{dt'} = -\frac{1}{a'(\lambda_n)}\int_{-\infty}^{+\infty} dx' \left\{\epsilon P[\phi]\psi_1(x',\lambda_n)\varphi_2(x',\lambda_n) + \epsilon^* P^*[\phi]\psi_2(x',\lambda_n)\varphi_1(x',\lambda_n)\right\}, \tag{7.70}$$

$$\frac{db_n}{dt'} = 4i\lambda_n^2 b_n + \frac{b_n}{a'(\lambda_n)}\int_{-\infty}^{+\infty} dx' \left\{\varepsilon P[\phi]Q_1(x',\lambda_n) + \varepsilon^* P^*[\phi]Q_2(x',\lambda_n)\right\}. \tag{7.71}$$

Here

$$Q_i(x',\lambda_n) \equiv \frac{\partial}{\partial\lambda}\left[\varphi_i(x',\lambda_n)\psi_i(x',\lambda) - \varphi_i(x',\lambda)\psi_i(x',\lambda_n)\right]_{\lambda=\lambda_n}, \quad i = 1,2, \tag{7.72}$$

and $a'(\lambda_n)$ has been defined in (7.66). The general evolution equations for the parameters for the one soliton solution of the nonlinear Schrödinger equation (4.56) are

$$\frac{d\eta}{dt'} = -\frac{1}{3}\Re\left\{\epsilon\int_{-\infty}^{+\infty} d\tilde{y} P[\phi_s(\tilde{y})] e^{2i\xi' x' + i\theta}\,\mathrm{sech}\tilde{y}\right\}, \tag{7.73}$$

$$\frac{d\xi'}{dt'} = \frac{1}{2}\Im\left\{\epsilon\int_{-\infty}^{+\infty} d\tilde{y} P[\phi_s(\tilde{y})]\frac{\tanh\tilde{y}}{\cosh\tilde{y}} e^{2i\xi' x' + i\theta}\right\}, \tag{7.74}$$

$$\frac{d\zeta'}{dt'} = -4\xi' - \frac{1}{4\eta^2}\Re\left\{\epsilon\int_{-\infty}^{+\infty} d\tilde{y}\frac{\tilde{y}P[\phi_s(\tilde{y})]}{\cosh\tilde{y}} e^{2i\xi' x' - i\theta}\right\}, \tag{7.75}$$

$$\frac{d\theta}{dt} = 4(\xi'^2 - \eta^2) + \frac{1}{2\eta}\Im\left\{\epsilon\int_{-\infty}^{+\infty} d\tilde{y}\frac{1 - 2\eta x'\tanh\tilde{y}}{\cosh\tilde{y}} P[\phi_s(\tilde{y})] e^{2i\xi' x' - i\theta}\right\}, \tag{7.76}$$

where $\zeta' = -4\xi' t + x_0'$, $\theta = 4(\xi'^2 - \eta^2)t + \theta_0$. Applying these formulae, we can obtain the time-evolutions of the scattering data and the corresponding solutions of the equation (7.55) only if the perturbation $P[\phi]$ is known. For the dissipative perturbations,

$$\varepsilon P[\phi] = i\alpha_1\phi, \quad \varepsilon P[\phi] = i\alpha_2\phi_{x'x'} - i\alpha_3|\phi|^2\phi. \tag{7.77}$$

Ott *et al.*, Karpman *et al.*, and Nozaki *et al.* obtained the evolution equations for the amplitude η and velocity $v = -4\xi'$ of the soliton

$$\frac{d\eta}{dt} = -2\eta\left(\alpha_1 + \frac{4}{3}\alpha_1\eta^2 + \frac{8}{3}\alpha_3\eta^2\right) - \frac{1}{2}\alpha_2\eta^2 v^2,$$
$$\frac{dv}{dt} = -\frac{16}{3}\alpha_2\eta^2 v. \qquad (7.78)$$

In contrast to the unperturbed case, in which the soliton's amplitude η and velocity $-4\xi'$ may take arbitrary values, the stationary values in the presence of this perturbation are uniquely determined from $\eta^2 = 3\alpha_1/4(\alpha_2+2\alpha_3)$ and $\xi' = 0$. However, it is clear that the corresponding soliton is unstable, since as $|x'| \to \infty$, $\phi(x', t')$ has the same asymptotic behavior as the unstable trivial solution $\phi = 0$. When a metastable trivial solution is unstable against the following finite perturbations,

$$P(\phi) = -i\alpha_1\phi + i\alpha_2\phi_{x'x'} + i\alpha_3|\phi|^2 - i\alpha_4|\phi|^4\phi, \qquad (7.79)$$

where all the coefficients α_1 to α_4 are positive. Petviashvili and Sergeev proved that a hard excitation does occur under the condition $\alpha_3 \geq 2\sqrt{\alpha_1\alpha_4}$, the solution (4.53) or (7.53) may exist with zero velocity. The last term in (7.79) is necessary to provide global stability. The evolution equation for its amplitude can be obtained using the first order perturbation theory as

$$\frac{d\eta}{dt'} = 2\eta\left(-\alpha_1 - \frac{4}{3}\alpha_2\eta^2 + \frac{8}{3}\alpha_3\eta^2 - \frac{128}{15}\alpha_4\eta^4\right).$$

From this equation, Malomed showed that the stationary amplitudes are given by

$$\eta^2 = \frac{5}{64\alpha_2}\left\{(2\alpha_3 - \alpha_2) \pm \left[(2\alpha_3 - \alpha_2)^2 - \frac{96}{5}\alpha_1\alpha_4\right]\right\}^{1/2}$$

where the upper and lower signs correspond, respectively, to stable and unstable solitons. The condition, $\alpha_3 \geq \alpha_2/2 + \sqrt{(6/5)\alpha_1\alpha_4}$, for the existence of the soliton is more restrictive than the above condition for hard excitation. The coefficient α_2 in (7.77) may be negative. In this case, $\alpha_1 = 0$ is required. Thus it is necessary to supplement the perturbation (7.77) by an additional stabilizing term $-i\alpha_4\phi_{x'x'x'x'}(\alpha_4 > 0)$. Pismen demonstrated that in this case the perturbation may support a stationary soliton solution with a nonzero velocity. However, if α_j is changed, then bifurcation between the quiescent and steadily moving solitons occurs.

For the following nonlinear Schrödinger equation with a dissipative perturbation

$$i\phi'_t + \phi_{x'x'} + 2|\phi|^2\phi = \epsilon\phi\int_{-\infty}^{+\infty}\frac{|\phi(\tilde{x})|^2}{x' - \tilde{x}}d\tilde{x}, \qquad (7.80)$$

where ϵ is a small real parameter that describes the nonlinear Landau damping of the microscopic particle. Ichikawa derived the following evolution equation for the soliton velocity and amplitude using the adiabatic approximation in such a case

$$\frac{dv}{dt'} = q\epsilon\eta^2, \quad \frac{d\eta}{dt'} = 0,$$

where

$$q \equiv \frac{192}{\pi^3}\zeta(3) \approx 7.44.$$

Thus the nonlinear Landau damping acts upon a microscopic particle (soliton) as a constant force. Similar results were obtained by Kodama and Hasegawa for the equation

$$i\phi_{t'} + \phi_{x'x'} + 2|\phi|^2\phi = -\frac{1}{2}\epsilon|\phi_{x'}|^2\phi. \tag{7.81}$$

This equation resembles that for the evolution of the envelope of an electromagnetic wave in a nonlinear one-mode waveguide with regard to nonlinear dissipation due to induced Raman scattering. The corresponding evolution equations for the soliton's parameters are

$$\frac{dv}{dt'} = \frac{128}{15}\epsilon\eta^4, \quad \frac{d\eta}{dt'} = 0. \tag{7.82}$$

In the presence of both dissipation and an external periodic force, $P(\phi) = -i\alpha\phi + \varepsilon\exp(i\Omega t')$, where Ω is the frequency of the external force, a soliton may be able to survive. Kaup *et al.* obtained a perturbed solution for a quiescent soliton in the form of

$$\phi(x', t') = 2i\eta(t')\frac{\exp[i\Omega t' - i\theta'(t')]}{\cosh[2\eta(t')x']}, \tag{7.83}$$

where $\theta'(t')$ is a real phase. Equations (7.73) and (7.75) reduce to the autonomous dynamical system

$$\frac{d\eta}{dt'} = -2\alpha\eta + \frac{1}{2}\pi\epsilon\sin\theta, \quad \frac{d\theta'}{dt'} = \Omega - 4\eta^2, \tag{7.84}$$

in such a case. They possess two stable stationary points which are actually equivalent to each other

$$\eta_n = \frac{1}{2}(-1)^n\sqrt{\Omega}, \quad \theta'_n = n\pi + \theta'_0, \quad (n = 0, 1), \tag{7.85}$$

where $\sin\theta'_0 = 2\alpha\sqrt{\Omega}/\pi\epsilon$. It follows from the expression for θ'_0 that a stable soliton, with its frequency-locked to the external ac force, exists provided that the amplitude of the force exceeds the threshold value $f_{\text{thr}} = 2\alpha\sqrt{\Omega}/\pi$.

Kivshar *et al.* investigated the effect of the perturbation

$$\epsilon P(\phi) = \epsilon\phi^* e^{i\omega_0 t'} - i\alpha\phi. \tag{7.86}$$

In this case, the soliton is taken to be the same form as (7.83), and the equations analogous to (7.84) are

$$\frac{d\eta}{dt'} = -2\alpha\eta - 2\epsilon\eta\sin(2\theta'), \quad \frac{d\theta'}{dt'} = \omega_0 - 4\eta^2 - \epsilon\eta\cos(2\theta'). \tag{7.87}$$

They have the following stable stationary points

$$\eta_n = \frac{1}{2}(-1)^n\left(\omega_0 + \sqrt{\epsilon^2-\alpha^2}\right)^{1/2}, \quad \theta'_n = n\pi + \theta'_0 \tag{7.88}$$

where $n = 0,\ 1$, and $\sin(2\theta'_0) = -\alpha/\epsilon$. Clearly, a stable soliton exists for $\alpha < \epsilon < \sqrt{\omega_0^2+\alpha^2}$. In addition to the stationary points (7.88), equations in (7.87) possesses the trivial stationary point $\eta = 0,\ \theta' =$ constant, which is also stable under the condition $\epsilon < \sqrt{\omega_0^2+\alpha^2}$. This means that there is a separatrix on the phase plane (η, θ'), which is a boundary between attraction basins of the two types of stationary points.

Karpman and Maslov demonstrated how the perturbative approach can be applied to the following nonlinear Schrödinger equation with slowly varying coefficients

$$i\tilde{\phi}_{t'} + \alpha(\epsilon t')|\tilde{\phi}|^2\tilde{\phi} + \beta(\epsilon t')\tilde{\phi}_{x'x'} = 0. \tag{7.89}$$

The substitution

$$\tau = \int_0^\tau \beta(\epsilon t')dt', \quad \phi(x',\tau) = \sqrt{\frac{\sigma}{2}}\tilde{\phi}(x',t'), \quad (\sigma \equiv \alpha\beta)$$

transforms the above equation into the standard form (7.55) with the perturbation

$$P(\phi) = \frac{i\sigma_\tau}{2\sigma}\phi.$$

Grimshaw investigated the one-soliton dynamics governed by the same equation (7.89) with slowly varying coefficients by using a multiscale expansion technique.

7.5 Direct Perturbation Theory in Nonlinear Quantum Mechanics

There are many ways to solve the perturbed dynamic-equations in nonlinear quantum mechanics besides the inverse scattering transformation discussed above. In this section we will introduce a direct perturbation theory which is applicable to both perturbed integrable and nonintegrable systems.

7.5.1 *Method of Gorshkov and Ostrovsky*

We now consider a system described by the following perturbed nonlinear Klein-Gordon equation in natural unit system

$$\phi_{tt} - \phi_{xx} + F(\phi) = \varepsilon R(\phi). \tag{7.90}$$

Clearly, at $\varepsilon = 0$ and $F(\phi) = -\phi + \phi^3$, it is a nonintegrable ϕ^4-field equation which has kink ($\sigma = 1$) and antikink ($\sigma = -1$) solutions. We now solve the perturbed equations (7.90) by an asymptotic method based on a small-parameter expansion proposed by Gorshkov and Ostrovsky (see book by Abdullaev). This method yields analytic forms of evolution of single solitons under weak perturbations.

Consider a set of general differential field-equations in first order in time and space,

$$M(\phi, \phi_\tau, \nabla\phi, X) = 0, \tag{7.91}$$

where ϕ is a vector function, $\phi = \{\phi_1, \cdots, \phi_N\}$, $\tau = \varepsilon t$ and $X = \varepsilon x$ ($\varepsilon \ll 1$) are time and coordinates, respectively. We now consider the evolution of a particular steady-state solution of (7.91) which for $\varepsilon = 0$ has the form.

$$\phi = \phi^{(0)}(\tilde{\xi}, A), \tag{7.92}$$

where $A = \{A_1, \cdots, A_m\}$ are arbitrary constants, $\tilde{\xi} = x - vt$ for (7.90). We expect that the solutions have the following forms for large $|\tilde{\xi}|$

$$\phi^{(0)}(\tilde{\xi}, A) = \phi^{(\pm)(0)}, \quad \tilde{\xi} \to \pm\infty. \tag{7.93}$$

Following the procedure of the small-parameter method, we seek the solution of the following form

$$\phi(x,t) = \phi^{(0)}(\tilde{\xi}, A, X, \tau) + \sum_{n=1}^{N} \varepsilon^n \phi^{(n)}(\tilde{\xi}, X, \tau), \tag{7.94}$$

where n is the order of expansion. Note that A is a slowly varying function of X and τ. Inserting (7.94) into (7.91) and equating the terms of the same order in ε yield a set of linear equations in $\phi^{(n)}$,

$$\hat{K}\phi^{(n)} = G^{(n)}, \tag{7.95}$$

where

$$\hat{K} = \frac{\partial M^{(0)}}{\partial \phi_{\tilde{\xi}}} \frac{d}{d\tilde{\xi}} + \frac{\partial M^{(0)}}{\partial \phi}, \quad G^{(1)} = -\frac{\partial M}{\partial \varepsilon} - \frac{\partial M}{\partial \phi_\tau}\phi_\tau^{(0)} - \frac{\partial M}{\partial \nabla\phi}\nabla_X \phi_X^{(0)},$$

and

$$M^{(0)} = M(\varepsilon = 0, \phi = \phi^{(0)}).$$

We can infer from (7.95) that its solution, $\phi^{(n)}$, is of the form

$$\phi^{(n)} = W\left(D^{(n)} + \int_0^{\tilde{\xi}} d\tilde{\xi}' W^+ G^{(n)}\right), \tag{7.96}$$

where $D^{(n)}$ = constant, W is a matrix corresponding to the solution of the equation $\hat{K}W = 0$, and

$$W^{+} = W^{-1}\left[\frac{\partial M^{(0)}}{\partial \phi_{\tilde{\xi}}}\right]^{-1}. \tag{7.97}$$

The W_is are found from the known $\phi^{(0)}$ by varying it with respect to $\tilde{\xi}$ and A, *i.e.*,

$$W_1 = \phi^{(0)}_{\tilde{\xi}}, \quad W_i = \phi^{(0)}_{A_i}, \quad i = 1, 2, 3, \cdots, m+1. \tag{7.98}$$

For the expansion (7.94) to hold we have to impose the condition of boundedness in terms of this series that are derived from (7.96). Among the terms in (7.96), one can see an exponentially growing W_i. Thus the finiteness of the solutions $\phi^{(n)}$ can be achieved by two ways.

(1) For an odd function $W^{+}G^{(n)}$, the solution $\phi^{(n)}$ is finite if the constants $D^{(n)}$s satisfy the relation,

$$D^{(n)}_x = \int_0^\infty d\tilde{\xi} W_i^{+} G^{(n)}. \tag{7.99}$$

(2) For an even function $W^{+}G^{(n)}$, the condition

$$\int_0^\infty d\tilde{\xi} W_i^{+} G^{(n)} = 0$$

must be satisfied. A secular growth of $\phi^{(n)}$, due to finite solutions of (7.98) for $\tilde{\xi} \to \pm\infty$, is ruled out by the relations

$$\lim_{\tilde{\xi}\to\pm\infty} W_i^{+} G^{(n)} = 0, \quad i = l+1, l+2, \cdots, m+1.$$

The above is the direct perturbation theory of nonlinear equation (7.91).

We now consider the solutions of the perturbed nonlinear Klein-Gordon equation (7.90) following the same procedure. In (7.90), $R(\phi)$ is a nonlinear operator, and $R(\phi^{(0)}) \to 0$ when $\tilde{\xi} \to \pm\infty$. The soliton solution of (7.90) for $\varepsilon = 0$ is of the form

$$\phi^{(0)} = \phi\left(\frac{\tilde{\xi}}{\sqrt{1-v^2}}\right), \quad \tilde{\xi} = x - vt.$$

By means of the expansion (7.94), the linear operator $\hat{K}$ in (7.95) is found to be

$$\hat{K} = (v^2 - 1)\frac{d^2}{d\tilde{\xi}^2} - F'(\phi^{(0)}).$$

There exists a solution, $\Phi_1 = \phi^{(0)}(\tilde{\xi})$, of $\hat{K}\Phi = 0$, that decreases for $\tilde{\xi} \to \pm\infty$.

We deduce the solution for each order of the approximation

$$\phi^{(n)} = \Phi_1\left(D_1^{(n)} + \int_0^{\tilde{\xi}} d\tilde{\xi}'\Phi_2 G^{(n)}\right) + \Phi_2\left(D_2^{(n)} + \int_0^{\tilde{\xi}} d\tilde{\xi}'\Phi_2 G^{(n)}\right),$$

where

$$\Phi_2 = \phi_{\tilde{\xi}}^{(0)} \int_0^{\tilde{\xi}} d\tilde{\xi}' \left[\phi_{\tilde{\xi}'}^{(0)}\right]^{-2}.$$

For $\phi^{(n)}$ to be finite, it is necessary and sufficient that the orthogonality condition

$$\int_{-\infty}^{\infty} d\tilde{\xi} \phi_{\tilde{\xi}}^{(0)} G^{(n)} = 0 \tag{7.100}$$

be satisfied. Thus, we have

$$G^{(1)} = \phi_{\tilde{\xi}}^{(0)} \frac{dv}{dt} + 2v\phi_{\tilde{\xi}\tau}^{(0)} + \varepsilon R(\phi^{(0)})$$

for $n = 1$. Substitution of this into (7.100) results in the following equation for a soliton with velocity v

$$\frac{d}{dt}\left(\frac{v}{\sqrt{(1-v^2)}}\right) = -\frac{\varepsilon}{\langle \phi_{y'}^2 \rangle} \int dy' \phi_{y'} R(\phi^{(0)}),$$

where

$$y' = \frac{\tilde{\xi}}{\sqrt{(1-v^2)}}, \quad \langle \phi_{y'}^2 \rangle = \int_{-\infty}^{\infty} dy' \phi_{y'}^2.$$

In a similar way one can derive equations describing soliton evolution in the following two-component perturbed nonlinear Schrödinger equation system

$$i\phi_{t'} + \phi_{x'x'} - u\phi = \varepsilon R(\phi), \tag{7.101}$$

$$u_{t't'} - u_{x'x'} = (|\phi|^2)_{x'x'}. \tag{7.102}$$

If we assume that $\tilde{\xi} = x' - vt'$, $u = u(\tilde{\xi})$, and $\phi(x', t') = \varphi(\tilde{\xi}) \exp(i\alpha x' + i\beta t')$, then from (7.102), we get

$$u_0 = \tilde{u} - \frac{\varphi_0^2}{1-v^2},$$

where $\tilde{u}$ is an integral constant. In the presence of the perturbation, we let $\tilde{u} = \tilde{u}(x', t')$. According to the direct perturbation theory discussed above, solution of (7.101) and (7.102) can be written in the form of a series

$$\phi(x,t) = \left[\varphi_0(\tilde{\xi}, X, \tau) + \sum_{n=1} \varepsilon^n \varphi^{(n)}(\tilde{\xi}, X, \tau)\right] e^{i(v\tilde{\xi}/2+\theta)},$$

$$u = u_0(\tilde{\xi}, X, \tau) + \sum_{n=1} \varepsilon^n u^{(n)}(\tilde{\xi}, X, \tau), \tag{7.103}$$

where φ_0, u_0 are soliton solutions in the unperturbed case, $v = v(t)$ are slowly varying function of space and time, $X = \varepsilon x'$, $\tau = \varepsilon t'$, and ε is a small parameter which is proportional to the ratio of the soliton scale to the given low-frequency wave scale. Inserting (7.103) into (7.97), (7.95), (7.101) and (7.102), and equating coefficients with the same powers of ε, we obtain a set of linear equations

$$\hat{K}_1\Re\varphi^{(n)} = \left[\frac{d^2}{d\tilde{\xi}^2} - (\lambda'^2 + 3u_0)\right]\Re\varphi^{(n)} = \Re G^{(n)},$$
$$\hat{K}_2\Im\varphi^{(n)} = \left[\frac{d^2}{d\tilde{\xi}^2} - (\lambda'^2 + u_0)\right]\Im\varphi^{(n)} = \Im G^{(n)}.$$

In accordance with the method of Gorshkov and Ostrovsky, it is necessary for $G^{(n)}$ to be orthogonal to the eigenfunctions of $\hat{K}_1$ and $\hat{K}_2$ which decrease to zero as $\tilde{xi} \to \pm\infty$ in order to stop the growth of the corrections, *i.e.*,

$$\langle\varphi_{0\tilde{\xi}}\Re G^{(n)}\rangle = 0, \quad \langle\varphi_{0\tilde{\xi}}\Im G^{(n)}\rangle = 0,$$

where

$$\langle\cdots\rangle \equiv \int^{t'}(\cdots)d\tilde{\xi}.$$

From this, we obtain the following for the soliton parameters,

$$\frac{d}{dt'}\lambda'(1-v^2) = 0, \quad \lambda'(1-v^2)\left(\frac{dv}{dt'} + 2\frac{\partial\tilde{u}}{\partial x'}\right) + \frac{8}{3}\frac{d}{dt'}(v\lambda'^3) = 0.$$

Thus, this system may be rewritten as a second-order differential equation

$$\frac{d^2x'}{dt'^2} = -2(\tilde{u}_{x'})\left[1 + \frac{8}{3}m^2\frac{1+5\dot{x}'^2}{(1-\dot{x}'^2)^4}\right]^{-1},$$

where $\dot{x}' = dx'/dt'$, $m =$ constant is the "quanta number".

7.5.2 *Perturbation technique of Bishop*

For small perturbations, Bishop developed other direct perturbation techniques. For example, for the following perturbed Sine-Gordon equation in natural unit system,

$$\phi_{tt} - \phi_{xx} + \sin\phi = \varepsilon R(\phi), \tag{7.104}$$

where ε is a small positive parameter, $\varepsilon R(\phi)$ is a perturbation action. Bishop transformed this equation into a set of two first-order differential equations with respect to the time-derivative

$$\psi_t - \phi_{xx} + \sin\phi = \varepsilon R(\phi),$$
$$\psi = \phi_t. \tag{7.105}$$

Equations (7.105) may be obtained from the variation principle

$$\frac{\delta L}{\delta\phi} = \frac{\delta L}{\delta\psi} = 0,$$

from

$$L(\phi,\psi) = L_0(\phi,\psi) - \varepsilon\phi R(\phi), \tag{7.106}$$

where

$$L_0(\phi,\psi) = \frac{1}{2}\phi_x^2 + 1 - \cos\phi - \phi_t\psi + \frac{1}{2}\psi^2,$$

ϕ and ψ are considered to be independent.

Introducing the matrices

$$\Psi = \begin{pmatrix}\phi\\ \phi_t\end{pmatrix}, \quad R' = \begin{pmatrix}0\\ R\end{pmatrix}, \quad M = \begin{pmatrix}0 & 1\\ -\partial_{xx} & \sin\phi\end{pmatrix},$$

equation (7.105) can be written as

$$\frac{\partial\Psi}{\partial t} + M\Psi = \varepsilon R'(\Psi). \tag{7.107}$$

Obviously, solution of the corresponding unperturbed equation, (7.107) with $\varepsilon = 0$, is given by

$$\Psi_0 = \begin{pmatrix}\phi_0\\ \phi_{0,t}\end{pmatrix}, \tag{7.108}$$

where

$$\begin{aligned}\phi_0(x,t) &= 4\tan^{-1}\left\{\exp\left[\frac{\eta(x-x_0-vt)}{\sqrt{1-v^2}}\right]\right\},\\ \phi_{0,t}(x,t) &= -2\eta v\ \mathrm{sech}\left[\frac{x-x_0-vt}{\sqrt{1-v^2}}\right].\end{aligned} \tag{7.109}$$

When perturbed, the solution of (7.107) can be expressed as

$$\Psi = \Psi_0 + \varepsilon\Psi_1, \quad \Psi_1 = \begin{pmatrix}\phi_1\\ \phi_{1,t}\end{pmatrix}. \tag{7.110}$$

Inserting (7.110) into (7.107) and retaining only the first-order correction terms for ε and in time-dependence of the parameters v and $x_{0'}$, we obtain

$$\left[\begin{pmatrix}1\,0\\ 0\,1\end{pmatrix}\frac{\partial}{\partial t} + M\right]\Psi_1 = Q(\phi_0,\phi_{0,t}), \tag{7.111}$$

where

$$Q(\phi_0,\phi_{0,t}) \equiv R'(\Psi) - \frac{1}{\varepsilon}\sum_{g_i}\frac{\partial\Psi_0}{\partial g_i}\frac{dg_i}{d\tau}, \quad (i=1,2) \tag{7.112}$$

is the effective perturbation and $g_1 = v$, $g_2 = x_0$. In order for Ψ_1 not to contain secular terms which increase linearly with time, we have to require that the effective perturbation (7.112) to be orthogonal to the functions $I\partial\Psi_0/\partial g_i$, where

$$I = \begin{pmatrix} 0 & -1 \\ 1 & 0 \end{pmatrix}, \quad g_i = v, x_0.$$

From the orthogonality condition we obtain a set of two equations

$$\varepsilon\left[I\frac{\partial\Psi_0}{\partial g_i}, R'(\Psi_0)\right] = \sum_{g_k}\left(I\frac{\partial\Psi_0}{\partial g_i}, \frac{\partial\Psi_0}{\partial g_k}\right)\frac{dg_k}{dt}, \quad (i = 1, 2).$$

Because the energy corresponding to the unperturbed Sine-Gordon equation (7.105) with $\varepsilon = 0$ is

$$E = \frac{1}{2}\int_{-\infty}^{\infty}\left[\phi_{0t}^2 + \phi_{0x}^2 + 2(1 - \cos\phi_0)\right]dx = \frac{8}{\sqrt{1 - v^2}} = \text{constant},$$

where ϕ_0 is given in (7.108), we assume that the small perturbation leads only to a change of the velocity v with time and that the general form of the excitation (7.108) is conserved. Then to first order approximation, the following relation holds

$$\frac{dE}{dt} = \int \varepsilon R(\phi)\phi_t dx.$$

Using (7.108) in the above equation, the direct forms of the equations for dv/dt and dx_0/dt were obtained by Bishop and are given here,

$$\begin{aligned}\frac{dv}{dt} &= \frac{1}{4}\varepsilon\eta(1 - v^2)\int R(\phi_0)\,\mathrm{sech}[\Omega(x,t)]dx,\\ \frac{dx_0}{dt} &= -\frac{1}{4}\varepsilon\eta(1 - v^2)\int R(\phi_0)\Omega(x,t)\,\mathrm{sech}[\Omega(x,t)]dx,\end{aligned} \qquad (7.113)$$

where

$$\Omega(x,t) \equiv \eta\sqrt{1 - v^2(t)}[x - x_0(t) - v(t)t].$$

If $R(\phi_0)$ is not related to ϕ_0, but is only related to coordinate x, *i.e.*, $R(\phi_0) = R(x)$, then (7.113) becomes

$$\frac{dv}{dt} = \frac{1}{4}\varepsilon\eta(1 - v^2)\int R(x)\,\mathrm{sech}[\Omega(x,t)]dx.$$

7.6 Linear Perturbation Theory in Nonlinear Quantum Mechanics

Linear perturbation theory allows one to study the influence of small external perturbations on behaviors of microscopic particles described by the nonlinear Schrödinger equation and the Sine-Gordon equation. Many scientists such as Pang, Fogel, Trullinger, Bishop, and Krumhansl, contributed to the development of the linear perturbation theory. The theory is based on an expansion of the soliton wave function in terms of the complete set of eigenfunctions of the self-conjugate differential operator. Such an operator has, among its eigenfunctions, a single function representing the stationary wave propagating with the same velocity as the free microscopic particle (soliton). In this section, we describe the linear perturbation technique for the nonlinear Schrödinger equation (3.2) with $A(\phi) = 0$, as put forward by Pang and by Fogel *et al.*, respectively.

7.6.1 *Nonlinear Schrödinger equation*

We introduce a small quantity ε, to denote the small external perturbation potential $V(x)$. In terms of this, equation (3.2) can be written as

$$i\frac{\partial\phi}{\partial t'} + \frac{\partial^2\phi}{\partial x'^2} + b|\phi^2|\phi = \epsilon V(x')\phi, \tag{7.114}$$

where $t' = t/\hbar$, $x' = \sqrt{2m/\hbar^2 x}$. Pang assumed the following form for the solution of (7.114)

$$\phi = \phi_0 + \phi' = (f + \varepsilon F)e^{i\theta(x',t')}. \tag{7.115}$$

Here $\phi_0 = f(x',t')e^{i\theta(x',t')}$ is an unperturbed solution of (7.114) with $\varepsilon = 0$, and it is the same as (4.8). Inserting (7.115) into (7.114) and neglecting terms higher than the second order in ε, we obtain the following equation for F.

$$\begin{aligned} &i\frac{\partial F}{\partial t'} + \frac{\partial^2 F}{\partial X'^2} - v^2[1 - 4\,\text{sech}^2(vX')]F + 2v^2\,\text{sech}^2(vX')F^* \\ &\quad = v\sqrt{\frac{2}{b}}\,\text{sech}(vX')V(X' + v_e t'), \end{aligned} \tag{7.116}$$

where

$$X' = x' - v_e t', \quad v^2 = \frac{1}{4}(v_e^2 - 2v_e v_e) = \left(\frac{v_e}{2}\right)^2\delta, \quad \delta = 1 - \frac{2v_c}{v_e}. \tag{7.117}$$

Performing the transformations $vX' = y$ and $v^2 t' = \tau$, equation (7.116) becomes

$$\begin{aligned} &i\frac{\partial}{\partial\tau}F(y,\tau) + \frac{\partial^2}{\partial y^2}F(y,\tau) - (1 - 4\,\text{sech}^2 y)F(y,\tau) + 2\,\text{sech}^2 yF^*(y,\tau) \\ &\quad = \frac{1}{v}\sqrt{\frac{2}{b}}\,\text{sech}yV\left(\frac{y}{v} + \frac{v_e\tau}{v^2}\right). \end{aligned} \tag{7.118}$$

Now let

$$F(y,\tau) = F_1(y,\tau) + iF_2(y,\tau). \tag{7.119}$$

Substituting (7.119) into (7.118), we get

$$\frac{\partial F_1}{\partial\tau} + \frac{\partial^2 F_2}{\partial y^2} - (1 - 2\,\text{sech}^2 y)F_2 = 0, \tag{7.120}$$

$$-\frac{\partial F_2}{\partial\tau} + \frac{\partial^2 F_1}{\partial y^2} - (1 - 6\,\text{sech}^2 y)F_1 = \frac{1}{v}\sqrt{\frac{2}{b}}\,\text{sech}yV(y,\tau). \tag{7.121}$$

Differentiating (7.120) and (7.121) with respect to τ results in the following equations for F_1 and F_2.

$$\frac{\partial^2 F_2}{\partial \tau^2} + \hat{M}_2 F_2 = A_2(y,\tau),$$
$$\frac{\partial^2 F_1}{\partial \tau^2} + \hat{M}_1 F_1 = A_1(y,\tau),$$

where

$$A_2 = -\frac{v^2}{v_e}\sqrt{\frac{2}{b}}\,\text{sech}y\frac{\partial V}{\partial y},$$
$$A_1 = -\frac{1}{v}\sqrt{\frac{2}{b}}\left(2\,\text{sech}y\tan y\frac{\partial V}{\partial y} - \text{sech}y\frac{\partial^2 V}{\partial y^2}\right),$$
$$\hat{M}_1 = \frac{d^4}{dy^4} - 2(1-4\,\text{sech}^2 y)\frac{d^2}{dy^2} - 24\,\text{sech}^2 y\tanh y\frac{d}{dy} + (1+16\,\text{sech}^2 y - 24\,\text{sech}^4 y),$$
$$\hat{M}_2 = \frac{d^4}{dy^4} - 2(1-4\,\text{sech}^2 y)\frac{d^2}{dy^2} - 8\,\text{sech}^2 y\tanh y\frac{d}{dy} + 1.$$

Clearly, A_2 possesses the property of a force. The operators $\hat{M}_1$ and $\hat{M}_2$ satisfy the following eigenequations,

$$\hat{M}_2 g_2(vX') = \left(\frac{\omega}{v^2}\right)^2 g_2(vX'), \tag{7.122}$$
$$\hat{M}_1 g_1(vX') = \left(\frac{\omega}{v^2}\right)^2 g_1(vX'), \tag{7.123}$$

where $y = vX'$. We can show easily that $\hat{M}_1$ and $\hat{M}_2$ are not Hermitian operators, but are Hermite conjugate to each other, *i.e.*, $\hat{M}_1^+ = \hat{M}_2$, and $\hat{M}_1^+ = \hat{M}_1$. Their eigenfunctions are also, orthogonal to each other, *i.e.*,

$$\int g_2^*(X',\omega)g_1(X',\omega')dX' = \delta(\omega-\omega').$$

The eigenfunctions belong to the eigenvalues $\omega = \pm(vk^2+v^2)$ and $\omega = 0$ in (7.122) and (7.123) can be obtained easily. k can take any value, $-\infty < \kappa < \infty$, when $\omega = 0$. The eigenfunctions are

$$g_1(x') = c_1\,\text{sech}(vX')\tanh(vX'),$$
$$g_2(x') = c_2\,\text{sech}(vX'), \tag{7.124}$$

where

$$c_1 = c_2 = \sqrt{2v}.$$

Pang showed that (7.124) are only local solutions of (7.122) and (7.123) when $\omega = 0$.

We now seek for a power-series solution of $\hat{M}_2$. We consider first the asymptotic behavior of $\hat{M}_2$ when $y \to \infty$, *i.e.*,

$$\hat{M}_2 \underset{y\to\infty}{\longrightarrow} \frac{d^4}{dy^4} - 2\frac{d^2}{dy^2} + 1.$$

The characteristic equation of $\hat{M}_2$ is

$$-\frac{\omega^4}{\upsilon^4} + k^4 + 2k^2 + 1 = 0,$$

or

$$\omega^2 = (\upsilon^2 k^2 + \upsilon^2)^2.$$

Assume the following trial solution

$$g_2(y,k) = h_a \cos(ky) + h_b \sin(ky). \tag{7.125}$$

Substituting (7.125) into (7.123), and using the fact that $\cos(ky)$ and $\sin(ky)$ are linearly independent, we can get

$$\begin{aligned}
&\frac{d^4 h_a}{dy^4} - \left[6k^2 + 2(1 - 4\,\mathrm{sech}^2 y)\right]\frac{d^2 h_a}{dy^2} - 8\,\mathrm{sech}^2 y \tanh y \frac{dh_a}{dy} - 8k^2\,\mathrm{sech}^2 y h_a \\
&\qquad +4k\left[\frac{d^3 h_b}{dy^3} - (k^2 + 1 - 4\,\mathrm{sech}^2 y)\frac{dh_b}{dy} - 2\,\mathrm{sech}^2 y \tanh y h_b\right] = 0, \\
&\frac{d^4 h_b}{dy^4} - \left[6k^2 + 2(1 - 4\,\mathrm{sech}^2 y)\right]\frac{d^2 h_b}{dy^2} - 8\,\mathrm{sech}^2 y \tanh y \frac{dh_b}{dy} - 8\kappa^2\,\mathrm{sech}^2 y h_b \\
&\qquad -4k\left[\frac{d^3 h_a}{dy^3} - (k^2 + 1 - 4\,\mathrm{sech}^2 y)\frac{dh_a}{dy} - 2\,\mathrm{sech}^2 y \tanh y h_a\right] = 0.
\end{aligned}$$

A further transformation, $z = \tan y$, was made by Pang. Under this transformation, equation (7.123) becomes

$$\begin{pmatrix} N_1 & N_2 \\ -N_2 & N_1 \end{pmatrix}\begin{pmatrix} h_a \\ h_b \end{pmatrix} = 0, \tag{7.126}$$

where

$$\begin{aligned}
N_1 &= (1-z^2)^3\frac{d^4}{dz^4} - 12z(1-z^2)\frac{d^3}{dz^3} - 2(1-z^2)(3k^2+1-14z^2)\frac{d^2}{dz^2} \\
&\qquad -4z(3k^2-1)\frac{d}{dz} - 8k^2 \\
N_2 &= 4k\left[(1-z^2)^2\frac{d^3}{dz^3} - 6z(1-z^2)\frac{d^2}{dz^2} - (k^2-1-2z^2)\frac{d}{dz} - 2z\right].
\end{aligned}$$

With the following unitary transformation

$$S = \frac{1}{\sqrt{2}}\begin{pmatrix} 1 & i \\ i & 1 \end{pmatrix}, \quad S^{-1} = \frac{1}{\sqrt{2}}\begin{pmatrix} 1 & -i \\ -i & 1 \end{pmatrix},$$

equation (7.126) becomes

$$\begin{pmatrix} N_1 - iN_2 & 0 \\ 0 & N_1 + iN_2 \end{pmatrix} \begin{pmatrix} h_a + ih_b \\ i(h_a - ih_b) \end{pmatrix} = 0. \tag{7.127}$$

Equation (7.127) is equivalent to two equations. Their solutions can be assumed to be of the form

$$h_a = z^p \sum_{n=0}^{\infty} c_n z^n, \quad h_b = z^\gamma \sum_{m=0}^{\infty} b_m z^m. \tag{7.128}$$

Solution of (7.127) can be obtained by substituting (7.128) into it. Since we desire a finite solution in the complete range of interest, the series in (7.128) must be truncated. It can be shown that this truncation is unique. Besides a constant coefficient, it has a group of solutions of the following form in such a case

$$\begin{aligned} h_a(z) &= (k^2 - 1) + 2kz, \\ h_b(z) &= (k^2 - 1) - 2kz. \end{aligned}$$

The solutions g_1 and g_2 can be then be expressed as

$$\begin{aligned} g_1(x') &= c_1 \operatorname{sech}(vX') \tanh(vX'), \\ g_2(x') &= c_2 \operatorname{sech}(vX'), \end{aligned}$$

when $\omega = 0$, and

$$\begin{aligned} g_1(X', k) = \frac{1}{k^2+1}\sqrt{\frac{v}{2\pi}} &\left\{[k^2 - 1 + 2k\tanh(vX') + 2\operatorname{sech}^2(vX')]\cos(kvX')\right. \\ &\left. + [k^2 - 1 + 2k\tanh(vX') + 2\operatorname{sech}^2(vX')]\sin(kvX')\right\}, \\ g_2(X', k)\frac{1}{k^2+1}\sqrt{\frac{v}{2\pi}} &\left\{[k^2 - 1 + 2k\tanh(vX')]\cos(kvX')\right. \\ &\left. + [k^2 - 1 + 2k\tanh(vX')]\sin(kvX')\right\}, \end{aligned}$$

when $\omega = \pm(v^2k^2 + v^2)$, where g_1 and g_2 satisfy the following orthogonality and normalization conditions,

$$\begin{aligned}
&\int_{-\infty}^{\infty} g_1(X')g_2(X',\kappa)dX' = 0,\\
&\int_{-\infty}^{\infty} g_1(X',\kappa)g_2(X',\kappa)dX' = \delta(\kappa-\kappa'),\\
&\int_{-\infty}^{\infty} g_1^2(X')dX' = \frac{2}{3},\\
&\int_{-\infty}^{\infty} g_2(X')g_1(X',\kappa)dX' = 0,\\
&\int_{-\infty}^{\infty} g_1(X',\kappa)g_1(X')dX' = \frac{\pi\kappa}{3\sqrt{2}}\operatorname{sech}\left(\frac{\pi\kappa}{2}\right),\\
&\int_{-\infty}^{\infty} g_2^2(X')dX' = 2,\\
&\int_{-\infty}^{\infty} g_2(X',\kappa)g_2(X')dX' = -\sqrt{\frac{\pi}{2}}\operatorname{sech}\left(\frac{\pi\kappa}{2}\right).
\end{aligned} \tag{7.129}$$

Thus, F_1 and F_2 can be expanded using the eigenfunctions g_1 and g_2, *i.e.*,

$$\begin{aligned}
F_1(y,\tau) &= \varphi_{10}(\tau)g_1(y) + \int_{-\infty}^{\infty} dk\varphi_{1k}(\tau)g_1(y,k),\\
F_2(y,\tau) &= \varphi_{20}(\tau)g_2(y) + \int_{-\infty}^{\infty} dk\varphi_{2k}(\tau)g_2(y,k).
\end{aligned} \tag{7.130}$$

Applying the conditions in (7.129), we get

$$\begin{aligned}
\hbar\frac{d^2\varphi_{1k}(\tau)}{d\tau^2} + (k^2+1)^2\varphi_{1k}(\tau) &= A_1(\tau,k),\\
\hbar\frac{d^2\varphi_{2k}(\tau)}{d\tau^2} + (k^2+1)^2\varphi_{2k}(\tau) &= A_2(\tau,k),
\end{aligned} \tag{7.131}$$

and

$$\begin{aligned}
\frac{d^2\varphi_{10}(\tau)}{d\tau^2} &= \frac{3}{2}A_1(\tau) - \sqrt{\frac{\pi}{8}}\int_{-\infty}^{\infty} dkA_1(\tau,k)k\operatorname{sech}\left(\frac{\pi k}{2}\right) = H_1(\tau),\\
\frac{d^2\varphi_{20}(\tau)}{d\tau^2} &= \frac{1}{2}A_2(\tau) + \sqrt{\frac{\pi}{8}}\int_{-\infty}^{\infty} dkA_2(\tau,k)k\operatorname{sech}\left(\frac{\pi k}{2}\right) = H_2(\tau),
\end{aligned} \tag{7.132}$$

where

$$\begin{aligned}
A_1(\tau) &= \int_{-\infty}^{\infty} A_1(\tau,y)g_1(y)dy, \quad A_2(\tau) = \int_{-\infty}^{\infty} A_2(\tau,y)g_2(y)dy,\\
A_1(\tau,\kappa) &= \int_{-\infty}^{\infty} A_1(\tau,y)g_1(y)dy, \quad A_2(\tau,k) = \int_{-\infty}^{\infty} A_2(\tau,y)g_1(y,k)dy.
\end{aligned}$$

From (7.131) and (7.132), we can see that the amplitudes of the "classical modes" $\varphi_{10}(\tau)$ and $\varphi_{20}(\tau)$ in (7.132) which possess discrete eigenvalues, satisfy Newton-type of equations of motion, and the amplitudes of the "quantum modes", $\varphi_{1k}(\tau)$ and $\varphi_{2k}(\tau)$ in (7.131) which possess continuous eigenvalues, satisfy the wave equations. There is a gap, δ, between the "classical mode" and the "quantum mode", of size $\triangle E = v^2$. When the system is quantized, *i.e.*, $\delta \to 0$, the "quantum mode" is near the "classical mode", and the gap approaches zero. When $\delta = 0$, the "classical mode" disappears and only the "quantum mode" remains. However, when the classical condition is satisfied, *i.e.*, $v_e \to \infty$, the gap approaches infinity. The two "quantum modes" are separated at $\pm\infty$. Now there is only a "classical mode". In this case the system shows pure classical behavior, a Newton-type equation of motion is sufficient. Therefore, for a potential field $V(x')$, the solitons of the nonlinear Schrödinger equation possess not only classical mechanical properties, but also quantum mechanical properties.

Once the solutions in (7.131) and (7.132) are known, we can determine φ_{1k}, φ_{2k}, φ_{10} and φ_{20}. Inserting them into (7.130) gives us F_1 and F_2. From (7.119) and (7.115) we can then obtain ϕ, which clearly has the soliton characters.

7.6.2 *Sine-Gordon equation*

We now consider the following Sine-Gordon equation with a small perturbation $R(x)$ in natural unit system,

$$\phi_{tt} - \phi_{xx} + \sin\phi = R[x]. \tag{7.133}$$

Fogel *et al.* showed that its solutions can be expressed in the following form (see book by Davydoy)

$$\phi(x,\tau) = \phi_0(\pm x) + \phi_1(x,t), \tag{7.134}$$

where $\phi_1(x,\tau)$ is a small change from the unperturbed solution (7.22), $\bar{\xi}' = \gamma(x-vt)$, and $\gamma = (1-v^2)^{-1}$.

Inserting (7.134) into (7.133) and retaining only first-order terms in $\phi_1(x,\tau)$, one obtains

$$\left[\frac{\partial^2}{\partial t^2} - \frac{\partial^2}{\partial x^2} + 1 - 2\,\mathrm{sech}^2\bar{\xi}'\right]\phi_1(\bar{\xi}',t) = R[x]. \tag{7.135}$$

In the coordinate system $(\bar{\xi}',t)$ which moves with velocity v, equation (7.135) becomes

$$\left[\frac{\partial^2}{\partial t^2} - \frac{\partial^2}{\partial \bar{\xi}'^2} + 1 - 2\,\mathrm{sech}^2\bar{\xi}'\right]\phi_1(\bar{\xi}',t) = R[\gamma(\bar{\xi}'+vt)]. \tag{7.136}$$

To solve this inhomogeneous equation, we make use of the eigenfunctions and eigenvalues of the Schrödinger equation

$$\hat{K}' f_k(\bar{\xi}') = \Omega^2(k) f_k(\bar{\xi}'), \tag{7.137}$$

with the self-conjugate operator

$$\hat{K}' = -\frac{d^2}{d\bar{\xi}'^2} + V(\bar{\xi}'), \quad V(\bar{\xi}') = 1 - 2\,\mathrm{sech}^2\bar{\xi}'. \tag{7.138}$$

The eigenvalues of (7.137) contain a zero level $\Omega_0 = 0$ and a continuum spectrum $\Omega^2(k) = 1 + k^2$. The corresponding eigenfunctions are

$$f_0(\bar{\xi}') = 2\,\mathrm{sech}\bar{\xi}', \quad \text{and} \quad f_\kappa(\bar{\xi}') = \frac{1}{\sqrt{2\pi}}[k + i\tanh\bar{\xi}']e^{ik\bar{\xi}'}, \tag{7.139}$$

respectively. From (7.22) and (7.139), we can get

$$\pm f_0(\bar{\xi}') = \frac{d}{d\bar{\xi}'}\phi_0(\pm\bar{\xi}'). \tag{7.140}$$

The small displacement of $\bar{\xi}'$ in the function (7.22) is consistent with result of the first-order perturbation treatment, as shown in (4.39)

$$\phi_0[\pm(\bar{\xi}' + a)] = \phi_0(\pm\bar{\xi}') \pm a f_0(\bar{\xi}').$$

The term $af_0(\bar{\xi}')$ may be considered as an operator for a displacement a of the soliton coordinate $\bar{\xi}'$. Thus the function in (7.137) may be referred to as a translation mode.

The eigenfunctions (7.139) of the self-conjugate operator (7.138) satisfy the conditions of orthogonality

$$\int f_0^2(\bar{\xi}')d\bar{\xi}' = 8,$$

$$\int f_0(\bar{\xi}')f_k(\bar{\xi}')d\bar{\xi}' = 0,$$

$$\int f_k^*(\bar{\xi}')f_{k_1}(\bar{\xi}')d\bar{\xi}' = \delta(\bar{k} - \bar{k}_1),$$

and completeness

$$\frac{1}{8}f_0(\bar{\xi}')f_0(\bar{\xi}_1) + \int f_k(\bar{\xi}')f_k^*(\bar{\xi}_1)d\bar{k} = \delta(\bar{\xi}' - \bar{\xi}_1).$$

Similar to (7.130), we expand the unknown function $\phi_1(\bar{\xi}, t)$ in (7.136) using the complete set of basis functions given in (7.139)

$$\phi_1(\bar{\xi}', t) = \frac{1}{8}\psi(0, t)f_0(\bar{\xi}') + \int \psi(k, t)f_k(\bar{\xi}')dk. \tag{7.141}$$

Here the factor $\psi(0, t)/8$ in front of the translation mode describes the motion of the center-of-mass of the microscopic particle, and the coefficient $\psi(k, t)$ defines a change in its form. Inserting (7.141) into (7.136) and making use of the orthogonality conditions given above, we obtain

$$\frac{d^2\psi(0,t)}{dt^2} = \int R[\gamma(\bar{\xi}' + vt)]f_0(\bar{\xi}')d\bar{\xi}',$$
$$\frac{d^2\psi(k,t)}{dt^2} + \Omega_k^2\psi(k,t) = \int F[\gamma(\bar{\xi}' + vt)]f_k^*(\bar{\xi}')d\bar{\xi}', \qquad (7.142)$$

which determine the first-order correction to the motion of the microscopic particle (soliton) due to the perturbation $R(x)$. Once R$[\gamma(\bar{\xi}' + vt)]$ is known, we can determine $\psi(0,t)$ and $\psi(k,t)$. Thus, $\phi_1(\bar{\xi}', t)$ can also be determined.

7.7 Nonlinearly Variational Method for the Nonlinear Schrödinger Equation

The time-dependent variational principle proposed by Dirac is of great use in nonlinear quantum mechanics. Since the nonlinear Schrödinger equation is a Hamiltonian integrable system, the same variational principle can be applied to the nonlinear Schrödinger equation to obtain time-dependent approximate variational solutions. In this section, we introduce the Dirac variational method and apply it to the nonlinear Schrödinger equation, which were performed by Cooper *et al.*

As mentioned in Chapter 3, the action for the nonlinear Schrödinger equation, in natural unit system, is defined as

$$S = \int dtL, \qquad (7.143)$$

where

$$L = \frac{i}{2}\int d^dx(\phi^*\phi_t - \phi_t^*\phi) - H.$$

The Hamiltonian of the nonlinear Schrödinger equation in d-dimensions with nonlinearity parameter σ is of the form

$$H = \int d^dx\left[\nabla\phi^*\nabla\phi - \frac{b(\phi^*\phi)^{\sigma+1}}{\sigma+1}\right].$$

The nonlinear Schrödinger equation with a given nonlinearity follows from Dirac's variational principle. If S is stationary against variations in ϕ and $\phi^*(x,t)$, *i.e.*,

$$\frac{\delta S}{\delta\phi} = \frac{\delta S}{\delta\phi^*} = 0,$$

we can get

$$i\phi_t + \nabla^2\phi + b(\phi^*\phi)^\sigma\phi = 0,$$
$$i\phi_t^* - \nabla^2\phi^* - b(\phi^*\phi)^\sigma\phi^* = 0. \qquad (7.144)$$

The variational principle is a form of Hamilton's least-action principle. In the variational principle, ϕ is an arbitrary square integrable function subject to

$$\frac{d}{dt}\left[\int d^d x(\phi^*\phi)\right] = 0. \tag{7.145}$$

To obtain an approximate solution of the nonlinear Schrödinger equation, Cooper, Lucheron and Shepard considered a restricted class of $\phi(x,t) = \phi_\nu(x,t)$, here $\phi_\nu(x,t)$ is constrained but is able to capture the known behavior of ϕ for the problem at hand. They constructed only one simplest trial wave function which is suitable for Gaussian initial data and leads to a very simple Lagrangian dynamics for the variational parameters. For a one-dimensional system ($d = 1$), the trial wave function is written as

$$\phi_v(x,t) = \frac{\sqrt{N}}{[2\pi G(t)]^{1/4}} e^{-[x-q(t)^2][G^{-1}(t)/4 - iQ(t)] + ip(t)[x-q(t)]}, \tag{7.146}$$

which automatically satisfies the constraint (7.144). The normalized probability function is $\rho(x,t) = \phi^*\phi/N$, here N plays the role of the conserved "mass", M, of the nonlinear Schrödinger equation. The physical meaning of the variational parameters given above is very clear, *i.e.*,

$$\langle x\rangle = \int dx x\rho(x) = q(t), \quad \langle [x - q(t)]^2\rangle = G(t),$$
$$\langle -i\partial_x\rangle = p(t), \quad \langle i\partial_t\rangle = p\dot{q} - G\dot{Q}.$$

Thus G is the variance and p and Q are the canonical conjugates to q and G. The action for the trial wave function can be expressed as

$$S(q,p,G,Q) = N\int dt(p\dot{q} + Q\dot{G} - H_{\text{eff}}) \tag{7.147}$$
$$= N\int dt\left[p\dot{q} + Q\dot{G} - p^2 - 4QGQ - \frac{1}{4G} + \frac{bN^\sigma}{(2\pi G)^{\sigma/2}(1+\sigma)^{3/2}}\right].$$

Then, $p(t)$, the momentum conjugate to position $q(t)$, is conserved. The variational equations $\delta S/\delta q_i = \delta S/\delta p_i = 0$ yield $\dot{q} = 2p$, $\dot{p} = 0$, so that the Gaussian wave function moves with a constant velocity $q(t) = vt$, $v = 2p_0$, and

$$\dot{G} = 8QG, \quad \dot{Q} = \frac{1}{4G^2} - 4Q^2 - \frac{b\sigma N^\sigma}{\pi^{\sigma/2}(2G)^{\sigma/2+1}(1+\sigma)^{3/2}}. \tag{7.148}$$

The above equation has a first integral which gives the conservation of energy

$$\frac{\langle H\rangle}{N} = E = p_0^2 + 4QGQ + \frac{1}{4G} - \frac{bN^\sigma}{(2\pi G)^{\sigma/2}(1+\sigma)^{3/2}}.$$

Using (7.148), this can be rewritten as

$$\dot{G}^2 = 16G\hat{E} + \frac{16bN^\sigma G}{(2\pi G)^{\sigma/2}(1+\sigma)^{3/2}} - 4,$$

where $\hat{E} = E - p_0^2$ is the energy available to the generalized coordinate G. When $\sigma = 0$, which corresponds to the linear Schrödinger equation, we have

$$\dot{G}^2 = 16G(\hat{E} + b) - 4.$$

Thus

$$G(t) = G_0 + 4(\hat{E} + b)(t - t_0)^2 + (t - t_0)[16G_0(\hat{E} + b) - 4]^{1/2},$$

which shows the spreading of the Gaussian wave packet with time for an initial data G_0 in linear quantum mechanics.

When $\sigma = 1$ which corresponds to the general nonlinear Schrödinger equation, one has

$$\dot{G}^2 = 16G\hat{E} + 4bN\sqrt{\frac{G}{\pi}} - 4. \tag{7.149}$$

From (7.148), we see that there is an energy regime, $-b^2N^2/16\pi \leq \hat{E} < 0$, for which G, the width of the Gaussian, does not spread but oscillates between two bounds

$$G_{\pm} = \frac{b^2N^2}{8\hat{E}^2\pi}\left[1 \pm \left(1 - \frac{16\pi|\hat{E}|}{b^2N^2}\right)^{1/2}\right]^2.$$

The lower extreme $\hat{E} = -b^2N^2/16\pi$ is a fixed point of the equation where the width of the Gaussian is $2\sqrt{\pi}/bN$. Thus in this approximation there is a range of initial data where the Gaussian does not spread in time. This is a feature of the microscopic particle described by the nonlinear Schrödinger equation in nonlinear quantum mechanics. It demonstrates again that the microscopic particle has exactly the property of classical particle, *i.e.*, it can always retain the shape during its motion and cannot be dispersed with time.

Cooper *et al.* obtained also the following expression for $G(t)$ for $\sigma = 2$ which corresponds to the ϕ^6-nonlinear Schrödinger equation.

$$G(t) = \left[\sqrt{G_0 + \frac{f(N)}{16\hat{E}}} - 2\sqrt{\hat{E}}(t - t_0)\right]^2 - \frac{f(N)}{16\hat{E}},$$

where

$$f(N) = \frac{16bN^2}{2\pi \times 3^{3/2}} - 4.$$

For $d \neq 1$, it is sufficient to consider a spherically symmetric trial wave function that is centered at all times at the origin (*i.e.*, $p(t) = q(t) = 0$). We parameterize the trial wave function as follows,

$$\phi(r,t) = \sqrt{N}\left(\frac{2\alpha}{\pi}\right)^{d/4} e^{-r^2[\alpha(t) - iQ(t)]}. \tag{7.150}$$

The effective "mass" is given by

$$N = \Omega_d \int \phi^*(r)\phi(r)r^{d-1}dr,$$

where

$$\Omega_d = \frac{2\pi^{d/2}}{\Gamma(d/2)}.$$

With $\rho(r) = \phi^*\phi/N$, we have

$$\langle r^2 \rangle = G(t) = \frac{d}{4\alpha(t)}, \quad \langle i\partial_t \rangle = -G\dot{Q},$$

so that Q is still canonically conjugate to G. Thus the action for the trial wave function is of the form

$$S(G,Q) = N \int dt(Q\dot{G} - H_{\text{eff}}) \tag{7.151}$$

$$= \int dt \left[Q\dot{G} - 4QGQ - \frac{d^2}{4G} + \frac{bN^\sigma}{(1+\sigma)^{(d+2)/2}} \left(\frac{d}{2\pi G} \right)^{\sigma d/2} \right].$$

The variational condition $\delta S = 0$ leads to the following equations of motion

$$\dot{G} = 8QG, \quad \dot{Q} = \frac{d^2}{4G^2} - 4Q^2 - \frac{b\sigma d N^\sigma}{2G} \left(\frac{d}{2\pi G} \right)^{\sigma d/2} \frac{1}{(1+\sigma)^{(d+2)/2}}.$$

The first integral of the first equation gives the energy E

$$\frac{\langle H \rangle}{N} = E = \frac{\dot{G}^2}{16G} + \frac{d^2}{4G} - \frac{bN^\sigma}{(1+\sigma)^{(d+2)/2}} \left(\frac{d}{2\pi G} \right)^{\sigma d/2}.$$

The key equation is

$$\dot{G}^2 = 16GE - 4d^2 + \frac{16bGN^\sigma}{(1+\sigma)^{(d+2)/2}} \left(\frac{d}{2\pi G} \right)^{\sigma d/2}.$$

Using the same arguments as in the case of $d = 1$, we can find that $G \to 0$ as long as $\sigma d \geq 2$. At the critical point $\sigma d = 2$, we can obtain the "mass" N corresponding to the singularity. From the criterion and $\dot{G} < 0$ for small G we can determine the parameter which is given by

$$4d^2 \leq \frac{d}{2\pi} \frac{16bN^\sigma}{(1+\sigma)^{(d+2)/2}}.$$

In terms of $d = 2/\sigma$, we obtain

$$N \geq N_c = \left(\frac{\pi d}{2b} \right)^{d/2} \left(1 + \frac{2}{d} \right)^{d(d+2)/4} \tag{7.152}$$

for the critical "mass". Thus for $d = 1$, $b = 1$, $N_c = \sqrt{\pi/2}3^{3/4} = 2.8569$; for $b = 1$, $d = 2$, $N_c = 4\pi = 12.5664$; for $b = 1$, $d = 3$, $N_c = 69.4696$. These can be compared with the exact numerical results in one and two dimensions, $d = 1$, $b = 1$, $N_c = 2.7$; $d = 2$, $b = 1$, $N_c = 11.73$, obtained by Rose, Weinstein and Schochet. Therefore, equation (7.152) gives quite reasonable results.

Cooper, Shepard and Simmons solved approximately the nonlinear Schrödinger equation (7.144) by the perturbation δ-expansion proposed by Bender *et al.* which is an expansion in terms of the degree of nonlinearity of the equation. This technique consists of first replacing σ by a parameter δ and treating δ as a small perturbation parameter.

To improve the dependence of the above description on the coupling constant b, Cooper *et al.* further introduced two parameters, M and c, in (7.144), and rewrite (7.144) as

$$i\frac{\partial\phi}{\partial t} + \frac{\partial^2\phi}{\partial x^2} + bM^c\left|\frac{\phi\phi}{M}\right|^\delta \phi = 0,$$

which is equivalent to (7.144) when $c = \delta$. The δ-expansion is obtained by first expanding this equation as a power series in δ.

$$i\frac{\partial\phi}{\partial t} + \frac{\partial^2\phi}{\partial x^2} + bM^c\phi\sum_{n=0}^{\infty}\frac{(\delta\ln|\tilde{\phi}\phi/M|)^n}{n!} = 0. \tag{7.153}$$

Assuming a power series solution for ϕ in terms of

$$\delta\phi = \sum_{n=0}^{\infty}\phi_n(x,t,M)\delta^N,$$

one obtains a system of linear equations for the ϕ_n with known driving terms. For the exact solution, ϕ is independent of M when $c = \delta$. However, when we expand ϕ in δ this is no longer the case and M can be treated as a variational parameter. That is, we will assume that ϕ has an expansion up to the order N in δ,

$$\phi^{(N)} = \sum_{n=0}^{N}\phi_n(x,t,M)\delta^n, \tag{7.154}$$

and M is a function of x and t by the requirement of $\partial\phi^{(N)}/\partial M\,|_{c=0} = 0$, which is known as a scaling relation.

The linear delta expansion, proposed by Okopinska *et al.*, Duncan *et al.* and Tones *et al.*, consists of replacing (7.144) by

$$i\frac{\partial\phi}{\partial t} + \frac{\partial^2\phi}{\partial x^2} + b\lambda\phi + \delta(b|\tilde{\phi}\phi| - b\lambda)\phi = 0, \tag{7.155}$$

which contains two new parameters δ and λ. When $\delta = 1$, it reduces to (7.143). At any finite order in δ, however, the solution depends on λ and this dependence can be minimized by imposing the "principle of minimal sensitivity", which is to impose the condition $\partial\phi^{(N)}/\partial\lambda\,|_{\delta=1}=0$, in the finite order N expansion

$$\phi^{(N)}=\sum_{n=0}^{N}\phi_n(x,t,\lambda)\delta^n \tag{7.156}$$

where λ is a function of x and t.

To solve for ϕ in the δ-expansion, we assume that ϕ can be written as

$$\phi=\phi_0+\delta\phi_1+\delta^2\phi_2+\cdots \tag{7.157}$$

and $\phi_n=\Psi_n e^{igt}$, where $g=bM^c$. Substituting (7.157) into (7.153), we get

$$\begin{aligned}
i\frac{\partial\Psi_0}{\partial t}+\frac{\partial^2\Psi_0}{\partial x^2}&=0,\\
i\frac{\partial\Psi_1}{\partial t}+\frac{\partial^2\Psi_1}{\partial x^2}&=f_1=-g\Psi_0\ln\left(\frac{\Psi_0^*\Psi_0}{M}\right),\\
i\frac{\partial\Psi_2}{\partial t}+\frac{\partial^2\Psi_2}{\partial x^2}&=f_2=-g\frac{\Psi_1^*\Psi_0+\Psi_0^*\Psi_1}{\Psi_0^*}-g\Psi_1\ln\left(\frac{\Psi_0^*\Psi_0}{M}\right)-\\
&\qquad\frac{1}{2}g\Psi_0\ln^2\left(\frac{\Psi_0^*\Psi_0}{M}\right).
\end{aligned} \tag{7.158}$$

The corresponding boundary conditions are,

$$\phi(x)=\Psi(x)=h(x)=\Psi_0(x),\quad\text{and}\quad\Psi_n(x)=0,\quad\text{for } n\geq 1, \tag{7.159}$$

at $t=0$. For $n=0$, the solution of (7.158) is given by

$$\Psi_0(x,t)=\frac{e^{-i\pi/4}}{\sqrt{4\pi t}}\int_{-\infty}^{\infty}d\tilde{x}e^{i(x-\tilde{x})^2/4t}h(\tilde{x}). \tag{7.160}$$

For $n\geq 1$ the solution can be expressed in terms of the Green function as follows

$$\Psi_n(x,t)=\int_0^t d\tilde{t}\int_{-\infty}^{\infty}dx'G(x-\tilde{x},t-\tilde{t})f_n(\tilde{x},\tilde{t}), \tag{7.161}$$

with

$$G(x,t)=\frac{e^{-3i\pi/4}}{\sqrt{4\pi t}}\theta(t)e^{ix^2/4t},$$

and f_n given by the right hand side of (7.158).

For the linear δ-expansion, equation (7.157) still holds and $\phi_n=\Psi_n e^{igt}$, where $g=b\lambda$. The structure of the Ψ_n equation is exactly the same as that for the δ-expansion except for the right-hand side, f_n, which are now given by

$$\begin{aligned}
f_1&=b[\lambda\Psi_0-(\Psi_0^*\Psi_0)^\sigma\Psi_0],\\
f_2&=b[\lambda\Psi_1-(\sigma+1)(\Psi_0^*\Psi_0)^\sigma\Psi_1-\sigma(\Psi_0^*\Psi_0)^{\sigma-1}(\Psi_0^*\Psi_0)\Psi_0].
\end{aligned} \tag{7.162}$$

In order to show how the variational approximation works in a simple case, Cooper *et al.* considered the initial condition

$$\phi(x,t=0) = h(x) = C\delta(x). \tag{7.163}$$

Then from (7.160), we get

$$\Psi_0(x,t) = \frac{C}{\sqrt{4\pi t}} e^{ix^2/4t - i\pi/4}, \tag{7.164}$$

so that

$$\phi_0(x,t) = \frac{C}{\sqrt{4\pi t}} e^{ix^2/4t + igt - i\pi/4}. \tag{7.165}$$

For the functions f_n corresponding to the rest of the terms in the two perturbation expansions, we can obtain

$$f_n(x,t) = \Psi_0(x,t)T_n(t), \tag{7.166}$$

after some iterations, for the δ-function initial data. Using the properties of the Green function in (7.161), we can obtain

$$\Psi_n(x,t) = (-i)\Psi_0(x,t)\varphi_n(t), \quad \varphi_n(t) = \int_0^t T_n(t')dt'. \tag{7.167}$$

These perturbation expansions lead to the following form for solutions of all orders in δ.

$$\phi(x,t) = \phi_0(x,t)\alpha(t). \tag{7.168}$$

Inserting (7.168) into (7.144), the following can be obtained for $\alpha(t)$

$$i\frac{d\alpha(t)}{dt} - g\alpha(t) + \frac{C^{2\sigma}|\alpha^*\alpha|^\sigma \alpha(t)}{(4\pi t)^\sigma} = 0. \tag{7.169}$$

For the linear δ-expansion substituting (7.168) into (7.155), we can obtain

$$i\frac{d\alpha(t)}{dt} - \delta g\alpha(t) + \frac{\delta C^{2\sigma}|\alpha^*\alpha|^\sigma \alpha(t)}{(4\pi t)^\sigma} = 0. \tag{7.170}$$

The boundary conditions for (7.169) and (7.170) are $\alpha(t=0)=1$. Let

$$\alpha(t) = R(t)e^{i\gamma(t)}, \tag{7.171}$$

then $R(t)$ satisfies

$$-gR + i\frac{dR}{dt} - R\frac{d\gamma}{dt} + b\left(\frac{R^2C^2}{4\pi t}\right)^\sigma R = 0. \tag{7.172}$$

Using the boundary condition at $t=0$, one gets

$$R(t) = 1, \quad \gamma(t) = -gt + b\left(\frac{C^2}{4\pi}\right)^\sigma \int_{t_0}^t \frac{d\tilde{t}}{\tilde{t}^\sigma}. \tag{7.173}$$

Substituting (7.173) into (7.172) and (7.168), an exact solution of (7.144) for all σ can be obtained. For the linear δ-expansion

$$T_1 = b\left[\lambda - \frac{C^{2\sigma}}{(4\pi t')^{\kappa}}\right], \tag{7.174}$$

where σ is a continuous variable. Thus $\varphi_1(t)$ in (7.167) has different analytic behavior for $\sigma < 1$, $\sigma = 1$ and $\sigma > 1$, which are

$$\varphi_1(t) = b[\lambda t + \chi(\sigma, t)], \quad \chi(\sigma, t) = -\left(\frac{C^2}{4\pi}\right)^{\sigma} \int_{t_0}^{t} \frac{d\tilde{t}}{\tilde{t}^{\sigma}}. \tag{7.175}$$

Thus to the δth order, one has

$$\phi^{(1)}(x,t) = e^{ib\lambda t}\Psi_0 \left\{1 - i\delta b[\lambda t + \chi(\sigma, t)]\right\}.$$

By determining the variational parameter via

$$\left.\frac{\partial \phi}{\partial \lambda}\right|_{\delta=1} = 0,$$

Cooper *et al.* obtained $\lambda(\sigma,t)t = -\chi(\sigma,t)$, so that to this order

$$\phi(x,t) = \Psi_0 e^{-b\chi(\sigma,t)}.$$

This is the exact solution for the boundary conditions given above. Extending the calculation to the order of δ^2, they obtained

$$\phi(x,t) = e^{ib\lambda t}\Psi_0 \left\{1 - i\delta b[\lambda t + \chi(\sigma,t)] - \frac{1}{2}\delta^2 b^2 [\lambda t + \chi(\sigma,t)]^2 + \cdots\right\}$$

for $\sigma < 1$. This series in δ can be written as

$$\phi(x,t) = \Psi_0 e^{i[b\lambda t(1-\delta) - \delta b\chi(\sigma,t)]}. \tag{7.176}$$

This solution is independent of the variational parameter λ at $\delta = 1$ and by direct substitution it can be shown that it is a solution of (7.155). Thus the solution to the original nonlinear Schrödinger equation is

$$\phi(x,t) = \Psi_0 e^{-ib\chi(\sigma,t)}.$$

This solution satisfies the boundary condition $t = 0$ for $\sigma < 1$. For $\sigma = 1$ and $\sigma > 1$, one gets an additional infinite phase as $t \to 0$, unless we allow $t \to t_0$ first and then let $t = t_0 \to 0$.

For the δ-expansion, instead of (7.174), we have

$$T_1(t) = -g \ln\left(\frac{C^2}{4\pi t M}\right).$$

In accordance with the above method, one can obtain

$$\varphi_1(t) = -gt\left[1 - \ln\left(\frac{4\pi tM}{C^2}\right)\right],$$

so that

$$\Psi_1 = ig\Psi_0 t\left[1 + \ln\left(\frac{C^2}{4\pi tM}\right)\right].$$

Thus to the δth order,

$$\phi = \Psi_0 e^{igt}\left\{1 + igt\delta\left[1 + \ln\left(\frac{C^2}{4\pi tM}\right)\right]\right\}.$$

By taking the exponential of both sides of the above equation, Cooper *et al.* obtained

$$\phi = \Psi_0 \exp\left(igt\left\{1 + \delta\left[1 + \ln\left(\frac{C^2}{4\pi tM}\right)\right]\right\}\right)$$

Using the above scaling law, they obtained

$$1 + \ln\left(\frac{C^2}{4\pi tM}\right) = 0.$$

Thus, $M = C^2 e/4\pi t$ and the final result is

$$\phi = \Psi_0 \exp\left[ibt\left(\frac{C^2 e}{4\pi t}\right)^{\delta}\right]. \tag{7.177}$$

Comparing the above result with the exact solution, we find that the t-dependence is correct except at $\delta = 1$. However, the coefficient of the t-dependence should have a factor $(1-\delta)^{-1}$ instead of e^{δ}. That is,

$$\phi(x,t) = \Psi(x,t)\exp\left[\frac{ibt}{1-\delta}\left(\frac{C^2}{4\pi t}\right)^{\delta}\right],$$

which can be only obtained by (7.177) at $\delta < 1$.

Therefore, Cooper *et al.* obtained a series of variational approximations to the initial value problems for the nonlinear Schrödinger equation with arbitrary nonlinearity σ. For the Dirac delta-function initial conditions, the variational principle associated with the linear δ-expansion gives the exact result for arbitrary σ when keeping the first order terms in the expansion. The exact result changed its analytic structure at $\sigma = 1$, which is the integrable case. Study showed that the δ-expansion results are less accurate for this problem. The solution converges to the exact answer when $\sigma = \delta < 1$. In the δ-expansion we do not seem to be able to handle the case of $\delta \geq 1$.

7.8 D Operator and Hirota Method

Solutions of dynamic equations in the nonlinear quantum mechanics can also be obtained by the Hirota method of function transformation. In this method, a D operator is defined by the relation (7.2). Let's further define an operator D_z and a differential operator $\partial/\partial z$ as follows

$$\begin{aligned} D_z &= \delta D_t + \varepsilon D_x, \\ \frac{\partial}{\partial z} &= \delta\frac{\partial}{\partial t} + \varepsilon\frac{\partial}{\partial x}, \end{aligned} \tag{7.178}$$

where δ and ε are constants. These operators have the following properties:

(1) $D_z^m f \cdot g = (-1)^m D_z^m g \cdot f$;

(2) $D_z^n f \cdot 1 = \left(\frac{\partial}{\partial z}\right)^n f$;

(3) $D_z^m f \cdot g = D_z^{m-1}(f_z \cdot g - f \cdot g_z)$;

(4) $D_z^m e^{p_1 x} e^{p_2 x} = (p_1 - p_2)^m e^{(p_1-p_2)x}$;

(5) $e^{\varepsilon D_z} f(x) \cdot g(x) = f(x+\varepsilon)g(x-\varepsilon) = \left[\exp\left(\varepsilon\frac{\partial}{\partial z}\right)\right] f(x) \left[\exp\left(-\varepsilon\frac{\partial}{\partial z}\right)\right] g(x)$;

(6) $e^{\varepsilon D_z} fg \cdot hq = [e^{\varepsilon D_z} f \cdot g] \cdot [e^{\varepsilon D_z} g \cdot h]$;

(7) $\exp\left(\varepsilon\frac{\partial}{\partial z}\right)\left(\frac{f}{g}\right) = \frac{e^{\varepsilon D_z} f \cdot g}{\cosh(\varepsilon D_z)\, g \cdot g}$;

(8) $2\cosh\left(\varepsilon\frac{\partial}{\partial z}\right)\log f = \log[\cosh(\varepsilon D_z) f \cdot f]$;

(9) $e^{\varepsilon D_z} f \cdot g = \exp\left[2\cosh\left(\varepsilon\frac{\partial}{\partial z}\right)\log g\right]\left[\exp\left(\varepsilon\frac{\partial}{\partial z}\right)\cdot\left(\frac{f}{g}\right)\right]$;

(10) $e^{\varepsilon D_z} f \cdot g = \exp\left[\sinh\left(\varepsilon\frac{\partial}{\partial z}\right)\log\left(\frac{f}{g}\right) + \cosh\left(\varepsilon\frac{\partial}{\partial z}\right)\log(fg)\right]$;

(11) $D_x^n D_y^m \cdots D_z^\kappa f \cdot g = (-1)^{i+j+\cdots+\kappa} D_x^n D_y^m \cdots D_z^\kappa g \cdot f$;

(12) $\left(2\cosh\varepsilon\frac{\partial}{\partial z}\right)\ln f = \ln(f+\varepsilon) + \ln(f-\varepsilon) = \ln[\cosh(\varepsilon D_z) f]$,

where f, g, and q are functions of x and t.

For the nonlinear Schrödinger equation,

$$i\phi_{t'} + p\phi_{x'x'} + b|\phi|^2\phi = 0, \tag{7.179}$$

we assume $\phi = g/f$, where g is complex and f is real. Using the identity

$$\exp\left(\varepsilon\frac{\partial}{\partial z}\right)\left(\frac{g}{f}\right) = \frac{g(x'+\varepsilon)f(x'-\varepsilon)}{f(x'+\varepsilon)g(x'-\varepsilon)} = \frac{\exp(\varepsilon D_{x'})g \cdot f}{\exp(\varepsilon D_{x'})f \cdot f}$$

where $t' = t/\hbar$, $x' = x\sqrt{2m/\hbar^2}$, we find that g and f must satisfy

$$\frac{(iD_{t'} + D_{x'}^2)g \cdot f}{f^2} - \frac{g}{f}\left(\frac{pD_{x'}^2 f \cdot f - bgg^*}{f^2}\right) = 0.$$

Substituting $\phi = g/f$ into (7.179), we have

$$(iD_{t'} + pD_{x'}^2)g \cdot f = 0,$$
$$pD_{x'}^2 f \cdot f - bgg^* = 0, \tag{7.180}$$

where f and g satisfy (7.2). The one-envelope-soliton solution of the nonlinear Schrödinger equation is given by

$$g = Ae^{i\theta}e^{\eta}, \quad f = 1 + ae^{\eta+\eta^*}, \tag{7.181}$$

with

$$\theta = kx' - \omega t', \quad \eta = K'x' - \Omega t' - \eta^0, \quad a = \frac{bA^2}{2[p(k+k^*)^2]},$$

and the dispersion relation is

$$-i(\Omega + i\omega) + p(K + ik)^2 = 0.$$

Here we see that the decoupling of (7.180) does not result in a simple relation between g and f, and the Hirota form (7.179) is a coupled system (7.180) in g and f. To obtain solution of (7.180), we assume that

$$g = \varepsilon g_1 + \varepsilon^3 g_3 + \cdots,$$
$$f = 1 + \varepsilon^2 f_2 + \cdots. \tag{7.182}$$

We then find that

$$ig_{1t'} + g_{1x'x'} = 0,$$
$$f_{2xx} = \frac{1}{2}g_1 g_1^*,$$
$$ig_{3t} + g_{3x'x'} = -(iD_{t'} + pD_{x'}^2)g_1 \cdot f_2,$$
$$\cdots\cdots.$$

Thus these equations may be solved recursively until an inhomogeneity is found to vanish where upon the term defined by that equation and all subsequent ones may be take to be zero. Substituting (7.181) into $\phi = g/f$, we obtain the one-soliton solution, as given in (7.53) or (4.56). Continuing in this way, the N-soliton solution of the nonlinear Schrödinger equation can be constructed.

For the Sine-Gordon equation, (7.6) in natural unit system, if we require $\partial\phi/\partial x \to 0$ when $|x| \to \infty$, we can assume that

$$\phi(x,t) = 4\tan^{-1}\left[\frac{g(x,t)}{f(x,t)}\right], \tag{7.183}$$

where

$$f(x,t) = \sum_{n=0}^{N/2} \sum_{N_n} a(i_1, i_2 \cdots, i_n) e^{\eta_{i_1} + \eta_{i_2} + \cdots + \eta_{i_{2n}}},$$

$$g(x,t) = \sum_{m=0}^{|(N-1)/2|} \sum_{N_{2m+1}} a(j_1, j_2 \cdots, j_{2m+1}) e^{\eta_{j_1} + \eta_{j_2} + \cdots + \eta_{j_{2m+1}}}. \tag{7.184}$$

Here

$$a(i_1, i_2 \cdots, i_n) = \begin{cases} \prod_{\kappa<\ell}^{(n)} (i_k, i_\ell), & n \geq 2 \\ 1, & n = 0, 1 \end{cases}$$

$$a(i_k, i_\ell) = \frac{(P_{ik} - P_{i\ell})^2 - (\Omega_{ik} - \Omega_{i\ell})^2}{(P_{ik} + P_{i\ell})^2 - (\Omega_{ik} + \Omega_{i\ell})^2},$$

$$\eta_i = P_i x - \Omega_i t - \eta_i^0,$$

$$p_i^2 - \Omega_i^2 = 1.$$

Inserting (7.183) into (7.6), we get,

$$f g_{xx} - 2 f_x g_x + f_{xx} g - (f g_{tt} - 2 f_t g_t + f_{tt} g) = f g, \tag{7.185}$$

$$f_{xx} f - 2 f_x^2 - f f_{xx} - (f_{tt} f - 2 f_t^2 + f f_{tt}) = g_{xx} g - 2 g_x^2 - (g_{tt} g - 2 g_t^2 + g g_{tt}).$$

Substituting (7.184) into (7.185), we can determine $a(i_k, i_\ell)$ and $\eta_{i_{2n}}, \eta_{j_{2m+1}}$. We can thus finally obtain the solution of (7.183), and find out the N-soliton solution of (7.6). To obtain the one-soliton solution, we insert (7.183) into (7.6) to get

$$(D_x^2 - D_t^2)(f \cdot g) = f g,$$

$$(D_x^2 - D_t^2)(f \cdot f - g \cdot g) = 0. \tag{7.186}$$

Obviously, $g = 0$ and $f = 1$ is a solution. Let

$$f = 1 + \varepsilon^2 f^{(1)} + \varepsilon^4 f^{(2)} + \cdots,$$

$$g = \varepsilon g^{(1)} + \varepsilon^3 g^{(2)} + \cdots. \tag{7.187}$$

Substituting (7.187) into (7.186) and equating the coefficients of the ε terms, we get the following linear equation

$$g_{xx}^{(1)} - g_{tt}^{(1)} = g^{(1)}.$$

The solution is given by $g^{(1)} = \exp(kx - \omega t + \delta)$, where $k^2 = \omega^2 - 1$, or equivalently, $\phi = 4 \tan^{-1}[\exp(kx - \omega t + \delta)]$ which is similar to (6.14) or (7.22).

For the ϕ^4-field equation (3.5) with $A(\phi) = 0$ in natural unit system, we can assume $\phi = g/f$, and insert it into (3.5). Tajiri *et al.* obtained that

$$(D_t^2 - D_x^2 - \alpha)g \cdot f = 0,$$
$$(D_t^2 - D_x^2)f \cdot f - \beta|g|^2 = 0,$$

where

$$g = e^{\eta}, \quad f = 1 + ae^{\eta+\eta^*},$$
$$a = \frac{1}{2}\frac{\beta}{(\Omega + \Omega^*) \pm (P + \Omega^*)}.$$

Here β is a nonlinear coefficient and α is a linear coefficient in this equation, $\eta = Px - \Omega t \pm \eta_0$, and $P^2 - \Omega^2 = -\alpha$. Its one-soliton solution can be finally written as

$$\phi = \frac{P_r}{\sqrt{\beta}}\sqrt{2(v^2 \pm 1)}\exp\left\{i\left[\sqrt{\frac{\alpha}{v^2-1} - P_r^2}(x - vt) + \eta_{0r}\right]\right\}\operatorname{sech}[P_r(x - vt + \theta_r)],$$

where P_r, Ω_r, and η_{0r} are real parts of P, Ω and η_0,

$$v = \frac{\Omega_r}{P_r}, \quad \theta_r = \frac{1}{P_r}\left(\eta_{0r} + \frac{\ln a}{2}\right),$$

and $v^2 > 1$.

Tajiri *et al.* generalized the Hirota's method to two- and three-dimensional systems and to solve dynamic equations of the microscopic particle in these systems. For the three-dimensional nonlinear Schrödinger equation,

$$i\phi_{t'} + p\phi_{x'x'} + q'\phi_{y'y'} + r'\phi_{z'z'} + b|\phi|^2\phi = 0, \tag{7.188}$$

with $b > 0$ (for $p, q' > 0$ and $r' < 0$), we introduce the dependent variable transformation,

$$\phi = \frac{g(x', y', z', t')}{f(x', y', z', t')},$$

where $f = f^*$. Substituting it into (7.188), we have

$$\begin{cases} (iD_{t'} + pD_{x'}^2 + q'D_{y'}^2 + r'D_{z'}^2)g \cdot f = 0, \\ (pD_{x'}^2 + q'D_{y'}^2 + r'D_{z'}^2)f \cdot f - b|g|^2 = 0. \end{cases}$$

Here, the bilinear operators are defined by

$$D_{x'}^{\kappa}D_{y'}^{1}D_{z'}^{m}D_{t'}^{n}a(x', y', z', t') \cdot \tilde{b}(x', y', z', t')$$
$$\equiv \left(\frac{\partial}{\partial x'} - \frac{\partial}{\partial \tilde{x}'}\right)^{\kappa}\left(\frac{\partial}{\partial y'} - \frac{\partial}{\partial \tilde{y}'}\right)^{l}\left(\frac{\partial}{\partial z'} - \frac{\partial}{\partial \tilde{z}'}\right)^{m}\left(\frac{\partial}{\partial t'} - \frac{\partial}{\partial \tilde{t}'}\right)^{n}$$
$$\cdot a(x', y', z', t')\tilde{b}(\tilde{x}, \tilde{y}, \tilde{z}, \tilde{t})\Big|_{x'=\tilde{x}, y'=\tilde{y}, z'=\tilde{z}, t'=\tilde{t}}$$

The one-envelope-soliton solution of the 3D nonlinear Schrödinger equation is given by

$$g = Ae^{i\theta}e^{\eta}, \quad f = 1 + ae^{\eta+\eta^*}$$

where

$$\begin{aligned}\theta &= kx' + ly' + mz' - \omega t', \\ \eta &= Kx' + Ly' + Mz' - \Omega t' - \eta^0, \\ a &= \frac{bA^2}{2\left[p(K' + K^*)^2 + q'(L + L^*)^2 + r'(M + M^*)^2\right]},\end{aligned}$$

and the dispersion relation is given as follows

$$-i(\Omega + i\omega) + p(K + ik)^2 + q'(L + il)^2 + r'(M + im)^2 = 0.$$

7.9 Bäcklund Transformation Method

Applying the Bäcklund transformation (BT), we can obtain new soliton solutions from a given solution of the nonlinearly dynamic equations in nonlinear quantum mechanics. However, there are many different types of Bäcklund transformations in nonlinear quantum systems as described in Section 3.3. Therefore, it is important to choose an appropriate Bäcklund transformation in order to obtain the solutions of a given equation.

7.9.1 *Auto-Bäcklund transformation method*

The auto-Bäcklund transformation method was studied by Rao and Rangwala, and Rogers *et al.*. For a 2×2 eigenvalue problem

$$\begin{aligned}\psi_x &= \hat{Y}\psi, \\ \psi_t &= \hat{T}\psi,\end{aligned} \tag{7.189}$$

with

$$\psi = \begin{pmatrix} \psi_1 \\ \psi_2 \end{pmatrix}, \quad \hat{Y} = \begin{pmatrix} \lambda & q(x,t) \\ r(x,t) & -\lambda \end{pmatrix}, \quad \hat{T} = \begin{pmatrix} A(x,t,\lambda) & B'(x,t,\lambda) \\ C(x,t,\lambda) & -A(x,t,\lambda) \end{pmatrix}, \tag{7.190}$$

where ψ_1 and ψ_2 are eigenfunctions, while λ is an eigenvalue, the compatibility condition $\psi_{xt} = \psi_{tx}$ yields

$$\hat{Y}_t - \hat{T}_x = [\hat{Y}, \hat{T}] = \hat{Y}\hat{T} - \hat{T}\hat{Y}. \tag{7.191}$$

Equation (7.191) results in the following set of conditions on A, B' and C,

$$A_x = qC - rB', \quad B'_x - 2\lambda B' = q_t - 2Aq, \quad C_x + 2\lambda C = r_t + 2Ar. \tag{7.192}$$

If we set

$$\hat{L} = \begin{pmatrix} \frac{\partial}{\partial x} & -q \\ r & -\frac{\partial}{\partial x} \end{pmatrix}, \quad \hat{B} = \hat{T},$$

then the Lax system in (3.20) – (3.22), where $\hat{B}$ and $\hat{L}$ are linear operators subject to the compatibility condition $\hat{L}_t + [\hat{L}, \hat{B}] = 0$, can result in the Ablowitz-Kaup-Newell-Segur (AKNS) system. Specializations of A, B' and C in the AKNS system lead to nonlinear evolution equations amenable to the inverse scattering method for the solution of privileged initial value problems. For the Sine-Gordon equation (7.6), there are $r = -q = -\phi_x/2$ and

$$A = \frac{1}{4\lambda}\cos\phi, \quad B' = C = \frac{1}{4\lambda}\sin\phi. \tag{7.193}$$

For the nonlinear Schrödinger equation (4.40) with $\sqrt{2m/\hbar^2}x \to x'$, $t/\hbar \to t'$, there are $r = -q^* = -\phi^*$, and

$$A = 2i\lambda^2 + i|\phi|^2, \quad B' = i\phi_{x'} + 2i\lambda\phi, \quad C = i\phi^*_{x'} - 2i\lambda\phi^* \tag{7.194}$$

Auto-Bäcklund transformations for the above equations are constructed via the AKNS formalism. The key step in the derivation involves the introduction of $\Gamma = \Psi_1/\Psi_2$, so that the AKNS system (7.189) – (7.194) results in a pair of Riccati equations

$$\begin{aligned} \Gamma_x &= 2\lambda\Gamma + q - r\Gamma^2, \\ \Gamma_t &= 2A\Gamma + B' - C\Gamma^2. \end{aligned} \tag{7.195}$$

If $r = -q$ in the Sine-Gordon equation, the AKNS system yields

$$\begin{aligned} \psi_{xx} - \lambda^2\psi &= \phi\psi, \\ \psi^*_{xx} - \lambda^2\psi^* &= \phi^*\psi^*, \end{aligned} \tag{7.196}$$

where $\psi = \psi_1 + i\psi_2$ and $\phi = -iq_x - q^2$.

The system (7.196) is invariant under the Crum-type transformation

$$\begin{aligned} \psi' &= \frac{1}{\psi}, \quad \phi' = \phi + 2(\ln\psi')_{xx}, \\ \psi^{*\prime} &= \frac{1}{\psi}, \quad \phi^{*\prime} = \phi^* + 2(\ln\psi^{*\prime})_{xx}. \end{aligned} \tag{7.197}$$

Whence

$$w' = w - i\ln\left(\frac{\psi}{\psi^*}\right) = w + 2\tan^{-1}\left(\frac{\psi_2}{\psi_1}\right),$$

where w and w' are potentials given by $q = w_x$, $q' = w'_x$. Thus, under the transformation (7.197),

$$\Gamma = \cot\left[\frac{1}{2}(w' - w)\right]. \tag{7.198}$$

Substitution of (7.198) in the Riccati equations (7.195) provides a generic Bäcklund transformation for the present subclass of the AKNS system with $r = -q$, namely

$$\begin{aligned} w_t - w'_t &= 2A\sin(w' - w) - (B' + C)\cos(w' - w) + B' - C, \\ w_x + w'_x &= 2\lambda\sin(w - w'), \quad \tilde{x} = x, \quad \tilde{t} = t. \end{aligned} \tag{7.199}$$

If we set $\alpha' = \lambda/2$, $w = -\phi/2$, substitution of the specialization (7.193) in (7.199) produces the following auto-Bäcklund transformation for the Sine-Gordon equation (7.6)

$$\begin{aligned} \phi'_x &= \phi_x - 2\alpha'\sin\left(\frac{\phi + \phi'}{2}\right) = BT_1(\phi, \phi_x, \phi'), \\ \phi'_t &= -\phi_t + \frac{1}{\alpha'}\sin\left(\frac{\phi - \phi'}{2}\right) = BT_2(\phi, \phi_t, \phi'), \end{aligned} \tag{7.200}$$

where $x = x'$, $t = t'$, and α' is a non-zero Bäcklund parameter. The invariance can be easily proved. In fact, applying the integrability condition

$$\frac{\partial BT_1}{\partial t} - \frac{\partial BT_2}{\partial x} = 0,$$

we can get $\phi_{xt} = \sin\phi$. However, if we write (7.200) in the following form,

$$\begin{aligned} \phi_x &= \phi'_x - 2\alpha'\sin\left(\frac{\phi + \phi'}{2}\right) = B\Gamma'_1(\phi', \phi'_x, \phi), \\ \phi_t &= \phi'_t + \frac{2}{\alpha'}\sin\left(\frac{\phi - \phi'}{2}\right) = BT'_2(\phi', \phi'_t, \phi), \end{aligned}$$

from the integrability condition,

$$\frac{\partial BT'_1}{\partial t} - \frac{\partial BT'_2}{\partial x} = 0,$$

we can also get $\phi'_{xt} = \sin\phi'$. Thus the invariance of the Sine-Gordon equation is verified. In fact, the Bäcklund transformation (7.200) is the same as (3.45).

Setting $\phi' = 0$ in (7.200), we get a pair of equations

$$\phi_x = 2\alpha'\sin\left(\frac{\phi}{2}\right), \quad \phi_t = \frac{2}{\alpha'}\sin\left(\frac{\phi}{2}\right),$$

which yield a soliton solution (6.14), *i.e.*,

$$\phi = 4\tan^{-1}\left\{\exp\left[\pm\frac{x - x_\circ - vt}{\sqrt{1 - v^2}}\right]\right\}, \quad v = \frac{1 - \alpha'^2}{1 + \alpha'^2}.$$

For the nonlinear Schrödinger equation with $r = -q^*$, generic auto-Bäcklund transformations were generated by Konno and Wadati via Γ' and q' which leave the pair of Riccati equations (7.195) invariant. In fact, (7.194) can result in the auto-Bäcklund transformation

$$\phi_x + \phi'_x = (\phi - \phi')\sqrt{4\lambda^2 - |\phi + \phi'|^2}, \quad \tilde{x} = x, \quad \tilde{t} = t, \tag{7.201}$$
$$\phi_t + \phi'_t = i(\phi_x - \phi'_x)\sqrt{4\lambda^2 - |\phi + \phi'|^2} + \frac{i}{2}(\phi + \phi')\left[|\phi + \phi'|^2 + |\phi - \phi'|^2\right].$$

The corresponding nonlinear superposition principle can also be obtained as follows

$$(\phi_0 - \phi_1)\sqrt{4\lambda^2 - |\phi_0 + \phi_1|^2} - (\phi_0 - \phi_2)\sqrt{4\lambda^2 - |\phi_0 + \phi_2|^2} + \tag{7.202}$$
$$(\phi_2 - \phi_{12})\sqrt{4\lambda^2 - |\phi_2 + \phi_{12}|^2} - (\phi_1 - \phi_{12})\sqrt{4\lambda_2^2 - |\phi_1 + \phi_{12}|^2} = 0,$$

where $\phi_i = B_\lambda \phi_0$, $(i = 1, 2)$, $\phi_{12} = B_{\lambda_1} B_{\lambda 2} \phi_0 = B_{\lambda_2} B_{\lambda_1} \phi_0$ and ϕ_0 is a starting solution. Using (7.201) or (7.202), we can get the soliton solution of the nonlinear Schrödinger equation (4.40) as (4.8) or (4.56).

7.9.2 *Bäcklund transform of Hirota*

The Bäcklund transformation provides a mean of constructing new solution from known solutions of the dynamic equations. In the context of bilinear equations, the Bäcklund transformation method was introduced by Hirota to solve nonlinear dynamic equations in nonlinear quantum mechanics based on the D operator in Section 7.8.

For the nonlinear Schrödinger equation (7.179) and

$$\sqrt{\frac{2m}{\hbar^2}}x \rightarrow x', \quad \frac{t}{\hbar} \rightarrow t',$$

Hirota assumed that

$$\Omega = \left[\left(iD_{t'} + D_{x'}^2\right)(g \cdot f)\right] f'^2 - f^2 \left[\left(iD_{t'} + D_{x'}^2\right)(g' \cdot f')\right], \tag{7.203}$$

which vanishes provided that (g, f) and (g', f') satisfy the first of (7.180). By means of a pair of identities he rewrote (7.203) as

$$\Omega = iD_{\tilde{t}'}(g \cdot f' + f \cdot g')ff' - (gf' + fg')iD_{t'} f \cdot f' +$$
$$2D_{\tilde{x}'}[D_{\tilde{x}}(g \cdot f' + f \cdot g')] \cdot ff' + (g \cdot f' - f \cdot g')D_{\tilde{x}'}^2 f \cdot f' - D_{\tilde{x}'}^2(g \cdot f' - f \cdot g') \cdot ff' +$$
$$\frac{1}{2}\left[(gf' + fg')(gg'^* - g^* g') - (gf' - fg')(gg'^* + g^* g')\right], \tag{7.204}$$

where it is also assumed that (g, f) and (g', f') satisfy the second equation in (7.180). Considering the relations between successive "rungs of the soliton ladder", Hirota obtained

$$D_{\tilde{x}'}(g \cdot f' + f \cdot g') = \mu(gf' - fg').$$

Thus (7.204) decouples to give

$$iD_{\tilde{t}'}(g \cdot f' + f \cdot g') - (D_{\tilde{x}'}^2 + \lambda)(g \cdot f' - f \cdot g') = 0,$$
$$\left[iD_{\tilde{t}'} + 2\mu D_{\tilde{x}'} + D_{x'}^2 - (2\mu^2 - \lambda)\right] f \cdot f' = gg'^*. \tag{7.205}$$

Equation (7.205) constitutes a Bäcklund transformation between (g, f) and (g', f') satisfying (7.180). We can find (g', f') and the corresponding solution if (g, f) or initial value of $\phi(x,t)$ is given. In such a case we must choose $\mu = K_N - K_N^*$, $\lambda = 2(K_N^2 + K_N^{*2}) + (K_N^2 - K_N^*)^2$, where K_N is a constant introduced by integration of the first of (7.182).

Applying the Hirota's transformation, $\phi = 2i\log(g/f)$, to the Sine-Gordon equation, $\phi_{xt} = \sin\phi$, it becomes

$$D_x D_t f \cdot f = \frac{1}{2}(f^2 - g^2), \quad D_t D_x g \cdot g = \frac{1}{2}(g^2 - f^2).$$

The Bäcklund transformation of the Sine-Gordon equation is

$$D_x f' \cdot f = \frac{K}{2} g' \cdot g, \quad D_t f' \cdot g = \frac{1}{2K} g' \cdot f, \tag{7.206}$$

as well as their complex conjugates, where K is a real constant. In order to obtain a superposition formula, we rewrite (7.206) as

$$D_t f_1 \cdot g_0 = -\frac{1}{2K_1} g_1 \cdot f_0, \quad D_t \hat{f}_1 \cdot g_0 = \frac{1}{2K_2} \hat{g}_1 f_0,$$
$$D_t f_2 \cdot g_1 = -\frac{1}{2K_2} g_2 \cdot f_1, \quad D_t f_2 \cdot \hat{g}_1 = -\frac{1}{2K_1} g_2 \cdot \hat{f}_1, \tag{7.207}$$

where $\hat{f}_1$ and $\hat{g}_1$ are one-soliton solution with parameter K_2, K_1 and K_2 are the corresponding parameters of soliton solutions. Using the properties of the D operator,

$$D_x^2(g \cdot f)hq - gf(D_x^2 h \cdot q) = D_x\{(D_x g \cdot q) \cdot hf + gq \cdot (D_x h \cdot f)\},$$

Hirota obtained the following superposition formula for the soliton solution of the Sine-Gordon equation,

$$\frac{g_0/f_0}{g_2^*/f_2} = \frac{K_2(g_1^*/f_1) - K_1(\hat{g}_1/\hat{f}_1)}{K_2(\hat{g}_1/\hat{f}_1) - K_1(g_1/f_1)}. \tag{7.208}$$

Inserting $\phi = 2i\log(g/f)$ into (7.208) we obtain,

$$e^{i(\phi_2 - \phi_0)/2} = \frac{K_2 - K_1 e^{i(\phi_1 - \hat{\phi}_1)/2}}{-K_1 + K_2 e^{i(\phi_1 - \hat{\phi}_1)/2}},$$

which gives

$$\tan\left[\frac{1}{4}(\phi_2 - \phi_0)\right] = \frac{K_1 + K_2}{K_1 - K_2} \tan\left[\frac{1}{4}(\phi_1 - \hat{\phi}_1)\right]. \tag{7.209}$$

This is the same as (3.44) with $\alpha_1 = K_1$, $\alpha_2 = K_2$, *i.e.*, it is the superposition formula of the Sine-Gordon equation. Applying (7.206), or (7.208), or (7.209) we can get a new solution of the Sine-Gordon equation from a known solution.

7.10 Method of Separation of Variables

Lamb proposed a method of separation of the variables to solve nonlinearly dynamic equations. For the Sine-Gordon equation given in (6.13) in natural unit system, Lamb expressed its solution in the form

$$\tan\left[\frac{1}{4}\phi(x,t)\right] = g(x)F(t). \tag{7.210}$$

Inserting (7.210) into (6.13) results in two first order differential equations as follows

$$\begin{aligned} g_x^2(x) &= -C'g^4(x) + \ell g^2(x) + \beta', \\ F_t^2(t) &= -\beta' F^4(t) + (\ell - 1)F^2(t) + C', \end{aligned} \tag{7.211}$$

where C', β' and ℓ are integral constants. When $C' = \beta' = 0$, equation (7.211) become

$$g_x = n\sqrt{\ell}g, \quad F_t = n\sqrt{\ell - 1}F, \quad n = \pm 1.$$

The solutions of this set of equations can be easily obtained. Equation (7.210) can now be written as

$$\tan\left[\frac{1}{4}\phi(x,t)\right] = \Re\exp\left\{n\sqrt{\ell}\left[x - x_0 - \frac{\sqrt{\ell - 1}}{\ell}t\right]\right\}. \tag{7.212}$$

Thus there are three kinds of solutions corresponding to $\ell > 1$, $0 < \ell < 1$ and $\ell < 0$ in (7.212) respectively.

(1) In the case of $\ell > 1$, we assume that $\sqrt{\ell - 1}/\ell = v$, $\sqrt{\ell} = \gamma = 1/\sqrt{1 - v^2}$ with $0 < v^2 < 1$. In such a case, equation (7.212) is replaced by

$$\phi(x,t) = 4\tan^{-1}\left[e^{n\gamma(x - x_0 - vt)}\right]. \tag{7.213}$$

This is consistent with results given in (7.22).

(2) In the case of $0 \le \ell < 1$, we assume that $\sqrt{\ell - 1}/\ell = i\omega$, $\sqrt{\ell} = 1/\sqrt{1 + \omega^2}$ with $0 \le \omega^2 < 1$. Then (7.212) becomes

$$\phi(x,t) = 4\tan^{-1}\left[\exp\left(\frac{nx}{1 + \omega^2}\right)\cos\left(\frac{\omega t}{1 + \omega^2}\right)\right].$$

This represents a stable solitary wave with frequency $\omega/\sqrt{1 + \omega^2}$.

(3) In the case of $\ell < 0$, assuming $\sqrt{(\ell - 1)/\ell} = \tilde{v}$, $\tilde{\gamma} = 1/\sqrt{\tilde{v}^2 - 1}$, we get

$$\phi(x,t) = 4\tan^{-1}\{\cos[\tilde{\gamma}(x - \tilde{v}t)]\}.$$

This solution depicts a periodic wave with a velocity $\tilde{v} > 1$ and wavelength $\lambda = 2\pi/\tilde{\gamma} = 2\pi\sqrt{\tilde{v}^2 - 1}$.

When $C' = 0$, $\beta' = 1$, (7.211) becomes

$$g_x^2 = \ell g^2 + 1, \quad F_t^2 = -F^4 + (\ell - 1)F^2. \tag{7.214}$$

In the case of $\ell > 1$, there are two types of solutions. The first one is given by

$$g(x) = \sinh(\gamma x)/\sqrt{\ell}, \quad F(t) = \sqrt{\ell - 1}\mathrm{csch}(\gamma v t).$$

Equation (7.210) can now be written as

$$\phi_{k-k}(x,t) = 4\tan^{-1}\left[\frac{v\sinh(\gamma x)}{\cosh(\gamma v t)}\right]. \tag{7.215}$$

This solution describes two interacting kinks moving with velocities v and $-v$, respectively, as described in Section 6.2.

Equation (7.214) also has the solution

$$g(x) = i\sqrt{\ell}\cosh(\gamma x), \quad F(t) = -i\sqrt{\ell - 1}\ \mathrm{sech}(\gamma v t).$$

Thus

$$\phi_{K-A}(x,t) = 4\tan^{-1}\left[\frac{v\cosh(\gamma x)}{\sinh(\gamma v t)}\right]. \tag{7.216}$$

It describes the collision between a kink and an antikink in which they pass through each other, as described in Section 6.2.

When $C' = 0$ and $\beta' = 1$ and $0 < \ell < 1$, equation (7.214) has the following solutions

$$g(x) = i\cosh(\sqrt{\ell}\,x)/\sqrt{\ell}, \quad F(t) = i\sqrt{1-\ell}\,\mathrm{sech}(\sqrt{1-\ell}\,t).$$

Thus

$$\phi(x,t) = 4\tan^{-1}\left[\frac{\sin(wt/\sqrt{1+w^2})}{w\cosh(x/\sqrt{1+w^2})}\right], \quad w = \sqrt{\frac{1-\ell}{\ell}}. \tag{7.217}$$

This solution denotes a localized pulsated entity – the bound state of a kink and an antikink which is a breather or bion. It is also called $o\pi$-pulse because it satisfies the boundary condition $\phi(x,t) \to 0$ as $|x| \to \infty$. It has an internal frequency $\omega = w/\sqrt{1+w^2}$, spread over a region which is inversely proportional to $1/\sqrt{1+w^2}$.

The corresponding traveling breather can be obtained by the Lorentz transformation

$$t \to \gamma(t - x/v), \quad x \to \gamma(x - vt),$$

where $\gamma = 1/\sqrt{1-v^2}$. The solution can be written as

$$\phi(x,t) = 4\tan^{-1}\left\{\frac{\tan\mu' \sin[(t-x/v)\gamma\cos\mu']}{\cosh[\gamma(x-vt)]\sin\mu'}\right\}, \tag{7.218}$$

with $\mu' = \tan^{-1}(1/w)$. The frequency of the pulsation in the moving breather is $\omega' = \gamma\cos\mu' = \gamma w'/\sqrt{1+w^2}$. According to

$$E = \frac{1}{2}\int_{-\infty}^{\infty} dx\left[\phi_x^2 + \phi_t^2 + 2(1-\cos\phi)\right], \quad \text{and} \quad P = -\int_{-\infty}^{\infty} dx\phi_t\phi_x,$$

we can obtain the energy and momentum of the moving breather

$$E_{br} = \frac{16\sin\mu'}{\sqrt{1-v^2}}, \quad \text{and} \quad P_{br} = \frac{16v\sin\mu'}{\sqrt{1-v^2}}. \tag{7.219}$$

We then obtain the relation

$$E_{br}^2 = P_{br}^2 + M_{br}^2,$$

where $M_{br} = 16\sin\mu'$ is the mass of a motionless breather. Since the total energy of a free kink or an antikink is equal to 16γ, the binding energy of the breather is $E_B = 16\gamma(1-\sin\mu')$.

In the case of $C' \neq 0$ and $\beta' \neq 0$, Lamb and Davydov assumed that

$$g(x) = \tilde{g}(x)\sqrt{\frac{\beta'}{C'}}, \quad F(t) = \tilde{F}(t)\left(\frac{C'}{\beta'}\right)^{1/4},$$
$$\alpha' = \sqrt{C'\beta'}, \quad g(x)F(t) = \tilde{g}(x)\tilde{F}(t).$$

Equation (7.211) then becomes

$$\tilde{g}_x^2 = -\alpha'\tilde{g}^4 + \ell\tilde{g}^2 + \alpha', \quad \tilde{F}_t^2 = -\alpha'\tilde{F}^4 + (\ell-1)\tilde{F}^2 + \alpha'.$$

When $\alpha' > 0$, we can denote

$$\tilde{g}_x^2 = \alpha'(\tilde{g}_0^2 - \tilde{g}^2)(\tilde{g}^2 - g_1^2),$$

where

$$g_0^2 = -\frac{1}{2\alpha'}\left[\ell + \sqrt{\ell^2 + 4\alpha'}\right], \quad g_1^2 = \frac{1}{2\alpha'}\left[\sqrt{\ell^2+4\alpha'} - \ell\right].$$

Thus

$$\hat{\phi}(x) = \tilde{g}_0 \operatorname{cn}(u,K), \quad \tilde{F}(t) = \tilde{F}_0 \operatorname{cn}\left[\frac{\sqrt{\alpha'}\tilde{F}_0 t}{J}, J\right],$$

where

$$u = \alpha' K^{-1} g_0 (x - x_0), \quad K^2 = \frac{\ell + \sqrt{\ell^2 + 4\alpha'}}{2\sqrt{\ell^2 + 4\alpha'}}$$

$$\tilde{F}_0 = \frac{1}{2\alpha'} \left\{ \ell - 1 + \sqrt{(\ell - 1)^2 + 4\alpha'} \right\},$$

$$J^2 = \frac{\ell - 1 + \sqrt{(\ell - 1)^2 + 4\alpha'}}{2\sqrt{(\ell - 1)^2 + 4\alpha'}}.$$

The solution of the Sine-Gordon equation in such a case is

$$\tan\left[\frac{\phi(x,t)}{4}\right] = \tilde{g}_0 \tilde{F}_0 \operatorname{cn}\left[\frac{g_0\sqrt{\alpha}}{K}(x - x_0), K\right] \operatorname{cn}\left[\frac{\sqrt{\alpha}\tilde{F}_0}{J}t, J\right]. \tag{7.220}$$

This solution represents "plasma vibrations" of standing periodic waves.

If β' in (7.210) is replaced by $-\beta'$ and $C' \neq 0$, $\beta' < 0$, $\beta'' = \sqrt{-\beta' C'} > 0$, in accordance with the above method, Lamb and Davydov obtained solution of the Sine-Gordon equation

$$\tan\left[\frac{1}{4}\phi(x,t)\right] = \tilde{g}_0 \tilde{F}_0 \operatorname{dn}(u, K) \operatorname{sn}(V, J'),$$

with

$$u = \tilde{g}_0 \sqrt{\beta''}(x - x_0), \quad V = \frac{\sqrt{\beta''}}{(-\beta)} \tilde{F}_0 t,$$

$$(-J')^2 = \frac{1 - \ell - \sqrt{(1 - \ell)^2 - 4\beta''}}{1 - \ell + \sqrt{(1 - \ell)^2 - 4\beta''}},$$

$$\tilde{g}_0^2 = \frac{\ell + \sqrt{\ell^2 - 4\beta''}}{2\beta''}, \quad \tilde{F}_0^2 = \frac{1 - \ell + \sqrt{(1 - \ell)^2 - 4\beta''}}{4\beta''}.$$

This solution describes breather oscillations.

7.11 Solving Higher-Dimensional Equations by Reduction

In the previous section, we solved dynamic equations in one-dimensional systems. In this section, we discuss methods for solving higher-dimensional, for example, two and three-dimensional, dynamic equations in the nonlinear quantum mechanics. We introduce a method of reduction from higher to lower dimensional equations proposed by Hirota and Tajiri. As a matter of fact, for the two-dimensional case, the solutions for some similarity variables have been studied by Nakamura *et al.* They obtained an explosion-decay mode solution by generalizing the similarity type plane wave solution derived by Redekopp. Tajiri investigated the similarity solutions of one- and two-dimensional cases using Lie's method.

(I) For the two-dimensional nonlinear Schrödinger equation with $r' = 0$ in (7.188), *i.e.*,

$$i\phi_{t'} + p\phi_{x'x'} + q'\phi_{y'y'} + b|\phi|^2\phi = 0, \tag{7.221}$$

where p, q, b are constants, Tajiri considered an infinitesimal one-parameter (ε) Lie's group in the (x', y', t', ϕ) space

$$\begin{aligned} \bar{x} &= x' + \varepsilon X(x', y', t', \phi) + O(\varepsilon^2), \\ \bar{y} &= y' + \varepsilon Y(x', y', t', \phi) + O(\varepsilon^2), \\ \tilde{t} &= t' + \varepsilon T(x', y', t', \phi) + O(\varepsilon^2), \\ \tilde{\phi} &= \phi + \varepsilon U(x', y', t', \phi) + O(\varepsilon^2). \end{aligned} \tag{7.222}$$

Then

$$\begin{aligned} \bar{\phi}_{\bar{t}} &= \phi_t + \varepsilon[U_t] + O(\varepsilon^2), \\ \bar{\phi}_{\bar{x}\bar{x}} &= \phi_{x'x'} + \varepsilon[U_{x'x'}] + O(\varepsilon^2), \\ \bar{\phi}_{\bar{y}\bar{y}} &= \phi_{y'y'} + \varepsilon[U_{x'x'}] + O(\varepsilon^2), \\ \frac{\partial\bar{\phi}}{\partial\bar{t}} &= \varphi_t + \varepsilon[U_t], \end{aligned} \tag{7.223}$$

where

$$[U_{t'}] = \frac{\partial U}{\partial t'} + \left(\frac{\partial U}{\partial\phi} - \frac{\partial T}{\partial t'}\right)\varphi_{t'} - \frac{\partial Y}{\partial t'}\varphi_{y'} - \frac{\partial X}{\partial t'}\varphi_{x'} - \frac{\partial T}{\partial\phi}\varphi_{t'}^2 - \frac{\partial Y}{\partial\phi}\varphi_{t'}\varphi_{y'} - \frac{\partial X}{\partial\phi}\varphi_{t'}\varphi_{x'}.$$

Similarly

$$\begin{aligned} [U_{x'x'}] = {} & \frac{\partial^2 U}{\partial x'^2} + \left(2\frac{\partial^2 U}{\partial x'\partial\phi} - \frac{\partial^2 X}{\partial x'^2}\right)\varphi_{x'} - \frac{\partial^2 Y}{\partial x'^2}\varphi_{y'} - \frac{\partial^2 T}{\partial x'^2}\varphi_{t'} \\ & + \left(\frac{\partial^2 U}{\partial\phi^2} - 2\frac{\partial^2 X}{\partial x'\partial\phi}\right)\varphi_{x'}^2 - 2\frac{\partial^2 Y}{\partial x'\partial\phi}\varphi_{x'}\varphi_{y'} - 2\frac{\partial^2 T}{\partial x'\partial\phi}\varphi_{x'}\varphi_{t'} \\ & - \frac{\partial^2 X}{\partial\phi^2}\varphi_{x'}^3 - \frac{\partial^2 Y}{\partial\phi^2}\varphi_{x'}^2\varphi_{y'} - \frac{\partial^2 T}{\partial\phi^2}\varphi_{x'}^2\varphi_{t'} + \left(\frac{\partial U}{\partial\phi} - 2\frac{\partial X}{\partial x'}\right)\varphi_{x'x'} \\ & - 2\frac{\partial Y}{\partial x'}\varphi_{x'y'} - 2\frac{\partial T}{\partial x'}\varphi_{x't'} - 3\frac{\partial X}{\partial\phi}\varphi_{x'}\varphi_{x'x'} - \frac{\partial Y}{\partial\phi}\varphi_{y'}\varphi_{x'x'} \\ & - \frac{\partial T}{\partial\phi}\varphi_{t'}\varphi_{x'x'} - 2\frac{\partial Y}{\partial\phi}\varphi_{x'}\varphi_{x'y'} - 2\frac{\partial T}{\partial\phi}\varphi_{x'}\varphi_{x't'}, \end{aligned}$$

where

$$t' = \frac{t}{\hbar}, \quad , x' = x\sqrt{\frac{2m}{\hbar^2}}, \quad y' = y\sqrt{\frac{2m}{\hbar^2}}.$$

Assuming that the 2D-nonlinear Schrödinger equation (7.221) is invariant under the transformations (7.222) and (7.223), we get by comparing terms in first order of ε,

$$i[U_{t'}] + p[U_{x'x'}] + q'[U_{y'y'}] + b(2|\phi|^2 U + \phi^2 U^*) = 0. \tag{7.224}$$

The solutions of (7.224) gives the infinitesimal elements (X, Y, T, U) while leaving invariant (7.221). For (7.224), Tajiri obtained the following for the infinitesimal elements

$$\begin{aligned} X &= \alpha x' + kt'x' + \frac{\beta}{q'}y' + p\gamma t' + \theta_1, \\ Y &= \alpha y' + kt'y' - \frac{\beta}{q'}x' + p\delta t' + \theta_2, \\ T &= 2\alpha t' + kt'^2 + \theta_3, \\ U &= \left\{ i\omega - \alpha - \kappa\left(t - \frac{i}{4p}x'^2 - \frac{i}{4q'}y'^2\right) + \frac{i}{2}\gamma x' + \frac{i}{2}\delta y' \right\}\phi, \end{aligned} \tag{7.225}$$

where α, β, γ, δ, κ, θ_1, θ_2, θ_3, and ω are arbitrary constants. Thus, the similarity variables and form are given by solving the characteristic equations,

$$\frac{dx'}{X} = \frac{dy'}{Y} = \frac{dt'}{T} = \frac{d\phi}{U}. \tag{7.226}$$

The general solution of (7.226) involves three constants, two of them become new independent variables and the third constant plays the role of a new dependent variable. Note that different similarity variables and form can be obtained by integration of (7.226) for different choices for values of the constants α, β, γ, δ, κ, θ_1, θ_2, θ_3 and ω in (7.225).

Case (1), α, γ, δ, κ, θ_1, θ_2, θ_3 and $\omega \neq 0$ and $\beta = 0$. From the integrals of $dx'/X = dt'/T$, $dy'/Y = dy/T$ and $dt'/T = d\phi/U$, Tajiri obtained the new independent and dependent variables as follows

$$\begin{aligned} \xi &= \frac{x'}{\sqrt{|Q(t)|}} - \frac{(\theta_1\kappa - p\gamma\alpha)t + (\theta_1\alpha - p\gamma\theta_3)}{(\kappa\theta_3 - \alpha^2)\sqrt{|Q|}}, \\ \eta &= \frac{y}{\sqrt{|Q(t)|}} - \frac{(\theta_2\kappa - q'\delta\alpha)t + (\theta_2\alpha - q'\delta\theta_3)}{(\kappa\theta_3 - \alpha^2)\sqrt{|Q|}}, \end{aligned} \tag{7.227}$$

and

$$\phi = \frac{1}{\sqrt{|Q(t)|}} e^{iF(x',y',t')} w(\xi, \eta), \quad Q(t') = \kappa t'^2 + 2\alpha t' + \theta_3, \tag{7.228}$$

where

$$F(x',y',t') = \omega \int \frac{dt'}{Q(t')} + \sigma_1 \left\{ \frac{1}{2}(\gamma\xi + \delta\eta) \int \frac{dt'}{\sqrt{|Q|}} \right.$$
$$+\frac{1}{2}\int \frac{1}{\sqrt{|Q|}} \left[\int \frac{\gamma(p\gamma t' + \theta_1) + \delta(q'\delta t' + \theta_2)}{Q\sqrt{|Q|}} dt' \right] dt'$$
$$+\frac{\kappa}{4p}\int \left(\int \frac{p\gamma t' + \theta_1}{Q\sqrt{|Q|}} dt' \right)^2 dt' + \frac{\kappa}{4q'}\int \left(\int \frac{q'\delta t' + \theta_2}{Q\sqrt{|Q|}} dt' \right)^2 dt'$$
$$+\frac{\kappa\xi}{2p}\int \left(\int \frac{p\gamma t' + \theta_1}{Q\sqrt{|Q|}} dt' \right) dt' + \frac{\kappa\eta}{2q'}\int \left(\int \frac{q'\delta t' + \theta_2}{Q\sqrt{|Q|}} dt' \right) dt'$$
$$\left. +\frac{\kappa t' + \alpha}{4}\left(\frac{\xi^2}{p} + \frac{\eta^2}{q'} \right) \right\},$$

and in the above, σ_1 is a step function in Q, *i.e.*,

$$\sigma_1 = \begin{cases} 1 & \text{for } Q > 0, \\ -1 & \text{for } Q < 0. \end{cases}$$

Inserting (7.228) into (7.221), we have

$$pw_{\xi\xi} + q'w_{\eta\eta} + b|w|^2 w - \left\{ A\left(\frac{\xi^2}{p} + \frac{\eta^2}{q'} \right) + B \right\} w = 0, \qquad (7.229)$$

with

$$A = \frac{1}{4}(\kappa\theta_3 - \alpha^2),$$

$$B = \sigma_1 \left\{ \left[\omega - \frac{1}{4(\kappa\theta_3 - \alpha^2)} \right] \left[\frac{\kappa\theta_1^2}{p} + \frac{\kappa\theta_2^2}{q'} + (p\gamma^2 + q'\delta^2)\theta_3 - 2\alpha(\gamma\theta_1 + \delta\theta_2) \right] \right\}.$$

Following the same procedure as in case (1), we can get the similarity variables and forms for other cases. The results are summarized in the following.

Case (2) $\theta_1 \neq 0$ and $\theta_2 \neq 0$, the similarity variables and forms are $\xi = x' - \theta y'$, $\tau = t'$, $\phi = w(\xi, \tau)$, where $\theta = \theta_1/\theta_2$. The differential equation is

$$iw_\tau + (p + q'\theta^2)w_{\xi\xi} + b|w|^2 w = 0. \qquad (7.230)$$

Case (3) $\delta \neq 0$ (or $\gamma \neq 0$), the similarity variables and forms are $\xi = x'$ (or y'), $\tau = t'$, $\phi = \exp[iy'^2/(4q't')]w(\xi,\tau)$ or $\exp[ix'^2/(4pt')]w(\xi,\tau)$. The differential equation is

$$iw_\tau + \frac{i}{2\tau}w + pw_{\xi\xi} + b|w|^2 w = 0,$$

or

$$iw_\tau + \frac{i}{2\tau}w + q'w_{\xi\xi} + b|w|^2 w = 0.$$

Case (2) above is the reduction transformations for essentially one-dimensional propagation and symmetry. The reduction of case (3) was first found by Nakamura. As an example of case (1), we give the reductions of the 2D-nonlinear Schrödinger equation to the nonlinear Klein-Gordon equation or the ϕ^4-equation in the following.

If $\kappa = 1$, $\omega \neq 0$, $\alpha = \beta = \gamma = \delta = \theta_1 = \theta_2 = \theta_3 = 0$, $\xi = x'/t'$, $\eta = y'/t'$, then

$$\phi = \frac{1}{t'} \exp\left[i\left(\frac{x'^2}{4pt'} + \frac{y'^2}{4q't'} - \frac{\omega}{t'}\right)\right] w(\xi, t'), \tag{7.231}$$

and

$$pw_{\xi\xi} + q'w_{\eta\eta} - \omega w + b|w|^2 w = 0.$$

But if $\omega = 1$, $\theta_1 \neq 0$, $\theta_2 \neq 0$, $\theta_3 \neq 0$, $\alpha = \beta = \gamma = \delta = \kappa = 0$, $\xi = x' - (\theta_1/\theta_2)t'$, $\eta = y' - (\theta_2/\theta_3)t'$, then

$$\phi = \exp\left\{i\left[\frac{\theta_1}{2p\theta_3}x + \frac{\theta_2}{2q'\theta_3}y - \left(\frac{\theta_1^2}{2p\theta_3^2} + \frac{\theta_2^2}{2q'\theta_3^2} - \frac{1}{\theta_3}\right)t'\right]\right\} w(\xi, \eta),$$

and

$$pw_{\xi\xi} + q'w_{\eta\eta} - mw + b|w|^2 w = 0,$$

where

$$m = \frac{1}{\theta_3} - \frac{\theta_1^2}{4p\theta_3^2} - \frac{\theta_2^2}{4q'\theta_3^2}.$$

We note that the above equations are the nonlinear Klein-Gordon equation or the ϕ^4-equation when $pq' < 0$.

The solutions of the 1D-nonlinear Schrödinger equation and nonlinear Klein-Gordon equation can also be transformed into the solutions of the 2D-nonlinear Schrödinger equation through the similarity transformations. Substituting the soliton solutions of the 1D-nonlinear Schrödinger equation and the nonlinear Klein-Gordon equation into the transformations, we get the well known soliton solutions extended in the (x, y) plane. The solutions obtained by substituting the soliton solutions of the nonlinear Klein-Gordon equation into the transformation (7.231) are the self-similar soliton-like solutions,

$$\phi = \pm\frac{1}{t'} \exp\left[i\left(\frac{x'^2}{4pt'} + \frac{y'^2}{4q't'} - \frac{\omega}{t'}\right)\right] \times \tag{7.232}$$
$$\begin{cases} \sqrt{\dfrac{2\omega}{b}}\,\mathrm{sech}\sqrt{\dfrac{\omega}{p+qc_1^2}}\left(\dfrac{x'}{t'} - c_1\dfrac{y'}{t'} - \zeta_0\right), & \text{for } \dfrac{\omega}{p+q'c_1^2} > 0,\ \dfrac{b}{p+q'c_1^2} > 0, \\ \sqrt{\dfrac{\omega}{b}}\tanh\sqrt{\dfrac{-\omega}{2\left(p+q'c_1^2\right)}}\left(\dfrac{x'}{t'} - c_1\dfrac{y'}{t'} - \zeta_0\right), & \text{for } \dfrac{\omega}{p+q'c_1^2} < 0,\ \dfrac{r}{p+q'c_1^2} > 0, \end{cases}$$

which was first obtained by Nakamura.

(II) Tajiri considered further the solutions of the 3D-nonlinear Schrödinger equation (7.188) using the same reduction method. Assuming that the coefficients p, q' and r' are not all the same sign, for example, $p, q' > 0$ and $r' < 0$, and using the transformation,

$$X = \frac{1}{\sqrt{p}}(x' - \theta_1 t'), \quad Y = \frac{1}{\sqrt{q'}}(y' - \theta_2 t'), \quad Z = \frac{1}{\sqrt{-r'}}(z' - \theta_3 t'), \quad z' = z\sqrt{\frac{2m}{\hbar^2}},$$

$$\phi = \exp\left\{i\left[\frac{\theta_1}{2p}(x' - \theta_1 t') + \frac{\theta_2}{2q'}(y' - \theta_2 t') + \frac{\theta_3}{2r'}(z' - \theta_3 t') + \theta_4 t'\right]\right\} U(X, Y, Z), \tag{7.233}$$

Tajiri obtained

$$U_{ZZ} - U_{XX} - U_{YY} + cU - b|U|^2 U = 0, \tag{7.234}$$

which is formally the 2D-nonlinear Klein-Gordon equation, where θ_1, θ_2, θ_3, and θ_4 are arbitrary constants, and

$$c = \theta_4 - \frac{\theta_1^2}{4p} - \frac{\theta_2^2}{4q'} - \frac{\theta_3^2}{4r'}.$$

Making transformation once more

$$\begin{aligned}
\xi &= X + \alpha Y = \frac{x'}{\sqrt{p}} + \alpha\frac{y'}{\sqrt{q'}} - \left(\frac{\theta_1}{\sqrt{p}} + \alpha\frac{\theta_2}{\sqrt{q'}}\right)t', \\
\tau &= X - \frac{1}{\alpha}Y \mp \frac{\sqrt{1+\alpha^2}}{\alpha}Z \\
&= \frac{x'}{\sqrt{p}} - \frac{y'}{\alpha\sqrt{q'}} \mp \frac{\sqrt{1+\alpha^2}}{\alpha\sqrt{-rq'}}z' - \left(\frac{\theta_1}{\sqrt{p}} - \frac{\theta_2}{\alpha\sqrt{q'}} \mp \frac{\theta_3\sqrt{1+\alpha^2}}{\alpha\sqrt{-rq'}}\right)t', \\
U &= \exp\left\{i\left[\pm\frac{\alpha(a^2-c)}{2a\sqrt{1+\alpha^2}}\left(X - \frac{1}{\alpha}Y\right) + \frac{1}{2}\left(a + \frac{c}{a}\right)Z\right]\right\} G(\xi, \tau),
\end{aligned} \tag{7.235}$$

we get

$$iG_\tau \pm \frac{\alpha}{2a}\sqrt{1+\alpha^2}G_{\xi\xi} \pm \frac{\alpha b}{2a\sqrt{1+\alpha^2}}|G|^2 G = 0, \tag{7.236}$$

which is formally the 1D-nonlinear Schrödinger equation, where a and α are arbitrary constants. It is well known that the 1D-nonlinear Schrödinger equation can be reduced to the following equation

$$\frac{d^2F}{d\zeta^2} = 2F^3 + \zeta F, \tag{7.237}$$

by the similarity transformation. Using results in (7.233) – (7.235), equation (7.188) with $r < 0$ can be reduced to (7.237) by the following transformation,

$$\begin{aligned}
\zeta &= \left(\frac{a\beta}{\alpha\sqrt{1+\alpha^2}}\right)^{1/3}\left\{\pm\left[\frac{1}{\sqrt{p}}x' + \frac{\alpha}{\sqrt{q'}}y' - \left(\frac{\theta_1}{\sqrt{p}} + \frac{\alpha\theta_2}{\sqrt{q'}}\right)t'\right]\right\} - \frac{\alpha\sqrt{1+\alpha^2}}{4a}\beta \\
&\times\left\{\left[\frac{1}{\sqrt{p}}x' - \frac{1}{\alpha\sqrt{q'}}y' \mp \frac{\sqrt{1+\alpha^2}}{\alpha\sqrt{-r'}}z' - \left(\frac{\theta_1}{\sqrt{p}} + \frac{\theta_2}{\alpha\sqrt{q'}} \mp \frac{\sqrt{1+\alpha^2}}{\alpha\sqrt{-r'}}\theta_3\right)t'\right]^2\right\} \\
\phi &= \left(\frac{a\beta}{\alpha\sqrt{1+\alpha^2}}\right)^{1/3}\left(-\frac{2(1+\alpha^2)}{b}\right)^{1/2}\exp\left\{i\left[\frac{\theta_1}{2p}(x'-\theta_1 t') + \frac{\theta_2}{2q'}(y'-\theta_2 t')\right.\right. \\
&+\frac{\theta_3}{2r'}(z'-\theta_3 t') + \theta_4 t' \pm \frac{\alpha(a^2-c)}{2a\sqrt{1+\alpha^2}}\left(\frac{x'-\theta_1 t'}{\sqrt{p}} - \frac{y'-\theta_2 t'}{\alpha\sqrt{q'}}\right) \\
&+\frac{1}{2}\left(a+\frac{c}{a}\right)\frac{z'-\theta_3 t'}{\sqrt{-r'}} + \frac{\beta}{2}\left(\frac{x'-\theta_1 t'}{\sqrt{p}} + \frac{\alpha(y'-\theta_2 t')}{\sqrt{q'}}\right) \\
&\times\left(\frac{x'-\theta_1 t'}{\sqrt{p}} - \frac{y'-\theta_2 t'}{\alpha\sqrt{q'}} \mp \frac{\sqrt{1+\alpha^2}(z'-\theta_3 t')}{\alpha\sqrt{-r'}}\right) \\
&\left.\left.\mp\frac{\beta^2\alpha}{12a}\sqrt{1+\alpha^2}\left(\frac{x'-\theta_1 t'}{\sqrt{p}} - \frac{y'-\theta_2 t'}{\alpha\sqrt{q'}} \mp \frac{\sqrt{1+\alpha^2}\,(z'-\theta_3 t')}{\alpha\sqrt{-r'}}\right)^3\right]\right\}F(\zeta),
\end{aligned} \tag{7.238}$$

where β is an arbitrary constant. We note here that (7.236) with $b > 0$ is transformed into the equation with $b < 0$ by the substitution $\tau \to -\tau$ and $\xi \to i\xi$.

The 2D-nonlinear Schrödinger equation (7.221) can be easily reduced to the 1D-nonlinear Schrödinger equation by the similarity transformation

$$\xi = x' - \theta y', \quad \tau = t', \quad \phi = G(\xi, \tau) \tag{7.239}$$

where θ is an arbitrary constant. The similarity transformation for the reduction of (7.221) to (7.237) is given by

$$\begin{aligned}
\zeta &= \left[\frac{\beta}{2(p+q'\theta^2)}\right]^{1/3}\left[x' - \theta y' - \frac{1}{2}\beta(p+q'\theta^2)t'^2\right], \\
\phi &= \left[-\frac{2(p+q'\theta^2)}{b}\right]^{1/2}\left[\frac{\beta}{2(p+q'\theta^2)}\right]^{1/3} \\
&\times\exp\left\{i\frac{\beta}{2}\left[(x'-\theta y')t' - \frac{\beta}{3}(p+q'\theta^2)t'^3\right]\right\}F(\zeta).
\end{aligned} \tag{7.240}$$

We note here that there are some differences between the similarity variables (7.238) and (7.240). For the wave described by the similarity variable (7.238), the direction of wave propagation depends on β, where β is a parameter for the reduction of the 1D-nonlinear Schrödinger equation to (7.237). On the other hand, for the wave described by the variable (7.240), the direction is independent of β. It has been pointed out that there is a close connection between the Painleve equation

(7.237) and the completely integrable nonlinear evolution equations by Ablowitz *et al.*. They conjectured that when the ordinary differential equations obtained by the similarity reduction from a given partial differential equation are the Painleve-type, the partial differential equation will be integrable. According to this conjecture, it seems that the 2D-nonlinear Schrödinger equation (7.221) has a N-soliton solution, which however is a parallel propagating solution so that the propagation is effectively one-dimensional. On the other hand, the 3D-nonlinear Schrödinger equation has three-dimensional propagating N-soliton solution which is not parallel propagating. It is well known that the 1D-nonlinear Schrödinger equation has N-soliton solutions. Substituting the well-known N-soliton solutions of the 1D-nonlinear Schrödinger equation into the similarity transformations (7.236) and (7.239), we can get N-soliton solutions of the 3D-nonlinear Schrödinger equation and of the 2D-nonlinear Schrödinger equation, respectively. Then, we have three-dimensional propagating N-soliton solutions of the 3D-nonlinear Schrödinger equation and parallel propagating N-soliton solutions of the 2D-nonlinear Schrödinger equation.

Bibliography

Abdullaev, F. (1994). Theory of solitons in inhomogeneous media, Wiley and Sons, New York.

Ablowitz, M. J. and Segur, H. (1977). Phys. Rev. Lett. **38** (1103) .

Ablowitz, M. J., Kaup, D. J. Newell, A. C. and Segur, H. (1973). Phys. Rev. Lett. **31** (125) .

Ablowitz, M. J., Ramani, A. and Segur, H. (1980). J. Math. Phys. **21** (715) .

Baeriswyl, D. and Bishop, A. R. (1992). Solitons and polarons in solid state physics, World Scientific, Singapore.

Barut, A. D. (1978). Nonlinear equations in physics and mathematics, D. Reidel Publishing, Dordrecht.

Bender, G. M., Milton, K. A., Moshe, M., Pinsky, S and Simmons Jr. L. M. (1987). Phys. Rev. Lett. **58** (2615) .

Bender, G. M., Milton, K. A., Moshe, M., Pinsky, S and Simmons Jr. L. M. (1988). Phys. Rev. D **37** (1472) .

Binder, G. M., Milton, K. A., Pinsky, S. and Simmons Jr., L. M. (1989). J. Math. Phys. **30** (1447) .

Bishop, A. R. (1978). Soliton and physical perturbations, in Solitons in actions, eds. by K. Lonngren and A. C. Scott, Academic Press, New York, pp. 61-87.

Bullough, R. K. and Caudrey, P. J. (1980). Solitons, Springer-Verlag, Berlin.

Chen, H. H. (1978). Phys. Fluids **21** (377) .

Cooper, F., Lucheroni, C. and Shepard, H. (1992). Phys. Lett. A **170** (184) .

Cooper, F., Shepard, H. K. and Simmons Jr., L. M. (1991). Phys. Lett. A **156** (436) .

Davydov, A. S. (1985). Solitons in molecular systems, D. Reidel Publishing, Dordrecht.

Dirac, P. A. M. (1930). Proc. Cambridge Philos. Soc. **26** (376) .

Drazin, P. G. and Johnson, R. S. (1989). Solitons, an introduction, Cambridge Univ. Press, Cambridge.

Duncan, A. and Moshe, M. (1988). Phys. Lett. B **215** (352) .

Eilenberger, G. (1981). Solitons: mathematical methods for physicists, Springer, Berlin.

Fogel, M. B., Trullinger, S. E. and Bishop, A. R. (1976). Phys. Lett. A **59** (81) .

Fogel, M. B., Trullinger, S. E., Bishop, A. R. and Krumhansl, J. A. (1976). Phys. Rev. Lett. **36** (1411) .

Fogel, M. B., Trullinger, S. E., Bishop, A. R. and Krumhansl, J. A. (1977). Phys. Rev. B **15** (1578) .

Gardner, C. S., Greene, J. M., Kruskal, M. D. and Miure, R. M. (1967). Phys. Rev. Lett. **19** (1095) .

Gorshkov, K. A. and Ostrovsky, L. A. (1981). Physica D **3** (428) .

Grimshaw, R. (1979). Proc. R. Soc. London, Ser. A **368** (359) and 377.

Guo, Bai-lin and Pang Xiao-feng, (1987). Solitons, Chin. Science Press, Beijing.

Hirota, R. (1972). J. Phys. Soc. Japan **33** (551) ; Phys. Rev. Lett. **27** (1192) .

Hirota, R. (1973). J. Math. Phys. **14** (805) .

Hirota, R. (1974). Prog. Theor. Phys. **52** (1498) .

Hirota, R. (1977). Prog. Theor. Phys. **57** (797) .

Hirota, R. (1978). J. Phys. Soc. Japan **45** (174) .

Hirota, R. (1982). J. Phys. Soc. Japan **51** (323) .

Hui, W. H. (1979). J. Appl. Math. Phys. **30** (929) .

Ichikava, Y. (1979). Phys. Scr. **20** (296) .

Jones, H. F. and Moshe, M. (1990). Phys. Lett. B **234** (492) .

Karpman, V. I. (1977). JETP Lett. **25** (271) .

Karpman, V. I. and Maslov, E. M. (1977). Sov. Phys. JETP **46** (281) .

Karpman, V. I. and Maslov, E. M. (1978). Sov. Phys. JETP **48** (252) .

Karpman, V. I. and Maslov, E. M. (1982). Phys. Fluids **25** (1686) .

Kaup, D. J. (1976). SIAM. J. Appl. Math. **31** (121) .

Kaup, D. J. (1977). SIAM. Phys. Rev. A **16** (704) .

Kaup, D. J. and Newell, A. C. (1978). Phys. Rev. B **18** (5162) ; Proc. R. Soc. London, Ser A **361** (413) .

Kivshar, Yu. S. (1988). J. Phys. Soc. Japan **57** (4232) .

Kivshar, Yu. S. and Malomed, B. A. (1989). Rev. Mod. Phys. **61** (763) .

Kodama, Y. and Hasegawa, A. (1982). Opt. Lett. **7** (339) .

Konno, K. and Wadati, M. (1975). Prog. Theor. Phys. **53** (1652) .

Konno, K. and Wadati, M. (1983). J. Phys. Soc. Japan **52** (1) .

Kundu, A. (1987). Physica **D25** (399) .

Lamb, G. L. (1969). Phys. Lett. A **29** (507) .

Lamb, G. L. (1971). Rev. Mod. Phys. **43** (99) .

Lamb, G. L. (1980). Elements of soliton theory, Wiley, New York.

Lax, P. D. (1968). Comm. Pure and Appl. Math. **21** (467) .

Liu, S. K. and Liu, S. D. (2000). Nonlinear equations in physics, Beijing Univ. Press, Beijing.

Lonngren, K. E. and Scott, A. C. (1978). Solitons in action, Academic, New York, p. 153.

Malomed, B. A. (1984). Phys. Lett. A **102** (83) .

Malomed, B. A. (1985). Physica D **15** (374) and 385.

Malomed, B. A. (1987). Opt. Commun. **61** (192) .

Malomed, B. A. (1987). Physica D **24** (155) .

Malomed, B. A. (1987). Physica D **27** (113) .

Malomed, B. A. (1987). Phys. Lett. A **120** (28) .

Malomed, B. A. (1987). Phys. Lett. A **123** (459) and 494.

Malomed, B. A. (1988). J. Phys. C **21** (5163) .

Malomed, B. A. (1988). Physica D **32** (393) .
Malomed, B. A. (1988). Phys. Rev. B **38** (9242) .
Malomed, B. A. (1988). Phys. Scr. **38** (66) .
Malomed, B. A. (1989). Phys. Lett. A **136** (395) .
Nakamura, A. (1981). J. Phys. Soc. Japan **50** (2469) .
Nakamura, A. (1982). J. Math. Phys. **23** (417) .
Nimmo, J. J. C. (1983). Phys. Lett. A **99** (279) .
Nimmo, J. J. C. (1988). Symmetric functions and the KP hierarchy in linear evolutions, ed. by J. J. P. Leon, World Scientific, Singapore.
Nokajima, K., Onodera, Y. Nakamura, T. and Sato, R. (1974). J. Appl. Phys. **45** (4095) .
Nozaki, K. and Bekki, N. (1983). Phys. Rev. Lett. **50** (1226) .
Nozaki, K. and Bekki, N. (1984). Phys. Lett. A **102** (383) .
Nozaki, K. and Bekki, N. (1985). J. Phys. Soc. Japan **54** (2362) .
Okopinska, A. and Moshe, M. (1987). Phys. Rev. D **35** (1835) .
Ott, E. and Sudan, R. N. (1969). Phys. Fluids **12** (2388) .
Pang, Xiao-feng (1985). J. Low Temp. Phys. **58** (334) .
Pang, Xiao-feng (1994). Theory of nonlinear quantum mechanics, Chongqing Press, Chongqing.
Pang, Xiao-feng (2003). Soliton physics, Sichuan Sci. and Tech. Press, Chengdu.
Petviashvili, V. I. and Sergeev, A. M. (1984). Sov. Phys. Dokl. **29** (493) .
Pismen, L. M. (1987). Phys. Rev. A **35** (1873) .
Rao, T. A. and Rangwala, A. A. (1988). Bäcklund transformation and soliton wave equation, in Solitons, introduction and application, eds. M. Lakshwanan, Springer, Berlin, p. 176.
Rogers, C. (1985). Theor. Math. Phys. **26** (395) .
Rogers, C. (1986). J. Phys. A **19** (L496) .
Rogers, C. (1986). Phys. Scr. **33** (289) .
Rogers, C. (1989). J. Phys. A **18** (L105) .
Rogers, C. and Shadwick, W. F. (1982). Bäcklund transformations and their applications, Academic Press, New York.
Rose, H. A. and Weinstien, M. I. (1988). Physica D **30** (207) .
Schochet, S. and Weinstein, M. I. (1986). Commun. Math. Phys. **106** (569) .
Tajiri, M. (1983). J. Phys. Soc. Japan **52** (1908) and 2277.
Tajiri, M. (1984). J. Phys. Soc. Japan **53** (1634) .
Wai, P. K. A., Chen, H. and Lee, Y. C. (1990). Phys. Rev. A **41** (426) .
Zabusky, N. J. and Kruskal, M. D. (1965). Phys. Rev. Lett. **15** (240) .
Zakharov, V. E. and Shabat, A. B. (1972). Sov. Phys.-JETP **34** (62) .
Zakharov, V. E. and Shabat, A. B. (1973). Sov. Phys.-JETP **37** (823) .

Chapter 8

Microscopic Particles in Different Nonlinear Systems

In this chapter, we will discuss the states and properties of microscopic particles in various nonlinear systems. We will consider motion of microscopic particles in inhomogeneous media with linearly varying density, in dissipative and random media, and in the presence of external electromagnetic fields, traveling and standing acoustic waves, and other types of time-dependent potential fields. Many-body effects and collisions of microscopic particles in perturbed cases will be also discussed in this chapter.

8.1 Charged Microscopic Particles in an Electromagnetic Field

The two-deminsional nonlinear Schrödinger equation of a charged microscopic particle in an electromagnetic field is of the following form, in natural unit system,

$$i\frac{\partial}{\partial t}\phi(\vec{X},t) = \left\{-\frac{1}{2}\left[\nabla - \frac{1}{2}i\vec{B}(t)\times\vec{X}\right]^2 - \vec{E}(t)\cdot\vec{X} + b\mid\phi(\vec{X},t)\mid^2\right\}\phi(\vec{X},t), \quad (8.1)$$

where $\vec{X}' = (x,y,z)$ denotes the Cartesian coordinates, ∇ denotes the gradient with respect to $\vec{X}$, and $\vec{E}(t)$ and $\vec{B}(t)$ represent a time-dependent uniform electromagnetic field. (Such a field may be realized approximately in the middle of a solenoid. Strictly speaking, the electric field is given by $\vec{E} - (1/2)\vec{B}\times\dot{\vec{X}}$). The various constants have been absorbed into $\vec{X}$, t, $\vec{B}$, $\vec{E}$ and ϕ. Thus b can take the values ± 1, since the sign of the kinetic energy term can be changed, if ϕ is replaced by ϕ^*. Takagi *et al.* assumed $b = -1$ and

$$\vec{B} = (0,0,B), \quad \vec{E} = (E_1, E_2, 0), \quad (8.2)$$

and limited the solution to the following form

$$\phi(\vec{X},t) = \varphi(\vec{r},t)\exp\left(ikz - \frac{1}{2}ik^2 t\right),$$

where $\vec{r} = (x, y)$ and κ is a real parameter. Thus (8.1) becomes the following (1+2)-dimensional equation

$$i\frac{\partial\varphi}{\partial t} = \left[-\frac{1}{2}\triangle - \omega(t)D + \frac{1}{2}\omega^2(t)\vec{r}^2 - \vec{E}(t)\cdot\vec{r} + b|\varphi|^2\right]\varphi, \tag{8.3}$$

where $\triangle$ is the two-dimensional Laplacian,

$$D = -i\left(x\frac{\partial}{\partial y} - y\frac{\partial}{\partial x}\right),$$

$\vec{E} = (E_1, E_2)$ and $\omega = B/2$.

The electromagnetic field may be eliminated by the following three successive transformations to noninertial frames of reference. The first step is to transform to a rotating (or Larmor) frame by defining φ via

$$\varphi(x, y, t) = \varphi'[x\cos\theta(t) - y\sin\theta(t), x\sin\theta(t) + y\cos\theta(t), t], \tag{8.4}$$

where

$$\theta(t) = \int_0^t dt'\omega(t').$$

Thus (8.3) becomes

$$i\frac{\partial}{\partial t}\varphi'(\vec{r}, t) = \left[-\frac{1}{2}\triangle + \frac{1}{2}\omega^2(t)\vec{r}^2 - \vec{E}'(t)\cdot\vec{r} + b|\varphi(\vec{r}, t)|^2\right]\varphi(\vec{r}, t), \tag{8.5}$$

where

$$\vec{E}' = (E_1', E_2')$$

with

$$E_1' = E_1\cos\theta - E_2\sin\theta,$$
$$E_2' = E_1\sin\theta - E_2\cos\theta.$$

The second step is to introduce a scale factor $a(t)$ which is an arbitrary positive solution of the equation

$$\ddot{a} + \omega^2 a = 0, \tag{8.6}$$

with initial condition $a(0) = 1$. We then define the scaled time as

$$\tau(t) = \int_0^t \frac{d\tilde{t}'}{[a(\tilde{t}')]^2}. \tag{8.7}$$

and transform to the dilating frame by defining φ' via

$$\varphi'(\vec{r}, t) = \frac{1}{\sqrt{[a(t)]^d}}\psi'\left\{\frac{\vec{r}}{a(t)}, \tau(t)\right\}\exp\left\{\frac{i\dot{a}(t)\vec{r}^2}{2a(t)}\right\}, \tag{8.8}$$

here $d(=2)$ is the dimension of the space. This transformation can eliminate the harmonic potential in (8.5). It can also ensure that the cubic nonlinear term in the (1+2) dimensions is invariant under this transformation. Takagi found that

$$i\frac{\partial}{\partial t}\psi'(\vec{r},\tau) = \left[-\frac{1}{2}\triangle - \vec{E}_1(\tau)\cdot\vec{r} + b|\psi'(\vec{r},\tau)|^2\right]\psi'(\vec{r},\tau), \tag{8.9}$$

where

$$\vec{E}_1 = \{a[t(\tau)]\}^3\vec{E}'[t(\tau)],$$

and $t(\tau)$ is the inverse of (8.7).

Finally, introducing a time-dependent vector $\vec{R} = (R_1, R_2)$ as a particular solution of

$$\frac{d^2\vec{R}}{d\tau^2} = \vec{E}_1,$$

or

$$\vec{R}(\tau) = \int_0^\tau d\tau' \int_0^{\tau'} d\tau'' \vec{E}_1(\tau''),$$

and the velocity $w(\tau)$: $\vec{w} = (w_1, w_2) = d\vec{R}/d\tau$. Takagi further performed a time-dependent translation (or a generalized Galilean transformation) by defining ψ via

$$\psi'(\vec{r},\tau) = \psi\left[\vec{r} - \vec{R}(\tau), \tau\right]\exp\left[i\vec{w}(\tau)\cdot\vec{r} - \frac{i}{2}\int_0^\tau d\tau' w^2(\tau')\right], \tag{8.10}$$

and found that

$$i\frac{\partial}{\partial\tau}\psi(\vec{r},\tau) = \left[-\frac{1}{2}\triangle + b|\psi(\vec{r},\tau)|^2\right]\psi(\vec{r},\tau). \tag{8.11}$$

This is the (1+2)-dimensional nonlinear Schrödinger equation in the absence of the electromagnetic fields. The solution of (8.11) may be written as

$$\psi(\vec{r},\tau) = g(x,t)\exp\left(i\mu_2 y - \frac{1}{2}i\mu_2^2\tau\right),$$

where μ_2 is a known real parameter. We can obtain the single envelope-soliton solution following approach of Chiao *et al.* and Karpman in the case of an attractive self-interaction ($b=-1$),

$$g(x,\tau) = \eta'\mathrm{sech}[\eta'(x - x_0 - v_1\tau)]\exp\left[iv_1 x - \frac{1}{2}i\left(v_1^2 - \eta'^2\right)\tau\right],$$

where η', x_0 and v_1 are real parameters. Hence one obtains the solution

$$\varphi(\vec{r},t) = \frac{\eta'}{a(t)}\mathrm{sech}\left\{\frac{\eta'}{a(t)}\left[\vec{q}(t)\cdot\vec{r} - b'(t)\right]\right\}\exp\left[iQ(\vec{r},t)\right], \tag{8.12}$$

where

$$\vec{q}(t) = (\cos\theta, -\sin\theta),$$
$$b'(t) = [x_0 + R_1(\tau) + v_1\tau]a,$$
$$Q(\vec{r},t) = \frac{1}{a}\left(\frac{1}{2}\dot{a}\vec{r}^2 + \vec{k}\cdot\vec{r}\right) - \vec{v}\cdot\vec{R}(\tau) - \frac{1}{2}\left(\vec{v}^2 - \eta^2\right)\tau - \frac{1}{2}\int_0^\tau d\tau'\vec{w}^2(\tau').$$

Introducing

$$\vec{v} = (v_1, v_2),$$
$$\vec{k} = (u_1\cos\theta + u_2\sin\theta, -u_1\sin\theta + u_2\cos\theta),$$
$$(u_1, u_2) = \vec{v} + \vec{w}(\tau),$$

where θ, a, and τ are functions of t, defined in (8.4), (8.6) and (8.7), respectively. Equation (8.12) is the solution corresponding to the following initial condition

$$\varphi(\vec{r},0) = \eta\mathrm{sech}[\eta'(x - x_0)]\exp\left[\frac{1}{2}i\dot{a}(0)\vec{r}^2 + i\vec{v}\cdot\vec{r}\right]. \tag{8.13}$$

The direction $\vec{q}(t)$ of propagation of the envelope $|\varphi|$ rotates, and the instantaneous frequency of the rotation is the Larmor frequency $\omega(t)$ or half of the cyclotron frequency $B(t)$. The center $b'(t)$ (*i.e.* the radial distance from the origin) performs forced harmonic motion $\ddot{b}'(t) + \omega^2 b'(t) = E_1\cos\theta - E_2\sin\theta$. At the same time the width and the peak height are modulated by the scale factor a which itself executes harmonic motion (8.6).

In the case of $\vec{E} = 0$, $\vec{v} = 0$, and a constant magnetic field and a positive ω value, Takagi chose the solution $a = \cos\omega t$ of (8.6), so that the solution (8.12) becomes

$$\varphi(\vec{r},t) = \eta'\frac{\mathrm{sech}\,\{\eta'[(x - x_0) - y\tan(\omega t)]\}}{\cos(\omega t)}\exp\left[\frac{i}{2}\left(\frac{\eta'^2}{\omega} - \omega\vec{r}^2\right)\tan(\omega t)\right]. \tag{8.14}$$

Then under a constant magnetic field, the initial field configuration, $\varphi(\vec{r},0) = \eta'\mathrm{sech}[\eta'(x - x_0)]$, will collapse at time $\pi/2w$ and rotate by 90° within its lifetime. The solution cannot be extended beyond this time. In the case of a pulsed magnetic field imposed on the system which has a constant value for $0 < t < t_0$ and vanishes otherwise, where t_0 is a parameter less than $\pi/2w$, Takagi again chose (8.14) for $0 < t < t_0$ and $a = \cos(\omega t_0) - (t - t_0)\omega\sin(\omega t_0)$ for $t > t_0$ accordingly. Solution (8.12) then corresponds again to the initial condition (8.16) and collapses at time $t_0 + \omega^{-1}\cos(\omega t_0)$. However, if we choose

$$a = \begin{cases} \dfrac{\cos[\omega(t_0 - t)]}{\cos(\omega t_0)}, & 0 < t < t_0 \\ \dfrac{1}{\cos(\omega t_0)}, & t > t_0 \end{cases}, \tag{8.15}$$

then solution (8.12) corresponds to the initial condition

$$\varphi(\vec{r},0)=\eta'\,\mathrm{sech}[\eta(x-x_0)]\exp\left\{\frac{1}{2}\left[i\vec{r}^2\omega\tan(\omega t_0)\right]\right\}. \tag{8.16}$$

For $t>t_0$, this solution can be expressed as

$$\begin{aligned}\varphi(\vec{r},t)=\eta'\cos(\omega t_0)\mathrm{sech}\{\bar{\eta}[x\cos(\omega t_0)-y\sin(\omega t_0)]-\eta' x_0\}\\ \exp\left\{\frac{1}{2}i\bar{\eta}\left[t-t_0+\omega^{-1}\tan(\omega t_0)\right]\right\},\end{aligned} \tag{8.17}$$

where $\bar{\eta}=\eta'\cos(\omega t_0)$. If the initial phase factor is arranged as in (8.16), then the field configuration rotates by ωt_0 and its scale changes by the factor $\cos(\omega t_0)$, but the lifetime is infinite.

The generic solitons in the case of $b=+1$ are unstable in the presence of a magnetic field. But, within their lifetime, their motion may be controlled to a certain extent by application of time-dependent electromagnetic fields.

The above model may also be used to describe charged bosons interacting via short-range force. φ would then represent the quantized field operator, and the above result would generate a classical solution (*i.,e.,* coherent part of φ) on which quantum treatment may also be based.

8.2 Microscopic Particles Interacting with the Field of an External Traveling Wave

In the case of a microscopic particle interacting with the field of an external traveling wave, the nonlinear Schrödinger equation can be written as

$$i\frac{\partial\phi}{\partial t'}+\frac{\partial^2\phi}{\partial x'^2}+b|\phi^2|\phi=\varepsilon e^{ikx'-\omega t'}, \tag{8.18}$$

where $t'=t/\hbar$, $x'=x\sqrt{2m/\hbar^2}$, ε is the coupling strength which is small, *i.e.*, $\varepsilon\ll 1$, ω and k are frequency and wave number of the driving field, respectively. Obviously when $\varepsilon=0$, the solution of (8.18) is given by (4.56). At present, it is given by

$$\phi_s(x',t')=\frac{2\eta\exp\{i[\eta x'/2+(\eta^2-v^2)t'/4-\hat{\theta}_0-\pi/2]\}}{\sqrt{b}\cosh[-\eta(x'-vt'-2\hat{x}_0/\eta)/2]}, \tag{8.19}$$

where v, η, $\hat{\theta}_0$ and $\hat{x}_0$ are free parameters. η defines the amplitude and width $(1/\eta)$ of the microscopic particle (soliton), v is the speed of the microscopic particle, $2\hat{x}_o/\eta$ is the location of its center of mass, and $\hat{\theta}_0$ is its initial phase.

According to Faddeev and Takhtajan, the above four parameters of the single-soliton solution form a Hamiltonian dynamic system. η becomes the canonical momentum conjugate to the coordinate $\theta=\hat{\theta}_0-(\eta^2-v^2)t'/4$, and the velocity v

is the canonical momentum conjugate to $x_0' = \hat{x}_\circ - (\eta v/2)t'$. The corresponding Hamiltonian is of the form

$$H(\phi_s) = \int_{-\infty}^{\infty} \left(\left| \frac{\partial \phi_s}{\partial t'} \right|^2 - \frac{b}{2}|\phi|^4 \right) dx' = \frac{1}{4}\left(\eta v^2 - \frac{1}{3}\eta^3 \right) \tag{8.20}$$

when $\varepsilon \neq 0$. The perturbed system has the Hamiltonian form, equation (8.18) can be obtained from the Hamilton equation

$$i\frac{\partial \phi}{\partial t'} = \frac{\delta H}{\delta \phi^*},$$

where the Hamiltonian is given by

$$H[\phi] = \int_{-\infty}^{\infty} dx' \left\{ \left| \frac{\partial \phi}{\partial x'} \right|^2 - \frac{b}{2}|\phi|^4 + 2\varepsilon \Re \left[\phi^* e^{i(kx' - \omega t')} \right] \right\}. \tag{8.21}$$

Cohen employed the adiabatic approximation to study variations of the parameters of the microscopic particle (soliton) due to the external field. Assuming a weak coupling ($\varepsilon \ll 1$), he neglected the radiation effects and possible formation of other solitons. Inserting the unperturbed solution (8.19) into (8.21), one can get

$$\begin{aligned} H[\phi_s] &= H(\eta, v, \theta, x_0, t') \qquad (8.22) \\ &= \frac{1}{4}\left(\eta v^2 - \frac{1}{3}\eta^3 \right) + \varepsilon \pi \sin \left[\frac{(2k - v)x_0'}{\eta} - \omega t' + \theta \right] \operatorname{sech} \left[\frac{(2k - v)\pi}{2\eta} \right]. \end{aligned}$$

The Hamiltonian (8.22) describes a dynamical system with two and a half degrees of freedom, as it is nontrivially coupled to an explicitly time-dependent driving force.

It can be easily shown that the perturbed one-soliton Hamiltonian (8.22) has an additional integral of motion

$$\eta[v - 2k] = \text{const}. \tag{8.23}$$

This integral follows from the complete and irreduced system, (8.18), possessing an exact conservation law,

$$I(\psi) = \Im \int_{-\infty}^{\infty} \left(\psi \frac{\partial \psi *}{\partial x'} \right) dx' = \text{const}. \tag{8.24}$$

where $\psi = \phi \exp[-i(kx' - \omega t')]$. Equation (8.23) can be obtained immediately by inserting (8.19) for ϕ into (8.24).

Cohen chose the integral of motion (8.23) as a new momentum and made the corresponding canonical transformation in the one-soliton parameters' phase space, and exploited the above mentioned symmetry in the time-dependence of the one-soliton Hamiltonian to reconstructed a new Hamiltonian

$$H(q, P, \Omega, Q) = \frac{1}{4}\left[\frac{1}{q}(P + 2qk)^2 - \frac{1}{3}q^3 \right] - \omega q + \varepsilon \frac{\pi \cos \Omega}{\cosh(\pi P/2q^2)}, \tag{8.25}$$

where the newly introduced canonical coordinates are given by

$$q = \eta, \quad P = \eta(v - 2k), \quad Q = \frac{x_0'}{\eta}.$$
$$\Omega = \theta - \frac{v - 2k}{\eta} x_0' - \omega t' + \frac{\pi}{2}, \tag{8.26}$$

The new momentum P is a constant of motion of the system. Therefore, the Hamiltonian (8.25) represents a system with one effective degree of freedom and is thus integrable. This fact excludes any possibility of chaotic motion in the reduced one-soliton system. Obviously, there is a possible resonance between the external wave and the microscopic particle (soliton) in (8.25) which is related to a stable (elliptic) fixed point. To look for the fixed point, we set both $\partial H/\partial q$ and $\partial H/\partial \Omega$ to zero. This yields two conditions. The first is $\Omega_0 = n\pi$ $(n = 0, 1)$. The second condition is obtained by limiting ourselves to the zero order approximation in ε, that is,

$$q_0 = \pm \left[2(k^2 - \omega) \pm \sqrt{4(k^2 - \omega)^2 - P^2}\right]^{1/2}. \tag{8.27}$$

The nontrivial, multivalued resonance condition for the momentum q is caused by the nonstandard dependence of the Hamiltonian (8.25) on the momentum q. From (8.27) we can get two conditions for the existence of a resonance between the driving field and microscopic particle (soliton), *i.e.*, $k^2 > \omega$, which is a condition on the dispersion relation of the external wave alone, and $4(k^2 - \omega)^2 > P^2$, which is a condition on the parameters of the microscopic particle (soliton) and external wave. The linear stability of the fixed points and (8.27) are determined by the sign of the product UR', where $R' = \partial^2 H_0/\partial q^2$ $(q = q_0)$ [H_0 is obtained from the Hamiltonian (8.25) by setting ε to 0], and $U = \varepsilon\pi \cosh^{-1}(\pi P/2q_0^2)$. In the first order of ε, the resonance condition for q becomes

$$q^4 - 4(k^2 - \omega)q^2 + P^2 \mp \varepsilon \frac{4\pi^2 P \sinh(\pi P/2q_0^2)}{q_0 \cosh^2(\pi P/2q_0^2)} = 0, \tag{8.28}$$

where the values of q_0 should be taken from the zeroth order approximation (8.27). From (8.28), we can see that for a given P, two of the resonant values of q do not exist when

$$\frac{Pq_0 \cosh^2[(\pi/2)(P/q_0^2)]}{4\pi^2 P \sinh[(\pi/2)(P/2q_0^2)]} < \varepsilon,$$

which leads to a topological change in the phase plane.

The resonant condition in laboratory coordinates can be obtained from $\dot{\Omega} = 0$, *i.e.*,

$$\dot{\theta} - \frac{d}{dt'}\left[\left(\frac{v - 2k}{q}\right) x_0'\right] - \omega = 0,$$

or

$$\dot{\theta} - \frac{d}{dt'}\left(\frac{Px_0}{q^2}\right) - \omega = 0. \tag{8.29}$$

However, P is a constant of motion for the reduced system and $\dot{P} = \dot{q} = 0$ at resonance. Therefore (8.29) becomes

$$\dot{\theta} - \left(\frac{v-2k}{q}\right)\dot{x}_0' = \omega.$$

Now $\dot{x}_0' = qv_s/2$, and v_s is the velocity of the center of mass of the microscopic particle (soliton). Letting $k = v/2$ be the soliton internal wave number, and $\omega_c = \dot{\theta}$ (in the leading order in ε), the resonant condition can be written as

$$\omega_c - kv_s = \omega - kv_s. \tag{8.30}$$

This condition is the same as the Doppler shifted resonance between two waves, the external (pumping) wave with k and ω and the carrier wave of the microscopic particle (soliton) with k and ω_c. There are two Doppler shifts at the resonant condition. The first enters the right side of (8.30), and it is by the center-of-mass velocity of the soliton. The second, entering the left hand side of (8.30), is less intuitive, and is also by the center-of-mass velocity of the soliton, but it is related to wave number of the soliton carrier wave. This shows the particle-like feature of the microscopic particle (soliton). When $k \to 0$, the system reduces to the well-known resonance with a homogeneous ac drive. For $\omega \to 0$ and $P = 0$, the resonant values of the soliton amplitude are reduced to $\eta = q = \pm\sqrt{2\omega}$. For $\omega > 0$, from (8.27) we can find that there is no resonance for $k = 0$ due to $k^2 > \omega$, which means that the resonance is unique to coupling to the external traveling field. When the values of the wave vectors of the driving wave and the internal soliton wave are close each other, *i.e.*, $\kappa = v/2$, this translates to $P = 0$. From (8.28) we see that in this limiting case, the reduced system will be at the bifurcation point for all values of ε. For $k = 0$ and $v_s = 0$, the resonance (8.30) is reduced to a simple resonance.

8.3 Microscopic Particle in Time-dependent Quadratic Potential

In the following, we consider the states of a microscopic particle placed in a time-dependent quadratic potential,

$$V(x,t) = d_1(t)x + d_2(t)x^2.$$

The corresponding nonlinear Schrödinger equation for the microscopic particle is

$$i\phi_t = \frac{1}{2m}\phi_{xx} - b|\phi|^2\phi + V(x,t)\phi, \quad V(x,t) = \sum_{n=1}^{2} d_n(t)x^n, \tag{8.31}$$

where $\hbar$ is set to 1, $d_1(t)$ and $d_2(t)$ are arbitrary functions of t except that $d_2(t)$ satisfies certain conditions. Solutions of (8.31) were studied by Nogami and Toyama by transforming the coordinate from the "laboratory system" to the system which is fixed on moving body in which the center of the microscopic particle (soliton) is at rest at the origin, separating $V(x,t)$ into the center-of-mass motion and the internal structure of the soliton. The transformed equation determines the structure of the microscopic particle which turns out to be independent of its motion. This method is a generalization of Husimi's method for solving the linear Schrödinger equation with the same $V(x,t)$.

When $V(x,t)=0$, the solution of (8.31) is given by (4.53) or (8.19), which can now be written as

$$\phi(x,t) = B_0(x-vt)\exp\left\{t\left[mvx - \left(\epsilon_0 + \frac{1}{2}mv^2\right)t\right]\right\}, \tag{8.32}$$

where

$$B_0(x-vt) = \sqrt{\frac{k}{2}}\,\text{sech}[k(x-vt)],$$
$$\epsilon_0 = -\frac{k^2}{2m}, \quad k = \frac{1}{2}mb.$$

$B_0(x)$ is a bound state solution of the time-independent equation

$$-\frac{1}{2m}B_{0xx} - bB_0^3 = \epsilon_0 B_0. \tag{8.33}$$

For the external potential $V(x) = max$ (a is a constant), Chen and Liu showed that (8.31) is integrable. Its one-soliton solution behaves like a classical particle of mass m subject to the potential $V(x) = max$. The shape of the soliton remains the same as that of the free soliton given by B_0 of (8.33).

For the nonlinear Schrödinger equation with an external potential, Chen and Liu further showed, by employing a transformation as used by Husimi, that (8.31) with $V(x) = m\omega^2x^2/2$ has an one-soliton solution which again behaves like a classical particle. However, the structure of the soliton in this case is different from the free soliton. Husimi's transformation changes the space variable from x to $\tilde{x}' = x - \xi(t)$, which is the coordinate relative to the moving origin $\xi(t)$. Later $\xi(t)$ will be taken as the center of mass of the soliton. Let the solution $\phi(x,t)$ of (8.31) be of the form

$$\phi(x,t) = \psi(\tilde{x}',t)e^{im\dot{\xi}\tilde{x}'}, \tag{8.34}$$

where

$$\dot{\xi} = \frac{d\xi(t)}{dt}.$$

Inserting Husimi's transformation and (8.34) into (8.31), we get

$$\begin{aligned} i\psi_t = & -\frac{1}{2m}\psi_{\tilde{x}'\tilde{x}'} - b|\psi|^2\psi + V_2(\tilde{x}',t)\psi \\ & + \left[m\ddot{\xi} + V_\xi(\xi,t)\right]\tilde{x}'\psi - \left[\frac{1}{2}m\dot{\xi}^2 - V(\xi,t)\right]\psi. \end{aligned} \tag{8.35}$$

The relation

$$V(\tilde{x}'+\xi,t) = V_2(\tilde{x}',t) + V(\xi,t) + \tilde{x}'V_{\dot{\xi}}(\xi,t),$$

where

$$V_\xi(\xi,t) = \frac{\partial V(\xi,t)}{\partial \xi}$$

was used in obtaining (8.35). It can be demonstrated that the position of the microscopic particle (soliton) $\xi(t)$ satisfies the Newton's equation,

$$m\ddot{\xi} = -V_\xi(\xi,t). \tag{8.36}$$

In such a case, we can assume that

$$\phi(x,t) = g(\tilde{x}',t)\exp\left[im\dot{\xi}\tilde{x}' + i\int^t L(t')dt'\right], \tag{8.37}$$

where

$$\begin{aligned} L(t) &= \frac{1}{2}m\dot{\xi}^2 - V(\xi,t), \\ ig_t &= -\frac{1}{2m}g_{\tilde{x}'\tilde{x}'} - b|g|^2g + V_2(\tilde{x}',t)g. \end{aligned} \tag{8.38}$$

Equation (8.38) looks like (8.31). But there are two differences. (i) Equation (8.31) is written in the laboratory system, whereas, (8.38) refers to the moving system, with its origin at $x = \xi(t)$. (ii) The term V_1 is absent in (8.38) and thus the parity with respect to $\tilde{x}'$ is a good quantum number. The latter allows (8.38) to have a bound-state solution such that

$$\int_{-\infty}^{\infty} |g(\tilde{x}',t)|^2\tilde{x}'d\tilde{x}' = 0.$$

That is, this state is confined in a finite spatial region, and its density $|g(\tilde{x}',t)|^2$ does not have to be t-independent. Equation (8.38) determines the structure of the bound state, which is independent of $\xi(t)$. The center of mass of the bound state is at $\tilde{x}' = 0$, *i.e.*, $x = \xi(t)$.

Nogami and Toyama examined the compatibility between (8.31) and (8.33) based on the Ehrenfest's theorem and obtained that

$$m\ddot{\xi} = \int_{-\infty}^{\infty}(b\rho_x - V_x)\rho dx, \tag{8.39}$$

where

$$\rho(x,t) = |\phi(x,t)|^2.$$

The term with $\rho_x = \partial\rho/\partial x$ on the right-hand side of (8.39) vanishes because

$$\int_{-\infty}^{\infty} \rho_x \rho dx = \frac{1}{2}\left[\rho^2\right]_{-\infty}^{\infty} = 0.$$

Thus the Ehrenfest's theorem in the usual form can be obtained. For the $V(x,t)$ given in (8.31), the following holds

$$\int_{-\infty}^{\infty} V_x \rho dx = V_\xi(\xi, t).$$

Thus (8.39) reduces to (8.36) which shows the classical feature of the microscopic particle in the system. If $d_2(t)$ is t-independent, then g in (8.38) can be written as

$$g(\tilde{x}', t) = D(\tilde{x}')e^{-i\epsilon t},$$

where $D(\tilde{x}')$ is real. We take the solution of the ground-state (8.38) to represent the microscopic particle (soliton). Thus the solution of (8.31) can be

$$\phi(x,t) = D[x - \xi(t)]e^{iS(x,t)}, \tag{8.40}$$

where

$$S(x,t) = m\dot{\xi}(x-\xi) - \epsilon t + \int^t L(\tilde{t}')d\tilde{t}'.$$

and $D(x)$ satisfies

$$-\frac{1}{2m}D_{xx}(x) - bD^3(x) + V_2(x)D(x) = \varepsilon D(x). \tag{8.41}$$

In (8.41) we have set $\xi = 0$ so that $\tilde{x}' = x$ without losing generality. The soliton obtained in this way moves in the x-direction like a classical particle of mass m, and no "radiation" takes place. That is, the energy of the ϕ-field is completely carried by the soliton. The shape of the soliton, determined by $D(\tilde{x}') = D(x - \xi)$ of (8.41), remains the same throughout the course of motion. The solution is valid no matter how rapidly $g_1(t)$ varies with time.

For a linear potential, $V(x,t) = V_1(x,t) = m\alpha(t)x$, since $V_2 = 0$, we find that $D(x) = D_0(x)$ and $\epsilon = \epsilon_0$. $\xi(t)$ is determined by $m\ddot{\xi} = -\alpha(t)$. With this $\xi(t)$, $S(x,t)$ in (8.40) can be worked out, and the results are consistent with those of Chen and Liu, as mentioned in Chapter 4.

For the quadratic potential

$$V(x) = V_2(x) = \frac{1}{2}m\omega^2 x^2,$$

where ω is a constant, equation (8.41) becomes

$$-\frac{1}{2m}D_{xx} - bD^3 + \frac{1}{2}m\omega^2 x^2 D = \epsilon D. \tag{8.42}$$

If the relative strength of the external potential and the nonlinear self-interaction $\sqrt{m\omega}/\kappa \simeq 1$, where $\kappa = mb/2$, the contributions of the two interactions to the soliton binding are about the same. However, in the case of $\sqrt{m\omega}/\kappa \ll 1$, the solution of (8.42) cannot be found analytically. Nogami and Toyama solved the equation numerically, using 0.0005 for the strength of the quadratic potential, $m\omega^2/2$, or $\sqrt{m\omega}/\kappa \simeq 0.36$.

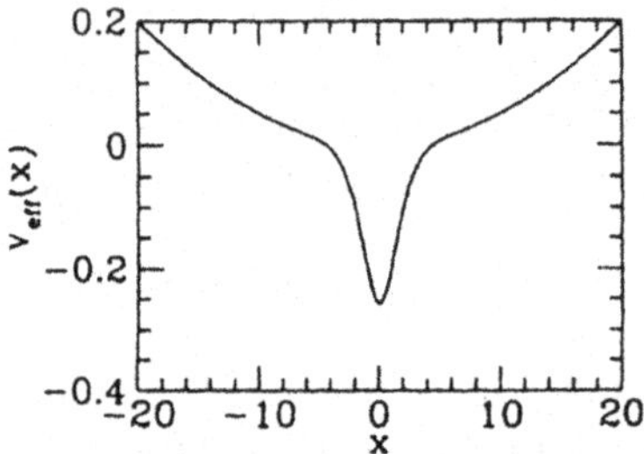

Fig. 8.1 The effective potential $V_{\text{eff}}(x)$ of (8.43) for the ground state of (8.42). The parameters of the model are $m = 1$, $b = 1$ and $m\omega^2/2 = 0.0005$. The x, t, and $V(x)$ can be taken as dimensionless quantities.

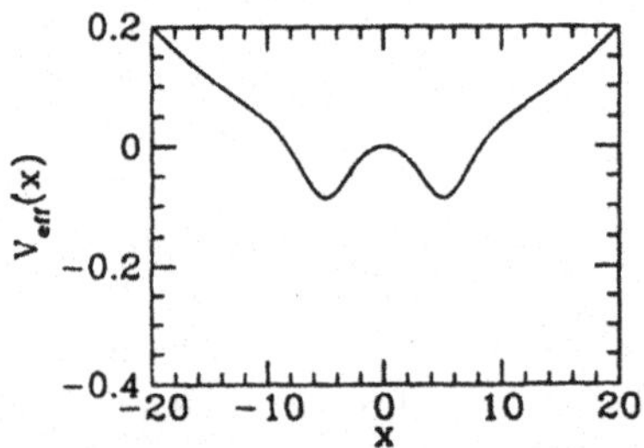

Fig. 8.2 The same as for Fig. 8.1, but for the first excited state.

Figures 8.1 and 8.2 show these effective potentials

$$V_{\text{eff}}(x) = V(x) - bD(x)^2 \tag{8.43}$$

for the ground state and the first excited state, respectively. The values of ϵ for these states are -0.1266 and -0.018, respectively. These can be compared with $\epsilon_0 = -k^2/2m = -0.125$ and $\omega/2 = 0.032$ obtained from $m\omega^2/2 = 0.0005$ and $\sqrt{m\omega}/k = 0.036$. Apart from the energy associated with the motion as a classical particle of mass m in the potential $V(\xi, t)$, the energy of the system is given by

$$\mathcal{E} = \int_{-\infty}^{\infty} \left[\frac{1}{2m}(D_x)^2 - \frac{1}{2}bD^4 + \frac{1}{2}m\omega^2 x^2 D^2 \right] dx. \tag{8.44}$$

$\mathcal{E}$ is not the same as ϵ. [$\mathcal{E}$ becomes ϵ, only if $-bD^4/2$ is replaced with $-bD^4$ in (8.44).] For the ground state and the first excited state, $\mathcal{E}$ are -0.0401 and 0.0169, respectively. D in (8.44) for the ground state approaches B_0 of (8.33), when (8.42) is compared with (8.33), while D for the first excited state is similar to that for the first excited state (of odd parity) of the quadratic potential without the nonlinear term. We take the soliton as the ground state of (8.42). The excited states are much more diffuse than the soliton state. The phase $S(x, t)$ in (8.40) can be worked out by using $\xi(t) = \xi_0 \cos(\omega t)$. This S, combined with the approximation $D \approx B_0$, can give the solution of Chen and Liu.

For the inverted quadratic potential,

$$V(x) = V_2(x) = -\frac{1}{2}m\omega^2 x^2,$$

Equation (8.41), with $\xi = 0$, becomes

$$-\frac{1}{2m}D_{xx}(x) - bD^3 - \frac{1}{2}m\omega^2 x^2 D = \epsilon D. \tag{8.45}$$

The nonlinear term $-bD^3$ ensures a bound state solution for (8.45). Consider a trial function D localized around the origin that can produce an effective attractive potential $-bD^2$. When it is delocalized, D and $-bD^2$ become less attractive and the energy of the system increases. If D is further delocalized away from the origin, the energy begins to decrease because D is more affected by $V(x)$ which is negative. The degree of localization of D can be measured by a parameter λ. Then the energy as a function of λ has a local minimum. This implies the existence of a bound state, which is denoted by

$$D(x, \lambda) = \sqrt{\frac{\lambda}{2}}\,\mathrm{sech}(\lambda x).$$

Thus, we can obtain $\mathcal{E}(D)$ from (8.44) by replacing $m\omega^2 x^2/2$ with $-m\omega^2 x^2/2$, *i.e.*,

$$\mathcal{E}(\lambda) = \frac{1}{6m}\left[\lambda^2 - mb\lambda - \frac{\pi^2}{4}\left(\frac{m\omega}{\lambda}\right)^2\right].$$

Setting $d\varepsilon(\lambda)/d\lambda$ to zero yields two real positive roots if

$$\frac{1}{2}m\omega^2 < \left(\frac{3}{4}\right)^3 \frac{1}{2\pi^2}\frac{\kappa^4}{m} \simeq 0.70,$$

where

$$\kappa = \frac{1}{2}mb.$$

The larger of the two roots corresponds to the minimum of $\mathcal{E}(\lambda)$.

Nogami and Toyama studied this problem numerically and found that a bound state exists when $m\omega^2/2 \leq 0.0007$, or $\sqrt{m\omega}/\kappa \leq 0.39$ and $m\omega^2/2 < 0.0013$. Figure 8.3 shows the effective potential $V_{\rm eff}(x)$ for the bound state obtained for $m\omega^2/2 = 0.0005$. They found that $\epsilon = -0.1230$ and $\mathcal{E} = -0.0434$ for this state. If $m\omega^2/2$ exceeds 0.0007, the bound state becomes unstable.

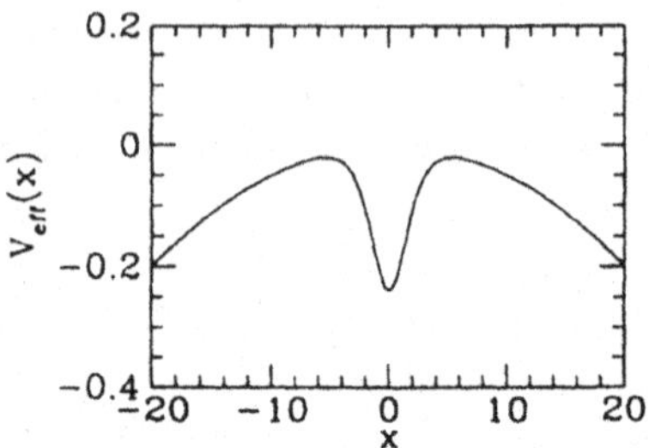

Fig. 8.3 Same as for Fig 8.1, but for $V_2(x) = -mx^2/2$.

For the linear-quadratic potential,

$$V(x,t) = \frac{1}{2}m\omega^2 x^2 - m\omega^2 G(t)x,$$

where $m\omega^2 G(t)x$ is an additional perturbation, Nogami and Toyama found, by solving the Newton's equation for $\xi(t)$ in (8.36), that

$$\xi(t) = \omega \int_{-\infty}^{t} G(\tilde{t}) \sin[\omega(t - \tilde{t})]d\tilde{t}. \tag{8.46}$$

A term in the form of $\xi_0 \cos(\omega t)$ may be added to the right-hand side of (8.46). The amplitude function $D(x - \xi)$ remains the same as D in (8.42). The phase $S(x,t)$ can be worked out by substituting (8.46) into (8.40). It is noted that $G(t)$ can be chosen arbitrarily as long as the integral in (8.46) is well defined.

Nogami and Toyama further considered the case of t-dependent $d_2(t)$ in (8.31). If the variation of $d_2(t)$ with time is sufficiently gentle, then an adiabatic approximation can be used for ϕ in (8.40). In this case, ϵ becomes t-dependent, and ϵt in S of (8.40) has to be replaced by $\int^t \epsilon(\tilde{t}')d\tilde{t}'$. On the other hand, if the variation of $d_2(t)$ becomes wild, then $g(x,t)$ which was in the ground state of (8.41) can get mixed with higher states and become diffuse. If $d_2(t) < 0$, $g(x,t)$ may escape to infinity.

Nogami and Toyama carried out a number of numerical experiments with (8.31). They observed that soliton states of a microscopic particle survives any rapid variation of $d_1(t)$. If $d_2(t)$ varies rapidly in the range of $-0.0007 < d_2(t) \leq 0.001$, [$d_2(t) = 0.001$ corresponds to $\sqrt{m\omega}/\kappa = 0.42$], it was found that the soliton does not break even when $d_2(t)$ changes suddenly which was surprising. No significant trace of "radiation" was found from the soliton.

In general, $V(x,t)$ in (8.31) has a constant curvature as a function of x. If the curvature varies with x, the above derivation does not apply. Hoverer, if the variation of the curvature is very small over the width of the microscopic particle (soliton) ($\simeq 1/\kappa$), the method may still be effective. When the microscopic particle is in a region where the curvature of $V(x,t)$ is negative, *i.e.*, $\partial^2 V(x,t)/\partial x^2 < 0$, the microscopic particle experiences a potential similar to $m\omega^2 x^2/2$. The soliton state of the microscopic particle can easily survive. The restriction on the curvature may be relaxed if $V(x,t)$ is bounded from below. For example, for the repulsive δ-function potential $V(x) = G'\delta(x)$, equation (8.41), with V_2 replaced by $G'\delta(x)$, is

$$-\frac{1}{2m}D_{xx} - bD^3 + G'\delta(x)D = \epsilon D.$$

It was shown by Nogami *et al.* and Foldy that this equation has a bound-state solution with

$$\epsilon = -\frac{\kappa'^2}{2m}, \quad \kappa' = \frac{1}{2}m(b - 2G'),$$

if $2G' < b$. By simulating the δ-function using a Gaussian function with a small width, Nogami and Toyama found that the curvature of this potential at $x = 0$ has a very large magnitude.

8.4 2D Time-dependent Parabolic Potential-field

The behaviors of a microscopic particle described by the nonlinear Schrödinger equation (3.2) in higher dimensional systems are of interest. Garcia-Ripoll *et al.* studied properties of the microscopic particle depicted by the following two-dimensional nonlinear Schrödinger equation with a time-dependent parabolic potential

$$i\frac{\partial\phi}{\partial t} = -\frac{1}{2}\Delta\phi + |\phi|^2\phi + \frac{1}{2}\varepsilon(t)(x^2 + y^2)\phi, \tag{8.47}$$

where

$$\Delta = \frac{\partial^2}{\partial x^2} + \frac{\partial^2}{\partial y^2},$$

$\hbar = m = b = 1$, and $A(\phi) = 0$ in (3.2), ε is a perturbation parameter. In the case of a radially symmetric system, the above authors represented the solution of (8.47) by $\phi(r,\theta,t) = \phi(r,t)e^{im\theta}$, in the polar coordinates, which includes both a typical radially symmetric problem corresponding to $m = 0$ and the vortex line solution with $m \neq 0$. The simplified equation for $\phi(r,t)$ is

$$i\frac{\partial\phi}{\partial t} = -\frac{1}{2r}\frac{\partial}{\partial r}\left(r\frac{\partial\phi}{\partial r}\right) + \left[\frac{m^2}{2r} + |\phi|^2 + \frac{\varepsilon(t)}{2}r^2\right]\phi. \tag{8.48}$$

This equation is nonintegrable and has no exact solution even in the case of $\varepsilon(t)$ being a constant. Its solutions are given by the stationary points of the action

$$S = \int_{t_2}^{t_1} \mathcal{L}(t),$$

where

$$\mathcal{L}(t) = \frac{i}{2}\int\left(\phi\frac{\partial\phi^*}{\partial t} - \phi^*\frac{\partial\phi}{\partial t}\right)d^2x + \int\frac{\varepsilon(t)}{2}r^2|\phi|^2d^2x + \frac{1}{2}\int\left(\left|\frac{\partial\phi^*}{\partial r}\right|^2 + \frac{m^2}{r^2}|\phi|^2 + |\phi|^4\right)d^2x. \quad (8.49)$$

We can define the following integral quantities for the system

$$\begin{aligned} I_1(t) &= \int|\phi|^2d^2x, \\ I_2(t) &= \int|\phi|^2r^2d^2x, \\ I_3(t) &= i\int\left(\phi\frac{\partial\phi^*}{\partial r} - \phi^*\frac{\partial\phi}{\partial r}\right)rd^2x, \\ I_4(t) &= \frac{1}{2}\int\left(|\nabla\phi|^2 + \frac{m^2}{r^2}|\phi|^2 + |\phi|^4\right)d^2x, \end{aligned} \quad (8.50)$$

where $d^2x = rdrd\theta$ and the integration with respect to θ yields a factor of 2π because of the symmetry. These quantities are related to the number, width, radial momentum and energy of the microscopic particles.

It is remarkable that the I_j satisfy the following simple and closed evolution laws

$$\frac{dI_1}{dt} = 0, \quad \frac{dI_2}{dt} = I_3, \quad \frac{dI_3}{dt} = -2\varepsilon(t)I_2 + 4I_4, \quad \frac{dI_4}{dt} = -\frac{1}{2}\varepsilon(t)I_3. \quad (8.51)$$

The first equation comes from the phase invariance of (8.48) under the global phase transformations and corresponds to the conservation of particle number. The other equations can also be obtained in connection with the invariance of the action under symmetry transformations. The equations (8.51) form a linear non-autonomous system for the unknowns, I_j $(j = 1,\cdots,4)$, that has several positive invariants under time evolution, of which the most important is

$$Q = 2I_4I_2 - \frac{I_3^2}{4} > 0.$$

Thus (8.51) can be reduced to a single equation for the parameter $I_2(t)$ as follows

$$\frac{d^2I_2}{dt^2} - \frac{1}{2I_2}\left(\frac{dI_2}{dt}\right)^2 + 2\varepsilon(t)I_2 = \frac{2Q}{I_2}. \quad (8.52)$$

If (8.52) can be solved, then the use of (8.51) would allow us to track the evolution of the other I_j's. By defining $R(t) = \sqrt{I_2}$, which is the width of the microscopic particle (soliton), and inserting it into (8.52), Garcia-Ripoll *et al.* obtained the following

$$\ddot{R} + \varepsilon(t)R = \frac{Q}{R^3}. \tag{8.53}$$

This is a singular (nonlinear) Hill equation. Following approach of Reid *et al.*, Garcia-Ripoll *et al.* gave the following general solution of (8.53)

$$R(t) = \left[\phi^2(t) + \frac{Q}{W^2}\varphi^2(t)\right]^{1/2}, \tag{8.54}$$

where $\phi(t)$ and $\varphi(t)$ are two solutions of the equation

$$\ddot{\Phi} + \varepsilon(t)\Phi = 0, \tag{8.55}$$

satisfying the initial conditions: $\phi(t_0) = R(t_0)$, $\dot{\phi}(t_0) = R'(t_0)$, $\varphi(t_0) = 0$, and $\varphi'(t_0) \neq 0$. W in the above equation is the Wronskian $W = \phi\dot{\varphi} - \varphi\dot{\phi} =$ constant. Equations (8.53) – (8.55) show that the particulars of the microscopic particle, for example, its width, are determined by the Hill equations. They are closely related to the solutions of (8.55) which is explicitly solvable only for particular choices of $\varepsilon(t)$, but its solutions are well characterized and its properties are known. A practical application of (8.54) is the design of $\varepsilon(t)$ starting from the desired properties of the wave packet.

Ripoll *et al.* assumed that $\varepsilon(t)$ depends on a parameter $\varepsilon(t) = 1 + \bar{\varepsilon}(t)$, where $\bar{\varepsilon}(t)$ is a periodic function of time with a zero mean value and a peak value of ϵ_0. There exists a complete theory which gives the intervals of ϵ_0 within which all solutions of (8.55) are bounded (stability intervals) and intervals within which all solutions are unbounded (instability intervals). Both types of intervals are ordered in a natural way.

In the case of $\varepsilon(t) = 1 + \epsilon_0\cos(\omega t)$, the regions where exact resonances occur can be determined by several methods. First, for a fixed ϵ_0, there always exists an infinite ordered series, $\{\omega_n\}$, $\{\omega_n'\}$, which approaches zero, and if ω belongs to this series, equation (8.55) describes a resonance. Second, when ω is fixed, resonances appear if ϵ_0 is large enough. Further, a stability diagram can be drawn in the ϵ_0-ω plane. The boundaries of these regions are called characteristic curves. If $B(\epsilon,\omega)$ is the discriminant of the equation, the characteristic curves can be obtained by solving the equations $B(\epsilon,\omega) = 2$ and $B(\epsilon,\omega) = -2$. In particular, instability regions start at frequencies $\omega = 2, 1, 1/2, \cdots, 2/n^2, \cdots$. The resonant behavior depends only on the mutual relation between the parameters but not on the initial data.

The above analysis on problems with the cylindrically symmetry is completely rigorous up to this point. In the following, we will use various approximations to discuss other related problems.

Notice that the above derivation is confined to the case of a special symmetry for the potential and the solution in order to obtain exact results. When we consider a general two- or three-dimensional model of the nonlinear Schrödinger equation (3.2) with $A(\phi) = 0$, and these constraints are removed, a corresponding set of coupled Hill's equations may still be possible with certain approximations, for example, the scaling laws or the variational ansatz. One such approach requires the wave function to have a quadratic form in the complex phase,

$$\phi = |\phi| \exp\left(i \sum_{jk} \alpha_{jk} x_j x_k \right), \quad (j, k = 1, 2, 3).$$

By introducing this trial function into (8.49), and using some transformations, Garcia-Ripoll *et al.* obtained the following from the Lagrange's equations for the parameters

$$\ddot{R}_i + \varepsilon_i(t) R_i = \frac{1}{R_i^3} + \frac{Q}{R_i R_1 R_2 R_3}. \tag{8.56}$$

Here R_i, $(i = 1, 2, 3)$, are the three root mean square radii of the solution for each spatial direction, *i.e.*, the extensions of the previous $R(t)$ to the three-dimensional problem. In this case, $\varepsilon_i(t)$ are not necessarily equal. Equations (8.56) are not integrable and form a six-dimensional non-autonomous dynamical system. Numerical study on this approximate model (8.56) produces an extended family of resonances which is more or less the Cartesian product of those of (8.54) with minor displacements due to the coupling. The numerical simulation of the nonlinear Schrödinger equation (3.2) with $A(\phi) = 0$ thus confirms the predictions of the simple model (8.56).

Although resonant behavior has been proven in a particular case, it is of interest to find out whether such behavior is also seen in non-symmetric problems. Garcia-Ripoll *et al.* thus looked for stationary solution $\phi(\vec{r}, t) = \psi(\vec{r}) e^{-i\omega t}$ for a stationary trap, $V = V(\vec{r})$. They wrote the nonlinear eigenvalue problem as

$$\omega\psi = -\frac{1}{2}\Delta\psi + |\psi|^2\psi + V(\vec{r})\psi. \tag{8.57}$$

Solutions of the above equation can be used to expand any solution of the non-stationary problem (3.2). By doing so one finds out that the energy absorption process in the time-dependent potential is determined by the separation between the eigenvalues, $\omega_i - \omega_j$, of any two different modes of (8.57). If these differences can be approximated by multiples of a fixed set of frequencies, then an appropriate parametric excitation will induce a sustained process of energy gain (and width growth) such as the one we discussed earlier.

Garcia-Ripoll *et al.* also studied the spectrum $\{\omega\}$ in an axial symmetric 3D potential $V(\vec{r}) = \varepsilon_r(x^2 + y^2) + \varepsilon_z z^2$, using a variant of a pseudospectral scheme and the harmonic oscillator. The results show that the spectrum exhibits an ordered structure, with different directions of uniformity that may be excited by the parametric perturbation. Results in 1D, 2D, and 3D cases without symmetry assumptions were also compared, and are similar despite the fact that the spectrum becomes more complex as dimensionality increases.

In conclusion, by using the moment technique for the cylindrically symmetric problem, a singular Hill equation was obtained which is reducible to a linear Hill equation. Existence of resonances can be verified analytically. For a periodic perturbation, there exist strong extended resonances for relevant parameters of the solution even when the solution is constrained by conservation laws. It is inferred using the parabolic ansatz, analysis of nonlinear spectrum, and numerical simulations that this behavior is also present in non-radially symmetric problems.

8.5 Microscopic Particle Subject to a Monochromatic Acoustic Wave

If the potential field $V(x,t)$ given in (3.2) is due to the following monochromatic acoustic wave,

$$V(x,t) = V_{\rm ac}(x,t) = A\cos(kt - kx + \theta_0)$$

where $k > 0$, and θ_0 is a phase shift of acoustic oscillations in the center of the microscopic particle ($x = 0$). Obviously, $V_{\rm ac}(x,t)$ satisfies the linear variational equation, $V_{{\rm ac}tt} - V_{{\rm ac}xx} = 0$. In such a case, equation (3.2) becomes

$$i\phi_{t'} + \phi_{x'x'} + 2|\phi|^2\phi = V_{\rm ac}(x',t')\phi, \tag{8.58}$$

where $t/\hbar \to t'$, and $x' \to \sqrt{2m/\hbar^2}\,x$. This equation describes the coupling between the microscopic particle (soliton) and the field of the linear acoustic wave. If the amplitude of the acoustic wave, A, is sufficiently small, *i.e.*, $A \ll \eta^2(1-v^2)$, where η and v are the amplitude and velocity of the soliton described by (4.8) or (4.56), respectively, which is a solution of (8.58) with $V_{\rm ac} = 0$, then it is natural to employ the inverse scattering method discussed earlier to solve (8.58) in the presence of the interaction term.

Inserting the perturbation in (8.58) into the general perturbation-induced evolution equation (7.70) for the radiation amplitudes $B(\lambda,t)$, and assuming the soliton to be quiescent, Kivshar *et al.* obtained

$$\frac{dB}{dt'} = \frac{i\pi}{8}\frac{Ak^2}{\lambda^2+\eta^2}\,\mathrm{sech}\left[\frac{\pi(k+2\lambda)}{4\eta}\right] e^{i\{[k-4(\lambda^2+\eta^2)]+\theta_0\}}. \tag{8.59}$$

Using (8.59), Kivshar and Malomed also obtained the emission intensity using the same approach as that used in studying emission problems described by the perturbed Sine-Gordon equation.

Here, we integrate (8.59) directly. We can multiply the right-hand side of (8.59) by $\exp(\alpha t')$, where α is an infinitesimal small but positive parameter. Physically, this is equivalent to turn on adiabatically a perturbation that was absent at $t' = -\infty$. Then the following can be obtained,

$$B^*(\lambda, t') = -\frac{\pi}{8}\frac{Ak^2}{\lambda^2+\eta^2}\text{sech}\left[\frac{\pi(k+2\lambda)}{4\eta}\right]\frac{e^{-i\{[k-4(\lambda^2+\eta^2)]t'+\theta_0\}}}{[k-4(\lambda^2+\eta^2)]+i\alpha}. \tag{8.60}$$

Substituting (8.59) and (8.60) into the equation

$$\frac{d}{dt}N'(\lambda) = \frac{2}{\pi}\Re\left(B^*\frac{dB}{dt}\right), \tag{8.61}$$

where

$$N'(\lambda) = \frac{1}{\pi}\ln\left|\frac{1}{a(\lambda)}\right|^2$$

is the spectral density of the microscopic particle number, $\lambda = \xi + i\eta$ is a spectral parameter, $v = -4\xi$, $a(\lambda)$ is defined in (7.63) in connection with

$$\lim_{\alpha\to 0}\frac{1}{F+i\alpha} \equiv P\left(\frac{1}{F}\right) - i\pi\delta(F),$$

where P represents the principal value, we find

$$\frac{d}{dt'}N'(\lambda) = \frac{1}{8}(\pi Ak)^2\,\text{sech}^2\left[\frac{\pi(k+2\lambda)}{4\eta}\right]\delta\left(\lambda^2+\eta^2-\frac{k}{4}\right). \tag{8.62}$$

The total emission rate of the microscopic particle number is given by

$$\frac{dN}{dt'} \equiv \int_{-\infty}^{\infty}\frac{d}{dt'}N'(\lambda)d\lambda = \left[\frac{dN}{dt'}\right]_+ + \left[\frac{dN}{dt'}\right]_-, \tag{8.63}$$

where

$$\left[\frac{dN}{dt'}\right]_\pm = \frac{(\pi Ak)^2}{8\sqrt{k-4\eta^2}}\,\text{sech}^2\left[\frac{\pi}{4\eta}\left(\sqrt{k-4\eta^2}\mp k\right)\right]. \tag{8.64}$$

From (8.62) – (8.64) we know that emission takes place provided that $k > k_0 \equiv 4\eta^2$, and it is concentrated at two points of the spectrum, $2\lambda_\pm = \pm\sqrt{k-4\eta^2}$. The group velocity of the envelope wave of high-frequency emission is $v_{\text{gr}} = -4\lambda$. That is, $\lambda < 0$ and $\lambda > 0$ correspond to waves emitted to the right [with intensity $(dN/dt')_+$] and to the left [with intensity $(dN/dt)_-$], respectively. As can be seen from (8.64), the intensity of the waves emitted forward, relative to the sound velocity, is greater than that of the waves emitted backwards. Using the conservation of the total microscopic particle number, we can find the decay rate of the soliton,

$d\eta/dt'$. The number of the microscopic particle (soliton) is $N_{\rm sol} = 4\eta$ (Zakharov *et al.*, 1980), so that

$$\frac{d\eta}{dt'} = -\frac{1}{4}\frac{dN}{dt'}. \tag{8.65}$$

According to the above condition, $\eta^2 < k/4$, it is natural to distinguish between "heavy" ($\eta \approx \sqrt{k}$) and "light" ($\eta \ll \sqrt{k}$) solitons. We can estimate the characteristic decay time τ_1 for a "heavy" soliton into a "light" soliton using (8.63) and (8.65). The result is given by

$$\tau_1 \approx \left[\frac{\eta}{d\eta/dt'}\right]_{\eta\sim\sqrt{k}} \approx \frac{1}{A^2 k}.$$

From (8.63) and (8.64), we see that the decay rate of the "light" soliton takes the following form

$$\frac{d\eta}{dt'} = -\frac{1}{2}\left(\frac{\pi A}{8}\right)^2 k^{3/2} \exp\left[-\frac{\pi(\sqrt{k}-k)}{4\eta}\right], \quad (0 < k \ll 1). \tag{8.66}$$

Integrating this equation results in the following expression for the soliton's decay rate

$$\eta \approx \frac{\pi}{4}\sqrt{k}\ln(kA^2t'). \tag{8.67}$$

Far from the soliton, the envelope wave of high frequency emission looks like the monochromatic wave $\phi = \alpha_\pm \exp[-i(4\lambda_\pm^2 t' - 2\lambda_\pm x')]$. By equating the microscopic particle number flux $j_\pm = -4\lambda_\pm\alpha_\pm^2$ carried by the wave (8.67) to the emission rate $(dN/dt')_\pm$ [see (8.63) and (8.64)], Kivshar and Malomed obtained a general expression for the amplitude of the emitted wave

$$\alpha_\pm^2 = \frac{4}{|\lambda_\pm|}\left[\frac{dN}{dt'}\right]_\pm.$$

Kivshar and Malomed also studied the influence of the random acoustic wave field on a microscopic particle (soliton). The field is defined by a random initial conditions, $V_{\rm ac}(x', t'=0) = V_0(x')$, $V_{{\rm act}'}(x', t'=0) = \tilde{V_0}(x')$, subject to the Gaussian correlation, $\langle V_0(x)\rangle = \langle\tilde{V_0}(x)\rangle = \langle V_0(x)\tilde{V}(x)\rangle = 0$, $\langle V_0(x)V_0(\tilde{x})\rangle = V_0^2\delta(x-\tilde{x})$, $\langle\tilde{V_0}(x)\tilde{V_0}(\tilde{x})\rangle = \tilde{V_0}^2\delta(x-\tilde{x})$, where $\tilde{V}_0^2 \ll \eta^7$, $V_0^2 \ll \eta^3$. The random initial conditions give rise to acoustic wave packets with both possible values of the group velocities $c_{1,2} = \pm 1$.

The potential for the random acoustic wave field in this case can be written as

$$V_{\rm ac}(x', t') = \sum_{j=1}^{2}\int_{-\infty}^{+\infty} dq A_j(q) e^{i(c_j q t' - q x')}.$$

It can be shown that the following correlations hold for the spectral amplitudes $A_j(q)$,

$$\langle A_j(q)\rangle = 0,$$
$$\langle A_1(q)A_2(q')\rangle = \frac{\pi}{2}\left(V_0^2 - \frac{\tilde{V}_0^2}{q^2}\right)\delta(q-q'), \tag{8.68}$$
$$\langle A_1(q)A_1(q')\rangle = \langle A_2(q)A_2(q')\rangle = \frac{\pi}{2}\left(V_0^2 - \frac{\tilde{V}_0^2}{q^2}\right)\delta(q-q').$$

The expression for dB/dt corresponding to (8.68) is given by

$$\frac{dB}{dt'} = i\frac{\pi}{4(\lambda^2+\eta^2)}\int_{-\infty}^{+\infty} dq q^2 \operatorname{sech}\left[\frac{\pi(q+2\lambda)}{4\eta}\right]\times \tag{8.69}$$
$$\sum_{j=1}^{2} A_j(q)\exp\{i[c_j q - 4(\lambda^2+\eta^2)]t'\}.$$

Integrating (8.69) in the same way as it was done earlier, we get

$$B^*(\lambda) = \frac{\pi}{4(\lambda^2+\eta^2)}\int_{-\infty}^{+\infty} dq q^2 \operatorname{sech}\left[\frac{\pi(q+2\lambda)}{4\eta}\right]\times \tag{8.70}$$
$$\sum_{j=1}^{2} A_j^*(q)\frac{e^{-i[c_j q - 4(\lambda^2+\eta^2)]t'}}{[c_j q - 4(\lambda^2+\eta^2)] + i\alpha}.$$

Multiplying (8.69) by (8.70) and averaging the product according to (8.68), we can obtain the average emission rate of spectral density,

$$\left\langle \frac{d}{dt'}N'(\lambda)\right\rangle \equiv \frac{2}{\pi}\Re\left\langle B^*(\lambda)\frac{dB(\lambda)}{dt'}\right\rangle \tag{8.71}$$
$$= \frac{16\pi^3}{(\lambda^2+\eta^2)^2}\left[V_0^2 + \frac{\tilde{V}_0^2}{16(\lambda^2+\eta^2)^2}\right]\sum_{j=1}^{2}\operatorname{sech}^2\left\{\frac{\pi}{2\eta}\left[\lambda + 2c_j(\lambda^2+\eta^2)\right]\right\}.$$

The emission induced by a random acoustic field is described by the smooth spectral density given in (8.71). When $\eta \geq 1$, the spectral density is smeared over a broad spectral range $\lambda^2 \leq \eta^2$. If $\eta \ll 1$, equation (8.71) has exponentially sharp maxima at the points $\lambda_j^{(1)} = -2c_j\eta^2$ and $\lambda_j^{(2)} = -c_j/2$. The latter lies outside the range of applicability of (8.58), while in the vicinity of the former, (8.71) takes the form

$$\left\langle \frac{d}{dt'}N'(\lambda)\right\rangle \approx \pi^3(\tilde{V}_0^2 + 16\eta^4 V_0^2)\sum_{j=1}^{2}\operatorname{sech}^2\left[\frac{\pi(\lambda+2c_j\eta^2)}{2\eta}\right]. \tag{8.72}$$

Using (8.72) and the conservation of the total number of microscopic particles,

Kivshar *et al.* derived the average decay rate of the microscopic particles,

$$\frac{d\eta}{dt'} = -\frac{1}{4}\int_{-\infty}^{+\infty} d\lambda \left\langle \frac{d}{dt'}N'(\lambda)\right\rangle = -2\pi^2(\tilde{V}_0^2 + 16\eta^4 V_0^2)\eta. \quad (8.73)$$

That is, at the late stage (when $\eta^2 \ll \tilde{V}_0/4V_0$), the microscopic particle decays according to $\eta(t') \approx \eta_0 \exp(-2\pi^2\tilde{V}_0^2 t')$, which is much faster than that described by (8.67).

The feature that distinguishes (8.73) from (8.65) is the absence of an exponentially small factor. This is due to the fact that there exist long-wave components in the random wave field with wave numbers q for which the exponential-smallness condition $\eta^2 \ll q$ [see (8.64)] does not hold. However, we should remember that (8.73) describes soliton decay induced by long-wave component of the random acoustic wave field [see (8.70) –(8.72)]. The main range of the acoustic wave numbers is $|q| - \eta^2 \ll 1$]. Thus contribution from short-wave component cannot be covered by (8.58).

The above discussion can be applied to other problems, *e.g.* emission of acoustic waves by colliding microscopic particles which was observed in numerical simulation by Degtyarev *et al.* (1974). Within the framework of perturbation theory, explicit evaluations of the total number of emitted particles and its spectral density are possible if the two colliding microscopic particles (solitons) have the same amplitudes, *i.e.*, $\eta_1 = \eta_2 \equiv \eta$, and their velocities $v_1 = -v_2 \equiv v$ lying in the interval $\eta_2 \ll v^2 \ll 1$.

8.6 Effect of Energy Dissipation on Microscopic Particles

As it is known, the energy of a microscopic particle is dissipated, when it is moving in a medium. The dissipation results in damping of its motion or radiation of the microscopic particle. It was proposed by Malomed *et al.* that linear dissipation effect can be adequately accounted for by the following two-term potential

$$P_{\text{diss}} = -\gamma_0\phi + \gamma_1\phi_{xx},$$

where γ_0 and γ_1 are dissipation constants, and satisfy $0 < \gamma_0, \gamma_1 \ll \eta$, with η being the amplitude of the soliton, ϕ_s in (4.56), of the nonlinear Schrödinger equation with $V(x,t) = P_{\text{diss}} = 0$, which may be regarded as a perturbation. The corresponding equation is given by

$$i\phi_{t'} + \phi_{x'x'} + 2|\phi|^2\phi = P_{\text{diss}} = -\gamma_0\phi + \gamma_1\phi_{x'x'}. \quad (8.74)$$

In the absence of the dissipation, the basic emission equation (7.70) may be written in the general form [see (8.59)]

$$\frac{dB(\lambda)}{dt'} = \epsilon f(\lambda) e^{i(\Omega - 4\lambda^2)t'}, \quad (8.75)$$

where Ω is the frequency of emission, $f(\lambda)$ is an arbitrary function of λ.

When dissipation occurs, equation (8.75) is modified as the following

$$\frac{dB(\lambda)}{dt'} = -\gamma B(\lambda) + \epsilon f(\lambda) e^{i(\Omega - 4\lambda^2)t'}, \tag{8.76}$$

where $\gamma \equiv \gamma_0 + 4\eta^2\gamma_1$. Let $B(\lambda)e^{\gamma t} \equiv \bar{B}(\lambda)$, then (8.76) can be written as

$$\frac{d\tilde{B}^*(\lambda)}{dt'} = \epsilon f^*(\lambda) e^{\gamma t'} e^{-i(\Omega - 4\lambda^2)t'}. \tag{8.77}$$

Multiplying (8.77) by the evident solution of this equation,

$$\tilde{B}(\lambda) = \frac{\epsilon f(\lambda)}{i(\Omega - 4\lambda^2) + \gamma} e^{\gamma t'} e^{i(\Omega - 4\lambda^2)t'},$$

Kivshar *et al.* obtained

$$\frac{1}{2}\frac{d}{dt'}|\tilde{B}(\lambda)|^2 \equiv \Re\left[\frac{d\tilde{B}^*(\lambda)}{dt'} B(\lambda)\right] = \frac{\gamma\epsilon^2 |f(\lambda)|^2 e^{2\gamma t'}}{\gamma^2 + (\Omega - 4\gamma^2)^2}. \tag{8.78}$$

Changing from $\bar{B}(\lambda)$ back to $B(\lambda)$, and comparing (8.78) with the non-dissipative equation (8.62), we can infer that the delta function in the expression for the spectral density of the radiation rate are "smeared" as $\delta(\Omega - 4\lambda^2) \to \pi^{-1}\gamma/[\gamma^2 + (\Omega - 4\lambda^2)^2]$. This phenomenon may be called the Lorenz broadening. Its physical meaning is that, due to dissipative absorption, the amplitude of the emitted wave decreases further away from the soliton. The wave is thus not exactly monochromatic.

Kodama *et al.* and Malomed *et al.* further considered the compensating effect to the dissipative effect by introducing a periodic pumping action. In such a case, the dynamic equation of the microscopic particle can be written as

$$i\phi_{t'} - \phi_{x'x'} - 2|\phi^2|\phi = \gamma\phi\left[-1 + \sum_{n=-\infty}^{\infty} \delta(t' - \pi\tau)\right], \tag{8.79}$$

where γ is the dissipation constant, τ is the spatial period of the pumping, which may be regarded as a small perturbation *i.e.*, $\tau\gamma \ll 1$ and $\gamma/\eta^2 \ll 1$. The coefficients of the two terms on the right-hand side of (8.79) are chosen such that full compensation of the dissipation is achieved for all possible values of the amplitude η and the velocity v of the microscopic particle (soliton). However, under the action of the pumping pulses, the microscopic particle generates radiation in the form of small-amplitude waves, which can be considered as "noise". Thus a stationary level of radiation occurs in the case of one soliton propagating in the medium, which was studied by Malomed.

Applying the general inverse scattering transformation evolution equation (7.70) to the radiation amplitudes of the perturbed system, Malomed found an increased amplitude of the radiation field $b(\lambda)$ generated by the microscopic particle (soliton) under the action of the nth pumping pulse

$$\triangle_n b(\lambda) = -\pi\gamma\tau \operatorname{sech}\left[\frac{\pi(\lambda+v/4)}{2\eta}\right] \exp\left[-4i\left(\eta^2 + \frac{1}{16}v^2 + \lambda\frac{1}{2}v\right)n\tau\right]. \tag{8.80}$$

Because of dissipation, $b(\lambda)$ evolves according to

$$b(\lambda, t') = b(\lambda, 0)e^{-\gamma t' + 4i\lambda^2 t'}. \tag{8.81}$$

Form (8.80) and (8.81), the amplitude $b(\lambda, t')$ can be found for $n\tau < t' < (n+1)\tau$, as the sum of an infinite geometrical progression,

$$b(\lambda, t') = \sum_{m=0}^{\infty} \Delta_{n-m} b(\lambda) \exp[(-\gamma + 4i\lambda^2)(t' - (n-m)\tau)] \tag{8.82}$$

$$= \pi\gamma\tau \operatorname{sech}\left[\frac{\pi(\lambda+v/4)}{2\eta}\right] \frac{\exp[-\gamma + 4i\lambda^2(t'-n\tau) - 4i(\eta^2 + v^2/16 + v\lambda/2)n\tau]}{\exp\{-\gamma\tau + 4i[\eta^2 + (\lambda+v/4)^2]\tau\}}.$$

Under the condition $\gamma\tau \ll 1$, it follows from (8.82) that in an equilibrium state the occupation numbers of the radiation modes (the spectral density of the number of quanta) take the stationary values according to (8.61), *i.e.*,

$$N'(\lambda) = \frac{\pi}{4}(\gamma\tau)^2 \frac{\operatorname{sech}^2[\pi(\lambda+v/4)/2\eta]}{\sin^2\{2[\eta^2 + (\lambda+v/4)^2]\tau\}}, \tag{8.83}$$

provided that

$$\left|\sin\left\{2\eta^2 + 2\tau\left(\lambda + \frac{v}{4}\right)^2\right\}\right| \gg \gamma\tau.$$

If this condition does not hold, *i.e.*,

$$2\left[\eta^2 + \left(\lambda + \frac{v}{4}\right)^2\right]\tau = \pi j + \frac{\theta}{2}, \tag{8.84}$$

where $|\theta| \ll \gamma\tau$ and j is an integer, it follows from (8.82) that

$$N'(\lambda) = \frac{\pi(\gamma\tau)^2}{(\gamma\tau)^2 + \theta^2} \operatorname{sech}^2\left[\frac{\pi\sqrt{\pi j - 2\tau\eta^2}}{2\eta\sqrt{2\tau}}\right]. \tag{8.85}$$

In the regime of the soliton's motion with sufficiently low pumping frequency, *i.e.*, with $\tau\eta^2 \gg 1$, (8.83) and (8.85) describe a spectral wave packet whose center lies at the point $\lambda = -v^2/4$ and whose width is $|\lambda + v/4| \approx \eta$. As can be seen from (8.84), inside the packet the spectral density $n(\lambda)$ has a large number ($\approx \tau\eta^2$) of sharp maxima separated by intervals $\Delta\lambda \approx (\eta\tau)^{-1}$; the width of each maximum is $\delta\lambda \approx \gamma/\eta \ll \Delta\lambda$ due to the underlying condition $\gamma\tau \ll 1$.

Adding up the occupation numbers, one can get

$$\tilde{N}'(\lambda) = \frac{1}{\pi}\sum_{m=0}^{\infty}|\triangle_{n-m}b(\lambda)|^2 e^{-2\gamma\tau m} = \frac{\pi}{2}\gamma\tau\,\mathrm{sech}^2\left[\frac{\pi(\lambda+v/4)}{2\eta}\right]. \tag{8.86}$$

This describes the smoothed spectral structure of the wave packet. In calculation of the total number of quanta scattered in the radiation field,

$$N_{rad} \equiv \int_{-\infty}^{+\infty} N'(\lambda)d\lambda,$$

the smoothed spectral density (8.86) yields the same results as (8.83) – (8.85). In particular, $N_{\rm rad} = 2\gamma\tau\eta$. Comparing this to $N_{\rm sol} = 4\eta$ of the quanta bound for the soliton, we discover that in the stationary regime $N_{\rm rad}$ is indeed much smaller than $N_{\rm sol}$, *i.e.*, $N_{\rm rad}/N_{\rm sol} = \gamma\tau/2 \ll 1$.

Kodama and Hasegawa gave the soliton-like solution of the following nonlinear Schrödinger equation with dissipation and high-order dispersion effects

$$i\phi_{t'} + \frac{1}{2}\phi_{x'x'} + |\phi^2|\phi = -i\gamma\phi + i\beta\phi_{x'x'x'}.$$

The soliton-like solution is

$$\begin{aligned}\phi(x',t') = \varphi(x',t')&\left\{1+\beta\left[2\eta^2 t' - 3\eta\tanh(\eta(x'-2\beta\sigma))\right] + \frac{i}{2}\gamma x'^2\right\}e^{i\theta}\\ &+O\left(\gamma^2,\beta^2,\gamma\beta\right),\end{aligned}$$

where $|t'| \ll O(\gamma^{-1/2})$, and

$$\varphi(x',t') = \eta\,\mathrm{sech}[\eta(x'-2\beta\sigma)],\quad \eta = \phi_0 e^{-i\gamma x'},\quad \sigma = \frac{\phi_0^2}{8\gamma}\left(1-e^{-4\gamma t'}\right).$$

Here ϕ_0 is the initial amplitude of the microscopic particle (soliton). The above results show that the dissipation effect of the medium reduces the amplitude and increases the width of the microscopic particle (soliton) at the rate of $e^{-2\gamma t'}$ and $e^{2\gamma t'}$, respectively. However, since the area (the amplitude time the width) is conserved, the features of the microscopic particle are maintained even in the presence of the dissipation loss, but the dispersion distorts the group velocity of the microscopic particle by $2\beta\sigma$.

Nozaki and Bekki gave the soliton-like solution of the nonlinear Schrödinger equation with dissipation and a driving field

$$i\phi_{t'} + \phi_{x'x'} + 2|\phi^2|\phi = -i\gamma\phi - i\varepsilon e^{i\omega t'}.$$

The solution is of the form

$$\begin{aligned}\phi(x',t') &= 2\eta\,\mathrm{sech}(2\eta x')e^{-2i\sigma - i\pi/2}\\ &-\frac{A_0}{\pi}\left[\tilde{\rho}e^{-4i\sigma}\mathrm{sech}^2(2\eta x') + \tilde{\rho}^*\tanh^2(2\eta x')\right] + c(t'),\end{aligned}$$

where

$$\tilde{\rho}_{t'} = i\omega\tilde{\rho} + \delta i\eta|c|e^{2ix'} + i\pi\gamma e^{2\chi} - \gamma\tilde{\rho},$$
$$\chi = 2\sigma - \arg(c), \quad c(t) = ce^{i\omega t},$$
$$\tilde{\rho} = \bar{\rho}[i\arg(c)], \quad \bar{\rho} = \rho(\xi = 0),$$

at $|x| \to \infty$. A_0 (< 0) is a constant weight-factor.

8.7 Motion of Microscopic Particles in Disordered Systems

Motion of microscopic particles in disordered media is an interesting problem and was extensively studied because the competition between disorder and nonlinearity can result in some new and complex effects. Disorder is the origin of exponential decay of the transmission coefficient (Anderson localization). Study shows that weak nonlinearity acting against the disorder changes the exponential length dependence of the transmission coefficient into a power-law dependence. However, a strong nonlinearity can cause a localization decay for the microscopic particle. Moreover, there exists a threshold transmission value for microscopic particles in the media. This problem was studied by Kivshar *et al.* and can be described by the following equation

$$i\phi_{t'} + \phi_{x'x'} + 2|\phi^2|\phi = \varepsilon(x')\phi, \tag{8.87}$$

where $\varepsilon(x') = \varepsilon\sum_n \delta(x' - x'_n)$ describes point impurities with equal intensities ε at random positions x'_n. It can also represent structural disorder of associated systems. If the last term on the left-hand side of (8.87) is neglected, then the equation represents propagation of monochromatic waves in a random inhomogeneous medium which leads to the stochastic dynamic equation,

$$-\phi_{x'x'} + \varepsilon(x')\phi = k^2\phi,$$

where k is the wave number, and to the phenomenon of localization of states by random inhomogeneities due to scattering. Localization means that the transmission coefficient T decays exponentially with the system length L. If $\varepsilon(x)$ is a stationary ergodic random process, then a positive finite number for the localization exists. The localization length $\lambda(k)$ is defined by $L^{-1}\ln T(k) \approx -\lambda^{-1}(k)$. For large L ($\gg \lambda(k)$), a very little transmission is allowed.

Kivshar *et al.* considered the scattering of microscopic particles (solitons) by a random distribution of point impurities with equal intensities ϵ. Once a soliton is incident on the disordered layer, assuming from the left, it will decompose into reflected (r) and transmitted (t) parts. After passing through each impurity the wave packet will reorganize itself and become again a soliton plus some waves of small-amplitudes. In such a case, there are two integrals of motion, the energy E

and the "number of particles" N, defined by

$$E = \int_{-\infty}^{+\infty} dx' \left[|\phi_{x'}|^2 + \varepsilon(x')|\phi|^2 - |\phi|^4 \right], \quad N = \int_{-\infty}^{+\infty} dx' |\phi|^2. \tag{8.88}$$

The total energy transmission coefficient is given by $T^{(E)} = E_t/E_i$ where E_t is the transmitted energy, E_i the incident energy. The transmission coefficient corresponding to the "number-of-particles" is given by $T^{(N)} = N_t/N_i$. Obviously, the constraints, $E_i = E_t + E_r = \text{const.}$, and $N_i = N_t + N_r = \text{const.}$ hold. When the impurity concentration ρ is low, the average distance between two nearby impurities may be larger than the soliton size. In this limit, the scattering can be treated as that by many independent impurities. Then $T \approx \Pi_j T_j$, where T_j is the transmission coefficient of the jth impurity. This approach is still used in disordered systems through the mean transmission coefficients. The transmitted soliton for the jth impurity is then the incident soliton for the $(j+1)$th scatterer. Thus we have (see chapter 4)

$$\begin{aligned} E_{j+1} &= E_j T_j^{(E)}(E_j, N_j), \\ N_{j+1} &= N_j T_j^{(N)}(E_j, N_j), \end{aligned} \tag{8.89}$$

and

$$\begin{aligned} \Delta E_{j+1} &= E_{j+1} - E_{j=} - E_j R_j^{(E)}(E_j, N_j), \\ \Delta N_{j+1} &= N_{j+1} - N_j = -N_j R_j^{(N)}(E_j, N_j), \end{aligned} \tag{8.90}$$

where $R_j^{(E)} = 1 - T_j^{(E)}$ and $R_j^{(N)} = 1 - T_j^{(N)}$ are the reflection coefficients of energy and number-of-particles, respectively. These coefficients can be calculated for $\epsilon \ll 1$, by employing the perturbation theory based on the inverse scattering method. The results are

$$\begin{aligned} R^{(N)} &= \frac{\pi\epsilon^2}{64Nv} \int_0^{+\infty} dy F(y, \alpha), \\ R^{(E)} &= \frac{\pi\epsilon^2 v}{256E} \int_0^{+\infty} dy y^2 F(y, \alpha), \end{aligned} \tag{8.91}$$

where

$$F(y, \alpha) = \frac{[(y+1)^2 + \alpha^2]^2}{\cosh^2[(\pi/4\alpha)(y^2 + \alpha^2 - 1)]}, \tag{8.92}$$

and $\alpha \equiv N/v$. These results, obtained within the Born approximation, are valid if $\epsilon \ll 1$ and $v^2 \gg |\epsilon|\alpha$. Using the soliton solution ϕ_s (4.8) or (4.56) in (8.87) with $\varepsilon(x') = 0$, and considering that there are $(\Delta x')\rho$ impurities in the interval $\Delta x'$ and that energy and number of solitons are function of 2η and v, given by $N = 4\eta$ and

$E = N(v^2 - N^2/3)/4$, we can derive the following equations from (8.90)

$$\frac{dN}{dz} = -\frac{1}{v}\int_0^{+\infty} dyF(y,\alpha), \tag{8.93}$$

$$\frac{dv}{dz} = -\frac{1}{2N}\int_0^{+\infty} dy(y^2-1)F(y,\alpha) - \frac{N}{2v^2}\int_0^{+\infty} dyF(y,\alpha), \tag{8.94}$$

where the distance is measured in units of $x_0' = 64/\pi\rho\epsilon^2$, *i.e.*, $z = x'/x_0'$. In the linear limit, $\alpha \ll 1$, equations (8.93) and (8.94) can be solved analytically which yields $v(x) = v(0) = \text{constant}$. Thus

$$T^{(N,E)} = \frac{N(x)}{N(0)} = \frac{E(x)}{E(0)} = e^{-x'/\lambda_0}, \tag{8.95}$$

where $\lambda_0 \equiv v^2(0)/\rho\epsilon^2 = 1/\rho R_1$, and R_1 is the reflection coefficient of one impurity. This indicates that the transmission coefficient decays exponentially, which is the same as that of the corresponding linear problem, where $\lambda_0 = \lambda(k_0)$, and k_0 represents the wave number of the carrier packet.

Kivshar *et al.* studied numerically these quantities by employing the usual rectangle method for estimating the integrals, and Euler's procedure to integrate the equations. Results in some cases were verified with a leap-frog scheme. In this case, they found that the asymptotic change in $T^{(N,E)}(z)$ depends essentially on the value of the parameter $\alpha(0) = N(0)/v(0)$ that is related to the nonlinearity of the incoming wave. The greater the α, the larger the number of quasiparticles in the soliton becomes, and the smaller its spatial extension. If α is small, the wave is similar to a linear wave packet. It can be proved by computing the derivative of $\alpha(z)$ that the solution α_c of the transcendental equation $\alpha_c^2 - 2 + G(\alpha_c) = 0$, with

$$G(\alpha) \equiv \frac{\displaystyle\int_0^{+\infty} dy(y^2-1)F(y,\alpha)}{\displaystyle\int_0^{+\infty} dyF(y,\alpha)},$$

is such that $\alpha(0) = \alpha_c$ implies $\alpha(z) = \alpha_c$ along the whole disordered layer. Solutions of (8.83) and (8.94) confirm that $\alpha(z)$ is monotonically increasing (decreasing) if $\alpha(0) > \alpha_c$ ($\alpha(0) < \alpha_c$). By solving the equation approximately and integrating (8.93) and (8.94), Kivshar *et al.* found that $\alpha_c \approx 1.28505$. Hence, for the initial conditions $\alpha(0) \ll \alpha_c$, the system evolves, in agreement with the analytical result of (8.95), to a final state in which N approaches zero exponentially while v approaches a constant positive value, satisfying $\alpha(\infty) = 0$ as required (see Fig. 8.4). If $1 \approx \alpha < \alpha_c$, the decay consists of an initial, slow transient after which a fast exponential behavior appears. When $\alpha(0) > \alpha_c$, it leads to a situation in which both N and v become practically constant and so does $\alpha(\infty)$, having some limit value around $\alpha \approx 10$. The dependence of the transmission coefficients on z is determined not only by α but also by the values of $N(0)$ and $v(0)$. In fact, the smaller the $N(0)$ and

$v(0)$, the smaller the interval needed to reach the asymptotic regime, and the initial slope can be roughly 10^5 for $N(0), v(0) \approx 0.1$. But both transmission coefficients approach their asymptotic constant, nonzero values.

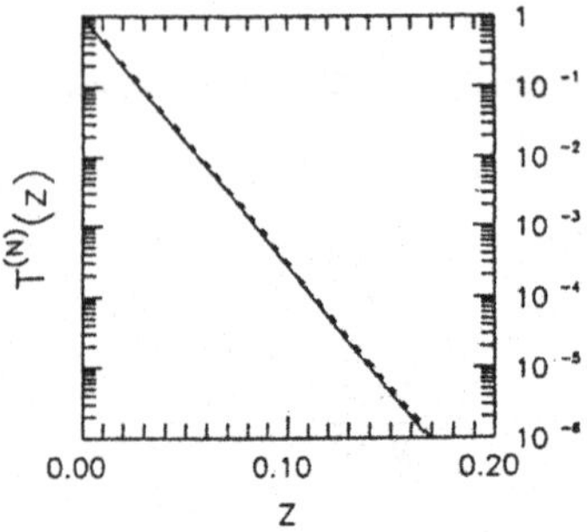

Fig. 8.4 Transmission coefficient $T^{(N)} = Nv/N(0)$ vs. z corresponding to the initial conditions $N(0) = 0.01$, $V(0) = 0.5$, and $\alpha(0) = 0.02$ (The solid line represents numerical results while the dashed line shows the analytical results).

The above discussions show that strong nonlinearity can completely inhibit the localization effects stipulated by the disorder. This effect appears over a threshold nonlinearity. Below this threshold, the transmission coefficient approaches zero as the size of the system increases, either exponentially (Fig. 8.4) or exponentially after a short transient. Above the threshold value there is an undistorted motion of the microscopic particle along the disordered system, *i.e.*, the transmission coefficient does not decay and localization does not happen in the system anymore, probably because of the small soliton width for large values of α.

8.8 Dynamics of Microscopic Particles in Inhomogeneous Systems

Next, we consider the dynamics of microscopic particles in a weakly inhomogeneous medium without dissipation, from the solutions of the nonlinear Schrödinger equation, (3.2). We assume that the medium has a linearly changing density and a linearly changing transverse component of the refraction index. In such a case, the potential $V(x) = \varepsilon\beta(t)x$ is a slowly varying function. It can thus be treated as a perturbation. The perturbed nonlinear Schrödinger equation then takes the following form,

$$i\phi_{t'} + \phi_{x'x'} + 2|\phi|^2\phi = \varepsilon\beta(t')x'\phi \tag{8.96}$$

where $\sqrt{2m/\hbar^2}x \rightarrow x'$, $t/\hbar \rightarrow t'$. Abdullaev proposed the following form for a solitonic solution of this equation

$$\phi = 2\eta\text{sech}\{2\eta[x' - \xi(t')]\}e^{iA(t')x'+iB(t')}, \tag{8.97}$$

with

$$\xi(t') = -2\varepsilon \int_0^{t'} \int_0^{t''} \beta(t''')dt'''dt''',$$

$$\xi A(t') = \varepsilon \int_0^{t'} \beta(t'')dt'',$$

$$B(t') = 4\eta^2 t' - \int_0^{t'} A^2(t'')dt''.$$

It shows that the inhomogeneity leads only to the modulation of the soliton velocity. At a constant gradient ($\beta < 0$), the microscopic particle is decelerated, and then reverses its direction of motion. If we insert (8.97) into (8.96), we can determine the relation among $\xi(t')$, A and B. A change of variables turns the perturbed equation into an ordinary integrable nonlinear Schrödinger equation. This may also be applied to describe the propagation of a soliton in a periodically modulated medium. The equation

$$i\phi_{t'} + \phi_{x'x'} + 2|\phi|^2\phi = a\cos\left(\frac{t'}{t'_0}\right)x'\phi,$$

describes behavior of a charged microscopic particle in an *ac* electric field. Its soliton solution can be written as

$$|\phi(x',t')|^2 = 4\eta^2 \operatorname{sech}^2\left\{2\eta\left[x' - x'_0 - vt' + 4at'^2 \sin\left(\frac{t'}{2t'_0}\right)\right]\right\}. \quad (8.98)$$

Abdullaev also considered properties of microscopic particles in a non-stationary medium described by the following nonlinear Schrödinger equation with time-dependent coefficients

$$i\phi_{t'} + \frac{1}{2}\phi_{x'x'} + |\phi|^2\phi = i\varepsilon\gamma(\varepsilon t')\phi. \quad (8.99)$$

Its solution is given in (8.97).

From the above, we see that the influence of non-stationary perturbed potential on the microscopic particle in the nonlinear quantum mechanics can lead to an effective damping or amplification of the amplitude of the microscopic particle. To the first order in ε, application of perturbation theory yields the following amplitude of the microscopic particle,

$$\eta = 2\eta_0\varepsilon \int \gamma(\varepsilon t')dt'$$

for slowly varying inhomogeneities.

For a δ-like impurity inhomogeneity, $V(x') = -\varepsilon\delta(x' - x'_0)$, we get from (4.20) – (4.22) that the effective potential of the system is of the form

$$V(x'_0) = 8\varepsilon\eta^3 \operatorname{sech}^2(2\eta x'_0).$$

The possibility of transmission and reflection of microscopic particle (soliton), and of oscillating regimes, depend on the initial energy of the microscopic particle. Integration of (4.25) yields the soliton trajectory for $\varepsilon < 0$,

$$X'(t') = \sinh^{-1}\left\{\sqrt{\frac{2|v_0|}{|E|}} - \sin[2|E|(t' - t'_0)]\right\} \tag{8.100}$$

where

$$v_0 = -4\varepsilon\eta^3, \quad X' = 2\eta x'_0,$$

and E is the energy of the soliton.

Phase plane trajectories of (4.25) can be shown when $E < 0$, which corresponds to oscillations of microscopic particle around the impurities. $E \geq 0$ results in unbounded motion of the microscopic particles. To show the oscillations of microscopic particle in the δ-like potential, Abdullaev used an action angle variable (S, θ). The action is given by

$$S = \frac{\pi}{2}\oint pdq = \sqrt{2|v_0|} - \sqrt{2|E|}. \tag{8.101}$$

The value of the action upon the separatrix t_0 equals $S = S_s - \sqrt{2|E|}$. Using (8.101), Abdullaev obtained the cyclic frequency of soliton oscillations,

$$\omega = \frac{\partial E}{\partial S} = \sqrt{2|E|} = (S_s - S).$$

When $S \to S_s$, $\omega(S) \to 0$.

We now consider motion of microscopic particles in a weakly inhomogeneous medium. The inhomogeneity can be caused by space-dependent density fluctuation or gradient, $\delta n(x) = (n(x) - n_0)/n_0$ which is the normalized density deviation from the uniform density n_0. In such a case, the nonlinear Schrödinger equation becomes

$$i\phi_{t'} + \phi_{x'x'} + 2|\phi|^2\phi = \delta n(x')\phi. \tag{8.102}$$

For an arbitrary density profile $\delta n(x)$, equation (8.102) is difficult to solve in general. However, in the case where the inhomogeneity is sufficiently gentle so that the density scale length is large compared with the width of the microscopic particle (soliton) under consideration, $\delta n(x)$ can always be approximated by a linear local density gradient. On the other hand, the microscopic particle undergoes nonuniform accelerations in an arbitrary density. To follow the microscopic particle, a time-dependent density gradient, instead of a constant one, can be used. Chen and Liu adopted an adiabatic approximation to study the dynamics of microscopic particles in such a system. They expanded the density profile $\delta n(x')$ around the position of the microscopic particle $\bar{x}_0(t')$ as

$$\begin{aligned}\delta n(x') &= \delta n[x' - \bar{x}_0(t') + \bar{x}_0(t')] \\ &\simeq \delta n[\bar{x}_0(t')] + [x' - \bar{x}_0(t')]\delta n'[\bar{x}_0(t')] + \cdots .\end{aligned} \tag{8.103}$$

When $2|\delta n'/\delta n''|_{x'=\bar{x}} \gg (x' - \bar{x}_0(t'))$, or $2L \gg d$, where L is the density scale length and d is the width of the microscopic particle, we can neglect the higher order terms. Then the arbitrary density profile is approximated by a linear and time-dependent density profile. If $\delta n = \delta n' = 0$, we also need the condition

$$[x' - \bar{x}_0(t')]^2 \delta n''(\bar{x}_0(t')) \ll 4|\phi|^2_{\max}.$$

Combining (8.102) and (8.103), we get the following nonlinear Schrödinger equation for a linear, time-dependent density gradient

$$i\phi_{t'} + \phi_{x'x'} + [2|\phi|^2 - B(t') - c(t')]\phi = 0, \tag{8.104}$$

where $B(t') = \delta n'[\bar{x}_0(t')]$ and $c(t') = \delta n[\bar{x}_0(t')] - \bar{x}_0(t')\delta n'[x'(t')]$. Applying the transformations

$$\begin{aligned} x' &\to X = x' - I(t'), \\ t' &\to \tau = t', \\ \phi &\to \phi' = \phi e^{iXp(t') + i\theta(t')}, \end{aligned} \tag{8.105}$$

with

$$\begin{aligned} p(t') &= \int_0^{t'} B(z)dz, \\ I(t') &= 2\int_0^{t'} p(z)dZ, \\ \theta(t') &= \int_0^{t'} [p^2(z) + B(z)I(z) + c(z)]dz, \end{aligned}$$

equation (8.104) can be reduced to a standard nonlinear Schrödinger equation,

$$i\phi'_\tau + \phi'_{XX} + 2|\phi'|^2\phi' = 0. \tag{8.106}$$

Thus the solution of (8.104) can be easily obtained. The direct correspondence between the two equations indicates the existence of multi-soliton solutions to (8.104). They are now non-uniformly accelerated. The acceleration is given by $a(t') = -d^2I(t')/dt'^2 = -2B(t')$.

The above equations describe the general propagations of microscopic particles in an arbitrary density profile in the absence of any source. If some physical process is responsible for the building up of a large amplitude of an initially localized wave packet, its subsequent time-evolution with the source switched off is then described by (8.102). Since the initial wave packet is sufficiently localized, we can approximate the density profile at the position of the initial wave packet by a linear density gradient. Chen and Liu studied a parabolic density profile, $\delta n(x') = -\alpha^2 x'^2$. In this case, (8.102) becomes

$$i\phi_{t'} + \phi_{x'x'} + 2|\phi|^2\phi - \alpha^2 x'^2 \phi = 0. \tag{8.107}$$

In fact, the one-soliton solutions of this equation can be obtained directly by assuming $\phi = Ae^{i\theta}$. Then, equation (8.107) becomes

$$\begin{aligned} (A^2)_{t'} + 2(A^2\theta_{x'})_{x'} &= 0, \\ -(\alpha^2 x'^2 + \theta_{t'} + \theta_{x'}^2)A + A_{x'x'} + 2A^3 &= 0. \end{aligned} \tag{8.108}$$

From (4.33), we know that the microscopic particle in this case has a speed given by $\dot{\bar{x}}_0 = 4\xi\cos[2\alpha(t'-t_0')]$ at position $\bar{x}_0 = (2\xi/\alpha)\sin[2\alpha(t'-t_0')]$, where ξ is an integration constant, and is related to the initial soliton speed by $4\xi = \dot{\bar{x}}_0 \mid_{t=t_0}$. It is then easily seen that the amplitude A must be a function of $Y = \alpha x' - 2\xi\sin[2\alpha(t' - t_0')]$ alone. We then have

$$\begin{aligned} A_{t'} &= -4\xi\cos[2\alpha(t'-t_0')]A_{x'}, \\ \theta_x &= 2\xi\cos[2\alpha(t'-t_0')], \\ \theta_t &= -4\xi\alpha\sin[2\alpha(t'-t_0')x' + h(t'), \end{aligned} \tag{8.109}$$

$$\theta = 2\xi x'\cos[2\alpha(t'-t_0')] + \int_0^{t'} h(\tau)d\tau + \theta_0. \tag{8.110}$$

Substituting these equations into (8.108), we get

$$h(t') = -4\xi^2\cos[4\alpha(t'-t_0')] + 4\eta^2, \tag{8.111}$$

and

$$-(Y + 4\eta^2)A + \alpha^2 A_{x'x'} + 2A^3 = 0. \tag{8.112}$$

Thus

$$A \cong \begin{cases} 2\eta\,\mathrm{sech}\left(\dfrac{2\eta Y}{\alpha}\right), & \text{for } Y \ll 2\eta, \\ 0, & \text{for } Y \gg 2\eta. \end{cases} \tag{8.113}$$

Combination of (8.108) and (8.113) gives the soliton solution (8.114), *i.e.*,

$$\begin{aligned} \phi &= 2\eta\,\mathrm{sech}\left\{2\eta\left[x' - \frac{2\xi}{\alpha}\sin[2\alpha(t'-t_0')]\right]\right\} \\ &\times \exp\left\{i\left[2\xi x'\cos[2\alpha(t'-t_0')] - \frac{\xi^2}{\alpha}\sin[4\alpha(t'-t_0')] + 4\eta^2(t'-t_0') + \theta_0\right]\right\}. \end{aligned} \tag{8.114}$$

This represents a localized object oscillating with frequency 2α and amplitude $2\xi/\alpha$ in the harmonic potential well, same as a classical particle. The solution (8.114) can also be obtained using the adiabatic approximation. Expanding $\delta n(x') = \alpha^2 x'^2$ according to (8.103), we get

$$\begin{aligned}\delta n(x') &= [(\alpha x' - \alpha \bar{x_0}) + \alpha \bar{x}_0]^2 \cong (a\bar{x_0})^2 + 2(\alpha x' - a\bar{x}_0)\alpha\bar{x_0} \\ &= 2\alpha^2 \bar{x}_0 x' - \alpha^2 \bar{x}_0^2 = 4\alpha\xi \sin[2\alpha(t' - t_0')]x' - 4\xi^2 \sin^2[2\alpha(t' - t_0')].\end{aligned}$$

Comparing this with (8.104), we obtain

$$\begin{aligned}B(t') &= 4\alpha\xi \sin[2\alpha(t' - t_0')], \\ c(t') &= -4\xi^2 \sin^2[2\alpha(t' - t_0')].\end{aligned} \tag{8.115}$$

The one-soliton solution given in (8.113) and (8.115) is then obtained from (8.105). It turns out to be identical to (8.114) which is obtained by direct integration. In fact, (8.107) is a special case of (8.31), but the way of solving the equation and their physical meanings are different.

8.9 Dynamic Properties of Microscopic Particles in a Random Inhomogeneous Media

It is well known that there always exist fluctuations in physical parameters in any medium. Such fluctuations in external fields and initial conditions result in stochastic dynamics of microscopic particles and other nonlinear waves in the system. Apart from the stochastization of the soliton parameters, processes related to soliton-radiated random waves, stochastic soliton and breather decay, as well as "dynamic chaos" of nonlinear wave due to strong instability waves under periodic perturbations, *etc.* might also exist. In this section, we study the dynamic properties of microscopic particles in such random inhomogeneous media using the mean field method, the adiabatic approximation and the statistical Born approximation, respectively, as it was done by Abdullaev *et al.*

8.9.1 *Mean field method*

We begin with a brief introduction to the mean field method. Consider a nonlinear wave equation

$$L\phi = \varepsilon(x,t)M\phi + Q\phi^2, \tag{8.116}$$

where Q is a deterministic linear operator, L and M are linear integro-differential operators, $\varepsilon(x,t)$ is a random function with prescribed statistics. The wave field will be written as the sum of a mean field $\langle\phi\rangle$ and a scattered term, $\delta\phi$, *i.e.*,

$$\phi = \langle\phi\rangle + \delta\phi, \quad \langle\delta\phi\rangle = 0. \tag{8.117}$$

Substituting (8.117) into (8.116), and averaging over all the realizations of ε, we have

$$L\langle\phi\rangle = \langle\varepsilon M\delta\phi\rangle + Q\langle\phi\rangle^2 + Q\langle\delta\phi^2\rangle,$$
$$L\delta\phi = \varepsilon M\langle\phi\rangle + \varepsilon M\delta\phi - \langle\varepsilon M\delta\phi\rangle + Q(\delta\phi^2 - \langle\delta\phi^2\rangle) + 2Q\langle\phi\rangle. \quad (8.118)$$

In the case of $\varepsilon \ll 1$, $\delta\phi$ is found to be

$$\delta\phi = \frac{1}{L}\varepsilon M\langle\phi\rangle, \quad (8.119)$$

and the equation satisfied by the mean field becomes

$$L\langle\phi\rangle = N\langle\phi\rangle + Q\langle\phi\rangle^2. \quad (8.120)$$

This equation is closed with respect to $\langle\phi\rangle$, and can be solved.

For a stochastic nonlinear Schrödinger equation, which is of the form

$$i\phi_{t'} + \frac{1}{2}\phi_{x'x'} + |\phi|^2\phi = -\varepsilon(x',t')\phi, \quad (8.121)$$

the mean field satisfies

$$i\langle\phi\rangle_{t'} + \frac{1}{2}\langle\phi\rangle_{x'x'} + \langle|\phi|^2\phi\rangle + \langle\varepsilon(x',t')\phi\rangle = 0. \quad (8.122)$$

According to the mean field method, we can write

$$\langle|\phi|^2\phi\rangle \cong |\langle\phi\rangle^2|\langle\phi\rangle.$$

Klyatzkin used the Furutzu-Novikov formula and $\langle\varepsilon(t,x)\rangle = 0$ to decouple the mean $\langle\varepsilon\phi\rangle$ by

$$\langle\varepsilon(x',t')\phi(x',t')\rangle = \int_0^{t'} dt'' \int_{-\infty}^{\infty} dx'' \langle\varepsilon(x'',t'')\varepsilon(x',t')\rangle x' \left\langle \frac{\delta\phi(x',t')}{\delta\varepsilon(x'',t'')} \right\rangle. \quad (8.123)$$

The causality condition requires that

$$\frac{\delta\phi(x',t')}{\delta\varepsilon(x'',t'')} = 0, \quad t'' < t'.$$

Integrating (8.122) between zero and t' and calculating again the variational derivative $\delta\phi/\delta\varepsilon(x'',t'')$, we can arrive at

$$\lim \frac{\delta\phi(x',t')}{\delta\varepsilon(x'',t'')} = 2iA_\varepsilon(0)\langle\phi(x',t')\rangle\delta(x'-x''), \quad (8.124)$$

where the correlator $\langle\varepsilon(x',t')\varepsilon(x'',t'')\rangle = 2A_\varepsilon(x''-x')\delta(t''-t')$. Inserting (8.124) into (8.122) and (8.123), we get

$$i\langle\phi\rangle_{t'} + \frac{1}{2}\langle\phi\rangle_{x'x'} + |\langle\phi\rangle|^2\langle\phi\rangle = -iA_\varepsilon(0)\langle\phi\rangle. \quad (8.125)$$

This is the nonlinear Schrödinger equation with damping, and can be solved analytically for $A_\varepsilon(0) \ll 1$ using the above approach.

8.9.2 *Statistical adiabatic approximation*

We consider the stochastic nonlinear wave equation

$$\hat{N}L\phi = R[\varepsilon(x', t', \phi, \phi_{t'}, \phi_{x'}, \cdots)], \tag{8.126}$$

where $\hat{N}L$ is a nonlinear operator for the nonlinear Schrödinger equation,

$$\hat{N}L = i\partial_{t'} + \partial_{x'x'} + b|\phi|^2,$$

R is a perturbation operator, $\varepsilon(x,t)$ is a random function which satisfies the Gaussian distribution with zero mean $\langle\varepsilon\rangle = 0$, and $\langle\varepsilon(x',t')\varepsilon(y',t')\rangle = B_{\ell_\varepsilon\tau_\varepsilon}(x'-y', t'-\tau)$, where ℓ_ε is a correlation length, τ_ε is a correlation time. For $\ell_\varepsilon, \tau_\varepsilon \to 0$, $B(x'-y', t'-\tau) \approx 2\sigma_\varepsilon^2\delta(x'-y')\delta(t'-\tau)$ which corresponds to the δ-correlation random process. $\varepsilon(x',t')$ is a weak random perturbation, *i.e.*, $\varepsilon \ll 1$. We can therefore infer that the parameters of the microscopic particle (amplitude, velocity, phase, *etc.*) change smoothly with time, while their mutual correlations remain the same as in the case of a free particle.

For the one-particle stochastic nonlinear Schrödinger equation with damping γ and under a random field $\varepsilon(t)$

$$i\phi_{t'} + \frac{1}{2}\phi_{x'x'} + |\phi|^2\phi = -i\gamma\phi + \varepsilon(t'), \tag{8.127}$$

applying the perturbation theory described in Sections 7.4 and 8.6, its solution can be obtained as

$$\phi_s = 2\eta(t')\,\mathrm{sech}\{2\eta[x' - \xi(t')]\}\exp\left[i\frac{2\eta v(t')}{\eta(t')}(x' - \xi) + i\theta(t')\right], \tag{8.128}$$

where the soliton parameters satisfy the following equations which are obtained using the inverse scattering method, based on the adiabatic approximation,

$$\begin{aligned}
\frac{d\eta}{dt'} &= -2\gamma\eta - \frac{\pi\varepsilon(t')\sin\theta(t')}{2\cosh[\pi v(t')/2\eta(t')]},\\
\frac{d\xi}{dt'} &= 2v - \frac{\pi^2\varepsilon(t')\cos\theta(t')\sinh[\pi v(t')/2\eta(t')]}{8\eta^2\cosh^2[\pi v/2\eta]},\\
\frac{dv}{dt'} &= \frac{\pi\varepsilon v\sin\theta(t')}{4\eta\cosh[\pi v/2\eta]},\\
\frac{d\theta}{dt'} &= 2(v^2 + \eta^2) - \frac{\pi\varepsilon\cosh\theta(t')}{\pi\cosh[\pi v/2\eta]}.
\end{aligned} \tag{8.129}$$

Abdullaev *et al.* expanded these parameters in the forms of

$$\begin{aligned} v &= v_0 + v_1 + \cdots, \\ \theta &= \theta_0 + \theta_1 + \cdots, \\ \eta &= \eta_0 + \eta_1 + \cdots \\ \xi &= \xi_0 + \xi_1 + \cdots, \end{aligned}$$

where $v_1, \eta_1, \theta_1, \xi_1 \approx \varepsilon$, and

$$\begin{aligned} v_0 &= \text{constant}, \\ \eta_0 &= \eta_i e^{-2\gamma t'}, \\ \eta_i &= \eta_0(t'=0), \\ \theta_0 &= 2v_0^2 t' + \theta_i + \frac{\eta_i^2}{\gamma}(e^{-2\gamma t'} - 1), \\ \xi_0 &= 2v_0 t' + \xi_i. \end{aligned} \tag{8.130}$$

In the case of $\gamma = 0$, we have

$$\frac{d\eta}{dt'} = -\frac{\pi\varepsilon(t')\sin\theta_0(t')}{2\cosh[\pi v_0/2\eta_0]}. \tag{8.131}$$

From this, Abdullaev *et al.* obtained the fluctuation in the soliton's amplitude (for $\gamma = 0$) which is given by

$$\langle \eta_1^2 \rangle = \frac{\pi\sigma_\varepsilon^2 t'}{\cosh[\pi v_0/2\eta_0]}\left\{1 - \frac{\sin[4(v_0^2+\eta_0^2)t']}{4(v_0^2+\eta_0^2)t'}\right\}. \tag{8.132}$$

Obviously, $\tau_\varepsilon \neq 0$ which leads to small-parameter corrections $\tau_\varepsilon/t'_s \ll 1$. The stochastic growth in the energy of the microscopic particle is a result of interaction with the fluctuating field. The linear amplitude increases with t' is accompanied by oscillations of frequency $4(v_0^2 + \eta_0^2)$. For $4(v_0^2 + \eta_0^2)t' \gg 1$, we obtain the diffusion coefficient

$$D_\eta = \lim_{t'\to\infty} \frac{\langle \eta_1^2 \rangle}{t'} = \frac{\pi\sigma_\varepsilon^2}{4\cosh^2(\pi v_0/2\eta_0)}.$$

The mean values $\langle v_1^2 \rangle$, $\langle \xi_1^2 \rangle$, and $\langle \theta_1^2 \rangle$ could be also obtained in a similar way.

We now study the parametric interaction between a microscopic particle and a random field. The nonlinear Schrödinger equation, describing propagation of the microscopic particle in a medium with fluctuating parameters, has the following form in this case

$$i\phi_{t'} + \frac{1}{2}\phi_{x'x'} + |\phi^2|\phi = -\omega^2(t')x'^2\phi. \tag{8.133}$$

For $\omega = \omega_0$, the microscopic particle (soliton) performs periodic oscillations about $x' = 0$. In the adiabatic approximation the parameters can be denoted by

$$\frac{d\xi}{dt'} = 2v, \quad \frac{d\eta}{dt'} = 0, \quad \frac{dv}{dt'} = \frac{1}{2}\omega^2(t')\xi. \tag{8.134}$$

The coordinate of the center of mass of the microscopic particle (soliton) satisfies

$$\frac{d^2\xi}{dt'^2} + \omega^2(t')\xi = 0. \tag{8.135}$$

Let $\omega^2(t') = \omega_0^2[1 + \varepsilon(t')]$, where $\varepsilon(t')$ is a random function. Abdullaev *et al.* expressed the mean values $\langle\xi\rangle$, and $\langle y\rangle$ $(y = \dot{\xi})$ by

$$\frac{d\langle\xi\rangle}{dt'} = \langle y\rangle, \quad \frac{d\langle y\rangle}{dt'} = -\omega_0^2\langle\xi\rangle.$$

whose solution corresponding to the initial conditions $\xi(0) = 0$ and $y(0) = 1$ is

$$\langle\xi\rangle = \frac{1}{\omega_0}\sin(\omega_0 t'), \quad \langle y\rangle = \cos(\omega_0 t').$$

The second moments are given by

$$\begin{aligned} \langle\dot{\xi^2}\rangle &= 2\langle y\rangle, \\ \langle\dot{\xi y}\rangle &= \langle y\rangle^2 - \omega_0^2\langle\xi^2\rangle, \\ \langle\dot{y^2}\rangle &= -2\omega_0^2\langle\xi y\rangle + 2\sigma_\varepsilon^2\omega_0^4\langle\dot{\xi^2}\rangle, \end{aligned}$$

with

$$\langle\xi^2\rangle = \langle\xi y\rangle = 0, \quad \langle y^2\rangle = 1, \quad t' = 0.$$

The solution of this set of equations for $\sigma^2\omega_0$ is

$$\langle\xi^2(t')\rangle = \frac{1}{2\omega_0^2}\left\{e^{\sigma^2\omega_0^2 t'} - e^{-\sigma^2\omega_0^2 t'/2}\left[\cos(2\omega_0 t') + \frac{3}{4}\sigma^2\omega_0\sin(\omega_0 t')\right]\right\}.$$

When $t' \approx 1/\sigma_\varepsilon^2$, the microscopic particle is accelerated stochastically due to fluctuations in the system parameters. The time of acceleration is $t' \gg t_s$, where $t_s \approx 1/2\eta v_s$, $t' = 1/\sigma_\varepsilon^2\omega_0^2$.

To further illustrate the stochastic parametric soliton resonance, Abdullaev *et al.* studied the following nonlinear Schrödinger equation

$$i\phi_{t'} + \frac{1}{2}\phi_{x'x'} + |\phi|^2\phi + i\tilde{\gamma}\phi - i\varepsilon(x', t')e^{-2\gamma_1 x'}\phi = 0, \tag{8.136}$$

where γ_1, and $\tilde{\gamma}$ are increment parameters related to the damping effect of the systems, ε is a random function defined by the field of partially coherent pumping and medium properties. In the following, ε is assumed to be a function of the coordinate t' only. Then for $\tilde{\gamma} = \varepsilon = 0$, equation (8.136) can be converted into a nonlinear Schrödinger equation with one-soliton solution (8.128).

Now we assume that $\varepsilon(t') = \varepsilon_0 + \tilde{\varepsilon}(t')$, where $\tilde{\varepsilon}(t')$ is a Gaussian random function. Utilizing the adiabatic approximation in the perturbation theory, we obtain the following for the amplitude of the microscopic particle (soliton)

$$\frac{d\eta}{dt'} = 2\eta\left[\varepsilon_0 e^{-2\gamma_1 t'} - \tilde{\gamma} + \tilde{\varepsilon} e^{-2\gamma_1 t'}\right]. \tag{8.137}$$

Given the initial condition $\eta = \eta_0$, equation (8.137) can be solved to give

$$\begin{aligned} \eta &= \eta_0 e^{h_0 + h_1}, \\ h_0 &= \frac{\varepsilon}{\gamma_1}\left(1 - e^{-2\gamma_1 t'}\right) - 2\gamma_1 t', \\ h_1 &= 2\int_{-\infty}^{\infty} \varepsilon(t') e^{-2\gamma_1 t'} dt'. \end{aligned} \tag{8.138}$$

Averaging (8.138) over all possible random values of the mean soliton amplitude yields

$$\langle \eta(t') \rangle = \eta_0 \exp\left(h_0 + \frac{1}{2}\langle h_1^2 \rangle\right). \tag{8.139}$$

For a δ-correlated random process ε, we have

$$\eta(t') = \eta_0 \exp\left[2(2\sigma_\varepsilon^2 + \varepsilon_0 - t') - 2\gamma_1(\varepsilon_0 + 4\sigma_\varepsilon^2)t'^2\right].$$

Expression (8.139) indicates that $\langle \eta(t') \rangle$ grows as the particle propagates for a range of parameters because energy is pumped into the soliton from the random field, which leads to the stochastic parametric resonance. When $\varepsilon_0 = \tilde{\gamma}$, the dissipative energy loss is exactly compensated by the interaction with the regular part of the pumping wave. Ignoring the dissipation of the pumping wave, we have $\gamma_1 = 0$ in this case. From (8.139), Abdnllaev *et al.* showed that $\langle \eta \rangle = \eta_0 \exp(4\sigma_\varepsilon^2 t')$. This clearly shows that the growth of the mean amplitude is due to fluctuation of ε. In such a case, the probability density function of the microscopic particles (solitons), in terms of their amplitudes, is given by

$$\begin{aligned} \rho_{t'}(\eta) &= \frac{1}{\sqrt{2\pi}\langle h_1^2 \rangle \eta} \exp\left\{-\frac{[\ln(\eta/\eta_0) - h_0]^2}{2\langle h_1^2 \rangle}\right\}, \\ \langle h_1 \rangle = 0, \quad \langle h_1^2 \rangle &= \frac{2\sigma_\varepsilon^2}{\gamma_1}\left[1 - e^{-4\gamma_1 t')}\right] = 8\sigma_\varepsilon^2 t'(1 - 2\gamma_1 t'). \end{aligned} \tag{8.140}$$

Therefore, the amplitude of the microscopic particle has a log-normal distribution.

8.9.3 *Inverse-scattering transformation based statistical perturbation theory*

Radiation of microscopic particle in a random inhomogeneous medium was studied by Abdullaev *et al.* using the statistical perturbation method. The following form was chosen for the nonlinear Schrödinger equation

$$i\phi_{t'} + \phi_{x'x'} + 2|\phi|^2\phi = \varepsilon f(x',t')R(\phi), \tag{8.141}$$

where $R(\phi) \to 0$ when $|\phi| \to \infty$.

When $\varepsilon = 0$, the solution of (8.141) is given by (8.19). Here we will examine the statistical features of the radiation field for some special types of perturbation. The correlation functions for the random function $f(x',t')$ have the form

$$\langle f(\tilde{x},\tilde{t})f(x',t')\rangle = \sigma_1^2 B_1(\tilde{x}-x')D_1(\tilde{x}-x');$$
$$\langle f(\tilde{x},\tilde{t})f^*(x't')\rangle = \sigma_2^2 B_2(\tilde{x}-x')D_2(\tilde{x}-x').$$

The mean spectral density of the energy of the wave radiated by the microscopic particle per unit time can be expressed as

$$\langle\rho(\lambda)\rangle = \frac{8\lambda^2}{\pi}\Re\left\langle b(\lambda,t')\frac{\partial b(\lambda,t')}{\partial t'}\right\rangle. \tag{8.142}$$

The Jost coefficient in the inverse scattering method satisfies the equation

$$\frac{\partial b(\lambda,t')}{\partial t'} = -2i\lambda^2 b(\lambda,t') + \frac{i\varepsilon\exp[ib(t')] - 2i\lambda\xi A(\lambda,\xi,\eta)}{2\eta[(\lambda-\xi)^2+\eta^2]}. \tag{8.143}$$

Algebraic calculation then leads to the following

$$A(\lambda,\xi,\eta) = \int\left\{e^{-i(\lambda-\xi)z/\eta}\left[(\lambda-\xi-i\eta\tanh z)^2 R[\phi_s]e^{i\phi(z,t)}\right]\right.$$
$$\left. - \left[\frac{\eta^2}{\cosh^2 z}R^*[\phi_s]e^{i\Phi(z,t)}\right]\right\}dz.$$

The change in the number of quanta M is given by

$$\frac{dN}{dt'} = \sum_k \frac{dN_k}{dt'} + \frac{2}{\pi}\int_{-\infty}^{\infty} d\lambda I(\lambda),$$

where

$$I(\lambda) = \Re\left\langle b(\lambda,t')\frac{\partial b^*(\lambda,t')}{\partial t'}\right\rangle_{t'\gg 1} = \frac{1}{8\eta^2[(\lambda-\xi^2)+\eta^2]^2}\int_{-\infty}^{\infty}\frac{d\kappa}{\eta}$$
$$\times\left\{\sigma_2^2|\varepsilon|^2\eta^4|A_1|^4 b_2(\kappa)d_2(\omega_0^{(-)}) + \sigma_2^2|\varepsilon|^2|A_2(\kappa)|^2 b_2(\kappa)d_2(\omega_0^{(+)})\right. \tag{8.144}$$
$$\left. + \sigma_2^2|\varepsilon|^2|A_2(\kappa)|^2 b_2(\kappa)d_1[\omega_0^{(-)}] - 2\eta^2\sigma_1^2\Re[\varepsilon^2 b_1(\kappa)d_1(\omega_0^{(+)})A_1(-\kappa)A_2^*(\kappa)]\right\},$$

with

$$(\omega_0^{(\pm)}) = \pm 4[(\lambda-\xi)^2+\eta^2] - 4\xi x',$$
$$A_1 = \int_{-\infty}^{\infty} dz e^{-i(\lambda/\eta-\xi/\eta-\kappa/2\eta)z} R\,\mathrm{sech}^2 z,$$
$$A_2 = \int_{-\infty}^{\infty} dz e^{i(\lambda/\eta+\xi/\eta-\kappa/2\eta)z} R^*(\lambda - i\eta\tanh z)^2.$$

In the case of a δ-correlated random process, and $R = \phi$, Abdullaev *et al.* obtained

$$\langle\rho(\lambda)\rangle = \frac{\pi^3\varepsilon^2\sigma_1^2\lambda^2[(\lambda-\xi)^2+\eta^2]^2}{2^6\xi^5\cosh^2[\pi(\lambda^2-\xi^2+\eta^2)/4\eta\xi]}.$$

There are two maxima in the above spectral density, $\rho_1 \approx \varepsilon^2\sigma_1^2\xi$ and $\rho_2 \approx \varepsilon^2\sigma_1^2\eta^4/\xi^2$, located at $\lambda = \pm\lambda_m$, respectively, where $\lambda_m = \sqrt{(\xi^2-\eta^2)} \approx \xi$. Their width is of the order of η. The total mean power emitted by the microscopic particle (soliton) can be obtained and is given by

$$P' = \int_{-\infty}^{\infty}\rho(\lambda)d\lambda = 4\varepsilon^2\sigma_1^2\eta\xi, \quad (\eta \ll \xi),$$

or

$$P' = \varepsilon\sigma_1^2\sqrt{\frac{\xi^3\eta}{2}}\exp\left(-\frac{\pi\eta}{2\xi}\right), \quad \text{for } \eta \gg \xi.$$

Nozaki and Bekki investigated the stochastic behaviors of microscopic particles described by the following nonlinear Schrödinger equation, with radom phases in both time and space in the presence of a small external oscillatory field

$$i\phi_{t'} + \phi_{x'x'} + 2|\phi|^2\phi = i(\varepsilon_1 - \varepsilon_2|\phi|^2)\phi + i\varepsilon_3\phi_{x'x'} - \frac{i\varepsilon_0}{T}\sum_{n=-\infty}^{\infty} e^{in\omega_0 t'},$$

where ε_j $(i = 0, 1, 2, 3)$ are small positive constants. The results show emission of small-amplitude plane waves with random phases. Statistical properties of random phases give the energy spectrum of the microscopic particle (soliton) and plane waves in the systems.

8.10 Microscopic Particles in Interacting Many-particle Systems

In interacting many-particle systems, such as quantum liquids, solids, molecular and magnet as well as nuclear hydrodynamics, the states and properties of microscopic particles described by the nonlinear Schrödinger equation will be changed due to the interaction among them. Barashenkov and Makhankov gave the dynamic equation of the microscopic particle in this case as follows

$$i\phi_{t'} + \Delta\phi - \alpha_1\phi + b|\phi|^2\phi - \alpha_5|\phi|^4\phi = 0, \tag{8.145}$$

where $t' = t/\hbar$, $x_i' = x_i\sqrt{2m/\hbar^2}$, $(i = 1, 2, 3)$, and $\alpha_5|\phi|^4\phi$ $\alpha_5 > 0$ is a nonlinear interaction arising from the many-particle interaction. This equation was used in Chapter 4 to discuss multi-particle collision in nonlinear systems, but its soliton solutions were not given. The work of Barashenkov and Makhankov on this equation is introduced in the following.

In one dimension and for $\alpha_5 = b = 1$, we can make the following transformation,

$$\phi = \frac{\varphi}{\sqrt{2(2\rho_0 + A)/3}}, \quad t' \to t'\left[\frac{4}{3}(A + 2\rho_0)^2\right], \quad x' \to \frac{2}{\sqrt{3}}x'(A + 2\rho_0),$$

with $\rho_0 > 0$, $2\rho_0 + A > 0$, and

$$\rho_0 = \frac{1}{2}(1 + \sqrt{1 + 4\alpha_1}), \quad \frac{A_{1,2}}{\rho_0} = -2 + \frac{3}{4\alpha_1} \pm \frac{3\sqrt{1 + 4\alpha_1}}{4|\alpha_1|}.$$

Then, equation (8.145) becomes

$$i\varphi_{t'} + \varphi_{x'x'} - \rho_0(\rho_0 + 2A)\varphi + 2(2\rho_0 + A)|\varphi|^2\varphi - 3|\varphi|^4\varphi = 0. \tag{8.146}$$

Equation (8.146) is more convenient for studying condensate excitations. By introducing two parameters, A and ρ_0, in (8.146), the coupling can be eliminated. Here ρ_0 is arbitrary, even though A/ρ_0 is required to take certain value. Thus, equation (8.145) can be called the "drop" form, and (8.146) is the "condensate" form. The drop form can be reduced to be condensate form if $\alpha_1 \geq -1/4$. The term $\alpha_1\phi$ in (8.145) can be eliminated with the substitution $\phi(x', t') = e^{i\alpha_1 t'}\bar{\phi}(x', t')$.

For the drop boundary conditions of

$$\phi(x', t') \to 0, \quad \text{at } |x'| \to \infty, \tag{8.147}$$

the energy and particle number integrals are given by

$$E = \int_{-\infty}^{\infty} dx' \left\{|\phi_{x'}|^2 - \frac{1}{2}|\phi|^4 + \frac{1}{3}|\phi|^6\right\}, \quad N = \int_{-\infty}^{\infty} dx'|\phi|^2, \tag{8.148}$$

respectively. Equation (8.145), with the boundary condition (8.147) and $\phi_{t'} = 0$, can be integrated twice to give

$$\phi_{dr} = e^{i\theta_0}\left\{\frac{-4\alpha_1}{1 + \sqrt{1 + 16\alpha_1/3}\cosh[2\sqrt{-\alpha_1}(x' - x'_0)]}\right\}^{1/2}. \tag{8.149}$$

Thus a moving soliton can be obtained via the Gallilei conversion

$$\theta_0 \to \frac{v}{2}x' - \frac{v^2}{4}t' + \theta_0,$$
$$\cosh[2\sqrt{-\alpha_1}(x' - x'_0)] \to \cosh[2\sqrt{-\alpha_1}(x' - vt' - x'_0)]. \tag{8.150}$$

From (8.149), the soliton-like solution is known to depend on three parameters x'_0, θ_0, and v and the solution exists when $-3/16 < \alpha_1 < 0$. In such a case, Barashenkov and Makhankov gave N and E as

$$N(\alpha_1) = \sqrt{3}\,\text{arch}\left[\frac{1}{\sqrt{1 + 16\alpha_1/3}}\right] \to 4\sqrt{-\alpha_1}|_{\alpha_1 \to 0},$$
$$E(\alpha_1) = \left(v^2 - \frac{3}{4} - \alpha_1\right)N(\alpha_1) + \frac{3}{4}\sqrt{-\alpha_1}.$$

When $\alpha_1 \to -3/16$, N grows without limit with the width of the soliton, and its amplitude approaches $\phi_0 = \sqrt{3}/2$. Such behaviors of the solution of (8.149) is extremely proper, and implies that the particle density in the soliton increases to its peak value due to an attractive force, but the three-body repulsion begins to act at small distances. This compensates for the attractive force and the increase in particle density so that further increase in the number of particles N gives rise to a growth in soliton size, and the binding energy of the particle remains constant. This bound state of a large number of particles is naturally regarded as a drop of their condensed state ("fluid") as shown in Fig. 8.5.

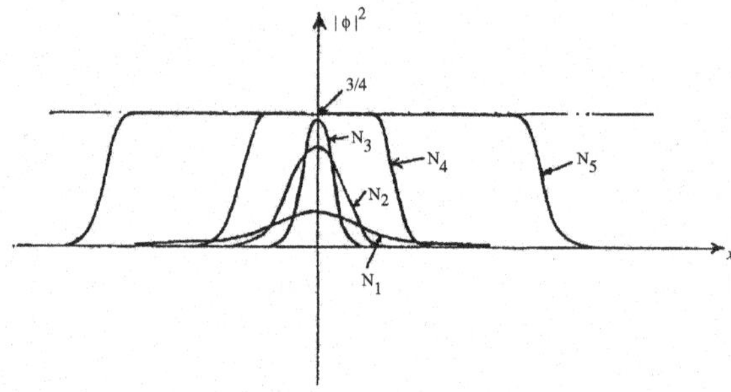

Fig. 8.5 The bound state of the nonlinear Schrödinger equation in the ϕ^4-ϕ^6 field for $\alpha_1 \in (-3/16, 0)$, and particle numbers $N_i = N(\alpha_i)$, $N_5 > N_4 > N_3 > N_2 > N_1$.

If $\alpha_1 = -3/16$, the solution (8.139) gives the condensate $\phi_c = e^{i\theta_0}\sqrt{3}/2$, *i.e.*, $N = \infty$, which corresponds to a transition from the drop state to the condensate state. Such a behavior of the model, along with the saturation effect, is one of the most remarkable features of this system.

Barashenkov and Makhankov studied the solutions of (8.146) under the nontrivial boundary condition $|\varphi|^2 \to \rho_0$. Integrals of motion of (8.148) are divergent in this case. They were thus rewritten as follows

$$E = \int_{-\infty}^{\infty} dx' \left[|\varphi_{x'}|^2 + (|\varphi|^2 - \rho_0)^2 (|\varphi|^2 - A) \right],$$
$$N = \int_{-\infty}^{\infty} dx' \left(|\phi|^2 - \rho_0 \right), \tag{8.151}$$

where A is a real constant. Barashenkov and Makhankov considered the dispersion of small oscillations of this condensate which can be expressed as

$$\varphi(x', t') = \sqrt{\rho_0} + \chi(x', t'),$$
$$\chi(x', t') = \eta_1 e^{i(\kappa x' - \omega t')} + \eta_2 e^{-i(\kappa x' - \omega t')}.$$

Thus η_1 and η_2 satisfy

$$\omega\eta_1 - \kappa^2\eta_1 - 2\rho_0(\rho_0 - A)\eta_1 - 2\varphi_0^2(\rho_0 - A)\bar{\eta}_2 = 0,$$
$$-\omega\eta_2 - \kappa^2\eta_2 - 2\rho_0(\rho_0 - A)\eta_2 - 2\varphi_0^2(\rho_0 - A)\bar{\eta}_1 = 0.$$

Setting the determinant of the coefficients of the above system to zero, one can get

$$\omega^2 = \kappa^2[\kappa^2 + 4\rho_0(\rho_0 - A)].$$

This is the Bogolubov dispersion, ignoring the change in the interaction due to variation of the sound velocity in such a condensate

$$v_{s0} = \lim_{\kappa\to 0}\frac{\omega}{\kappa} = 2\sqrt{\rho_0(\rho_0 - A)}.$$

For small amplitude nonlinear waves, φ has the form

$$\varphi(x', t') = \sqrt{\rho(x', t')}e^{i\theta(x',t')}. \tag{8.152}$$

Introducing $y = \theta_{x'}$, we get from (8.146) the following dynamics equations

$$\rho_{t'} = -2(y\rho)_{x'},$$
$$y_{t'} = \frac{1}{2}\left(\frac{\rho_{x'x'}}{\rho} - \frac{1}{2}\frac{\rho_{x'}^2}{\rho^2}\right)_{x'} - 2yy_{x'} + [(\rho_0 - \rho)(3\rho - \rho_0 - 2A)]_{x'}. \tag{8.153}$$

Choosing new coordinates, $\tau = \sqrt{\epsilon}t'$, $\xi = \sqrt{\epsilon}(x' - v_s t')$, and by means of "reductive perturbation theory", we expand ρ and y in powers of ϵ about the condensate, $\rho = \rho_0$, $y = 0$,

$$\rho = \rho_0 + \epsilon\rho_1 + \epsilon^2\rho_2 + \cdots,$$
$$y = \epsilon y_1 + \epsilon^2 y_2 + \cdots.$$

Substituting these into (8.153) and letting the coefficients of ϵ and ϵ^2 vanish, we get

$$\begin{aligned} v_{s0}\rho_1' &= 2\rho_0 y_1' = \partial_\xi, \\ v_{s0}y_1' &= 2(\rho_0 - A)\rho_1', \\ v_{s0}\rho_2' &= \dot{\rho}_1 + 2(y_1\rho_1)' + 2\rho_0 y_2' = \partial_\tau, \\ v_{s0}y_2' &= \dot{y}_1 + 2y_1y_1' - \frac{\rho_1'''}{2\rho_0} + 2(\rho_0 - A)\rho_2' + 3(\rho_1^2)', \end{aligned} \tag{8.154}$$

where the "$'$" and "$\cdot$" denotes differentiations with respect to ξ and t, respectively. The first (or third) of (8.154) gives $y_1' = v_{so}\rho_1'/2\rho_0$, while the second and the fourth reduce to the KdV equation

$$2\sqrt{\rho_0(\rho_0 - A)}\,\dot{\rho}_1 - \frac{1}{2}\rho_1''' + 3(2\rho_0 - A)(\rho_1^2)' = 0,$$

or

$$\partial_{\bar{\tau}}\rho_1 - \partial_\xi^3\rho_1 + 3(\rho_1^2)_\xi = 0, \tag{8.155}$$

by the scale transformation

$$\xi = \bar{\xi}\sqrt{|2(2\rho_0 - A)|}, \quad \tau = \bar{\tau}\frac{v_{s0}}{2\sqrt{(2\rho_0 - A)^3}}.$$

It can be shown that (8.155) has a one-soliton solution which is given by

$$\rho_1(\bar{\xi}, \bar{\tau}) = -\frac{\tilde{b}}{2}\text{sech}^2\left[\frac{\sqrt{\tilde{b}}}{2}(\bar{\xi} + \bar{b}\bar{\tau} - \bar{\xi}_0)\right], \tag{8.156}$$

or

$$\rho(x', t') = \left\{\rho_0 - \frac{b'}{2\rho_0 - A}\text{sech}^2\left[\sqrt{b'}\left(x' - v_{s0}t' + \frac{2b'}{v_{s0}}t' - x_0'\right)\right]\right\}^{1/2}, \tag{8.157}$$

where $b' = \tilde{b}/2(2\rho_0 - A) \geq 0$. Expressions (8.156) or (8.157) describes a localized rarefraction wave moving in the condensate at a velocity which is close, but less than the sound velocity. The quasisolitons and the background arise in an initial perturbation decay. However, the decay of a compression perturbation occurs in the solitonless part to give a dispersive packet of Bogolubov waves (some sound waves with small wave numbers). The rarefraction soliton (8.157) is accompanied by the emission of "linear" (Bogolubov type) waves moving in the same direction as the soliton, but faster, with the group velocity $v_g = 2(k^2 + v_{s0}^2/2)/\sqrt{k^2 + v_{s0}^2}$. These waves are usually called "foregoers".

Barashenkov and Makhankov obtained localized solutions of (8.146). Here, we use (8.152) and assume $\varphi(x', t') = \varphi(\xi')$, with $\xi' = x' - vt'$. Equation (8.146) then becomes

$$\rho\rho'' - \frac{1}{2}\rho'^2 + v\theta'\rho^2 - (\theta')^2\rho^2 - \rho^2(\rho - \rho_0)(3\rho - \rho_0 - 2A) = 0, \tag{8.158}$$

$$\rho\theta' + \frac{v'}{2}(\rho_0^2 - \rho) = 0.$$

Inserting the second equation into the first of (8.158), and integrating again, we get

$$2(\xi' - \xi_0') = \int d(\rho - \rho_0)^{-1}\left(\rho^2 - A\rho - \frac{v^2}{4}\right)^{-1/2}$$
$$= \int \frac{dq}{q}\sqrt{q^2 + (2\rho_0 - A)q + c}. \tag{8.159}$$

Here

$$q = \rho - \rho_0, \quad c = \rho_0^2 - A\rho_0 - \frac{v^2}{4} \equiv \frac{1}{4}(v_{s0}^2 - v^2).$$

There exists a soliton-like solution of (8.159) at $c > 0$ or $v^2 < v_{s0}^2$. It shows that the microscopic particle is localized in these systems in this case. Inverting the integral

(8.159), one can get

$$q_{\pm}(\xi') = \frac{\pm 2c}{\sqrt{A^2+v^2}\cosh[2\sqrt{c}(\xi'-\xi_0')] \pm (2\rho_0 - A)}. \tag{8.160}$$

The second of the above solutions, $q_-(\xi')$, is singular since $2\rho_0 > A$. The first is regular, and takes a different form in each of the following four regions, $A > \rho_0$, $\rho_0 > A > 0$, $-\rho_0/2 < A < 0$, and $A < -\rho_0/2$.

In the region $\rho_0 > A > 0$, the soliton is at rest and

$$\varphi_b = e^{i\theta_v}\frac{\sqrt{A}\cosh[v_{s0}(x'-x_0')/2]}{(1+A/\rho_0)\sinh^2[\sqrt{v_{s0}(x'-x_0')/2}]}. \tag{8.161}$$

φ_b is an even function of $(x'-x_0')$. It describes a bubble in the condensate, whose depth depends on A, the smaller the A, the greater the rarefaction in the bubble.

In the region $-\rho_0/2 < A < 0$ and $A < -\rho_0/2$, we have

$$\varphi_k = e^{i\theta_v}\frac{\sqrt{|A|}\sinh[v_{s0}(x'-x_0')/2]}{(1+|A|/\rho_0)\cosh^2[\sqrt{v_{s0}(x'-x_0')/2}]}. \tag{8.162}$$

Here φ_k is an odd function of $(x'-x_0')$ and has a kink form.

Phases of both solutions were found via (8.159) and (8.152) to be

$$\theta_{k,b} = \frac{\sqrt{2}e^{i\theta_v}\cosh(\nu - iu)}{\sqrt{(2\rho_0 - A)/\sqrt{A^2+v^2} + \cosh(2\nu)}}, \tag{8.163}$$

where $\nu = (1/2)\sqrt{v_{s0}^2 - v^2}(x' - vt' - x_0')$, $\cos(2u) = (A\rho_0 + v^2/2)/(\rho_0\sqrt{A^2+v^2})$, and the sign of A determines whether the solution is a kink or a bubble. For $A = 0$, or $\alpha_1 = -3/16$, Barashenkov and Makhankov gave a solution with mixed boundary conditions $\varphi(x',t') \to 0$ for $x' \to -\infty$, and $\varphi(x',t') \to \sqrt{\rho_0}e^{i\theta_\nu}$ for $x' \to +\infty$. In the case of finite energy, the solution is

$$\varphi_k(x') = \frac{e^{i\theta_\nu}\sqrt{\rho_0}}{\sqrt{1+2\rho_0\exp[\pm 2\rho_0(x'-x_0')]}}. \tag{8.164}$$

It connects two stable vacuum states, the condensate $|\varphi|^2 = \rho_0$ and the trivial state $|\varphi|^2 = 0$. When $\theta_\nu =$ constant, (8.164) describes a wave at rest. If $\theta_\nu = (v/2)x' - \omega t'$, the wave given in (8.164) moves together with the condensate.

Obviously, integrals of the hole number and energy for solutions (8.163) are

$$N_{k,b} = -\operatorname{arch}\left[\frac{2\rho_0 - A}{\sqrt{A^2+v^2}}\right],$$

$$E_{k,b} = \sqrt{c}\left(\frac{A}{2}+\rho_0\right) + \left[A\rho_0 + \frac{1}{4}(v^2 - A^2)\right]N_{k,b}.$$

For small amplitudes c (hence velocity v close to the sound velocity v_{s0}), the module of (8.163) is

$$\rho_{k,b}(x',t') = \left[\rho_0 - \frac{2c}{\sqrt{c^2+v^2}\cosh[2\sqrt{c}(\xi'-\xi_0')] + 2\rho_0 - A}\right]^{1/2}. \tag{8.165}$$

This coincides with the approximate solution (8.157) since

$$(2\rho_0 - A)^2 - (A^2 + v^2) = 4c,$$

or

$$\sqrt{A^2+v^2} \simeq 2\rho_0 - A - \frac{2c}{2\rho_0 - A}$$

at $c \ll |2\rho_0 - A|$. $\sqrt{c^2+v^2}$ in (8.165) is then replaced by $2\rho_0 - A$, and we can get

$$\rho_{k,b} = \left\{\rho_0 - \frac{2c/(2\rho_0 - A)}{1+\cosh[2\sqrt{c}(\xi'-\xi_0')]}\right\}^{1/2} \simeq \left\{\rho_0 - \frac{c/(2\rho_0 - A)}{\cosh^2[2\sqrt{c}(\xi'-\xi_0')]}\right\}^{1/2},$$

with $v = v_k - 2c/v_{s0}$, $\bar{b} = 2c/(2\rho_0 - A)$.

Besides the above solutions, the ϕ^6 model also has other localized solutions, for example, the rational solution. However, in the region of $A > \rho_0$, there are no localized solutions.

Barashenkov and Makhankov demonstrated the stabilities of the above hole-like excitation and the drop-like solitons by the spectral analysis and variational method. In the first case (particle-like solutions), the drop is stable. There are two types of localized excitations of the condensate: the bubbles (8.161) with $-1/4 < \alpha_1 < -3/16$, and the kinks (8.162) with $\alpha_1 > 3/16$. They are unstable. These solutions were also obtained by Cowan *et al.* through numerical simulation.

8.11 Effects of High-order Dispersion on Microscopic Particles

The effects of various high-order dispersions on the properties of microscopic particles were studied by Karpman, Kundu, and Pathria and Morris. Karpman obtained the solutions of the following nonlinear Schrödinger equation

$$i\phi_{t'} + \frac{1}{2}\phi_{x'x'} + |\phi|^2\phi = R\phi = -i\alpha_3\phi_{x'x'x'} - \alpha_4\phi_{x'x'x'x'}, \tag{8.166}$$

where the operator R is defined as

$$R = -i\alpha_3\frac{\partial^3}{\partial x'^3} - \alpha_4\frac{\partial^4}{\partial x^4},$$

with α_3 and α_4 being real and small coefficients. Karpman expressed the solution of (8.166) as

$$\phi = aF\left(t', x' - \int v dt'\right) \exp\left[\frac{1}{2} i \int a^2 dt' + i\theta(x', t')\right]. \tag{8.167}$$

where a and v are the amplitude and velocity of the microscopic particle, respectively. Variations of a and v with time are ignored. Inserting (8.167) into (8.166), Karpman obtained

$$\begin{aligned} \theta &= -\Omega t' + kx', \\ \Omega &= \tfrac{1}{2}k^2 - \alpha_3 k^3 - \alpha_4 k^4, \\ v &= k - 3\alpha_3 k^2 - 4\alpha_4 k^3, \end{aligned} \tag{8.168}$$

and

$$iF_{t'} + \frac{1}{2}a_2 F_{\xi\xi} + ia_3 F_{\xi\xi\xi} + a_4 F_{\xi\xi\xi\xi} + a^2\left(|F|^2 - \frac{1}{2}\right)F = 0, \tag{8.169}$$

with

$$\begin{aligned} \xi &= x' - \textstyle\int v dt', \\ a_2 &= 1 - 6\alpha_3 k - 12\alpha_4 k^2, \\ a_3 &= \alpha_3 + 4\alpha_4 k, \\ a_4 &= \alpha_4. \end{aligned}$$

For small α_3 and α_4, the solution of (8.169) can be written as

$$F(t', \xi) = \operatorname{sech}\left(\frac{a\xi}{\sqrt{a_2}}\right) + f(\xi, t'), \tag{8.170}$$

with $a_2(t') > 0$. The term $f(\xi, t')$, at sufficiently large ξ and t', describes the radiation field. Karpman gave an asymptotic expression for $f(\xi, t')$ which is

$$\begin{aligned} f(\xi, t') = \frac{\sqrt{a_2}}{a} \sum_{j=1}^{2} C_j k_j \exp\left(-\frac{\pi\sqrt{a_2}}{2a}|k_j|\right) \times \\ e^{ik_j\xi}\Theta\left(\left|\int dt v_j\right| - |\xi|\right)\Theta(v_j\xi). \end{aligned} \tag{8.171}$$

where $j\ (= 1, 2)$ is the number of the radiation modes, k_j are their wave numbers, given by $k_{1,2} \approx (-a_3 \pm \sqrt{a_3^2 + 2a_2a_4})/(2a_4)$, v_j are the corresponding group velocities in the reference frame specified by ξ and t', $v_j = a_2k_j - 3a_3k_j^2 - 4a_4k_j^3$, C_j are constants (of order 10), and $\Theta(Z)$ is the step function

$$\Theta(Z) = \begin{cases} 1 & (Z > 0), \\ 0 & (Z < 0). \end{cases}$$

The Θ-function in (8.171) states that the fronts of the modes with $k = k_{1,2}$ propagate with $v = v_{1,2}$ in the directions specified by the signs of $v_{1,2}$. The whole expression (8.171) is exponentially small, because $|k_{1,2}|a^{-1} \gg 1$.

When $a_3^2 \gg 2a_2|a_4|$, $k_1 \approx a_2/(2a_3)$, and $|k_1| \ll |k_2|$, (8.171) is approximately reduced to

$$f(\xi, t') \approx \frac{C\sqrt{a_2}}{a} k_1 \exp\left(-\frac{\pi\sqrt{a_2}}{2a}|k_1|\right) e^{ik_1\xi}\Theta\left(\left|\int dtv_1\right| - |\xi|\right)\Theta(v_1\xi), \quad (8.172)$$

where $C_1 = C$ and $v_1 \approx -a_2/(4a_3)$. Under such a condition the radiation field does not depend on α_4 in (8.166). Therefore, we essentially have a third-order nonlinear Schrödinger equation. When $a_3^2 \ll 2a_2|a_4|$, we have

$$k_1 \approx -k_2 \approx \sqrt{\frac{a_2}{2a_4}}, \quad v_1 \approx -v_2 \approx -\sqrt{\frac{a_2^3}{2a_4}}, \quad (8.173)$$

with $a_4 > 0$. In such a case, the third derivative term in (8.166) is insignificant and we arrive at the fourth-order nonlinear Schrödinger equation. The soliton radiation is in the form given above, with $a_4 > 0$, $C_1 = -C_2^* = C$. For $a_4 < 0$, the soliton does not radiate. Karpman showed that (8.172) may be valid even at $|k_1| < |k_2|$. However, if $|k_2| < |k_1|$ is sufficiently small, one should use the full expressions (8.171).

Based on conservations of particle number and momentum, *i.e.*,

$$\frac{dN}{dt'} = \frac{d}{dt'}\int_{-\infty}^{\infty} dx'|\phi(x', t'|^2 = 0,$$
$$\frac{dP}{dt'} = \frac{d}{dt'}\left[\frac{i}{2}\int_{-\infty}^{\infty} dx'(\phi^*\phi_{x'} - \phi\phi_{x'}^*)\right] = 0,$$

and the above solution, Karpman also obtained the variations of the amplitude a and the velocity v of the microscopic particle with time, which are caused by the radiation of the microscopic particle (soliton). They are

$$\frac{da(t')}{dt'} \approx \frac{\sqrt{2}|C|^2}{4\sqrt{\alpha_4^3}} \exp\left(-\frac{\pi}{\sqrt{2}\alpha_4 a}\right),$$
$$v(t') \approx k_1 \ln\left[\frac{a(t')}{a_0}\right],$$

where

$$a(t') = \begin{cases} a_0\left(1 + \dfrac{t'}{t'_{ch}\ell_0}\right) & (t' \ll t'_{ch}) \\ a_0\left[1 + \dfrac{1}{\ell_0}\ln\left(\dfrac{t}{t_{ch}}\right)\right] & (t' \gg t'_{ch}), \end{cases}$$

and

$$\ell_0 = \ell(0) = \frac{\pi\sqrt{\alpha_2}|\kappa_1|}{a_0}, \quad t'_{ch} = \frac{1}{|C_1|^2|\kappa v_1|_0\ell_0^2 e^{\ell_0}}, \quad a_0 = a(t=0).$$

Karpman further obtained solutions of (8.166) with $R = i\alpha_3\partial^3/\partial x^3$ and $\alpha_4\partial^4/\partial x^4$, respectively, and showed that the radiation of the microscopic particle (soliton) leads to a decrease in the amplitude of the microscopic particle. He also studied acceleration of the microscopic particles which arises due to conservation of momentum.

Karpman also studied resonant radiation effect of microscopic particles due to high order dispersion, described by the following nonlinear Schrödinger equation

$$i\phi_{t'} + \frac{1}{2}\phi_{x'x'} + b|\phi|^2\phi = -(i\alpha_1|\phi|^2\phi_{x'} + i\alpha_2\phi|\phi|^2)_{x'} + i\alpha_3\phi_{x'x'x'} + \alpha_4\phi_{x'x'x'x'},$$

with both linear and nonlinear (cubic) dispersion terms, third and fourth derivatives. He obtained the amplitudes and asymptotic form of the resonant radiations. According to him, the soliton-like solution of the above equation can be written as

$$\phi(x',t') = \frac{\exp\{i[\mu^2 t'/2 + \theta(x')]\}}{\sqrt{A_0A_1[A_0^2\sinh^2(\mu x') + A_1^2\cosh^2(\mu x')]}},$$

where

$$\theta(x') = -\frac{\alpha_1 + 2\alpha_2}{2C}\arctan\left[\frac{A_0}{A_1}\tanh(\mu x')\right],$$
$$C^2 = \frac{1}{12}[4(\alpha_1 + 2\alpha_2)\alpha_1 - (\alpha_1 + 2\alpha_2)^2],$$
$$A_1^2 = \frac{\sqrt{2C^2\mu^2 + q^2} + q}{2C^2},$$
$$A_0^2 = \frac{\sqrt{2C^2\mu^2 + q^2} - q}{2C^2},$$

at $\alpha_3 = \alpha_4$.

De Oliveira *et al.* obtained soliton-like solution of the nonlinear Schrödinger equation,

$$i\phi_{t'} + \frac{1}{2}\phi_{x'x'} + |\phi|^2\phi = +i\alpha_1|\phi|^2\phi_{x'} - i\varepsilon|\phi|^2\phi,$$

which is given by

$$\phi(x',t') = A(t')\text{sech}\left[\frac{x'}{\omega(t')}\right]e^{i\theta(x',t')},$$
$$\theta(t') = \theta_0 - \frac{1}{2}a_0^2 t' + \frac{2\mu'}{16\varepsilon}\ln\left(1 + \frac{8}{3}\varepsilon A_0'^2 t'\right),$$

where

$$A(t) = \frac{A_0'}{\sqrt{1+8\varepsilon A_0'^2 t'/3}}, \quad \mu' = 1 + a_0(\alpha_1 - 2a_2), \quad \omega(t') = \omega_0\sqrt{1+8\varepsilon A_0'^2 t'/3},$$

a_0 and a_2 are constants.

The influences of the nonlinear dispersion on microscopic particles described by the ϕ^6-nonlinear Schrödinger equation (Section 8.9) were investigated by Pathria and Morris. The generalized nonlinear Schrödinger equation is

$$i\phi_{t'} + \phi_{x'x'} + b|\phi|^2\phi + \alpha_1|\phi|^4\phi = -i\alpha_2(|\phi|^2)_{x'}\phi - i\alpha_3|\phi|^2\phi_{x'}. \tag{8.174}$$

Applying the following gauge transformation,

$$\phi(x', t') = \varphi(x', t')e^{i\theta(x', t')},$$

equation (8.174) becomes

$$i\varphi_{t'} + \varphi_{x'x'} + b|\varphi|^2\varphi + \beta_1|\varphi|^4\varphi + i\beta_2|\varphi_{x'}|^2\varphi + i\beta_3|\varphi|^2\varphi_{x'} = 0, \tag{8.175}$$

where $\beta_1 = \alpha_1 + 2\delta\alpha_2 - \delta\alpha_3 + 4\delta^2$, $\beta_2 = \alpha_2 + 4\delta$, $\beta_3 = \alpha_3$, and δ is an arbitrary constant.

Solutions of (8.174) which satisfy the integrability condition are constructed from the known solutions of the mixed nonlinear Schrödinger equation. For (8.174) with $4\alpha_1 = \alpha_2^2 - \alpha_2\alpha_3$, if $\alpha_3 = 0$ (so that $4\alpha_1 = \alpha_2^2$) and $b > 0$, the solitary wave solution corresponding to the one-hump solution of the cubic nonlinear Schrödinger equation is given by

$$\phi(x', t') = \sqrt{\frac{2\eta}{b}}\,\mathrm{sech}[\sqrt{\eta}(x' - vt') + x_0']e^{i\theta(x', t')}, \tag{8.176}$$

$$\theta(x', t') = \frac{v}{2}(x' - vt') - \frac{\sqrt{\eta}\alpha_2}{b}\tanh[\sqrt{\eta}(x' - vt) + x_0'] + \ell_1,$$

where $\eta = (v/2)(v/2 - c') > 0$, c', v, x_0 and ℓ_1 are arbitrary constants. Obviously, the solution (8.173) differs from the above soliton of cubic nonlinear Schrödinger equation only in its peculiar phase. The solitary waves given in (8.176) can exhibit also the same clean interactions as the solitons (4.53) of cubic nonlinear Schrödinger equation (4.37).

Choosing $\delta = -(2\alpha_2 + \alpha_3)/8$ and setting the forces $2\beta_2 + \beta_3 = 0$ in (8.175), we get an equation for φ

$$i\varphi_{t'} + \varphi_{x'x'} + b|\varphi|^2\varphi + \beta_1|\varphi|^4\varphi + i\beta_2|\varphi|^2_{x'}\varphi + i\beta_3|\varphi|^2\varphi_{x'} = 0, \tag{8.177}$$

where

$$\beta_1 = \alpha_1 - \frac{1}{16}(2\alpha_2 + \alpha_3)(2\alpha_2 - 3\alpha_3), \quad \beta_2 = -\frac{1}{2}a_2, \quad \beta_3 = \alpha_3.$$

Since $2\beta_2 + \beta_3 = 0$, equation (8.177) can be written in the following equivalent form

$$i\varphi_{t'} + \varphi_{x'x'} + F(\varphi)\varphi = 0, \tag{8.178}$$

where $F(\varphi)$ is a real function and is given by

$$F(\varphi) = b|\varphi|^2 + \beta_1|\varphi|^4 + \beta_3\Im(\varphi\bar{\varphi}_{x'}).$$

Pathria *et al.* obtained solutions of (8.174) using the transformed equation (8.178). Writing $\varphi(x', t')$ as

$$\varphi(x', t') = f(x' - vt')e^{ih(x'-cy)}, \tag{8.179}$$

where f and h are real functions with $h(y) = (v/2)y + \ell_1$, v, c and ℓ_1 are arbitrary constants. $\theta(x', t')$ can then be obtained and is given by

$$\theta(x', t') = 2\delta \int |\varphi|^2 dx = 2\delta \int f^2(x' - vt')d(x' - vt)$$
$$-\frac{1}{4}(2\alpha_2 + \alpha_3) \int f^2(x' - vt')d(x' - vt').$$

Inserting (8.179) into the transformed equation, (8.178), we get a cnoidal wave type equation for $g = f^2(x' - vt')$,

$$(g')^2 = -\frac{4\beta_1}{3}g^4 - (2b - v\beta_3)g^3 - (2vc - v^2)g^2 - 4C'g$$
$$\alpha(g - g_1)(g - g_2)g^2 = 0, \tag{8.180}$$

where C' is an arbitrary constant, $g' = \partial g/\partial x'$, g_1 and g_2 are determined by the solution of g, $\alpha = -4\beta_1/3$. Solution (8.179) may be represented in terms of the elliptic functions. If $C' = 0$, these solutions can be expressed in terms of elementary functions, including oscillatory, singular, phase jump, and solitary wave solutions. These are four-parameter families of solutions, with arbitrary constants v, c, ℓ_1 and x'_0. The parameters v and c representing, respectively, the speeds of the carrier and envelope waves of φ, partly determine the form of the solution. When $\beta_1 < 0$, and g_1 and g_2 are real and $g_1 > g_2 > 0$, the solitary wave solution of (8.174) is

$$\phi(x', t') = \left[\frac{g_1 g_2}{g_1 + (g_1 - g_2)\sin h^2(\Omega')}\right]^{1/2} e^{i\theta(x',t')}, \tag{8.181}$$

$$\theta(x', t') = -\frac{2\alpha_2 + \alpha_3}{4}\sqrt{-\frac{3}{\beta_1}}\tanh^{-1}\left[\sqrt{\frac{g_2}{g_1}}\tanh(\Omega')\right] + \frac{v}{2}(x' - ct') + \ell_1,$$

where

$$\Omega' = \sqrt{\frac{-g_1 g_2 \beta_3}{3}}(x' - vt') + \mu'.$$

When $\beta_1 > 0$, α in (8.170) is replaced by $-\alpha$. If g_1 and g_2 are real, and $g_1 > 0 > g_2$, a solitary wave solution of (8.174) can be obtained and is given by

$$\phi(x',t') = \left[\frac{g_1 g_2}{g_2 + (g_2 - g_1)\sinh^2(\Omega')}\right]^{1/2} e^{i\theta(x',t')},$$

$$\theta(x',t') = -\frac{2\alpha_2 + \alpha_3}{4}\sqrt{\frac{3}{\beta_3}}\tan^{-1}\left[\sqrt{-\frac{g_1}{g_2}}\tanh(\Omega')\right] + \frac{v}{2}(x' - ct') + \ell_1.$$

If g_1 and g_2 are real, and $g_1 > g_2 > 0$, then g is oscillatory and

$$\phi(x',t') = \left[\frac{g_1 g_2}{g_1 + (g_2 - g_1)\cos^2(\Omega')}\right]^{\frac{1}{2}} e^{i\theta(x',t')},$$

$$\theta(x',t') = -\frac{2\alpha_2 + \alpha_3}{4}\sqrt{\frac{3}{\beta_1}}\tan^{-1}\left[\sqrt{-\frac{g_2}{g_1}}\tanh(\Omega')\right] + \frac{v}{2}(x' - ct') + \ell_1.$$

Therefore, equation (8.174) has solitary wave solutions for both positive and negative values of β_1. Note that (8.174) is invariant under the Galilean transformation

$$\begin{aligned} x^* &\simeq A^2(x' + 2A^2Bt'), \\ t^* &= A^4 t', \\ B &= \frac{b}{\alpha_2}\left(1 - \frac{1}{A^2}\right), \\ \phi^* &= \frac{1}{A}\phi(x',t')e^{iA^2B(x'+A^2Bt')}, \end{aligned} \tag{8.182}$$

where A is an arbitrary nonzero real constant. Then, since

$$i\phi^*_{t^*} - \phi^*_{x^*x^*} - b|\phi^*|^2\phi^* - \alpha_1|\phi^*|^4\phi^* + i\alpha_2|\phi^*|^2_{x^*}\phi^* + i\alpha_3|\phi^*|^2\phi^*_{x^*} = 0,$$

we can construct new solutions for (8.174) using the above approach and (8.182).

The conservation laws of the system described by (8.174) were studied by Pathria *et al.*. If $\varphi(x',t')$ is a rapidly decreasing function of x' and $P(x')\partial_{x'}\varphi(x',t') \to 0$ as $x' \to \pm\infty$, for any polynomial $P(x')$, then the following are the conserving quantities for (8.175)

$$N_0 = \int |\varphi|^2 dx',$$

$$E_1 = \int_{-\infty}^{\infty} \left[4\Im(\varphi\varphi^*_{x'}) + \beta_3|\varphi|^4\right] dx',$$

$$E_2 = \int_{-\infty}^{\infty} \left[|\varphi_{x'}|^2 - \frac{1}{2}b|\varphi|^4 - \frac{1}{3}\beta_1|\varphi|^6\right] dx'.$$

Pathria *et al.* numerically solved the following equations using a pseudo-spectral split-step discretization,

$$i\phi_{t'} + \phi_{x'x'} + |\phi^2|\phi + |\phi|^4\phi - 2i|\phi|^2\phi = 0, \tag{8.183}$$

for the initial condition

$$\phi(x',0) = \frac{1}{\sqrt{2}}\operatorname{sech}\left(\frac{x'-15}{2}\right)\exp\left\{i\left[\frac{x'-15}{2}+\tanh\left(\frac{x'-15}{4}\right)\right]\right\}+$$

$$\frac{1}{2}\operatorname{sech}\left(\frac{x'-35}{2\sqrt{2}}\right)\times\exp\left\{i\left[-\frac{x'-35}{4}+\tanh\left(\frac{x'-35}{2\sqrt{2}}\right)\right]\right\}, \quad (8.184)$$

which corresponds to two initially well-spaced solitons of the form (8.176). The one initially on the right moves to left with speed of unity, and the one initially on the left moves to right with half the speed. The two solitary waves emerge from their encounter with their shape and velocities unchanged, although they may be displaced from the position they would have occupied had the collision not occurred. The elastic collision of the waves, which is shown in Fig. 8.6, demonstrates the stability of the microscopic particles (solitons).

Solution of the following nonlinear Schrödinger equation

$$i\phi_{t'} + \phi_{x'x'} - \frac{1}{2}|\phi|^2\phi - \frac{7}{4}|\phi|^4\phi - i|\phi|^2_{x'}\phi - 2i|\phi|^2\phi_{x'} = 0 \quad (8.185)$$

can be found and it is given by

$$\phi(x',t') = \left[\frac{4}{4+3\sinh^2(x'-2t'-15)}\right]^{1/2} e^{i\theta(x',t')},$$

$$\theta(x',t') = 2\tanh^{-1}\left[\frac{1}{2}\tanh(x'-2t'-15)\right] + x' - 15.$$

For the initial condition $\phi_0(x')$ given in (8.184), results of numerical simulation agree with theoretical prediction that a bell-shaped solitary wave propagates from left to right at a speed of 2, as shown in Fig. 8.7.

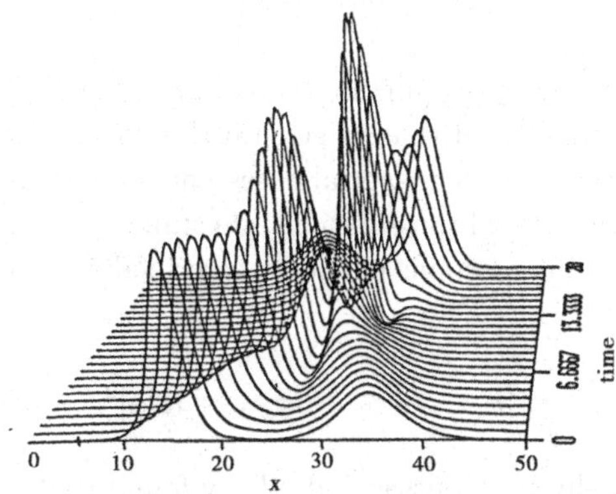

Fig. 8.6 Collision between microscopic particles described by (8.183).

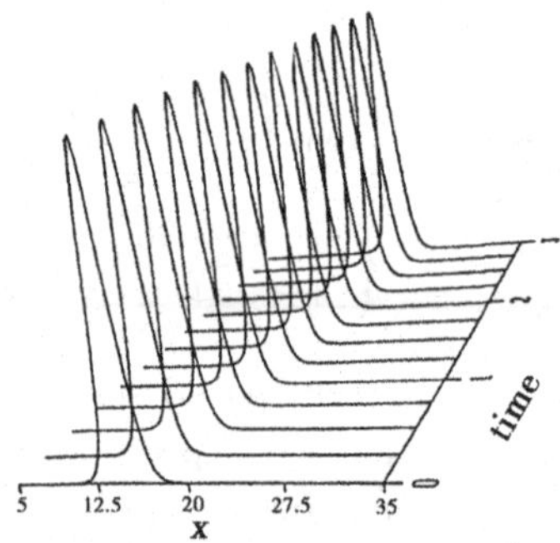

Fig. 8.7 Solitary wave solution of (8.185).

The properties of microscopic particles described by

$$i\phi_{t'} + \frac{1}{2}\phi_{x'x'} + |\phi^2|\phi = \gamma\phi - \alpha(|\phi|^2\phi)_{x'} \tag{8.186}$$

were investigated by Shchesnovish and Doktorov using an adiabatic approximation of perturbation-induced evolution of the soliton parameters based on the Riemann-Hilbert problem. γ and α in (8.186) are constants. The results show that the shape of the microscopic particle was distorted and the emission of linear waves occurs in such a case.

Crespe *et al.* studied the stability of soliton-like solution of the following nonlinear Schrödinger equation,

$$i\phi_{t'} + \frac{1}{2}\phi_{x'x'} + |\phi^2|\phi = i\delta\phi + i\varepsilon|\phi|^2\phi + i\beta\phi_{x'x'} + i\mu'|\phi|^2\phi - \nu'|\phi|^4\phi.$$

The soliton-like solution was found to be

$$\phi(x', t') = A(x')e^{i[\theta(x') - i\omega t']},$$

where $A(x')$ and $\theta(x')$ are some unknown functions which are determined by the original equation. It was found that in general the microscopic particles in such a case are unstable except in a few special cases and that there are some regions in the parameter space where soliton solutions are stable.

Various profiles of microscopic particles described by the following nonlinear Schrödinger equation,

$$i\phi_{t'} + \frac{1}{2}\phi_{x'x'} + |\phi|^2\phi = -ic\phi - i\beta\phi_{x'x'} + i\varepsilon|\phi|^2\phi - id|\phi|^2\phi. \tag{8.187}$$

were studied numerically by Moussa *et al.*. They found that soliton-like solution of this equation is of the form

$$\phi(x', t') = \eta'\text{sech}[\eta'(x' - \lambda)]e^{i(\eta'^2 t'/2 + \theta')},$$

where

$$\eta' = \frac{1}{16d}[5(2\varepsilon - \beta) \pm \sqrt{25(2\varepsilon - \beta)^2 - 480cd)}],$$

λ and θ' are arbitrary constants. The results reveal various shapes of the solitary solutions in such a case.

8.12 Interaction of Microscopic Particles and Its Radiation Effect in Perturbed Systems with Different Dispersions

We first consider collision between microscopic particles in a perturbed system which was investigated by Malomed. The microscopic particles are described by the following nonlinear Schrödinger equation with a nonlinear damping

$$i\phi_{t'} + \phi_{x'x'} + 2|\phi|^2\phi = \varepsilon\phi R(\phi), \tag{8.188}$$

where $R(\phi)$ is a perturbation, ε is real small number. Malomed considered

$$R(\phi) = (|\phi|^2)_{x'}, \tag{8.189}$$

and

$$R(\phi) = \frac{1}{\pi}P\int_{-\infty}^{\infty} \frac{|\phi(\tilde{x})|^2}{x' - \tilde{x}}dx'. \tag{8.190}$$

Here P stands for the principle value of the integral. $R(\phi)$ given in (8.189) specifies the dispersion effect, and that in (8.190) describes a nonlinear Landau damping effect. Obviously, when $\varepsilon = 0$, the soliton solution of (8.188) is given by (8.19). When $\varepsilon \neq 0$ the microscopic particle experiences an acceleration induced by the nonlinear damping. In the lowest order approximation, the equations of motion for the amplitude η and velocity v of the microscopic particle are $\eta_t = 0$, and $v_t = \delta'\varepsilon\eta^n$, with $\delta' = 128/13$ and $n = 4$ in the case of (8.189), or $\delta' \approx 7.443$ and $n = 3$ in the case of (8.190). The accelerated motion of the microscopic particle can generate radiation and self-damping. However, these effects are very weak, and depend on the amplitude of the microscopic particle.

Consider the collision between two microscopic particles and collision-induced changes in their amplitudes. Such changes are useful in understanding of energy transfer from the slower microscopic particle to the faster one. As it is known, the sum of amplitudes is conserved because the number of quanta bound in the soliton is $N_{\text{sol}} = 4\eta$ which is conserved for the soliton (8.19). We now examine the changes in the amplitude and velocity of the microscopic particle in the system described by (8.188), using the perturbation theory.

Assuming that the relative velocity v of the colliding microscopic particles (solitons) is large compared to their amplitude, Malomed considered the lowest order approximation and expressed the full wave field during the collision as a linear superposition of the unperturbed solutions (8.19),

$$\phi(x,t) = (\phi_{\text{sol}})_1 + (\phi_{\text{sol}})_2. \tag{8.191}$$

Inserting (8.191) into the right-hand side of (8.188), both (8.189) and (8.190) become polynomials in $(\phi_{\text{sol}})_1$ and $(\phi_{\text{sol}})_2$. The number of quanta is conserved in the system described by (8.188), *i.e.*,

$$N = \int_{-\infty}^{\infty} |\phi|^2 dx' = \text{constant}.$$

It can be shown that

$$\frac{d}{dt}(N_{\text{sol}})_1 \equiv \frac{d}{dt}\int_{-\infty}^{+\infty} |(\phi_{\text{sol}})_1|^2 dx = -i\int_{-\infty}^{+\infty} dx'\phi_1^*(x')P_1(x') + c.c, \tag{8.192}$$

where

$$P_1 = \epsilon(\phi_{\text{sol}})_1(\phi_{\text{sol}})_2[(\phi_{\text{sol}}^*)_2]_{x'} \tag{8.193}$$

Further analysis shows that evolution of the first soliton is dominated by the term of the polynomial given in (8.193) in the case of (8.189). Indeed, inserting the polynomial (8.193) into (8.192), one notes that the dominating term must contain one power of $(\phi_{\text{sol}})_2$ and one power of $(\phi_{\text{sol}}^*)_2$ (or their derivatives), lest the integral on the right-hand side of (8.192) will be exponentially small (the integrand will contain a rapidly oscillating exponent), and that either $(\phi_{\text{sol}})_2$ or $(\phi_{\text{sol}}^*)_2$ must be expressed by its derivative, which gives an additional large multiplier $\approx v$. Using $\eta = N_{\text{sol}}/4$ and (8.19), (8.193) and (8.192) in the following

$$\delta^{(1)}\eta_1 = \frac{1}{4}\int_{-\infty}^{+\infty} dt\frac{d}{dt}(N_{\text{sol}})_1,$$

we get

$$\delta^{(1)}\eta_{1,2} = 4\epsilon\eta_1\eta_2\text{sgn}(v_{1,2} - v_{2,1}). \tag{8.194}$$

The superscript (1) in (8.194) indicates result obtained in the first order of ϵ. Evidently, (8.194) satisfies the conservation law $\delta^{(1)}\eta_1 + \delta^{(1)}\eta_2 = 0$. For (8.190), the result is similar

$$\delta^{(1)}\eta_{1,2} = 8\epsilon\eta_1\eta_2(v_{1,2} - v_{2,1})^{-1}. \tag{8.195}$$

Malomed considered an ensemble of solitons with different initial amplitudes and velocities. Solitons with the largest initial amplitude will acquire the largest velocity according to $v_{t'} = \delta'\varepsilon\eta^n$, and from (8.194) and (8.195), their amplitudes will be further increased due to collisions with the slower solitons. Malomed estimated that if the mean distance between the microscopic particles (solitons) is L and a characteristic value of the initial amplitudes is η_0, then the time T during which the microscopic particles (solitons) will separate into "large" and "small" ones is about $T \approx \sqrt{L/\eta_0^5}/\epsilon$ for (8.189) and $T \approx L/(\epsilon\eta_0)$ for (8.190).

Using perturbed theory and numerical method, Malomed also investigated collision of microscopic particles described by (8.188). The collision-induced changes in the amplitudes of both microscopic particles and their radiation losses in the collision process were obtained for the case of relative velocity of the microscopic particles being much larger than their amplitudes. The collision between two microscopic particles of amplitudes η_1 and η_2 and the relative velocity $v(v^2 \gg \eta_1^2, \eta_2^2)$ results in a radiation loss which depends on the changes $\delta^{(2)}\eta_n$ $(n = 1, 2)$ in the amplitudes of the colliding microscopic particles, *i.e.*,

$$\delta^{(2)}\eta_{1,2} = \begin{cases} -4\varepsilon^2\eta_{1,2}\eta_{2,1}^2, & \text{for (8.189)}, \\ -\dfrac{16\varepsilon^2}{v^2}\eta_{1,2}\eta_{2,1}^2, & \text{for (8.190)}. \end{cases} \tag{8.196}$$

However, the changes in the amplitudes of the microscopic particles produced by the adiabatic (non-radiative) exchange should also be considered, which are first order effects in ε and are given by (8.194) and (8.195). Therefore, total changes in the amplitudes of the microscopic particles should be the sum of the above two effects.

Contribution of the collision to the change in the velocities of the solitons can be calculated. However, this effect, unlike the energy exchange and the radiative losses, is not of principal importance since the velocities of the solitons change continuously between collisions.

Malomed also investigated radiative losses in collision processes of microscopic particles obeying the following nonlinear Schrödinger equation, with high-order (third order) dispersions

$$i\phi_{t'} + \phi_{x'x'} + 2|\phi|^2\phi = i\varepsilon_1\phi_{x'x'x'} + i\varepsilon_2|\phi|^2\phi_{x'} + i\varepsilon_3\phi^2\phi_{x'}, \tag{8.197}$$

where ε_1, ε_2 and ε_3 are coefficients of the third-order linear and nonlinear dispersions, respectively, which are some real and small perturbation effects. Malomed calculated the radiative losses using perturbation theory based on the inverse scattering method. In his treatment of the collision process, he also assumed that the relative velocity of the two microscopic particles is much larger than their amplitudes, *i.e.*, $v = |v_1 - v_2| \gg \eta_1, \eta_2$. Hence, what we discussed in the above is still applicable here.

In the lowest order approximation, Malomed expressed the solution of (8.197) in the form of

$$\phi(x', t') = \sum_{n=1}^{2}\left[\phi_{\text{sol}}^{(n)}(x', t') + \delta\phi^{(n)}(x', t')\right], \tag{8.198}$$

where the correction $\delta\phi^{(n)}(x', t')$ are determined by the following linear equation

$$\delta\phi_{t'}^{(1)} + i\delta\phi_{x'x'}^{(1)} + 4|\phi_{\text{sol}}^{(1)}|^2\delta\phi^{(1)} + 2[\phi_{\text{sol}}^{(1)}]^2\delta\phi^{(1)*} = -4|\phi_{\text{sol}}^{(2)}|^2\phi^{(1)} + 2[\phi_{\text{sol}}^{(2)}]^2\phi^{(1)*}. \tag{8.199}$$

It is easily seen that the lowest-order approximate solution to (8.199) is given by keeping only the first term on either side of the equation, and the fully modified solitonic wave form can be expressed as

$$\tilde{\phi}_{\text{sol}}^{(1)} \equiv \phi_{\text{sol}}^{(1)} + \delta\phi_{\text{sol}}^{(1)} = \phi_{\text{sol}}^{(1)} e^{i\theta}, \tag{8.200}$$

where $v = v_1 - v_2$ is the relative velocity of the colliding microscopic particles, and

$$\theta = -\frac{8\eta_2}{v}\tanh[2\eta_2(x' - vt')].$$

The collision results in the following change in the phase of the first microscopic particle

$$\triangle\theta_1 \equiv \theta(x' - vt' = -\infty) - \theta(x' - vt' = +\infty) = 16\frac{\eta_2}{v}. \tag{8.201}$$

As discussed in Chapter 4, the expression of the collision-induced phase shift in the unperturbed nonlinear Schrödinger equation is

$$\triangle\theta_1 = 2\tan^{-1}\left[\frac{8\eta_2 v}{v^2 + 16(\eta_1^2 - \eta_2^2)}\right]. \tag{8.202}$$

Thus (8.201) is nothing but the lowest-order term in the expansion of (8.202) in powers of v^{-1}. Furthermore, the collision between the microscopic particles gives rise to a shift in the center of mass of the first soliton, and a corresponding expansion which starts from the term v^{-2}.

According to the inverse scattering method, the radiation part of the wave field is described by the spectral amplitudes (reflection coefficient) $B(\lambda)$ and the spectral parameter 2λ (radiation wave number). The basic ingredient of the perturbation theory is the general perturbation-induced evolution equation for $B(\lambda)$, which is given by

$$\frac{dB(\lambda)}{dt'} = -a(\lambda)e^{-4i\lambda^2 t'}\int_{-\infty}^{+\infty} dx' \left\{[\psi^{(1)*}(\lambda, x')]^2 P^*(x') + [\psi^{(2)*}(\lambda, x')]^2 P(x')\right\}, \tag{8.203}$$

where $P(x')$ denotes an emission-generating perturbing term on the right-hand side of (8.197), and $\psi^{(1,2)}(\lambda, x')$ are components of the Jost function, which are related to $\phi(x')$ by the Zakharov-Shabat equations (4.42) or (3.22). The quiescent ($v = 0$) one-soliton solution is given by (8.19). The transmission coefficient in (8.203) is $a(\lambda) = (\lambda - i\eta)/(\lambda + i\eta)$, and

$$\psi_{\text{sol}}^{(1)}(\lambda, x') = \frac{e^{i\lambda x'}}{\lambda + i\eta}[\lambda + i\eta\tanh(2\eta x')],$$
$$\psi_{\text{sol}}^{(2)}(\lambda, x') = \frac{i\eta}{\lambda + i\eta}e^{i\lambda x' + 4i\eta^2 t'}\operatorname{sech}(2\eta x'). \tag{8.204}$$

Since it is assumed that $|v_1 - v_2| \gg \eta_1, \eta_2$ in the lowest order approximation, a correction corresponding to the modified soliton wave form given by (8.200) can be found. Substituting (8.200) into the Zakharov-Shabat equation (4.42), the correction to $\psi(\lambda, x')$ can be found to be

$$\delta\psi = -i\theta \begin{pmatrix} 0 \\ \psi_{\text{sol}}^{(2)} \end{pmatrix}. \tag{8.205}$$

Inserting (8.200), (8.204) and (8.205) into (8.203) with the perturbing term $i\epsilon_1\phi_{x'x'x'}$, and integrating (8.203) over t', we obtain, within the lowest order approximation,

$$B_{\text{fin}}(\lambda) = \int_{-\infty}^{\infty} dt' \frac{dB(\lambda)}{dt'} = i\frac{16\pi\epsilon_1\eta_2\lambda}{15v}\frac{17\eta_1^2 + 3\lambda^2}{\eta_1^2 + \lambda^2}\,\text{sech}\left(\frac{\pi\lambda}{2\eta_1}\right). \tag{8.206}$$

Equation (8.197) (with $\epsilon_1 = \epsilon_3 = 0$) conserves the number of quanta, the momentum and the energy,

$$N = \int_{-\infty}^{+\infty} dx' |\phi(x')|^2,$$
$$P \equiv i\int_{-\infty}^{+\infty} dx' \phi\phi_{x'}^*,$$
$$E \equiv \int_{-\infty}^{+\infty} dx' \left(\frac{1}{2}|\phi_{x'}|^2 - |\phi|^4 + i\epsilon_1\phi_{x'}^*\phi_{x'x'}\right).$$

Thus the spectral densities of the radiation parts of these three conserved quantities are

$$N'(\lambda) = \frac{1}{\pi}|B_{\text{fin}}(\lambda)|^2,$$
$$P'(\lambda) = -\frac{2\lambda}{\pi}|B_{\text{fin}}(\lambda)|^2,$$
$$\mathcal{E}'(\lambda) = -\frac{4\lambda^2}{\pi}|B_{\text{fin}}(\lambda)|^2,$$

so that the net number of quanta, momentum, and energy carried by the radiation are given by

$$\begin{aligned} N_{\text{rad}} &= \int_{-\infty}^{+\infty} d\lambda N'(\lambda), \\ P_{\text{rad}} &= \int_{-\infty}^{+\infty} d\lambda P'(\lambda), \\ E_{\text{rad}} &= \int_{-\infty}^{+\infty} d\lambda E'(\lambda). \end{aligned} \tag{8.207}$$

The same quantities for the unperturbed soliton are

$$N_{\text{sol}} = 4\eta, \quad P_{\text{sol}} = 2\eta v, \quad E_{\text{sol}} = -\frac{16}{3}\eta^3 + \eta v^2.$$

Substituting (8.206) into (8.207) and integrating the spectral density, the total number of quanta emitted by the first microscopic particle (soliton) can be obtained and is given by

$$N_{\text{rad}}^{(1)} = \left(\frac{16\pi\eta_1\eta_2\epsilon_1}{15v}\right)^2 \eta_1 J, \tag{8.208}$$

where

$$J \equiv \int_{-\infty}^{+\infty} dz \left[\frac{z(17+3z^2)}{1+z^2}\right]^2 \operatorname{sech}^2\left(\frac{\pi z}{2}\right) \approx 50.43.$$

Using the conservation of the net number of quanta and (8.208), the collision-induced change in the amplitude of the microscopic particle (soliton) can be obtained as

$$\delta\eta_1 = -\frac{1}{4}N_{\text{rad.}}^{(1)} \tag{8.209}$$

For the second soliton, changes in the amplitudes of the microscopic particles are given by (8.208) and (8.209) with η_1 and η_2 replaced by their transposes. Using the momentum and energy conservations we can find the collision-induced changes in the velocities of the microscopic particles through the balance equations for the momentum and energy in the center-of-mass reference frame defined by $\eta_1 v_1 + \eta_2 v_2 = 0$. In fact, from (8.206) and the assumption of $|v_1 - v_2| \gg \eta_1, \eta_2$, one can conclude that the spectral wave packets emitted by the two microscopic particles are centered at $\lambda_1 = v_1/4$ and $\lambda_2 = v_2/4$, respectively, and the widths of the packets are given by $\delta\lambda_n \approx \eta_n$ $(n = 1, 2)$. According to (8.207), this implies that, in the first-order approximation, the emitted radiation carries zero momentum, and $\delta\eta_1/\delta\eta_2 = \eta_1/\eta_2$ in such a case. Based on these, Malomed obtained changes in the velocities of the particles δv_1 and δv_2 as

$$\eta_1\delta v_1 + \eta_2\delta v_2 = 0. \tag{8.210}$$

Substituting (8.207), (8.210) and $\eta_1 v_1 = -\eta_2 v_2$ into the energy-balance equation, Malomed determined δv_1,

$$\dot{v}_1\delta v_1 = \frac{8\eta_2}{\eta_1^2}(\eta_1^2 + \eta_2^2 - \eta_1\eta_2)\delta\eta_1.$$

For the effects of the other two perturbing terms involving ϵ_2 and ϵ_3 in (8.197), on the emission of radiation and on the amplitudes and velocities of the solitons, a tedious calculation cannot be avoided. As mentioned above, the particular cases of $\epsilon_1 = 0$, $\epsilon_2 = 2\epsilon_3$ and $\epsilon_3 = 0$, $\epsilon_2 = 6\epsilon_1$ can be integrated exactly. The soliton-soliton collisions are elastic in these cases. Thus one may infer that the contributions to

$B(\lambda)$ due to the second and third terms, respectively, in (8.197), or the first [see (8.206)] and second, may cancel each other. If this is true, all the above results are applicable to the general case, (8.197), if the parameter ϵ_1 in (8.206) and (8.208) is replaced by

$$\epsilon_{\text{eff}} \equiv \epsilon_1 - \frac{1}{6}\epsilon_2 + \frac{1}{3}\epsilon_3.$$

Besley *et al.* used multiscale perturbation theory in conjunction with the inverse scattering method to study the interaction of many microscopic particles (solitons) obeying the following nonlinear Schrödinger equation which has a small correction to the nonlinear potential,

$$i\phi_{t'} + \frac{1}{2}\phi_{x'x'} + |\phi|^2\phi + F[\phi, \phi^*] = 0,$$

subjecting to the initial condition $\phi(x', 0) = \phi_0(x')$ for certain initial field $\phi_0(x')$. In the limit of the perturbation term $F[\phi, \phi^*]$ being small, it can be assumed that the microscopic particles are all moving with the same velocity at the initial instant. This maximizes the effect each microscopic particle has on the others as a consequence of the perturbation. Over a long time, the amplitudes of the microscopic particles remain the same, while their centers of mass move according to the Newton's equation, with a force for which the above authors presented an integral formula.

Interaction of microscopic particles through a quintic perturbation was also studied by the above authors. They give more details since symmetries, which are related to the form of the perturbation and to the small number of particles involved, allow the problem to be reduced to a one-dimensional problem with a single parameter, the effective mass. They calculated the binding energy and oscillation frequency of nearby solitons in the stable case, when the perturbation is an attractive correlation to the potential. Numerical results verified accuracy of the perturbative calculation and revealed its range of validity.

8.13 Microscopic Particles in Three and Two Dimensional Nonlinear Media with Impurities

Properties of microscopic particles described by the nonlinear Schrödinger equation in two and three dimensional systems were studied by many scientists, for example, Gaididei *et al.*, Desyatnikov *et al.*, Infeld and Rowlands, Pokrovsky and Talopov, Germanschewski *et al.*, Konopelchenko, and so on. Here the work of Desyatnikov *et al.* and Gaididei *et al.* will be briefly discussed.

Gaididei *et al.* investigated dynamics of nonlinear excitation in two-dimensional systems with impurities. In their study, the following assumptions were made, the density of excitations is low, the impurity is located at site 0 of a lattice consisting of similar hosts, and the impurity substitution for the host does not distort the lattice significantly. The dynamic equation of the microscopic particle in the system is written in the form of the following nonlinear Schrödinger equation

$$i\hbar\phi_t + \frac{\hbar^2}{2m}\nabla^2\phi + b|\phi|^2\phi = E(\vec{r})\phi, \tag{8.211}$$

where

$$\nabla^2 = \frac{\partial^2}{\partial x^2} + \frac{\partial^2}{\partial y^2}, \quad \text{and} \quad \int d\vec{r}|\phi(\vec{r},t)|^2 = 1. \tag{8.212}$$

$E(\vec{r})$ in (8.211) is the continuum limit of the on-site excitation energy $E_{\vec{n}}$. It determines an energetic profile for an excitation in the vicinity of the impurity molecule. Gaididei *et al.* considered only some general properties of soliton dynamics in the vicinity of an impurity. $E(\vec{r})$ is given as an axially symmetric Gaussian function

$$E(\vec{r}) = Ee^{-(\vec{r}/r_0)^2}, \tag{8.213}$$

where E is the strength of the impurity and r_0 is its radius. Letting

$$\vec{\rho} = \frac{\vec{r}}{r_0}, \quad \tau = \frac{\hbar t}{2mr_0^2}, \quad \varphi = \sqrt{\frac{2mbr_0^2}{\hbar^2}}\phi, \quad \epsilon = \frac{2mEr_0^2}{\hbar^2},$$

equations (8.211) and (8.213) are replaced by

$$i\varphi_\tau + \nabla^2\varphi + |\varphi|^2\varphi = V(\rho)\varphi, \quad V(\rho) = \epsilon e^{-\rho^2}, \tag{8.214}$$

where ε characterizes the strength of the impurity and

$$\int d\vec{\rho}|\varphi|^2 = N. \tag{8.215}$$

In general, the two-dimensional nonlinear Schrödinger equation possesses bright vortex soliton solutions, *i.e.*, localized solutions with an internal velocity (spin), and some unstable solutions that may either disperse or collapse. These two types of solutions are separated by the so-called ground state solution whose width does not change in time. The ground state solution to (8.214) with $V = 0$ is approximately given by

$$\varphi_s = B_s \operatorname{sech}\left(\frac{\rho}{A_s}\right)e^{i\tau}, \tag{8.216}$$

if the particle is initially at rest and centered at $\rho = 0$. The ground state amplitude B_s and width A_s are given by

$$B_s = \sqrt{\frac{12\ln 2}{4\ln 2 - 1}}, \quad A_s = \sqrt{\frac{2\ln 2 + 1}{6\ln 2}}, \tag{8.217}$$

respectively. Substituting (8.216) into (8.215) yields $N = N_s = 11.7$. For the initial conditions, $\varphi = \varphi(\rho, 0)$ with N being larger (smaller) than N_s, the solution, $\varphi =$

$\varphi(\rho,\tau)$, of the two-dimensional nonlinear Schrödinger equation collapses (disperses) in finite time.

To study the motion of a microscopic particle (soliton) in the neighborhood of an impurity, Gaididei *et al.* assumed that the radius r_0 of the impurity is large compared with the width of the microscopic particle. In this case we can expand the impurity potential $V(\rho)$ in powers of ρ. Keeping only terms of the second order, *i.e.*, $V(\rho) \approx \epsilon(1-\rho^2)$, the problem is reduced to that of a microscopic particle moving in a two-dimensional parabolic potential, given by

$$V(\rho,\tau) = -\epsilon(\tau)\rho^2, \tag{8.218}$$

where $\epsilon(\tau)$ is an arbitrary function of time. Transforming to the noninertial frame of reference in which the center of mass of the excitation is at rest, we have

$$\varphi(\vec{\rho},\tau) = \Psi(\vec{\rho},\tau)\exp\left\{\frac{i}{2}\dot{\vec{R}}\cdot\vec{\rho} - \frac{i}{4}\int_0^\tau d\tilde{\tau}\left[\dot{\vec{R}}(\tilde{\tau})^2 + \epsilon(\tilde{\tau})\vec{R}^2(\tilde{\tau})\right]\right\}, \tag{8.219}$$

where

$$\vec{R} = \frac{1}{N}\int d\vec{r}\cdot\vec{r}\,|\varphi(\vec{r},\tau)|^2$$

is the position of the center of mass and $\vec{\rho}\,' = \vec{\rho} - \vec{R}(\tau)$ is the coordinate in the new frame of reference, $\dot{\vec{R}} = \partial\vec{R}/\partial\tau$. It can be shown that

$$\ddot{\vec{R}} - 4\epsilon(\tau)\vec{R} = \vec{0}, \tag{8.220}$$

and

$$i\Psi_\tau + \nabla^2_{\rho'}\Psi + |\Psi|^2\Psi + \epsilon(\tau)\rho'^2\Psi = 0. \tag{8.221}$$

Equations (8.220) and (8.221) show that in the parabolic potential, the external (position of center of mass, $\vec{R}$) and the internal $[\Psi(\vec{\rho},\tau)]$ degrees of freedom of the microscopic particle are separated. This is a unique property of the harmonic potential. Anharmonic terms in $V(\rho)$ would lead to coupling between the external and internal degrees of freedom of the microscopic particles. The center of mass of the microscopic particle behaves like a time-dependent oscillator. To consider the internal motion of the microscopic particle in the impurity field, we introduce a change of variables which is referred to as lens transformation

$$\Psi(\vec{\rho},\tau) = \frac{1}{a(\tau)}\Phi(\xi,T)\exp\left[i\left(T + \frac{\dot{a}}{4a}\rho'^2\right)\right], \tag{8.222}$$

where $a(\tau)$ is the width of the microscopic particle, and

$$\xi = \frac{|\rho'|}{a(\tau)}, \quad T = \int_0^\tau \frac{d\tilde{\tau}}{a^2(\tilde{\tau})} \quad \text{and} \quad \dot{a} = \frac{da}{dt}$$

are new space and time variables, respectively. Substituting (8.222) into (8.221) yields

$$i\Phi_T + \left[\frac{\partial^2}{\partial\xi^2} + \frac{1}{\xi}\frac{\partial}{\partial\xi}\right]\Phi + |\Phi|^2\Phi - \triangle\xi^2\Phi - \Phi = 0, \tag{8.223}$$

where

$$\triangle = \frac{1}{4}a^3\ddot{a} - \epsilon(\tau)a^4. \tag{8.224}$$

The same problem was also investigated by Karlsson *et al.* using the collective coordinate method. We are interested in localized solutions ($\Phi \to 0$ for $\xi \to \infty$) of (8.223). If $\triangle$ is a positive constant, we can set $\Phi_T = 0$ and (8.223) becomes

$$\left(\frac{\partial^2}{\partial\xi^2} + \frac{1}{\xi}\frac{\partial}{\partial\xi}\right)\Phi - \triangle\xi^2\Phi + |\Phi|^2\Phi - \Phi = 0. \tag{8.225}$$

In the case of $\triangle > 0$, equation (8.225) has localized solutions, and in this case, (8.222) describes a non-collapsing soliton. We solve (8.225) now.

If $\xi \to \xi/\triangle^{1/4}$, $\Phi \to \triangle^{1/4}\Phi$, Gaididei *et al.* obtained from (8.225)

$$\left(\frac{\partial^2}{\partial\xi^2} + \frac{1}{\xi}\frac{\partial}{\partial\xi}\right)\Phi + (|\Phi|^2 - \xi^2)\Phi = \frac{1}{\sqrt{\triangle}}\Phi. \tag{8.226}$$

The corresponding energy is given by

$$E = \int d\xi\xi\left(\left|\frac{\partial\Phi}{\partial\xi}\right|^2 - \frac{1}{2}|\Phi|^4 + \xi^2|\Phi|^2 + \frac{1}{\sqrt{\triangle}}|\Phi|^2\right). \tag{8.227}$$

The following form can be assumed for a trial function,

$$\Phi = \sqrt{\frac{N}{2\pi\ln 2}\frac{1}{\beta}}\mathrm{sech}\left(\frac{\xi}{\beta}\right), \tag{8.228}$$

where β is a variational parameter. Inserting (8.228) into (8.227) and minimizing E with respect to β, we obtain

$$\beta^4 = \triangle = \frac{4}{27}\left(1 - \frac{N}{N_s}\right)\frac{1+2\ln 2}{\zeta(3)},$$

where ζ denotes the Riemann zeta function and $N_s = 4\pi\ln 2(2\ln 2+1)/(4\ln 2-1) \simeq 11.7$ is the number of excitations in the ground state. For $N < N_s$, returning to the original variable, the soliton solution of (8.221) was found to be

$$\Psi = \frac{1}{a(\tau)}\sqrt{\frac{N}{2\pi\ln 2}}\,\mathrm{sech}\xi\exp\left[i\left(T + \frac{\dot{a}}{4a}\rho'^2\right)\right]. \tag{8.229}$$

The soliton dynamics in the parabolic potential (8.218) is governed by (8.220) for the center of mass and (8.224) for the width of the soliton. Equations (8.220) and (8.224) belong to the class of Ermakov-Pinney equations, the solution of which was obtained by Pinney and it is

$$a(t) = \left[u^2(t) + \frac{4\triangle}{W^2}v^2(t)\right]^{1/2}. \tag{8.230}$$

where $W = \dot{u}v - u\dot{v}$ is the Wronskian and (u,v) is the fundamental set of solutions of the respective linear equation $\ddot{y} - 4\epsilon(t)y = 0$, which coincides with (8.220) as far as the motion of the center of mass is concerned. If

$$\epsilon(\tau) = -\frac{\Omega_0^2}{4} \equiv -\frac{2m|E|r_0^2}{\hbar^2} = \text{const.} \tag{8.231}$$

then the impurity is an acceptor of excitations. Inserting (8.231) into (8.220) and (8.230), Gaididei *et al.* obtained

$$R(t) = R_0 \cos[\omega(t + t_0)],$$
$$a^2(t) = A^2 \cos^2[\omega(t + t_1)] + \frac{\hbar^2\triangle}{2m|E|r_0^2A^2}\sin^2[\omega(t + t_1)].$$

Here R_0, A, t_0 and t_1 are arbitrary constants, $\omega = \sqrt{2|E|/m}/r_0$ is the frequency of oscillation of the center of mass. The width $a(t)$ of the soliton oscillates at a frequency of 2ω. Here, ω depends on the depth $|E|$ of the impurity, the radius r_0, and the mass of the microscopic particle. The above approach is valid if the radius of the impurity r_0 is larger than the width of the soliton (*i.e.*, $a < 1$).

If

$$\epsilon(\tau) = -\frac{\Omega_0^2}{4}[1 + \lambda\cos(\Omega\tau)], \tag{8.232}$$

then the acceptor molecule oscillates around its equilibrium position with the frequency Ω. The acceptor can perform either translational or vibrational motion. The parameter λ is proportional to the amplitude of the impurity oscillations. Inserting (8.232) into (8.220), we get

$$\ddot{R} + \Omega_0^2[1 + \lambda\cos(\Omega\tau)]R = 0, \tag{8.233}$$

which is the Mathieu's equation. The (Ω_0, λ) plane can be divided into a stability region [$R(\tau)$ remains finite as τ approaches to infinity] and an instability region (parametric resonance). For the first resonant range of $|\Omega - 2\Omega_0| < \Omega_0\lambda/2$, Gaididei, *et al.* obtained an approximate solution of (8.233) which is given by

$$R(\tau) = R_0\{e^{\gamma\tau}\cos[\Omega(\tau + \tau_0)] + e^{-\gamma\tau}\sin[\Omega(\tau + \tau_0)]\}, \tag{8.234}$$

where R_0 and τ_0 are constants that depend on the initial position of the microscopic particle and its initial velocity. γ is given by

$$\gamma = \frac{1}{2}\left[\left(\frac{\Omega_0\lambda}{2}\right)^2 - (\Omega - 2\Omega_0)^2\right]^{1/2}. \tag{8.235}$$

We can see from (8.234) – (8.235) that when an acceptor oscillates about its equilibrium position with a frequency in the range given, it cannot trap the microscopic particle. The amplitude of the oscillations of the microscopic particle increases exponentially with the amplification coefficient γ.

Gaididei *et al.* also obtained numerical solutions of (8.223) using the split-step Fourier method. The numerical results are consistent with the analytical results given above.

Desyatnikov *et al.* investigated solutions of high-dimensional ϕ^6-nonlinear Schrödinger equation

$$i\phi_{t'} + \nabla^2\phi + |\phi|^2\phi - |\phi|^4\phi = 0, \tag{8.236}$$

where

$$\nabla^2 = \frac{\partial^2}{\partial x'^2} + \frac{\partial^2}{\partial y'^2} + \frac{\partial^2}{\partial z'^2},$$

and $t' = t/\hbar$, $x' = x\sqrt{2m}/\hbar$, $y' = y\sqrt{2m}/\hbar$, $z' = z\sqrt{2m}/\hbar$. In the case of two-dimension, Quiroga-Teixeiro and Michinel found that there exists a 2D vortex soliton of (8.236) which was shown to be stable by means of numerical simulation. However, Kruglov *et al.* found that the helical vortex soliton of this model, with an amplitude periodically modulated along the propagation direction, is unstable. Meanwhile, this 2D vortex soliton with the quadratic nonlinearity, although exists as a stationary solution, is subject to a strong azimuthal instability. This was demonstrated numerically by Firth and Skryabin, and Petrov *et al.*, and observed experimentally in optical fiber by Petrov *et al.*. A similar strong azimuthal instability of the 2D spinning soliton has been predicted by numerical simulation with saturable nonlinearity by Petrov *et al.*. However, in the 3D case, numerical simulation showed that there is also a 3D soliton solution to (8.236). Desyatnikov *et al.* first obtained for radially symmetric soliton solution without the spin. Their work will be introduced in the following.

Let $r = \sqrt{x'^2 + y'^2 + z'^2}$ and assume the solution of (8.236) has the following form

$$\phi(x', y', z', t) = e^{ikt'}\psi(\vec{r}).$$

Substituting the above into (8.236), we get

$$\psi_{rr}(\vec{r}) + \frac{2}{r}\psi_r(\vec{r}) - k\psi(\vec{r}) + \psi^3(\vec{r}) - \psi^5(\vec{r}) = 0, \tag{8.237}$$

The boundary condition is defined by means of the asymptotic expression,

$$\begin{aligned} &\psi(\vec{r}) \approx \alpha(k)(1 + cr^2), \quad \text{at } r \to 0; \\ &\psi(\vec{r}) \approx \frac{B(k)}{r}e^{-\sqrt{k}r}, \quad \text{at } r \to \infty. \end{aligned} \tag{8.238}$$

where $c = (\alpha^4 - \alpha^2 + k)/6$, $\alpha(k)$ and $B(k)$ are functions of k.

Desyatnikov *et al.* obtained numerically the solution of (8.237) for different k by the shooting method as shown in Fig. 8.8. From this figure we can see that the effective size of the soliton increases with k, which results in a flatter distribution of the field at the center of the soliton. This corresponds to a decrease in the curvature parameter c in (8.238). When $c = 0$, the amplitude of the soliton $\alpha(k)$ at $r = 0$ approaches to two limiting values: $\alpha_{1,2}(k) = \pm\sqrt{1+\sqrt{1-4k}/2}$.

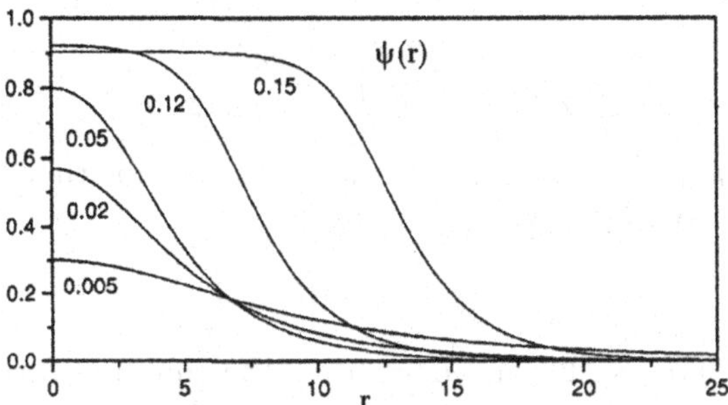

Fig. 8.8 The zero spin 3D soliton solutions of (8.237). The values of the propagation constant k are indicated near the curves.

In any dimension, there is a similar upper boundary for the values of k at which solitons exist. In 1D, one has, instead of (8.237), an equation

$$\psi'' - k\psi + \psi^3 - \psi^5 = 0,$$

which has the well-known exact soliton solution

$$\psi^2(x') = \frac{4k}{1+\sqrt{1-(16/3)k}\cosh[2\sqrt{k}(x'-x'_0)]},$$

which is the same as (8.149) except that $-\alpha_1$ is now replaced by k. Obviously, this solution exists if $k < k_{\max}^{(1D)} = 3/16$. In 2D, it was found, using the same shooting method, that $k_{\max}^{(2D)} \approx 0.18$. If $k_{\max}^{(3D)} < k_{\max}^{(2D)} < k_{\max}^{(1D)}$, the upper boundary for a soliton solution of (8.237) to exist decreases as the space dimension increases. The most important physical characteristic of the 3D soliton is its energy,

$$E(k) = 4\pi \int_0^\infty \psi^2(\vec{r};k)\vec{r}^2 d\vec{r}.$$

In 2D, the soliton is interpreted as a spatial cylindrical beam. Its energy reaches a finite value $E_{\min}^{(2D)} \approx 11.75$ at $k = 0$. In 1D, the energy of the soliton vanishes as $k \to 0$. The divergence of the energy of the $3D$ soliton at $k \to 0$ is a consequence of a minimum energy required the $3D$ zero-spin solitons, whose numerical value is $E_{\min}/4\pi \approx 15$. Therefore, there is a U-shaped dependence of $E(k)$ on k in $3D$. This means that there are two soliton solutions, with different values of k, at each energy value $E > E_{\min}$. This is a distinctive feature of the $3D$ case, which is also known in the saturable model.

The spherical spatiotemporal coordinates can be used to construct solutions to (8.237) in the form of 3D solitons with an integer spin $m \neq 0$. *et al.* searched for solutions of the form

$$\phi(x', y', z', t') = e^{ikt' + im\tilde{\varphi}}\psi(r', \theta), \tag{8.239}$$

where $\cos\theta \equiv z/r$, and $\tilde{\varphi}$ is the usual angular coordinate in the transverse plane (x, y). Then, equation (8.237) is transformed into

$$\frac{1}{r^2}\frac{\partial}{\partial r}\left(r^2\frac{\partial\psi}{\partial r}\right) + \frac{1}{r^2\sin\theta}\frac{\partial}{\partial\theta}\left(\sin\theta\frac{\partial\psi}{\partial\theta}\right) - \frac{m^2\psi}{r^2\sin^2\theta} - k\psi + \psi^3 - \psi^5 = 0. \tag{8.240}$$

A variational approximation was developed to describe the 3D spinning solitons and the following trial function was used

$$\psi(r, \theta) = \Psi(r)\sin\theta. \tag{8.241}$$

Inserting (8.241) into (8.240) and integrating the equation with respect to θ, but keeping an arbitrary dependence of $\Psi(r)$ on r, the following can be obtained.

$$\frac{d^2\Psi}{dr^2} + \frac{2}{r}\frac{d\Psi}{dr} - 2\frac{\Psi}{r^2} - k\Psi + \frac{4}{5}\Psi^3 - \frac{24}{35}\Psi^5 = 0. \tag{8.242}$$

Solutions to this equation corresponding to different values of k are displayed in Fig. 8.9. This is a spinning soliton. These solutions were found numerically by means of the shooting method adjusted to the obvious boundary conditions that $\Psi(r)$ must vanish linearly at $r \to 0$ and exponentially at $r \to \infty$. Similar to that in the $2D$ case obtained by Teixeiro and Michinel, it was found that the slope of the function $\Psi(r)$ at $r = 0$ increases with k up to a maximum value at $k = 0.09$, and then decreases after this (in the 2D case, a maximum was reached at $k = 0.145$). The energy of the spinning soliton (8.239) is given by

$$E = 2\pi\int_0^\infty r^2 dr\int_0^\pi \psi^2(r, \theta)\sin\theta d\theta,$$

or

$$E(k) = \frac{8\pi}{3}\int_0^\infty \Psi^2(r, k)r^2 dr.$$

The energy given by the latter expression is displayed as a function of the propagation constant k. The minimum of the soliton energy is $E_{\min}/4\pi \approx 62.6$, which is located at $k = k_{cr} \approx 0.033$. This suggests that the spinning soliton with $m = 1$ can be stable at $k > k_{cr}$. It was not possible to have soliton solutions to (8.242) when k exceeds some maximum value, which was shown to coincide with $k_{\max}^{(3D)} \approx 0.15$, the upper bound of the region where the zero-spin solitons exists. Thus, the upper bound for multidimensional solitons does not depend strongly on the value of the spin, but on the spatial dimension. Desyatnikov *et al.* verified that the upper bound for 2D solitons with $m = 0$ almost coincides with that of $m = 1$ ($k_{\max}^{(2D)} \approx 0.18$ for

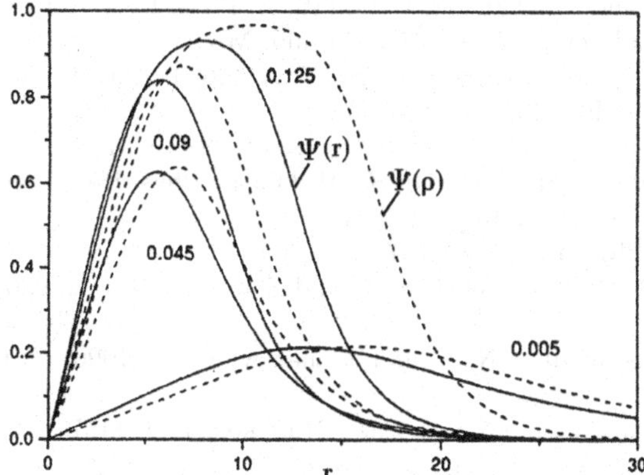

Fig. 8.9 3D spinning solitons with $m = 1$. The solid and dashed curves show, respectively, the functions $\Psi(r)$ and $\Psi(\rho)$ [see (8.241) and (8.243)] with the same values of k, as indicated.

both $m = 0$ and $m = 1$). The ratio of the minimum energy of the $m = 0$ and $m = 1$ soliton in the 2D case (which are $E^{2D}_{\min} = 50$ for the soliton with $m = 1$ and $E^{2D}_{\min} = 11.75$ for the zero-spin soliton) is $50/11.75 \simeq 4.26$. Comparing this to the same ratio in the $3D$ case, $62.6/15 \simeq 4.17$, we find that, in any dimension, formation of a spinning soliton requires energy which is roughly four times that necessary for the formation of a spinless soliton. Thus, experimental generation of the spinning soliton is expected to be harder than that of zero-spin, but not impossible.

Desyatnikov *et al.* also gave the solution of (8.237) in the form of (8.239) in cylindrical coordinate system $(\rho, \tilde{\varphi}, \tau)$. $\psi(r', \theta)$ in (8.239) is replaced by

$$\psi(\rho, \tau) = \Psi(\rho)\text{sech}(\mu t'), \tag{8.243}$$

The form of $\Psi(\rho)$ is shown in Fig. 8.9. The soliton solution found in such a case has same features as that in the spherical coordinates, as discussed above.

Bibliography

Abdullaev, F. (1982). Lebedel Institute Reports on Physics **10** (3) .

Abdullaev, F. (1982). Theor. Math. Phys. **51** (454) (in Russian).

Abdullaev, F. (1994). Theory of solitons in inhomogeneous media, Wiley and Sons, New York.

Abdullaev, F., Abrarev, R. M. and Darmanyan, S. A. (1989). Opt. Lett. **14** (131) .

Abdullaev, F., Darmanyan, S. and Umarev, B. (1985). Phys. Lett. A **108** (7) .

Abdullaev, F., Darmanyan, S. and Khabibullaev, P. (1990). Optical solitons, Springer, Berlin.

Abdullaev, F.,*et al.* (1985). Contributions of III Inter. symposium on some problems of statistical mechanics, Dubna, JINR, 3-7.

Ablowitz, M. J. and Segur, H. (1981). Solitons and the inverse scattering transform, SIAM, Philadelphia.

Barashenkov, I. V. and Makhankov, V. G. (1988). Phys. Lett. A **128** (52) .
Benjanian, T. B. and Feir, J. E. (1967). J. Fluid. Mech. A **21** (1439) .
Besley, T. A., Miller, P. D. and Akhmedier, N. N. (2000). Phys. Rev. E **61** (7121) .
Chen, H. H. (1978). Phys. Fluids **21** (377) .
Chen, H. H. and Liu, C. S. (1976). Phys. Rev. Lett. **37** (693) .
Chiao, R. Y., Garmire, E. and Townes, C. H. (1964). Phys. Rev. Lett. **13** (479) .
Chirikov, B. V. (1979). Phys. Rep. **52** (263) .
Cohen, G. (2000). Phys. Rev. E **61** (874) .
Cowan, S., Enn, R. H., Rangnekar, S. S. and Sanghera, S. S. (1986). Can. J. Phys. **64** (311) .
Crespe, J. M. S., Akhmediev, N. N. and Afanasjev, V. V. (1996). J. Opt. Soc. Am. B **13** (1409) .
Degtyarev, L. M., Makhankov, V. G. and Rudakov, L. I. (1974). Sov. Phys. JETP **40** (264) .
de Oliveira, J. M., Cavalcanti, S. B., Cerda, S. C. and Hickmann, J. M. (1998). Phys. Lett. A **247** (294) .
Desyatnikov, A., Maimistov, A., Malomed, B. (2000). Phys. Rev. E **61** (3107) .
Faddeev, L. D. and Takhtajan, L. A. (1987). Hamiltonian methods in the theory of solitons, Springer, Berlin.
Firth, W. J. and Skryabin, D. V. (1997). Phys. Rev. Lett. **79** (245) .
Foldy, L. L. (1976). Am. J. Phys. **44** (1196) .
Gaididei, Yu. B., Rasimussen, K. ϕ. and Christiansen, P. L. (1995). Phys. Rev. E **52** (2951) .
Garcia-Ripoll, J. J. and Perez-Garcia, V. M. (1999). Phys. Rev. A **60** (4864) .
Garcia-Ripoll, J. J. and Perez-Garcia, V. M. (2004). E-print, http://xxx.lanl.gov/abs/Patt-sol/9904006.
Garcia-Ripoll, J. J., Perez-Garcia, V. M. and Torres, P. (1999). Phys. Rev. Lett. **83** (1715) .
Germanschewski, K., Grauer, R., Berge, L., Mezentsev, V. K. and Rasmussen, J. J. (2001). Physica **D151** (175) .
Husimi, K. (1953). Prog. Theor. Phys. **9** (381) .
Infeld, E. and Rowlands, G. (1990). Nonlinear waves, solitons and chaos, Cambridge Univ. Press. Cambridge.
Karlsson, M., Anderson, D. and Desaix, M. (1992). J. Opt. Soc. Am. **17** (22) .
Karpman, V. I. (1967). JETP Lett. **6** (277) .
Karpman, V. I. (1991). Phys. Lett. A **160** (531) .
Karpman, V. I. (1993). Phys. Lett. A **181** (211) ; Phys. Rev. E **47** (2073) .
Karpman, V. I. (1994). Phys. Lett. A **193** (355) .
Karpman, V. I. (1998). Phys. Lett. A **244** (394) and 397.
Karpman, V. I. (2000). Phys. Rev. E **62** (5678) .
Karpman, V. I. and Shagalov, A. G. (1991). Phys. Lett. A **160** (538) .
Karpman, V. I. and Shagalov, A. G. (1999). Phys. Lett. A **254** (319) .
Kerner, E. H. (1958). Can. J. Phys. **36** (371) .
Kivshar, Yu. S. and Gredeskul, S. A. (1990). Phys. Rev. Lett. **64** (1693) .
Kivshar, Yu. S. and Malomed, B. A. (1989). Rev. Mod. Phys. **61** (763) .
Klyatzkin, V. I. (1980). Stochastic equations and waves in randomly inhomogeneous media, Nauka, Moscow (in Russian).
Kodama, Y. (1985). J. Stat. Phys. **39** (597) .
Kodama, Y. (1985). Physica D **16** (14) .
Kodama, Y. (1985). Phys. Lett. A **107** (245) .

Kodama, Y. (1985). Phys. Lett. A **112** (193) .
Kodama, Y. and Hasegawa, A. (1982). Opt. Lett. **7** (339) .
Konopelchenko, B. G. (1995). Solitons in multidimensions, World Scientific, Singapore.
Kruglov, V. I., Logvin, Yu. A. and Volkov, V. M. (1992). J. Mod. Opt. **39** (2277) .
Kundu, A. (1987). Physica **D25** (399) .
Makhankov, V. G. (1990). Soliton phenomenology, Kluwer, Amsterdam.
Malomed, B. A. (1987). Sov. J. Plasma Phys. **13** (360) .
Malomed, B. A. (1988). Phys. Scr. **38** (66) .
Malomed, B. A. (1990). Phys. Rev. A **41** (4538) .
Malomed, B. A. (1991). Phys. Rev. A **43** (3114) .
Malomed, B. A. (1991). Phys. Rev. A **44** (1413) .
Menyuk, C. R. (1986). Phys. Rev. A **33** (4367) .
Moussa, R., Goumri-said, S. and Aourag, H. (2000). Phys. Lett. A **266** (173) .
Nogami, Y. (1991). Am. J. Phys. **59** (64) .
Nogami, Y. and Toyama, F. M. (1994). Phys. Lett. A **184** (245) .
Nogami, Y. and Toyama, F. M. (1994). Phys. Rev. E **49** (4497) .
Nogami, Y., Vallieres, M. and Van Dijk, W. (1976). Am. J. Phys. **44** (886) .
Novikov, S. P., Manakov, S. V., Pitaevsky, L. P. and Zakharov, V. E. (1984). Theory of solitons, the inverse scatteaing method, Consultants Bureau, New York.
Nozaki, K. and Bekki, N. (1983). Phys. Rev. Lett. **50** (1226) .
Nozaki, K. and Bekki, N. (1986). Physica D **21** (381) .
Pathria, D. and Morris, J. L. (1989). Phys. Scr. **39** (673) .
Petrov, D. V. and Torner, L. (1997). Opt. Quantum Electron **29** (1037) .
Petrov, D. V., Torner, L., Martorell, J., Valaseca, R., Torres, J. P. and Cojocaru, C. (1998). Opt. Lett. **23** (1444) .
Pinney, E. (1950). Proc. Am. Math. Soc. **1** (681) .
Pokrovsky, V. L., Talopov, A. L. (1986). Thermodynamics of two dimensional soliton systems in soliton, ed. Trallingen, S. Zakharov, V. E. and Pokrovsky, V. L., North-Holland, Amsterdam, p. 73.
Quiroga-Teixeiro, M. and Michinel, H. (1997). J. Opt. Soc. Am. **B14** (2004) .
Reid, J. L. and Ray, J. R. (1984). Z. Angew. Math. Mech. **64** (365) .
Shchesnovish, V. S. and Doktorov, E. V. (1999). Physica D **129** (115) .
Stevenson, P. M. and Roditi, I. (1986). Phys. Rev. D **33** (2305) .
Takagi, S. (1991). Phys. Lett. A **160** (251) .
Zakharov, V. E., Manako, S. V., Noviko, S. P. and Pitayersky, I. P. (1984). Theory of solitons, Plenum, New York.

Chapter 9

Nonlinear Quantum-Mechanical Properties of Excitons and Phonons

In the remaining two chapters, we will continue to discuss applications of the nonlinear quantum mechanics. We will discuss nonlinear properties of exciton, phonon, proton, polaron and magnon. Through these discussions, and what we already know about electrons and helium atom from Chapter 2, we can get further understanding on properties and motion of microscopic particles in these nonlinear systems. These discussions will also establish the nonlinear quantum mechanics theory presented in Chapter 3 as the correct theory for describing properties and motion of microscopic particles in nonlinear systems. We will begin with motion of exciton and phonon in a molecular crystal.

9.1 Excitons in Molecular Crystals

Let's first consider a particular molecular crystal, the acetanilide ($CH_3COHNC_6H_5)_x$ or ACN. Two close chains of hydrogen-bonded amide-I groups which consists of atoms of carbon, oxygen, nitrogen and hydrogen (CONH) run through the acetanilide crystal. Its crystal structure has been determined and a unit cell of ACN is shown in Fig. 9.1. The space group is D_{2h}^{15} ($Pbca$) and the unit cell or factor group is D_{2h} for this crystal. The average lattice constants are $a = 1.9640$ nm, $b = 0.9483$ nm, and $c = 0.7979$ nm. There are eight molecules in an unit cell and at the amide-I frequency, each of these has one degree of freedom (d.f.). Thus, there are three infrared-active modes (B_{1u}, B_{2u}, and B_{3u}), four Raman-active modes (A_g, B_{1g}, B_{2g}, and B_{3g}), and one inactive mode (A_u). However, at low frequency (< 200 cm^{-1}), each molecule exhibits 6 d.f. (three translations and three rotations. This gives 48 low-frequency modes: 24 Raman active modes ($6A_g+6B_{1g}+6B_{2g}+6B_{3g}$), 18 infrared-active modes ($6B_{1u}+6B_{2u}+6B_{3u}$) and six (A_u) modes corresponding to the acoustic modes of translation and rotation). All of these active modes are seen in infrared absorption and Raman experiments. ACN is an interesting system because the nearly planar amide-I groups have bond lengths which are close to those found in polypeptide (see Fig.9.2). Since physical properties of such a hydrogen bonded amide-I system are very sensitive to bond lengths, study of ACN revealed some new phenomena. For example, in the experiments of infrared absorption and Raman scattering, a new amide-I band red-shifted from the main peak at 1666 cm^{-1} by about 16 cm^{-1} when the crystalline acetanilide is cooled from 320 to 10 K. No other major changes occur from 4000 to 800 cm^{-1}. The intensity of this new band increases steadily from room temperature till 70 K. The band at 1650 cm^{-1} is not present in amorphous materials or ACN methylated at the position where hydrogen-bonded distances occur, but it is recovered after annealing. Similar phenomena can be observed in Raman scattering experiments (Alexander 1985, Alexander and Krumbansl 1986, Benkui and John 1998, Careri *et al.* 1983, 1984, 1985, 1998, Davydov 1968, 1973, 1975, 1977, 1979, 1982, 1985, Eilbeck *et al.* 1984, Kenkre *et al.* 1994). Pope and Swenberg 1982, Scott 1990, 1992, 1998, Scott *et al.* 1985, 1989, Silinsh and Capek 1994, Tekec *et al.* 1998, Xiao 1998).

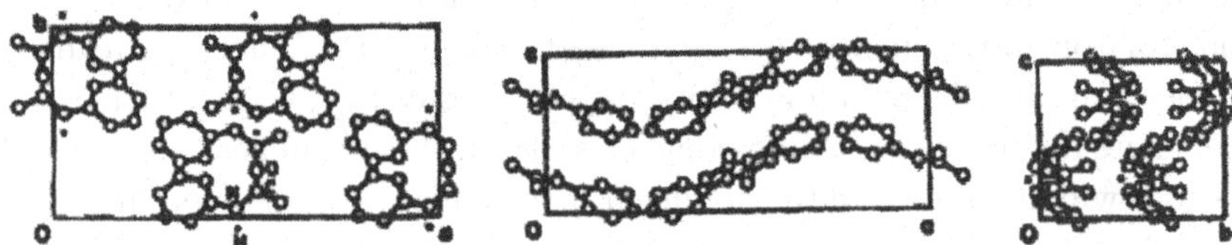

Fig. 9.1 Various views of the unit cell of ACN, with cell parameters $a = 19.640$ Å, $b = 9.483$ Å, and $c = 7.979$ Å.

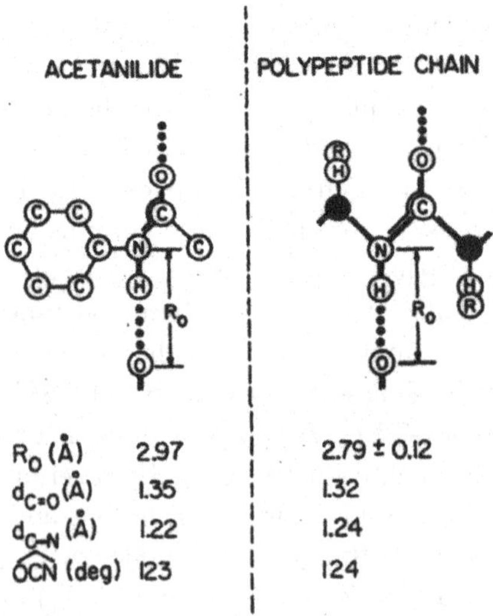

Fig. 9.2 Comparison between peptide groups of ACN and a protein molecule.

As it is known, the characteristic feature of the amide-I group, CONH, in polypeptides is the amide-I mode which mainly involves stretching of the C=O bond. This mode is observed as an infrared absorption peak at 1666 cm^{-1} in ACN and near this value in a wide variety of materials, including the amide-I groups. The corresponding spectroscopic evidence of the new band at 1650 cm^{-1} has been mentioned earlier, but detailed measurements of the crystal structure and specific heat as a function of temperature preclude assignment of the new band to (1) a conventional amide-I mode, (2) crystal defect states, (3) Fermi resonance or (4) frozen kinetics between two different subsystems. The correct assignment is the self-trapping of the amide-I vibrational energy. This is based on the following experimental facts: (1) the ^{15}N substitution induces a small shift to the amide-I at 1666 cm^{-1}, and the new band is also shifted by the same amount; (2) deuterium substitution at the NH position strongly affects both the amide-I and the new band in a complicated way; (3) upon cooling a decrease in the integrated absorption of the normal amide-I and a corresponding increase in the integrated absorption of the 1650 cm^{-1} band are observed; (4) the 1650 cm^{-1} band and the amide-I band show the same dichroism over the temperature range integrated; (5) the measurements of specific heat, the dielectric constant and the volume expansion as a function of temperature rule out the occurrence of rotational isomerism or of a polymorphic transition which would affect some other infrared and Raman absorption bands, but not the new band. The self-trapping mechanism of the amide-I vibrational energy used by Scott and Eilbeck *et al.* comes from the Davydov model of vibration energy transport in alpha-helix protein which was described in Section 5.6. They have given a good account of the properties of first excited state in this model. Scott and co-workers, Alexander and Krumbansl have also obtained an exponential dependence of the absorption intensity on temperature, $\exp(-\beta T^2)$, and explained the experimentally observed intensity change of the new band with decreasing temperature in terms of a complementary polaron or soliton of the self-trapping state of exciton on the basis of the Davydov model (Davydov referred to the intramolecular excitation occurred in the systems as exciton in such a case). However, the red shift of a few cm^{-1} given by this model is much smaller than the experimental value of 16 cm^{-1}. This indicates that these models need further improvement and development.

Pang and co-workers also proposed that the new band of amide-I is caused by the self-trapping of amide-I vibrational quanta (exciton) in their soliton model (the vibrational quantum is referred to as exciton, in accordance with the exciton theory by the Davydov). But this model is different from Davydov's soliton model or Alexander's complementary polaron model. Here it is the nonlinear interaction between the localized amide-I vibrational-quantum and low frequency vibrations of the lattice (phonons) that results in the self-trapping of exciton as a soliton. The mechanism of this soliton model can be described as follows. Through their intrinsic nonlinear interaction, an amide-I vibrational quantum acts as a source of low-frequency phonons and causes shifts in the average positions of the lattice molecules in the ground state. These shifts, or lattice distortions, in turn react, through the same nonlinear interaction, as a potential well to trap the amide-I vibrational quantum and prevent the energy dispersion of the amide-I vibrational quantum via the dipole-dipole interaction that occurs in neighboring peptides with certain electric moments, resulting in the soliton. Evidently, the soliton is formed by self-trapping of excitons interacting with low frequency lattice phonons. This is a dynamic self-sustaining entity that propagates together with the lattice deformation along the molecular chains. The main properties of the soliton is that it can move over macroscopic distances with velocity v and retain its wave form, energy, momentum, and other quasi-particle properties. Using this model, Pang and coworkers also gave an exponential dependence of the infrared-absorption intensity on temperature which is consistent with the experimental result (Pang 1987, 1989a, 1990, 1992a-c, 1993a-k, 1994b-c, 1996, 1997, 1999, 2000a, 2001a-g, 2004a, Pang and Chen 2001, 2002a).

From these studies, we know that collective excitations can result from localized fluctuation of the amide-I vibrational quanta and structural deformation of the molecular chains due to photoexcitation in the molecular crystals. The Hamiltonian of the systems given by Pang is as follows

$$\begin{aligned} H &= H_{ex} + H_{ph} + H_{\text{int}} \\ &= \frac{1}{2m}\sum_n p_n^2 + \frac{m\omega_0^2}{2}\sum_n r_n^2 + \frac{m\omega_1^2}{2}\sum_n r_n r_{n+1} \\ &\quad + \frac{1}{2M}\sum_n P_n^2 + \frac{\beta}{2}\sum_n (R_n - R_{n-1})^2 + m\chi_1 \sum_n (R_{n+1} - R_{n-1}) r_n^2 \\ &\quad + m\chi_2 \sum_n (R_{n+1} - R_n) r_n r_{n+1} - m\chi_2 \sum_n (R_{n-1} - R_n) r_n r_{n-1}. \end{aligned} \tag{9.1}$$

This Hamiltonian includes vibrational excitation of amide-I caused by localized fluctuation, vibration of lattice molecules caused by the structural deformation of the molecular chains, and the interaction between the two modes of motion in the crystals, respectively. Here m is the mass of the amide-I vibrational quantum (exciton), ω_0 and ω_1 are the diagonal and off-diagonal elements of the dynamic matrix of the vibrational quantum, ω_0 is also the Einstein vibrational frequency of the exciton, $m\omega_1^2 r_n r_{n+1}/2$ the interaction between the nearest neighboring excitons caused by the dipole-dipole interaction in the molecular chains, r_n and $p_n = m(r_n)_t$ the normal coordinates of the ith excitons and its canonical conjugate momentum, respectively, M the mass of a molecule or peptide group in the unit cell, $2\chi_1 = \partial\omega_0^2/\partial R_n$ and $2\chi_2 = \partial\omega_1^2/\partial R_n$ the change of energy of exciton and of coupling interaction between the excitons for an unit extension of molecular chain, respectively, R_n and $P_n = M(R_n)_t$ the canonically conjugate operators of displacement and the momentum of the molecule and β the elastic constant of the molecular chains. The term H_{ex} in H is the Hamiltonian of harmonic vibration of amide-I including the off-diagonal factor, H_{ph} is the Hamiltonian of harmonic vibration of the molecular chain and $H_{\rm int}$ is the interaction Hamiltonian between the two modes of motion. This Hamiltonian is significantly different from that of the Davydov model used by Scott and co-workers. As far as the vibration of amide-I is concerned, we adopt a harmonic oscillator model with optical vibration that includes an off-diagonal factor, which comes from the interaction between the excitons, and interaction with the lattice phonon. Thus, the vibrational frequencies of amide-I are related to the displacements of lattice molecules, which shows the occurrence of the interaction between the amide-I vibrations and displacement of lattice molecules. Their relationship is described by

$$\omega_0^2(R_n) \approx \omega_0^2 + \frac{\partial \omega_0^2}{\partial R_n}(R_n - R_{n-1}) = \omega_0^2 + \chi_1(R_n - R_{n-1}),$$
$$\omega_1^2(R_n) \approx \omega_1^2 + \chi_2(R_n - R_{n-1}).$$

Inserting the above into the vibrational Hamiltonian of amide-I and taking into account the effect of the neighboring lattice molecules on both sides of the excitons, (9.1) can be obtained. Therefore, the Hamiltonian given above has high symmetry and a one-to-one correspondence. On the other hand, the Hamiltonian in the Davydov model does not have such a one-to-one correspondence. In other words, there is no interaction between the amide-I vibrational quantum and displacement of the lattice molecule resulting from the interaction between the neighboring excitons in the Davydov model (see Section 5.6). Thus, the Davydov model has encountered many difficulties in studying dynamics of energy transport in the systems. The Hamiltonian given above includes not only the optical vibration of amide-I, but also the resonant interaction caused by the dipole-dipole interaction between neighboring excitons; and it also takes into account both the change of the relative displacement of the neighboring lattice molecules resulting from the vibration of amide-I and the correlation interaction between the neighboring excitons. Therefore, the Hamiltonian given above has higher symmetry and one-to-one correspondence in H_p and $H_{\rm int}$. It can give a more complete description on the dynamics of the systems compared to the Davydov model and other models.

Since the amide-I vibration and the vibration of molecular chains are all quantized, we introduce the following canonical second quantization transformation,

$$r_n = \sqrt{\frac{\hbar}{2m\omega_0}}(b_n^+ + b_n), \quad p_n = -i\sqrt{\frac{\hbar m\omega_0}{2}}(b_n - b_n^+), \tag{9.2}$$

$$R_n = \sum_q \sqrt{\frac{\hbar}{2MN\omega_q}}(a_q + a_{-q}^+)e^{ir_0qn}, \quad P_n = i\sum_q \sqrt{\frac{N\hbar\omega_q}{2M}}(a_{-q}^+ - a_q)e^{ir_0qn}, \tag{9.3}$$

where $i = \sqrt{-1}$, $\omega_q = 2\sqrt{\beta/M}\sin(r_0q/2)$ is the frequency of the phonon with wave vector q, N the number of unit cells in the molecular chain, r_0 the distance between the molecules, $b_n^+(b_n)$ and $a_q^+(a_q)$ are the creation (annihilation) operators of the exciton and phonon, respectively.

Using (9.2) and (9.3), (9.1) becomes

$$\begin{aligned} H = &\sum_n \varepsilon_0 \left(b_n^+ b_n + \frac{1}{2} \right) - J \sum_n (b_n^+ b_{n+1} + b_n b_{n+1}^+) + \sum_q \hbar\omega_q \left(a_q^+ a_q + \frac{1}{2} \right) \\ &+ \sum_{n,q} \left[g(q)(b_n^+ b_n + b_n b_n^+) + g_1(q)(b_n^+ b_{n+1} + b_n b_{n+1}^+) \right. \\ &\left. - g_2(q)(b_n^+ b_{n-1} + b_n b_{n-1}^+) \right] (a_q + a_{-q}^+) e^{inr_0 q}, \end{aligned} \tag{9.4}$$

where $\varepsilon_0 = \hbar\omega_0$, $J = \hbar\omega_1^2/4\omega_0$, and

$$g(q) = \sqrt{\frac{\hbar}{2NM\omega_q}} \left(\frac{\hbar\chi_1}{2\omega_0} \right) \left[e^{ir_0 q} - e^{-ir_0 q} \right],$$

$$g_1(q) = \sqrt{\frac{\hbar}{2NM\omega_q}} \left(\frac{\hbar\chi_2}{2\omega_0} \right) \left[e^{ir_0 q} - 1 \right],$$

$$g_2(q) = \sqrt{\frac{\hbar}{2NM\omega_q}} \left(\frac{\hbar\chi_2}{2\omega_0} \right) \left[e^{-ir_0 q} - 1 \right].$$

Due to the fact that the collective excitations generated by the localized fluctuation of the excitons and structural deformation of the molecular chains are coherent, Pang proposed the following form for the trial wave function of the systems

$$|\Phi\rangle = |\varphi(t)\rangle U(t)|0\rangle_{ph} = \frac{1}{\lambda'} \left(1 + \sum_n \varphi_n(t) b_n^+ \right) |0\rangle_{ex} \cdot U(t)|0\rangle_{ph}, \tag{9.5}$$

with

$$U(t) = \exp\left\{ \sum_n \frac{1}{i\hbar} [u_n(t) P_n - \pi_n(t) R_n] \right\}, \tag{9.6}$$

or

$$U(t) = \exp\left\{ \sum_q [\alpha_q(t) a_q^+ - \alpha_q^*(t) a_q] \right\},$$

where $|0\rangle_{ex}$ and $|0\rangle_{ph}$ are the excitonic and phononic vacuum-states, respectively, λ' the normalization factor. We assume hereafter that $\lambda' = 1$ for convenience of calculation unless it is explicitly mentioned. $\varphi_n(t)$, $u_n(t) = \langle\Phi|R_n|\Phi\rangle$ and $\pi_n(t) = \langle\Phi|P_n|\Phi\rangle$ are three sets of unknown functions. Here,

$$|\varphi\rangle = \frac{1}{\lambda'} \left(1 + \sum_n \varphi_n(t) b_n^+ |0\rangle_{ex} \right)$$

is not the Davydov wave function (neither $|D_1\rangle$ in Section 9.5 nor $|D_2\rangle$ in (5.72)) which is given by $|\varphi_D\rangle = \sum_n \varphi_n(t) b_n^+ |0\rangle_{ex}$. It is not an excitation state of a single particle, but rather a coherent state, or more accurately, a quasi-coherent state. It can be approximately written as

$$|\varphi\rangle \sim \frac{1}{\lambda'} \exp\left[\sum_n \varphi_n(t) b_n^+\right] |0\rangle_{ex} = \frac{1}{\lambda'} \exp\left\{\sum_n \left[\varphi_n(t) b_n^+ - \varphi_n^*(t) b_n\right]\right\} |0\rangle_{ex}.$$

The final representation is a standard coherent state. The above derivation is mathematically justified in the case of small $\varphi_n(t)$ (*i.e.*, $\varphi_n(t) \ll 1$). Therefore, the wave function can be viewed as an effective truncation of a standard coherent state, which is referred to as a quasi-coherent state. However, it is not an eigenstate of the number operator $\hat{N} = \sum_n b_n^+ b_n$ since $\hat{N}|\varphi\rangle = \sum_n \varphi_n(t) b_n^+ |0\rangle_{ex} = |\varphi\rangle - |0\rangle_{ex}$. Thus, $|\varphi\rangle$ indeed represents a coherent superposition of the one-quantum amide-I vibrational state and the ground state of the exciton. But the number of quantum are determinate in this state and it contains only one exciton because

$$N = \langle\varphi|\hat{N}|\varphi\rangle = \sum_n \langle\varphi|b_n^+ b_n|\varphi\rangle = \sum_n |\varphi_n(t)|^2 = 1.$$

Therefore, $|\varphi\rangle$ given in (9.5) not only exhibits coherent feature of collective excitations of the excitons, caused by the nonlinear interaction generated by the exciton-phonon interaction, which makes the wave functions of the states of the exciton and phonon (the wave function of the phonon is also standard coherent sate) symmetrical, but it can also make the number of the excitons conserved in the Hamiltonian (9.4) as well. Meanwhile, the wave function given above has another advantage. The equation of motion of the soliton can be obtained from the following Heisenberg equations for the creation and annihilation operators of the exciton and phonon by using (9.4) – (9.6)

$$i\hbar\frac{\partial}{\partial t}\langle\Phi|b_n|\Phi\rangle = \langle\Phi|b_n, H|\Phi\rangle, \quad i\hbar\frac{\partial}{\partial t}\langle\Phi|a_q|\Phi\rangle = \langle\Phi|a_q, H|\Phi\rangle. \tag{9.7}$$

The equations of motion for $\varphi_n(t)$, $\alpha_q(t)$ and $\alpha_{-q}^*(t)$ are

$$\begin{aligned} i\hbar\dot{\varphi}_n = {} & \hbar\omega_0\varphi_n - J(\varphi_{n+1} + \varphi_{n-1}) + \sum_q \{2g(q)(\alpha_q + \alpha_{-q}^*)\varphi_n \\ & + (g_1(q) - g_2(q))\left[(\alpha_q + \alpha_{-q}^*)\varphi_{n+1} + (\alpha_q + \alpha_{-q}^*)\varphi_{n-1}\right]\} e^{ir_0qn}, \end{aligned} \tag{9.8}$$

$$\begin{aligned} i\hbar\dot{\alpha}_q = {} & \hbar\omega_q\alpha_q + \sum_n \{2g(q)|\varphi_n(t)|^2 \\ & + [g_1(q) - g_2(q)](\varphi_n^*\varphi_{n+1} + \varphi_n^*\varphi_{n-1})\} e^{-ir_0qn}, \end{aligned} \tag{9.9}$$

$$\begin{aligned} i\hbar\dot{\alpha}_{-q}^* = {} & -\hbar\omega_q\alpha_q^* - \sum_n \{2g(q)|\varphi_n(t)|^2 \\ & + [g_1(q) - g_2(q)](\varphi_n^*\varphi_{n+1} + \varphi_n^*\varphi_{n-1})\} e^{-ir_0qn}. \end{aligned} \tag{9.10}$$

From (9.9) and (9.10), we can get

$$\ddot{\alpha}_{-q}^* + \ddot{\alpha}_q = -\omega_q^2(\alpha_q + \alpha_{-q}^*) - \sum_n \left\{ \frac{4g(q)\omega_q}{\hbar} |\varphi_i|^2 + \frac{2g_1(q) - 2g_2(q)}{\hbar}\omega_q(\varphi_n^*\varphi_{n+1} + \varphi_n^*\varphi_{n-1}) \right\} e^{-ir_0qn}.$$

From (9.3), we get the Fourier transformation of the variable $u_n = \langle\Phi|R_n|\Phi\rangle$,

$$u_n = \frac{1}{\sqrt{N}}\sum_q u_q(t)e^{iqx},$$

where $x = nr_0$ and

$$u_q(t) = \sqrt{\frac{\hbar}{2M\omega_q}}(\alpha_q + \alpha_{-q}^*).$$

Thus we get in the long-wave length approximation

$$\ddot{u}_q + v_0^2q^2u_q = \frac{1}{\sqrt{N}}\sum_n \left\{ \frac{2iq\hbar r_0\chi_1}{M\omega_0}|\varphi_n|^2 + \frac{iq\hbar r_0\chi_2}{M\omega_0}(\varphi_n^*\varphi_{n+1} + \varphi_n^*\varphi_{n-1}) \right\} e^{-ir_0qn},$$

where $\omega_q \approx v_0q$, $v_0 = r_0\sqrt{\beta/M}$. Now multiplying the above equation by $\exp(ir_0qn)/\sqrt{N}$, summing over the wave number, q, and making the continuum approximation, we get

$$\frac{\partial^2u(x,t)}{\partial t^2} - v_0^2\frac{\partial^2u(x,t)}{\partial x^2} = \frac{2\hbar r_0(\chi_1+\chi_2)}{M\omega_0}\frac{\partial|\varphi(x,t)|^2}{\partial x}. \tag{9.11}$$

At the same time, (9.7) becomes

$$i\hbar\frac{\partial\varphi(x,t)}{\partial t} = (\hbar\omega_0 - 2J)\varphi(x,t) - Jr_0^2\frac{\partial^2\varphi(x,t)}{\partial x^2} + \frac{2\hbar r_0(\chi_1+\chi_2)}{\omega_0}\varphi(x,t)\frac{\partial u(x,t)}{\partial x}. \tag{9.12}$$

Inserting (9.11) into (9.12), we can get the nonlinear Schrödinger equation for motion of the exciton

$$i\hbar\frac{\partial\varphi(x,t)}{\partial t} = (\hbar\omega_0 - 2J)\varphi(x,t) - Jr_0^2\frac{\partial\varphi(x,t)}{\partial x^2} + \frac{4\hbar^2r_0^2(\chi_1+\chi_2)^2}{M\omega_0(v^2-v_0^2)}|\varphi(x,t)|^2\varphi(x,t), \tag{9.13}$$

where

$$\frac{\hbar^2}{2m} = Jr_0^2.$$

Here m is the effective mass of the exciton. This shows clearly that the excitation and motion of the excitons in the ACN is a nonlinear problem, and the excitons in the laws of nonlinear quantum-mechanical systems satisfy or obey truly the nonlinear quantum mechanics, its motion can be exactly described by the nonlinear Schrödinger equation. An exciton in such a case has the properties of microscopic particles as described in Chapters 3 – 6. Obviously, the nonlinear interaction energy (G) is provided by the coupling between the excitons and phonons, and the exciton self-trapes as a soliton by this nonlinear interaction. The soliton-solution of (9.13) is of the form

$$\varphi(x,t) = \sqrt{\frac{\mu}{2}}\mathrm{sech}[\mu'(x-x_0-vt)]\exp\left\{\frac{i}{\hbar}\left[\frac{\hbar^2 v(x-x_0)}{2Jr_0^2} - E_{\mathrm{sol}}t\right]\right\}, \quad (9.14)$$

$$u(x,t) = -\frac{\hbar(x_1+x_2)}{2\omega_0\beta(1-s^2)}\tanh[\mu'(x-x_0-vt)], \quad (9.15)$$

where

$$\mu = \frac{G}{4J}, \quad G = \frac{4\hbar^2 r_0^2(\chi_1+\chi_2)^2}{Mv_0^2\omega_0^2(1-s^2)}, \quad \mu' = \frac{\mu}{r_0},$$
$$s = \frac{v}{v_0}, \quad \varepsilon = \hbar\omega_0 - 2J, \quad \varepsilon_0 = \hbar\omega_0, \quad (9.16)$$

Using

$$\alpha_q = \sqrt{\frac{M\omega_q}{2\hbar}}u_q + i\sqrt{\frac{1}{2M\hbar\omega_q}}\pi_q,$$

and the results given above, we found that $\alpha_q(t)$ in (9.3) is given by

$$\alpha_q(t) = \frac{1}{\sqrt{N}}\sum_n \alpha_n(t)e^{-iqx}$$
$$= \frac{i\hbar(\chi_1+\chi_2)\pi M(\omega_q+vq)}{2Mv_0^2\omega_0(1-s^2)\hbar N\omega_q}\sinh\left(\frac{\pi r_0 q}{2\mu}\right)e^{iqvt} = \alpha_q e^{iqvt}. \quad (9.17)$$

The energy of the soliton can be found by using (9.14) and (9.15) and is given by

$$E_{\mathrm{sol}} = \int_{-\infty}^{\infty} H d\zeta = \varepsilon_0 - 2J + \frac{\hbar^2 v^2}{4Jr_0^2} - \frac{\mu^2 J}{3} = E_0 + \frac{M_{sol}v^2}{2},$$

where

$$E_0 = \varepsilon_0 - 2J - \frac{\hbar^4(\chi_1+\chi_2)^4}{3\beta^2 J\omega_0^4}$$

is rest energy of the soliton. The mass of the soliton is given by

$$M_{\mathrm{sol}} = m + \frac{4\hbar^4(\chi_1+\chi_2)^4(1-3s^2/2-s^4/2)}{3\beta^2 J\omega_0^4 v_0^2(1-s^2)^2} > m.$$

Therefore, the energy and the rest energy of the soliton is about $\hbar^4(\chi_1+\chi_2)^4/3\beta^2 J\omega_0^4(1-s^2)$ and $\hbar^4(\chi_1+\chi_2)^4/3\beta^2 J\omega_0^4$ lower than those of the exciton,

$E' = \varepsilon_0 - 2J + \frac{1}{2}mv^2$ and $E'_0 = \varepsilon_0 - 2J$, respectively, but the mass of the soliton is greater than that of the exciton, m. The exciton-soliton is thus stable.

Pang *et al.* also studied the influence of anharmonic vibration of the molecules on the soliton. The phonon's Hamiltonian in (9.1) is replaced by

$$H'_{ph} = \frac{1}{2M}\sum_n P_n^2 + \frac{\beta}{2}\sum_n (R_n - R_{n-1})^2 + \frac{1}{3}\lambda_1 (R_n - R_{n-1})^3. \tag{9.18}$$

It was found that only terms such as $(\hbar\omega_0 - 2J)\varphi(x,t)$ in (9.13) are changed. Thus only the phase, velocity, energy, amplitude and width of the soliton are affected.

9.2 Raman Scattering from Nonlinear Motion of Excitons

In the following three sections, we will discuss physical phenomena arising from nonlinear excitations and motions of the excitons, *i.e.*, Raman scattering, Mössbauer effect and infrared absorption, which were studied by Pang.

In order to study the Raman scattering, we first diagonalize the soliton Hamiltonian (9.4) following procedures of Eremko *et al.*. Then we transform it to the intrinsic reference frame which moves with the soliton at velocity v. In this case, (9.4) is replaced by

$$\bar{H} = H - \sum_k \hbar k v (b_k^+ b_k + a_k^+ a_k), \tag{9.19}$$

where

$$b_k^+ = \frac{1}{\sqrt{N}}\sum_n e^{iknr_0} b_n^+, \quad b_k = (b_k^+)^+.$$

Since the soliton excitation is always connected with the deformation of molecular chain or intramolecular distances, it is necessary take this deformation into consideration in (9.19). In the light of Eremko *et al.*, such a transition is realized by the replacement $(R_i, P_i) \to (a_q^+, a_q) \to (A_q^+, A_q)$ according to the relations: $A_q^+ = a_q^+ - \alpha_q^*/\sqrt{N}$, $A_q = a_q - a_q/\sqrt{N}$, where A_q^+ (A_q) is the creation (annihilation) operator of the new phonon. The coherent phonon state (lattice distortion) then becomes the vacuum state of the new phonon, $|\bar{0}\rangle_{ph} = \exp[a_q(t)a_q^+ - a_q^*(t)a_q]|0\rangle_{ph}$, where $A_q|\bar{0}\rangle_{ph} = 0$. We can now carry out the canonical transformation for the partial diagonalization of the soliton Hamiltonian by the following transformation

$$B_\lambda^+ = \sum_n \varphi_\lambda(t) b_n^+, \quad B_\lambda = (B_\lambda^+)^+,$$

$$\sum_\lambda \varphi_\lambda^*(n)\varphi_\lambda(n') = \delta_{nn'}, \quad \sum_n \varphi_\lambda^*(n)\varphi_{\lambda'}(n) = \delta_{\lambda\lambda'}.$$

The so-called partial diagonalization of the Hamiltonian means diagonalization of the part of the Hamiltonian which does not contain the creation and annihilation operators of the new phonons. In the continuum approximation, the condition imposed onto the function $\varphi_\lambda(n)$ to realize such a diagonalization is equivalent to the following problem on the eigenfunction $\varphi(n/\lambda)$ and the eigenvalues E_λ,

$$\left[-J\frac{\partial^2}{\partial n^2} - \frac{i\hbar v}{r_0}\frac{\partial}{\partial n} - 2\mu^2 J r_0^2 \text{sech}^2(\mu n) + \varepsilon_0 - 2J\right]\varphi\begin{pmatrix} n \\ \lambda \end{pmatrix} = E_\lambda \varphi\begin{pmatrix} n \\ \lambda \end{pmatrix}.$$

It has a unique bound state solution,

$$\varphi\begin{pmatrix} n \\ s \end{pmatrix} = \sqrt{\frac{\mu}{2}}\text{sech}(\mu n)\exp\left(\frac{i\hbar v n}{2Jr_0}\right),$$
$$E_s = \varepsilon_0 - 2J - \frac{\hbar^2 v^2}{4Jr_0^2} - \mu^2 J, \tag{9.20}$$

and an unbound state solution given by

$$\varphi\begin{pmatrix} n \\ k \end{pmatrix} = \frac{\mu \cdot \tanh(\mu n) - ikr_0}{\sqrt{2}\pi(\mu - ikr_0)}\exp\left(\frac{i\hbar v n}{2Jr_0} + ikr_0 n\right),$$
$$E_k = \varepsilon_0 - 2J - \frac{\hbar^2 v^2}{4Jr_0^2} + (kr_0)^2 J. \tag{9.21}$$

E_s is less than that of lowest unbound state $E_\lambda^0 = \varepsilon_0 - 2J$ by $\mu^2 J$ in such a case. Thus, the partially diagonalized Hamiltonian in (9.19) becomes

$$\begin{aligned}\bar{H} = {} & E_s B_s^+ B_s + \sum_k E_k B_k^+ B_k + \sum_q \hbar(\omega_q - vq)A_q^+ A_q + \frac{G^2}{48J} \\ & + \frac{1}{\sqrt{N}}\sum_q \hbar(\omega_q - vq)(A_q^+ \alpha_q + A_q^+ \alpha_q^*)(1 - B_s^+ B_s) \\ & + \frac{1}{N}\sum_{qk} \tilde{F}(k,q)(B_s^+ B_{-k} + B_k^+ B_s)(A_{-q}^+ + A_q) \\ & + \frac{1}{\sqrt{N}}\sum_{qk} F(k,q)(B_{k+q}^+ B_k)(A_q + A_{-q}^+),\end{aligned} \tag{9.22}$$

where

$$\begin{aligned}\tilde{F}(k,q) = {} & \frac{2ir_0 qJ\pi\hbar}{\omega_0(G + 4Jir_0 k)}\sqrt{\frac{\hbar J}{GM\omega_q}}\left[\chi_1(e^{iqr_0} - e^{-iqr_0}) + \chi_2(e^{iqr_0} - e^{-iqr_0})\right] \\ & \times \text{sech}\left[\frac{2\pi r_0 J(q-k)}{G}\right],\end{aligned}$$

$$\begin{aligned}F(k,q) = {} & \sqrt{\frac{\hbar^3}{2M\omega_0^2\omega_q}}\left[\chi_1(e^{iqr_0} - e^{-iqr_0}) + \chi_2(e^{iqr_0} - e^{-iqr_0})\right] \\ & \times \left\{1 - \frac{4iGqr_0}{(G - 4ikr_0 J)[G + 4i(l+q)r_0 J]}\right\}\end{aligned}$$

are the coupling constants of the bound state and the unbound state of the new phonons, respectively. Obviously, the distribution amplitude of the amide-I vibrational excitations, (9.20) and (9.21), can be obtained if an exact deformation potential is given a priori. This treatment allows us to see that the Hamiltonian (9.22) or (9.19) describing the molecular chain in the reference frame moving with velocity v has a soliton state given by (9.20), with energy

$$E_{\rm sol} = E_s + W = \varepsilon_0 - 2J - \frac{\hbar^2 v^2}{4Jr_0^2} - \frac{G^2}{48J},$$

and a delocalized state corresponding to the exciton in the undeformed chain with energy E_k given in (9.21). The former is a localized coherent structure with a size of the order of $2\pi r_0/\mu$ that propagates with velocity v and can transfer energy $E_{\rm sol} < \varepsilon_0$. Unlike a bare exciton that can be scattered by interaction with phonons, the soliton state describes a quasi-particle consisting of the exciton plus the lattice deformation and hence it already includes the interaction with the acoustic phonons. Therefore, the soliton is not scattered and affected by this interaction, and it can maintain its form, energy, momentum and other quasi-particle properties over a macroscopic distance in the absence of external fields and/or impurities.

This result shows that the soliton energy is lower than that of the lowest exciton state ($k = 0$) by $G^2/48J$. Thus, an energy gap between the soliton state and the exciton state occurs which corresponds to the binding energy of the soliton. This is due to the fact that the soliton, unlike an exciton, is localized, and is a dynamically self-sustaining entity as a result of self-trapping of the exciton through interaction with the deformation of the molecular chain. Because of the self-trapping, the energy of the exciton drops by about $G^2/24J$ which is the same as the deformation energy of the lattice,

$$W = \frac{1}{N}\sum_q \hbar(\omega_q - vq)|\alpha_q|^2 = \frac{G^2}{24J}.$$

The energy of the soliton formed by this mechanism is about $G^2/48J$ less than that of the exciton. Thus, the energy gap between them is $G^2/48J$. The energy gap can result in a red shift from the main peak at 1666 cm^{-1} in the infrared absorption and Raman spectra. This red shift of about 16 cm^{-1} has been observed experimentally in ACN, as mentioned earlier.

Using the following values for the physical parameters of ACN, $J = (3-4)\text{cm}^{-1}$, $\chi = \hbar\chi_1/2\omega_0 = (56-62)PN$, $\beta = (4.8-13)\text{N/m}$, $M = 2.25\times10^{-25}$ kg, $\chi' = \hbar\chi_2/2\omega_0 = (6-8)PN$, $r_0 = (2-4.5)$ Å, $\omega_0 = (2-4)\times10^{14}$ s, $\omega_1 = (3-6)\times10^{13}$ s, Pang obtained the binding energy of the soliton or the energy gap to be $18.1-33$ cm^{-1} which is consistent with the experimental value of 16 cm^{-1}. In contrast, the binding energy of the soliton in the Davydov model is about $3-7$ cm^{-1}, which is too small compared to the experimental value. Obviously, the soliton or the localization of the exciton is formed due to the nonlinear interaction energy $G =$

$(1.5-3.2)\times10^{-21}$ J which strongly suppresses its dispersive energy $J = 0.795\times10^{-22}$ J. This indicates clearly that the observed red shift is caused by the formation of the soliton. Thus it is experimentally confirmed that the exciton is actually localized in ACN.

Identification of the corresponding Stokes component in the Raman scattering spectrum can be one of the ways of experimentally determining the energy gap between the soliton and exciton, and hence the solitons themselves. It is, therefore, necessary to calculate the differential cross-section of the Raman scattering due to the solitons. We assume here that the Raman scattering process is activated by some intermediate states of the molecular chains associated with the soliton, e.g., electronic excitation whose Hamiltonian is

$$H_e = \sum_{kn} \varepsilon_n(k) D^+_{kn} D_{kn},$$

where D^+_{kn} (D_{kn}) is the creation (annihilation) operator of the electronic excitation of the nth band with wave vector k and energy $\varepsilon_n(k)$. At the same time, let the incident light wave be quantized in the volume $V' = Nr_0S'$ and be denoted by the Hamiltonian

$$H_Q = \sum_{Q\sigma} \hbar\omega_Q C^+_{Q\sigma} C_{Q\sigma},$$

where $C^+_{Q\sigma}$ $(C_{Q\sigma})$ is the creation (annihilation) operator of the photon with wave vector Q, energy $\hbar\omega_Q$, and unit polarization vector $\vec{e}_\sigma(Q)$. In such a case, the interaction Hamiltonian among the soliton, electronic excitation and light wave leading to the Raman scattering can be denoted by

$$\begin{aligned}\bar{H}_{\text{int}} &= \frac{1}{N}\sum_{kk'}\sum_{q} X_{nn}(q) D^+_{K+qn} D_{kn'} B^+_{k'-q} B_{k'} + \\ \frac{1}{N}&\sum_{kk'}\sum_{q} \bar{X}_{nn'}(q) D^+_{kn} D_{kn'} B^+_{k'-q} B_k (B^+ B_{k'} - B^+_{k'} B) + \\ &\sqrt{N}\sum_{Q\sigma n} U_{n\sigma}(Q)(C^+_{n\sigma} + C_{-Q\sigma})(D^+_{Qn} - D_{-Qn}),\end{aligned} \tag{9.23}$$

where $X_{nn'}$ and $\bar{X}_{nn'}$ are the interaction coefficients among the excitons and solitons and electronic excitations with different wave vectors, respectively, $U_{n\sigma}(Q)$ is the coupling constant between the electronic excitations and the light wave, *i.e.*,

$$U_{n\sigma}(Q) = -i\varepsilon_n(0)\sqrt{\frac{2\pi}{\hbar\omega_Q V'}}\vec{e}_\sigma(Q)\cdot\vec{d}_n,$$

where $\vec{d}_n$ is the dipole moment of the transition from the ground state of electron into the state of the nth band, the z-axis is directed along the molecular chain.

Pang calculated the transition probability per unit time of the electron-light-soliton (exciton) system from the initial state,

$$|m\rangle = C^+_{Q_0\sigma_0}|0\rangle_Q B^+_s|0\rangle_{ex}\prod_q \frac{1}{\sqrt{\tilde{n}_q!}}(A^+_q)^{\tilde{n}_q}|\tilde{O}\rangle_{ph}|0\rangle_e,$$

to the final state,

$$|fk\rangle = C^+_{Q\sigma}|0\rangle_Q B^+_k|0\rangle_{ex}\prod_q \frac{1}{\sqrt{\tilde{n}_q!}}(a^+_q)^{n_q}|0\rangle_{ph}|0\rangle_e,$$

caused by the perturbation potential, (9.23). Here the initial state consists of the photon with wave vector $\vec{Q}_0$ and polarization $\vec{e}_{\sigma_0}(\vec{Q}_0)$, the soliton moving with the velocity v, the new phonon and the electronic excitation, while the final state consists of the photon with the wave vector $\vec{Q}$ and polarization $\vec{e}_\sigma(\vec{Q})$, the exciton and a certain number of ordinary phonons and electronic excitations. In the above, $|0\rangle_Q$ and $|0\rangle_e$ are the vacuum states of the photons and the electronic excitation states, respectively. According to the quantum mechanical perturbation theory, the probability of transition from the initial state to the final state in the present case can be determined by

$$\frac{d}{dt}\bar{W}(Q_0,\sigma_0 - Q,\sigma) = \frac{d}{dt}\left[\frac{1}{\hbar^6}\sum_{kf} P^{(ph)}\right.$$
$$\left.\left|\int_{-\infty}^t dt_1\int_{-\infty}^{t_1} dt_2\int_{-\infty}^{t_2} dt_3\langle fk|\bar{H}_{\text{int}}(t_1)\bar{H}_{\text{int}}(t_2)\bar{H}_{\text{int}}(t_3)|m\rangle\right|^2\right], \tag{9.24}$$

where

$$\bar{H}_{\text{int}}(t) = e^{iH_0t/\hbar}\bar{H}_{\text{int}}(t)e^{-iH_0t/\hbar},$$

$$H_0 = H + H_Q + H_e - \hbar\sum_{kn} kvD^+_{kn}D_{kn} - \hbar\sum_{Q\sigma}(\vec{Q}\cdot\vec{v})C^+_{Q\sigma}C_{Q\sigma}.$$

However, we are more interested in the long-time behavior of $d\bar{W}/dt$. By straightforward calculation, the final transition probability can be obtained approximately,

$$\lim_{t\to\infty}\frac{d\bar{W}}{dt} \approx \frac{\pi^2 r_0^2}{4V'^2}\frac{\gamma\sqrt{J}}{\mu^2 J + \triangle_{ks}}\left(\frac{\gamma}{2.42\omega_a}\right)^{g_0} \tag{9.25}$$

$$\times\left\{2[(\triangle_{ks}(0) - W)^2 + (\hbar\gamma)^2][\sqrt{(\triangle_{ks}(0) - W^2) + (\hbar\gamma)^2} - \sqrt{(\triangle_{ks}(0) - W^2)^2}]\right\}^{-1/2}$$

$$\left|\sum_{nn_0}\frac{(Q_{0z} - Q_z)U_{\sigma_0 n_0}(Q_0)[X_{nn_0}(Q_{0z} - Q_z) + \tilde{X}_{nn_0}(Q_{0z} - Q_z)]U^*_{\sigma n}(Q)}{[\varepsilon_{n_0}(Q_{0z}) - \hbar\omega Q_n(\varepsilon_n(-Q_z) + \hbar\omega_Q)]}\right.$$

$$\left. + (Q_z - Q_{0z})U_{\sigma_0 n_0}(Q_0)\frac{X_{nn_0}(Q_z - Q_{0z}) + \bar{X}_{nn_0}(Q_z - Q_{0z})U^*_{\sigma n}(Q)}{(\varepsilon_{n_0}(-Q_{0z}) + \hbar\omega_{Q\sigma})(\varepsilon_n(-Q_z) + \hbar\omega_Q)}\right|^2$$

$$\times\left\{2 + \tanh^2\left[\frac{\pi}{2\mu}\left(\sqrt{\frac{\triangle_{ks}}{J}} - (Q_{0z} - Q_z)r_0\right)\right]\right.$$

$$\left. + \tanh^2\left[\frac{\pi}{2\mu}\left(\sqrt{\frac{\triangle_{ks}}{J}} + (Q_{0z} - Q_z)r_0\right)\right]\right\},$$

where

$$\gamma = \frac{2J\mu r_0 K_B T}{\hbar^2 v_0}, \quad \triangle_{ks} = E_k - E_s - \frac{1}{3}\mu^2 J + \hbar(\omega_Q - \omega_{Q_0}).$$

The cross-section of Raman scattering is closely related to the transition probability, (9.25). This transition probability has a sharp peak when the frequency of the scattered light is in the vicinity of the frequency of the incident light, ω_{Q_0}. Naturally, we are of interest only in the cross-section of Raman scattering in such a case which is

$$\frac{1}{N}\frac{dN}{ds} = \frac{d^2\delta}{d\omega_Q d\Omega} = \frac{V'^2\omega_{Q0}^2}{4\pi^2}\frac{d\bar{W}}{dt} \approx \frac{r_0^2\omega_{Q0}^2\omega_Q^2\hbar\gamma\sqrt{J}}{16\pi^3\mu^3 J h^2 C^6}\left(\frac{\gamma}{2.42\omega_0}\right)^{g_0}$$

$$\times\left\{2\left[(\triangle_{ks}(0) - W)^2 + (\hbar\gamma)^2\right]\left[\sqrt{(\triangle_{ks}(0) - W)^2 + (\hbar\gamma)^2} - \sqrt{(\triangle_{ks}(0) - W)^2}\right]\right\}^{-1/2}$$

$$\times\sum_k\left|\sum_{nn_0}\frac{(Q_{0z} - Q_z)U_{\sigma_0 n_0}(Q_0)[X_{nn_0}(Q_{0z} - Q_z) + \tilde{X}_{nn_0}(Q_{0z} - Q_z)]U^*_{\sigma n}(Q)}{[\varepsilon_{n_0}(Q_{0z}) - \hbar\omega Q_n(\varepsilon_n(-Q_z) - \hbar\omega_Q)]}\right.$$

$$\left. + \frac{(Q_z - Q_{0z})U_{\sigma_0 n_0}(Q_0)[X_{nn_0}(Q_z - Q_{0z}) + \bar{X}_{nn_0}(Q_z - Q_{0z})U^*_{\sigma n}(Q)]}{[(\varepsilon_{n_0}(-Q_{0z}) + \hbar\omega_{Q\sigma})(\varepsilon_n(-Q_z) + \hbar\omega_Q)]}\right|^2$$

$$\times\operatorname{sech}^2\left\{\frac{\pi\omega_0 r_0}{2\mu c}\left[\sin\theta_0 - \frac{\omega_Q}{\omega_{Q_0}}(\cos\theta_0\cos\theta + \sin\theta_0\sin\theta\cos\varphi)\right]\right\}$$

$$\times\left[\cos\theta_0 - \frac{\omega_0}{\omega_{Q_0}}(\sin\theta_0\sin\theta\cos\varphi + \cos\theta_0\sin\theta)\right]^2, \tag{9.26}$$

where $d\Omega = d\theta d\varphi\sin\theta$ is the element of solid angle in the direction of propagation of the scattered light, θ_0 is the angle between the incident light $\vec{Q}_0$ and the z-axis, θ is the angle between $\vec{Q}_0$ and the scattered light $\vec{Q}$, φ is the angle between the projection of the vector $\vec{Q}$ on to the xy-plane and the X-axis. Obviously, the cross-section of the Raman scattering is directly related to the size of the energy gap, the direction and intensity of incident light, the dipole moment and its orientation corresponding to the transition from the ground state to the resonance electronic level. The distribution and intensity of the scattered light are different for different frequency of the incident light. If the frequency of the incident light is in the following region,

$$0 < \hbar\omega_{Q_0} < \mu^2 J - \varepsilon_n(0),$$

or

$$\mu^2 J + \varepsilon_n(0) < \hbar\omega_{Q_0} < 2\varepsilon_n(0),$$

then (9.26) represents a resonant Raman scattering. The inequalities ensure not only the smoothness of the cross-section as a function of ω_{Q_0} but also the applicability of the above perturbation theory to the calculation of the cross-section. However, when the frequency of the incident light satisfies $\mu^2 J \ll \hbar\omega \ll \varepsilon_n(0)$ for all n, then (9.26) represents a nonresonant Raman scattering. In such a case, part of the energy of the incident light is converted into the excitation energy of the exciton-soliton. Thus, the angular-dependent part in (9.26) may be approximately treated as a constant.

Detailed examination of (9.26) show that the Raman scattering has the following properties. (1) When the frequencies of the scattered light is close to that of the incident light, the scattering cross-section is the largest. (2) For a fixed direction of the scattered light, the cross-section and the shape of the spectral line are mainly determined by the factor,

$$\frac{\hbar\gamma\sqrt{J}(2.42\omega_a/\gamma)^{-g_0}}{\{2[(\triangle_{ks}(0)-W)^2+(\hbar\gamma)^2][\sqrt{(\triangle_{ks}(0)-W)^2+(\hbar\gamma)^2}-(\triangle_{ks}(0)-W)]\}^{1/2}},$$

regardless it is a resonant or a nonresonant Raman scattering. (3) In general, the scattering indicatrices (angular dependence of the scattering cross-section) are different for the resonant scattering and nonresonant scattering. For the former the scattering indicatrix depends strongly on the orientation of the dipole moment of the transition from the ground state to the resonance electronic levels. For the latter, when the frequencies of the scattered light ω_Q are close to that of the Raman peak, its scattering indicatrix is independent of the coupling strength, and has a simple form. (4) The intensity of the scattered light reaches a maximum when the frequency ω_Q approaches $\omega_{Q_0} + 2J - G^2/48J - \varepsilon_0$. This maximum scattering intensity is associated with the transition of the system from the soliton state which is accompanied by lattice deformation, to the delocalized exciton state without lattice deformation. Therefore, the soliton becomes unstable as the incident light gets scattered. These characteristics of Raman scattering and the asymmetry exhibited in the spectral line shape (sharper on the shor wave side and smoother on the long wave side) which is connected to the corresponding energy dependence of the density of the delocalized states, allow us to confirm the existence of soliton state of excitons in the organic molecular crystal ACN.

9.3 Infrared Absorption of Exciton-Solitons in Molecular Crystals

In the light of Hamilton theory in the nonlinear quantum mechanics described in Chapter 3, Pang expressed the effective Hamiltonian corresponding to (9.13) as

$$H_{\text{eff}} = \frac{1}{r_0}\int\left(\frac{1}{2}Jr_0^2|\varphi_x|^2 + \frac{1}{2}\varepsilon|\varphi|^2 - \frac{1}{4}G|\varphi|^4\right)dx. \tag{9.27}$$

The corresponding descrete form is

$$H_{\text{eff}} = \frac{J}{2}\sum_n(\varphi_n\varphi_{n+1}^* + \varphi_{n-1}\varphi_n^*) + \sum_n\left(\frac{1}{2}\varepsilon|\varphi_n|^2 - \frac{1}{4}G|\varphi_n|^4\right). \tag{9.28}$$

In earlier discussion, we considered only properties of exciton-soliton which is formed through coupling of amide-I oscillators (excitons) with longitudinal acoustic phonons. To explain the temperature dependence of intensity of the infrared absorption of the exciton-soliton, one has to take the interaction between the soliton and optical phonons in ACN into account. It is believed that this phenomenon is related to the zero-phonon line spectrum arising from random thermal modulations of the exciton-solitons by the optical phonons. In this case, the Hamiltonian of this system can be written as

$$H = H_{\text{eff}} + H_{\text{op}} + H' \tag{9.29}$$

where H_{eff} is the Hamiltonian of the effective amide-I oscillator system given by (9.28), which satisfies $H_{\text{eff}}|m\rangle = \hbar\omega(m)|m\rangle$. Here the eigenstate and the energy eigenvalue are collectively denoted by $|m\rangle$ and $\hbar\omega(m)$, respectively, in terms of a quantum number m. H_{op} is the Hamiltonian for the optical-phonon system which is taken to be a set of harmonic oscillators,

$$H_{\text{op}} = \sum_n h_n(Q_n) = \sum_n\left[-\frac{\hbar^2}{2M^*}\left(\frac{\partial^2}{\partial Q_n^2}\right) + \frac{1}{2}M^*\omega_n^2Q_n^2\right], \tag{9.30}$$

with

$$h_n(Q_n)\Phi_{\kappa_n}(Q_n) = \hbar\omega_n\left(\kappa_n + \frac{1}{2}\right)\Phi_{\kappa_n}(Q_n) = \hbar\omega_n(\kappa_n)\Phi_{\kappa_n}(Q_n), \quad \kappa_n = 0, 1, 2, \cdots,$$

$$\omega(N) = \sum_n \omega_n(\kappa_n),$$

where h_n, Q_n and ω_n are the Hamiltonian, the normal coordinate and the eigenfrequency of the nth harmonic oscillator, respectively, in which M^* is the effective mass, and $\Phi_{\kappa_n}(Q_n)$ is an eigenfunction of h_n characterized by the integer κ_n. H'

in (9.29) describes the interaction between the amide-I oscillator system and the optical phonon system which we take to be linear in the Q'_ns, *i.e.*,

$$H' = \sum_{n'} V_{n'}(\varphi_{n'})Q'_{n'}. \tag{9.31}$$

We can find that

$$V_{n'}(\varphi_{n'}) = \frac{1}{\sqrt{N}}\sum_{n''} \lambda_{n''}\varphi^2_{n'}\vec{e}(\vec{k},\eta)e^{i\vec{k}\cdot\vec{u}_{n'}}\sin(\vec{k}\cdot\vec{u}_{n''}), \tag{9.32}$$

where $\vec{e}(\vec{k},\eta)$ is a polarization vector associated with the wave vector $\vec{k}$ and branch η ($\eta = 1,2,3$, two transverse and one longitudinal), $\lambda_{n''}$ is an interaction constant in which the index n'' indicates the range of the oscillator-phonon interaction.

In order to study the properties of the entire system given by (9.29), we use an adiabatic approximation, *i.e.*, the phonon state depends adiabatically on the amide-I oscillator state. This can be justified on the ground that the characteristic frequency of about 1650 cm^{-1} of the amide-I oscillator system is much higher than the maximum frequency of the phonon system which is about 200 cm^{-1}. In order to obtain the adiabatic potential U_m associated with the state $|m\rangle$ of the amide-I oscillator system, we neglect the change in wave function of the amide-I oscillator system caused by the perturbation H'. Thus the effective Hamiltonian H_{eff} of this system for the state $|m\rangle$ can be written in terms of a set of the Hamiltonians of the displaced harmonic oscillators of the phonon system

$$\begin{aligned} H_{\text{eff}}(m) &= \hbar\omega(m) + \sum_{n'}\frac{1}{2}M^*\omega^2_{n'}Q^2_{n'} - \sum_{n'}\frac{\hbar^2}{2M^*}\frac{\partial^2}{\partial Q^2_{n'}} + \sum_{n'} q_{n'}(m)Q_{n'} \\ &= U_m - \sum_{n'}\frac{\hbar^2}{2M^*}\frac{\partial^2}{\partial Q^2_{n'}} = \sum_{n'} h_{n'}[Q_{n'} - Q_{0n'}(m)] + \text{const.}, \end{aligned} \tag{9.33}$$

where

$$q_{n'}(m) = \langle m|V_{n'}(m)|m\rangle, \quad Q_{0n'}(m) = -\frac{q_{n'}(m)}{M\omega^2_{n'}}.$$

Using this Hamiltonian, we can find the infrared absorption coefficient arising from the soliton motion. In accordance with the first order Born approximation in quantum radiation and theory of absorption, the absorption coefficient per unit volume of the amide-I oscillator system described by (9.33) can be expressed as

$$\begin{aligned} B_{fn'(\omega)} = \frac{4\pi\omega}{\varepsilon' chV_0}\int dte^{-i\omega t}\Bigg\langle \sum_{m_f}\langle m_{n'}|P(t)|m_f\rangle\langle m_f|P|m_{n'}\rangle \\ \prod_{n'}\sum_{k'_n}\Gamma(\kappa_{n'},\kappa'_n,m_n,m_f)^2 e^{i[\omega(\kappa'_n)-\omega(\kappa_{n'})]t}\Bigg\rangle_T, \end{aligned} \tag{9.34}$$

where

$$P(t) = e^{iH_{\text{eff}}t/\hbar} P e^{-iH_{\text{eff}}t/\hbar},$$

$$\Gamma(\kappa_{n'}, \kappa'_{n'}, m_n, m_f) = \int_{-\infty}^{\infty} d\zeta_{n'} \Phi_{\kappa_{n'}}[\zeta_{n'} - \zeta_{0n'}(m_n)] \Phi_{\kappa'_{n'}}[\zeta_{n'} - \zeta_{on'}(m_f)],$$

in which

$$\zeta_{n'} = Q_{n'} \sqrt{\frac{M^* \omega_{n'}}{\hbar}},$$

$$\zeta_{0n'}(n') = Q_{0n'}(m) \sqrt{\frac{M^* \omega_{n'}}{\hbar}} = -q_{n'}(m) \sqrt{\frac{\hbar}{M^* \omega_{n'}}} \frac{1}{\hbar \omega_{n'}} \ll 1.$$

In the above equations, $m_{n'}(\kappa_{n'})$ and $m_f(\kappa'_{n'})$ with $\omega(N_n) = \sum_{n'} \omega_{n'}(\kappa_{n'})$ and $\omega(N_f) = \sum_{n'} \omega_{n'}(\kappa'_{n'})$ are the quantum numbers denoting the initial and the final states of the amide-I oscillator system (the nth harmonic oscillator), respectively. The quantities, c, P, ε' are the speed of light, the effective dipole moment of the amide-I oscillator and the refractive index, respectively, V_0 is the volume of the total system, $\langle \cdots \rangle_T$ denotes the average over all initial states in the canonical equilibrium. Following procedure of Takeno, we get

$$\Gamma(\kappa_{n'}, \kappa'_{n'}, m_n, m_f) = \begin{cases} 1 - \frac{1}{2}\left(\kappa_{n'} + \frac{1}{2}\right) \zeta_{0n'}^2(m_n, m_f), & \text{for } \kappa'_{n'} = \kappa_{n'}, \\ \sqrt{\frac{\kappa_{n'} + 1}{2}} \zeta_{0n'}(m_n, m_f), & \text{for } \kappa'_{n'} = \kappa_{n'} + 1, \\ -\sqrt{\frac{\kappa_{n'}}{2}} \zeta_{0n'}(m_n, m_f), & \text{for } \kappa'_{n'} = \kappa_{n'} - 1, \\ 0, & \text{for } \kappa'_{n'} \neq \kappa_{n'}, \kappa_{n'} \pm 1, \end{cases}$$

and

$$\prod_{n'} \sum_{\kappa'_{n'}} \Gamma(\kappa_{n'}, \kappa'_{n'}, m_n, m_f) e^{i[\omega(\kappa'_{n'}) - \omega(\kappa_{n'})]t} \cong \prod_{n'} [1 + Z_{n'}(t, m_n, m_f)],$$

where

$$Z_{n'}(t, m_n, m_f) = \frac{1}{2} \zeta_{0n'}^2(m_n, m_f) \left[(\kappa_{n'} + 1) e^{-i\omega_{n'}t} + \kappa_{n'} e^{-i\omega_{n'}t} - 2\kappa_{n'} + 1 \right],$$

$$\zeta_{0n'}(m_n, m_f) = \zeta_{0n'}(m_n) - \zeta_{0n'}(m_f) \ll 1.$$

In accordance with the above approximation procedure, the thermal average in the integrand of (9.34) can be expressed separately for the amide-I oscillator state and the phonon state, which are denoted by the angular bracket with subscript sT

and aT, respectively. Taking first the average over the phonon state, we get

$$\left\langle \prod_{n'} [1 + Z_{n'}(t, m_{n'}, m_f)] \right\rangle_{aT} = \left\langle \exp_A \left[\sum_{n'} Z_{n'}(t, m_n, m_f) \right] \right\rangle_{aT} \tag{9.35}$$

$$= \exp \left\{ \left\langle \exp_A \left[\sum_{n'} Z_{n'}(t, m_n, m_f) \right] - 1 \right\rangle_{aT}^{D} \right\} \cong \exp \left[\left\langle \sum_{n'} Z'_n(t, m_n, m_f) \right\rangle_{aT} \right],$$

where the symbol $\exp_A(\sum_{n'} Z_{n'})$ is a leveled exponential function that levels off the product $Z_{n'}$s by neglecting terms which contain higher orders than unity in any of the $Z_{n'}$s, and the superscript (D) attached to the bracket symbol indicates a cumulative average, as proposed by Kabo. Using (9.34), equation (9.35) can be written as

$$B_{fn'(\omega)} = \frac{4\pi\omega}{\varepsilon' ch V_0} \int_{-\infty}^{\infty} dt e^{-i\omega t} \left\langle \sum_{m_f} e^{-Zo(T,m_n,m_f)} e^{-Z_1(T,t',m_n,m_f)} \right\rangle$$

$$\times \langle m_n | P(t) | m_f \rangle \langle m_f | P | m_n \rangle_{sT},$$

$$Z_0(T, m_n, m_f) = \sum_{n'} \left\langle \zeta_{0n'}^2(m_n, m_f) \left(\kappa_{n'} + \frac{1}{2} \right) \right\rangle_{aT},$$

$$Z_1(T, t, m_i, m_f) = \frac{1}{2} \sum_{n'} \langle \zeta_{0n'}^2(m_{n'}, m_f) \left[(\kappa_{n'} + 1) e^{i\omega_{n'} t} + \kappa_{n'} e^{-i\omega_{n'} t} \right] \rangle_{aT}.$$

The function $\exp[Z_1(T, t, m_n, m_f)]$ can be expanded as

$$\exp[Z_1(T, t, m_n, m_f)] = 1 + Z_1(T, t, m_n, m_f) + \frac{1}{2!} Z_1^2(T, t, m_n, m_f) + \cdots.$$

The first, the second and the third terms on the right-hand side are contributions to the zero, one and two-phonon processes. In the general case, one and multi-phonon processes give weak side-band effects. Therefore, we may pay attention exclusively to the zero-phonon process and the main absorption line. Then we get

$$B_{nf}(\omega) = \frac{4\pi\omega}{\varepsilon' ch V_0} \int_{-\infty}^{\infty} dt e^{-i\omega t}$$

$$\times \left\langle \sum_{m_f} e^{-Zo(T,m_n,m_f)} \langle m_n | P(t) | m_f \rangle \langle m_f | P | m_n \rangle \right\rangle_{sT}. \tag{9.36}$$

In order to determine $Z_0(T, m_n, m_f)$, we have to find out $\zeta_{0n}(m_n, m_f)$. Evaluating the matrix element of H' in (9.31) with respect to $|m\rangle$ for low frequency phonons and in the adiabatic approximation, $\zeta_{0n'}$ is separated into two parts, phonon and soliton,

$$\zeta_{0n'}(m_n, m_f) \cong \zeta_{0n'}^{(1)} \zeta_s^{(2)}(m_n, m_f), \tag{9.37}$$

where

$$\zeta_{0n'}^{(1)} \cong \sqrt{2}\sum_l \lambda_l \vec{e}(n')\frac{\mu_{n'}}{v_{0n'}}\sqrt{\frac{1}{\hbar M^*\omega_{n'}}} = \frac{F_{n'}}{\sqrt{\omega_{n'}}},$$

$$\zeta_s^{(2)}(m_n, m_f) \cong \frac{1}{\sqrt{N}}\sum_i \left(\langle m_n|Q_iQ_i^* + Q_i^*Q_i|m_{n_i}\rangle - \langle m_f|Q_iQ_i^* + Q_i^*Q_i|m_f\rangle\right).$$

Here

$$F_{n'} = \sum_l \sqrt{2}\lambda_l\vec{e}(n')\left(\frac{\mu_{n'}}{v_{0n'}}\right)\frac{1}{\sqrt{\hbar M^*}}, \quad \omega_{n'} \cong v_{0n'}k,$$

$v_{0n'}$ is the speed of the nth phonon mode. Since the n'-dependence of $\vec{e}(n')$ and $v_{0n'}$ comes mainly from $\omega_{n'}$, we can neglect the n'-dependence of $F_{n'}$ and rewrite it as F. Inserting the above formula into (9.36), and using the Debye's approximation for the frequency spectrum of phonons, Pang finally obtained

$$\begin{aligned} Z_0(T, m_n, m_f) = Z_0(T) &= \left[\zeta_s^{(2)}\right]^2 \sum_{n'} \frac{1}{\omega_{n'}}\left\langle F_{n'}^2\left(\kappa_{n'} + \frac{1}{2}\right)\right\rangle_{aT} \\ &\approx \left[\zeta_s^{(2)}\right]^2 F^2 \sum_{n'}\frac{1}{\omega_{n'}}\left(\langle\kappa_{n'}\rangle_{aT} + \frac{1}{2}\right). \end{aligned} \tag{9.38}$$

where C and λ, of which the explicit expressions are omitted here, are constants, and θ is the Debye temperature of the ACN. If $T \ll \theta$, we have $Z_0(T) = C + \lambda T^2$.

Until now, no specification has been given for the quantum number m. It may be divided into two parts, one being associated with the bound states or the soliton states of the microscopic particle, the other with continuum or scattering states of excitons, denoted by m_s and m_e, respectively. Correspondingly, the absorption coefficient can be divided into the soliton absorption line $B_{sf(\omega)}$ and the exciton absorption line $B_{ex(\omega)}$, appearing at about 1650 cm^{-1} and 1665 cm^{-1}, respectively, *i.e.*,

$$B_{n'f(\omega)} = B_{sf(\omega)} + B_{ex(\omega)}.$$

From (9.37) and (9.38), the following relation can be derived

$$\langle m_s|Q_{n'}Q_{n'}^* + Q_{n'}^*Q_{n'}|m_s\rangle \gg \langle m_e|Q_{n'}Q_{n'}^* + Q_{n'}^*Q_{n'}|m_e\rangle.$$

Thus we have,

$$\zeta_{on}(m_{sn}, m_{sf}) \gg \zeta_{on}(m_{en}, m_{ef}),$$

or

$$Z_0(T, m_{sn}, m_{sf}) \gg Z_0(T, m_{en}, m_{ef}).$$

Since the mean square displacement of the exciton system associated with the state $|m_e\rangle$ is a sum of a large number of oscillatory functions with respect to the space variable n', the temperature dependence of $B_{ex(\omega)}$ is very weak as compared to that of $B_{sf(\omega)}$. When the soliton density is so small that their mutual interactions are negligible, we can consider only a single excited state corresponding to the first excited state of single-soliton states. Pang then approximately obtained

$$B_{ex}(\omega) = \frac{4\pi\omega}{\varepsilon' chV_0} \int_{-\infty}^{\infty} dte^{-i\omega t} \langle P_1(t)P_1 \rangle_{m_eT}, \tag{9.39}$$

$$B_{sf(\omega)} = \frac{4\pi N_s\omega}{\varepsilon' chV_0} e^{-Z_0(T)} \int_{-\infty}^{\infty} dte^{-i\omega t} \langle P_1(t)P_1 \rangle_{m_sT} \approx e^{-(C+\lambda T^2)} B_s^{(0)}(\omega),$$

where

$$B_s^{(0)}(\omega) = \frac{AP_1^2}{\pi} \frac{W_s\omega}{(\omega^2 - \omega_s^2)^2 + W_s^2\omega^2},$$

and N_s and P_1 are the soliton number and an effective dipole moment associated with a single soliton state, respectively, ω_s is the eigenfrequency of the soliton, W_s and A are, respectively, a natural width of the single-soliton absorption line and a constant. Since the soliton energy is much higher than that of phonons, $B_s^{(0)}(\omega)$ does not depend on temperature in the above approximation.

From (9.39), we see that the soliton absorption line spectrum $B_{sf}(\omega)$ is a product of the phonon-free soliton absorption line spectrum $B_s^{(0)}(\omega)$ and the Debye-Waller factor $\exp[-(c + \alpha T^2)]$. This indicates that the intensity of infrared absorption due to the nonlinear excitation of the excitons in ACN depends on the temperature exponentially, which is consistent with results obtained by Alexander and Krumbansl, *i.e.*, $\exp(-\beta T^2)$, using a polaron model. This temperature dependence of the infrared absorption line at about 1650 cm^{-1} is analogous to that of localized or resonant harmonic phonon modes due to impurities in alkali halide crystals, where the impurity-induced infrared absorption spectra have a characteristic temperature dependence in the form of $\exp(-\alpha T^2)$ (α = constant) at low temperatures. Although the physical origins of the localized modes are different in these two cases, the Debye-Waller factor results from the same origin, *i.e.*, random thermal modulations of low frequency optical phonons. Furthermore, the temperature dependence of the exciton absorption line, if any, is quite different from that of the soliton in the absence of the Debye-Waller factor in $B_{ex}(\omega)$. If we identify $B_{sf(\omega)}$ and $B_{ex(\omega)}$ with the absorption spectra of the ACN at about 1650 cm^{-1} and 1665 cm^{-1}, respectively, which were observed experimentally, the experimental results in ACN can be explained. Therefore, it is firmly believed that soliton can exist in ACN and the above theory gives a correct description of it.

Pang *et al.* used again the exciton-soliton model and other methods for the interaction of the amide-I vibrational quanta (exciton) with optical phonons to study infrared absorption at finite temperature. They used an improved theory for color centers in solid state and the nonlinear quantum perturbation method. Their numerical result for the exponential temperature dependence of the intensity of the anomalous band at 1650 cm^{-1} is basically consistent with the experimental data.

9.4 Finite Temperature Excitonic Mössbauer Effect

From the above discussion, we can see that there can be subsonic soliton ($v < v_0$) in the system if linear harmonic vibration of the molecular chains is taken into consideration. The subsonic soliton is formed by self-trapping of excitons through interaction with localized deformation of lattice. In other words, the lattice must be distorted for the soliton to occur. This means that positions of atoms in the lattice will shift. Thus, the states of atoms will be changed, and the nuclei at the lattice sites will be excited or activated in such a case. γ-quantum emission of the active nuclei would then follow. If the γ-quantum is exactly equal to the molecular excitation energy, *i.e.*, when the molecular chain does not change its state in the course of emission, an observable Mössbauer effect will occur in the system. In fact, when an active nucleus emits a γ-ray, the transition energy, $\triangle E_{nm}$, in principle, may be distributed between the γ-photon, the nucleus that emitted the photon, and the chain as whole, and eventually between vibrations of the chains (in the case being considered, instead of pure phonon vibrations, we have localized self-trapped solitonic state which is a result of the exciton-phonon interaction). The energy needed for a nucleus to leave its site in the chain is at least 10 eV. But the recoil energy which a molecular chain can receive as a whole is small ($N \gg 1$) and does not exceed several tenths of an electronvolt. As a result, an atom whose nucleus has emitted a γ-ray cannot change its position in the lattice, so that it may be neglected. Thus the transition energy can only be distributed between the γ-quantum and the solitons. A Mössbauer transition occurs if the solitonic state of the chain is not changed, and the γ-photon receives the entire energy of the transition. Therefore, solitonic excitation and motion in acetanilide could cause changes in Mössbauer effect, compared to the case in which pure phonon modes spread over the whole chain. In the latter, the lattice remains unchanged after the γ-quantum emission of an active nucleus. We can thus prove the existence of soliton by experimentally observing the changes in Mössbauer effect in ACN.

The purpose of this section is to discuss the properties and changes of Mössbauer effect arising from this mechanism and to calculate the corresponding γ-radiative Mössbauer transition probability by means of the above theory, (9.1) – (9.6). This discussion is also useful in understanding and clarifying the natures of the soliton in such systems and in facilitating development of soliton experiments. Since the domain of the lattice where the soliton is present is subject to a deformation which travels along the molecular chain with a velocity $v < v_0$. In the context of our analysis, the molecular crystal is in equilibrium with a thermostat having $T \neq 0$, so that we can use a model that couples the soliton with thermal phonons.

For the system with a finite temperature, the collective excitations and motions of the excitons and phonons will be changed due to the thermal effect. A straightforward result is the thermal excitation of the phonons. Therefore, $U(t)|0\rangle_{ph}$ in (9.5) must be replaced by the following

$$|\alpha_\nu\rangle = U_n^+|\nu\rangle = \exp\left\{\sum_{qn}[\alpha_{nq}(t)a_q^+ - \alpha_{nq}^*(t)a_q]\right\}\prod_q \frac{1}{\sqrt{\nu_q'!}}(a_q^+)^{\nu_q'}|0\rangle_{ph}, \quad (9.40)$$

where

$$|\nu\rangle = \prod_q \frac{1}{\sqrt{\nu_q'!}}(a_q^+)^{\nu_q'}|0\rangle_{ph}$$

form a complete phononic set which represents the elementary excitation of single phonon due to finite temperature $T \neq 0$ K. Using (9.4) – (9.7) and (9.40), we can obtain the equations of motion for the exciton and phonon in the system. However, because the molecular crystal being studied is in contact with a thermostat at finite temperature $T \neq 0$ K, similar to the calculation of the expectation value of (9.7), we can calculate the thermal mean values of these quantities using

$$\bar{Y} = \langle Y\rangle = Tr(\rho_{\nu\nu})_{ph}Y = \sum_\nu \langle\nu|\rho|\nu\rangle_{ph}\langle\Phi|Y|\Phi\rangle, \quad (9.41)$$

where the density matrix, $\sum_\nu\langle\nu|\rho|\nu\rangle$ is given by

$$\begin{aligned}\langle\nu|\rho|\nu\rangle_{ph} = (\rho_{\nu\nu})_{ph} &= \frac{\langle\nu|\exp(-H_{ph}/K_BT)|\nu\rangle}{\sum_\nu\langle\nu|\exp(-H_{ph}/K_BT)|\nu\rangle}\\ &= \frac{\langle\nu|\exp[-\sum_q \hbar\omega_q(a_q^+a_q)/K_BT]|\nu\rangle}{\sum_\nu\langle\nu|\exp[-\sum_q \hbar\omega_q(a_q^+a_q)/K_BT]|\nu\rangle}.\end{aligned} \quad (9.42)$$

Here the diagonal matrix elements of the Hamiltonian are

$$\langle\Phi|H|\Phi\rangle = \langle\Phi|H_{ex} + H_{\rm int}|\Phi\rangle + \langle\alpha_\nu|H_{ph}|\alpha_\nu\rangle.$$

Making use of the method discussed earlier and the following relationships

$$\begin{aligned}&\langle\alpha_\nu|(a_q + a_q^+)|\alpha_\nu\rangle = -[\alpha_{nq}(t) + \alpha_{n-q}^*(t)]; \quad \langle\alpha_\nu|a_q^+a_q|\alpha_\nu\rangle = (\nu_q + |\alpha_{n-q}|^2);\\ &\sum_\nu \exp\left(-\frac{\hbar\omega_q\nu_q'}{K_BT}\right)\langle\nu|\exp(\alpha_{nq}^*a_q)\exp(-\alpha_{nq}a_q^+)|\nu\rangle = (\nu_q+1)\exp[-|\alpha_{nq}|^2(\nu_q+1)];\\ &\nu_q = \left[\exp\left(\frac{\hbar\omega_q}{K_BT} - 1\right)\right]^{-1}; \quad \exp(-\bar{W}_{n,n\pm1}) = \langle\alpha_{nq}|\alpha_{n\pm1q}\rangle;\\ &\bar{W}_{nn\pm1} = \exp\left\{\sum_q\left[\alpha_{n\pm q}^*\alpha_{nq} - \frac{1}{2}(|a_{n+1q}|^2 + |a_{nq}|^2)\right]\right\}\\ &\qquad\times\prod_q\left\{\sum_{m=1}^n \frac{(-1)^m|a_{nq} + a_{n+1q}|^2\nu_q'!}{(m!)(\nu_q'-m)!}\right\},\end{aligned} \quad (9.43)$$

we get, after some tedious calculations, the following equations of motion for the exciton and phonon in the molecular crystals at finite temperature

$$\frac{\partial^2 u}{\partial t^2} - v_0^2 \frac{\partial^2 u}{\partial x^2} = \frac{\hbar r_0(\chi_1+\chi_2)}{2M\omega_0} \frac{\partial |\varphi|^2}{\partial x}, \tag{9.44}$$

$$\begin{aligned} i\hbar \frac{\partial \varphi(x,t)}{\partial t} &= [\varepsilon_0 - 2JB(T,q)]\varphi(x,t) - Jr_0^2 B(T,q)\frac{\partial^2 \varphi(x,t)}{\partial x^2} \\ &+ \frac{\hbar(\chi_1+\chi_2)r_0}{\omega_0}\frac{\partial u(x,t)}{\partial x}\varphi(x,t). \end{aligned} \tag{9.45}$$

Thus the motion of the exciton in the nonlinear quantum systems at finite temperature still satisfies the following nonlinear Schrödinger equation

$$\begin{aligned} i\hbar \frac{\partial \varphi(x,t)}{\partial t} &= [\varepsilon_0 - 2JB(T,q)]\varphi(x,t) - Jr_0^2 B(T,q)\frac{\partial^2 \varphi(x,t)}{\partial x^2} \\ &+ G|\varphi(x,t)|^2\varphi(x,t). \end{aligned} \tag{9.46}$$

Its soliton solution is

$$\begin{aligned} \varphi(x,t) &= \sqrt{\frac{\mu(T)}{2}}\operatorname{sech}\left[\frac{\mu(T)}{r_0}(x-x_0-vt)\right] e^{i(\bar{k}x-\hbar\omega_{\rm sol}t)} \\ &= A(T)\operatorname{sech}\left[\frac{\mu(T)}{r_0}(x-x_0-vt)\right] e^{i(\bar{k}x-\hbar\omega_{\rm sol}t)}. \end{aligned} \tag{9.47}$$

It can then be obtained that

$$u(x,t) = \frac{-\hbar(\chi_1+\chi_2)}{2\omega_0\beta(1-s^2)}\tanh\left[\frac{\mu(T)}{r_0}(x-x_0-vt)\right], \tag{9.48}$$

and

$$\begin{aligned} G &= \frac{\hbar^2 r_0^2(\chi_1+\chi_2)^2}{\omega_0^2 M v_0^2(1-s^2)}, \\ \mu(T) &= \frac{c\bar{m}(T)}{2\hbar^2}, \quad \bar{m}(T) = m\exp[\bar{W}_n], \quad \bar{k} = \frac{\bar{m}(T)v}{\hbar}, \\ \frac{\hbar^2}{2\bar{m}(T)r_0^2} &= JB(T,q) \approx \frac{J}{4\hbar\beta(1-s^2)}\left[4\hbar\beta(1-s^2) - \frac{\hbar^2 r_0(\chi_1+\chi_2)^2}{4\omega_0^2 v_0}F(T,q)\right]\times \\ &\exp\left[\frac{\hbar^5 r_0(\chi_1+\chi_2)^6}{128\beta^3J^2(1-s^2)^3\omega_0^6 v_0}F(T,q)\right], \end{aligned} \tag{9.49}$$

$$\bar{\varepsilon}_0(T) = \varepsilon_0 - \frac{\hbar}{mr_0^2}e^{\bar{W}_n} = \varepsilon_0 - 2JB(T,q), \quad \hbar\omega_{sol} = \varepsilon_0(T) + \frac{\hbar^2[k^2-\mu^2(T)]}{2m},$$

$$F(q,T) = \frac{r_0}{\pi N}\sum_q |q|(1+\nu_q) = \begin{cases} \dfrac{2r_0k_BT}{\pi\hbar v_0}, & \text{for } k_BT \gg \hbar\omega_q, \\ \dfrac{1}{2} + \dfrac{k_B^2T^2r_0^2}{3\hbar v_0^2}, & \text{for } k_BT << \hbar\omega_q. \end{cases}$$

(9.47) represents a subsonic soliton ($v < v_0$, $s = v/v_0 < 1$). It can be shown that this soliton is stable because its rest energy is lower than the bottom of the exciton band at about $1/3J\mu^2(T)$. The presence of the energy gap in the spectrum of the excited state of the molecular chain is one of the reasons for the better stability of the soliton. This soliton state describes a quasiparticle consisting of the exciton and lattice deformation. It thus already includes interactions with the acoustic phonons. Therefore, destruction of the soliton requires the removal of the lattice deformation. That is, in order to split the soliton into an exciton and an undeformed lattice it is necessary to expend sufficient energy to allow a transition from the soliton state to a free exciton state. The transition probability to a lattice state with no distortion is proportional to the Frank-Condon factor which is negligibly small for a long molecular chains. Otherwise, since solitons always move with velocity less than the velocity of the longitudinal sound in the chain, they do not emit phonons. In other words, their kinetic energy cannot be transformed into energy of thermal motion. The soliton is thus stable.

From the above discussion, we see that when the subsonic soliton is formed by self-trapping of the exciton through interaction with the acoustic phonon in the molecular crystals at finite temperature $T \neq 0$ K, the lattice molecules or atoms will have a displacement given by (9.48), *i.e.*, the part of the crystal lattice where the soliton is present is subjected to a deformation. The states and positions of atoms and nuclei in this region of the molecular chains will be changed. The active nuclei may emit γ-photons in such a case and the Mössbauer effect could occur in the crystals, as mentioned above. In the following, we evaluate the probability of the γ-radiative Mössbauer transition resulting from the active nuclei which is assumed to be located at node n_0 in the molecular chain.

Following the approach of Gashi *et al.* and Ivic *et al.*, the interaction potential due to the emission mentioned above can be expressed as

$$V_{\text{int}} = A'\pi(x_L, P_L, \sigma_L)\Gamma(P, u_{n_0}), \tag{9.50}$$

where $\pi(x_L, P_L, \sigma_L)$ is an operator which describes the internal state of the active nucleus located at node n_0 and depends on its coordinate x_L, momentum P_L and spin σ_L, $\Gamma(P, u_{n_0})$ is an operator related to the emitted γ-photon, P is the momentum of the emitted γ-quantum, u_{n_0} is the position vector of the center of mass of lattice node. Obviously, $u_{n_0} = n_0 r_0 + R_{n_0}$, in the emitting process, where R_{n_0} denotes the recoil displacement of the emitted nucleus from its node. In general, $\Gamma(P, u_{n_0})$ can be represented in term of a periodical potential or a plane wave, *i.e.*, $\Gamma(P, u_{n_0}) = d\exp(i\vec{P}\cdot\vec{u}_{n_0}/\hbar)$. Thus (9.50) becomes

$$V_{\text{int}} = \bar{A}\pi(x_L, P_L, \sigma_L)e^{i\vec{P}\cdot n_0\vec{r}_0/\hbar}e^{i\vec{P}\cdot\vec{R}_{n_0}/\hbar}, \tag{9.51}$$

where d and A' are constants related to the characteristics of the molecular chain and nucleus, and we can set $\bar{A} = dA'$. In accordance with the quantum theory of radiation, the transition matrix element due to the change of the state in such a case can be written as

$$T_{n\to m} = \langle m\Phi^m|V_{\text{int}}|n\Phi^n\rangle = \bar{A}\langle\Phi^m|e^{i\vec{P}\cdot\vec{R}_{n_0}/\hbar}e^{i\vec{P}\cdot n_0\vec{r}_0/\hbar}|\Phi^n\rangle\cdot \langle m|\pi(x_L, P_L, \sigma_L)|n\rangle. \tag{9.52}$$

Since $\pi(x_L, P_L, \sigma_L)$ depends only on the fixed numbers of internal degrees of freedom of the nucleus, it is unnecessary to know its explicit form if we are only interested in the emission of the molecular crystals. Thus the matrix element, $\langle m|\pi(x_L, P_L, \sigma_L)|n\rangle$, can be treated as a constant value when the change of inner state of the nucleus is small, and can be absorbed into $\bar{A}$. Then the relative transition probability can be determined when the same soliton is found in the molecular crystals before and after the emission, and is given by

$$T_{n\to m} = \bar{A}\langle\Phi|e^{i\vec{P}\cdot\vec{R}_{n_0}/\hbar}|\Phi\rangle e^{i\vec{P}\cdot n_0\vec{r}_0/\hbar}. \tag{9.53}$$

$|\Phi\rangle$ in (9.5) and (9.40) represents the amplitude of the soliton. Following the procedures of Ivic *et al.* and Gashi *et al.*, Pang inserted (9.5), (9.40), and

$$R_{n_0} = \sum_q \sqrt{\frac{\hbar}{2NM\omega_q}}\vec{e}_q(a_q + a^+_{-q})e^{inr_0q}$$

into (9.53), and obtained

$$T_{n\to m} = \bar{A}\sum_n (1+|\varphi_n|^2)T_{\nu\nu}(n)e^{in_0\vec{r}_0\cdot\vec{P}/\hbar},$$

where

$$T_{\nu\nu}(n) = \prod_q \exp\left\{\beta_q^*\alpha_{n-q}^* - \beta_q\alpha_{nq} + \frac{|\beta_q|^2}{2}\right\}\sum_{n_q=1}^{\infty}\frac{(-1)^{n_q}|\beta_q|^{2n_q}(n_q+\nu_q')!}{(n_q!)^2\nu_q!}, \tag{9.54}$$

with

$$\beta_q = \frac{i}{\hbar}\sqrt{\frac{\hbar}{2MN\omega_q}}(\vec{P}\cdot\vec{e}_q)e^{in_0r_0q} \quad \text{and} \quad \beta_{-q} = -\beta_q^*.$$

Since the molecular crystal being considered is in contact with a thermostat at temperature $T \neq 0$ K, we should take the thermal average over the phonon states for the matrix element (9.53) using the density matrix (9.42). Equation (9.53) then becomes

$$\begin{aligned}\langle T_{n\to m}\rangle_{ph} &= \bar{A}\sum_n \left\langle(1+|\varphi_n|^2)T_{\nu\nu}(n)e^{in_0\vec{r}_0\cdot\vec{P}/\hbar}\right\rangle_{ph}\\ &= \bar{A}\sum_n (1+|\varphi_n|^2)\langle T_{\nu\nu}(n)\rangle_{ph}e^{in_0\vec{r}_0\cdot\vec{P}/\hbar},\end{aligned}$$

$$\langle T_{\nu\nu}(n)\rangle_{ph} = \sum_\nu \rho_{\nu\nu}T_{\nu\nu}(n) = \langle T_{sol}\rangle_{ph}\exp\left\{-\sum_q\frac{(\vec{P}\cdot\vec{e}_q)^2}{2MN\omega_q}\left(\nu_q+\frac{1}{2}\right)\right\},$$

where $\langle T_{sol}\rangle_{ph}$ contains the matrix element for the solitonic part. Inserting α_{nq} and α^*_{n-q}, which are related to the soliton solution (9.47), obtained from previous equations of motion into (9.54), the soliton part of the matrix element, $\langle T_{sol}\rangle_{ph}$, can be written as

$$\langle T_{sol}\rangle_{ph} = \exp\left\{-\frac{1}{N}\sum_q \frac{(\chi_1+\chi_2)(\vec{P}\cdot\vec{e}_q)qr_0|\varphi_n|^2}{\omega_0 M v_0^2(1-s^2)|q|^2} e^{iqr_0(n_0-n)}\right\}.$$

After tedious calculations, Pang finally obtained

$$\begin{aligned}\langle T_{n\to m}\rangle_{ph} &= \bar{A}\exp\left\{-\sum_q \frac{(\vec{P}\cdot\vec{e}_q)^2}{2MN\hbar\omega_q}\left(\nu_q+\frac{1}{2}\right)\right\}\Re\left\{\frac{1}{r_0}\int_{-\infty}^{\infty}dx(1+|\varphi|^2)e^{i\Omega x}\right\}\\ &= \bar{A}\exp\left\{-\sum_q \frac{(\vec{P}\cdot\vec{e}_q)^2}{2MN\hbar\omega_q}\left(\nu_q+\frac{1}{2}\right)\right\}\Re\left\{1+\frac{\pi\Omega r_0}{2\mu(T)}\mathrm{csch}\left[\frac{\pi\Omega r_0}{2\mu(T)}\right]e^{i\Omega vt}\right\},\end{aligned}$$

where

$$\Omega = \frac{(\chi_1+\chi_2)r_0\mu(T)}{2Mv_0^2(1-s^2)N\omega_0}\sum_q(\vec{P}\cdot\vec{e}_q).$$

The Mössbauer transition probability is given by

$$\begin{aligned} f = \langle W_{n\to m}\rangle = |\langle T_{n\to m}\rangle_{ph}|^2 &= \bar{A}^2\exp\left\{-\sum_q \frac{(\vec{P}\cdot\vec{e}_q)^2}{2MN\hbar\omega_q}(\nu_q+\frac{1}{2})\right\}\\ &\times\Re\left\{1+\frac{\pi\Omega r_0}{2\mu(T)}\mathrm{csch}\left[\frac{\pi\Omega r_0}{2\mu(T)}\right]e^{i\Omega vt}\right\},\end{aligned} \tag{9.55}$$

where

$$\begin{aligned}\mu(T) &= \frac{\mu(0)}{4\beta\hbar(1-s^2)}\left[4\hbar\beta(1-s^2) - \frac{\hbar^2 r_0(\chi_1+\chi_2)^2}{4\omega_0^2 v_0}F(T,q)\right]\\ &\times\exp\left\{\frac{\mu^3(0)\pi r_0 J}{4\hbar v_0}F(T,q)\right\},\\ F(T,q) &= \frac{2r_0K_BT}{\hbar v_0\pi}, \quad \mu(0) = \frac{\hbar^2(\chi_1+\chi_2)^2}{4\beta J(1-s^2)\omega_0^2}.\end{aligned}$$

From (9.55), we can obtain the following properties of the Mössbauer effect.

(1) The transition probability is a product of the subsonic soliton part related to the temperature and the phononic part. Generally speaking, if the molecular chain is populated exclusively by solitons, the Mössbauer effect is different from that in the case where pure phonon modes are spread over the whole chain. The latter leads to a probability with a factor

$$\exp\left\{-\sum_q \frac{(\vec{P}\cdot\vec{e}_q)^2}{2MN\hbar\omega_q}\coth\left(\frac{\hbar\omega_q}{2K_BT}\right)\right\}$$

in (9.55). In the former the soliton transition probability is manifested by the factor in the curly brackets in (9.55), $\Re\{\cdots\}$, which is to be multiplied by the factor corresponding to the pure phonon states.

(2) The transition probability given above depends on temperature T. Temperature influences the probability through the following factors, the amplitude of the soliton, $A(T)$, $\mu(T)$ and

$$\nu_q = \frac{1}{\exp(\hbar\omega_q/K_B T) - 1}$$

in the case of pure phonon states. Knowledge of these parameters allows us to estimate the value of the transition probability. In general, although an exact calculation for the probability is very difficult, an approximate numerical estimation of (9.55) is possible by using generally accepted values for the parameters given earlier. In Fig. 9.3, the average of the relative transition probability $\bar{f}/\bar{A}^2$ is shown as a function of temperature T at $v = 0$ or $s = 0$. Here $\lambda = 2.25 \times 10^{-11}$ m is used for the average wave-length of γ-photon.

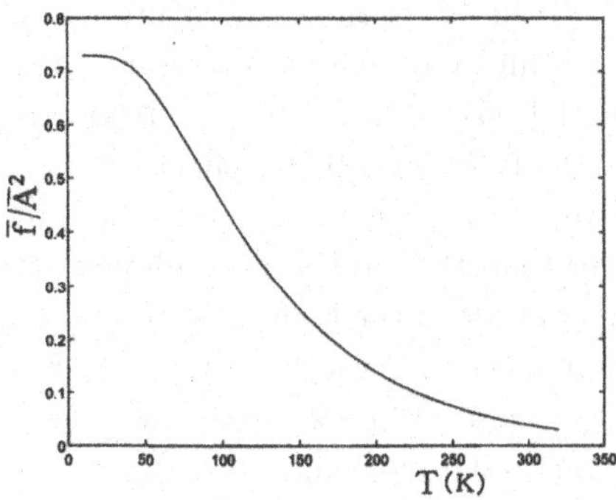

Fig. 9.3 The average relative transition probability $\bar{f}/\bar{A}^2$ is shown as a function of temperature T for $s = 0$ and $\lambda = 2.25 \times 10^{-11}$ m.

From this figure, we see that the transition probability overall decreases exponentially as the temperature increases, except at very low temperature. The effect of temperature on the probability comes from the phononic part. It can be shown that $\bar{f}/\bar{A}^2$ is $3 - 3.5\%$ at $T = 300$ K and $s = 0$. Therefore, the transition probability is very small at 300 K.

The average value of the soliton part of the transition probability in (9.55),

$$\frac{\bar{f}_s}{\bar{A}^2} = \Re\left[1 + \frac{\pi\Omega r_0}{2\mu(T)} \operatorname{csch}\left(\frac{\pi\Omega r_0}{2\mu(T)}\right) e^{i\Omega vt}\right]^2,$$

can be evaluated at high and low temperature limits using the parameters given in Section 9.2, and the results are $\bar{f}_s/\bar{A}^2 = 1.1738$ at $T = 300$ K and $\bar{f}_s/\bar{A}^2 = 1.1857$ at $T = 10$ K, respectively. As far as the influence of temperature on the transition probability is concerned, the value of the soliton part at low temperature is larger than that at high temperature, but the difference between them is very small in the long-wave length approximation. This demonstrates that the effect of the temperature on the soliton part of the transition probability of the Mössbauer effect is also very small. Furthermore, it can be shown that the contribution of the soliton state to the transition probability is larger than that of the pure phonon modes in this model, particularly at high temperature.

(3) The transition probability depends closely on the strength of the coupling coefficient, $(\chi_1 + \chi_2)$ and $\mu(0)$. From (9.55), we see that $\bar{f}/\bar{A}^2$ decreases with increasing $(\chi_1 + \chi_2)$ and $\mu(0)$ at $s = 0$, $T = 300$ K and $\lambda = 2.25 \times 10^{-11}$ m, respectively, and that the transition probability is pure phononic, when $(\chi_1+\chi_2) = 0$ or $\mu(0) = 0$. This shows clearly that the Mössbauer effect and the corresponding transition probability result from the motions of the solitons and thermal phonons and the interaction between them in the molecular crystals.

(4) The transition probability decreases with increasing soliton velocity, v, since the parameter $\mu(T)$ is related to v. When the velocity of the soliton approaches the sound speed v_0, the Mössbauer effect transition probability decreases obviously. We can similarly calculate the effect of temperature on the transition probability of the soliton part at high and low temperatures and at $s = 0.5$. The results are 1.0399 at $T = 300$ K and 1.05126 at $T = 10$ K, respectively. This shows clearly that the transition probability of Mössbauer effect decreases as the velocity of the soliton increases, compared to that in the case of $s = 0$, but its contributions to the probability are still larger than that in the pure phonon state in such a case.

(5) In general, the magnitude of the transition probability depends on the inherent characteristics of the molecular crystals, $(M, r_0, v_0, N, \omega_0$ and $\omega_1)$, the properties of the quasiparticles generated in collective excitation, $(\omega_q, q, v$, and $m)$, and the environment condition of the molecular crystals, *i.e.*, temperature. Therefore, the transition probability of different molecular crystals would be different.

Pang also calculated Mössbauer transition probability when the anharmonic vibrations of the molecules (9.18) arising from the temperature is taken into consideration in the Hamiltonian. In such a case, the exciton becomes a supersonic soliton. Apply the above method, the supersonic exciton-soliton can be approximately represented by

$$\varphi(x,t) = A(T)\,\mathrm{sech}\left[\frac{1}{r_0}\left(\frac{v^2}{2v_0^2} - \frac{1}{2}\right)^{-1/2}(x - vt)\right] e^{i(\tilde{K}x - \omega_{\rm sol}t)}, \tag{9.56}$$

where $v > v_0$, $A(T)$ is the amplitude of the supersonic soliton which depends on the temperature. Thus the corresponding Mössbauer transition probability is obviously different from (9.55). The features of the Mössbauer effect caused by the nonlinear excitation of the excitons are discussed in details in Pang 1987, 1989a, 1990, 1993e, 1999, 2001c.

9.5 Nonlinear Excitation of Excitons in Protein

We now move on to discuss properties of nonlinear excitations and motions of excitons in protein molecules on the basis of an improved Davydov theory, proposed by Pang *et al.* (Pang 1987, 1989a, 1990, 1992b, 1993b, 1993e-h, 1999, 2001c, 2001e, Pang and Chen 2001).

As it is known, many biological processes are associated with bioenergy transport through protein molecules, where energy is released by the hydrolysis of adenosine triphosphate (ATP). Understanding the mechanism of bioenergy transport in such systems has been a long-standing problem that remains of great interest today. As an alternative to electronic mechanisms, one can assume that the energy is stored as vibrational energy in the C=O stretching mode (amide-I) of a polypeptide chain. Based on the idea of Davydov (see Section 5.6), one can take into account the coupling between the amide-I vibrational quantum (exciton) and the acoustic phonon in the amino acid lattice. Through the coupling, nonlinear interaction occurs in the motion of exciton, which results in a self-trapped state of the exciton. The latter, together with lattice deformation of the amino acid molecules can travel over macroscopic distances along the molecular chains, retaining the wave shape, energy and momentum. In this way, bioenergy can be transported as a localized exciton or soliton. This model of the bioenergy transport was first proposed by Davydov in the 1970s.

Davydov's idea yields a compelling picture for the mechanism of bioenergy transport in protein molecules and consequently has been the subject of a large number of works (Bolterauer and Opper 1991, Borwn 1988 Brizhik and Davydov 1969, 1983, Brown, *et al.* 1986, 1987, 1988, 1989, Christiansen 1990, Cottingham and Schweitzer 1989, Cruzeiro *et al.* 1985, Cruzeiro-Hansson 1992, 1993, 1994 Cruzeiro-Hansson and Scott 1994, Cruzeiro-Hansson and Takeno 1997, Davydov 1980, 1981, 1982, 1991, Davydov and Kislukha 1973, 1977, Förner 1991, 1992, 1993, 1994, 1996, Förner and Ladik 1991, Hyman *et al.* 1981, Ivic 1998, Ivic and Brown 1989, Kerr and Lomdahl, 1989, 1991, Lawrence *et al.* 1986, Lomdahl and Kerr 1985, 1991, Macneil and Scott 1984, Mechtly and Shaw 1988, Motschman *et al.* 1989, Pang 1986, Schweitzer 1992, Schweitzer Cottingham 1991, Scott 1982, 1983, 1984, Skrinjar *et al.* 1984, 1988, Takeno 1984, 1985, 1986, Tekec *et al.* 1998, Wang *et al.* 1988, 1989, 1991, Zekovic and Ivic 1999). Problems related to the Davydov model (see Section 5.6), including the foundation and the accuracy of the theory, the quantum and classical properties, and the thermal stability and lifetimes of the Davydov soliton, have been extensively studied. However, considerable controversy has arisen regarding whether the Davydov soliton is sufficiently stable in the region of biological temperature to provide a viable explanation for bioenergy transport. Many numerical simulations have been based essentially on classical equations of motion and are subject to the criticism that they are likely to yield unreliable estimates for the stability of the soliton since the dynamics of the soliton is not being determined by the nonlinear Schrödinger equation. Simulations based on the $|D_2\rangle$ state, *i.e.*, $|\Phi(t)\rangle$ in (5.70) in which $B_n^+(B_n)$ is the creation (annihilation) operator of exciton at site n, u_n is the displacement operator of the amino acid molecule at site n, P_n is its conjugate momentum operator, generally agree that the stability of the soliton decreases with increasing temperature and that the soliton is not sufficiently stable in the region of biological temperature (300 K). Since the dynamical equations used in the simulations are not equivalent to the nonlinear Schrödinger equation, the stability of the soliton obtained by these numerical simulations is unreliable. On the other hand, simulations based on the $|D_1\rangle$ state, *i.e.*,

$$|D_1\rangle = \sum_n \varphi_n(t) B_n^+ |0\rangle_{ex} \exp\left\{\sum_q \left[\alpha_{nq}(t) a_q^+ - \alpha_{nq}^*(t) a_q\right]\right\} |0\rangle,$$

where $|0\rangle = |0\rangle_{ex}|0\rangle_{ph}$, $a_q^+(a_q)$ is the creation (annihilation) operator of the lattice phonon, and $\alpha_{nq}(t)$ and $\alpha_{nq}^*(t)$ are some undetermined functions, with Davydov's thermal treatment, where the equations of motion are derived from a thermally averaged Hamiltonian, yielded a surprising result that stability of the soliton can be enhanced with increasing temperature. Evidently, this conclusion is not reliable because the Davydov procedure in which one constructs an equation of motion for an average dynamical state from an average Hamiltonian, corresponding to the Hamiltonian averaged over a thermal distribution of phonons, is inconsistent with standard concepts of quantum-statistical mechanics in which a density matrix must be used to describe the system. There has been no exact fully quantum-mechanical treatment for the numerical simulation of the Davydov soliton. However, for its thermal equilibrium properties, a quantum Monte Carlo simulation had been carried out by Wang *et al.*. In this simulation, correlation characteristic of solitonlike quasiparticles occur only at low temperatures, $T < 10$ K, for widely accepted parameter values. This is consistent at a qualitative level with the result of Cottingham and Schweitzer. The latter is a straightforward quantum-mechanical perturbation calculation. The lifetime of the Davydov soliton obtained by using this method is too small (about $10^{-12} - 10^{-13}$ sec.) to be useful in biological processes. This indicates clearly that the Davydov solution is not a true wave function of the systems. A systematic study with different parameters, different types of disorder, different thermalization schemes, different wave functions, and different associated dynamics by Föner *et al.* lead to a very complicated picture for the Davydov model. These results did not completely rule out Davydov's theory, but they did not eliminate the possibility of another wave function and a more sophisticated Hamiltonian of the system which might result in longer soliton lifetimes and good thermal stability.

Indeed, the question of lifetime of soliton in protein molecules is twofold. In the Langevin dynamics, the problem consists of uncontrolled effects arising from the semiclassical approximation. In quantum treatments, the lacking of an exact wave function for the soliton has been the issue. The exact wave function of the fully quantum mechanical Davydov model has not been known up to now. Different wave functions have been used to describe the states of fully quantum-mechanical systems. Although some of these wave functions lead to exact quantum states and exact quantum dynamics for the $J = 0$ state, they share the same problem as the original Davydov wave function, namely the degree of approximation included when $J \neq 0$ is not known. Davydov's wave function thus has to be modified. It was thought that the soliton with a multiquantum state (n$\geq$ 2), such as coherent states of Brown *et al.*, multiquantum state of Kerr and Lomdahl, and Schweitzer *et al.*, two-quantum states of Cruzeiro-Hansson and Förner, would be thermally stable in the regime of biological temperature and could provide a realistic mechanism for bioenergy transport in protein molecules. However, the assumption of the standard coherent state is unsuitable or impossible for biological protein molecules because there are innumerable particles in this state and the particle number cannot be conserved. The assumption of a multiquantum state ($n > 2$) along with a coherent state is also inconsistent with the fact that the energy released in ATP hydrolyses (about 0.43 eV) can excite only two quanta of amide-I vibration. On the other hand, numerical result obtained using the two-quantum model by Förner reveals considerable differences from one-quantum dynamics, *i.e.*, the soliton with a two-quantum state is more stable than that with a one-quantum state. Cruzeiro-Hansson considered that Förner's two-quantum state is not exact in the semiclassical case. Therefore, he constructed an "exact" two-quantum state for the semiclassical Davydov system, given by

$$|\varphi(t)\rangle = \sum_{n,m=1}^{N} \varphi_{nm}(\{u_l\},\{P_l\},t)B_n^+ B_{m,}^+|0\rangle_{ex}.$$

However, further investigation by Pang *et al.* concluded that the Cruzeiro-Hamsson wave function does not represent exactly the two-quantum state. An energy release of about 0.43 eV in the ATP hydrolysis does not support the model.

On the basis of the work by Cruzeiro-Hansson, Förner, and others, Pang improved and extended the Davydov model by modifying simultaneously the Hamiltonian and the wave function of the systems. He included the coupling interaction between acoustic phonons and the amide-I vibrational modes in the original Davydov Hamiltonian, and replaced the one-quantum exciton state in Davydov's wave function by a quasicoherent two-quantum state. This resulted in an equation of motion and properties of the soliton completely different from those in the Davydov model. It is believed that this model is able to resolve the controversy concerning the thermal stability and lifetime of soliton in protein molecules.

The Davydov model is indeed too simple, both in its wave function and Hamiltonian. It simply cannot properly describe the basic properties of collective excitations in protein molecules. Davydov's theory was applied to protein molecules as an exciton-soliton model in an one-dimensional molecular chain. Although the molecular structure of the α-helix protein is analogous to some molecular crystals such as ACN, the α-helix protein molecule consists of three peptide channels.

Many properties and functions of the protein molecules are completely different from those of ACN. Protein molecules act as a kind of soft condensed matter as well as bio-self-organization with active functions, for instance, self-assembly and self-renewing. The physical concepts of coherence, order, collective effects, and mutual correlation are very important in protein molecules compared with other more general molecular systems.

There is an obvious asymmetry in the Davydov wave function given above since the phononic part is a coherent state while the excitonic part is only an excitation state of a single particle. It is thus unreasonable for the same nonlinear interaction generated by the coupling between the excitons and phonons to produce different states for the phonon and exciton. Taking these into consideration, Pang modified the Davydov wave function according to the following,

$$|\Phi(t)\rangle = |\varphi(t)\rangle|\beta(t)\rangle = \frac{1}{\lambda}\left[1+\sum_n \varphi(t)B_n^+ + \frac{1}{2!}\left(\sum_n \varphi(t)B_n^+\right)^2\right]|0\rangle_{ex}$$
$$\times \exp\left\{-\frac{i}{\hbar}\sum_n [\beta_n(t)P_n - \pi_n(t)u_n]\right\}|0\rangle_{ph}, \tag{9.57}$$

where u_n and P_n are the displacement and momentum operators of the amino acid molecule at site n, $\varphi(t)$, $\beta_n(t) = \langle\Phi(t)|u_n|\Phi(t)\rangle$ and $\pi_n(t) = \langle\Phi(t)|P_n|\Phi(t)\rangle$ are three sets of unknown functions, λ is a normalization constant. We assume thereafter $\lambda = 1$ for convenience of calculation.

Since the resonant dipole-dipole interaction between the neighboring amide-I vibrational quanta (excitons) in neighboring amino acids with electrical moment of about 0.8-3.5D has been taken into account in the Davydov Hamiltonian, we may as well consider the changes of relative displacement of the neighboring amino acids arising from this interaction. Therefore, it is reasonable to include the interaction term $\chi^2(u_{n+1} - u_n)(B_{n+1}^+B_n + B_m^+B_{n+1})$ in Davydov's Hamiltonian to represent correlations between collective excitations in protein molecules. Although the dipole-dipole interaction is small as compared to the energy of the amide-I vibrational quantum, the resulting change in relative displacement of neighboring peptide groups cannot be ignored due to the sensitive dependence of dipole-dipole interaction on the distance between amino acids in the protein molecules. The modified Davydov's Hamiltonian given by Pang is as follows.

$$\begin{aligned} H &= H_{ex} + H_{ph} + H_{\text{int}} \qquad (9.58)\\ &= \sum_n \left[\varepsilon_0 B_n^+B_n - J(B_n^+B_{n+1} + B_nB_{n+1}^+)\right] + \sum_n\left[\frac{P_n^2}{2M} + \frac{1}{2}w(u_n - u_{n-1})^2\right]\\ &\quad + \sum_n \left[\chi_1(u_{n+1} - u_{n-1})B_n^+B_n + \chi_2(u_{n+1} - u_n)(B_{n+1}^+B_n + B_n^+B_{n+1})\right], \end{aligned}$$

where $\varepsilon_0 = \hbar\omega_0 = 1665\ \text{cm}^{-1}$ is the energy of the exciton (the C=O stretching mode), χ_1 and χ_2 are the nonlinear coupling constants which represent the modulations of the on-site energy and resonant (or dipole-dipole) interaction energy, respectively, of excitons produced by molecular displacements, M is the mass of an amino acid molecule and w is the elastic constant of the protein molecular chains, J is the dipole-dipole interaction energy between neighboring sites. This Hamiltonian has better symmetry and describes correctly the mutual correlations between the collective excitations in protein molecules.

Obviously, the exciton wave function given in (9.57) is not an excitation state of a single particle, but rather a coherent state, because

$$|\varphi(t)\rangle \approx \exp\left\{\sum_n [\varphi(t)B_n^+ - \varphi(t)^* B_n]\right\} |0\rangle_{ex}.$$

More precisely, the new wave function is a truncated standard coherent state and retains only three terms in the expansion of the latter in the case of small $\varphi_n(t)$ [*i.e.*, $|\varphi_n(t)| \ll 1$]. $|\varphi(t)\rangle$ is thus referred to as a quasi-coherent state. It is not an eigenstate of the particle number operator, $\hat{N} = \sum_n B_n^+ B_n$, since $\hat{N}|\varphi(t)\rangle = 2|\varphi(t)\rangle - [2 + \sum_n \varphi_n(t) B_n^+]|0\rangle_{ex}$. But the number of quanta in this state is determinate, and there are two excitons *i.e.*,

$$N = \langle\varphi(t)|\hat{N}|\varphi(t)\rangle = \sum_n |\varphi_n(t)|^2 \left[1 + \sum_n |\varphi_n(t)|^2\right] = 2,$$

where

$$\sum_n |\varphi_n(t)|^2 = 1, \quad [B_n, B_n^+] = \delta_{nm}.$$

Thus, $|\varphi(t)\rangle$ indeed represents a coherent superposition of multiquantum states. Therefore, this wave function is different from the one-quantum state proposed by Davydov, the two-quanta states proposed by Förner and Cruzeiro-Hansson, the standard coherent state proposed by Brown *et al.*, and the multi-quantum states proposed by Kerr *et al.* and Schweitzer *et al.*. The wave function given in (9.57) not only gives the coherent property of collective excitations of exctions and phonons due to the nonlinear exciton-phonon interaction, which results in a symmetric wave function for the states of the system, but also agrees with the fact that the energy released in ATP hydrolysis (about 0.43 eV) is only sufficient to create two excitons. The number of excitons in the Hamiltonian (9.58) is thus conserved.

Using (9.57) and (9.58), and the Heisenberg equations for the expectation values of the operators, u_n and P_n, in $|\Phi(t)\rangle$

$$\begin{aligned} i\hbar\frac{\partial}{\partial t}\langle\Phi(t)|u_n|\Phi(t)\rangle &= \langle\Phi(t)|[u_n, H]\Phi(t)\rangle, \\ i\hbar\frac{\partial}{\partial t}\langle\Phi(t)|P_n|\Phi(t)\rangle &= \langle\Phi(t)|[P_n, H]|\Phi(t)\rangle, \end{aligned} \tag{9.59}$$

we can obtain the equation of motion for $\beta_n(t)$

$$M\ddot{\beta}_n = \omega(\beta_{n+1} - 2\beta_n + \beta_{n-1}) + 2\chi_1(|\varphi_{n+1}|^2 - |\varphi_{n-1}|^2) \\ +2\chi_2[\varphi_n^*(\varphi_{n+1} - \varphi_{n-1}) + \varphi_n(\varphi_{n+1}^* - \varphi_{n+1}^*)]. \tag{9.60}$$

For the equation of motion for $\varphi(t)$, a basic assumption is that $|\Phi(t)\rangle$ is a solution of the time-dependent Schrödinger equation

$$i\hbar\frac{\partial}{\partial t}|\Phi(t)\rangle = H|\Phi(t)\rangle.$$

Using (9.57) and (9.58), we can, after some tedious calculations, get

$$i\hbar\frac{\partial\varphi_n}{\partial t} = \varepsilon_0\varphi_n - J(\varphi_{n+1} - \varphi_{n-1}) + \chi_1(\beta_{n+1} - \beta_{n-1})\varphi_n \tag{9.61}$$

$$+\chi_2(\beta_{n+1} - \beta_n)(\varphi_{n+1} - \varphi_{n-1}) + \frac{5}{2}\left[W(t) - \frac{1}{2}\sum_m(\dot{\beta}_m\pi_m - \dot{\pi}_m\beta_m)\right]\varphi_n,$$

where

$$W(t) = \langle\beta(t)|H_{ph}|\beta(t)\rangle = \sum_n\left[\frac{\pi_n^2}{2M} + \frac{1}{2}\omega(\beta_n - \beta_{n-1})^2\right] + \sum_q\frac{1}{2}\hbar\omega_q.$$

In the continuum approximation, Pang obtained from (9.60) and (9.61) that

$$i\hbar\frac{\partial\varphi(x,t)}{\partial t} = R(t)\varphi(x,t) - Jr_0^2\frac{\partial^2\varphi(x,t)}{\partial x^2} - G_P|\varphi(x,t)|^2\varphi(x,t), \tag{9.62}$$

and

$$\frac{\partial\beta}{\partial x} = -\frac{4(\chi_1 + \chi_2)}{\omega(1 - s^2)r_0}|\varphi(x,t)|^2, \tag{9.63}$$

where

$$R(t) = \varepsilon_0 - 2J + \frac{5}{2}\left[W(t) - \frac{1}{2}\sum_m(\dot{\beta}_m\pi_m - \dot{\pi}_m\beta_m)\right],$$

$s = v/v_0$, and $v_0 = r_0\sqrt{w/M}$ is the sound speed in the protein molecular chains. Equation (9.62) is a standard nonlinear Schrödinger equation. It shows that the protein molecules are nonlinear quantum systems, in which exciton becomes a soliton that has the properties of microscopic particle in nonlinear quantum systems described in Chapters 3-6. The soliton solution of (9.62) then is

$$\varphi(x,t) = \sqrt{\frac{\mu_P}{2}}\,\mathrm{sech}\left[\frac{\mu_P}{r_0}(x - x_0 - vt)\right]\exp\left\{i\left[\frac{\hbar v}{2Jr_0^2}(x - x_0) - E_v\frac{t}{\hbar}\right]\right\}, \tag{9.64}$$

with

$$\mu_P = \frac{2(\chi_1 + \chi_2)^2}{\omega(1 - s^2)J}, \quad G_P = \frac{8(\chi_1 + \chi_2)^2}{\omega(1 - s^2)}, \quad A = \varepsilon_0 - 2J. \tag{9.65}$$

Although the forms of above equation of motion and the corresponding soliton solution, (9.62) – (9.65), are similar to those of the Davydov soliton (see Section 5.6), they are significantly different because the parameters in the equation of motion and their representations have obviously different meanings. Immediate results of this newly proposed model include an increase of the nonlinear interaction energy G_P which is given by

$$G_P = 2G_D \left[1 + 2\left(\frac{\chi_2}{\chi_1}\right) + \left(\frac{\chi_2}{\chi_1}\right)^2\right],$$

a corresponding increase in the amplitude of the soliton, and a decrease in its width due to an increase of μ_P,

$$\mu_P = 2\mu_D \left[1 + 2\left(\frac{\chi_2}{\chi_1}\right) + \left(\frac{\chi_2}{\chi_1}\right)^2\right],$$

compared to the Davydov soliton. Here

$$\mu_D = \frac{\chi_1^2 J}{w(1-s^2)} \quad \text{and} \quad G_D = \frac{4\chi_1^2}{w(1-s^2)}$$

are the corresponding values in the Davydov model. The localized feature of the soliton is enhanced, and its stability against quantum fluctuation and thermal perturbations increased considerably as compared with the Davydov soliton.

The energy of the soliton in the improved model becomes

$$E = \langle\Phi(t)|H|\Phi(t)\rangle = 2\left[(\varepsilon_0 - 2J) + \frac{\hbar^2 v^2}{2Jr_0^2} - \frac{2\mu_P^2}{3}J\right] = E_0 + \frac{1}{2}M_{sol}v^2. \quad (9.66)$$

The rest energy of the soliton is given by

$$E_0 = 2(\varepsilon_0 - 2J) - \frac{8(\chi_1+\chi_2)^4}{3w^2 J} = E_s^0 + W, \quad (9.67)$$

where

$$W = \frac{2(\chi_1+\chi_2)^4}{3w^2}J$$

is the deformation energy of the lattice. The effective mass of the soliton is

$$M_{mol} = 2m_{ex} + \frac{8(\chi_1+\chi_2)^4(9s^2+2-3s^4)}{3w^2 J(1-s^2)^3 v_0^2}. \quad (9.68)$$

In such a case, the binding energy of the soliton can be obtained and is given by

$$\begin{aligned} E_{BP} &= -\frac{8(\chi_1+\chi_2)^4}{3Jw^2} \\ &= 8E_{BD}\left[1 + 4\left(\frac{\chi_2}{\chi_1}\right) + 6\left(\frac{\chi_2}{\chi_1}\right)^2 + 4\left(\frac{\chi_2}{\chi_1}\right)^3 + \left(\frac{\chi_2}{\chi_1}\right)^4\right]. \end{aligned} \quad (9.69)$$

It is obvious that E_{BP} is larger than that of the Davydov soliton, $E_{BD} = -\chi_1^4/3Jw^2$.

The binding energy of the soliton given by the improved model is about several ten times larger than that of the Davydov soliton. The increase in the soliton binding energy is due to the two-exciton nature of the improved model and the additional interaction term $\sum_n x_2(\mu_{n+1} - \mu_n)(B_{n+1}^+ B_n + B_n^+ B_{n+1})$ in the Hamiltonian, (9.58). However, we can see from (9.69) that the two-exciton nature plays the main role in the increase of the binding energy and the enhancement of thermal stability for the soliton, compared to the additional interaction term, $\chi_2 < \chi_1$. The increase in binding energy results in significant changes of properties of the soliton. To compare the various correlations, it is useful to consider them as functions of a composite coupling parameter, such as that given by Young *et al.* and Scott, that can be written as

$$4\pi\sigma_P = \frac{(\chi_1 + \chi_2)^2}{2w\hbar\omega_D},$$

where $\omega_D = \sqrt{w/M}$ is the band edge for acoustic phonons (Debye frequency). If $4\pi\sigma_P \gg 1$, the coupling is said to be strong, and if $4\pi\sigma_P \ll 1$, it is said to be weak. Using the following widely accepted values for various parameters for the α-helix protein molecule, *i.e.*, $J = 1.55 \times 10^{-22}$ J, $w = 13 - 19.5$ N/m, $M = (11.7 - 19.1) \times 10^{-25}$ kg, $\chi_1 = 62 \times 10^{-12}$ N, $\chi_2 = (10 - 18) \times 10^{-12}$ N, $r_0 = 4.5 \times 10^{-10}$ m, it can be estimated that the coupling constant lies in the range of $4\pi\alpha_p = 0.11 - 0.273$. The corresponding range in the Davydov model is $4\pi\sigma_P = 0.036 - 0.045$. Therefore, the improved model is not a weakly coupled theory. Using the notation of Venzel and Fischer, Nagy *et al.*, it is convenient to define another composite parameter $\gamma = J/2\hbar\omega_D$. In terms of these two composite parameters, $4\pi\sigma_P$ and γ, the soliton binding energy given by the improved model can be written as

$$\frac{E_{BP}}{J} = \frac{8}{3}\left(\frac{4\pi\sigma_P}{\gamma}\right)^2, \quad M_s = 2m_{ex}\left[1 + \frac{32}{3}\left(\frac{4\pi\sigma_P}{\gamma}\right)^2\right]. \tag{9.70}$$

Using the above values for the various parameters, we can get $\gamma = 0.08$. E_{BP}/J calculated using the above parameters is shown in Fig. 9.4 as a function of $4\pi\sigma_P$. For the Davydov model, it can be shown that

$$\frac{E_{BD}}{J} = \frac{1}{3}\left(\frac{4\pi\sigma_D}{\gamma}\right)^2, \quad M'_{\text{sol}} = m_{ex}\left[1 + \frac{2}{3}(4\pi\sigma_D)^2\right], \quad 4\pi\sigma_D = \frac{\chi_1^2}{2w\hbar\omega_0}.$$

E_{BD}/J of the Davydov model is also shown in Fig. 9.4 versus $4\pi\sigma_D$. From this figure, we can see that the difference between the soliton binding energies of the two models becomes larger with increasing $4\pi\alpha$. Therefore, localization of the soliton is enhanced in the improved model due to the increased nonlinear interaction. The soliton binding energy increases due to the increase in exciton-phonon interaction in the improved model. This leads to the enhanced stability of the soliton against quantum and thermal fluctuations.

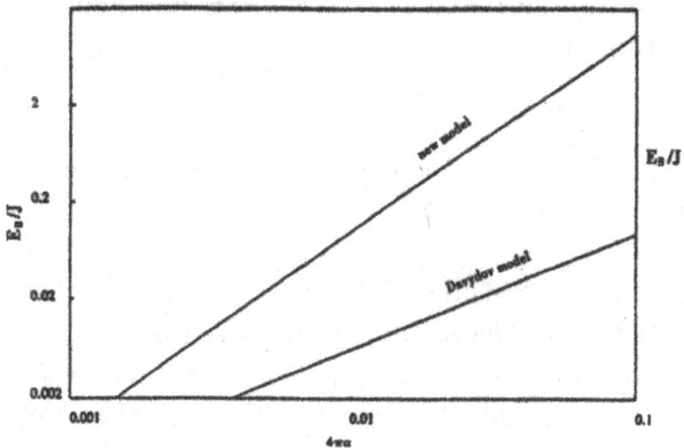

Fig. 9.4 Soliton binding energy (E_B) obtained using the improved model and that of the Davydov model are shown as functions of the coupling constant, $4\pi\alpha$. The binding energy is given in terms of dipole-dipole interaction energy J.

As a matter of fact, the nonlinear interaction energy which is responsible for the formation of the soliton is

$$G_P = \frac{8(\chi_1 + \chi_2)^2}{(1 - s^2)} w = 3.8 \times 10^{-21}\,\text{J},$$

which is larger than the linear dispersion energy, $J = 1.55 \times 10^{-22}$ J. This means that the nonlinear interaction is so large that it suppresses the linear dispersion effect in the equation of motion, leading to the exceptional stability of the soliton. On the other hand, the nonlinear interaction energy in the Davydov model is only

$$G_D = \frac{4\chi_1^2}{1 - s^2} w = 1.18 \times 10^{-21}\,\text{J},$$

which is about three to four times smaller than G_P. The Davydov soliton is thus less stable compared to the soliton in the improved model. Furthermore, the binding energy of the soliton in the improved model, $E_{BP} = (4.16 - 4.3) \times 10^{-21}$ J in (9.66), is somewhat larger than the thermal perturbation energy, $K_B T = 4.13 \times 10^{-21}$ J, at 300 K and about four times larger than the Debye energy, $K_B \Theta = \hbar\omega_D = 1.2 \times 10^{-21}$ J, (Here ω_D is the Debye frequency). Therefore, transition of the soliton to a delocalized state can be suppressed by the large energy difference between the initial (solitonic) state and the final (delocalized exciton) state, which is very difficult to be compensated by the energy of the absorbed phonon. The soliton is thus robust against quantum and thermal fluctuations. It has a large lifetime and good thermal stability in the biological temperature range. On the other hand, the binding energy of the Davydov soliton, $E_{BD} = \chi_1^4/3w^2 = 0.188 \times 10^{-21}$ J, is about 23 times smaller than that of the soliton in the improved model, about 22 times smaller than $K_B T$, and about 6 times smaller than $K_B \Theta$, respectively. Therefore, the Davydov soliton is easily destroyed by thermal perturbation and quantum transition effects, which is why the Davydov soliton has a very short lifetime, and it is unstable at the biological temperature of 300 K.

9.6 Thermal Stability and Lifetime of Exciton-Soliton at Biological Temperature

The thermal stability and lifetime of the exciton-soliton at 300 K in the protein molecules is crucial to the validity of the improved model. It is directly related to whether soliton can provide the mechanism for bioenergy transport in protein molecules. In this section, we will examine the thermal stability and lifetime of exciton-soliton at biological temperature in more details.

Making use of (9.3), (9.58) can be written as

$$H = \sum_n \left[\varepsilon_0 B_n^+ B_n - J\left(B_n^+ B_{n+1} + B_{n+1}^+ B_n\right)\right] + \sum_q \hbar\omega_q \left(a_q^+ a_q + \frac{1}{2}\right) \quad (9.71)$$

$$+\frac{1}{\sqrt{N}} \sum_{q,n} \left[g_1(q) B_n^+ B_n + g_2(q)\left(B_n^+ B_{n+1} + B_n^+ B_{n+1}\right)\right](a_q + a_{-q}^+) e^{inr_0 q},$$

where

$$g_1(q) = 2\chi_1 i \sqrt{\frac{\hbar}{2M\omega_q}} \sin(r_0 q); \quad g_2(q) = \chi_2 \sqrt{\frac{\hbar}{2M\omega_q}} \left(e^{ir_0 q} - 1\right). \quad (9.72)$$

We now partially diagonalize the model Hamiltonian (9.71) and calculate the lifetime of the soliton, (9.64), following procedure given earlier and results in (9.19) – (9.22).

Since we are interested in the case where a soliton is initially moving with a velocity v along the chains, it is convenient to carry out the analysis in a frame of reference in which the soliton is at rest. We thus consider the Hamiltonian in this rest frame of the soliton, $\tilde{H} = H - vP$, where P is the total momentum, and is given by

$$P = \sum_q \hbar q \left(a_q^+ a_q - B_q^+ B_q\right),$$

where

$$B_q^+ = \frac{1}{\sqrt{N}} \sum_n e^{iqnr_0} B_n^+.$$

In order to obtain a simple analytical expression, we make the usual continuum approximation corresponding to (9.71). This gives

$$\tilde{H} = \int_0^L dx 2\left\{(\varepsilon_0 - 2J)\varphi^+(x)\varphi(x) + Jr_0^2\frac{\partial\varphi^+}{\partial x}\frac{\partial\varphi}{\partial x}\right.$$
$$\left.-\frac{i\hbar v}{2}\left[\frac{\partial\varphi^+}{\partial x}\varphi(x) - \varphi^+(x)\frac{\partial\varphi}{\partial x}\right]\right\} + \sum_q \hbar(\omega_q - qv)a_q^+a_q +$$
$$\frac{1}{\sqrt{N}}\sum_q 2\left[g_1(q) + 2g_2(q)\right]\left(a_{-q}^+ + a_q\right)\int_0^L dx e^{ikx}\varphi^+(x)\varphi(x), \qquad (9.73)$$

where $\varphi(x)$ now represents the field operator corresponding to B_n in the continuum limit (whereas it only indicated a numerical value previously). Here $L = Nr_0$, $-\pi < kr_0 < \pi$, and $\omega_q \approx \sqrt{w/M}r_0|q|$, $x = nr_0$. Since the soliton excitation is connected with the deformation of intermolecular spacing, it is necessary to take this deformation into account in transforming the phonons between the reference frames. Such a transformation can be realized by means of the following transformation on the phonon operators

$$b_q = a_q - \frac{1}{\sqrt{N}}\alpha_q, \quad b_q^+ = a_q^+ - \frac{1}{\sqrt{N}}\alpha_q^*, \qquad (9.74)$$

which describe phonons relative to a chain with a particular deformation. Here b_q (b_q^+) is the annihilation (creation) operator of the new phonon. The vacuum state of the new phonons is

$$|\tilde{0}\rangle_{ph} = \exp\left\{\frac{1}{\sqrt{N}}\sum_q\left[\alpha_q(t)a_q^+ - \alpha_q^*(t)a_q\right]\right\}|0\rangle_{ph},$$

which is a coherent phonon state, *i.e.*, $b_q|\tilde{0}\rangle_{ph} = 0$. Performing the canonical transformation,

$$\varphi(x) = \sum_j A_jC_j(x), \quad \varphi^+(x) = \sum_j C_j^*(x)A_j^+, \qquad (9.75)$$

where

$$\int C_1^*(x)C_j(x)dx = \delta_{1j}, \quad \sum_j C_j^*(x')C_j(x) = \delta(x - x'), \quad \int dx|C_j(x)|^2 = 1,$$

and the operators A_s^+ and A_k^+ are the creation operators for the bound states $C_s(x)$ and delocalized state $C_k(x)$, respectively. Pang partially diagonalized the

Hamiltonian (9.73) using (9.74) – (9.75), to get

$$\tilde{H} = W + E_s A_s^+ A_s + \sum_k E_k A_k^+ A_k + \sum_q \hbar(\omega_q - qv) b_q^+ b_q$$
$$+ \frac{1}{\sqrt{N}} \sum_q \hbar(\omega_q - qv)(b_q^+ \alpha_q + \alpha_q^* b_q)(1 - A_s^+ A_s)$$
$$+ \frac{1}{\sqrt{N}} \sum_{kk'q} F(k, k', q)(b_{-q}^+ + b_q) A_{k'}^+ A_k$$
$$- \frac{1}{\sqrt{N}} \sum_{kq} \tilde{F}(k, q)(b_{-q}^+ + b_q)(A_s^+ A_{-k} - A_s^+ A_s), \tag{9.76}$$

and

$$C_s(x) = \sqrt{\frac{\mu_P}{2r_0}} \operatorname{sech}\left(\frac{\mu_P x}{r_0}\right) \exp\left(\frac{i\hbar x v}{2Jr_0^2}\right), \tag{9.77}$$

$$C_k(x) = \frac{\mu_P \tanh(\mu_P x / r_0) - ikr_0}{\sqrt{N r_0}(\mu_P - ikr_0)} \exp\left(ikx + \frac{i\hbar v x}{2Jr_0^2}\right), \tag{9.78}$$

with

$$E_s = 2\left(\varepsilon_0 - 2J - \frac{\hbar^2 V^2}{2Jr_0^2} - \mu_P J\right),$$
$$E_k = 2\left[\varepsilon_0 - 2J - \frac{\hbar^2 V^2}{2Jr_0^2} - J(kr_0)^2\right],$$

respectively. In (9.76),

$$F(k, k', q) = 2[g_1(q) + 2g_2(q)] \int_0^L dx e^{iqx} C_{k'}^*(x) C_k(x)$$
$$\approx 2[g_1(q) + 2g_2(q)] \left\{1 - \frac{i\mu_P q r_0}{[\mu_P + i(k+q)r_0][\mu_P - ikr_0]}\right\}$$
$$\approx F[k, (k+q), q] \delta_{k'k+q},$$

$$\tilde{F}(k, q) = 2[g_1(q) + 2g_2(q)] \int_0^L dx e^{iqx} C_{k'}^*(x) C_s(x)$$
$$= \frac{2\pi}{\sqrt{2\mu_P}} [g_1(q) + 2g_2(q)] \left(1 - \frac{iqr_0}{\mu_P - ikr_0}\right) \operatorname{sech}\left[\frac{\pi(k-q)r_0}{2\mu_P}\right],$$

where α_q is determined by the condition,

$$(\omega_q - vq)\alpha_q = (\omega_q + qv)\alpha_q^*,$$

which is required in order to determine the factor $(1 - A_s^+ A_s)$ in $\tilde{H}$ in (9.76). Thus we get

$$\alpha_q = \frac{i\pi(\chi_1 + \chi_2)}{w\mu_P(1 - s^2)} \sqrt{\frac{M}{2\hbar\omega_q}} (\omega_q + qv) \operatorname{csch}\left(\frac{\pi q r_0}{2\mu_P}\right),$$

and

$$W = \frac{2}{3}\mu_P^2 J.$$

For this α_q, $|\tilde{0}\rangle_{ph}$ is the coherent phonon state. However, unlike $C_k(x)$ in (9.78) which is an unbound state, the bound state $C_s(x)$ in (9.77) is self-consistent with the deformation. Such a self-consistent state of the intramolecular excitation and deformation forms a soliton which is stationary in its intrinsic reference frame.

For the soliton described by the state

$$|\Psi\rangle = \frac{1}{\sqrt{2!}}(A_s^+)^2 |0\rangle_{ex} |\tilde{0}\rangle_{ph},$$

the average energy is given by

$$\langle \Psi | \tilde{H} | \Psi \rangle = 2\left(\varepsilon_0 - 2J - \frac{\hbar^2 v^2}{4 J r_0^2}\right) - \frac{4}{3} J \mu_P^2. \tag{9.79}$$

Evidently, the average energy in the soliton state $|\Psi\rangle$, (9.79), is equal to the soliton energy E_{sol} given above, or the sum of the energy of the bound state in (9.77), E_s, and the deformation energy of the lattice, W, *i.e.*,

$$\langle \Psi | \tilde{H} | \Psi \rangle = E_{sol} = E_s + W.$$

This is an interesting result. It shows clearly that the quasi-coherent soliton formed by this mechanism is a self-trapping state of the two excitons plus the corresponding deformation of the lattice. However, it should be noted that $|\Psi\rangle$ is not an exact eigenstate of $\tilde{H}$, due to the presence of $A_k^+ A_s$ and $A_s^+ A_{-k}$ in $\tilde{H}$.

For discussion of the decay rate and lifetime of the soliton state, it is convenient to write $\tilde{H}$ in (9.76) as $H_0 + V_1 + V_2$, as suggested by Cottingham and Schweitzer, where

$$H_0 = W + E_s A_s^+ A_s + \sum_k E_k A_k^+ A_k \sum_q \hbar(\omega_q - vq) b_q^+ b_q + \\ + \frac{1}{\sqrt{N}} \sum_q \hbar(\omega_q - vq)(\alpha_q b_q^+ + \alpha_q^* b_q)(1 - A_s^+ A_s), \tag{9.80}$$

$$V_1 = \frac{1}{\sqrt{N}} \sum_{kk'q} F(k, k+q, q)(b_{-q}^+ + b_q) A_{k'}^+ A_k, \tag{9.81}$$

$$V_2 = \frac{1}{N} \sum_{kq} \tilde{F}(k, q)(b_{-q}^+ + b_q)(A_s^+ A_k - A_s^+ A_{-k}). \tag{9.82}$$

Here, H_0 describes the relevant quasi-particle excitations in the protein. This is a soliton together with phonons relative to the distorted lattice. The resulting delocalized excitation belongs to an exciton-like band with phonons relative to a uniform lattice. The bottom of the band of the latter is at the energy $4J\mu_p^2/3$ relative to the soliton, in which the topological stability associated with removing the lattice distortion is included.

We now calculate the decay rate of the soliton by using (9.80) and V_2 in (9.82), following the quantum perturbation of Schweitzer *et al.*. We first derive a formula for the decay rate of a soliton containing n quanta in a general system in which the three terms in (9.57) are replaced by the $(n+1)$ terms of a coherent state,

$$\exp\left[\sum_n \varphi_n(t)B_n^+\right]|0\rangle_{ex}.$$

After that, we obtain again the decay rate of the soliton with two-quanta. Thus, the above choice for H_0 is such that $|n\rangle$ is the ground state of H_0, with energy $W+nE_s'$, in the subspace of n excitations, *i.e.*,

$$\left\langle n\left|\sum_i B_i^+B_i\right|n\right\rangle = \left\langle n\left|\left(A_s^+A_s+\sum_k A_k^+A_k\right)\right|n\right\rangle = n.$$

In this subspace the eigenstates have the simple form

$$|n-m,k_1,k_2,\cdots,k_m,\{n_q\}\rangle\frac{1}{\sqrt{(n-m)!}}(A_s^+)^{n-m}A_{k_1}^+A_{k_2}^+\cdots A_{k_m}^+|0\rangle_{ex}\times \prod_q\frac{(d_q^+)^{n_q}}{\sqrt{n_q!}}|\tilde{0}\rangle_{ph}^{n-m},$$

where

$$d_q=b_q+\frac{m}{n}\frac{1}{\sqrt{N}}\alpha_q=a_q-\frac{n-m}{n}\frac{1}{\sqrt{N}}\alpha_q,\quad d_q|\tilde{0}\rangle_{ph}^{n-m}=0,$$

n and m $(m\le n)$ are integers. The corresponding energy of the systems is

$$E^{(0)}_{n-m;k_1,\cdots,k_{m1};\{n_q\}}=\left[1-\left(\frac{m}{n}\right)^2\right]W+(n-m)E_s'+\sum_{j=1}^m E_{k_j}'+\sum_q\hbar(\omega_q-vq)n_q.$$

Here E_s' is the energy of a bound state with one exciton, E_k' is the energy of the an unbound (delocalized) state with one exciton. When $m=0$, the excitation state is a n-type soliton plus phonons related to the chain with a deformation corresponding to the n-type soliton. For $m=n$, the excited states are delocalized and the phonons are relative to a chain without any deformation. Furthermore, except for small k, the delocalized states are approximations of ordinary excitons. Thus the decay of the soliton is nothing but a transition from the initial state with the n-type soliton plus the phonons,

$$|n\rangle = \frac{1}{\sqrt{n!}} \prod_q \frac{(b_q^+)^{n_q}}{\sqrt{n_q!}} (A_s^+)^n |0\rangle_{ex} |\tilde{0}\rangle_{ph}, \tag{9.83}$$

with energy

$$E_s\{n_q\} = W + nE_s' + \sum_q \hbar(\omega_q - vq)n_q,$$

to the final state with delocalized excitons and the same phonons,

$$|\alpha k\rangle = \prod_q \frac{(a_q^+)^{n_q}}{\sqrt{n_q!}} |0\rangle_{ph} (A_k^+)^n |0\rangle_{ex}, \tag{9.84}$$

with energy

$$E_k\{n_q\} = nE_k' + \sum_q \hbar(\omega_q - vq)n_q$$

due to V_2, in the perturbation interaction $V = V_1 + V_2$. In this case, the initial phonon distribution is taken to be at thermal equilibrium. In the lowest order perturbation theory, the probability of the above transition is given by

$$\bar{W} = \frac{1}{\hbar^2} \int_0^t dt' \int_0^t dt'' \sum_{\alpha k'} \sum_l P_l^{(ph)} \left\langle n \left| \exp\left(\frac{iH_0 t''}{\hbar}\right) V_2 \exp\left(\frac{iH_0 t''}{\hbar}\right) \right| \alpha k' \right\rangle \times \left\langle \alpha k' \left| \exp\left(\frac{iH_0 t'}{\hbar}\right) V_2 \exp\left(\frac{-iH_0 t'}{h}\right) \right| n \right\rangle. \tag{9.85}$$

We can calculate the transition probability of the soliton resulting from the perturbation potential, $(V_1 + V_2)$, using the first-order perturbation theory. Following the procedure of Cottingham and Schweitzer, we estimate only probability for the transition from the soliton state to the delocalized exciton states due to the potential V_2, which can be treated satisfactorily by perturbation theory since the coefficient $\tilde{F}(k, q)$ defined in (9.76) is proportional to an integral over the product of the localized state and a delocalized state, and therefore is of order $1/\sqrt{N}$. The V_1 term in the Hamiltonian represents interaction between the delocalized excitons and the phonons. The main effect of V_1 is to modify the spectrum of the delocalized excitations in the weak coupling limit ($J\mu_p/K_B T_0 \ll 1$, T_0 is defined below). This results in a shift in the energies of delocalized excitons and phonons, and a finite lifetime. These effects are ignored in our calculation since they are only of second order in V_1.

The sum over l in (9.85) is over an initial set of occupation numbers for phonons relative to the distorted lattice with probability distribution P_1^{ph}, which is taken to be the thermal equilibrium distribution for a given temperature T. Using (9.80) and (9.82), we can get

$$\bar{W} = \frac{1}{\hbar^2}\frac{\pi^2}{2n\mu_1 N^2}\sum_k\sum_{k'}\sum_{k''}\left[g_1^*(k)+2g_2^*(k)\right]\left[g_1(k'')+2g_2(k'')\right]$$
$$\times\frac{(kr_0)(k''r_0)}{(n\mu_1)^2+(k'r_0)^2}\,\mathrm{sech}\left[\frac{\pi r_0}{2n\mu_1}(k-k')\right]\mathrm{sech}\left[\frac{\pi r_0}{2n\mu_1}(k''-k')\right]$$
$$\times\int_0^t dt'\int_0^t dt''\exp\left\{-\frac{i}{\hbar}\left[n\left(n^2-\frac{2}{3}n\right)\mu_1^2 J+nJ(k'r_0)^2\right](t'-t'')\right.$$
$$\times\left\langle\!\!\left\langle\exp\left[i\sum_q(\omega_q-qv)b_q^+b_q(t'-t'')\right](b_k^+ + b_{-k})\right.\right.$$
$$\left.\left.\left.\times\exp\left[i\sum_q(\omega_q-qv)a_q^+a_q(t'-t'')\right](b_{-k'}^+ + b_{k'})\right\rangle\!\!\right\rangle\right\},$$

where

$$g_1(k)+2g_2(k) = 2\chi_1\sqrt{\frac{\hbar}{2M\omega_k}}\left[A(\cos(r_0k)-1)+i(A+1)\sin(r_0k)\right]$$
$$\approx 2i(A+1)(r_0k)\chi_1\sqrt{\frac{\hbar}{2M\omega_k}},$$
$$\mu_1 = \frac{\chi_1^2(1+A^2)}{\omega(1-s^2)J},\quad A = \frac{\chi_2}{\chi_1},$$

with

$$\langle\!\langle\tilde{A}\rangle\!\rangle = \frac{\mathrm{Tr}\left\{\tilde{A}\exp\left[-\beta\sum_q\hbar(\omega_a-qv)b_q^+b_q\right]\right\}}{\mathrm{Tr}\left\{\exp\left[-\beta\sum_q\hbar(\omega_a-qv)b_q^+b_q\right]\right\}}$$

Here $\beta = 1/k_BT$, A is the ratio of the new nonlinear interaction term to that in the Davydov model.

To estimate the soliton lifetime, we are interested in the long-time behavior of $d\bar{w}/dt$. By straightforward calculation, the average transition probability or decay rate of the soliton was obtained by Pang and it is given by

$$\Gamma_n \lim_{t\to\infty}\frac{d\bar{W}}{dt} = \frac{2}{n\mu_1\hbar^2}\frac{\pi^2}{N^2}\sum_{kk'}|g_1(k)+2g_2(k)|^2\frac{(r_0k)^2\mathrm{sech}^2[\pi(k-k')r_0/2n\mu_1]}{(n\mu_1)^2+(k'r_0)^2}$$
$$\times\Re\int_0^\infty dt\exp\left\{-i\left[nJ(k'r_0)^2+n\left(n^2-\frac{2}{3}n\right)\frac{\mu_1^2Jt}{\hbar}+R_n(t)+\xi_n(t)\right]\right\}$$
$$\times\frac{\exp[i(\omega_k-kv)t]}{\exp[\beta\hbar(\omega_k-kv)]-1}, \tag{9.86}$$

where

$$R_n(t) = -\frac{1}{n^2 N}\sum_k |\alpha_k|^2 \left[i - e^{-i(\omega_k - kv)t}\right],$$

$$\xi_n(t) = -\frac{4}{n^2 N}\sum_k \frac{|\alpha_k|^2 \sin^2[(\omega_k - kv)t/2]}{\exp[\beta\hbar(\omega_k - kv) - 1]}.$$

This is a general analytical expression for the decay rate of a soliton containing n quanta at any temperature within the lowest order perturbation theory.

From (9.86), we can see that Γ_n, $R_n(t)$, $\xi_n(t)$ and $\mu = n\mu_1$ given above all change with increasing number of quanta, n. Thus, Γ_n is different for different n. In the following, we derive an explicit expression for the decay rate of the soliton with two-quanta ($n = 2$). In this case, the integral in the decay rate expression can be evaluated explicitly, and $R_2(t)$ and $\xi_2(t)$ in (9.86) can be obtained using approximation method. The results are for $v \to 0$ and $\omega_q \to \sqrt{w/M}$ are given in terms of the digamma function. We are interested in the long-time steady behavior. Thus we take the limit $t \to \infty$. Then

$$R_2(t) = -R_0\left[\ln\left(\frac{1}{2}\omega_\alpha t\right) + 1.578 + \frac{1}{2}i\pi\right],$$

$$\xi_2(t) \approx -\frac{\pi R_0 k_B T t}{\hbar},$$

where

$$R_0 = \frac{4(\chi_1 + \chi_2)^2}{\pi\hbar w}\sqrt{\frac{M}{w}} = \frac{2J\mu_P r_0}{\pi\hbar v_0}, \quad \omega_\alpha = \frac{2\mu_P}{\pi}\sqrt{\frac{w}{M}}, \quad T_0 = \frac{\hbar\omega_\alpha}{K_B}.$$

We have used $\coth(\omega_\theta t/2) \approx 1$ in obtaining the above results. That is,

$$\lim_{t\to\infty} \xi_2(t) = -\eta t, \quad \eta = \frac{\pi R_0}{\beta\hbar} = \frac{\pi R_0 k_B T}{\hbar}.$$

For protein molecules, we have $R_0 < 1$, $T_0 < T$ and $R_0 T/T_0 < 1$. In this case, the decay rate of the soliton with two quanta was derived by Pang and it is given by

$$\Gamma_2 = \lim_{t\to\infty}\frac{d\bar{W}}{dt} = \frac{2}{\mu_P}\left(\frac{\pi}{N}\right)^2 \sum_{kk'} \frac{(kr_0)^2|g_1(k) + 2g_2(k)|^2 \text{sech}^2[(\pi r_0/2\mu_P)(k-k')]}{[\mu_P^2 + (k'r_0)^2][\exp(\beta\hbar\omega_k) - 1]}$$
$$\times \frac{1}{(2.43\omega_\alpha)^{R_0}} \frac{\{\eta^2 + [4\mu_P^2 J/3 + 2(k'r_0)^2 J - \hbar\omega_k]^2/\hbar^2\}^{(1+R_0)/2}}{\hbar^2\eta^2 + [4\mu_P^2 J/3 + 2(k'r_0)^2 J - \hbar\omega_k]^2}$$
$$\times \left\{1 - \frac{1}{2}\left[\frac{R_0\pi}{2} + (1 - R_0)\frac{4\mu_P^2 J/3 + 2(k'r_0)J - \hbar\omega_k}{\hbar\eta}\right]^2\right\}. \quad (9.87)$$

This is evidently different from that of the Davydov model with one quantum ($n =$

1) obtained by Cottingham and Schweitzer, which is of the form

$$\Gamma_D = \frac{2\pi\chi_1^2}{N\hbar\mu_D}\sum_{kk'}\frac{(kr_0)^2\sin^2(kr_0)\mathrm{sech}^2[(\pi r_0/2\mu_D)(k-k')]}{2M\omega_k[\mu_D^2+(k'r_0)^2][\exp(\beta\hbar\omega_k)-1]}$$
$$\times\left(\frac{\eta_D}{\omega_\alpha^D}\right)^{R_0^D}\frac{\hbar^2\eta_D}{\hbar^2\eta_D^2+[J\mu_D^2/3+J(k'r_0)^2-\hbar\omega_k]}, \tag{9.88}$$

where

$$\eta_D = \frac{\pi R_0^D K_B T}{\hbar}, \quad R_0^D = \frac{2\chi_1^2}{\pi\hbar w}\sqrt{\frac{M}{w}}, \quad \omega_\alpha^D = \frac{2\mu_D}{\pi}\sqrt{\frac{M}{w}}.$$

Equations (9.87) and (9.88) are different not only in their values, but also in the factors they contain. In (9.87), there is an additional factor,

$$1-\frac{1}{2}\left\{\frac{R_0\pi}{2}+\frac{1-R_0}{\hbar\eta}\left[\left(\frac{4}{3}\mu_P^2J+2(k'r_0)^2J-\hbar\omega_k\right)\right]\right\}^2,$$

and the factor $(\eta_D/\omega_\alpha)^{R_0^D}\eta_D$ in (9.88) is replaced by

$$\frac{1}{(2.43\omega_d)^{R_0}}\left\{\eta^2+\frac{1}{\hbar^2}\left[\frac{4}{3}\mu_P^2J+2(k'r_0)^2J-\hbar\omega_k\right]^2\right\}^{(1+R_0)/2}$$

in (9.87), as a result of the two-quanta nature of the wavefunction and the additional interaction term in the Hamiltonian in the improved model. Unlike that in the Davydov model, η, R_0 and T_0 in (9.87) are not necessarily small. Using the parameters of protein molecules given in Section 9.5, we find that $\eta = 6.527\times10^{13}$ s^{-1}, $R_0 = 0.529$ and $T_0 = 294$ K for the soliton in the improved model. When compared to $R_0 = 0.16$, $T_0 = 95$ K, and $\eta_D = 2.096\times10^{13}$ s^{-1} for the Davydov soliton at $T = 300$ K, we see that the values of η, R_0 and T_0 in the improved model are about 3 times larger than the corresponding values in the Davydov model due to the increases of μ_P and of the nonlinear interaction coefficient G_P.

Using (9.87) and the commonly accepted values for the various parameters of protein given in Section 9.5, we can numerically compute the decay rate, Γ_2, and the lifetime of the soliton, $\tau = 1/\Gamma_2$, for the α-helical protein molecules. For wavevectors in the Brillouin zone and $v = 0.2v_0$, values of Γ_2 fall between 1.54×10^{10} s^{-1} and 1.89×10^{10} s^{-1}. This corresponds to a soliton lifetime, τ, between 0.53×10^{-10} s and 0.65×10^{-10} s at $T = 300$ K, or $\tau/\tau_0 = 510-630$, where $\tau_0 = r_0/v_0$ is the time for the soliton to travel a distance of one lattice spacing at the speed of sound, which is equal to $\sqrt{M/w} = 0.96\times10^{-13}$ s. In this period, the soliton, traveling at two tenths of the speed of sound in the chain, would travel several hundreds of lattice spacings. That is several hundred times of that of the Davydov soliton for which $\tau/\tau_0 < 10$ at 300 K. In other words, the Davydov soliton travels at one half of the sound speed can cover less than 10 lattice spacings during its lifetime. The soliton excitation in the improved model has a sufficiently long lifetime to be a successful carrier of bio-energy. Therefore, the quasi-coherent soliton is a viable mechanism for bio-energy transport at biological temperature for the above ranges of parameters.

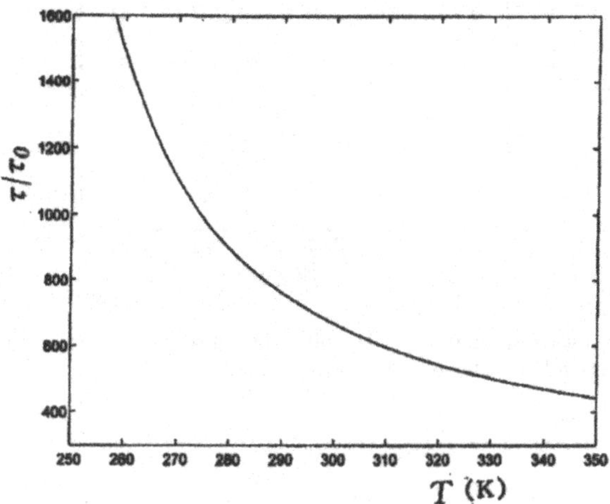

Fig. 9.5 Soliton lifetime τ, relative to τ_0, as a function of temperature T using parameters characteristic of the α-helical molecule.

We are interested in the relation between the lifetime of the quasi-coherent soliton and temperature. Figure 9.5 shows the relative lifetime τ/τ_0 of the soliton versus temperature T for a set of widely accepted values for the various parameters given in Section 9.5. Since it is assumed that $v < v_0$, the soliton will not travel the entire length of the chain unless τ/τ_0 is large compared with L/r_0, where $L = Nr_0$ is a typical length of protein molecular chains. Hence for $L/r_0 \approx 100$, $\tau/\tau_0 > 500$ is a reasonable criterion for the soliton to be a possible mechanism of bio-energy transport in protein molecules. The lifetime of the quasi-coherent soliton shown in Fig. 9.5 decreases rapidly as temperature increases. But below $T = 310$ K, it is still large enough to fulfill the requirement. Thus the soliton can play an important role in biological processes.

For comparison, we show simultaneously, in Fig. 9.6, $\log(\tau/\tau_0)$ versus temperature for the Davydov soliton and the soliton with a quasi-coherent two-quanta state. We find that τ/τ_0 in the two models are very different. The τ/τ_0 of the Davydov soliton is too small, and it can only travel a distance of less than ten lattice spacings at half the speed of sound in protein molecules. Thus the Davydov soliton is ineffective for biological processes.

Dependency of the soliton lifetime on other parameters can also be studied by using (9.87). Using parameters around the accepted values given earlier, it was found that the lifetime of the soliton increases with increasing $(\chi_1 + \chi_2)$ (Pang 1987, 1989a, 1990, 1993e, 1999, 2001c).

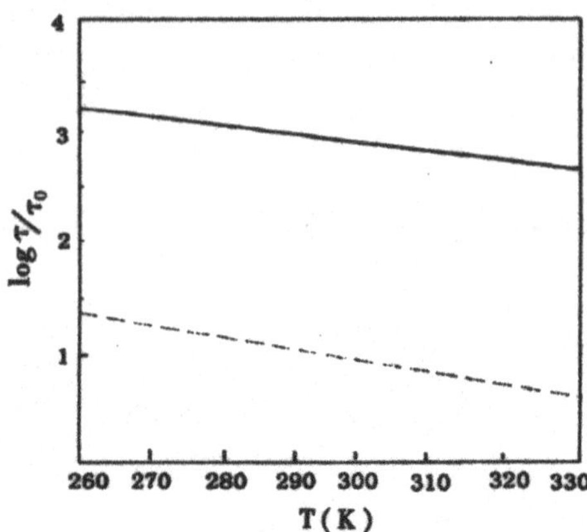

Fig. 9.6 $\log(\tau/\tau_0)$ versus temperature. The solid line represents result of the new model, while the dashed line corresponds to that of the Davydov model.

9.7 Effects of Structural Disorder and Heart Bath on Exciton Localization

The results discussed so far were all obtained analytically, in which the protein molecules were regarded as a periodic system and some approximate methods such as long-wavelength approximation, continuum approximation, or long-time approximation, were used. Some average values as given in Section 9.5 were used for various parameters of the protein molecules in the calculations. However, in reality, a protein is not a periodic system, but rather it consists of 20 different amino acid residues with molecular weights ranging from 75 m_p (glycine) to 204 m_p (tryptophane). This corresponds to a variation between $0.67\bar{M}$ and $1.80\bar{M}$, where m_p is the mass of the proton, $\bar{M}$ is the average mass of the amino acid molecules. Hence, there exists structure disorder in proteins. Careri *et al.* demonstrated that a relatively small amounts of disorder in amorphous film of acetanilide (ACN), a protein-like crystal, is enough to destroy the spectral signature of a "soliton". Therefore, it is necessary to investigate the influences of structural disorder on the stability of soliton in proteins at biological temperature. This will be the topic of this section. A numerical simulation based on the Runge-Kutta method will be used obtain related results in which these disorder effects of the protein molecules are considered, without using approximation.

The transformation, $a_n(t) \to a_n(t)\exp[iR(t)t/h]$, enables us to eliminate the term of $R(t)a_n(t)$ in (9.61), where $a_n(t) = \varphi_n(t)$ is a complex function of time t, and can be represented by $a_n(t) = a(t)r_n + ia(t)i_n$. Here $a(t)r_n$ and $a(t)i_n$ are real and imaginary parts of $a_n(t)$. Equations (9.60) and (9.61) can then be approximately written as

$$\hbar \dot{a r}_n = -J(ai_{n+1} + ai_{n-1}) + \chi_1(q_{n+1} - q_{n-1})ai_n$$
$$+\chi_2(q_{n+1} - q_n)(ai_{n+1} + ai_{n-1}), \tag{9.89}$$
$$-\hbar \dot{a i}_n = -J(ar_{n+1} + ar_{n-1}) + \chi_1(q_{n+1} - q_{n-1})ar_n$$
$$+\chi_2(q_{n+1} - q_n)(ar_{n+1} + ar_{n-1}), \tag{9.90}$$
$$\dot{q}_n = \frac{y_n}{M}, \tag{9.91}$$
$$\dot{y}_n = w(q_{n+1} - 2q_n + q_{n-1}) + 2\chi_1(ar_{n+1}^2 + ai_{n+1}^2 - ar_{n-1}^2 - ai_{n-1}^2)$$
$$+4\chi_2[ar_n(ar_{n+1} - ar_{n-1}) + ai_n(ai_{n+1} - ai_{n-1})], \tag{9.92}$$

where $|a_n|^2 = |ar_n|^2 + |ai_n|^2$ and $q_n = \beta_n$. Equations (9.89) – (9.92) are the equations of motion which determine the states of the soliton. We will obtain their solutions numerically using the fourth-order Runge-Kutta method. Obviously, there are four equations for every peptide group. Therefore, for a protein molecule consisting of N amino acids, we have to solve a system of $4N$ equations. To use the Runge-Kutta method, we need first to discretize the equations, in which the time is discretized and denoted by j, and the step size of the space variable is denoted by h.

In the numerical simulation, Pang *et al.* used eV as the unit of energy, Å for length and ps for time. The following are details of the numerical simulation: a time step size of 0.0195 ps, the total energy $E = \langle\Phi(t)|H|\Phi(t)\rangle$ was conserved to within 0.0012 eV, (a possible imaginary part of the energy can be developed during the simulation due to numerical inaccuracy which was kept to below 0.001 eV), the norm was conserved up to 0.3 pp (parts per million). A fixed chain of N units was used and an initial excitation in the form of $a_n(0) = A\,\mathrm{sech}[(n-n_0)(\chi_1+\chi_2)^2/4JW]$, where A is a normalization constant, was assumed. Pang *et al.* used $N = 50$ in the simulation. For the lattice, $q_n(0) = \pi_n(0) = 0$ was applied. The simulation was performed using a data parallel algorithm and the MATLAB software.

9.7.1 *Effects of structural disorder*

Using average values of the various parameters given earlier in Section 9.5, Pang *et al.* first computed the solution of the above equations (9.89) – (9.92) in the case of a free and uniform periodic chain. The result is shown in Fig. 9.7. From this figure, we see that the amplitude and energy of the solution remain constant throughout the course of motion. The collision behavior of two solitons, setting up from opposite ends of the chain, can be investigated and the results are shown in Fig. 4.4. From this figure, we see that the two solitons go through each other without scattering. Therefore, the solution of the above equations are truly "soliton". We can thus confirm that the exciton described by the above nonlinear Schrödinger equation is indeed localized as a soliton in the case of a free uniform periodic protein molecules.

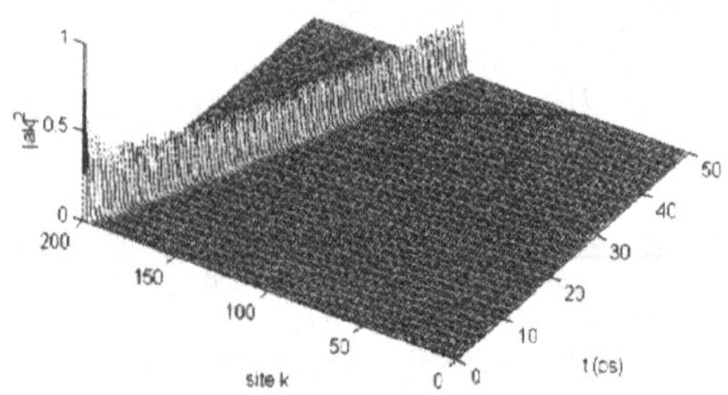

Fig. 9.7 Soliton in a free and uniform periodic chain.

However, what would be the behaviors of the soliton when there are "structural disorders" in the protein molecules? To answer this question, we introduce structural disorder into the systems and carry out the simulation. A random number generator can be used to produce random sequences for the various parameters along the protein molecular chains.

We first consider the influence of a disorder in mass sequence on the stability of the soliton. This disorder is modeled by a nonuniformity at given sites or a deviation from the uniform mass distribution $\bar{M} = 114m_p = (1.17 - 1.91) \times 10^{-25}$ kg per amino acid residue. In a first series of calculations we study the effect of the mass nonuniformity at given sites on the soliton. This mass disorder may be caused by amino acid side groups and local geometric distortion of molecular chains, imported impurity, or some other molecules bound to the protein (reactive centers such as heme groups). As an example, only the mass at site 49 was increased, all other masses were kept at $\bar{M}$. For $M_{49} = 100\bar{M}$, the soliton is still very stable. Very surprisingly, up to $100000\bar{M}$, no obvious perturbations and decays appear in the motion of the soliton. These results are shown in Fig. 9.8. From these results, one can conclude that general disorder of mass sequence of the amino acid residues at a given site does not disturb the soliton at all. This, however, may not be true in the case of vast impurity or under the influences of other disorders.

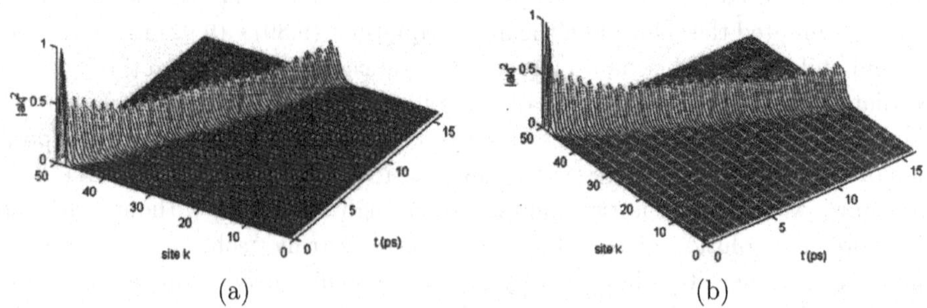

Fig. 9.8 Soliton for (a) $M_{49} = 100\bar{M}$ and (b) $M_{49} = 100000\bar{M}$.

The influence of a random series of masses distributed along the whole chain can be studied. Here we introduce a small parameter α_k to denote the variation of mass at each point in the molecular chain, *i.e.*, $M_k = \alpha_k \bar{M}$. The values of α_k is randomly generated with equal probability within a prescribed interval. The aperiodicity due to α_k, for example, $0.67 \leq \alpha_k \leq 400$, does not affect the stability of the soliton. However, when the mass variation is sufficiently large, for example, $0.67 \leq \alpha_k \leq 700$, the vibrational energy becomes dispersed, as shown in Fig. 9.9. The interval $0.67 \leq \alpha_k \leq 400$ over which the soliton can move unperturbed is evidently larger than the variation of masses of the natural amino acids ($0.67 \leq \alpha_k \leq 1.80$). Therefore, the soliton is very robust against mass disorder of natural amino acids.

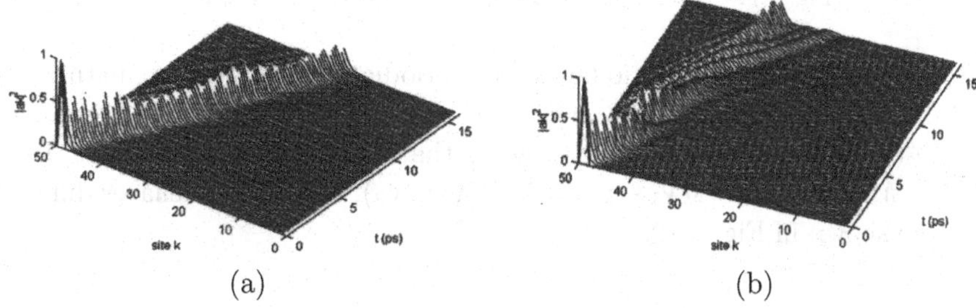

(a) (b)

Fig. 9.9 Soliton with (a) $0.67 \leq \alpha_k \leq 400$, and (b) $0.67 \leq \alpha_k \leq 700$.

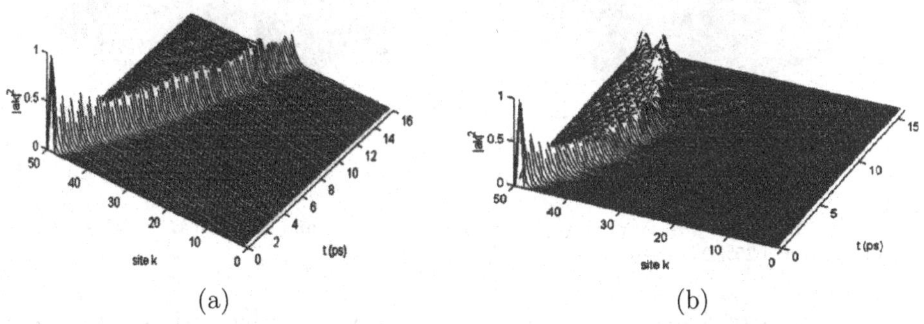

(a) (b)

Fig. 9.10 Soliton with (a) $\Delta w = \pm 45\% \bar{w}$ and (b) $\Delta w = 75\% \bar{w}$.

We can also investigate the influence of fluctuations in the spring constant w, arising from the structural disorder, on the stability of the soliton. For a random variation up to $\pm 45\% \bar{w}$, no change is found in the dynamics of the soliton. For $\pm 55\% \bar{w}$, the soliton velocity is only slightly reduced compared with the case of $\bar{w}$. Finally, for $\pm 75\% \bar{w}$, the soliton disperses slowly and its propagation is irregular. The results are shown in Fig. 9.10. If in addition w is aperiodic, the soliton is stable up to $\pm 40\% \bar{w}$, and then at $45\% \bar{w}$ the slow dispersive phenomenon occurs.

The soliton, however, is more sensitive to the variation in J, resulting also from structural disorder. For variation in J alone the soliton is stable for a change up to $9\%\bar{J}$, but it disperses at $\triangle J = \pm 15\%\bar{J}$, as shown in Fig. 9.11.

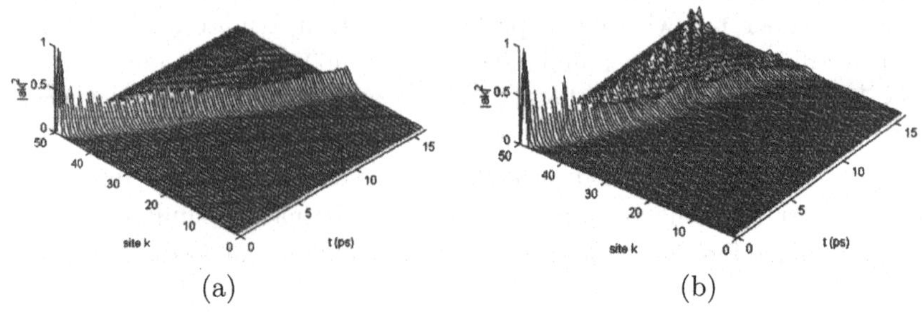

Fig. 9.11 Soliton at (a) $\triangle J = \pm 9\%\bar{J}$, and (b) $\triangle J = \pm 15\%\bar{J}$.

If $(\chi_1 + \chi_2)$ alone is aperiodic or is aperiodic together with a natural mass variation, the states of the soliton will be changed. $(\chi_1 + \chi_2)$ can be varied up to $\pm 25\%(\bar{\chi}_1 + \bar{\chi}_2)$ without any change in the dynamics of the soliton. But for $0.67\bar{M} \leq M < 2\bar{M}$ and $\triangle(\chi_1 + \chi_2) = 25\%(\bar{\chi}_1 + \bar{\chi}_2)$, the soliton behaves differently. These are shown in Fig. 9.12.

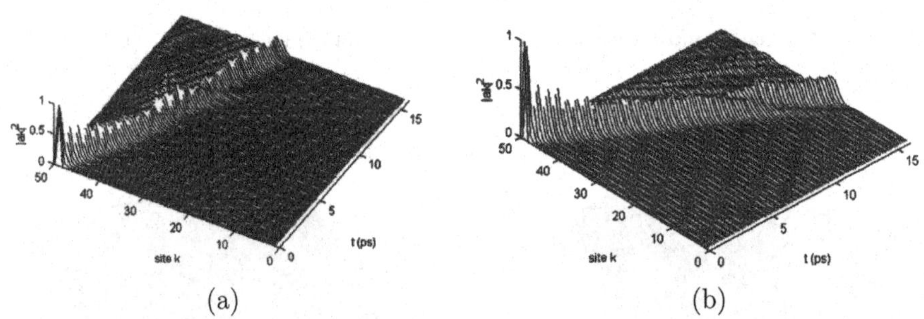

Fig. 9.12 Soliton at (a) $\triangle(\chi_1 + \chi_2) = \pm 25\%\triangle(\bar{\chi}_1 + \bar{\chi}_2)$ and (b) $0.67\bar{M} < M < 2\bar{M}$ and $\triangle(\chi_1 + \chi_2) = \pm 25\%\triangle(\bar{\chi}_1 + \bar{\chi}_2)$.

In the case of changing the ground state energy $\triangle\varepsilon_0$, which arises from different amino acid side groups and corresponding local geometric distortions due to imported impurities, it can be found that for an isolated impurity in the middle of the chain which results in a change of the ground state energy, $\triangle\varepsilon_0 = \varepsilon\delta_n$, the soliton can pass the impurity only if $\varepsilon < 1$ meV. In other cases, it is reflected or dispersed. In the case of a random fluctuation but $\triangle\varepsilon_0 = \varepsilon|\beta_n|$, $\varepsilon < 1$ meV, $|\beta_n| \leq 0.5$, the soliton can pass the chain. The results are shown in Fig. 9.13. For higher values of ε, the excitation disperses.

Finally, if all types of disorder, variation of mass, fluctuations in the spring constant, dipole-dipole interaction constant, coupling constant, and ground state energy, are taken into account, for example, $\triangle w = \pm 10\%\bar{w}$, $\triangle J = \pm 5\%\bar{J}$, $\triangle(\chi_1 +$

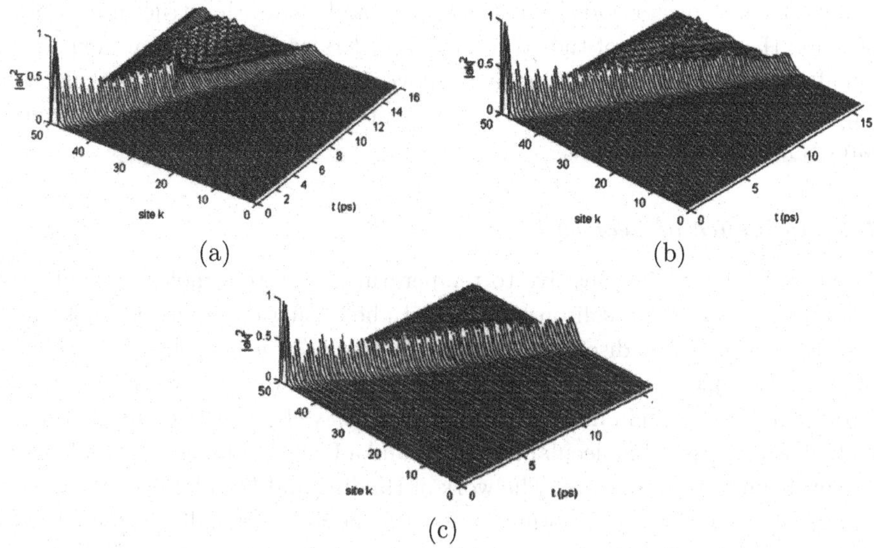

Fig. 9.13 The soliton for (a) $\triangle\varepsilon_0 = \varepsilon\delta_n$, $\varepsilon = 0.5$ meV, (b) $\triangle\varepsilon_0 = \varepsilon|\beta_n|$, $\varepsilon < 1$ meV, $|\beta_n| \leq 0.5$, and (c) $\triangle w = \pm 10\%\bar{w}$, $\triangle J = \pm 5\%\bar{J}$, $\triangle(\chi_1+\chi_2) = \pm 5\%(\bar{\chi}_1+\bar{\chi}_2)$, $0.67\bar{M} < M < 2\bar{M}$, $\triangle\varepsilon_0 = \varepsilon|\beta_n|$, $\varepsilon = 0.4$ meV, $|\beta_n| \leq 0.5$.

$\chi_2) = \pm 5\%(\bar{\chi}_1+\bar{\chi}_2)$, $0.67\bar{M} < M < 2\bar{M}$ and $\varepsilon = 0.4$ meV ($\triangle\varepsilon_0 = \varepsilon|\beta_n|$) occur simultaneously, it is observed that the soliton is still stable, as show in Fig. 9.13(c). Therefore the soliton can be said very robust against these structural disorders.

We have seen that the soliton is stable even when the structure disorders are large, for example, for a mass change at a given site up to $M_n = 100000\bar{M}$, a random mass nonuniformity in the range of $0.67 < \alpha_k < 400$, a change in the spring constant up to $\pm 40\%\bar{w}$, fluctuations in the interaction constant J, coupling constant $(\chi_1+\chi_2)$ and ground state energy ε_0 up to $10\%\bar{J}$, $\pm 25\%(\bar{\chi}_1+\bar{\chi}_2)$ and $\triangle\varepsilon_0 = \varepsilon|\beta_n|$, $\varepsilon = 1$ meV, $|\beta_n| \leq 0.5$ and $\triangle\varepsilon_0 = \varepsilon\delta_n$, $\varepsilon = 0.5$ meV, respectively; or a combination of fluctuations with $\triangle w = \pm 10\%\bar{w}$, $\triangle J = \pm 5\%\bar{J}$, $\triangle(\chi_1+\chi_2) = \pm 5\%(\bar{\chi}_1+\bar{\chi}_2)$, $0.67\bar{M} < M < 2\bar{M}$, $\triangle\varepsilon_0 = \varepsilon|\beta_n|$, $\varepsilon = 0.4$ meV, $|\beta_n| \leq 0.5$. Therefore, the new soliton is very robust against these structural disorders in the protein molecules. However, the actual degree of disorder in protein molecules is unknown. Also protein molecules are bio-self-organizations with high-order and coherent feature, which are necessary condition for a protein to perform its biological functions. A large structural disorder in a biological protein means a degeneration of the structure and the disability of the functions of the protein molecules which lead to disease of living systems. Therefore, larger structural disorders are not likely to occur in protein molecules. Then, discussing effects of large disorder in all parameters on the stability of soliton may not be relevant and practical in normal biological protein molecules. Since proteins are not simply random heterpolymers. The amino acids are not free particles, but are covalently bonded to the main polypeptide chains.

Its nonuniformity or aperiodicity is thus small, and so are the structural disorders. Therefore, the influences of the structural disorders on the soliton are expected to be small. Nevertheless, the results given above indicate that soliton is stable in case of larger structural disorders, and the new soliton is robust against such structure disorders in protein molecules.

9.7.2 *Influence of heat bath*

Soliton are known to be sensitive to temperature in protein molecules. The thermal stability of Davydov soliton has been studied using different methods. In the following, we focus our discussion on the effects of (1) a head bath to which the proteins are coupled to; and (2) of structural disorder due to a non-uniform mass distribution in the protein molecule. Both effects have been studied by Halding and Lomdahl using classical molecular dynamics and a Lennard-Jones potential between the peptide units of an α-helix. The work of Halding and Lomdahl is an example of the application of a classical thermalisation scheme to a classically described lattice. Lomdahl and Kerr, Lawrence and co-workers studied the stability of the Davydov soliton at finite temperature and found that at 300 K the Davydov soliton is destroyed. The method of Lomdahl and Kerr will be used here to treat the influence of the heat bath on the new soliton.

In accordance with Lomdahl and Kerr, Pang *et al.* considered a random-noise force $F_n(t)$ and a dissipation force, $M\Gamma\dot{q}_n$, resulting from the interaction between the heat bath at temperature T and the protein molecules, and included these forces in the displacement equation of amino acid, (9.60), which now can be written as

$$\begin{aligned} M\ddot{q}_n(t) &= w[q_{n+1}(t) - 2q_n(t) + q_{n-1}(t)] + 2\chi_1\left[|a_{n+1}|^2 - |a_{n-1}|^2\right] \\ &\quad +2\chi_2\left\{a_n^*(t)[a_{n+1}(t) - a_{n-1}(t)] + a_n(t)[a_{n+1}^*(t) - a_{n-1}^*(t)]\right\} \\ &\quad -M\Gamma\dot{q}_n + F_n(t), \end{aligned} \tag{9.93}$$

where Γ is a vibrational dissipation coefficient of amino acids. The correlation function of the random force can be represented by

$$\langle F(x,t)F(0,0)\rangle = 2M\Gamma K_B T\frac{\delta(x)\delta(t)}{\tau}.$$

It is assumed that the random deviations obey the normal distribution with a standard deviation of $\sqrt{\sigma}$ and a zero expectation value. That is,

$$N(F_n) = \frac{1}{\sqrt{2\pi\sigma}}e^{-F_n^2/2\sigma},$$

where $\sigma = 2MK_BT\Gamma/\tau$ and τ is a time constant, Γ is the reciprocal time constant of the heat bath. It can be shown that

$$F_n(t) = \sqrt{6}\sum_{r=1}^{L}\left[X_{nr}(t) - \frac{1}{2}\right].$$

Here $X_{nr}(t)$ is a random number between 0 and 1. If we choose $L = 12$, then the deviation of $[X_{ur}(t) - 1/2]$ is $1/12$, and the standard deviation of $F_n(t)$ is $\sqrt{\sigma}$. The domain of random noise force is $|F_n(t)| \leq 6\sqrt{\sigma}$.

Meanwhile, we can verify numerically that over a sufficiently long time the mean kinetic energy is given by

$$\left\langle \sum_n \frac{1}{2} M \dot{q}_n(t) \right\rangle = \frac{1}{2} N K_B T.$$

Even in the presence of damping and noise force, equations (9.89) – (9.92) have a conserved quantity $\sum_n |a_n(t)^2| = \text{constant}$.

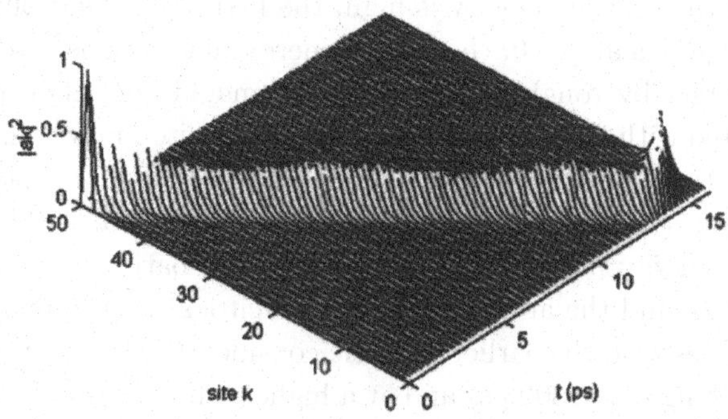

Fig. 9.14 The state of the soliton at 300 K.

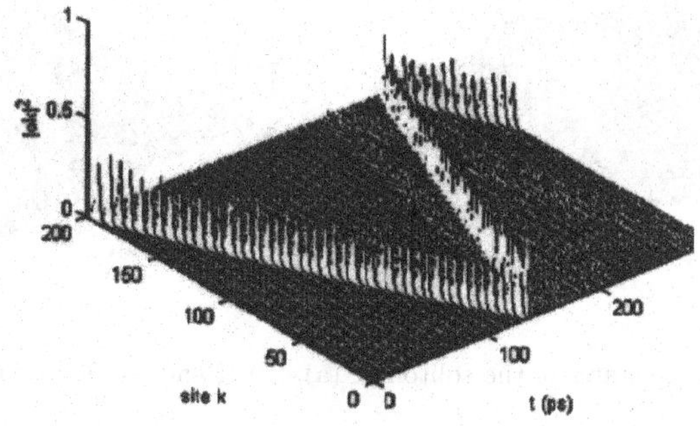

Fig. 9.15 The state of the soliton at 300 K for a time of 300 ps.

Applying the formulae given above, we can study the influences of dissipation force and random noise force, arising from the heat bath, on the new soliton. However, since the proteins now are in contact with a heat bath of about 300 K, one has to find out whether the soliton motion will be destroyed by the thermal motion of the lattice. For a C=O vibrational energy of $\varepsilon_0 = 3.28 \times 10^{-20} J$ and a temperature of 300 K, the Boltzmann factor is 3×10^{-20}, *i.e.*, only three out of 10,000 oscillators are excited in thermal equilibrium. Therefore, it can be safely assumed that the heat bath affects the soliton motion only via the lattice, which exhibits a quasi-continuum of vibrational states. Through the coupling between the protein and the heat bath, the heat bath also affects the oscillator system. Prior to the soliton commencement, the system is in equilibrium with the heat bath. The oscillator system is in its ground state, while the lattice is in thermal motion, which can be described by a linear combination of its normal modes. With the commencement of the soliton a non-equilibrium state is created. We can consider two extreme cases. In the first, the time that the soliton takes to travel through the protein is small compared to the time it takes the heat bath to re-establish equilibrium with the system, and in the second, the soliton velocity is small compared to the velocity of equilibration. If the soliton velocity is high, the first case would be more realistic.

If the soliton starts at $t = 0$, the lattice energy fluctuations associated with the heat bath are larger by roughly three orders of magnitudes than the local lattice energies associated with the soliton motion, but the soliton moves through the chain completely undisturbed at the biological temperature (300 K) as shown in Fig. 9.14. Here average values of the parameters of proteins were used. Therefore, despite the large lattice energy fluctuations due to the heat bath, the nonlinear coupling between the lattice and the amide oscillators (excitons) is still able to stabilize the soliton. This agrees with the earlier analytic consideration. The state of the soliton in the cases of a long time (300 ps) and at a higher temperature of 310 K are shown in Fig. 9.15 and Fig. 9.16(a), respectively. However, at high temperature of 320 K, the soliton starts to disperse as shown in Fig. 9.16(b).

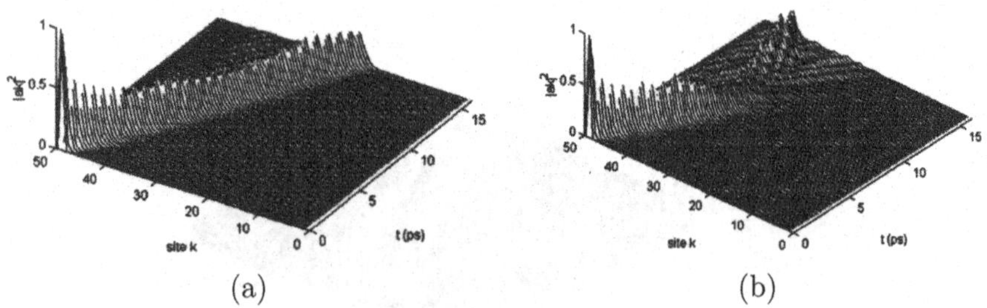

Fig. 9.16 The states of the soliton at (a) 310 K and (b) 320 K, respectively.

These results show clearly and sufficiently that the soliton in the improved model is thermally stable at 300 K, and the lifetime of the soliton is at least 300 ps in which the soliton can travel over thousands of lattice spacings. The critical temperature of the soliton is about 320 K. These are in agreement with the analytic results, and it not only shows that the analytic model is valid, but also establishes the soliton as a "possible" carrier of bio-energy transport in protein molecules.

The Davydov soliton does not have such properties. For comparison we show

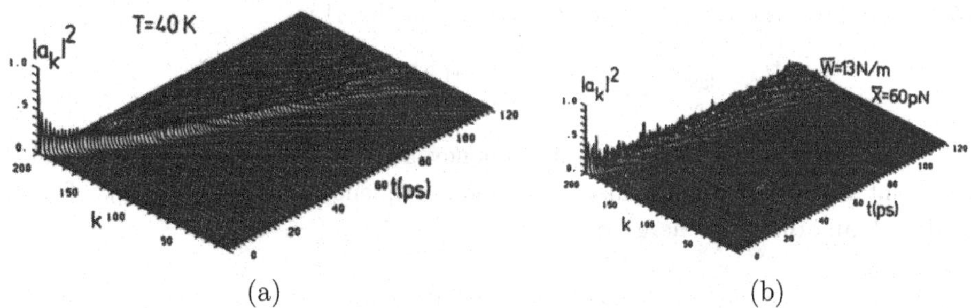

(a) (b)

Fig. 9.17 The states of the Davydov soliton at (a) 40 K, and (b) 300 K, respectively.

in Fig. 9.17 the results of the Davydov soliton obtained by Förner using the same values for the various physical parameters. The temperatures are $T = 40$ K and 300 K, respectively. We see clearly from Fig. 9.17 that the Davydov soliton is not thermally stable at 300 K, and its critical temperature is about 40 K.

Finally, Pang *et al.* studied the thermal stability of the new soliton at 300 K under the influences of structural disorders. The results show that the new soliton is still stable at this temperature against the changes of parameters as those used in Fig. 9.13 (c).

9.8 Eigenenergy Spectra of Nonlinear Excitations of Excitons

In this section, we calculate the eigenenergy spectra of the nonlinear excitations of excitons, or quantum vibrational energy spectra of amide-I in protein, based on the Hamiltonian and dynamic equations in the theory of bio-energy transport discussed in previous sections. From the earlier discussion, we know that both Davydov's Hamiltonian and that in the improved model, (5.64) and (9.58), respectively are quantum mechanical by construction. Both analysis are based on the coherent state ansatz and are therefore as "classical" as a quantum-mechanical treatment can be. An alternative approach would be to start from purely classical equations, and then subsequently apply the semi-classical or Bohr-Sommerfeld quantization. This has been done by Scott *et al.*, Pang, and Chen *et al.*. The advantage of this method is that it allows for an arbitrary number of amide-I quanta (excitons).

The starting point is to define normal displacement and momentum coordinates Q_n and P_n for amide-I (exciton), for which the Hamiltonian is given by $\sum_n(P_n^2 + Q_n^2)$. The motion of the exciton can then be described in terms of a complex amplitude

$$A_n \equiv \omega_0^{1/2}(P_n - iQ_n).$$

In terms of A_n, the classical analogue of H_{ex} in (5.64) is

$$H_{\text{ex-clas}} = \sum_n \left[\omega_0|A_n|^2 - J(A_{n+1}^* A_n + A_n^* A_{n+1})\right], \tag{9.94}$$

where ω_0 is the amide-I vibrational frequency. It is then assumed that the amide-I mode interacts with some (unspecified) low-frequency phonon mode q_n of amino acid with an adiabatic energy of

$$H_{\text{ph-clas}} = \frac{1}{2} w_{\text{clas}} \sum_n q_n^2.$$

The interaction energy can be taken as

$$H_{\text{int-clas}} = 2\chi_{\text{clas}} \sum_n q_n |A_n|^2.$$

The total classical Hamiltonian is then

$$H_{\text{clas}} = H_{\text{ex-clas}} + H_{\text{ph-clas}} + H_{\text{int-clas}}. \tag{9.95}$$

Minimizing (9.95) with respect to q_n yields

$$q_n = -\frac{\chi_{\text{clas}}}{w_{\text{clas}}} |A_n|^2.$$

In the continuum approximation, (9.95) reduces to

$$H_{\text{clas}} = \sum_n \left[\omega_0|A_n|^2 - J(A_{n+1}^* A_n + A_n^* A_{n+1}) - \frac{1}{2}\gamma|A_n|^4\right], \tag{9.96}$$

where $\gamma = \chi_{\text{clas}}^2 / w_{\text{clas}}$ is the nonlinear interaction constant. The equation of motion corresponding to (9.96) is

$$\left(i\frac{d}{dt} - \omega_0\right)|A_n|^2 + J(A_{n+1} + A_{n-1}) + \frac{1}{2}\gamma(|A_n|^2)A_n = 0. \tag{9.97}$$

In addition to the energy H_{clas}, equation (9.97) has another constant of motion, *i.e.* the particle number,

$$N = \sum_n |A_n|^2.$$

Equation (9.97) happens to be the discrete nonlinear Schrödinger equation for one-chain in the protein molecules. Therefore, the vibration of protein molecules is indeed a nonlinear problem as can be seen from the above. In Pang's improved model, χ_{clas}^2 in γ is approximately replaced by $(\chi_1 + \chi_2)_{\text{clas}}^2$.

Equation (9.97) is actually a special case of the following more general discrete self-trapping dynamic (DSTD) equation

$$i\frac{dA}{dt} + \bar{M}A + \gamma D(|A|^2)A = 0,$$

where $A = \{A_\alpha\}$ is a complex α-vector, α corresponds to number of the modes of motion, $D(|A|^2)$ denotes the diagonal matrix, diag. $(|A_1|^2, |A_2|^2, \cdots, |A_\alpha|^2)$, and $\bar{M}$ is a real symmetric matrix. This equation has been extensively studied. For the α-helix protein molecules consisting of the three chains, the DSTD equation, or (9.97), becomes

$$\left(i\frac{d}{dt} - \omega_0\right) A + \bar{M}A + \gamma D(|A|^2)A = 0 \tag{9.98}$$

where $A =$ col.$(A_1, A_2, A_3, A_4, A_5, A_6, A_7, A_8, A_9)$ is a complex 9-vectors, $D(|A|^2)$ denotes the diagonal matrix, diag. $(|A_1|^2, |A_2|^2, |A_3|^2, |A_4|^2, |A_5|^2, |A_6|^2, |A_7|^2, |A_8|^2, |A_9|^2)$ and $\bar{M}$ is a real symmetric matrix of order 9, which represents various interactions between neighboring amide-Is, *i.e.*,

$$\bar{M} = \begin{bmatrix} 0 & J & \varepsilon_1 & \varepsilon_2 & \varepsilon_3 & \varepsilon_3 & \varepsilon_2 & \varepsilon_1 & J \\ J & 0 & J & \varepsilon_1 & \varepsilon_2 & \varepsilon_3 & \varepsilon_3 & \varepsilon_2 & \varepsilon_1 \\ \varepsilon_1 & J & 0 & J & \varepsilon_1 & \varepsilon_2 & \varepsilon_3 & \varepsilon_3 & \varepsilon_2 \\ \varepsilon_2 & \varepsilon_1 & J & 0 & J & \varepsilon_1 & \varepsilon_2 & \varepsilon_3 & \varepsilon_3 \\ \varepsilon_3 & \varepsilon_2 & \varepsilon_1 & J & 0 & J & \varepsilon_1 & \varepsilon_2 & \varepsilon_3 \\ \varepsilon_3 & \varepsilon_3 & \varepsilon_2 & \varepsilon_1 & J & 0 & J & \varepsilon_1 & \varepsilon_2 \\ \varepsilon_2 & \varepsilon_3 & \varepsilon_3 & \varepsilon_2 & \varepsilon_1 & J & 0 & J & \varepsilon_1 \\ \varepsilon_1 & \varepsilon_2 & \varepsilon_3 & \varepsilon_3 & \varepsilon_2 & \varepsilon_1 & J & 0 & J \\ J & \varepsilon_1 & \varepsilon_2 & \varepsilon_3 & \varepsilon_3 & \varepsilon_2 & \varepsilon_1 & J & 0 \end{bmatrix}. \tag{9.99}$$

The dynamic equation of amide-I vibration quantum (exciton), with a order-9 matrix (9.99), obtained here is similar to that obtained by Scott *et al.* from the following equations for the three channels of α-helix protein.

$$\begin{aligned} i\frac{dA_{nl}}{dt} &= -J(A_{n+1} - 2A_{nl} + A_{n-1,l}) - K_1(R_{n+1,l} - R_{n-1,l})A_{nl} + LF_L \\ &\quad + K_2[(R_{n-1,l}A_{n-1,l} - R_{n+1,l}A_{n+1,l}) + R_{nl}(A_{n+1,l} - A_{n-1,l})] \\ &\quad - 2JA_{nl} + NF_N + PF_p + QF_Q + SF_S + TF_T \\ &\quad + UF_U + VF_V + XF_X + ZF_Z, \end{aligned}$$

$$\begin{aligned} \frac{d^2R_{n,l}}{dt^2} - w(R_{n+1,l} - R_{nl} + R_{n-1,l}) &= K_3(|A_{n+1,l}|^2 - |A_{n+1,l}|^2) \\ &\quad + K_4[A^*_{nl}(A_{n+1,l} - A_{n-1,l}) + A_{nl}(A^*_{n+1l} - A^*_{n-1l})], \end{aligned}$$

where

$$K_1 = \frac{1}{2 \times 10^{11}\hbar}\sqrt{\frac{M}{w}}\chi_1, \quad K_2 = \frac{1}{2 \times 10^{11}\hbar}\sqrt{\frac{M}{w}}\chi_2,$$
$$K_3 = \frac{10^{11}}{2w}\chi_1, \quad K_4 = \frac{10^{11}}{2w}\chi_2$$

represent the nonlinear coupling between vibrations of the amide-I and displacements of amino acids. Here M is one third of the mass of a unit cell, w is the linear force constant of the molecular chains, χ_1 is the change in the vibrational energy of the amide-I per unit extension of an amino acid, and χ_2 is the change in the longitudinal dipole-dipole coupling energy per unit extension or displacement of amino acid. In these equations, J, L, N, P, Q, R, S, R, U, V, X, Y and Z are various dipole-dipole interaction energies between pairs of amid-I bonds. Their values are 7.8, 12.358, 3.873, 1.592, 1.0082, 0.637, 0.472, 0.387, 0.196, 0.159, 0.116, 0.09 cm^{-1}, respectively. The subscript j indicates unit cells along the helix, while l specifies one of the three spines. The symbol "LF_L" is a shorthand for the dipole-dipole interactions between laterally adjacent amide-I bonds. Thus

$$\begin{aligned} i\dot{A}_{n,1} &= L(A_{n,2} + A_{n-1,2}) + \cdots, \\ i\dot{A}_{n,2} &= L(A_{n3} + A_{n1}) + \cdots, \\ i\dot{A}_{n3} &= L(A_{n,2} + A_{n+1,1}) + \cdots. \end{aligned}$$

$NF_N, \cdots, ZF_Z$ are defined similarly. In our discussion, the influences of small dipole-dipole interactions, $(L, N, P, Q, R, S, T, U, V, X, Z)$, on the matrix $\bar{M}$ are included in ε_1, ε_2 and ε_3. Therefore, equation (9.99) can indeed represent the dynamic properties of the exciton, interacting with displacements of amino acid residues, in the α-helical protein molecules.

Upon quantization using the method discussed in Chapter 6, the complex amplitudes are replaced by the creation and annihilation operators of a harmonic oscillator, $\hat{B}_\alpha^+$ and $\hat{B}_\alpha$, in the second quantization with the following properties,

$$\hat{B}_\alpha^+|m_\alpha\rangle = \sqrt{(m_\alpha + 1)}|m_\alpha + 1\rangle, \quad \hat{B}_\alpha|m_\alpha\rangle = \sqrt{m_\alpha}|m_\alpha - 1\rangle,$$

and

$$[B_\alpha, \hat{B}_j^+] = \delta_{\alpha j}.$$

The particle number and the energy corresponding to (9.99) then are represented by the operators

$$\hat{N} = \sum_\alpha^9 \left(\hat{B}_\alpha^+ \hat{B}_\alpha + \frac{1}{2}\right), \tag{9.100}$$

$$\begin{aligned} \hat{H} = \left(\omega_0 - \frac{1}{2}\gamma\right)\hat{N} - J\sum_{\alpha \neq j}^9 \hat{B}_\alpha^+ \hat{B}_j - \varepsilon_1 \sum_{\alpha \neq j_1 \neq j}^9 \hat{B}_\alpha^+ \hat{B}_{j_1} - \varepsilon_2 \sum_{\alpha \neq j_2 \neq j_1 \neq j}^9 \hat{B}_\alpha^+ \hat{B}_{j_2} \\ -\varepsilon_3 \sum_{\alpha \neq j_2 \neq j_1 \neq j \neq j_3}^9 \hat{B}_\alpha^+ \hat{B}_{j_3} - \frac{1}{2}\gamma \sum_\alpha^9 \hat{B}_\alpha^+ \hat{B}_\alpha \hat{B}_\alpha^+ \hat{B}_\alpha, \end{aligned} \tag{9.101}$$

where $\hbar$ is taken to be 1 and both frequencies and energies are measured in the same unit (cm^{-1}).

Choosing a set of basic function, $|m, k\rangle = |m_1\rangle|m_2\rangle|m_3\rangle|m_4\rangle|m_5\rangle|m_6\rangle|m_7\rangle \cdot |m_8\rangle|m_9\rangle$, where m $(= m_1 + m_2 + m_3 + m_4 + m_5 + m_6 + m_7 + m_8 + m_9)$ is the total quantum number, $k = 1, 2, \cdots, d(m)$, and $d(m)$ is equal to the number of different ways that m particles can be distributed in the nine states. We are interested in the common eigenfunction, $|\Psi_m\rangle$, of both operators. If we take

$$|\Psi_m\rangle = C_1|m,1\rangle + C_2|m,2\rangle + \cdots + C_d(m)|m,d(m)\rangle, \tag{9.102}$$

then the Hamiltonian equation $\hat{H}|\Psi_m\rangle = E_m|\Psi_m\rangle$ reduces to an algebra equation on the column vector, $\bar{C}_m = \text{Col.}(C_1, C_2, \cdots, C_{d(m)})$,

$$H_m\bar{C}_m = E_m\bar{C}_m, \tag{9.103}$$

where H_m is a $d(m) \times d(m)$, real and symmetrical matrix with diagonal elements, $H_{mr} = \langle m,r|\hat{H}|m,r\rangle$ and the off-diagonal elements $H_{rs} = \langle m,r|\hat{H}|m,s\rangle$. Here $1 \le r \le d(m)$, $1 \le s \le d(m)$, and $r \ne s$.

Using (9.103), we can numerically obtain eigenenergy spectra of the nonlinear excitations of the exciton in protein molecule, using the following accepted values for the various parameters for the three channel α-helix protein molecules

Parameter	Value
β	(39 – 58.5) N/m
M	$(3.51 - 5.73) \times 10^{-25}$ kg
χ	(56 – 62) PN
J	7.8 cm^{-1}
χ_1	(10 – 15) PN
r_0	4.5 Å
ε_1	16.231 cm^{-1}
ε_2	13.951 cm^{-1}
ε_3	15.363 cm^{-1}
ε_0	(0.205 – 0.210) eV

We can then obtain the parameters needed in our calculations, which are $\omega_0^1 = 1693.98$ cm^{-1}, $J^1 = 7.80$ cm^{-1}, $\gamma^1 = 40.68$ cm^{-1} for the Davydov model and $\omega_0^2 = 1712.08$ cm^{-1}, $J^2 = 7.8$ cm^{-1}, $\gamma^2 = 49.73$ cm^{-1} for the improved model. The calculated eigenenergies for $m = 3$ for both models are listed in Table 9.1.

From Table 9.1, we see that even though the energy spectra of the protein is quite complicated, its distribution shows the following trends. (1) The vibrational energy spectra consists of a series of manifolds or energy-bands, *i.e.* there are several energy levels corresponding to each vibrational quantum-number m. For example, the first three excitation states, $m = 1$ from $1610 - 1678$ cm^{-1}, $m = 2$ from $3179 - 3358$ cm^{-1} and $m = 3$ from $4735 - 5001$ cm^{-1}, consist of 8, 44 and 164 energy levels, respectively. Hence, as m increases, the gap between energy levels, $\triangle E$, decreases gradually. For instance, $\triangle E$ ranges from 6 to 23 cm^{-1} for $m = 1$, while it varies in the ranges of $0 - 14$ and $0 - 11$ cm^{-1} for $m = 2$ and $m = 3$, respectively. (2) The vibrational spectra have strong local mode characteristics, *i.e.*, the gap between the energy-levels depends strongly on the nonlinear interaction, γ,

Table 9.1 The vibrational energy spectra of protein molecules with three channels in cm^{-1}.

M	Exp^a	Cal^b	Cal^c	M	Exp^a	Cal^b	Cal^c
1		1611.01	1610.42	1		1612.95	1612.01
1		1628.35	1627.64	1		1631.61	1630.11
1	1650	1654.37	1653.81	1	1662	1662.95	1661.98
1	1666	1668.23	1667.65	1		1679.27	1678.73
2	3150	3206.33	3179.40	2		3212.17	3203.19
2	3205	3213.60	3204.71	2		3224.25	3211.85
2		3225.39	3212.95	2		3226.57	3213.21
2	3216	3233.34	3216.84	2		3234.71	3218.19
2		3246.71	3242.48	2		3248.75	3242.45
2	3250	3252.57	3249.68	2		3259.67	3258.78
2		3260.85	3259.87	2		3263.57	3261.77
2		3264.66	3260.95	2		3265.73	3262.97
2		3267.91	3263.67	2	3267	3269.99	3267.39
2		3270.45	3269.43	2		3278.57	3277.71
2	3279	3279.97	3278.89	2	3280	3282.18	3280.21
2		3283.91	3282.84	2		3284.75	3283.97
2		3286.54	3285.44	2		3287.56	3286.49
2		3288.24	3287.44	2		3293.14	3290.49
2		3299.61	3298.96	2		3300.81	3300.09
2		3301.73	3301.15	2		3304.95	3302.13
2		3310.54	3309.47	2		3311.27	3310.21
2		3313.24	3312.91	2		3314.73	3313.37
2		3322.27	3321.54	2		3323.29	3322.49
2		3325.11	3323.56	2		3328.47	3327.96
2		3331.54	3329.16	2		3338.04	3333.91
2		3319.17	3345.11	2		3360.61	3358.58
3		4782.91	4735.46	3		4783.15	4735.96
3		4787.16	4736.91	3		4787.51	4737.08
3		4788.17	4737.54	3		4788.40	4737.64
3		4788.57	4738.21	3		4789.14	4738.54
3	4752	4789.68	4749.26	3	4803	4819.76	4805.31
3		4819.95	4805.30	3		4823.84	4806.84
3		4825.25	4808.56	3		4826.45	4809.18
3		4829.18	4812.86	3	4813	4829.88	4813.76
3		4841.69	4824.16	3		4842.36	4824.96
3		4842.91	4825.51	3		4843.02	4826.33
3		4846.16	4830.52	3		4847.97	4831.91
3		4848.71	4832.61	3		4848.86	4833.15
3		4850.95	4834.67	3		4851.66	4836.24
3		4852.92	4836.46	3		4855.38	4837.21
3		4857.12	4839.73	3	4841	4858.92	4841.94
3		4859.24	4842.45	3		4860.42	4843.91
3		4861.44	4846.93	3		4862.66	4847.94
3		4862.97	4849.86	3		4864.35	4851.62
3		4866.61	4852.32	3		4868.62	4852:76

Table 9.1 continued.

M	Exp[a]	Cal[b]	Cal[c]	M	Exp[a]	Cal[b]	Cal[c]
3		4869.92	4854.47	3		4870.62	4855.86
3		4872.21	4855.96	3		4873.12	4857.09
3		4873.57	4857.94	3		4875.32	4858.36
3		4875.66	4859.72	3		4876.24	4860.12
3		4877.01	4861.55	3		4877.65	4863.25
3		4879.12	4866.28	3		4879.49	4866.31
3		4879.92	4868.07	3		4880.73	4869.29
3		4881.04	4869.45	3		4881.36	4869.69
3		4882.37	4970.11	3		4883.71	4870.82
3		4887.31	4871.84	3		4887.63	4874.31
3		4888.61	4874.81	3		4889.36	4876.25
3		4889.71	4875.91	3		4890.12	4876.83
3		4891.82	4877.52	3		4892.56	4877.91
3		4892.72	4878.72	3		4893.32	4878.96
3		4894.24	4879.31	3		4895.27	4879.61
3		4895.76	4881.73	3		4896.62	4884.66
3		4897.96	4885.56	3		4898.96	4886.57
3		4899.32	4886.93	3		4900.70	4887.13
3		4901.09	4887.29	3		4901.27	4887.79
3		4902.32	4888.42	3		4902.67	4889.92
3		4903.24	4890.49	3		4903.42	4891.82
3		4904.26	4893.42	3		4905.54	4894.32
3		4905.86	4996.62	3		4916.84	4897.83
3		4907.12	4898.56	3		4907.92	4899.37
3		4909.81	4899.81	3		4910.62	4901.83
3		4911.52	4905.21	3		4915.72	4906.32
3		4916.62	4907.52	3		4918.83	4909.51
3		4919.31	4913.52	3		4920.52	4914.72
3		4921.73	4914.89	3		4922.42	4915.67
3		4922.83	4916.08	3		4923.53	4916.21
3		4923.76	4918.42	3		4924.84	4918.93
3		4927.06	4919.92	3		4927.26	4921.83
3		4928.23	4923.66	3		4928.61	4924.26
3		4929.12	4924.96	3		4929.76	4925.36
3		4930.66	4926.92	3		4933.46	4927.30

[a]Experimental results from Careri *et al.* 1983, 1984, 1998, Careri and Eilbeck 1985, Eilbeck *et al.* 1984, Scott 1990, 1992, 1998, Scott *et al.* 1985, 1989.
[b]Results of the Davydov model.
[c]Results of the improved model.

which can be seen from (9.101) and (9.103). This is due to the fact that γ is much greater than J, ε_1, ε_2 and ε_3. (3) The local-mode degeneracies of energy levels appear at higher-lying vibrational states which begin to occur at $m \geq 2$. More precisely, there are degeneracies at 3242 cm^{-1} and 3259 cm^{-1} for $m = 2$, and for $m = 3$, there are degeneracies at 4735, 4737, 4738, 4737, 4857, 4859, 4866, 4869, 4870, 4877, 4878, 4879, 4887, 4916, 4918, 4924 cm^{-1}. Therefore, the degeneracy increases with increasing m.

A major advantage of the above method, (9.97) – (9.103) for computing energy levels is that it allows for treatment of more than one amide-I quanta (excitons), and its validity (in the zero dispersion limit) can be directly verified through (9.103). It is clear from Table 9.1 that theoretically calculated values are in good agreement with the experimental data obtained from infrared absorption and Raman scattering experiments in protein and ACN, and results obtained using the improved theory are even closer to the experimental values compared with those of Davydov's theory, suggesting that the improved model is more appropriate for the protein molecules. Pang further studied the properties of infrared absorption of the protein molecules, using data given in Table 9.1 (see Pang 1992a, 1993c, 1993d, 1994b, 1997, 2000a, 2001d).

9.9 Experimental Evidences of Exciton-Soliton State in Molecular Crystals and Protein Molecules

Energy transport in protein molecules is a very fundamental problem in life science. Because of the form of the exciton-soliton associated with the vibrational change of the C=O stretching and deformation of amino acids or peptide groups, existence and motion of the exciton-soliton can be observed and determined by infrared absorption and Raman spectra, and so on, in ACN and proteins. A lot of experiments were carried out on ACN, protein, cell and animal tissue to verify the existence of exciton-solitons in these systems. These experiments include infrared absorption and Raman scattering in acetanilide and α-helical collagen proteins, specific heat of acetanilide and protein, measurement of lifetime of thermal pulse or ground-state recovery time in acetanilide, and infrared emission of human tissue, *etc.* Based on these experiments, it can be concluded that the exciton-soliton could exist and is thermally stable at biological temperature 300 K.

9.9.1 *Experimental data in acetanilide*

In the early 1970s, Careri carried out a novel experimental investigation on protein dynamics. His intention was to circumvent problems arising in the study of real protein by looking at hydrogen bonded crystals that might be regarded as "model proteins". The basic idea can be understood by comparing the molecular structure of acetanilide and the structure of a typical polypeptide chain in natural protein as shown in Fig. 9.3. The similarity of bond lengths and angles of the peptide group (HNCO) and the same peptide chain of hydrogen bond $\cdots$ H-N-C=O-$\cdots$ H-N-C=O-$\cdots$ H-N-C=O$\cdots$ suggests that the dynamic properties of ACN might provide clues to the corresponding properties of natural protein molecules. This is also the reason that we first discuss the behaviors of ACN.

9.9.1.1 *Infrared absorption and Raman spectra*

In the measurement of Careri *et al.*, the infrared spectra were obtained using three different infrared spectro-photometers. For the study of temperature dependence of the amide-I region, a Nicolet model 7000 Fourier-transform infrared spectrophotometer was used. Spectra were collected for 100 scans using a band width of 0.5 cm^{-1}. The sample was thermostated using a closed-cycle helium refrigerator. Far-infrared absorption spectra were measured using a Michelson interferometer (model 720) equipped with a Golay cell. Samples consist of pellets obtained from a mixture of grounded ACN and polyethylene power. Pure polyethylene pellets were used to measure background transmission. Raman spectra were excited by a coherent radiation model 52 argon ion laser operating at 4880 Å or 5145 Å, with stabilized output power of 20-200 mW. Incident light was filtered by proper choice of interference filter and its intensity was monitored using a beam-splitter and a silicon photocell, scattered light was analyzed by a Jarrel-Ash model 25-300 Raman spectrometer and detected by an ITT model Fw-130 cooled photomultiplier using photon counting electronics. They measured the infrared spectra of amide-I (1600-1700 cm^{-1}), amide-II (1500-1600 cm^{-1}), amide-III (1300-1500 cm^{-1}), amide-IV or VI (500-700 cm^{-1}) and amide-V (700-800 cm^{-1}). The infrared spectra of amide-I in 1600-1700 cm^{-1} and 2500-3500 cm^{-1}, 4750-4850 cm^{-1} and 6200-6400 cm^{-1} are shown in Fig. 9.18. The absorption intensity vs. temperature is shown in Fig. 9.19. The Raman spectra of amide-I in 1630- 1700 cm^{-1} and low-frequency modes at 300 K and 50 K are shown in Fig. 9.20 and Fig. 9.21, respectively.

According to analysis in Section 9.1, the eigenenergy spectra of amide-I are 1665 cm^{-1}, 1662 cm^{-1} and 1659 cm^{-1}, which are assigned to the B_{2u}, B_{1u}, B_{3u} modes, respectively. The 1650 cm^{-1} mode in Fig. 9.18(a) and Fig. 9.20 should be assigned to the exciton-soliton. Thus, the exciton-soliton excitation, predicted by the above theory indeed exists in these systems. From Fig. 9.19, we know that the absorption intensity decreases in the form of $e^{-\beta T^2}$ with increasing temperature. This is consistent with the theoretical result of $e^{-(c+rT^2)}$ given in Section 9.4. The comparison between experimental (denoted by "*") and the theoretical values (solid curve) is shown in Fig. 9.19(b). From Figs. 9.18 – 9.20, we see that the peak values of the infrared absorption of ACN agree with the nonlinear eigenenergy spectra of the exciton-soliton given in Table 9.1, obtained from the above bioenergy transport theory. These interesting results provide experimental evidence to the existence of exciton-soliton in ACN and confirm the validity of above bioenergy-transport theory.

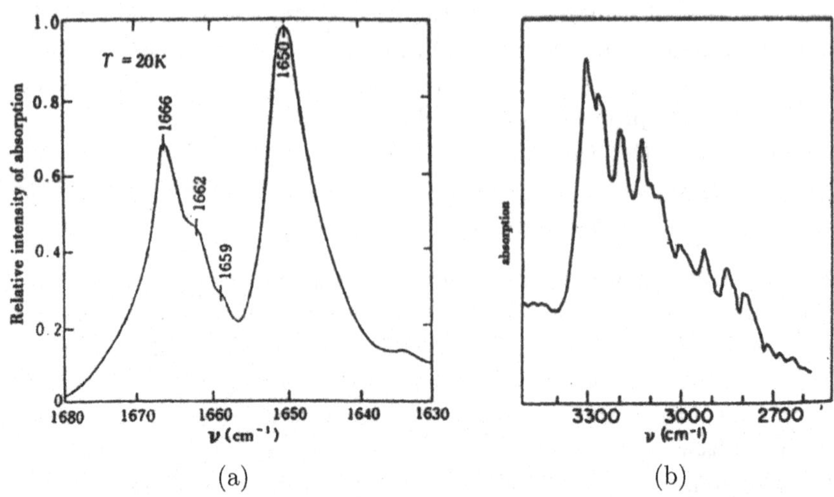

(a) (b)

Fig. 9.18 Infrared absorption spectra of ACN in (a) 1600-1700 cm^{-1} and (b) 2500-3500 cm^{-1}. (see Careri *et al.* 1983, 1998, Eilbeck *et al.* 1984, Scott 1990, 1992, 1998, Scott *et al.* 1985)

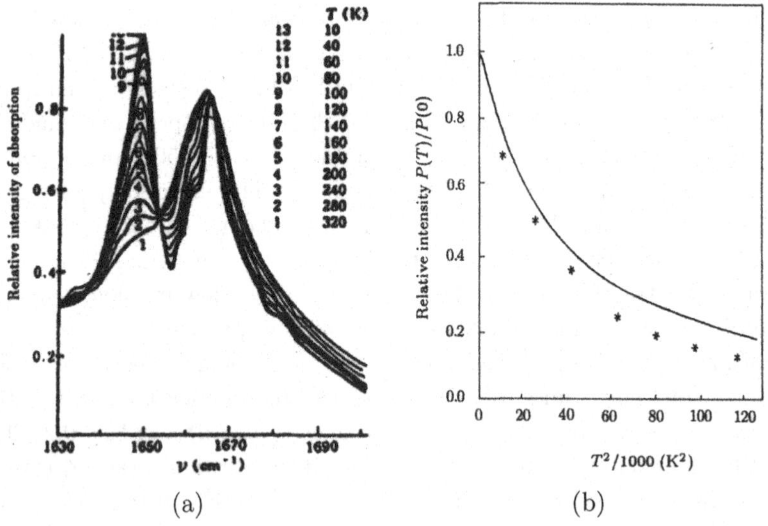

(a) (b)

Fig. 9.19 (a) The absorption intensity vs. temperature for ACN, and (b) comparison of experimental and theoretical values.

9.9.1.2 *Dynamic test of soliton excitation in acetanilide*

The experimental measurement of the exciton-soliton dynamics is of interest. It is useful to resolve the lifetime of the soliton since lifetime and mobility of soliton are important criteria for its usefulness as a means of energy transport. Fann *et al.* measured the relaxation time of the vibrational excitation of the 1650 cm^{-1} band

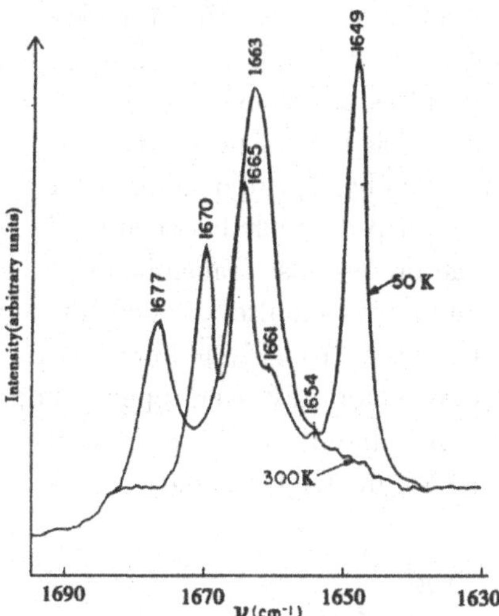

Fig. 9.20 Raman spectra in 1630-1700 cm^{-1} for ACN at 300 K and 50 K, respectively.

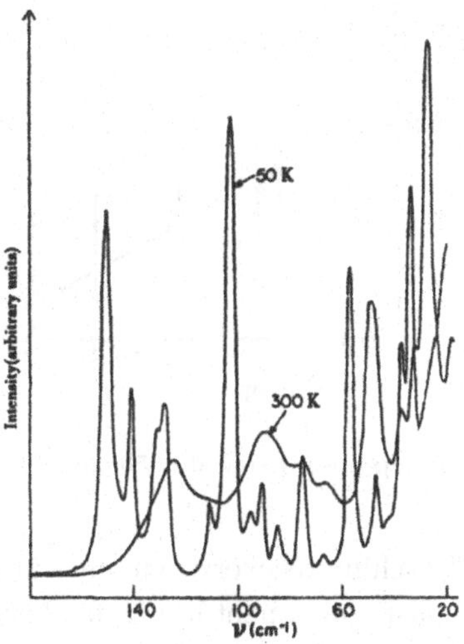

Fig. 9.21 Low frequency Raman spectra of ACN at 300 K and 50 K, respectively.

by transient-infrared-bleaching experiments which can set limits on the excitation lifetime. The source required for such measurements must be tunable around 6 μm (1650 cm^{-1}) with adequate spectral resolution to selectively excite the band (10 cm^{-1}). In addition, pulses with picosecond duration and sufficient intensity to bleach the transition are necessary to observe the bleaching recovery with adequate temporal resolution and signal-to-noise ratio.

The source Fann *et al.* choose to meet these criteria is the Mark free-electron laser (FEL) which produces pulses in bursts (macropulses) about 1.3 μs in duration. The Mark III laser output is divided into pump (95%) and probe (5%) beams by a CaF_2 wedge plate. Both pulses are focused by a single lens to a spot of 300-μm in diameter on the sample which is held in a closed-cycle refrigerator with CaF_2 windows. The ACN was purified by repeated sublimation, crystalline domains were grown by slow cooling from an ACN melt between CaF_2 disks of 2mm thickness. A substantial increase in the probe-pulse transmission at the 1650 cm^{-1} band was observed when the pump pulse was simultaneous with the probe pulse. The transmission increases due to the pump roughly follows the 1650 cm^{-1} band shape at 0 ps delay (pump and probe temporally overlapped), indicating that the effect is associated with the 1650 cm^{-1} band. This show clearly that the exciton-soliton excitation occurs in such a case in the acetanilide.

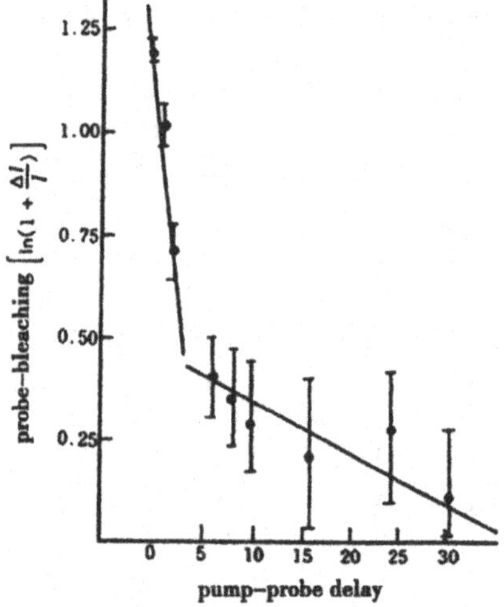

Fig. 9.22 Probe-bleaching vs. pump-probe delay at 1650 cm $^{-1}$ and 80 K ambient temperature.

Figure 9.22 shows the bleaching recovery dynamics at the band center as assembled from many scans. The fast component in Fig. 9.22 is limited by pulse-duration, while the slow component has a decay time of 15 ± 5 ps which can be associated with saturation recovery due to repopulation of the ground state of the 1650 cm^{-1} mode. The relaxation time or lifetime of 15 ± 5 ps indicates that the 1650 cm^{-1} band is strongly coupled to the lattice, and the 15 ± 5 ps is a typical relaxation value for a vibration line in the ACN. Fann *et al.* also measured the transmission recovery of the sample at 100 ps in which the probe decays to the "cold" base line. This is an important fact since it was thought that the 1650 cm^{-1} band should disappear on deposition of energy into the 130 cm^{-1} optical phonons. Loss of the 1650 cm^{-1} band due to such a heating should result in an increase in transmission approximately by a factor of 2 at 100 ps, which, however, was not observed. Therefore, the time scale of such vibrational relaxations observed is inconsistent with the lifetime of 10^{-10} s predicted by the theory of Pang *et al.*.

How could this happen? Further investigation is required in order to clarify this issue. In practice, if the rate of energy flow into the 130 cm^{-1} optical phonon is much slower than in other materials, or the energy flows of 1650 cm^{-1} into the 150-200 cm^{-1} optical phonon, instead of 130 cm^{-1} phonon, the ideal time scale of 10^{-10} s could be observed in this experiments because the soliton-optical phonon coupling can accelerate the cooling speed. Therefore, the above experiment needs to be improved.

It was believed that a soliton should travel a distance of more than 100 amino acids in its lifetime, as mentioned in Section 9.6. Therefore, it was thought that the lifetime of 15 ps for the exciton-soliton obtained by Fann *et al.* is too short to verify the usefulness of soliton in biological processes because it can only travel a distance of several tens of amino acids at subsonic speed in its lifetime. However, we should remember that the acetanilide Fann *et al.* used was at room temperature. According to the result in Section 9.4, the exciton-soliton travels at supersonic speed in such a case, as shown in (9.55). Thus the solution can actually travel at least over 150 amino acids during its lifetime of 15 ps. Then we can say that the experimental result of Fann *et al.* also supports the soliton theory of biological transport in ACN.

9.9.2 *Infrared and Raman spectra of collagen, E. coli. and human tissue*

9.9.2.1 *Infrared spectra of collagen proteins*

Collagen is a helical protein with three channels which is similar to the α-helix protein molecules. Recently, Xiao, and Pang *et al.* measured the infrared absorption spectra of the collagen. The sample was purchased from Sigma chemical Co. Ltd. and was used without further purification. It was placed between KBr windows and transferred to a temperature cell in the spectrometer. Infrared spectrum was recorded on a Perkin Elmer spectrum GX·FT-IR spectrometer equipped with a DTGS detector. The measurements were performed at a resolution of 4 cm^{-1} in the range of 400-4000 cm^{-1}. To obtain an acceptable signal-to-noise ratio, 16 scans were accumulated. Spectra were recorded in a variable-temperature cell with a reported accuracy of ± 1°C between 15°C and 95°C in intervals of 10°C.

The results obtained at 25°C in the this infrared absorption experiment are shown in Fig. 9.23. At high frequencies the spectrum is dominated by the amide where there are obviously two vibrational modes of amide-I, 1666.01 and 1650 cm^{-1}. Other amide bands are found above 1000 cm^{-1}. For example, the amide II occurs at 1542.09 cm^{-1}, amide-III at 1455.89 cm^{-1}, 1404.47 cm^{-1}, amide-IV at 1335.8 cm^{-1} and 1243.49 cm^{-1}, and amide-V at 1081 cm^{-1}. There are also rich spectral lines in the range of 2800 – 4000 cm^{-1}. For example, 3209.01 cm^{-1}, 3225.70 cm^{-1}, 3244.04 cm^{-1}, 3218.19 cm^{-1} and 3286.33 cm^{-1}, etc., which were not observed

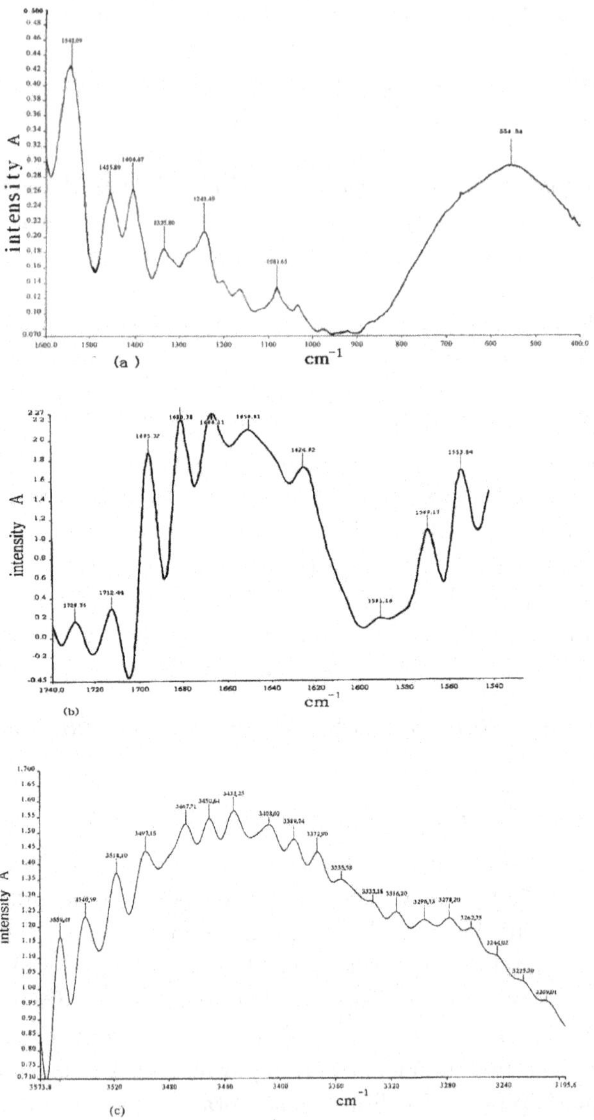

Fig. 9.23 Infrared absorption spectra of collagen proteins from 400 to 4000 cm^{-1} at 25°C.

in other experiments. Results of a detailed study of temperature-dependence of absorption intensity of the amide-I region of the collagen in the temperature range of 5 – 95°C are reported in Figs. 9.24 and 9.25. We can see from these figures that the intensity of the band at 1650 cm^{-1} increases on cooling without apparent change in frequency and shape, but it is weaken at 95°C. On the other hand, the amide-I absorption at 1666.11 cm^{-1} decreases on cooling. The peak intensities of the 1650 cm^{-1} and 1666.1 cm^{-1} bands as a function of temperature are shown in

Fig. 9.25. Clearly different temperature-dependence of the intensity can be seen for the two bands. It is surprising that the absorption intensity of the band at 1666.1 cm^{-1} increases linearly with temperature, but the intensity of the band at 1650 cm^{-1} decreases exponentially with temperature, especially in the range of 5 – 45°C which can be approximately expressed as $\exp[-(0.437 + 8.987 \times 10^{-6}T^2)]$. This exponential behavior, $\exp[-(0.437 + 8.98^{-6}T^2)]$ for the temperature-dependence of the intensity of the infrared absorption of the band at 1650 cm^{-1} is especially interesting. Alexander *et al.*, Scott *el al.* and Pang obtained an exponential decrease of the intensity of the 1650 cm^{-1} band with temperature, $I \propto \exp[-(a + bT^2)]$, in a low temperature range (10-280 K), where T is temperature of the system, a and b are constants, and a linear increase of the intensity of the 1666 cm^{-1} band with temperature, by using the soliton theory in ACN and the improved Davydov theory in the α-helical protein, respectively. Interestingly, the theoretical results based on the soliton theory for the α-helical protein resemble closely the experimental resutls of the collagen shown in Fig. 9.25. This serves as an additional experimental evidence to verify the existence of the nonlinear or soliton excitation in collagen.

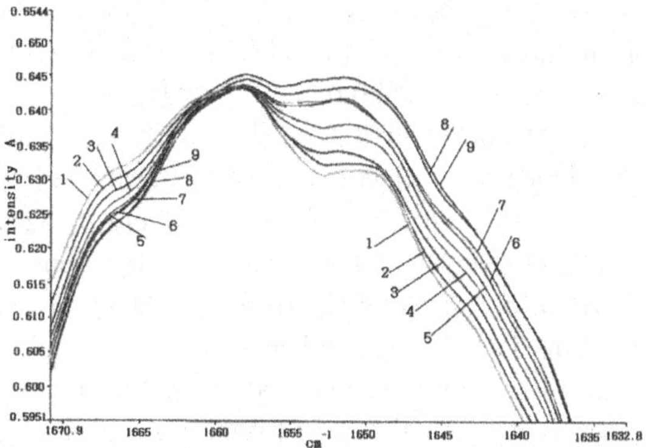

Fig. 9.24 The infrared absorption intensities of the collagen in the region of the amide-I mode at different the temperatures, (1) 95°C, (2) 85°C, (3) 75°C, (4) 65°C, (5) 55°C, (6) 45°C, (7) 35°C, (8) 25°C, and (9) 15°C.

The following conclusions can be drawn from these results. (1) Infrared absorption spectra of the collagen are basically similar to that of ACN. (2) The 1666 cm^{-1}, 1671 cm^{-1} and 1650 cm^{-1} lines are always present in the infrared absorption of both collagen and ACN. The 1666 cm^{-1} or the 1671 cm^{-1} line is the vibrational frequency of the amide-I, 1650 cm^{-1} is its anomalously new band. According to the analytic results on ACN obtained by Careri *et al.*, Eilbeck *et al.*, and Scott, respectively, given in Section 9.1, we can conclude that the 1666 cm^{-1} or the 1671^{-1} spectral line represents the vibrational excitation of amide-I (or exciton), but the 1650 cm^{-1} line should be assigned to the soliton excitation in these systems, *i.e.*, there are soliton motion in collagen. (3) The dependence of intensity of the infrared absorption on

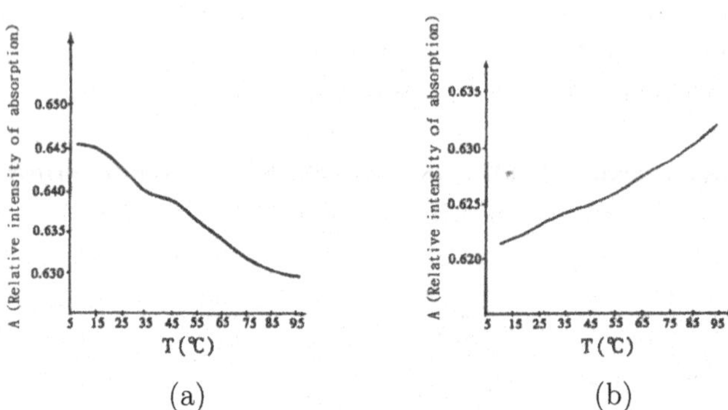

Fig. 9.25 Temperature dependence of strengths of (a) 1650 cm^{-1} and (b) 1666 cm^{-1} spectral lines in collagen in the region of 5 – 95°C.

temperature shown in Fig. 9.25 is basically the same as that obtained in ACN and protein molecules by Alexander and Krumbansl, Scott *el al.*, and Pang based on the soliton theory. Therefore, results shown in Fig. 9.25 are direct verification of the existence of soliton in these systems. (4) There are bands at 1680.31 cm^{-1}, 1666 cm^{-1}, 1650 cm^{-1}, 1624.94 cm^{-1}, 3209.01 cm^{-1}, 3225.7 cm^{-1}, 3244.04 cm^{-1}, 3262 cm^{-1}, 3278 cm^{-1}, 3296.33 cm^{-1}, 3316.2 cm^{-1}, 3333.18 cm^{-1} and 3355.58 cm^{-1} in Fig. 9.23 for collagen which are basically consistent with those at 1667 cm^{-1}, 1662 cm^{-1}, 1653 cm^{-1}, 1627 cm^{-1}, 3204.71 cm^{-1}, 3218.19 cm^{-1}, 3242.48 cm^{-1}, 3261.77 cm^{-1}, 3278.89 cm^{-1}, 3298.96 cm^{-1}, 3313.37 cm^{-1}, 3333.91 cm^{-1} and 3358.58 cm^{-1} in Table 9.1, respectively, obtained by the improved Davydov theory of bio-energy transport by Pang. This is not a coincidence. It shows that there exists soliton excitation in these systems and the improved theory proposed by Pang correctly describe bio-energy transport in α-helical protein molecules.

9.9.2.2 *Raman spectrum of collagen*

Cai *et al.* measured the laser-Raman spectrum from acidity-I type fiber collagen. This protein is extracted from lungs of killed mouse with a weight of about 200 g. The samples of acidity-I type is obtained using the Kathryn's method and purified. In the test, they put the purified collagen in the lower part of a microscope of SpeX1430-type laser-Raman instrument. The width of seam of Argon ion laser is about 800 μm. The observed room temperature Raman spectrum of the collagen was excited by a laser with a wavelength of 5145 Å and 250 – 800 mW power. The experimental result is shown in Fig. 9.26 in the region of 1000 – 1800 cm^{-1} and 2800 – 3000 cm^{-1}, respectively. From this figure, we see clearly that the collagen can emit infrared lights at wavelengths 1670, 1650, 1562 cm^{-1} and so on, which are basically consistent with that in ACN in 1600 – 1700 cm^{-1}. As it is known, the line at 1670 cm^{-1} is an eigenfrequency of amide-I, whereas 1650 cm^{-1} should be assigned to the exciton soliton in the collagens. Furthermore, these peak values of the Raman spectra approach the theoretical results in Table 9.1. It shows again the credibility of the theory of bioenergy transport.

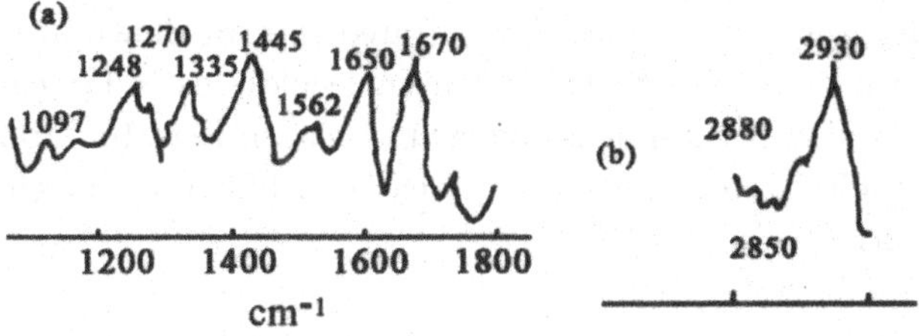

Fig. 9.26 Laser Raman spectra of collagen.

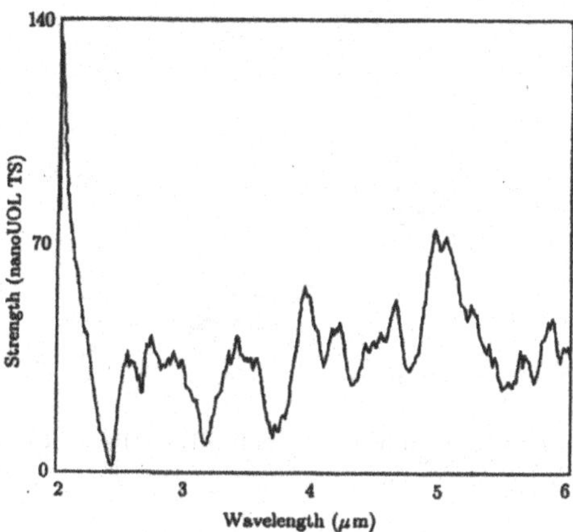

Fig. 9.27 Infrared spectra of tissue from a human finger.

9.9.3 *Infrared radiation spectrum of human tissue and Raman spectrum of E. col.*

Pang *et al.* and Chi *et al.* measured the infrared radiation of tissue of human body at room temperature by optical multichannel analyzer (OMA) with multichannels infrared probe systems and a single-photon counter method, respectively. It was found in the experiments that the intensities of the emitted infrared lights of different tissues of human bodies are different and decrease in the following order from the strongest to the weakest, fingers, palms, cheeks, fore arms, upper arms, chest and abdomen. Fig. 9.27 shows the measured result for a tissue of a human finger by the OMA and infrared probe systems in the wavelength of $2-6$ μm. The width of the seam of the OMA is modulated to 0.5 mm. In this experiment, the spectra of the substrate of the instrument was first measured which was followed by the radiation spectra of the tissue of a human finger. The finger tissue was placed on the seam. The environment light was screened during the measurement. The experimental result shown in Fig. 9.27 were obtained by 11 scans, each lasting about 0.99 seconds. It can been seen in Fig. 9.27 that human body can radiate infrared lights at wavelengths of $1 - 6\ \mu m$. The infrared spectra shown in Fig. 9.27 should be assigned to the vibrations of the protein molecules, instead of chemical reaction or other biomolecules, because it is similar to those in Figs. 9.18, 9.20, 9.23 – 9.26, and that given in Table 9.1. Therefore we can conclude that the energy spectra in Table 9.1 are correct.

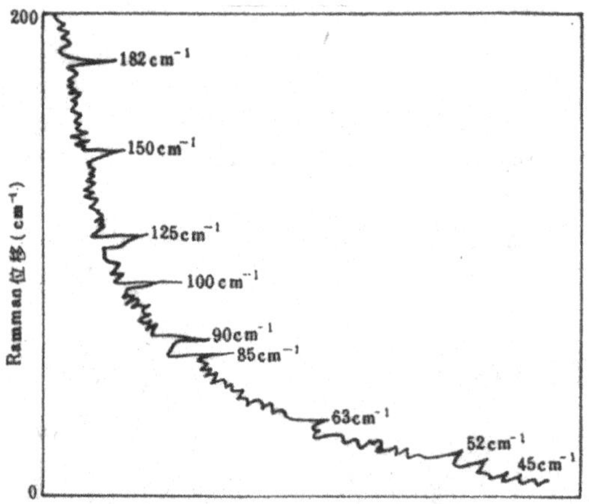

Fig. 9.28 Laser-Raman Spectrum of metabolically active E. Coli. courtesy of Webb

In 1980, Webb measured laser-Raman spectrum of the metabolically active E. Coli at low temperature. Results of this measurement are shown in Fig. 9.28, where nine lines can be clearly seen.

Spectra radiated from protein molecules can also be obtained if the energy spectra given in Table 9.1, which is shown in Fig. 9.29 for $m = 2$, is used. The nine lines observed here are consistent with the experimental data of the laser-Raman spectrum of E. Coli., given in Fig. 9.28, and of the Raman spectrum of ACN at 50 K, given in Fig. 9.21. This again confirms the correctness of the energy spectra given in Table 9.1, obtained by the theory of bioenergy transport.

Scott suggested that the energy would vary periodically between the three spins of the α-helix protein with a phase shift of 120°, much like the voltage on a standard three-phase power line. The frequency of alternation ν is related to the transverse dipole-dipole coupling energy, L, by $\nu = L/h = 3.85 \times 10^{11}$ Hz. Considering the overtones and the interactions of the moving soliton with a discrete lattice, a set of internal resonances $\{\nu_i(th)\}$ was suggested. In Webb's experiment on E. Coli, there was also a set of spectral lines, denoted by $\{\nu_i(\exp)\}$, due to internal resonances in the far infrared range between 10^{11} and 10^{12} Hz. It was interesting to notice

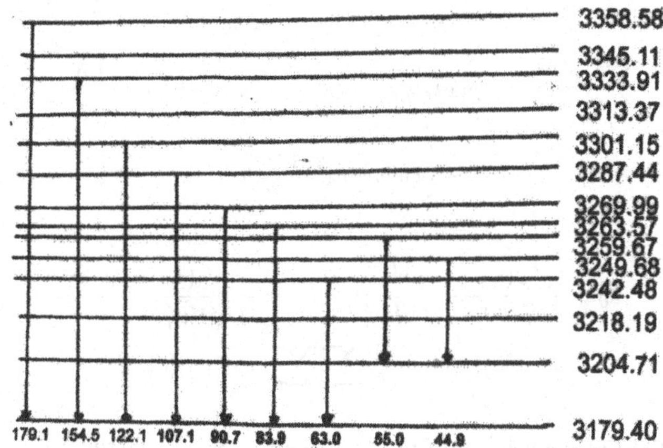

Fig. 9.29 The distribution of energy-levels for $m = 2$ and corresponding Raman spectra of proteins.

that $\{\nu_i(th)\} \approx \{\nu_i(\exp)\}$. This agreement seemed especially significant because $\{\nu_i(\exp)\}$ were measured without knowing the existence of the exciton-soliton, and the theoretical model was constructed with with the knowledge of the molecular structure of α-helix protein. Shortly thereafter it was suggested that Mie scattering from cell density fluctuations (clumps cells) could provide an explanation for Webb's observations. In order to settle this issue an experimental program was organized at the Los Alamos National Laboratory to measure again the laser Raman scattering from metabolically active cells using the OMA to record the spectra in 1980s. But the result is not shown in Fig. 9.28. Therefore, this experimental result given in Fig. 9.28 has yet to be reconfirmed.

9.9.4 *Specific heat of ACN and protein*

Careri *et al.* measured the specific heat of powdered crystalline ACN from liquid nitrogen to room temperature. The measured data were then fitted to the formula

$$C(T) = 4.59 \times 10^{-3}T + 1.505 \tag{9.104}$$

in units of $\mathrm{Jg^{-1}k^{-1}}$, where T is temperature in °C.

This temperature dependence of specific heat can be understood from the exciton-soliton theory of bioenergy transport in proteins and ACN. As a matter of fact, using the Davydov approximation, the quantity $\bar{W}$ contained in the factor

$$JB(T,q) = \frac{\hbar^2}{2mr_0^2}e^{-\bar{W}_n}$$

in (9.46) is given by the following

$$\bar{W} \approx W_{nn\pm 1} = \sum_q (1+\nu_q)\beta_{qn\pm 1}\beta_{qn}^* - \nu_q\beta_{qn\pm 1}^*\beta_{qn} + \left(\nu_q + \frac{1}{2}\right)(|\beta_{qn}|^2 + |\beta_{qn\pm 1}|^2)$$
$$\approx \frac{1}{2} r_0 \delta_0 |\varphi_n|^2 B' F(q,\tau),$$

where

$$\delta_0 = \frac{2(\chi_1+\chi_2)^2}{8r_0\beta(1-s^2)J\omega_0^2}, \quad B' = \frac{\pi r_0(\chi_1+\chi_2)^2}{4\omega_0^2\beta v_0^2\hbar(1-s^2)},$$
$$F(q,\tau) = \frac{r_0}{\pi N}\sum_q \sum (1+\nu_q)|q|.$$

In the first order approximation, (9.46) can now be written as

$$i\hbar\frac{\partial}{\partial t}\varphi(x,t) = (\varepsilon_0 - 2J)\varphi(x,t) - Jr_0^2\frac{\partial^2}{\partial x^2}\varphi(x,t) - G'(T)|\varphi|^2\varphi(x,t).$$

Its soliton solution is of the form

$$\varphi(x,t) = \sqrt{\frac{G'(T)}{8J}}\,\mathrm{sech}\left[\frac{G'(T)}{2Jr_0}(x-x_0-vt)\right]\exp\left[\frac{i\hbar v}{2Jr_0^2}(x-x_0) - \frac{iE_{\mathrm{sol}}t}{\hbar}\right],$$

where

$$G'(T) = \frac{\hbar^2(\chi_1+\chi_2)^2 r_0^2}{M\omega_0^2 v_0^2(1-s^2)}\left[1 - \frac{1}{4}B'F(q,T)\right].$$

The corresponding soliton energy is given by

$$E_{\mathrm{sol}} = (\varepsilon_0 - 2J) + \frac{1}{2}mv^2 - \frac{\hbar^4(\chi_1+\chi_2)^4(1-5s^2)}{24\omega_0^4\beta^2 J(1-s^2)^3}\left[1 - \frac{B'}{4}F(q,T)\right]^2.$$

Thus the specific heat arising from the motion of the exciton-soliton is

$$C_v = \frac{dE}{dT} = \frac{\hbar^4(\chi_1+\chi_2)^4 B'(1-5s^2)}{48\omega_0^4\beta^2 J(1-s^2)^3}\left[1 - \frac{1}{4}B'F(q,T)\right]\frac{dF(q,T)}{dT}.$$

At high temperature, $K_B T \gg \hbar\omega_q$, we have

$$F(q,T) = \frac{2K_B T}{\hbar v_0 \pi},$$

Thus

$$C_v = K_B(a + bT). \tag{9.105}$$

where T is the absolute temperature. Using the accepted values of parameters for ACN, we can obtain $a = 5.15$, $b = 0.0199\ \mathrm{K}^{-1}$. Thus (9.105) is consistent with experimental data given in (9.104). Meanwhile, Mrevlishvil and Goldanskill *et al.* measured the linear specific heat of various biopolymers including proteins and DNA. They obtained $C_v = K_B(a' + b'T)$, which resembles (9.105). The consistency between the experimental and the theoretical results for the specific heat again demonstrates that the above bio-energy transport theory is correct.

In short, from the experimental results in the ACN, collagen, E. Coli. and human tissues discussed above, we can conclude that exciton-soliton, described by the nonlinear Schrödinger equation, can exist, or can be localized in nonlinear systems such as ACN and α-helix protein molecules, and that the bioenergy transport theory and the nonlinear quantum mechanical theory established by Pang is correct.

9.10 Properties of Nonlinear Excitations of Phonons

From previous sections in this chapter we know that exciton-soliton or localization of exciton occurs due to exciton-phonon interaction. In such a case, the phonon is also localized at the same time and moves as soliton in the systems. This can be verified from (9.11) – (9.12) or (9.60) – (9.63). To do so, we first note that from (9.14) – (9.15) and

$$\mu = \frac{(\chi_1+\chi_2)^2\hbar^2}{Mv_0^2\omega_0^2(1-s^2)J},$$

where $s = v/v_0$, we can get

$$\frac{\partial}{\partial x}|\varphi(x,t)^2| = \frac{2(\chi_1+\chi_2)^3\hbar^3}{r_0^3\beta J^2(1-s^2)\omega_0^3}\left[u - \frac{4\omega_0^2(1-s^2)^2\beta^2}{\hbar^2(\chi_1+\chi_2)^2}u^3\right]. \tag{9.106}$$

Substituting (9.106) into (9.11), we obtain

$$u_{tt} - v_0^2 u_{xx} = \frac{4(\chi_1+\chi_2)^4\hbar^4}{M\omega_0^4 r_0^2 J^2\beta(1-s^2)}\left[u - \frac{4\omega_0^2(1-s^2)^2\beta^2}{\hbar^2(\chi_1+\chi_2)^2}u^3\right]. \tag{9.107}$$

This is a ϕ^4-equation for $u(x,t)$. It has a soliton solution, *i.e.*, (9.15). We can conclude that the phonon has been localized. Obviously, the localization of the phonon is due to nonlinear exciton-phonon interaction in ACN and protein molecules because (9.107) reduces to a linear wave equation, $u_{tt} - v_0^2 u_{xx} = 0$, if $\chi_1 = \chi_2 = 0$, here the lattice molecule performs harmonic vibration, and the solution, $u = A\sin(\omega t - \theta_0)$, is simply a plane wave. Localization or nonlinear excitation of the phonons occurs only in nonlinear systems. The phonon is "self-trapped" as a soliton through interaction with the exciton. Its motion is described by the nonlinear quantum mechanics.

If the lattice molecule or amino acid molecule is in an anharmonic state, its Hamiltonian is given by (9.18). In such a case, it can be shown that the phonon also moves as a soliton, or, the phonon can also be localized. We will discuss properties of this kind of phonon localization based on (9.18).

Inserting (9.3) into (9.18), we obtain

$$H_{ph} = \sum_q \hbar\omega_q \left(a_q^+ a_q + \frac{1}{2}\right) \tag{9.108}$$
$$+\sum_{q_1 q_2} F(q_1, q_2)(a_{q_1} + a_{-q_1}^+)(a_{q_2} + a_{-q_2}^+)(a_{(q_1+q_2)} + a_{-(q_1+q_2)}^+),$$

where

$$F(q_1, q_2) = \frac{8}{3} i\lambda_1 \sqrt{\frac{\hbar}{2M}\omega_{q_1}\omega_{q_2}\omega_{(q_1+q_2)}}$$
$$\times \sin\left(\frac{1}{2}r_0 q_1\right) \sin\left(\frac{1}{2}r_0 q_2\right) \sin\left[\frac{1}{2}r_0(q_1+q_2)\right]. \tag{9.109}$$

From (9.108) and the phonon part, $U|0\rangle_{ph}$ in (9.5), and the Heisenberg equations (9.7), we can get that

$$i\hbar\dot{\alpha}_q = \hbar\omega_q\alpha_q - \sum_k F(k-q)(\alpha_k + \alpha_{-k}^*)(\alpha_{k-q}^* + \alpha_{q-k}),$$
$$i\hbar\dot{\alpha}_{-q}^* = -\hbar\omega_q\alpha_{-q}^* + \sum_k F(k-q)(\alpha_k + \alpha_{-k}^*)(\alpha_{k-q}^* + \alpha_{q-k}).$$

From these two equations, we can further get

$$(\ddot{\alpha}_{-q}^* + \ddot{\alpha}_q) = -\omega_q^2(\alpha_{-q}^* + \alpha_q) + \frac{2\omega_q}{\hbar}\sum_k F(k-q)(\alpha_q + \alpha_{-q}^*)(\alpha_{k-q}^* + \alpha_{q-k}). \tag{9.110}$$

Because

$$u_n(t) =_{ph} \langle 0|U|R_n|U|0\rangle_{ph} = \sqrt{1}\sqrt{N}\sum_q u_q(t)e^{iqx},$$

where $(x = nr_0)$ and

$$u_q(t) = (\frac{\hbar}{2M\omega_q})^{1/2}(\alpha_q + \alpha_{-q}^*),$$

equation (9.110) becomes

$$(u_q)_{tt} + \omega_q^2 u_q = -\frac{8i\lambda_1}{3M}\sin\left(\frac{1}{2}r_0 q\right)\sum_k \sin\left(\frac{1}{2}r_0 k\right) \times$$
$$\sin\left[\frac{1}{2}r_0(q-k)\right] u_k u_{q-k} e^{-i(k-q)r_0}. \tag{9.111}$$

Taking into account the dispersion relation, then in the long wavelength and the second order approximation

$$\omega_q \approx \frac{1}{2}\sqrt{\frac{\beta}{M}}\left[\frac{1}{2}r_0 q - \frac{1}{3!}\left(\frac{1}{2}r_0 q\right)^3\right] = v_0 q\left(1 - \frac{1}{24}r_0^2 q^2\right),$$

(9.111) can be written as

$$(u_q)_{tt} + v_0^2 q^2 u_q - \frac{1}{12} v_0^2 r_0^2 q^4 u_q = -\frac{i\lambda_1 r_0^3 q}{3M} \sum_k (q-k) k u_k u_{q-k}.$$

In the continuum approximation, applying the following relation,

$$\frac{\partial u(x,t)}{\partial x} = \frac{1}{\sqrt{N}} \sum_q (iq) u_q(t) e^{iqx}$$

results in

$$\frac{\partial^2 u}{\partial t^2} - v_0^2 \frac{\partial^2 u}{\partial x^2} - \frac{v_0^2 r_0^2}{12} \frac{\partial^4 u}{\partial x^4} = \frac{\lambda_1 r_0^3}{3M} \frac{\partial}{\partial x} \left| \frac{\partial u}{\partial x} \right|^2. \tag{9.112}$$

Letting $Q(x,t) = -\partial u/\partial x$, then the above equation can be rewritten as

$$Q_{tt}(t) - v_0^2 \frac{\partial^2 Q(x,t)}{\partial x^2} - \frac{v_0^2 r_0^2}{12} \frac{\partial^4 Q(x,t)}{\partial x^4} = \frac{\lambda_1 r_0^3}{3M} \frac{\partial^2}{\partial x^2} |Q(x,t)|^2. \tag{9.113}$$

This is a nonlinear equation. It shows again that the phonon is a nonlinear microscopic particle in such a case. Applying the boundary conditions, $u(\pm\infty) = Q(\pm\infty) = 0$, it has the following soliton solution

$$Q(x,t) = \frac{3M(v^2 - v_0^2)}{4\lambda_1 r_0^3} \operatorname{sech}^2 \left[\frac{\sqrt{2}(x - vt)}{r_0 \sqrt{v^2/v_0^2 - 1}} \right], \quad (v > v_0). \tag{9.114}$$

Thus

$$u(x,t) = -\frac{3M(v^2 - v_0^2)^{3/2}}{4\sqrt{2}\lambda_1 r_0^2} \tanh \left[\frac{\sqrt{2}(x - vt)}{r_0 \sqrt{v^2/v_0^2 - 1}} \right], \quad (v > v_0) \tag{9.115}$$

where v is the velocity of the phonon-soliton. The shape of this soliton is similar to the soliton solution (9.15), but (9.115) is a supersonic soliton ($v > v_0$) which satisfies the nonlinear equation (9.113), while (9.15) is a subsonic soliton ($v < v_0$), which satisfies the ϕ^4-equation, (9.107). Therefore, the mechanism and properties of localization of the phonons in the two cases, (9.107) and (9.113), are obviously different.

Bibliography

Alexander, D. M. (1985). Phys. Rev. Lett. **60** (138) .

Alexander, D. M. and Krumbansl, J. A. (1986). Phys. Rev. B **33** (7172) .

Atkinson, K. E. (1987). An Introduction to Numerical Analysis, Wiley, New York.

Benkui, T. and John, P. B. (1998). Phys. Lett. A **240** (282) .

Bolterauer, H. (1991). in Davydov's soliton revisited: Self-trapping of vibrational energy, eds. Christiansen, P. L. and Scott, A. C., Plenum, New York, p. 99.

Bolterauer, H. and Opper, M. (1991). Z. Phys. **B82** (95) .

Brizhik, L. S. and Davydov, A. S. (1969). Phys. Stat. Sol. (b) **36** (11) .

Brizhik, L. S. and Davydov, A. S. (1983). Phys. Stat. Sol. (b) **115** (615) .
Brown, D. W. (1988). Phys. Rev. A **37** (5010) .
Brown, D. W. and Ivic, Z. (1989). Phys. Rev. B **40** (9876) .
Brown, D. W., Lindenberg, K. and West, B. J. (1986). J. Chem. Phys. **84** (1574) ; Phys. Rev. Lett. **57** (234) .
Brown, D. W., Lindenberg, K. and West, B. J. (1987). J. Chem. Phys. **87** (6700) ; Phys. Rev. B **35** (6169) .
Brown, D. W., Lindenberg, K. and West, B. J. (1988). Phys. Rev. B **37** (2946) .
Brown, D. W., West, B. J. and Lindenberg, K. (1986). Phys. Rev. A **33** (4104) and 4110.
Bullough, R. K. and Caudrey, P. J. (1982). Solitons, Springer-Verlag, New York, pp. 80-160.
Cai, G. P., Chen, L. L. and Yang, Q. N. (1992). Chin. J. Sickness of Labour-health **10** (129) .
Careri, G., Buontempo, U., Caeta, F., Gratton, E. and Scott, A. C. (1983). Phys. Rev. Lett. **51** (304) .
Careri, G., Buontempo, U., Galluzzi, F., Scott, A. C., Gratton, E. and Shyamsunder, E. (1984). Phys. Rev. B **30** (4689) .
Careri, G., Gransanti, A. and Ruple, J. A. (1998). Phys. Rev. A **37** (2703) .
Careri, G., Gratton, E. and Shyamsunder, E. (1998). Phys. Rev. A **37** (4048) .
Carr, J. and Eilbeck, J. C. (1985). Phys. Lett. A **109** (201) .
Caspi, S. and Ben-Jacob, E. (2000). Phys. Lett. A **272** (124) .
Chen, X. R., Gou, Q. Q. and Pang, X. F. (1996). Chin. Phys. Lett. **13** (660) .
Chen, X. R., Gou, Q. Q. and Pang, X. F. (1997). Chin. J. Atom. Mol. Phys. **14** (393) ; Chin. J. Chem. Phys. **10** (145) .
Chen, X. R., Gou, Q. Q. and Pang, X. F. (1998). Acta Phys. Sin. **7** (329) ; J. Sichuan University (nature) Sin. **35** (362) .
Chen, X. R., Gou, Q. Q. and Pang, X. F. (1999). Acta Phys. Sin. **8** (1313) ; Chin. J. Chem. Phys. **11** (240) ; Chin. J. Comput. Phys. **16** (346) ; Commun. Theor. Phys. **31** (169) .
Chen, Y. J., Lin, L. M. and Ma, C. L. (1993). Chin. J. Biophys. **9** (673) .
Chi, X. S., Pang, D. B., Li, Y. M., Yang, H. Y., Ye, D. O. and Zhu, G. X. (1994). Chin. J. Biophys. **10** (163) .
Christiansen, P. L. and Scott, A. C. (1990). Self-trapping of vibrational energy, Plenum Press, New York.
Cottingham, J. P. and Schweitzer, J. W. (1989). Phys. Rev. Lett. **62** (1792) .
Cruzeiro-Hansson, L. (1992). Phys. Rev. A **45** (4111) .
Cruzeiro-Hansson, L. (1993). Physica **68D** (65) .
Cruzeiro-Hansson, L. (1994). Phys. Rev. Lett. **73** (2927) .
Cruzeiro-Hansson, L. and Kenkre, V. M. (1994). Phys. Lett. A **190** (59) .
Cruzeiro-Hansson, L. and Takeno, S. (1997). Phys. Rev. E **56** (894) .
Cruzeiro-Hansson, L., Christiansen, P. C. and Scott, A. C. (1990). in Davydov's soliton revisited: Self-trapping of vibrational energy, eds. Christiansen, P. L. and Scott, A. C., Plenum, New York, p. 325.
Cruzeiro, L., Halding, J., Christiansen, P. L., Skovgard, O. and Scott, A. C. (1985). Phys. Rev. A **37** (703) .
Davydov, A. S. (1973). J. Theor. Biol. **38** (559) .
Davydov, A. S. (1975). Theory of molecular exciton, 2nd ed. Plenum Press, New York.
Davydov, A. S. (1977). J. Theor. Biol. **66** (379) .
Davydov, A. S. (1979). Phys. Scr. **20** (387) .
Davydov, A. S. (1980). Zh. Eksp. Theor. Fiz. **78** (789) ; Sov. Phys. JETP **51** (397) .

Davydov, A. S. (1981). Physica **3D** (1) .
Davydov, A. S. (1982). Sov. Phys. USP. **25** (898) ; Biology and quantum mechanics, Pergamon, New York.
Davydov, A. S. (1985). Solitons in molecular systems, D. Reidel Publishing, Dordrecht.
Davydov, A. S. (1991). in Davydov's soliton revisited: Self-trapping of vibrational energy, eds. Christiansen, P. L. and Scott, A. C., Plenum, New York, p. 11.
Davydov, A. S. and Kislukha, N. I. (1973). Phys. Stat. Sol. (b) **59** (465) .
Davydov, A. S. and Kislukha, N. I. (1976). Phys. Stat. Sol. (b) **75** (735) .
Eilbeck, J. C., Lomdahl, P. S. and Scott, A. C. (1984). Phys. Rev. B **30** (4703) .
Eremko, A. A., Gaididel, Yu. B. and Vaknenko, A. A. (1985). Phys. Stat. Sol. (b) **127** (703) .
Fann, W., Rothberg, L., Roberso, M., Benson, S., Madey, J., Etemad, S. and Austin, R. (1990). Phys. Rev. Lett. **64** (607) .
Fohlich, H. (1980). Adv. Electron. Electron Phys. **53** (86) .
Fohlich, H.,*et al.* (1983). Coherent excitation in biology, Springer, Berlin.
Förner, W. (1991). Phys. Rev. A **44** (2694) ; J. Phys.: Condensed Matter **3** (1915) and 3235.
Förner, W. (1992). J. Phys.: Condensed Matter **4** (4333) ; J. Comput. Chem. **13** (275) .
Förner, W. (1993). J. Phys.: Condensed Matter **5** (823) , 883, 3883, 3897; Physica **68D** (68) .
Förner, W. (1994). J. Phys.: Condensed Matter **6** (9089) .
Förner, W. (1996). Phys. Rev. B **53** (6291) .
Förner, W. (1998). J. Phys.: Condensed Matter **10** (2631) .
Förner, W. and Ladik, J. (1991). in Davydov's soliton revisited: Self-trapping of vibrational energy, eds. Christiansen, P. L. and Scott, A. C., Plenum, New York, p. 11.
Frushour, B. G.,*et al.* (1975). Biopolymers **14** (379) .
Gashi, F., Gashi, R., Stepancic, B. and Zakula, R. (1986). Physica **135A** (446) .
Glanber, R. J. (1963). Phys. Rev. **13** (2766) .
Goldanskill, V. I., Krupyanskii, Yu. F. and Flerov, V. N. (1983). Dokl. Akad. Nauk (SSSR) **272** (978) .
Guo, Bai-lin and Pang Xiao-feng, (1987). Solitons, Chin. Science Press, Beijing.
Halding, J. and Lomdahl, P. S. (1987). Phys. Lett. A **124** (37) .
Ho, M. W., Popp, F. A. and Warnke, U. (1994). Bioelectrodynamics and Biocommunication, World Scientific, Singapore.
Hyman, J. M., Mclaughlin, D. W. and Scott, A. C. (1981). Physica **D3** (23) .
Ivic, Z. (1998). Physica **D113** (218) .
Ivic, Z. and Brown, D. W. (1989). Phys. Rev. Lett. **63** (426) .
Ivic, Z., Sataric, M., Shemsedini, Z. and Zakula, R. (1988). Phys. Scr. **37** (546) .
Kabo, R. (1962). J. Phys. Soc. Japan **17** (206) .
Kenkre, V. M., Raghvan, S. and Cruzeiro-Hansson, L. (1994). Phys. Rev. B **49** (9511) .
Kerr, W. C. and Lomdahl, P. S. (1989). Phys. Rev. B **35** (3629) .
Kerr, W. C. and Lomdahl, P. S. (1991). in Davydov's soliton revisited: Self-trapping of vibrational energy, eds. Christiansen, P. L. and Scott, A. C., Plenum, New York, p. 23.
Knox, R. S., Maiti, S. and Mu, P. (1990). Search for remote transfer of vibrational energy in proteins, in Davydov's soliton revisited, eds. Christiansen, P. L. and Scott, A. C., Plenum Press, New York, p. 401.
Koeming, J. L. (1972). J. Polym. Sci-PD **59** (177) .
Konev, S. V. (1965). Excited states of biopolymers, Science and technique, Minsk.
Lawrence, A. F., McDaniel, J. C., Chang, D. B., Pierce, B. M. and Brirge, R. R. (1986).

Phys. Rev. A **33** (1188) .
Lipkin, H. I. (1973). Quantum mechamics, North-Holland, Amsterdam.
Lomdahl, P. S. and Kerr, W. C. (1991). in Davydov's soliton revisited: Self-trapping of vibrational energy, eds. Christiansen, P. L. and Scott, A. C., Plenum, New York, p. 259.
London, R. (1986). The quantum theory of light, 2nd ed., Oxford University Press, Oxford.
Macneil, L. and Scott, A. C. (1984). Phys, Scr. **29** (284) .
Mechtly, B. and Shaw, P. B. (1988). Phys. Rev. B **38** (3075) .
Motschman, H., Förner, W. and Ladik, J. (1989). J. Phys.: Condensed Matter **1** (5083) .
Mouritsen, O. G. (1980). Phys. Rev. B **22** (1127) .
Mrevlislivil, G. M. (1979). Usp. Fiz. Nauk **128** (273) [Sov. Phys. USP. **22** (433)].
Nagle, J. F., Mille, M. and Morowitz, H. J. (1980). J. Chem. Phys. **72** (3959) .
Pang, X. F. and Chen, X. R. (2000a). Chin. Phys. **9** (108) .
Pang, X. F. and Chen, X. R. (2000b). Commun. Theor. Phys. **32** (437) .
Pang, X. F. and Chen, X. R. (2001). J. Phys. Chem. Solids **62** (793) .
Pang, X. F. and Chen, X. R. (2002a). Commun. Theor. Phys. **37** (715) .
Pang, X. F. and Chen, X. R. (2002b). Phys. Stat. Sol. (b) **229** (1397) .
Pang, Xiao-feng (1986a). Chin. Acta Biochem. BioPhys. **18** (1) .
Pang, Xiao-feng (1986b). Chin. J. Appl. Math. **10** (278) .
Pang, Xiao-feng (1986c). Chin. J. Atom. Mol. Phys. **4** (275) .
Pang, Xiao-feng (1987). Chin. J. Atom. Mol. Phys. **5** (383) .
Pang, Xiao-feng (1989a). Chin. J. Atom. Mol. Phys. **7** (1235) .
Pang, Xiao-feng (1989b). Chin. J. Low Temp. Supercond. **10** (612) .
Pang, Xiao-feng (1990). J. Phys. Condens. Matter **2** (9541) .
Pang, Xiao-feng (1992a). Chin. J. Light Scattering **4** (125) .
Pang, Xiao-feng (1992b). J. Nature Sin. **15** (915) .
Pang, Xiao-feng (1992c). J. Sichuan Univ. (Nature) Sin. **29** (491) .
Pang, Xiao-feng (1993a). Acta Math. Sci. **13** (437) .
Pang, Xiao-feng (1993b). Acta Phys. Sin. **42** (1856) .
Pang, Xiao-feng (1993c). Chin. J. Biophys. **9** (631) .
Pang, Xiao-feng (1993d). Chin. J. Infrared Millimeter Wave **12** (377) .
Pang, Xiao-feng (1993e). Chin. Phys. **22** (612) .
Pang, Xiao-feng (1993f). Chin. Phys. Lett. **10** (381) .
Pang, Xiao-feng (1993g). Chin. Phys. Lett. **10** (437) .
Pang, Xiao-feng (1993h). Chin. Phys. Lett. **10** (517) .
Pang, Xiao-feng (1993i). Chin. Sci. Bull. **38** (1572) .
Pang, Xiao-feng (1993j). Chin. Sci. Bull. **38** (1665) .
Pang, Xiao-feng (1993k). J. Sichuan Univ. (Nature) Sin. **30** (48) .
Pang, Xiao-feng (1994a). Acta Phys. Sin. **43** (1987) .
Pang, Xiao-feng (1994b). Chin. J. Biophys. **10** (133) .
Pang, Xiao-feng (1994c). Phys. Rev. E **49** (4747) .
Pang, Xiao-feng (1994d). Theory of nonlinear quantum mechanics, Chongqing Press, Chongqing.
Pang, Xiao-feng (1995). Chin. J. Phys. Chem. **12** (1102) .
Pang, Xiao-feng (1996). Acta Math. Sci. (Suppl.) **16** (1) .
Pang, Xiao-feng (1997). Chin. J. Infrared Millimeter Wave **16** (288) .
Pang, Xiao-feng (1999). European Phys. J. B **10** (415) .
Pang, Xiao-feng (2000a). Chin. Phys. **9** (86) and 108.
Pang, Xiao-feng (2000c). J. Phys. Condens. Matter **12** (885) .
Pang, Xiao-feng (2001a). Commun. Theor. Phys. **33** (323) .

Pang, Xiao-feng (2001b). Commun. Theor. Phys. **35** (763) .
Pang, Xiao-feng (2001c). European Phys. J. B **19** (297) .
Pang, Xiao-feng (2001d). Inter. J. Infr. Mill. Waves **22** (277) and 291.
Pang, Xiao-feng (2001e). Physica D **154** (138) .
Pang, Xiao-feng (2001f). Phys. Rev. E **62** (6989) .
Pang, Xiao-feng (2001g). Phys. Sin. **28** (143) .
Pang, X. F. (Pang Xiao-feng), (2002). Chin. J. Atom. Mol. **19** (319) and 417.
Pang, Xiao-feng (2003). Soliton physics, Sichuan Sci. and Tech. Press, Chengdu.
Pang, Xiao-feng (2004a). Commun. Theor. Phys. **41** (165) and 561.
Pang, Xiao-feng (2004b). Commun. Theor. Phys. **42** (458) .
Pope, M. and Swenberg, C. E. (1982). Electronic Processes in Organic Crystals, Clarendon Press/Oxford University Press, Oxford/New York, 1982.
Popp, F. A., Li, K. H. and Gu, Q. (1993). Recent advances in biophoton research and its application, World Scientific, Singapore.
Potter, J. (1970). Quantum mechamics, North-Holland, Amsterdam.
Sataric, M. and Zakula, R. (1984). II Nuovo Cimento **D3** (1053) .
Sataric, M., Ivic, Z. and Zakula, R. (1986). Phys. Scr. **34** (283) .
Schweitzer, J. W. (1992). Phys. Rev. A **45** (8914) .
Schweitzer, J. W. and Cottingham, J. P. (1991). in Davydov's soliton revisited: Self-trapping of vibrational energy, eds. Christiansen, P. L. and Scott, A. C., Plenum, New York, p. 285.
Scott, A. C. (1982). Phys. Rev. A **26** (578) .
Scott, A. C. (1982). Phys. Scr. **25** (651) .
Scott, A. C. (1983). Phys. Rev. A **27** (2767) .
Scott, A. C. (1984). Phys. Scr. **29** (279) .
Scott, A. C. (1990). Physica D **51** (333) .
Scott, A. C. (1992). Phys. Rep. **217** (67) .
Scott, A. C. (1998). Phys. Lett. A **86** (603) .
Scott, A. C., Bigio, I. J. and Johnston, C. T. (1989). Phys. Rev. B **39** (12883) .
Scott, A. C., Gratton, E., Shyamsunder, E. and Careri, G. (1985). Phys. Rev. B **32** (5551) .
Silinsh, E. A. and Capek, V. (1994). Organic Molecular Crystals, AIP Press, New York.
Skrinjar, M. J., Kapor, D. W. and Stojanovic, S. D. (1988). Phys. Lett. A **133** (489) .
Skrinjar, M. J., Kapor, D. W. and Stojanovic, S. D. (1988). Phys. Rev. A **38** (6402) .
Skrinjar, M. J., Kapor, D. W. and Stojanovic, S. D. (1988). Phys. Scr. **39** (658) .
Skrinjar, M. J., Kapor, D. W. and Stojanovic, S. D. (1989). Phys. Rev. B **40** (1984) .
Spatschek, K. H. and Mertens, F. G. (1994). Nonlinear coherent structures in physics and Biology, Plenum Press, New York.
Stiefel, J. (1965). Einfuhrung in die Numerische Mathematik, Teubner Verlag, Stuttgart.
Takeno, S. (1985). Prog. Theor. Phys. **73** (853) .
Takeno, S. (1986). Prog. Theor. Phys. **71** (395) .
Takeno, S. (1986). Prog. Theor. Phys. **75** (1) .
Takeno, S. (1991). in Davydov's soliton revisited: Self-trapping of vibrational energy, eds. Christiansen, P. L. and Scott, A. C., Plenum, New York, p. 56.
Tan, Benkui and Boyd, J. P. (1998). Phys. Lett. A **240** (282) .
Tekec, J., Ivic, Z. and Priulj, Z. (1998). J Phys.: Condensed Matter **10** (1487) .
Vaconcellos, A. R. and Luzzi, R. (1993). Phys. Rev. E **48** (2246) .
Venzel, G. and Fischer, S. F. (1984). J. Phys. Chem. **81** (6090) .
Vladiminov, Y. A. (1965). Photochemistry and luminescence of protein, Science, Moskova.
Wanger, J. and Kongeter, A. (1989). J. Chem. Phys. **91** (3036) .

Wang, X., Brown, D. W., Lindenberg, K. (1988). Phys. Rev. A **37** (3357) .
Wang, X., Brown, D. W., Lindenberg, K. (1989). J. Mol. Liq. **4** (123) .
Wang, X., Brown, D. W., Lindenberg, K. (1989). Phys. Rev. B **39** (5366) .
Wang, X., Brown, D. W., Lindenberg, K. (1989). Phys. Rev. Lett. **62** (1792) .
Wang, X., Brown, D. W., Lindenberg, K. (1991). in Davydov's soliton revisited: Self-trapping of vibrational energy, eds. Christiansen, P. L. and Scott, A. C., Plenum, New York, p. 83.
Webb, S. J. (1980). Phys. Rep. **60** (201) .
Webb, S. J. (1985). The crystal properties of living cells ans seen by millinmeter microwaves and Raman spectroscopy, in the living state-II, ed. Mishra, R. K., World Scientific, Singapore, pp. 367-403.
Xiao, He-Lan, Cai Guo-ping, Sun Su-qin and Pang Xiao-feng (2003). Chin. Atom. Mol. Phys. **20** (211) .
Xie, A., van der Meer, A. F. G. and Austin, R. H. (2002). Phys. Rev. Lett. **88** (018102) .
Young, E., Shaw, P. B. and Whitfield, G. (1979). Phys. Rev. B **19** (1225) .
Zekovic, S. and Ivic, Z. (1999). Bioclectrochem. Bioenergetics **48** (297) .

Chapter 10

Properties of Nonlinear Excitations and Motions of Protons, Polarons and Magnons in Different Systems

In this chapter, we continue to discuss properties of excitations and motions of three types of microparticles, proton, polaron and magnon, in various nonlinear systems. We will see that their motions also obey nonlinear quantum mechanics.

10.1 Model of Excitation and Proton Transfer in Hydrogen-bonded Systems

There are many examples of hydrogen-bonded systems, which consist of a series of hydrogen bonds, in condensed matters and living systems, such as ice, solid alcohol, carbon hydrates and proteins. These systems exhibit a considerably large electrical conductivity even though electron transport through the systems is hardly supported. Through long period of studies, we have known now that this phenomenon is caused by proton transfer in these systems. However, understanding of proton transfer in such systems is a long-standing problem. Nonlinear dynamics and soliton motion provide a possibility to resolve this issue. In view of their close connection with phenomena related to proton transfer across biological membranes, study of such systems becomes even more important and is expected to provide insights to some fundamental processes of life.

In the studies of proton transfer processes in hydrogen-bonded systems, it is suffice to consider one-dimensional chains which are referred to as Bernal-Fowler filaments. In the normal state of a chain, each proton is linked to a heavy ion (or oxygen atom in ice) by a covalent bond on one side, and a hydrogen bond on the other. Therefore, there can be two types of arrangements of hydrogen bonded states in these systems, namely the X–H$\cdots$X–H$\cdots$X–H$\cdots$X–H$\cdots$ type and the H–X$\cdots$H–X$\cdots$H–X$\cdots$H–X$\cdots$H–X$\cdots$ type. Obviously the two systems should have the same energy. The potential experienced by the proton in such a system can be modeled by a double-well potential in which the two minima correspond to the two equilibrium positions of the proton between the two neighboring heavy ions (or oxygen atoms), as shown in Fig. 10.1. The barrier which separates the minima has a height which is in general of the order of the oscillation energy in a covalent X-H bond and is approximately 20 times larger than that in a hydrogen bond. Small amplitude harmonic vibration about their equilibrium positions is assumed for the protons in the hydrogen bonds.

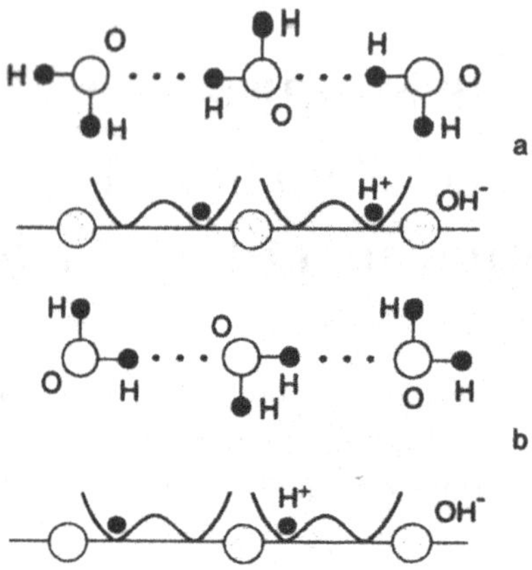

Fig. 10.1 The double-well potential in the hydrogen-bonded system with H^+ in one well (a) or the other (b) of the potential.

If the protons in such systems are perturbed by an externally applied field such as light or energy released by adenosine triphosphate (ATP) hydrolysis in protein molecules, then localized fluctuations of the protons occur, which result in changes in positions of the protons. Protons move and deviate from their equilibrium positions, for instance, by a translation, a jump, a shift, or by hopping in the interbonds and intrabonds. This phenomenon of proton transfer along the hydrogen-bonded chains was observed experimentally, and the protonic conductivity along the chains was found to be about $10^3 - 10^4$ times larger than that in the perpendicular direction. The motion of the protons may result in ionic and bonding (orientational or Bjerrum) defects which correspond to the exchange and rotation of bonds, respectively. Thus, transfer of protons along the hydrogen-bonded chain is a result of the transport of the two types of defects as shown in Figs. 10.2 and 10.3, respectively. It is possible that protons are transferred by jumping from one water molecule to another along the hydrogen-bonded chain, and that a migration of hydroxonium and hydroxyl ionic defects takes place in the intrabonds. When a proton moves from one end of a molecule to the other end, it may form a covalent bond with the molecule, for example, with an oxygen atom in ice, and the original proton moves to a neighboring molecule. Repetition of this process results in a continued motion of the proton along the chain. However, proton transfer cannot occur in one direction, but can be achieved with a re-orientation of OH groups by the second defect mechanism, the Bjerrum defect. The motion of an orientation defect contains simple successive rotations of OH groups, starting at one end of the chain and ending at the other. As a result of these rotations, a pair of D and L defects (see Fig. 10.3) can be created and they can move to different ends in any internal part of the chain. A sequence of rotations of all of the molecules in a filament returns the chain to its original state. It follows that the motion of another proton can occur only after the passage of a Bjerrum defect.

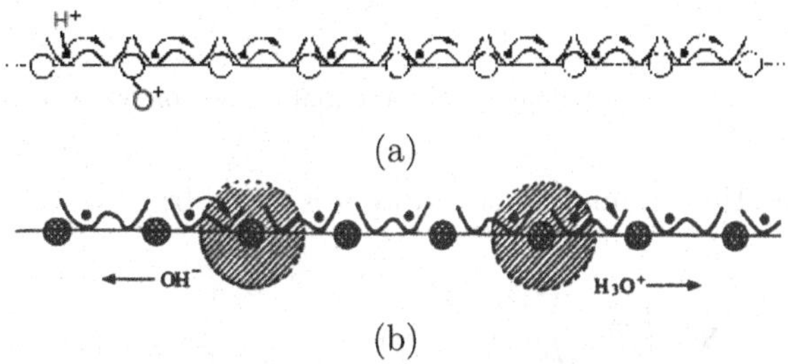

Fig. 10.2 Two ionic defects in a hydrogen-bonded system (ice).

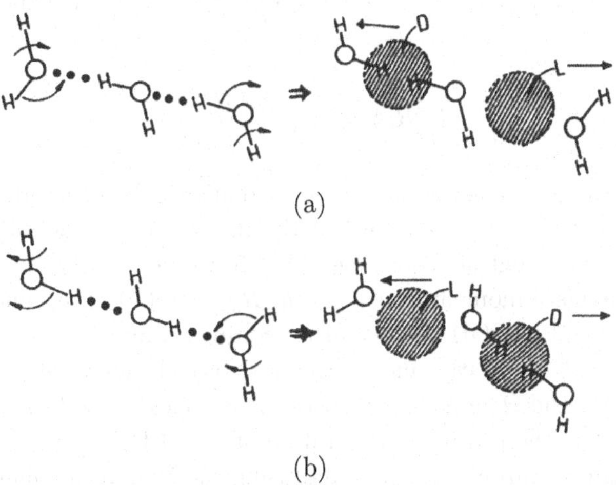

Fig. 10.3 The two Bjerrum defects in a hydrogen-bonded system (ice).

Potential models of the positive ionic and bonding defects in such systems are shown in Fig. 10.4 (a) and (b), respectively.

The soliton excitation model of proton transfer was first proposed by Antonchenko, Davydov and Zolotaryuk (the ADZ model) for crystal ice (Antonchenko

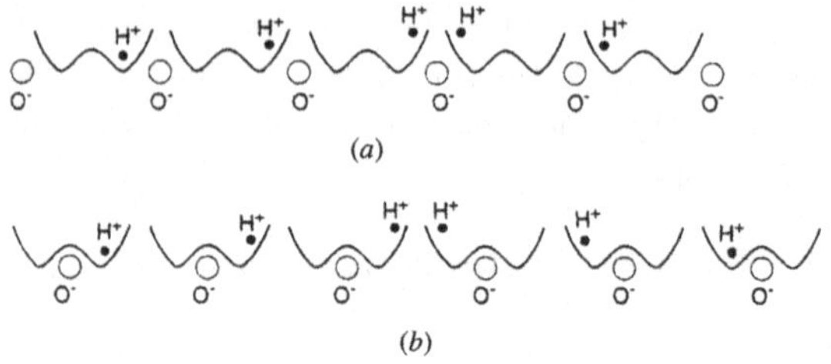

(a)

(b)

Fig. 10.4 The model with double-well potential curves for the two types of defect.

et al. 1983). The Hamiltonian of the system is given by

$$\begin{aligned} H &= H_p + H_{OH} + H_{\text{int}} \\ &= \sum_n \left\{ \frac{m}{2} \left[(R_n)_t^2 + \omega_1^2 (R_{n+1} - R_n)^2 \right] + U(R_n) \right\} \\ &\quad + \sum_n \frac{M}{2} \left[(u_n)_t^2 + \Omega_0^2 u_n^2 + \Omega_1^2 (u_{n+1} - u_n)^2 \right] \\ &\quad + \sum_n \chi u_n (R_n^2 - R_0^2), \end{aligned} \tag{10.1}$$

where the double-well potential for the proton, due to the neighboring OH^- ions, is given by

$$U(R_n) = U_0 \left(1 - \frac{R_n^2}{R_0^2} \right)^2,$$

with U_0 being the height of the barrier of the double-well potential, R_0 the distance between the local maximum and one of the minima of the double-well potential, and R_n the displacement of proton measured from the top of the barrier. The corresponding conjugate momentum is $p_n = m(R_n)_t$. In (10.1), u_n and $P_n = M\,(u_n)_t$ are displacement of the OH^- ion, and its conjugate momentum, respectively, M and m are the masses of OH^- and proton, respectively, ω_1 is vibrational frequency of the proton, Ω_0 and Ω_1 are characteristic frequencies of the OH^-, χ is a coupling coefficient between the proton and vibration of the OH^-.

In the continuum approximation, and using the Hamilton equation in the nonlinear quantum mechanics, we can get the following equations of motion for the proton and OH^- ion from (10.1),

$$R_{tt} - C_0^2 R_{xx} - \omega_0^2 \left(1 - \frac{R^2}{R_0^2} \right) R + 2\frac{\chi}{m} u R = 0, \tag{10.2}$$

$$u_{tt} - \Omega_0^2 u - v_1^2 u_{xx} + \frac{\chi}{M} \left(R^2 - R_0^2 \right) = 0, \tag{10.3}$$

respectively. Obviously, transfer of the protons and ionic motion are nonlinear problems. In the case of $\chi = 0$, the following analytic solution can be obtained.

$$R = \mp R_0 \tanh \left[\frac{\mu (x - x_0 - vt)}{r_0} \right], \quad \mu = \frac{2U_0}{m\omega_1^2 R_0^2 (1 - s^2)},$$

where $s = v/C_0$, $C_0 = \omega_1 r_0$, and r_0 is the lattice constant. When $v = v_1$, equations (10.2) – (10.3) have a soliton solution which is given by

$$R = \mp R_0 \tanh \left[\frac{\mu_0 (x - x_0 - vt)}{r_0} \right], \quad u = u_0 \operatorname{sech}^2 \left[\frac{\mu_0 (x - x_0 - vt)}{r_0} \right],$$

where

$$\mu_0 = \mu(R_0) = \sqrt{\frac{1}{m\omega_1^2 (1 - s_0^2)} \left(\frac{2U_0}{R_0^2} - \frac{\chi^2 R_0^2}{M\Omega_0^2} \right)},$$

$$s_0 = \frac{v_1}{C_0}, \quad u_0 = \frac{\chi R_0^2}{M\Omega_0^2}, \quad v_1 = \Omega_1 r_0.$$

If $\chi \neq 0$, and $v \neq v$, equations (10.2) – (10.3) do not have analytic solution. The solutions can only be found using approximation method or numerically.

Further investigations on the ADZ model lead to solutions for a far greater range of velocity values. The problem was addressed in a number of publications and a variety of theoretical extensions, including the one-component protonic chain with a new two-parameter, double periodic, one-site potential proposed by Pnevmatikos *et al.*, have been developed. Shortcoming of the ADZ model is that the coupling between the protons and oxygen atoms (or heavy ions) is only one mechanism of reducing the potential barrier, which protons have to overcome to move from one molecule to another. This is included in the ADZ model by coupling the proton motion with an optical mode of the heavy ionic sublattice. The nature of the proton sublattice depends on the systems to be studied. In this model, an ionic defect appears as a kink or solitary wave in the proton sublattice, propagating together with a localized contraction of the relative distance between neighboring oxygens. This excitation is referred to as a two-component solitary wave. In the ADZ model, the proton potential with the double-well ansatz plays an essential role in the description of the motion of ion defects as topological solitary waves. The nonlinear interaction generated by the coupling between the protons and oxygen atoms plays only a secondary role by reducing the height of the barrier. Therefore, the properties of the solitons are mainly determined by the double-well potential. Thus, this model is only effective for explaining the transfer of ion defects. The equations involved are also very difficult to solve, and exact analytical solutions cannot be obtained. Furthermore, if realistic values for the parameters of hydrogen-bonded systems are considered, the continuum approximation fails due to the narrowing of the domains of validity of the solutions with respect to the lattice spacing. This is the case in ice in which the H_3O^+ or OH^- ions become almost point defects. It is also very difficult to accept the one-component model proposed by Phevmatikos *et al.* because the influence of the heavy ionic sublattice on the protons in such a model was not considered in details. Therefore, the proton transfer in the hydrogen-bonded systems is still an open problem. A complete theoretical description of the combined effect of the transfers of both types of defects had not been possible until a model was proposed by Pang and Miller.

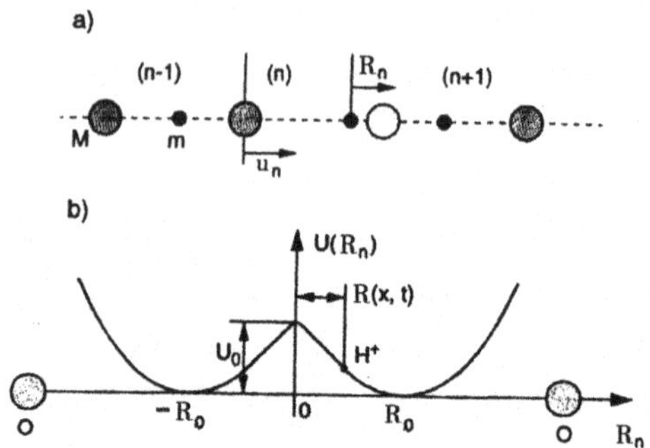

Fig. 10.5 The one-dimensional lattice model for a hydrogen-bonded quasi-diatomic chain.

Pang and Miller proposed a model to study the dynamic properties of proton transfer resulting from the localized fluctuations of protons and structural deformation of the heavy ionic sublattice due to the displacement of the protons. In this model, it is assumed that the hydrogen-bonded chain consists of two interacting sublattices of harmonically coupled protons of mass m and heavy ions (hydroxyl groups of ice, or complex negative ions, of mass M, as shown in Fig. 10.5. Each proton lies between a pair of heavy ions, usually referred to as 'oxygens'. The proton is connected by one covalent bond and one hydrogen bond to the two neighboring oxygens. Therefore, the potential energy of the proton in each hydrogen bond has the form of a double-well potential with the two minima corresponding to the two equilibrium positions of the proton. Obviously the double-well potential is motivated, physically, by the simultaneous electromagnetic interaction of the two neighboring oxygens with the proton.

If the proton can overcome the central barrier of the double-well potential and can move from one well to the other, the relative positions of the proton and the two neighboring oxygens have changed, and the positions of the covalent and hydrogen bonds are exchanged. Thus, ionic defects occur in the system. In such a case, the position of the proton in the hydrogen bond is mainly determined by the double-well potential. The proton displacement is controlled by the elastic interaction involved in the model. However, when the proton approaches the neighboring oxygen the coupling interaction between the proton and the oxygen will be greatly enhanced in the intrinsically nonlinear system. Thus, the relative position between the proton and the oxygen will also be significantly changed, *i.e.* the migration of the proton as well as the deformation of the heavy ionic sublattice by stretching and compression are enhanced. This phenomenon may result in change of direction of the covalent bond between the proton and oxygen, *i.e.* rotation of the bond. In fact, as far as the covalent bond is concerned, bond rotation in chemistry is simply carried out by relative displacements of the proton and the oxygens with charges. Thus the Bjerrum defect occurs due to the coupling interaction. Therefore, the mechanism for formation of this defect is different from that of the ionic defect mentioned above, although both are produced by changes in the relative positions of the protons and oxygens.

An electromagnetic interaction between neighboring protons, for example, the dipole-dipole interaction and resonant interaction, was included in the Pang-Miller model, besides the above double-well potential and the elastic interaction due to the covalent interaction and related actions. Thus, it is also natural to take into account the changes in the relative positions of the neighboring heavy ions resulting from this interaction. Assuming again the harmonic model with acoustic vibrations of low frequency for the heavy ionic sublattice, the Hamiltonian of system can be written as

$$\begin{aligned} H &= H_\rho + H_{ion} + H_{\text{int}} \\ &= \sum_n \left[\frac{p_n^2}{2m} + \frac{1}{2}m\omega_0^2 R_n^2 - \frac{1}{2}m\omega_1^2 R_n R_{n+1} + U(R_n)\right] \\ &\quad + \sum_n \left[\frac{P_n^2}{2M} + \frac{1}{2}\beta(u_n - u_{n-1})^2\right] \\ &\quad + \sum_n \left[\frac{1}{2}\chi_1 m\left(u_{n+1} - u_{n-1}\right) R_n^2 + m\chi_2\left(u_{n+1} - u_n\right) R_n R_{n+1}\right], \end{aligned} \tag{10.4}$$

where

$$U\left(R_n\right) = U_0\left[1 - \left(\frac{R_n}{R_0}\right)^2\right]^2.$$

In (10.4), R_n and $p_n = m(R_n)_t$ are the proton displacements and momenta, respectively, with R_n being defined relative to the mid-point between the nth and the $(n+1)th$ heavy ions or OH's in the static case. R_0 is the distance between the central maximum and one of the minima of the double well, U_0 is the height of the potential barrier. Similarly, u_n and $P_n = M(u_n)_t$ are the displacement of the nth heavy ion from its equilibrium position and its conjugate momentum, respectively. Furthermore, $\chi_1 = \partial\omega_0^2/\partial u$ and $\chi_2 = \partial\omega_1^2/\partial u$ are coupling constants between the protons and the heavy ionic sublattice, representing the changes in the vibrational energy of the protons and of the coupling energy between neighboring protons due to a unit extension of the heavy ion sublattice, respectively. ω_0 is the frequency of harmonic vibration of the proton. The quantity $(1/2)m\omega_1^2 R_n R_{n+1}$ is the correlation interaction between neighboring protons caused by the dipole-dipole interactions. ω_0 and ω_1 are the diagonal and off-diagonal elements of the dynamical matrix of the proton, respectively. β is the linear elastic constant of the heavy ionic sublattice. m and M are the masses of the proton and heavy ion, respectively. H_p in (10.4) is the Hamiltonian of the protonic sublattice with an on-site double-well potential $U(R_n)$, H_{ion} is the Hamiltonian of the heavy ionic sublattice with low-frequency harmonic vibration and H_{int} is the interaction Hamiltonian between the protonic and the heavy ionic sublattices.

The Pang-Miller model is still based on coupling of two oscillators (proton and heavy ion), and in that sense, it is similar to the ADZ model. However, it differs from the ADZ model in the following. (i) Due to the large mass and the large number of atoms or atomic groups, motion of the heavy ion is simply harmonic, with low-frequency acoustic vibration. In contrast, both acoustic and optical vibrations are allowed for heavy ions in the ADZ model, but the physical origins of this model are rather vague. Generally, it is believed that optical and acoustic vibrations are two different forms of vibration in nature. Therefore, the Pang-Miller model for heavy ion is more appropriate than the ADZ model. (ii) For protons lying in the double-well potential, Pang and Miller adopted a harmonic oscillator model with optical vibration, but it includes an off-diagonal factor, which comes from the interaction between neighboring protons, and interaction with the heavy ions. Thus the vibrational frequencies of the protons are related to displacements of the heavy ions. Therefore, the Hamiltonian in the Pang-Miller model has high symmetry and a one-to-one corresponding relation for these interactions. However, in the ADZ model, the vibration of the proton is acoustic. This is not reasonable because the vibration frequency of the proton is very high compared to the heavy ion due to its small mass and strong interaction. Therefore the Pang-Miller model is more appropriate than the ADZ model. Moreover, the relation between the protonic and interactional Hamiltonians in the ADZ model does not have a one-to-one correspondence, the physical meaning of the interaction Hamiltonian is also very vague or difficult to understand. It does not have a strict analytic solution and therefore it is difficult to investigate the properties and mechanism of proton transfer in the systems in the ADZ model. Because the Hamiltonian in the Pang-Miller model not only includes the optical vibration of the protons, but also the resonant interaction between the protons caused by the electromagnetic interactions between neighboring protons, and it also takes into account both the change of the relative displacement of the neighboring heavy ions resulting from the vibration of the proton and the correlation interaction between the neighboring protons, we can expect that the Pang-Miller model can reveal some new results compared to the ADZ model or other models.

10.2 Theory of Proton Transferring in Hydrogen Bonded Systems

When we use quantum theory to study the transfer of the protons, u_n, P_n and R_n, p_n should be regarded as operators. Following (9.2), we make a standard transformation in the second quantization representation

$$R_n = \sqrt{\frac{2m\omega_0}{\hbar}}(a_n + a_n^+), \quad p_n = \sqrt{\frac{\hbar m\omega_0}{2}}(-i)(a_n - a_n^+), \tag{10.5}$$

where $i = \sqrt{-1}$, a_n^+ (a_n) is the creation (annihilation) operator of the proton. Equation (10.4) becomes

$$\begin{aligned} H = \sum_n \Bigg\{ & \hbar\omega_0 a_n^+ a_n - \left[\frac{\hbar\omega_1^2}{4\omega_0} - \frac{\hbar\chi_2}{2\omega_0}(u_{n+1} - u_n)\right](a_{n+1}^+ + a_{n+1})(a_n^+ + a_n) + \\ & U_0\left[1 - \frac{\hbar}{m\omega_0 R_0^2}(a_n^+ + a_n)^2 + \frac{\hbar^2}{4m^2\omega_0^2 R_0^4}(a_n^+ + a_n)^4\right] + \\ & \frac{\hbar\chi_1}{4\omega_0}(u_{n+1} - u_{n-1})(a_n^+ + a_n)^2 + \left[\frac{M}{2}\dot{u}_n^2 + \frac{\beta}{2}(u_{n+1} - u_n)^2\right]\Bigg\}. \end{aligned} \tag{10.6}$$

From (10.6), we see that the motion of proton is a nonlinear problem because there is a term in the fourth power of the operator a_n^+ (or a_n) arising from the double-well potential and coupling interaction between proton and vibrational quantum of the heavy ions. Therefore, the characteristics of the proton is changed when compared with a bare proton. Because here we do not consider the spin of the proton, the wave function describing the collective excitations arising from the localized fluctuation of the protons and the structural deformation of the heavy ionic sublattice can be written as

$$\begin{aligned} |\Phi\rangle = |\varphi\rangle|\beta\rangle = & \frac{1}{\lambda'}\left[1 + \sum_n \varphi_n(t) a_n^+\right]|0\rangle_{pr} \times \\ & \exp\left\{\sum_n \frac{1}{i\hbar}\left[\beta_n(t)P_n - \pi_n(t)R_n\right]\right\}|0\rangle_{ph}, \end{aligned} \tag{10.7}$$

where $|0\rangle_{pr}$ and $|0\rangle_{ph}$ are the ground states of the proton and the vibrational excitation of the heavy ionic sublattice (phonon), respectively. $\varphi_n(t)$, $\beta_n(t) = \langle\Phi|R_n|\Phi\rangle$ and $\pi_n(t) = \langle\Phi|P_n|\Phi\rangle$ are three sets of unknown functions. We assume hereafter that $\lambda' = 1$ for convenience of calculation except when explicitly mentioned. The wave function of the proton,

$$|\varphi\rangle = \frac{1}{\lambda'}\left[1 + \sum_n \varphi_n(t) a_n^+\right]|0\rangle_{pr},$$

is not an excitation state of a single particle, but rather a coherent state, and contains only one proton. From (10.6) and (9.59) we obtained

$$\begin{aligned} (\beta_n)_{tt}(t) = & \beta(\beta_{n+1} + \beta_{n-1} - 2\beta_n) + \frac{\hbar\chi_1}{2\omega_0}(|\varphi_{n+1}|^2 - |\varphi_{n-1}|^2) \\ & - \frac{\hbar\chi_n}{2\omega_0}(\varphi_{n-1}^*\varphi_n + \varphi_n^*\varphi_{n-1} - \varphi_n\varphi_{n+1}^* - \varphi_n^*\varphi_{n+1}). \end{aligned} \tag{10.8}$$

From (10.6) and (9.7), we get approximately

$$\begin{aligned}-(\varphi_n)_{tt} \approx \omega_0^2\varphi_n - \frac{1}{2}\omega_1^2(\varphi_{n+1}+\varphi_{n-1}) + \left[\chi_1(\beta_{n+1}-\beta_{n-1}) - \frac{4U_0}{mR_0^2}\right]\varphi_n \\ +\chi_2[(\beta_{n+1}-\beta_n)\varphi_{n+1} + (\beta_n - \beta_{n-1})\varphi_{n-1}] \\ +\frac{8\hbar U_0}{m^2R_0^4\omega_0^2}|\varphi_n|^2\varphi_n + \frac{6\hbar U_0}{m^2R_0^4\omega_0^2}\varphi_n. \end{aligned} \tag{10.9}$$

In the continuum and long-wave length approximations, we can obtain the following from (10.8) and (10.9).

$$\begin{aligned}\varphi_{tt} = \varepsilon'\varphi + \frac{1}{2}\omega_1^2 r_0^2\varphi_{xx} - 2(\chi_1+\chi_2)r_0\beta_x(x,t)\varphi \\ -\frac{8\hbar U_0}{m^2R_0^4\omega_0}|\varphi|^2\varphi = f_1(\varphi,\beta(x,t)),\end{aligned} \tag{10.10}$$

$$M\frac{\partial^2\beta(x,t)}{\partial t^2} = \beta r_0^2\frac{\partial^2\beta(x,t)}{\partial x^2} + \hbar r_0\frac{(\chi_1+\chi_2)}{\omega_0}\frac{\partial|\varphi(x,t)|^2}{\partial x} = f_2(\varphi,\beta(x,t)), \tag{10.11}$$

where r_0 is lattice constant of the heavy ion sublattice and

$$\varepsilon' = \omega_1^2 - \omega_0^2 + \frac{4U_0}{mR_0^2}\left(1 - \frac{3\hbar}{mR_0^2\omega_0}\right).$$

Obviously, the equations of motion for the proton and heavy ion of the systems obtained in the second quantization representation, equations (10.10) are ϕ^4-equations. It is thus a nonlinear quantum mechanical problem and its motion can be described by the nonlinear quantum mechanics.

Assuming that $\xi = x - vt$, again from (10.10) and (10.11), we can get

$$\beta_x = -\frac{(\chi_1+\chi_2)\hbar r_0}{MC_0^2(1-s^2)\omega_0}|\varphi(x,t)|^2 + A, \tag{10.12}$$

$$\varphi_{tt} = \varepsilon\varphi + v_1^2\varphi_{xx} - g|\varphi|^2\varphi, \tag{10.13}$$

with

$$\begin{aligned}\varepsilon = \varepsilon' - 2A\frac{\chi_1+\chi_2}{r_0}, \quad g = \frac{\hbar U_0}{m^2R_0^2\omega_0} - \frac{2r_0^2(\chi_1+\chi_2)^2}{MC_0^2(1-s^2)\omega_0}, \\ v_1^2 = \frac{1}{2}\omega_1^2r_0^2, \quad s = \frac{v}{C_0}, \quad C_0 = r_0\sqrt{\frac{\beta}{M}},\end{aligned} \tag{10.14}$$

where A is an integral constant, C_0 is the sound velocity in the heavy ionic sublattice. From (10.12) and (10.13), we see clearly that the proton moves in the form of soliton in the nonlinear systems, and there are two nonlinear interactions in this model, the double-well potential and the coupling between the proton and the heavy ion. The competition and balance between the two nonlinear interactions can result in two different kinds of soliton solutions, corresponding to two different kinds of defects (ionic and bonded defects) in the systems. This competition between the two interactions is mainly controlled by the coupling interaction between the protons and heavy ion, $(\chi_1 + \chi_2)$, and proton velocity, v. When the coupling is weak, *i.e.*, when the distance between the proton and the heavy ion is large, the double-well potential dominates, namely, $g > 0$, $\varepsilon > 0$. In this case, the soliton solutions of (10.12) and (10.13) at $0 < v < v_1$ and $v < C_0$ can be written as

$$\varphi(x,t) = \pm\sqrt{\frac{\varepsilon}{g}}\tanh\zeta, \tag{10.15}$$

$$\beta(x,t) = \mp\frac{\sqrt{2}(\chi_1+\chi_2)\hbar r_0}{MC_0^2(1-s^2)g\omega_0}\sqrt{\varepsilon(v_1^2-v^2)}\tanh\zeta, \tag{10.16}$$

where

$$\zeta = \sqrt{\frac{\varepsilon}{2(v_1^2-v^2)}}(x-vt).$$

Equation (10.16) can be written as

$$\beta(x,t) = B\varphi(x,t), \tag{10.17}$$

where

$$B = \frac{\sqrt{2}(\chi_1+\chi_2)\hbar r_0}{MC_0^2(1-s^2)\omega_0}\sqrt{\frac{v_1^2-v^2}{g}}.$$

When $g < 0$ and $\varepsilon < 0$, $0 < v_1 < v$, or $v > C_0$, the solutions of (10.13) are still given by (10.15) and (10.16).

On the other hand, if the coupling interaction dominates relative to the double-well potential, *i.e.*, $g < 0$, $\varepsilon < 0$ and $0 < v < v_1$, $v < C_0$, the solutions of (10.12) and (10.13) are

$$\varphi(x,t) = \mp\sqrt{\frac{2|\varepsilon|}{|g|}}\text{sech}\zeta', \tag{10.18}$$

$$\beta(x,t) = \mp\frac{2(\chi_1+\chi_2)\hbar r_0}{MC_0^2(1-s^2)|g|\omega_0}\sqrt{|\varepsilon|(v_1^2-v^2)}\tanh\zeta', \tag{10.19}$$

where

$$\zeta' = \sqrt{\frac{|\varepsilon|}{2(v_1^2-v^2)}}(x-vt).$$

When $g < 0$, $\varepsilon < 0$ and $v_1 < v$, or $C_0 < v$, solutions of (10.12) and (10.13) are still given by (10.18) and (10.19). Therefore, there are different solutions for different parameters. This indicates that properties of the protons depend mainly on the coupling interaction between the proton and the heavy ions and on the velocity of the proton. Different $(\chi_1 + \chi_2)$ and v can lead to different values of ϵ and g. Thus, we have different forms and properties of soliton solutions in different systems and for different states of the proton.

For crystal ice, typical values of various physical parameters are $r_0 = 2.67$ Å, $R_0 = 1$ Å, $U_0 = 0.22$ eV, $\chi = \hbar\chi_1/2\omega_0 = 0.10$ eV/Å, $\chi' = \hbar\chi_2/2\omega_0 = 0.011$ eV/Å, $C_0 = 2\times10^4$ m/s, $v_1 = (7-9.5)\times10^3$ m/s, $m = m_p$, $M = 17m_p$, $\omega_0 = (1-1.5)\times10^{14}$ s^{-1}, $\omega_1 = (4-5)\times10^{13}$ s^{-1}. Using these values, the amplitude of the soliton in (10.15) for ice can be calculated and the result is

$$\varphi_m = \left\{\left[2\left(\omega_1^2 - \omega_0^2 + \frac{4U_0}{mR_0^2}\right)\left(1 - \frac{3\hbar}{mR_0^2\omega_0}\right) - 2A(\chi_1+\chi_2)r_0\right]\right.$$
$$\left.\left[\frac{8\hbar U_0}{m^2R_0^2\omega_0} - \frac{2\hbar r_0^2(\chi_1+\chi_2)^2}{MC_0^2(1-s^2)\omega_0}\right]^{-1}\right\}^{1/2} \approx 0.9.$$

Its width is given by

$$W_k = \pi\left\{\frac{2(v_1^2 - v^2)}{\omega_1^2 - \omega_0^2 + (4U_0/mR_0^2)(1 - 3\hbar/mR_0^2\omega_0) - 2A(\chi_1+\chi_2)r_0}\right\}^{1/2} \approx 8.1r_0,$$

where

$$A = \frac{\hbar r_0(\chi_1+\chi_2)}{MC_0^2(1-s^2)\omega_0}\sqrt{\frac{\varepsilon}{g}}.$$

This is an indication that the continuous approximation used in the calculation is appropriate to the ice.

To further understand the behaviors of the soliton solutions, we need to study carefully the effective potential $U(\varphi)$ of the systems corresponding to (10.12) and (10.13), because the properties of the protons are determined by the effective potential,

$$U(\varphi) = \frac{\hbar}{\omega_0}\left(-\frac{1}{2}\varepsilon\varphi^2 + \frac{1}{4}g\varphi^4\right) + U_0. \tag{10.20}$$

Obviously, the effective potential consists of a double-well potential and a nonlinear coupling interaction between the proton and heavy ion, and depends on g and ε.

(I) If $\varepsilon > 0$ and $g > 0$, the double-well potential plays a main role in determining properties of the protons. The protons and heavy ions become kink-antikink pairs, (10.15) – (10.16), which can cross over the intrabond barrier, to move from one well to the other of the double-well potential. The coupling interaction between the proton and heavy ion is only of secondary importance, which only reduces the height of the barrier, to allow the proton to easily cross over. This can be understood from the following. Consider the case of $\chi_1 = \chi_2 = 0$. The values of ε and g may change, but as long as $\varepsilon > 0$ and $g > 0$, it does not change the nature of the system. In this case, the effective potential has two degenerate minima

$$U_{\min} = -\frac{\varepsilon^2}{4g} + U_0 \tag{10.21}$$

at

$$\bar{\varphi}_0(\xi) = \varphi_{\min}(\xi) = \pm\sqrt{\frac{\varepsilon}{g}},$$

respectively. Therefore, $U_{\min}$ and its "location", $\bar{\varphi}_0(\xi)$, the barrier height, $\bar{U}_0^*$, of the effective potential depend on g/ε, or, on $(\chi_1+\chi_2)$, ω_0, ω_1, v and U_0. When $\chi_1 = \chi_2 = 0$ and $\omega_1 = \omega_0 = 0$, $U_{\min} \to 0$, $\bar{\varphi}_0(\xi) = \varphi(R_0)$. In such a case, the effective potential of the system and $U(\varphi)$ in (10.20) consist of only the double-well potential, $U(R_n)$ in (10.4), with $U(\varphi)$ given by

$$U(\varphi) = -\frac{2\hbar U_0}{m\omega_0 R_0^2}\left(1 - \frac{3\hbar}{2mR_0^2\omega_0}\right)\varphi^2 + \frac{2\hbar^2 U_0}{m^2R_0^2\omega_0^2}\varphi^4 + U_0,$$

which is not exactly the same as $U(R_n)$ given in (10.4) because the coefficients are different from those in $U(R_n)$, due to quantum effect and the approximate method used in the above derivation.

We can see from (10.20) and (10.21) that the $\bar{\varphi}_0(\xi)$ increases and $\bar{U}_0^*$ decreases with increasing ε and decreasing g, *i.e.*, with increasing coupling constants. Thus, the height of the barrier decreases and the positions of the minima of the potential-wells become further separated in such a case. This means that the possibility for the protons to cross over the barrier is increased with increasing coupling constants. The decrease of the barrier height, $\Delta U_0 = U_0 - \bar{U}_0^*$, and the change in the equilibrium "position", $\Delta\bar{\varphi}_0(\xi) = \varphi(R_0) - \bar{\varphi}_0(\xi)$, can be approximately given by

$$\begin{aligned}\Delta U_0 &\approx U_0\left[\frac{1}{8} - 1 + (2z - z^2)(y+1) - y\right] > 0,\\ \Delta\bar{\varphi}_0(\xi) &\approx \left[\frac{z}{2}\left(1+\frac{y}{2}\right) - \frac{y}{2}\right]\varphi(R_0) > 0,\end{aligned} \tag{10.22}$$

respectively, where

$$\varphi(R_0) = \sqrt{\frac{MR_0^2\omega_0}{8\hbar}},\quad y = \frac{(\chi_1+\chi_2)^2 m^2 R_0^4 r_0^2}{4U_0 M C_0^2(1-s^2)} > 0,$$
$$z = \frac{8\hbar}{R_0^2\omega_0 m} + \frac{mR_0^2}{U_0}\left[2A(\chi_1+\chi_2)r_0 + (\omega_0^2 - \omega_1^2)\right],$$

Equation (10.22) shows that the values of the minima of the potential changes from zero at $\bar{\varphi}_0(\xi) = \varphi(R_0)$ to negative at $\bar{\varphi}(\xi) > \bar{\varphi}_0(\xi)$, *i.e.*, the larger the $(\chi_1+\chi_2)$, the smaller $\bar{\varphi}_0(\xi)$, the lower the height of the barrier, and the more negative the values of the minima of the potential energy. Thus the possibility for the protons to jump over the barrier is enhanced greatly. It clearly shows that the solitons, given by (10.15) and (10.16), in the case of $\varepsilon > 0$ and $g > 0$ describes exactly the motion of the proton crossing over the barrier of the interbond double-well potential to jump from one molecule to another, resulting in ionic defects in the systems. The motions of proton-kinks are accompanied by compression or rarefaction of the heavy ion sublattice around the ionic defects. The physical meaning of the soliton $\varphi(x,t)$

with the plus sign in (10.15) is a localized reduction in the protonic density, (*i.e.*, expansion of the proton sublattice), arising from the motion of the kink-soliton. This effect leads to the creation of a negatively charged carrier and an extended ionic defect moving with a velocity v which is less than the speed of sound C_0. Therefore, the soliton corresponds to the OH^- ionic defect which appears in the Bernal-Fowler filaments. The other soliton solution with the minus sign in $\varphi(x,t)$ in (10.15) represents compression of the protonic sublattice and an increase of the localized proton density which leads to the creation of a positively charged carrier and an extended ionic defect. It corresponds to the H_3O^+ ionic defect. The solutions (10.15) and (10.16) thus represent proton transfer in the form of interbond ionic defects accompanied by a localized deformation of the heavy ionic sublattice. The soliton is referred to as a kink I soliton.

It is also obvious that $U_{\rm min}$ and $\bar{\varphi}_0(\xi)$ in (10.21) decrease with increasing velocity of the proton. It implies that the proton will be further separated from the original heavy ions when its velocity is increased.

(II) In the case of $\varepsilon < 0$ and $g < 0$ which corresponds to the soliton solutions (10.18) and (10.19), the coupling interaction between the protons and heavy ions plays the main role in determining the properties of the protons. The protons become another kind of solitons which shift over the intrabond barriers by the quasi-self-trapping mechanism. In this case, the double-well potential plays a minor role because the soliton solutions of (10.10) and (10.11) are still given by (10.18)-(10.19) when $U_0 = 0$. The above effective potential is still twofold degenerate in such a case and its minima are

$$U'_{\rm min} = \frac{|\varepsilon|^2}{4|g|} - U_0 \approx \frac{MC_0^2(1-s^2)\omega_0^4}{8r_0^2(\chi_1+\chi_2)^2}(1-z')^2(1+y') - U_0, \tag{10.23}$$

(here $A = 0$), located at

$$\bar{\varphi}'_0(\xi) = \pm\sqrt{\frac{|\varepsilon|}{|g|}} \approx \pm\varphi'_0\left(1-\frac{1}{2}z'\right)\left(1+\frac{1}{2}y'\right), \tag{10.24}$$

where

$$\varphi'_0 = \sqrt{\frac{MC_0^2(1-s^2)\omega_0^3}{2r_0^2(\chi_1+\chi_2)^2}},$$

$$z' = \frac{4U_0}{m\omega_0^2R_0^2}\left(1-\frac{3\hbar}{2mR_0^2\omega_0}\right) - \frac{\omega_1^2}{\omega_0^2} > 0, \quad y' = \frac{1}{y} > 0.$$

From (10.23) and (10.24), it is obvious that when $(\chi_1+\chi_2)$ and v increase, $\bar{\varphi}'_0(\xi)$ decreases. Therefore, strong coupling interaction and higher velocity of the proton bring the protons closer to the heavy ions, so that the distance between the proton and heavy ions decreases appreciably. On the contrary, $\bar{\varphi}'_0(\xi)$ increases, when $(\chi_1+\chi_2)$ and v decrease. This means that the coupling interaction between them and the velocity of the proton decrease with increasing distance between the proton and the heavy ions. Since there are two new equilibrium positions of the proton in such a case, the proton can shift from one side to the other side of the heavy ion by means of quasi-self-trapping and attraction interaction between the proton and the residual negative charge of the heavy ion, which leads to bond rotation or Bjerrum defect. Therefore, the solitons (10.18) and (10.19) in the case of $\varepsilon < 0$ and $g < 0$ represent a hopping motion of the proton over the interbond barrier of the heavy ion, *i.e.*, it represents the Bjerrum defect produced by the rotation of the X-H bond, arising from changes of the relative positions of the protons and the heavy ions. In such a case, there can be two protons between the two oxygen atoms and positive effective charge (the D Bjerrum defect), or no proton between them and negative effective charge (the L Bjerrum defect). This type of soliton is referred to as kink II. The plus sign of $\varphi(x,t)$ in (10.18) applies to the L Bjerrum defect, which leads to the creation of a negative effective charge, while the minus sign in $\varphi(x,t)$ in (10.18) applies to the case of the D Bjerrum defect which results in the creation of a positive effective charge. Thus, the Pang-Miller model supports two types of defects that occur in hydrogen bonded systems, *i.e.*, kink I $\rightarrow$ I^- ionic defect, and kink II $\rightarrow$ L Bjerrum defect and anti-kink I $\rightarrow$ I^+ ionic defect, anti-kink II $\rightarrow$ D Bjcrrum defect. The ionic defect is mainly produced by the double-well potential through the mechanism of proton crossing over the interbond barriers in the manner of translation, but the Bjerrum defect is a result of the coupling interaction through quasi-self-trapping in the manner of lattice deformation and relative intrabond shift of positions of the proton and the heavy ion.

In the transfer process, the protons cross over the interbond barriers in the form of kink solitons, and jump over the intrabond barriers in another soliton form. The coupling interaction also changes depending on the relative positions between the proton and the heavy ion. When the protons cross the interbond barriers, the coupling interaction is small due to their long separation from the heavy ions, and it plays only a secondary role in determining the properties of the protons. When the protons are near the heavy ions and when they shift over the intrabond barriers, the coupling interaction becomes so large that their positions relative to those of the heavy ions change considerably by means of quasi-self-trapping. In such a case the coupling interaction determines the properties of the protons, which transforms into another soliton form in the intrabonds. However, the above changes in the forms of proton transfer are not very sudden, but take place gradually. As a matter of fact, we see from (10.21) that the minima of the potential energy become more and more negative with increasing $(\chi_1+\chi_2)$. When the latter becomes so large that the coupling effect is greater than that of the double-well potential, the potential energy minima reduce to that of (10.22) and (10.23). In this process, the changes of the velocity of the proton transfer in different regions affect the potential energy of the system. It can also influence the form of the proton transfer as discussed above. Therefore, the Pang-Miller model gives a clear description of the proton transfer process in hydrogen bonded systems.

10.3 Thermodynamic Properties and Conductivity of Proton Transfer

The theory for proton transfer given in the previous section must be validated by experiments. To do so, it is necessary to study the thermodynamic properties of proton transfer and calculate its conductivity. Pang was the first one who tackled this problem.

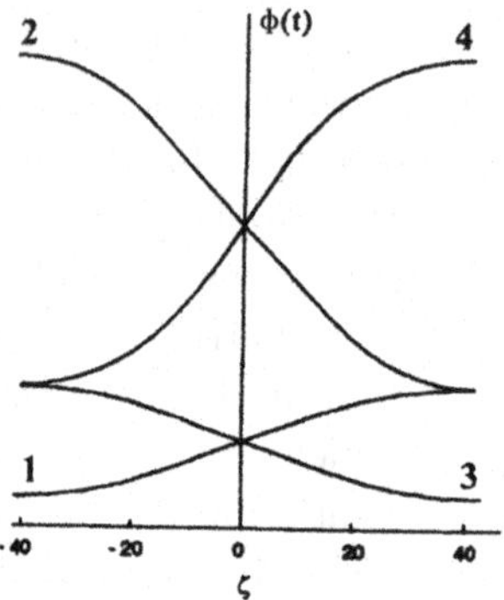

Fig. 10.6 The kink-anti-kink solutions in hydrogen bonded systems.

From (10.15) and (10.16), we can see that if the nonlinear autolocalized excitation in the protonic sublattice is a kink (or anti-kink), there is also an anti-kink (or kink) soliton in the heavy ionic sublattice, which is a "shadow" of the kink (or anti-kink) as shown in Fig. 10.6. They propagate together along the hydrogen-bonded chains in pairs with the same velocity. In Fig. 10.6, curve 1 (3) corresponds to the kink (antikink) soliton in the protonic sublattice and curve 2 (4) corresponds to the antikink (kink) soliton in the heavy ionic sublattice. The momentum of the kink-antikink pair can be obtained from

$$P = \frac{1}{r_0}\int\left(\frac{\hbar}{\omega_0}\varphi_x\varphi_t + M\beta_x\beta_t\right)dx = P_K + P_{ak} = M_{sol}v, \tag{10.25}$$

where $M_{sol} = m_k^* + m_{ak}^*$, with m_k^* and m_{ak}^* being the effective masses of the kink and antikink, respectively, which can be obtained by inserting (10.15) and (10.16) or (10.18) and (10.19) into (10.25), v is the velocity of the kink-antikink pair. Also in (10.25),

$$P_K = \frac{1}{r_0}\int\frac{\hbar}{\omega_0}\varphi_x\varphi_t dx, \quad P_{ak} = \frac{M}{r_0}\int\beta_x\beta_t dx$$

are the momenta of the kink and antikink, respectively. The mobility and conductivity of the proton can be obtained using (10.25).

Since proton transfer is associated with motions of ionic and bonded defects which are charged, when the proton-soliton is formed, a charge deviation from the regular protonic charge distribution occurs in the systems. The soliton charge is related to the quantity $d = R(\infty) - R(-\infty)$, which is equal to $\pm 2R_0$ for positive-negative ionic-defects, and $\pm(4\pi - R_0)$ for positive-negative charge bonded-defects, respectively. Since the proton transfer is caused by the combined transition of the ionic and bonded defects, we have $q = q_I + q_B$, where q_I and q_B are the partial charges of an ionic and bonded defect, respectively. According to the results of Pnevmatikos *et al.*, $q_i = -\alpha_i d_i$ (where i is I or B). When $\alpha_I = \alpha_B = \alpha$, the coefficient is found to be $\alpha = q/4\pi$. In most systems, $\alpha_I \neq \alpha_B$, and q_i is expected to be dependent on the dynamics of the heavy ions. Thus an electric current can occur due to motion of the charged proton-soliton along one direction under an externally electric-field. If we can obtain the corresponding electric conductivity, we can judge the validity of the theory given above by comparing the theoretical and experimental results.

Having this in mind, we consider the conduction of the charged kink-antikink soliton pair described above under a constant external force. Even though the same force is applied to both the protons and the heavy ions, the responses by the two sublattices may be different. Thus, we represent the effects by different fields F_1 and F_2. Considering the dissipation effects, due to influence of environment, on the motions of the proton and heavy ion, the equations of motion, (10.10) and (10.11), are replaced by

$$\varphi_{tt} = f_1(\varphi, \beta) - \Gamma_1 \varphi_t - \sqrt{\frac{m\omega_0}{\hbar}}\frac{F_1}{m},$$
$$\beta_{tt} = \frac{1}{M} f_2(\varphi, \beta) - \Gamma_2 \beta - \frac{F_2}{M}, \tag{10.26}$$

where F_1 and F_2 are the external forces on the proton and heavy ion, respectively, Γ_1 and Γ_2 are the damping coefficients for the motions of the proton and heavy ion, respectively. We assume that the effects of the external forces and dissipation effects to the kink-antikink pair of (10.15) and (10.16) are small so that they lead only to a small change to the velocity of the kink-antikink pair, but not to the waveform. Furthermore, we assume that the forces F_1 and F_2 are function of time only. Applying the boundary conditions for $\varphi(x,t)$ and $u(x,t)$, we can obtain, from (10.25) and (10.26), the following equation of motion for the proton-soliton

$$\frac{dv}{dt} + \gamma v = \frac{3\sqrt{2g}(v_1^2 - v^2)F}{2\varepsilon(m + B^2 M m\omega_0/\hbar)}, \tag{10.27}$$

where

$$\gamma = \frac{m\Gamma_1 + B^2\Gamma_2 M(m\omega_0/\hbar)}{m + B^2 M(m\omega_0/\hbar)} + 2v\frac{dv}{dt}, \quad F = \sqrt{\frac{m\omega_0}{\hbar}}(F_1 + BF_2).$$

Since we assumed that the velocity of the kink-antikink pair is smaller than the sound speeds of the two sublattices, *i.e.*, $v \ll C_0$ and $v \ll v_1$, we can neglect the terms containing v, and regard g, $\bar{\varphi}_0$, ε, B and γ as constants, *i.e.*, $g = g_0 = \text{const.}$,

$\varphi_0 = \bar{\varphi}_0^0 = \text{const.}$, $\varepsilon = \varepsilon_0 = \text{const.}$, $B = B_0 = \text{const.}$, $\gamma = \gamma_0 = \text{const.}$ Equation (10.27) then becomes,

$$\frac{dv}{dt} + \gamma_0 v = \frac{3\sqrt{2g_0 v}}{2\varepsilon_0}\left[m + B_0^2 M\left(\frac{m\omega_0}{\hbar}\right)\right]^{-1} F, \tag{10.28}$$

which is an analogue of the equation of motion of a macroscopic particle with damping in classical physics. It shows that the kink-antikink soliton pair behaves like a classical particle.

Since the kink-antikink pair is charged, the electric field force acting on the pair due to a constant electric field $\vec{E}$ can be written as

$$\vec{F} = q^*\vec{E} = \left(q_1 + B_0\sqrt{\frac{m\omega_0}{\hbar}}q_2\right)\vec{E},$$

where q_1 and q_2 are the effective charges of the kink and the antikink. We can obtain solution of (10.28) in the case of a steady current by approximating the mobility of the kink-antikink pair with

$$\mu = \frac{v}{|\vec{E}|} = \frac{3(q_1 + B_0\sqrt{m\omega_0/\hbar}q_2)v_1\sqrt{m\omega_0 g/\hbar}}{\sqrt{2}\varepsilon_0[m\Gamma_1 + B_0^2(m\omega_0/\hbar)M\Gamma_2]}. \tag{10.29}$$

The electrical conductivity of the kink-antikink pair in hydrogen bonded systems can be obtained and it is given by

$$\sigma = q^* n_0\mu = \frac{3n_0(q_1 + B_0\sqrt{m\omega_0/\hbar}q_2)^2 v_1\sqrt{m\omega_0 g/\hbar}}{\sqrt{2}\varepsilon_0[m\Gamma_1 + B_0^2(m\omega_0/\hbar)M\Gamma_2]}, \tag{10.30}$$

where n_0 is the number of protons in an unit volume. Using parameters given earlier and $\Gamma_1 \sim (0.6 - 0.7)\times 10^{14}\ \text{s}^{-1}$, $\Gamma_2 \sim (9.1 - 13)\times 10^{14}\ \text{s}^{-1}$, $q_1 = 0.68e$, $q_2 = 0.32e$, $n_0 = 10^{22}\ \text{mol}^{-1}$ for ice (Onsager 1969, Bjerrum 1952, Schmidt *et al.* 1971, Homilton and Ibers 1969, Bell 1973, Eisenberg and Kanzmann 1969, Eigen *et al.* 1962, Whalley *et al.* 1973, Weiner and Asker 1970, Pauling 1960, Peyrard 1995), it can be found that $\mu = (6.6-6.9)\times 10^{-6}\ \text{m}^2/\text{V·s}$ and $\sigma = (7.6-8.1)\times 10^{-3}$ $(\Omega\cdot\text{m})^{-1}$.

Experimentally, Eigen and Maeyer reported mobility as high as $(10-20)\times 10^{-6}$ $\text{m}^2/\text{V·s}$ for proton transfer in ice, while measurement by Nagle *et al.* yielded a lower value of $(5-10)\times 10^{-7}\ \text{m}^2/\text{V·s}$. The theoretical result based on the above model falls between these experimental values.

In order to further justify this soliton model of proton transfer in ice, we consider the temperature dependence of mobility using the above soliton model, following the work of Nylund and Tsironis. In their work, the ice is placed in a heat reservoir and a constant electric field is applied to accelerate the proton. The damping effect and a Langevin-type δ-correlated Gaussian stochastic force are considered in the dynamic equation for the oxygen atom. Following the approach of Nylund and Tsironis, we integrate numerically the dynamic equations for the proton and oxygen atom using the fourth-order Runge-Kutta method. We can then obtain the mobility (or velocity) of the thermal kink soliton as a function of inverse temperature. The results for two different field values are shown in Fig. 10.7 using solid lines. Also shown in Fig. 10.7 (dashed line) are the results of Nylund and Tsironis which was obtained using the ADZ soliton model. The most distinct feature of the mobility-temperature curves is the presence of two transition temperatures $T_{\text{max}} = 191$ K and $T_{\text{min}} = 210$ K, *i. e.*, the mobility first rises as the temperature increases, until it reaches a peak at 191 K, and it drops subsequently with further increase in temperature, and reaches a minimum at 210 K, before rising again. The up-down-up trend in this range of temperature seems a general feature in the temperature dependence of the soliton mobility and can also be observed for other values of electric fields and different barrier heights. This behavior is in qualitative agreement with experimental results obtained by Hobbs and Engelheart *et al.* in the same temperature range for crystal ice (see inset in Fig. 10.7). Therefore, temperature can enhance mobility at low and high temperatures, but a sharp drop occurs in the intermediate temperature region. In addition to the similarity of temperature dependences of the conductivity (experimental data) and the velocity (soliton model), another remarkable feature is the coincidence of the transition temperatures in the experimental and theoretical results. This coincidence provided evidence for the existence of soliton in crystal ice. It supports the soliton model of proton transfer given above as well as the treatment of the electric properties of ice.

Next, we discuss the thermodynamic properties of the systems. The effective Hamiltonian of the protons corresponding to (10.13) can be written as

$$H_{\text{eff}} = \int dx \left(\frac{1}{2}\varphi_t^2 + \frac{1}{2}v_1^2\varphi_z^2 + \frac{g}{4}\varphi^4 - \frac{\varepsilon}{2}\varphi^2 \right). \tag{10.31}$$

The thermodynamics of the systems can be analyzed using a transfer integral technique and statistical physical method. The classical partition function corresponding to (10.31) in terms of the field variables $\varphi(x,t)$ and its conjugate $\varphi^*(x,t)$ can be written as

$$Z(\beta, L) = \int dp \int du e^{-\beta H_{\text{eff}}} = Z_p Z_\varphi, \tag{10.32}$$

where L is the length of the system, T is its temperature, K_B is Boltzmann constant and $\beta = 1/K_B T$. Z_p and Z_φ are are given by

$$Z_p = \int dp e^{-\beta E_p} = (2\pi K_B T)^{N/2}, \quad Z_\varphi = \int du e^{-\beta E_\varphi}, \tag{10.33}$$

respectively, where

$$E_p = \int_0^L dz \frac{p^2}{2m}, \quad E_\varphi = \int_0^L dz \left(\frac{1}{2}v_1^2\varphi_x^2 + \frac{1}{4}g\varphi^4 - \frac{1}{2}\varepsilon\varphi^2 \right),$$

and

$$E_p + E_\varphi = H_{\text{eff}}.$$

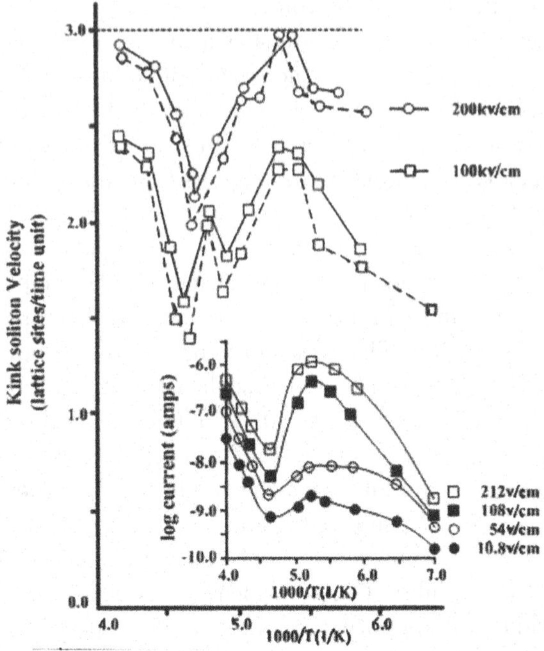

Fig. 10.7 Kink soliton velocity as a function of inverse temperature for two electric-field values. The solid lines represent results obtained using the present soliton model. The dashed lines are results of Nylund and Tsironis. The inset shows the experimental results.

The integrals given above can be evaluated exactly in the thermodynamic limit of a large system containing N heavy ions ($N \to \infty$) by using the eigenfunctions and eigenvalues of a transfer integral operator

$$\int dz_{i-1} e^{-\beta\varepsilon F_\varphi(\varphi_i,\varphi_{i-1})}\psi_i(\varphi_{i-1}) = e^{-\beta\varepsilon\varepsilon_i}\psi_i(\varphi_i), \tag{10.34}$$

where $F(\varphi_i, \varphi_{i-1})$ relates the potential energy components E_φ. This calculation is performed using the method of Krumhansl and Schrieffer, and Schneider and Stoll. $\psi_i(\varphi)$ and ε_i in (10.34) satisfy the following equation

$$\left(-\varphi^2 + \sqrt{\frac{g}{\varepsilon}}\varphi^4 - \frac{R_0^2}{2\beta^2\varepsilon v_1^2}\frac{d^2}{d\varphi^2}\right)\psi_i(\varphi) = (\varepsilon_i - q_0)\psi_i(\varphi), \tag{10.35}$$

with

$$q_0 = \frac{1}{2}\varepsilon\beta \ln\left(\frac{v_1 R_0\sqrt{\varepsilon\beta}}{2\pi}\right).$$

We have assumed that

$$m^* = \frac{\varepsilon v_1^2}{R_0^2 K_B^2 T^2}.$$

Since the single-site potential in (10.31) is bounded from below, so is the eigenspectrum, and we denote the lowest eigenvalue of the above Schrödinger equation by ε_0. Then, in the thermodynamic limit, we can approximately get

$$Z_u = e^{-\beta N \varepsilon \varepsilon_0}. \tag{10.36}$$

The free energy per particle can be obtained which is

$$f = \frac{K_B T}{2} \ln(2\pi K_B T) + \varepsilon \varepsilon_0, \tag{10.37}$$

where ε_0 is determined from (10.35).

Finally, we can find the internal energy per particle and the related specific heat at constant volume. The results are

$$\begin{aligned} e' &= f - T\frac{\partial f}{\partial T} \cong K_B T + \frac{4\varepsilon^2}{15g}\left(\frac{K_B T}{E_k^0}\right)^2 + O\left[(K_B T)^3\right], \\ c_v &= \frac{\partial e'}{\partial T} = K_B + \frac{8}{15}\mu\varepsilon^2 K_B T + O\left[(K_B T)^2\right], \end{aligned} \tag{10.38}$$

respectively, where $E_k^0 = 2\sqrt{2\varepsilon^3 v_1 m}/R_0 g$. Equation (10.38) gives the specific heat of hydrogen-bonded systems, including ice. It is linearly dependent on temperature, which is similar to (9.104) – (9.105) for the ACN and proteins.

10.4 Properties of Proton Collective Excitation in Liquid Water

In this section, we discuss collective excitation of protons in liquid water. We will apply the theory presented in earlier section to water, a hydrogen bonded system, to reveal the mechanism of magnetization of liquid water.

Water is familiar because it is closely related to growth and living of human beings, animals and plants. We can say that there would be no life without water in the world. However, the properties of water are still not fully understood, although it has been studied for several hundreds years. Magnetization of liquid water is a prototypical example. It has been known that liquid water can be magnetized, *i.e.*, when the water is exposed to a magnetic field, its optical, electromagnetic, thermodynamic and mechanical properties, such as density, surface tension, viscosity, melting temperature, solidification temperature, dielectric constant, conductivity, refractive index, and spectra (ultraviolet, infrared and laser-Raman), will be changed compared to non-magnetized water. These changes are due to magnetization of the liquid water.

Magnetized water has been widely used in industry, agriculture and medicine. For example, it can be used to expedite digestion of food and removing dirts in industrial boilers, and so on. Even though many models, for instance, the resonant model by Ke Laxin, the model of hydrogen bond destruction, and physical-chemistry reaction of water molecules with ions, were proposed to understand this process, the mechanism of magnetization has not been satisfactorily explained. In the resonant model, Ke Laxin proposed that the magnetization of water is caused by a resonant effect among vibrations of the components including water molecules and hydration water, chelate and impurities under an externally applied magnetic field with appropriate frequency, which leads to distortion of the hydrogen bonds, and changes in structures and properties of the water molecules. On the other hand, Jiang *et al.* proposed that the water molecules form hydration ions with some ions in liquid water. Thus the distribution and polarization and dynamic features of water molecules are changed. The externally applied magnetic field destroys the structure and distribution of the ionic-hydrations through the Lorenz force. The re-arrangements of the ionic-hydrations result in magnetization of water. In this process, destruction of some hydrogen bonds are believed to be necessary.

However, these models are only qualitative, and cannot fully explain the magnetization of water and its related properties. New ideas from nonlinear dynamics have provided a solution to this important problem. In this section, we will review the nonlinear theory of motion of protons and look into the mechanism of magnetization of liquid water. The model is based on the theory of proton conductivity in crystal ice discussed in Sections 10.1 and 10.2. In other words, it is a generalization of the theory of proton conductivity in the ice, since liquid water is expected to become ice at 0°C through first-order phase transition.

10.4.1 *States and properties of molecules in liquid water*

States and distribution of molecules in liquid water are different from that in ice and vapor phases of water. It is known that the molecules in liquid water are polarized, which can be easily demonstrated experimentally. For example, when a charged plastic pencil (by static electricity generated by friction contact) is brought to a water stream from a tap, the water deviates from its original flowing direction and bends towards the pencil. This is a simple and direct demonstration that the water molecules are polarized. Each molecule possesses a large dipole-moment. Thus, it can be expected that there exist many hydrogen-bonded chains formed by dipole-dipole interaction between molecules in the system.

The polarization of water molecules can be easily understood from the properties of the first-order phase transition. It is well known that when ice becomes water through a first-order phase transition at the melting point, 0°C, a certain amount of latent heat is released, due to changes of volume and entropy. As mentioned earlier, ice is a hexagonal structure formed by hydrogen bonding between water molecules and there is a large number of linear hydrogen-bonded chains in the system below 4°C, in which the ice lattice is extremely porous and contains many "vacancies" since the average number of nearest neighbor water molecules of each molecule (coordination number) is only about four. On melting, the ice lattice is partially destroyed, and at the same time, some vacancies are filled and the density of water becomes greater than that of ice. This is one of the principal anomalies of water. With further heating, up to 4°C, the condensation process continues. When the temperature reaches above 4°C, the amplitude of anharmonic vibrations increases, and these molecules no longer vibrate around their equilibrium positions and the ice lattice is completely destroyed. However, each molecule still has four neighbors and hydrogen bonding still exist. The contribution of hydrogen bonding to the total energy of intermolecular interaction (11.6 Kcal/mole) is about 69 Kcal/mole in such a state. However, because there are a lot of hydrogen bonds, the melting point (0°C) and the boiling point (100°C) of water are significantly higher than those of other molecular liquids which are bound together by van der Waals' forces. For example, the melting point and boiling point of methane (CH_4) are −186°C and $-161°C$, respectively. Obviously, the composition and structure of the intermolecular complexes and the number of associated hydrogen-bonded molecules in the complexes (clusters), as well as density of water depend on temperature, and the compositions decrease with increasing temperature due to disorder of thermal motion of water molecules. Rough estimates give about 240 molecules in a cluster at room temperature, 150 at 37°C and 120 at 45°C.

Raman spectrum of liquid water can give us more information about the structures and distribution of water molecules. Experiments by Jiang *et al.* and Pang *et al.* showed that there are three Raman peaks in the range of 300 – 3700 cm^{-1}, located at 300 – 900 cm^{-1} and 1600 – 1900 cm^{-1} and 2900 – 3800 cm^{-1}, respectively. The peak in the range of 300 – 900 cm^{-1} is associated with the libration of water molecule. The peak at 1600 – 1900 cm^{-1} is a narrow hand, arising from bending vibrations of HOH bonds. The peak at 2900 – 3800 cm^{-1} is a wide band, containing four peaks. The two peaks at 3241 cm^{-1} and 3415 cm^{-1} correspond to symmetric and antisymmetric stretching-vibrations of hydrogen-bonded OH bonds, and the two peaks at 3540 cm^{-1} and 3617 cm^{-1} represent symmetric and antisymmetric stretching-vibrations of free OH bonds. Since the Raman spectra show changes in the number of hydrogen bonds, the shape and intensity of these peaks in the 2900 – 3800 cm^{-1} band can be used to investigate the nature of the hydrogen bonded chains in liquid water. It was confirmed by Raman spectral study in this region that (1) there are a lot of hydrogen-bonded chains of water molecules; (2) the number of the hydrogen-bonded chains and the number of water molecules in the chains decrease with increasing temperature; (3) there is a red-shift of frequencies in the Raman spectra and infrared absorption spectra, when the water molecules are hydrogen bonded with neighboring molecules.

10.4.2 *Properties of hydrogen-bonded closed chains in liquid water*

From the above discussion, we know that there are hydrogen-bonded chains in liquid water. Considering the special properties of liquid water, such as long-range disorder in the motion of molecules and the uncertainty of equilibrium positions of the water molecules, it is reasonable to assume that some chains form closed loops by linking head and tail of different linear chains. These loop configurations can consist of various number of water molecules as shown in Fig. 10.8.

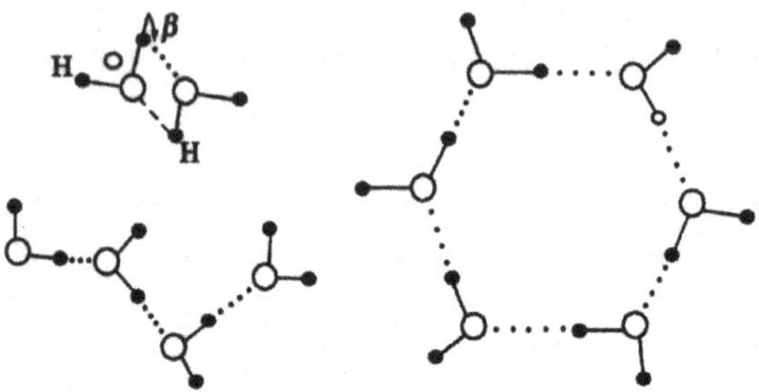

Fig. 10.8 Hydrogen bonds in a dimer ($\beta = 110°$), a linear hydrogen bonded ($\beta = 0$), and a cyclic hydrogen bonded water molecule.

Let β be the angle between the OH bond and hydrogen bond in the closed loops, as shown in Fig. 10.8. Its value depends on the number of water molecules or the number of hydrogen bonds in the closed loop. For instance, in a dimer which consists of 2 water molecules, β is 110 degrees, and in the five membered ring it is approximately 10 degrees.

Interactions between water molecules in the closed loops will cause further frequency shift of the OH stretching vibrations, compared to that in free water molecules. The maximum shift was observed experimentally in the case of $\beta = 0$ (linear hydrogen bond), the hydrogen-bonding energy of which is also a maximum. For 3540 cm^{-1}, the maximum shift is about 400 cm^{-1}, while for 3617 cm^{-1}, it is about 340 cm^{-1}. Therefore, the value of β can be used as an indicator for formation of closed hydrogen bonded loops in water. If a nonzero value of β is obtained experimentally, we can conclude that closed loops exist in the liquid water.

Although the magnitude of the frequency shift depends on the number of water molecules and the angle (β) between the OH bond and the hydrogen-bond in the closed loops, the functional relationship among them is very complicated. But it can be derived based on the theory of proton transfer in hydrogen-bonded chains in ice discussed in Sections 10.1 and 10.2, and making use of the Badger-Bauer rule which states that the energy of a hydrogen bond is proportional to the frequency shift, $\triangle\nu$, of the valency infrared vibrations of the OH group in a closed loop with respect to the vibrational frequency in the free molecule. From the structural properties of the closed hydrogen bonded systems, and the interaction and motion of proton illustrated in (10.4) and Fig. 10.5, we can obtain the ratio of the frequency shift $\triangle\nu = \nu_0 - \nu(\beta)$ at a given β to the maximum shift $(\triangle\nu)_{\max} = \nu_0 - \nu(0)$, as a function of β in the closed loops, where ν_0, $\nu(\beta)$ and $\nu(0)$ are the frequencies of intrinsic vibration, the valency infrared vibrations of the OH group for $\beta \neq 0$

(closed loop) and $\beta = 0$ (linear chain), respectively. This ratio can be approximately written as

$$\frac{\triangle\nu}{(\triangle\nu)_{\max}} = \frac{\nu_0 - \nu(0)\sqrt{1 + (1-\cos\beta)[\nu_0^2 - \nu^2(0)]/\nu(0)^2}}{(\triangle\nu)_{\max}}.$$

The variation of $\Delta\nu/(\Delta\nu)_{\max}$ with β, given by the above equation is shown in Fig. 10.9 (solid line), together with experimental data (symbol).

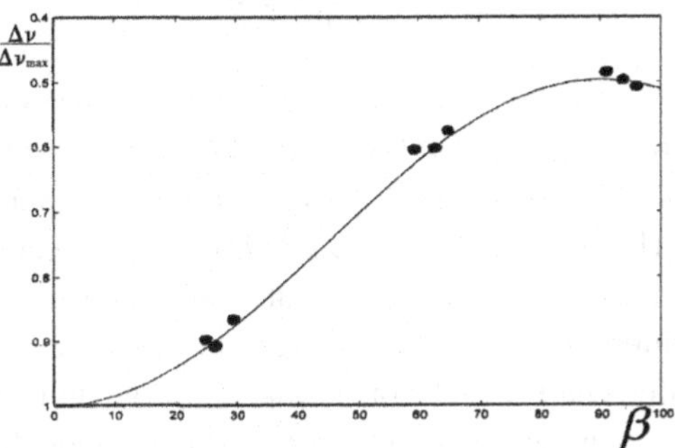

Fig. 10.9 The ratio of the frequency shift $\triangle\nu = \nu_0 - \nu(\beta)$ to the maximum shift $(\triangle\nu)_{\max} = \nu_0 - \nu(0)$ as a function of β for a ring chain.

This graph also reflects the dependence of the energy of the hydrogen bond on the angle β. It is clear that the theoretical result is in good agreement with the experimental data (Davydov 1982). It provides experimental confirmation that the theory developed by Pang *et al.* gives a fairly good description for the closed hydrogen bonded chains in liquid water.

10.4.3 *Ring electric current and mechanism of magnetization of water*

The hydrogen-bonded closed chain structure of liquid water is crucial for its magnetization. The closed chains or loops have many different forms and types. It may consist of polymerizations of two, three, four, five, $\cdots$, or n water molecules, as shown in Fig. 10.9. The form and type depend on the temperature of water. The numbers of water molecules in the closed chains decrease gradually with increase of water temerature. Since the closed hydrogen-bonded chains are also channels of proton transfer, we can generalize the proton transfer theory given in Section 10.4 to proton transfer in such closed loops. We assume that charged proton-solitons move along the closed loops. Then an electric current can be produced by this motion of the proton-soliton around the closed loops of hydrogen bonds, when a liquid water containing many closed hydrogen-bonded chains is exposed to an external magnetic field, H_{ex}. This is because the charged proton-solitons move around the hydrogen-bonded chains due to the action of the Lorenz force, $\vec{F} = q\vec{v} \times \vec{H}_{ex}/c$, where $\vec{v}$

is the velocity of the proton-soliton, c is the speed of light, q is the charge of the proton-soliton. The ring electric current $\vec{J} = qn\vec{v}$ further induces a magnetic field $H_{in}(\nabla \times \vec{H}_{in} = 4\pi\vec{J}/c)$ in the closed loops, where n is number of the proton-solitons per mole. Here, the velocity $\vec{v}$ is still determined by (10.28), while the force F in (10.28) is the Lorenz force. For a closed loop with a diameter a, the magnitude of the induced magnetic field on the symmetrical axis is given by

$$H_{in} = \frac{2\pi a^2 J}{c(a^2 + Z^2)},$$

where Z is the distance to a point on the symmetry axis from the center of the ring. Therefore, if the external magnetic field and the structure of the closed loops are known, we can determine the current J and the induced magnetic field H_{in} from (10.29). Obviously, there is an induced magnetic field in each ring. But they can differ in magnitudes and directions due to different conformations, different number of the protons and different number of hydrogen-bonds contained in the loops. Thus, these hydrogen-bonded chains, each consisting of a large number of water molecules, will arrange orderly via magnetic interaction, and form a locally ordered state. This leads to an ordered distribution of water molecules and magnetization of liquid water. The arrangement of these closed loops depends on the interactions between the induced magnetic fields of the loops and between the induced magnetic field and the external magnetic field. When the external field is very strong, all closed loops will be aligned along the direction of the external field, and the maximum magnetization is obtained. Based on the theory discussed earlier, the distribution and degree of the localized order of the magnetic-arrangements or magnetization of the water molecules can be determined, if we know the number of water molecules contained in a loop, the numbers and distribution of loops, and the magnitude and direction of the external magnetic field H_{ex}. The effect of magnetization of water can then be quantitatively determined. However, the calculation is very complicated and can only be carried out numerically for certain distributions of the loops in liquid water.

Therefore, magnetization of water is nothing but a magnetic ordering of water molecules in the closed hydrogen-bonded chains, in the presence of an external magnetic field through magnetic interaction. This is a nonlinear and local ordering phenomena, and it is a collective effect of a large number of molecules, rather than behaviors of individual molecules. The physical fundamentals for magnetization of water is the presence of a large number of closed hydrogen-bonded chains and ring proton currents in the liquid water. Its theoretical foundation is the theory of proton conduction which is a generalization of the proton transfer theory in ice. A direct effect of magnetization of water is the change of states and distribution of water molecules in the liquid state and the density of the water molecules, which lead to changes in other physical properties such as optical, electrical, mechanical and thermodynamic properties. For example, the dielectric constant, susceptibility, magnetoconductivity, refractive index, pH value, Raman spectrum of liquid water can be affected. Such changes in physical properties of the liquid water have been confirmed experimentally. For example, it was found that (Xie 1983, Joshi and Kamat 1965, Higashitani 1993, Ke Laxin 1982, Muller 1970, Liemeza 1976, Li 1976, Jiang *et al.* 1992, Pan and Xun 1985, Song 1997, Cao 1993, Binder 1984, Myneni *et al.* 2002, Davydov 1982, Koutselose 1995, Evans 1982, Evans 1983, Evans 1987, Mouritsen 1978, Chikazumi 1981, Kusalik 1995, Dwicki 1997) the dielectric constant, susceptibility, pH value, and magnetoconductivity all increase when liquid water is magnetized, and changes were also found in the Raman spectrum and the infrared absorption. Pang *et al.* and Jiang *et al.* demonstrated that molecular structure remains the same, the hydrogen bonds remain intact, the peak positions of Raman spectrum and those of infrared absorption are unchanged, but their intensities are changed when liquid water is magnetized, as shown in Figs. 10.10 and 10.11. The magnetization effect depends on the magnitude, direction and time of action of the external field, the larger the magnitude and longer the time in which the field is applied, the large the magnetization effect. Pang measured the refractive index and dielectric constant of magnetized water. For pure water, its refractive index is 1.3336 at 25 °C. For magnetized water, its refractive indeces are 1.3339, 1.3342 and 1.3346 at applied magnetic field H_{ex} = 1.8 T, 2.4 T, and 3.0 T, respectively. Therefore, the refractive index of magnetized water increases slowly with increasing applied magnetic fields. Accordingly, the dielectric constant of magnetized water increased by about 1.4% when the applied magnetic field wass increased from 1.8 T to 3.0 T.

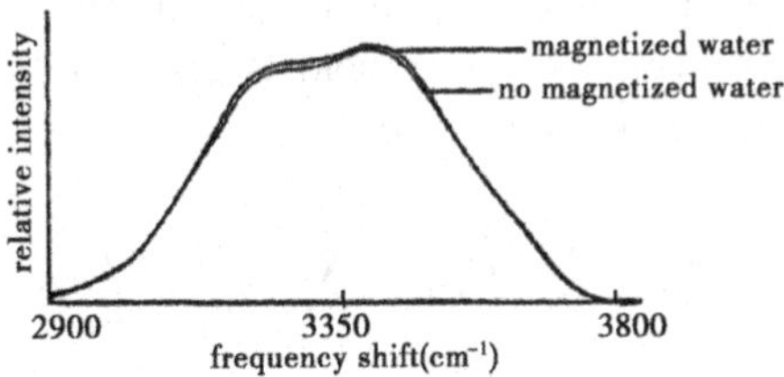

Fig. 10.10 Comparison of Raman spectra of magnetized and nonmagnetized water.

Similar to any other magnetic systems, the magnetization in liquid water can be saturated, in which the magnetization no longer change with further increase of the external magnetic field once it reaches a certain value. In order to validate this conclusion, Pang measured infrared absorption of magnetized water with an increasing applied magnetic field, H_{ex}, by a Nicolet Nexus 670 FT-IR spectrometer. The results are shown in Fig. 10.12. From this figure we see clearly that the intensity of absorption increases with increasing H_{ex}. After the magnetic field had been applied for five hours, the absorption reached the maximum value, confirming the saturation effect of magnetized water. The memory effect of magnetization in

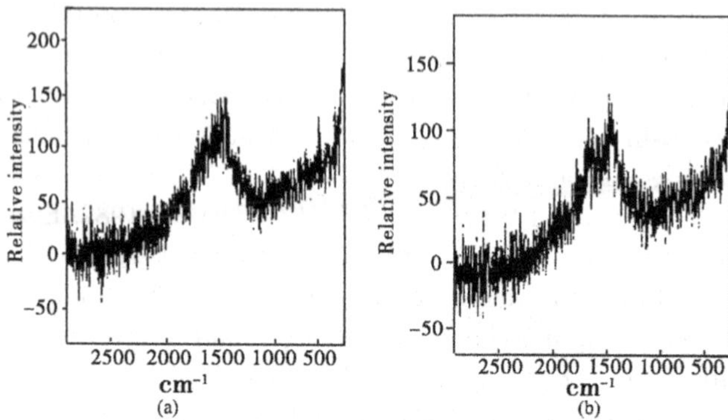

Fig. 10.11 Raman spectra of magnetized (a) and nonmagnetized (b) water.

which the magnetization remains when the applied field is removed is also observed in liquid water. In Pang's experiment, the intensity of the infrared absorption decreased gradually when the applied magnetic field was removed after the saturation magnetization of water. The magnetized water can restore the normal state after about four hours, indicating that the magnetized water has a memory effect. These phenomena can all be explained using the same theory because magnetization of water is a magnetic ordering process of water molecules in these closed loops under the action of the applied field. Evidently, this arrangement is directly related to the magnitude, direction and time of action of the external field. When the applied field is so strong that all induced magnetization of the closed loops can be orderly arranged along the direction of the applied field, the magnetization is saturated. In such a case the saturation magnetic field is given by the sum of the induced magnetic fields of all the closed loops, *i.e.*,

$$H_{sa} = \sum_{k=1}^{N} H'_{\text{kin}} = \sum_{k=1}^{N} \frac{2\pi a_k^2 J_k}{c(a_k^2 + Z_k^2)},$$

where N is the number of closed loops in the liquid water, the parameters a_k, J_k and Z_k are the diameter, induced current and magnetic field of the kth closed loop, respectively.

When the applied field is removed, the magnetization cannot immediately disappear due to interactions among these induced magnetization, which is the origin of the memory effect of the magnetized water. However, given sufficient time, the memory effect will eventually disappear. The duration of the memory effect can be obtained from (10.27). In fact, once the applied field is removed from the liquid water, we know from (10.27) that the velocity of the protons in the hydrogen-bonded systems will be changed according to

$$\frac{dv}{dt} = -\gamma v.$$

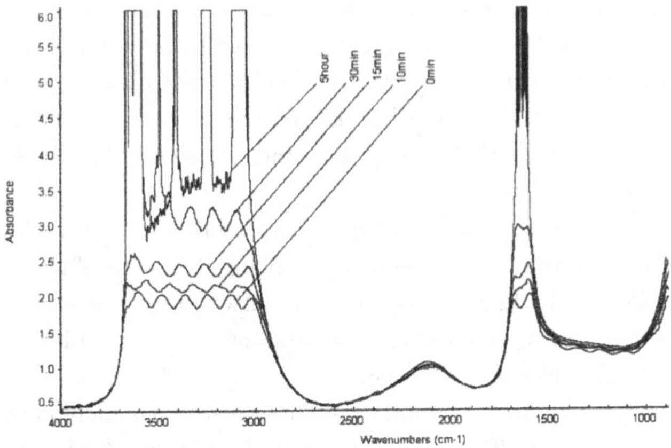

Fig. 10.12 The saturation effect of magnetized water. The effective strength of the applied magnetic field, H_{ex}, is denoted by the time exposed. The curve labeled "0 min" represents that of pure water (without magnetic field).

The solution of this equation is

$$v(t) = v(0)e^{-\gamma t} = v(0)e^{-t/\tau},$$

where $\tau = 1/\gamma$ is the damping or relaxation time of the velocity of the proton. It shows that the velocity approaches to zero, when $t > \tau = 1/\gamma$. On the other hand, we know from the above that the induced electric current J as well as the induced magnetic-field H'_{in} are proportional to the velocity of the protons, v. Thus J and H'_{in} will also approach zero, when the v approaches zero, and the interactions among these induced magnetic fields disappear gradually after a time of $\tau = 1/\gamma$. Therefore, $\tau = 1/\gamma$ is the lifetime of the induced electric current J and induced magnetic field H'_{in} in the absence of the external field. In other words, it is the memory time of the magnetized water. Evidently, this time is inversely proportional to the damping coefficients, Γ_1 and Γ_2 of the medium (see (10.27)). In short, magnetization related phenomena in liquid water can be satisfactorily explained by the proton transfer theory, indicating that the theory correctly describes the mechanism of magnetization of water.

Based on this mechanism, we can obtain the following properties of magnetized water. (1) The basic requirement of magnetization of liquid water is the existence of a large number hydrogen-bonded rings which consist of many molecules. The conductive carriers of the ring electric current in the rings are protons, instead of electrons. Because the mass of an electron is about 2000 times smaller than that of a proton, the velocity of protons is much smaller, and there are less number of protons than electron, the proton current in the rings is small. Furthermore, only a fraction of water molecules form ring structures, the magnetization effect of water is typically small.

(2) The external magnetic field H_{ex} results only in ring proton current and

induced magnetic moments in the closed hydrogen-bonded chains. It does not change the numbers of hydrogen-bonded water molecules. The hydrogen-bonds of water molecules are thus not destroyed due to magnetization, and the positions of the Raman and infrared absorption peaks remain the same. This is in agreement with experimental results of Jiang *et al.*, and Pang *et al.* which are shown in Figs. 10.10 and 10.11, respectively.

(3) Increasing the temperature of liquid water will increase the kinetic energy of the disordered motion of water molecules, resulting in breaks of the hydrogen bonds of water molecules, if the kinetic energy of the disordered motion is larger than the hydrogen-bonding energy. Therefore, the number of water molecules contained in the hydrogen bonded closed loops decreases with increase of temperature and the number of disrupted hydrogen bonds increases also with increasing temperature. Thus, the magnetization effect of liquid water is very small at high temperature, and it disappears at about $T = 100°\text{C}$.

10.5 Nonlinear Excitation of Polarons and its Properties

The concept of polaron was first proposed by Landau to describe the motion of an additional electron in a crystal lattice. It was suggested that the electron would polarize the crystal and the crystal energy would be lowered in the process. The localized electron and the lattice together are referred to as a polaron. Landau's idea was discussed in details by Pekar, Fröhlech and Holstein, and many others. However, many questions about polaron remain open.

We now take a look at the behavior of an additional electron in an one-dimensional polar crystal in the framework of Holstein model which takes into consideration possible appearance of soliton-like excitations of the additional electron in such systems. Assume that the system consists of N identical diatomic molecules, with atomic masses of m_1 and m_2, respectively, and located at their equilibrium positions, $x_n^{(1)}$ and $x_n^{(2)}$ of the nth molecular cell. We will consider only the variations of the interatomic distance due to the longitudinal optical oscillation with frequency ω_0, and ignore the longitudinal acoustic oscillations. The Hamiltonian operator for vibration of lattice molecules is given by

$$H_{ph} = -\frac{\hbar}{2M}\sum_n \frac{\partial^2}{\partial u_n^2} + \frac{1}{2}M\omega_0^2\sum_n u_n^2 + \frac{1}{2}M\omega_1^2\sum_n u_n u_{n+1}, \tag{10.39}$$

with

$$M = \frac{m_1 m_2}{m_1 + m_2}, \quad u_n = x_n^{(2)} - x_n^{(1)} = a.$$

The last term in (10.39) represents the nondiagonal dispersion effect of the optical phonons, with $\omega_1^2 \ll \omega_0^2$. According to Fedyanin *et al.*, the Hamiltonian of an electron interacting with the lattice is given by

$$H_e + H_{\text{int}} = (\varepsilon_0 + W)\sum_n N_n - J\sum_n \left(b_n^+ b_{n+1} + b_{n+1}^+ b_n\right) + V\sum_n N_n N_{n+1} \quad (10.40)$$

in the second quantization representation, where W is the shift of the energy level ε_0 of the electron at the nth site due to the influence of the electron at other sites (we assume there is only one electron level ε associated with a possible bound state). Also in (10.40), J is the overlap integral ($J > 0$) and V is the strength of the electron-phonon interaction (having the dimensions [energy] × [length]$^{-1}$), b_n^+ (b_n) is the creation (annihilation) operator of the additional electron $N_n = b_n^+ b_n$.

For small V, the Hamiltonian $H_{ph} + H_e + H_{\text{int}}$ can be transformed to the well-known Pekar-Fröhlich type which allows us to take into account the long-range interaction between the electron and the polarization field caused by the lattice deformation, whereas the Holstein Hamiltonian contains only short-range interaction. Applying the Bogolyubov variational principle, Fedyanin *et al.* transformed the Hamiltonian $H = H_{ph} + H_e + H_{\text{int}}$ into the following form, with "separation" of the phonon and electron subsystems

$$H = \tilde{H}_{ph} + \tilde{H}_e,$$

where

$$\tilde{H}_{ph} = \hbar\sum_q \omega_q \left(a_q^+ a_q + \frac{1}{2}\right),$$

$$\tilde{H}_e = \left(\varepsilon_0 + W - \frac{V^2}{2M\omega_0^2}\right)\sum_n N_n - Je^{-F_\theta}\sum_n \left(b_n^+ b_{n+1} + b_{n+1}^+ b_n\right) + \frac{V^2}{2M\omega_0^2}\sum_n N_n N_{n+1}, \quad (10.41)$$

with

$$\left[a_q, a_{q'}^+\right] = \delta_{qq'}, \quad \left[b_n, b_{n'}^+\right]_+ = \delta_{nn'}, \quad \left[\tilde{H}_e, \tilde{H}_{ph}\right]_- = 0,$$

$$N_n = b_n^+ b_n, \quad \omega_q^2 = \omega_0^2 + \omega_1^2 \cos qr_0,$$

$$F_\theta = \frac{V^2}{4MN\hbar\omega_0^4}\sum_q \omega_q |S_q|^2 \coth\left(\frac{\hbar\omega_q}{2K_BT}\right), \quad (\theta = K_BT).$$

In the above,

$$S_q = \omega_0^2 \sum_n \frac{\Delta_{n+1}^m - \Delta_n^m}{\Delta} e^{iqmr_0},$$

$$\Delta_n^m = \left(-\frac{\omega_1^2}{2\omega_0^2}\right)^{|n-m|} \frac{\Delta}{\omega_1^2}\left(1 - 2\frac{\omega_1^2}{\omega_0^2}\right)^{-2}.$$

Here $\triangle$ is the determinant of the matrix constituted of ω_1^2 and ω_0^2, $\triangle_n^m$ is its minor obtained by eliminating the mth column and the nth row.

For $\theta \to 0$, the basic contribution to $\triangle_n^m$ comes from the diagonal terms ($n = m$) of

$$S_\theta = \frac{V^2\theta}{2M\hbar^2\omega_0^4} + O\left(\frac{\omega_1^2}{\omega_0^2}\right).$$

Thus, the parameters corresponding to the phonon subsystem ω_q and V renormalize the Hamiltonian H_e due to the application of the Bogolyubov variational principle. Obviously, the Hamiltonian given in (10.41) is a nonlinear operator. Therefore, the polaron undergoes a change from a linear elementary excitation to a nonlinear soliton-excitation through the electron-phonon interaction.

Using the Heisenberg equation (9.7), Makhankov *et al.* showed that

$$\begin{aligned} i\hbar(\varphi_n)_t = &\left(\varepsilon_0 + W - \frac{V^2}{2M\omega_0^2}\right)\varphi_n(t) - Je^{-F_\theta}\left[\varphi_{n+1}(t) + \varphi_{n-1}(t)\right] \\ &+\frac{V^2}{M\omega_0^2}\left[1 - \varphi_{n+1}^*(t)\varphi_{n+1}(t) + \varphi_{n-1}^*(t)\varphi_{n-1}(t)\right]\varphi_n(t). \end{aligned} \tag{10.42}$$

In this case, the trial wave function can be written as

$$|\varphi(t)\rangle = U|0\rangle = \sum_n \varphi_n(t) b_n^+|0\rangle_{po}, \tag{10.43}$$

where $|0\rangle_{po}$ is the ground state of the polaron. In the continuum approximation, Makhankov and Fedyanin obtained the following from (10.44) by neglecting terms involving derivatives higher than the second.

$$\begin{aligned} i\hbar\varphi_t = &\left(\varepsilon_0 + W - \frac{V^2}{2M\omega_0^2}\right)\varphi(x,t) - Je^{-F_\theta}\left[2\varphi + r_0^2\varphi_{xx}\right] \\ &+\frac{V^2}{M\omega_0^2}\left[1 - |\varphi|^2 - r_0^2|\varphi_x|^2 - \frac{r_0^2}{4}|\varphi_{xx}|^2 - \frac{r_0^2}{2}\left(\varphi_{xx}^*\varphi + \varphi_{xx}\varphi^*\right)\right]\varphi(x,t). \end{aligned} \tag{10.44}$$

If we further retain only terms with orders less than the fourth with respect to dispersion and nonlinearity in (10.44), we can get

$$\begin{aligned} i\hbar\varphi_t(x,t) = &\left(\varepsilon_0 + W - \frac{V^2}{2M\omega_0^2} - 2Je^{-F_\theta} + \frac{V^2}{M\omega_0^2}\right)\varphi(x,t) \\ &-Jr_0^2e^{-F_\theta}\varphi_{xx} - \frac{V^2}{M\omega_0^2}|\varphi|^2\varphi(x,t). \end{aligned} \tag{10.45}$$

This is a standard nonlinear Schrödinger equation. Therefore, polarons can be described by nonlinear quantum mechanics. The soliton solution of (10.45) is given by

$$\varphi(x,t) = \frac{r_0\exp[i\,(kx - \omega t + \theta_0)]}{2W_s\cosh[(x - vt - x_0)/W_s]}, \tag{10.46}$$

and its energy is

$$E_s = \varepsilon_0 + W + \frac{V^2}{2M\omega_0^2} - 2Je^{-F_\theta} + \frac{M_s v^2}{2} - \frac{eF_\theta}{24J}\left(\frac{V^2}{M\omega_0^2}\right)^2, \tag{10.47}$$

where

$$M_s = \frac{\hbar^2 e^{F_\theta}}{2Jr_0^2}, \quad W_s = \frac{4r_0 JM\omega_0^2}{V^2 e^{F_\theta}}.$$

The polaron-soliton (10.46) is localized in the region $\approx W_s$. The polaron effect is defined by the last two terms in (10.47). The third term describes the renormalization of the electron level shift due to its translation jumps. Note that there arises a temperature renormalization of the conductive electron mass $m_e = \hbar^2/2r_0^2 J$ due to J is replaced by Je^{-F_θ} at present. Therefore, the polaron now gets heavier, *i.e.*, $M_s(\theta) > M_s(0)$, and the effective mass becomes temperature dependent in such a case. If $\theta \to 0$, then $M_s(\theta) > M_s(0)(1 + 3\chi\theta/8\omega_0)$, where χ is a coupling constant between electron and phonon and is given by $\chi = 4V^2/3M\hbar^2\omega_0^3$. The same effect, *i.e.* increase of mass, was observed for $\theta \neq 0$ in cyclotron resonance experiments by Rodriges and Fedyanin. Such a correspondence is naturally associated with the similar structures of the Pekar-Frölich and Holstein Hamiltonians.

For a polaron with a small radius, a canonical transformation method has been well developed. By means of this transformation, Makhankov and Fedyanin examined the main contribution to the electron mass renormalization defined by its strong interaction with the lattice vibration and obtained the Hamiltonian of a "residual" (weak) polaron lattice vibration interaction. In this case, the lattice Hamiltonian (10.39) is rewritten in terms of the normal coordinates Q_q and ξ_q via the unitary operator $U = \exp(-iL)$, where

$$L = i\sum_q \xi_q^0 \frac{\partial}{\partial \xi_q}, \tag{10.48}$$

with

$$\xi_q^0 = -\frac{V}{M\omega_q^2}\sqrt{\frac{2}{N}}\sum_n N_n \sin\left(qn + \frac{\pi}{4}\right).$$

The U transformations of the normal coordinates ξ_q viz, $\bar{\xi}_q = U\xi_q U^{-1} = \xi_q + \xi_q^1$ are such that ξ_q^1 may be regarded as the operator of lattice deformation. The overall Hamiltonian, $\tilde{H}$, is then divided into three parts: $\tilde{H}_{ph}$ which describes harmonic oscillations of lattice sites about their equilibrium positions, the polaron Hamiltonian

$\tilde{H}_p$,

$$\tilde{H}_p = (\varepsilon_0 + W)\sum_k N_k - Je^{-F_T}\sum_n \left(b_{n+1}^+ b_n + b_{n+1}^+ b_{n+1}\right) + V\sqrt{\frac{2}{N}}\sum_n N_n \sum_q \xi_q^0 \sin\left(qn + \frac{\pi}{4}\right) + \sum_q \frac{M\omega_q^2}{2}(\xi_q^0)^2, \quad (10.49)$$

where

$$F_T = \frac{2V^2}{4\hbar MN}\sum_q \frac{1}{\omega_q^3}\left\{\sin\left(qn + \frac{\pi}{4}\right)\sin\left[q(n+1) + \frac{\pi}{4}\right]\right\}\coth\left(\frac{\hbar\omega}{2K_BT}\right),$$

and the Hamiltonian $\bar{H}_{\rm int}$ of the residual interaction of the polaron with the oscillation of the deformed-lattice. The polaron motion described by (10.49) is adiabatic such that the lattice reconstructs completely with oscillations about the new equilibrium positions. $\bar{H}_{\rm int}$ describes the "friction process", *i.e.*, the polaron motion is accompanied by a change of the phonon number, and the polaron loses its energy and slows down. Such effects decrease the energy gain due to soliton formation.

Using $|\varphi(t)\rangle$ in (10.43), we now replaced the lattice deformation operator ξ_q^0 by

$$\bar{\xi}_q^0 = \langle\varphi(t)|\xi_q^0|\varphi(t)\rangle = -\frac{V}{M\omega_q^2}\sqrt{\frac{2}{N}}\sum_n |\varphi_n(t)|^2 \sin\left(qn + \frac{\pi}{4}\right).$$

Then $\tilde{H}_p$ is transformed into $\tilde{H}_p'$ which satisfies the following equation

$$i\hbar\frac{\partial}{\partial t}|\varphi(t)\rangle = \bar{H}_p'|\varphi(t)\rangle. \quad (10.50)$$

From (10.50) and the relation $\bar{H}_p' = \bar{H}_p(\bar{\xi}_q^0)$, we can obtain the equation satisfied by $\varphi(t)$ in the continuum approximation. Up to $O(\omega_1^2/\omega_0^2)$, the equation is

$$i\hbar\varphi_t(x,t) = (\varepsilon_0 + W + \tilde{\varepsilon}_0 - 2Je^{-F_T})\varphi(x,t) - Je^{-F_T}r_0^2\varphi_{xx} - \frac{V^2}{M\omega_0^2}|\varphi(x,t)|^2\varphi(x,t), \quad (10.51)$$

where

$$\tilde{\varepsilon}_0 = \frac{V^2}{2Mr_0^2\omega_0^2}\int |\varphi(x,t)|^2 dx.$$

Equation (10.51) is similar to (10.45), and therefore, has similar soliton solution as (10.46) – (10.47).

The canonical transformation adopted here is valid for small V so that the renormalization of the band width remains small, *i.e.* $Je^{-F_T} \gg V^2/M\omega_0^2$ (This is the condition for continuum approximation). If $e^{-F_T} \ll 1$, we can neglect the electron translational motion and arrive at the small radius polaron. For the electron-acoustic phonon interaction, Whitfield and Shaw showed that the polaron state becomes a soliton using the deformation potential technique.

Using the adiabatic approximation, Fedyanin *et al.* studied the properties of polaron in the case where the kinetic energy of the lattice oscillation can be neglected. In this approximation, the electron is assumed to move rapidly in the slowly varying field of the lattice deformation, which may take place if the time between two successive translation leaps of the electron, $\tau_1 \approx \hbar/J$, is much less than the characteristic relaxation time of the lattice deformation, $\tau_2 \approx \omega_0^{-1}$, *i.e.*, $\hbar\omega_0 \ll J$. This is reasonable for a sufficiently wide electron band J. In this case, the self-consistent electron amplitude, $\varphi(x,t)$ in (10.51), satisfies the following equation

$$i\hbar\varphi_t(x,t) = (\varepsilon_0 + W - 2J)\varphi(x,t) - Jr_0^2\varphi_{xx} - \frac{V^2|\varphi|^2\varphi}{M\omega_0^2}, \tag{10.52}$$

in the continuum limit.

Equation (10.52) is similar to (10.45), but the coefficients are constant. According to the above approach, (10.45) – (10.47), the total energy of the electron and the lattice in the adiabatic approximation is given by

$$E_s = \frac{M_s v^2}{2} - \frac{1}{48J}\left(\frac{V^2}{M\omega_0^2}\right)^2,$$

where $M_s = \hbar^2/2Jr_0^2$ is the soliton mass which is the same as that of the electron in conduction band in the given approximation, *i.e.*, there are no temperature renormalization of the energy shift due to the translation leaps $(-2J)$ and the soliton localization area $W_s^0 = (2Jr_0/V^2)M\omega_0^2$. This implies the electron being weakly connected with the lattice, since $V \to 0$ $(W_s^0 \to \infty)$, so that we have $\hbar\omega = \varepsilon_0 + W - 2J + \frac{1}{2}m_z v^2$ and $\varphi(x,t) \approx \exp[i(\omega t - kx)$, *i.e.*, the soliton wavefunction degenerates into a plane wave, in other words, the soliton is delocalized as an electron.

In their study, Fedyanin *et al.* used the normal coordinate ξ_q and its conjugate momentum P_q, and introduced an energy functional $E = \langle\varphi(t)|H|\varphi(t)\rangle$, with $|\varphi(t)\rangle$ given by (10.43), which is a function of ξ_q and P_q for a given $\varphi_n(t)$. The equation satisfied by ξ_q determines the motion of a conventional oscillator under a constant external force. The solution of the equation in such a case can be simply written as

$$\xi_q(t) = \xi_q(0)\cos(\omega_q t) - \frac{V^2}{M\omega_q^2}\sqrt{\frac{2}{N}}\sum_n |\varphi_n(t)|^2.$$

Inserting the above into the equation

$$\frac{\partial}{\partial t}|\varphi(t)\rangle = \hat{H}|\varphi(t)\rangle$$

yields, in the continuum approximating, the following equation for $\varphi(x,t)$,

$$i\hbar\varphi_t(x,t) = (\varepsilon_0 + B - 2J)\varphi(x,t) - Jr_0^2\varphi_{xx}(x,t) - \frac{V^2}{M\omega_0^2|\varphi|^2}\varphi(x,t), \qquad (10.53)$$

where

$$B = \sum_q \frac{V^2}{MN\omega_q^2}\left[\sum_n \frac{\partial}{\partial t}|\varphi_n(t)^2|\sin\left(qn + \frac{\pi}{4}\right)\right]^2$$

is the kinetic energy of lattice deformation consistent with the electron motion, and

$$\varepsilon_0 = \sum_q \frac{V^2}{MN\omega_q^2}\left[\sum_n |\varphi_n(t)^2|\sin\left(qn + \frac{\pi}{4}\right)\right]^2$$

is the elastic energy of the lattice. Note that in the above adiabatic approximation, we have neglected the energy of harmonic oscillations about the shifted equilibrium positions

$$E' = T' + V' = \sum_q \frac{1}{2}M\omega_q^2\xi_q^2,$$

which would lead to a constant correction to the excitation dispersion. Equation (10.53) is similar to (10.45) and (10.52), and thus has soliton solution similar to (10.46). It indicates that polaron in nonlinear system is a soliton and satisfies the nonlinear Schrödinger equation. It describes a quasiparticle of mass $M_s = me^{F_\theta}$, localized in the region of $W_s = W_s^0 e^{-F_\theta}$. The energy of the quasiparticle, $\varepsilon + W - 2Je^{-F_\theta} + V^2/(2M\omega_0^2)$, is less than that of the electron level, $\varepsilon + W + V^2/2M\omega_q^2$. Here

$$m = \frac{\hbar^2}{2Jr_0^2}, \quad W_s^0 = \frac{4JM\omega_0^2}{v^2}, \quad F_\theta = \frac{V^2\theta}{2M\hbar^2\omega_0^4} + O\left(\frac{\omega_1^2}{\omega_0^2}\right).$$

These properties of the polaron are independent of the method of calculation. For soliton satisfying the nonlinear Schrödinger equation (10.53), its "kinetic energy" is

$$E = \frac{1}{2}(m + \Delta m)v^2,$$

where

$$m = \frac{\hbar^2}{2Jr_0^2}, \quad \text{and} \quad \Delta m = \frac{4}{15}\frac{V^2}{M\omega_q^2}\left(\frac{r_0}{W_s^0}\right),$$

with the latter being the temperature independent renormalization of the polaron mass. Using $W_s^0 = 4JM\omega_0^2/V^3$, Δm is given by

$$\Delta m = \frac{(V^2/M\omega_0^2)^4}{240J^3r_0^2\omega_0^2}.$$

10.6 Nonlinear Localization of Small Polarons

Brown *et al.* proposed a theory of polaron formation based on the Fröhlich Hamiltonian in the transportless limit. It was found that two related processes are involved in this formation. In the first process, energy interpretable as the polaron binding energy is released by the exciton system into the normal modes of the lattice. Another process results in the dispersion of this energy throughout the lattice and results in the development of a persistent deformation around the region occupied by the exciton. This indicates that a polaron cannot be exist until the polaron binding energy is dispersed and the lattice deformation is completed. The amount of energy released into the lattice during polaron formation and the time dependence of this energy transfer are independent of the detail with which the initial bare exciton is distributed. Ivic and Brown further studied the nonlinear localization of the small polarons.

It is well known that a foreign particle or excitation in a deformable solid often causes local distortions of the host which can significantly affect the character and dynamics of the quasiparticles. For example, interstitial hydrogen isotopes in metal hydrides can cause large volume dilations; mobilities of photo injected charge carriers in organic molecular crystals such as anthracene and naphthalene exhibit novel temperature dependences; excitonic spectra in organic molecular crystals such as pyrene and α-perylene are significantly affected by local distortions; and certain features in Raman spectra of biological materials such as l-alanine have been identified with local vibrational modes. The quasiparticle transport in these deformable media was investigated by Ivic and Brown using the Fröhlich Hamiltonian

$$H = \sum_n E_n b_n^+ b_n - \sum_{nm} J_{mn} b_m^+ b_n + \sum_q \hbar\omega_q a_q^+ a_q + \\ \sum_{qn} \hbar\omega_q (\chi_n^q a_q^+ + \chi_n^{q*} a_q) b_n^+ b_n, \tag{10.54}$$

where J_{nm} is a transfer integral between different sites in the medium. For the translationally invariant acoustic chain model, they choose the phonon dispersion relation ω_q and the dimensionless coupling function χ_n^q to be

$$\omega_q = \omega_B \sin\left(\frac{1}{2}|qr_0|\right), \quad \chi_n^q = \chi^q e^{-iqR_n} + \frac{2i\chi\sin(qr_0)}{\sqrt{2NM\hbar\omega_q^3}} e^{-iqR_n},$$

respectively, where r_0 is the lattice constant, b_n and a_q are the annihilation operators of the small-polaron and phonon with wave vector q and frequency ω_q, respectively.

Exact solutions corresponding to the Fröhlich Hamiltonian are known only in the limit of infinite effective mass, wherein the transfer integrals between different sites in the medium are all zero. In general the solutions are found by the most approximate approaches which are either small-polaron theories, usually employing perturbation methods based on a set of polaron basis states, or large-polaron theories, usually employing variational methods to obtain nonlinear evolution equations. Among the latter are theories which describe quasiparticle transport in terms of envelope solitons. Ivic *et al.* used the latter to display the unification of polaron and soliton theories of quasiparticle transport in the small-polaron regime based on (10.54).

Following the notation adopted by Ivic *et al.*, we denote operators related to small-polaron by a tilde,

$$\begin{aligned}\langle|\tilde{D}(t)\rangle &\equiv \sum_{a} \tilde{\alpha}_n(t)\tilde{b}_q^+|0\rangle \otimes |\tilde{\beta}(t)\rangle,\\ \langle|\tilde{\beta}(t)\rangle &\equiv \exp\left\{\sum_{n} \tilde{\beta}_q(t)\tilde{a}_q^+ - \tilde{\beta}_q^*(t)\tilde{a}_q^+\right\}|0\rangle,\end{aligned} \tag{10.55}$$

where $\tilde{a}_n \equiv Ua_nU^+$ and $\tilde{b}_n \equiv Ub_nU^+$ are annihilation operators of the dressed-phonon and the small-polaron, respectively, with

$$U \equiv \exp\left[-\sum_{qm}\left(\chi_n^{q*}a_q^+ - \chi_n^{q*}a_q\right)b_m^+b_m\right]. \tag{10.56}$$

These $\tilde{D}$ states contain the exact small-polaron states in the infinite-mass limit ($\tilde{\beta}_q(t)=0$).

The Hamiltonian (10.54) is not diagonal in the small-polaron states when $J \neq 0$. However, the interaction between small polarons $(\tilde{b},\tilde{b}^+)$ and the dressed phonons $(\tilde{a},\tilde{a}^+)$ is weaker than the interaction between the bare quasiparticles (b,b^+) and bare phonons (a,a^+). The microscopic dynamics can be well described by a factored polaron state $|\tilde{D}(t)\rangle$. Since the phase mixing part of $\beta_{qn}(t)$ is chosen to be time independent, Ivic and Brown did not considered soliton, or polaron, formation from bare states, they considered instead the dressed polarons or solitons. To determine the equations of motion of polarons, Ivic *et al.* applied the variational principle of time-dependent quantum mechanics,

$$\delta\int_{t_1}^{t_2} dt\left\langle\psi(t)\left|i\hbar\frac{d}{dt} - H\right|\psi(t)\right\rangle = 0,$$

which was also used by others (Skrinjar, Kapor, and Stojanovic, Zhang, Romero-Rochin) and obtained the following evolution equations,

$$i\hbar\frac{d\tilde{\alpha}_n(t)}{dt} + i\hbar\sum_{q}\left[\chi_n^q\tilde{\beta}_q^*(t) - \chi_n^{q*}\tilde{\beta}_q(t)\right]\tilde{\alpha}_n(t) = \frac{\partial\langle H\rangle}{\partial\tilde{\alpha}_n^*(t)}, \tag{10.57}$$

$$i\hbar\frac{d\tilde{\beta}_q(t)}{dt} - i\hbar\sum_{n}\chi_n^q\frac{d}{dt}|\tilde{\alpha}_n(t)|^2 = \frac{\partial\langle H\rangle}{\partial\tilde{\beta}_q^*(t)} \tag{10.58}$$

where $\langle H\rangle$ is the expectation value of the Hamiltonian, (10.54), in the state $|\tilde{D}\rangle$. From (10.57) and (10.58), we can eliminate the explicit phonon variables in the latter, to obtain an integro-differential equation for the small-polaron probability amplitudes,

$$i\hbar\frac{d\tilde{\alpha}_n(t)}{dt} = \left(E - \sum_q |\chi^q|^2\hbar\omega_q\right)\tilde{\alpha}_n(t) - \sum_m \tilde{J}[\tilde{\alpha}_{n+1}(t) + \tilde{\alpha}_{n-1}(t)] + \int_0^t d\tau \sum_j G_{nj}(t-\tau)\frac{d}{dt}|\tilde{\alpha}_j(\tau)|^2\tilde{\alpha}_n(t) + g_n(t)\tilde{\alpha}_n(t), \qquad (10.59)$$

where

$$\tilde{J} = J\exp\left[-4\sum_q |\chi^q|^2 \sin^2\left(\frac{1}{2}qr_0\right)\right]$$

is the renormalized tunneling matrix element, obviously it is reduced from the bare value J by interactions with phonons, and

$$G_{nj}(t) = 2\sum_q \chi_n^q \chi_j^{q*}\hbar\omega_q\cos(\omega_q t),$$

$$g_n(t) = \sum_q \hbar\omega_q\left[\chi_n^q\tilde{\beta}_n^*(0)e^{i\omega_q t} + \chi_n^{q*}\tilde{\beta}_q(0)e^{-i\omega_q t}\right].$$

The kernels $G_{nj}(t)$ are related to the thermal part of the fluctuations $g_n(t)$, as discussed by Wang *et al.*. Ivic *et al.* expressed a small-polaron Bloch state by

$$|\psi(t)\rangle = \sum_n \tilde{u}(k)\exp\left\{i\left[kR_n - \frac{\bar{E}(k)}{\hbar}\right]t\right\}\tilde{b}_n^+|0\rangle \qquad (10.60)$$

with

$$\bar{E}(k) = \left(E - \sum_q |\chi^q|^2\hbar\omega_q\right) - 2\tilde{J}\cos(kr_0). \qquad (10.61)$$

They chose

$$\tilde{\alpha}_n(t) = \tilde{u}(k)\exp\left\{ikR_n - \frac{i\tilde{E}(k)t}{\hbar}\right\}$$

and $\tilde{\beta}_q(0) = 0$. Then (10.59) is linearized, indicating that the small-polaron, Bloch states are exact stationary solutions of the nonlinear evolution equation. In the continuum limit, Ivic *et al.* obtained the nonlinear Schrödinger equation given below by searching for solutions with the D'Alembert property $|\alpha(y,\tau)|^2 = |\alpha(y - v\tau)|^2$, and choosing $\beta_q(0)$ corresponding to a dressed soliton at zero temperature,

$$i\hbar\frac{d\tilde{\alpha}(x,t)}{dt} = -\frac{\hbar^2}{2\tilde{m}}\frac{\partial^2}{\partial x^2}\tilde{\alpha}(x,t) + \tilde{E}(0)\tilde{\alpha}(x,t) - \frac{v^2}{v_0^2}G(v)|\tilde{\alpha}(x,t)|^2\tilde{\alpha}(x,t), \qquad (10.62)$$

where $\tilde{m} = \hbar^2/2\tilde{J}r_0^2$, with $\tilde{J} = Je^{-\delta^2 F}$,

$$F = 4\sum_q |\chi^q|^2 \sin\left(\frac{r_0 q}{2}\right)$$

is the effective mass of the small-polaron,

$$\tilde{E}(0) = E - \delta(2-\delta)E_0 - 2Je^{-\delta^2 F}$$

is the bottom of the energy band of the small-polaron, as given by (10.61), and

$$G(v) = \frac{4\chi^2 r_0}{w}\frac{v_0^2}{v_0^2 - v^2}$$

is a velocity-dependent nonlinearity parameter. In the above, δ is defined by

$$\beta_{nq}(t) = -\delta\chi_n^q + \tilde{\beta}_q(t) \quad \text{and} \quad \tilde{\beta}_{nq}(t) = \delta\chi_n^q.$$

The above result shows that the small polaron is a soliton which satisfies the nonlinear Schrödinger equation. This soliton solution is

$$\alpha(x,t) = \sqrt{\frac{\mu}{2}}\frac{e^{-i[\tilde{E}(0)/\hbar]t}e^{i(kx-\omega t)}}{\cosh[\mu(x-vt)]}, \tag{10.63}$$

where

$$\hbar k \equiv \tilde{m}v, \quad \hbar\mu = \frac{\tilde{m}G(v)}{2\hbar}\frac{v^2}{v_0^2}, \quad \hbar\omega \equiv \frac{1}{2}\tilde{m}v^2 - \frac{v^4}{v_0^2}\frac{\tilde{m}G(v)^2}{8\hbar^2}.$$

The plane-wave solutions (10.60) of this equation are not bare Bloch states, but are small-polaron Bloch states with a larger effective mass ($\tilde{m} > m$) and a lower energy $[\tilde{E}(0) < E(0)]$ compared to a bare Block state, due to the dressing of the quasiparticle.

10.7 Nonlinear Excitation of Electrons in Coupled Electron-Electron and Electron-Phonon Systems

A coupled electron-electron or an electron-phonon system is often described by the modified Hubbard model. The one-dimensional modified Hubbard model was studied in detail by Lindner and Fedyanin, and Makhankov *et al.* The model Hamiltonian is written in the form of

$$H = H_{ph} + H_e + H_{\text{int}} \tag{10.64}$$

where H_{ph} and H_e are the Hamiltonians of the phonon and electron subsystems, respectively, while H_{int} describes the interaction between them. H_e corresponds to the Hubbard model in the nearest-neighbor approximation. We assume that the ground state is antiferromagnetic and that the lattice is divided into two equivalent ferromagnetic sublattices A and B such that the resulting magnetic moment is zero in the absence of an external magnetic field. Based on this, Makhankov *et al.* proposed the following for H_e,

$$H_e = \sum_{n\in A,j\in B;\sigma} J_{nj}(b^{A+}_{n,\sigma}b^{B}_{n,\sigma} + b^{B+}_{n,\sigma}b^{A}_{n,\sigma}) + \frac{V}{2}\sum_{n\in A,B;\sigma} N_{n,\sigma}N_{n,-\sigma} - \mu\sum_{n,\sigma} N_{n,\sigma}, \quad (10.65)$$

where

$$J_{nj} = \begin{cases} \bar{J}, & (j = n \pm 1), \\ 0, & (j \neq n \pm 1), \end{cases}$$

$b^+_{n,\sigma}$ $(b_{n,\sigma})$ is the creation (annihilation) operator for an electron with spin σ in a Wannier state at site n, $N_{n,\sigma} = b^+_{n,\sigma}b_{n,\sigma}$ is the number operator, J the transition amplitude between nearest neighbors, V the repulsive interaction between electrons in the same atom, μ denotes the chemical potential. The lattice part of the Hamiltonian, H_{ph}, is given in the harmonic approximation by

$$H_{ph} = \frac{M}{2}\sum_n [(R_n)_t]^2 + \frac{w}{2}\sum_n (R_{n+1} - R_n)^2, \quad (10.66)$$

here M is the atomic mass, w denotes the force constant of the system, $R_n \equiv R_n(t) = \bar{R}_n + u_n(t)$ is the position of the nth atom and $\bar{R}_n$ is its equilibrium position. In the linear approximation with respect to small deviations $u_n(t)$, the interaction between the electron subsystem and the lattice in (10.64) can be written as

$$\begin{aligned} H_{\text{int}} = I\sum_{n,\sigma}(u_{n,\sigma} - u_{n+1,\sigma})\,&\left[b^{B+}_{n,\sigma}b^{A}_{n+1,\sigma} + b^{A+}_{n+1,\sigma}b^{B}_{n,\sigma}\right. \\ &\left. + b^{B+}_{n+1,\sigma}b^{A}_{n,\sigma} + b^{A+}_{n,\sigma}b^{B}_{n+1,\sigma}\right], \end{aligned} \quad (10.67)$$

where

$$I = \left\langle (n+1) \left| \frac{\partial V(x - \bar{R}_n)}{\partial x} \right| n \right\rangle$$

denotes the matrix element of the local "force" between Wannier states of neighboring atoms, corresponding to the local potential $V = V(x - \bar{R}_n)$. Effects of anharmonicity in the lattice vibrations and nonlinearity in the expansion of electron integration force in this model have been studied and many interesting features of the microscopic particles (soliton) were revealed.

Since the ground state is antiferromagnetic, we are interested in the low-lying excitations, and in equations describing the probability amplitudes of spin up or down states $(\phi_{n\uparrow}, \phi_{n\downarrow})$ at the nth lattice site. We can eliminate terms of the type $(u_{n+1} - u_n)C^{B+}_{n,\sigma}C^{A}_{n-1,\delta}$, *etc.* by the analogic method discussed in Section 10.5. To simplify the calculation, Makhankov *et al.* employed a method similar to the adiabatic approach related to the ground state, and examined the change of the state

$|0\rangle$ under the action of the evolution operator, which is defined on single-particle states. Thus Lindner *et al.* and Makhankov *et al.* defined

$$\phi^p_{n,\sigma}(t) = \langle\varphi(t)|b^p_{n,\sigma}(t)|\varphi(t)\rangle = \langle 0|b^p_{n,\sigma}(t)|0\rangle = \langle 0|Ub_{n,\sigma}U|0\rangle. \tag{10.68}$$

Making use of (10.66) – (10.68), we obtain the following equation for $b^A_{n,\sigma}(t)$,

$$\begin{aligned} i(b^A_{n,\sigma}(t))_t = [b^A_{n,\sigma}, H]_- = Vb^A_{n,\sigma}(t)n_{n-\sigma} - \mu b^A_{n,\sigma}(t) \\ +[\bar{J} + I(u_n - u_{n-1})]b^B_{n+1,\sigma}(t) + [\bar{J} + (u_{n-1} - u_n)]b^B_{n-1,\sigma}(t). \end{aligned} \tag{10.69}$$

The corresponding equation for $b^B_{n,\sigma}$ is the same as (10.69) except that A and B are interchanged. Applying (10.68) and (10.69), we can get

$$\begin{aligned} i(\phi^A_{n,\sigma}(t))_t = V\langle 0|b^A_{n,\sigma}(t)n_{n,-\sigma}|0\rangle - \mu\phi^A_{n,\sigma}(t) \\ +[\bar{J} + I(u_n - u_{n-1})]\phi^A_{n+1,\sigma}(t) + [\bar{J} + I(u_{n-1} - u_n)]\phi^B_{n-1,\sigma}(t) \end{aligned} \tag{10.70}$$

and a similar equation for $\phi^B_{N,\sigma}(t)$, which is the same as the above except that A and B are interchanged. Applying the following decoupling procedure to (10.70),

$$\begin{aligned} \langle 0|b^p_{n,\sigma}(t)b^{p+}_{n,-\sigma}(t)b^p_{\eta,-\sigma}(t)|0\rangle \to \langle 0|b^p_{n,\sigma}(t)|0\rangle\langle 0|b^{p+}_{n,-\sigma}(t)|0\rangle\langle 0|b^p_{\eta,-\sigma}(t)|0\rangle \\ = \phi^p_{n,\sigma}(t)\phi^{p*}_{n,-\sigma}(t)\phi^p_{n,-\sigma}(t), \end{aligned} \tag{10.71}$$

where $p = A, B$, we finally get a system of nonlinear equations for the four complex function $\phi^P_{n\alpha}$ and the real function u_n. For the "pure" Hubbard model ($I = 0$), similar nonlinear equations can be obtained for $\phi_{n,\sigma}(t)$ as well.

We now discuss the influence of spin ordering on the lattice vibrations. Makhankov *et al.* constructed the functional,

$$\begin{aligned} F(\phi^0_{n,\sigma}, u_n) = \langle\varphi(t)|H|\varphi(t)\rangle = -\mu \sum_{n\in A\cup B;\sigma} |\phi_{n,\sigma}|^2 - \sum_{n,\sigma}[\bar{J} - I(u_n - u_{n+1})] \\ \times(\phi^{A*}_{n,\sigma}\phi^B_{n+1,\sigma} + \phi^{B*}_{n+1,\sigma}\phi^A_{n,\sigma} + \phi^{B*}_{n,\sigma}\phi^A_{n+1,\sigma} + \phi^{A*}_{n+1,\sigma}\phi^B_{n,\sigma}) \\ -\frac{1}{2}\sum_{n\in A\cup B} |\phi_{n,\sigma}|^2|\phi_{n,-\sigma}|^2 + \frac{M}{2}\sum_n[(R_n)_t]^2 + \frac{w}{2}\sum_n(R_{n+1} - R_n)^2. \end{aligned}$$

From the Hamilton equations,

$$M(R_n)_{tt} = P_n, \quad P_n = -\frac{\delta F}{\delta R_n},$$

it follows that

$$\begin{aligned} M(R_n(t))_{tt} = w[R_{n+1}(t) - 2R_n(t) + R_{n-1}(t)] \\ -I\sum_\sigma \{\phi^{A*}_{n,\sigma}(t)[\phi^B_{n+1,\sigma}(t) - \phi^B_{n-1,\sigma}(t)] - [\phi^{B*}_{n+1,\sigma}(t) - \phi^{B*}_{n-1,\sigma}(t)]\phi^A_{n,\sigma}(t) \\ -\phi^{B*}_{n,\sigma}(t)[\phi^A_{n+1,\sigma}(t) - \phi^A_{n-1,\sigma}(t)] + [\phi^{A*}_{n+1,\sigma}(t) - \phi^{A*}_{n-1,\sigma}(t)]\phi^B_{n,\sigma}(t)\}. \end{aligned} \tag{10.72}$$

This equation is a discrete analogue of the harmonic oscillator equation under the action of an external force which does not vanish if variation in the initial spin ordering takes place. The usual condition for antiferromagnetism $\langle N^A_{n,\sigma}\rangle = \langle N^B_{n\pm1,\sigma}\rangle$ now becomes $\phi^A_{n-\sigma} = \phi^B_{n\pm1,\sigma}$ which causes the last term in (10.72) $(-I)$ to vanish.

If we assume that the variation of magnetic order is described by a sufficiently smooth function of coordinate along the chain, we can use the continuum approximation

$$u_n(t) = R_n(t) - \bar{R}_n \to (\chi(\xi,t) - \xi)r_0, \quad \phi^p_{n,\sigma}(t) \to \phi^p_\sigma(\xi,t).$$

Then, expanding up to the second derivative and taking into account up to the cubic nonlinear terms, we have

$$R_{n+1}(t) \to r_0\left(\chi + \chi_\xi + \frac{1}{2}\chi_{\xi\xi} + \cdots\right),$$

$$\phi^B_{n\pm1,\sigma}(t) \to \phi^A_{-\sigma}(\xi,t) \pm \phi^{A'}_{-\sigma}(\xi,t) + \frac{1}{2}\phi^{A''}_{-\sigma}(\xi,t) + \cdots,$$

$$\phi^A_{n\pm1,\sigma}(t) \to \phi^B_{-\sigma}(\xi,t) \pm \phi^{B'}_{-\sigma}(\xi,t) + \frac{1}{2}\phi^{B''}_{-\sigma}(\xi,t) + \cdots,$$

where $\phi^B_\sigma(\xi,t) = 0$ if $n \in A$, and similarly $\phi^A_\sigma = 0$, if $n \in B$. Makhankov *et al.* finally obtained the following equations for $\phi_\sigma(x,t)$ and $\chi(\xi,t)$

$$i(\phi_\sigma(\xi,t))_t = [(\phi^n_{-\sigma}(\xi,t))_{\xi\xi} + 2\phi_{-\sigma}(\xi,t)][\tilde{J} + r_0 I(1 - \chi_\xi(\xi,t))] + I|\phi_{-\sigma}|^2\phi_\sigma - \mu\phi_\sigma, \tag{10.73}$$

$$\ddot{\chi}(\xi,t) = v_0^2\chi_{\xi\xi}(\xi,t) + \frac{2I}{r_0 M}\sum_\sigma \frac{\partial}{\partial\xi}(\phi^*_\sigma\phi_{-\sigma}), \tag{10.74}$$

respectively, where $v_0^2 = \frac{w}{M}r_0^2$.

In the quasi-static approximation (or setting $\chi(\xi,t) = \chi(\xi - vt)$), we can integrate (10.74) to obtain

$$\chi_\xi(\xi,t) = \frac{\tilde{I}}{2Ir_0}\sum_\sigma \phi^*_\sigma\phi_\sigma + K, \tag{10.75}$$

where

$$\tilde{I} = \frac{(2I)^2}{M(v^2 - v_0^2)}$$

with $v^2 \ll v_0^2$ and the constant K depending on initial conditions. Using (10.75), we can get from (10.73)

$$i(\phi_\sigma(\xi,t))_t = T(\phi^n_{-\sigma}(\xi,t))_{\xi\xi} + \left[2T - \tilde{I}\sum_\sigma \phi^*_\sigma(\xi,t)\phi_{-\sigma}(\xi,t)\right]\phi_{-\sigma}(\xi,t) - \mu\phi_\sigma(\xi,t) + V|\phi_{-\sigma}(\xi,t)|^2\phi_\sigma(\xi,t), \tag{10.76}$$

where

$$T = \bar{J} - r_0 I(1 - K).$$

In terms of even and odd combinations of $\phi_\sigma = \{\phi_\uparrow, \phi_\downarrow\}$,

$$\varphi^{(1)} = \phi_\uparrow + \phi_\downarrow, \quad \varphi^{(2)} = \phi_\uparrow - \phi_\downarrow,$$

where

$$\phi_\uparrow = \frac{\varphi^{(1)} + \varphi^{(2)}}{2}, \quad \phi_\downarrow = \frac{\varphi^{(1)} - \varphi^{(2)}}{2},$$

one gets

$$\begin{aligned} i\varphi_t^{(i)} = T\varphi_{\xi\xi}^{(i)} &- \left\{\delta_{i,1}2T - \delta_{i,2}2T - \mu - \frac{V-\hat{I}}{2}\left[|\varphi^{(1)}|^2 + |\varphi^{(2)}|^2 - 2|\varphi^{(i)}|^2\right]\right\}\varphi^{(i)} \\ &-\tilde{I}\left\{-\frac{1}{2}\left[|\varphi^{(1)}|^2 + |\varphi^{(2)}|^2\right]\varphi^{(i)} + \frac{3}{4}|\varphi^{(i)}|^2\varphi^{(i)} \right. \\ &\left. +\frac{1}{4}|\varphi^{(i)}|^2\left[\delta_{i,1}\varphi^{*(2)} + \delta_{i,2}\varphi^{*(1)}\right]\right\}, \end{aligned} \tag{10.77}$$

where $i = 1, 2$. Assuming that the average occupation number $n = |\phi_\uparrow|^2 + |\phi_\downarrow|^2$ is constant and neglecting the term $\phi_\sigma^*\phi_{-\sigma}\phi_{-\sigma}$ compared to $|\phi_{-\sigma}|^2\phi_\sigma$, we get two decoupled nonlinear Schrödinger equations for $\varphi^{(1)}$ and $\varphi^{(2)}$, respectively, and their soliton solutions are

$$\varphi^{(i)}(\xi, t) = \sqrt{2n}\,\mathrm{sech}\left[\sqrt{\frac{n(V-\hat{I})}{T}}(\xi - vt)\right]\exp\left(\frac{iv\xi}{T} - i\omega_i t\right), \quad (i = 1, 2),$$

where

$$\omega_1 = 2T - \mu - \frac{v^2}{4T}, \quad \omega_2 = -2T - \mu + (V - \hat{I}) + \frac{v^2}{4T},$$

and v is the soliton velocity. Therefore, electron in the Hubbard model with electron-electron and electron-phonon interactions satisfies the nonlinear Schrödinger equation or the laws of nonlinear quantum mechanics. The soliton consists of a "drop" associated with a "bubble", and is localized in the region

$$W_s \approx \sqrt{\frac{T}{n}(V - \hat{I})},$$

which is accompanied by a lattice distortion wave that is also bounded in the same region, and $\chi(\xi, t)$ in (10.74) or (10.75) can be represented approximately by $\chi \approx \tanh[(\xi - vt)/W_s]$ under appropriate initial conditions. However, for the pure Hubbard model, $\tilde{I} = 0$, $T = \tilde{J}$, and a half-filled band, *i.e.*, $n = 1$, it follows that $\omega_1 = \omega_2$ for $\mu = V/2$.

In the more general case, it is difficult to find the solution of (10.77). But if we neglect the last term on the right-hand side of (10.77), it becomes

$$i\varphi_t^{(i)} + \delta_{i,1}T\varphi_{xx}^{(i)} - \delta_{i,2}T\varphi_{xx}^{(i)} - \frac{1}{2}V\left(\eta|\varphi^{(i)}|^2 - |\varphi^{(i+1)}|^2\right)\varphi^{(i)} = 0, \tag{10.78}$$

where $i = 1, 2$ and

$$\eta = 1 - \frac{\hat{I}}{2V}, \quad \varphi^{(2)} = \varphi^{(1)}.$$

This is a vector nonlinear Schrödinger equation. In the case of $\eta = 1$ (or $\hat{I} = 0$) which corresponds to the situation when the electron-phonon interaction is "switched off", Makhankov *et al.* proved that (10.78) is integrable. They were able to obtain the Lax pair, examine its symmetry property, and construct the soliton solutions for given values of η.

10.8 Nonlinear Excitation of Magnon in Ferromagnetic Systems

Presence of magnons in ferromagnetic crystals can change the potential energy of interaction between atoms. In the case of a ferromagnetic chain, a deformation is created by excitation of the magnons which can result in localization of magnon through interaction between the spin deviation and the deformation of the lattice. Thus the magnon becomes a soliton in such a case, it can move without changing its shape. Nonlinear excitations in ferromagnetic chains were studied by Pushkarov and Pushkarov. They used the following Hamiltonian.

$$H = T + V - \frac{1}{4}\sum_{n\delta} J(u_n - u_{n+\delta})(S_n^+ S_{n+\delta}^- + S_n^- S_{n+\delta}^+)$$
$$-\frac{1}{2}\sum_{n\delta} \tilde{J}(u_n - u_{n+\delta}) S_n^z S_{n+\delta}^z. \tag{10.79}$$

Here u_n is the position of atom n in the chain, δ runs over the nearest neighbors of a given atom, $S_n^\pm$ are the cyclic spin components given by the relations $S_n^\pm = S_n^x \pm i S_n$, $J(u_n - u_{n+\delta})$ and $\tilde{J}(u_n - u_{n+\delta})$ are exchange integrals which are functions of $|u_n - u_{n+\delta}|$,

$$T = \frac{m}{2}\sum_m [(u_n)_t]^2$$

is the kinetic energy of the atoms of mass m. The harmonic approximation is assumed for interatomic interaction and the potential energy is given by

$$V = \frac{mv_0^2}{2r_0}\sum_n (u_{u+1} - u_n - r_0)^2,$$

where v_0 is the sound velocity and r_0 the lattice constant.

At low temperature when only a few spin waves are excited, the operators $S_n^\pm$ and S_n^z can be expressed in terms of the Bose operators a_n and a_n^+ as,

$$S_n^+ = \sqrt{2S} a_n, \quad S_n^- = \sqrt{2S} a_n^+, \quad S_n^z = S - a_n^+ a_n. \tag{10.80}$$

In the bilinear approximation with respect to the Bose operators, we can obtain the following by inserting (10.80) into (10.79).

$$H = T + V - \frac{S^2}{2}\sum_{n\delta}\tilde{J}(u_n - u_{n+\delta}) - \frac{S}{2}\sum_{n\delta}\Big\{\tilde{J}(u_n - u_{n+\delta})$$

$$\times\left[a_n^+ a_{n+\delta} + a_n a_{n+\delta}^+\right] - J(u_n - u_{n+\delta})\left[a_n^+ a_n + a_{n+\delta}^+ a_{n+\delta}\right]\Big\}. \quad (10.81)$$

From

$$i\hbar\frac{\partial}{\partial t}|\varphi\rangle = H|\varphi\rangle$$

and (10.43), we obtain

$$i\hbar\frac{\partial\varphi_n}{\partial t} = \left[T + V - \frac{S^2}{2}\sum_{n\delta}\tilde{J}(u_n - u_{n+\delta})\right]\varphi_n$$

$$-S\sum_{\delta}J(u_n - u_{n+\delta})\varphi_{n+\delta} + S\sum_{\delta}\tilde{J}(u_n - u_{n+\delta})\varphi_n. \quad (10.82)$$

We have assumed in (10.82) that

$$J(|u_n - u_{n+1}|) + J(|u_n - u_{n-1}|) \approx 2J,$$

$$J(|u_n - u_{n+1}|) + \tilde{J}(|u_n - u_{n-1}|) \approx 2\tilde{J}_0 - \tilde{J}_1(u_{n+1} - u_{n-1}),$$

$$\sum_{n\delta}\tilde{J}(u_n - u_{n+\delta}) \approx 2\tilde{J}_0 N - \tilde{J}_1\sum_{n}(u_{n+1} - u_{n-1}),$$

where

$$\tilde{J}_1 = -\frac{\partial\tilde{J}}{\partial u_n} > 0.$$

Equation (10.82) can then be written in the following form

$$i\hbar\frac{\partial\varphi_n}{\partial t} = \left[T + V - \tilde{J}_0 S^2 N + \tilde{J}_1 S\sum_{n}(u_{n+1} - u_{n-1}) + 2(\tilde{J}_0 - J)S\right]\varphi_n$$

$$-J(\varphi_{n+1} + \varphi_{n-1} - 2\varphi_n) - \tilde{J}_1 S(u_{n+1} - u_{n-1}). \quad (10.83)$$

Taking into account the explicit expressions of T and V as well as the Hamilton equations for u_n and $P_n = m(u_n)_t$, we get

$$m(u_n)_{tt} = mv_0^2(u_{n+1} + u_{n-1} - 2u_n) - \tilde{J}_1 S(|\varphi_{n+1}|^2 - |\varphi_{n-1}|^2). \quad (10.84)$$

In the long-wave and continuum approximations, equation (10.84) takes the form

$$\frac{\partial^2 u(x,t)}{\partial t^2} = v_0^2\frac{\partial^2 u(x,t)}{\partial x^2} - \frac{2\tilde{J}_1 S}{m}\frac{\partial}{\partial x}|\varphi(x,t)|^2. \quad (10.85)$$

In the continuum approximation (10.83) becomes

$$i\hbar\frac{\partial\varphi(x,t)}{\partial t} = \left[T+V-\tilde{J}_0S^2N+\frac{(2\tilde{J}_1)^2S}{m(v_0^2-v^2)}+2(\tilde{J}_0-J)S\right]\varphi(x,t)$$
$$-JS\frac{\partial^2\varphi(x,t)}{\partial x^2}-\frac{(2\tilde{J}_1S)^2}{m(v_0^2-v^2)}|\varphi(x,t)|^2\varphi(x.t), \tag{10.86}$$

with

$$T+V=\frac{m}{2}\left[\int_{-\infty}^{\infty}\left(\frac{\partial u}{\partial t}\right)^2dx+v_0^2\int_{-\infty}^{\infty}\left(\frac{\partial u}{\partial x}\right)^2dx\right]$$
$$=\frac{(2\tilde{J}_1S)^2}{2mv_0^2}\frac{1+s^2}{(1-s^2)^2}\int_{-\infty}^{\infty}|\varphi(x,t)|^4dx,$$

where $s=v/v_0$. Therefore, the magnon is a soliton in nonlinear systems, which satisfies the nonlinear Schrödinger equation, and it can be described by the nonlinear quantum mechanics. The soliton solution of (10.82) is

$$\varphi(x,t)=\frac{1}{\sqrt{2\bar{\mu}}}e^{i[\alpha x-\theta-\omega t]}\text{sech}\left(\frac{x-x_0-vt}{\bar{\mu}}\right),$$

where

$$\alpha=\frac{\hbar v}{2JS},\quad \bar{\mu}=\frac{Jmv_0^2}{\tilde{J}_1^2S}(1-s^2),$$
$$T+V=\frac{2(\tilde{J}_1S)^4}{3JS(mv_0^2)^2}\frac{1+s^2}{(1-s^2)^3},$$
$$\hbar\omega=T+V-\tilde{J}_0S^2N+\frac{(2\tilde{J}_1)^2S^3}{mv_0^2(1-s^2)}+2(\tilde{J}_0-J)S+\frac{\hbar^2v^2}{4JS}.$$

Here x_0 and x' are arbitrary constants. x_0 appears as a result of the translation invariance of the problem and together with the velocity v can be determined by the initial conditions. θ depends on the initial phase of the wave function.

The energy of the system with a magnon-soliton is given by

$$\hbar\omega=\hbar\omega_0+\frac{\hbar^2v^2}{4JS}+\frac{(\tilde{J}_1S)^4}{JS(mv_0^2)^2}\frac{(2+3s^2-s^4)s^2}{(1-s^2)^3}, \tag{10.87}$$

where

$$\hbar\omega_0=-\tilde{J}_0SN^2+2(\tilde{J}_0-J)S-\frac{(\tilde{J}_1S)^4}{JS(mv_0^2)^2}.$$

In (10.87), the total energy $\hbar\omega$ is given as a sum of the rest energy $\hbar\omega_0$ and kinetic energy. The first two terms in $\hbar\omega_0$ represent the ground state energy of the anisotropy spin system being considered and the last one describes the rest energy due to the solitary excitation. In the case of small velocities, ($v^2 \ll v_0^2$), the soliton moves as a particle with an effective mass $m^* = m_0^*(1 + \triangle m^*/m_0^*)$, where $m_0^* = \hbar^2/2JS$ is the effective mass of a free magnon and

$$\triangle m^* = \frac{4(\tilde{J}_1 S)^4}{m\hbar^2 v_0^4} m_0^*$$

is the change in mass due to localization of the magnon, *i.e.*, the transformation of the magnon into a "magnon-soliton". If the velocity v equals to zero, the soliton energy has the absolute minimum,

$$E_{\min} = -\tilde{J}_0 S^2 N + 2(\tilde{J}_0 - J)S - \frac{(\tilde{J}_1 S)^4}{JS(mv_0^2)^2}.$$

We see that the energy of the system with a soliton at rest is smaller than the energy of free magnon. Hence such a state is more favorable. As the velocity v increases, the energy also increases. At $v > v/\sqrt{5}$, the contribution of the last term in (10.87) predominates over the last one in $\hbar\omega_0$.

Chain elongation $\triangle$ occurs due to spin deviation. From (10.85), we obtain that

$$\triangle = \int_{-\infty}^{\infty} \frac{\partial u}{\partial x} dx = \frac{2\tilde{J}_1 S}{mv_0^2(1 - s^2)}.$$

Thus, $\triangle$ increases with velocity v. In the case of a stationary soliton, we have $\triangle = 2\tilde{J}_1 S/mv_0^2$.

Pushkarov and Pushkarov studied nonlinear excitations of magnons in ferromagnetic systems with a biquadratic exchange and anharmonic lattice deformation using the same method, and obtained many useful results. Davydov and Kislukha examined low-lying excitations of anisotropic Heisenberg ferromagnet in one-dimensional systems described by the general Hamiltonian

$$H = -\mu\tilde{H}_z \sum_n S_n^z + 2J \sum_n \left(\xi S_n^x S_{n+1}^x + \eta S_n^x S_{n+1}^y + S_n^z S_{n+1}^z - \frac{1}{4}\right), \tag{10.88}$$

where $B = (0, 0, \tilde{H}_z)$, $S = 1/2$. For a system of N spins which are rigidly fixed at lattice sites, if $-1 \leq (\xi, \eta) \leq 1$, *i.e.*, for any possible combination of S_j^x and S_j^y (compared to S_n^z), we have the so-called "easy axis" $0z$ model. When $\xi = \eta$, there is isotropy in the base (x-y) plane. If $\eta = 1$, $\xi \neq \eta$ (or alternatively $\xi = 1$, $\xi \neq \eta$), we have the "easy plane" model (y-z or x-z, respectively). Finally, setting $\xi = 0$ (or $\eta = 0$) gives rise to the Ising model in a transverse field. The simplest variant of (10.88) is an exact "Ising model" when $\xi = \eta = 0$. $\xi = \eta = 1$ yields the isotropic Heisenberg model, which includes also the case of $\xi \gg 1$ (or $\eta \gg 1$), if the last term $S_n^z S_{n+1}^z$ in (10.88) can be treated as a small perturbation. However, in the study of low-lying excitations related to nonlinearity, this small term in (10.88) must be included.

Applying the Jordan-Wigner transformation for $S_j^{\pm}$ and S_j^z

$$\begin{aligned} a_n^+ &= (-2)^{n-1} S_1^z S_2^z \cdots S_{n-1}^z S_n^+, \quad S_n^z = S_n^+ S_n^- - \frac{1}{2}, \\ a_n &= (-2)^{n-1} S_1^z S_2^z \cdots S_{n-1}^2 S_n^-, \quad S_n^{\pm} = S_n^x \pm i S_n^y, \end{aligned} \tag{10.89}$$

where a_n^+ and a_n are the Fermi operators $\{a_n, a_m^-\} = \delta_{mn}$, $\{a_n, a_n\} = \{a_n^+, a_n^+\} = 0$, we obtain, instead of (10.88), the Hamiltonian in terms of the Fermi operators

$$\begin{aligned} H = \frac{1}{2}\mu N\tilde{H}_z - (\mu\tilde{H}_z + 2J)\sum_k N_k + \frac{1}{2}J(\xi+\eta)\sum_n (a_n^+ a_{n+1} + a_{n+1}^+ a_n) - \\ \frac{1}{2}J(\xi-\eta)\sum_n (a_{n-1}^+ a_n^+ - a_{n-1}a_n) + 2J\sum_n N_n N_{n+1}, \end{aligned} \tag{10.90}$$

where $N_n = a_n^+ a_n$. The nonlinearity is now due to the term $S_n^z S_{n+1}^z$. The anomalous term $-(1/2)J(\xi-\eta)\sum_n (a_{n-1}^+ a_n^+ - a_{n+1}a_n)$ is defined by the difference between the x and y components. Note that model (10.88), like (10.90), corresponds to a system of rigidly fixed spins.

From the Heisenberg equation and (10.90) and (10.43), one can obtain

$$i\hbar\varphi_t(x,t) = -\Delta\varphi(x,t) + \frac{1}{2}J(\xi+\eta)r_0^2\varphi_{xx}(x,t) + \frac{1}{2}J(\xi-\eta)r_0\varphi_x^* + 2J|\varphi(x,t)|^2\varphi(x,t),$$

where

$$\Delta = 2J\left(1 - \frac{\mu\tilde{H}_z}{2J} - \frac{\xi-\eta}{2}\right) \equiv 2Jd.$$

Note that Δ can be positive (e.g. for $J > 0$, $|\xi| < 1$, and $|\eta| < 1$) or negative ($\xi \gg 1$, $\eta \approx 1$). In the case of $J > 0$ and $0 \le \xi \le 1$, $0 \le \eta \le 1$, $\Delta > 0$, $d > 0$, letting $\hbar = r_0 = 1$ and introducing new variables, $\phi = \varphi(x,t)/\sqrt{d}$, $t' = t\Delta$, and $x' = \sqrt{4d/(\xi+\eta)}x$, we have

$$i\phi_{t'} - \phi + \phi_{x'x'} + \alpha\phi_{x'}^* + |\phi|^2\phi = 0, \tag{10.91}$$

where

$$\alpha = \frac{\xi - \eta}{\sqrt{4d(\xi+\eta)}}.$$

Equation (10.91) is still a nonlinear Schrödinger equation, and should have a soliton solution. If $\alpha < 1$, $\alpha\phi^*$ could be treated as a small perturbation. Based on this equation, Makhankov *et al.* discussed structural stability of microscopic particles (soliton) described by the nonlinear Schrödinger equation given in Chapter 4. Fedyanin *et al.* also obtained an approximate solution of (10.91) using an alternative approach. They assumed $\phi = \sqrt{2}\exp(-it')F(x',t')$ and derived the following equation by inserting it into (10.91).

$$iF_{t'} + \frac{1}{2}F_{x'x'} + |F|^2F = -\frac{1}{2}\alpha F^*_{x'}\exp(it'). \tag{10.92}$$

Equation (10.92) has the following solution at $\alpha = 0$,

$$F_0 = 2\nu_0 \exp\left[i\left(\frac{\mu_0}{\nu_0}z_0 + \theta_0\right)\right]\text{sech}z_0,$$

where $z_0 = 2\nu_0(x' - \bar{x}'_0)$, with $\bar{x}'_0 = 4\mu t'$. If $\alpha \neq 0$ but very small, we can assume that the parameters, ν_0, μ_0, $\bar{x}'_0$ and θ_0, change only due to this perturbation. Following the perturbative approach of Karpman and Maslov, Fedyanin *et al.* obtained the first order correction in α to the time-dependences of the variables, ν, μ, $\bar{x}'$ and θ which are given by

$$\mu_{t'} = \frac{\alpha}{2}\Im\int_{-\infty}^{\infty}\frac{\sinh z}{\cosh^2 z}(iF^*_z)Q(z)dz,$$

$$\nu_{t'} = \frac{\alpha}{2}\Re\int_{-\infty}^{\infty}\frac{iF^*_z}{\cosh z}Q(z)dz,$$

$$\bar{x}'_{t'} = 4\mu + \frac{\alpha}{4\nu^2}\Re\int_{-\infty}^{\infty}\frac{z}{\cosh z}(iF^*_z)Q(z)dz,$$

$$\theta_{t'} = 2\mu\bar{x}'_{t'} - 4(\mu^2 - \nu^2) + \frac{\alpha}{2\nu}\Im\int_{-\infty}^{\infty}\left(\cosh^{-1}z - \frac{z\sinh^2 z}{\cosh^2 z}\right)(iF^*_z)Q(z)dz,$$

where

$$Q(z) = \exp\left[i\left(-\frac{\mu}{\nu}z - \theta + 2t'\right)\right].$$

However, note that

$$v_{t'} \approx \alpha\Re\left[e^{2it}\int_{-\infty}^{\infty}iF^*F^*_z dz\right] = 0.$$

We then have

$$\mu \simeq -\frac{2}{3}\alpha\nu_0^2\sin 2t', \quad \bar{x}' = -\frac{\alpha}{2}\cos 2t', \quad \theta = 4\nu_0^2 t'.$$

The perturbed solution of (10.91) can be written as

$$\phi(x,t) = A_0\text{sech}\left\{\frac{A_0}{\sqrt{2}}[x' - \cos(2t')]\right\}\exp\left\{-i\left(1 - \frac{A_0^2}{2}\right)t + \frac{\alpha A_0^2 x'}{6}\sin(2t')\right],$$

where $A_0 = 2\sqrt{1-\omega}$. This indicates that the magnon is still a soliton in this case. The soliton solution of (10.91) can be obtained using other approaches.

From the above discussion, we can conclude that the magnon is localized, as a soliton, due to nonlinear interactions in nonlinear ferromagnetic systems. The motion of the magnon-soliton can be very well described by the nonlinear quantum mechanics. Existence of the magnon-solitons has been verified by neutron scattering and NMR experiments in $CuGeO_3$ by Ronnow *et al.*, Enderle *et al.*, and Revarat *et al.*, respectively. The readers are referred to the paper by Mikeska and Steiner (1991) for further details on magnetic soliton.

10.9 Collective Excitations of Magnons in Antiferromagnetic Systems

Collective excitation and motion of magnons due to magnon-phonon interactions or magnon-magnon interaction in Heisenberg antiferromagnetic systems have been extensively studied by Pang and co-workers and many other scientists. The results show that the characteristics of the collective excitation in these systems are quite different from those in ferromagnetic systems. In this section, we present some results obtained by Pang *et al.* on collective excitation in anisotropic Heisenberg antiferromagnets with magnon-phonon and magnon-magnon interactions.

Assuming the double-sublattice model, the Hamiltonian of the Heisenberg antiferromagnet can be expressed as

$$H = T + V + \frac{1}{2}\sum_{n}^{A}\sum_{\delta}^{A}\left[\xi_{n,n+\delta}S_n^x S_{n+\delta}^x + \eta_{n,n+\delta}S_n^y S_{n+\delta}^y + J_{n,n+\delta}S_n^z S_{n+\delta}^z\right]$$
$$+\frac{1}{2}\sum_{j}^{B}\sum_{\delta}^{B}\left[\xi_{j,j+\delta}S_j^x S_{j+\delta}^x + \eta_{j,j+\delta}S_j^y S_{j+\delta}^y + J_{j,j+\delta}S_j^z S_{j+\delta}^z\right], \qquad (10.93)$$

where

$$T = \frac{m}{2}\sum_n \dot{u}_n^2, \quad V = \frac{mv_0^2}{r_0}\sum_n (u_{n+1} - u_n - u_0)^2$$

are the kinetic and potential energies of lattice oscillations, respectively, with m being the "mass" of a spin, r_0 the lattice constant, and v_0 the sound velocity in the crystal which we set equal to unity in subsequent calculations, $S^{\kappa}_{n(j)}$ $(\kappa = x, y, z)$ is the spin component at site $n(j)$ in the k-direction. Applying the transformation, $S^+_{n(j)} = (S^x_{n(j)} + iS^y_{n(j)})$ and making use of the Dyson-Maleev representation of the spin operators

$$S_a^+ = \sqrt{2S}\left(1 - \frac{a^+a}{4S}\right)a, \quad S_a^- = \sqrt{2S}a^+\left(1 - \frac{a^+a}{4S}\right), \quad S_a^z = S - a^+a,$$
$$S_b^+ = \sqrt{2S}b^+\left(1 - \frac{b^+b}{4S}\right), \quad S_b^- = \sqrt{2S}\left(1 - \frac{b^+b}{4S}\right)b, \quad S_a^z = b^+b - S,$$

where a^+ (a) and b^+ (b) are the creation (annihilation) operators of the Heisenberg magnon field on the two sublattices, respectively. Taking into account the symmetry of the sublattices A and B and the fact that the A and B sublattices are neighbors of each other, the Hamiltonian (10.93) can be approximately written as

$$H \approx T + V - J_0 N S^2 + S \sum_n^A \sum_\delta^A J_{n,n+\delta} a_n^+ a_n + S \sum_n^B \sum_\delta^B J_{n,n+\delta} b_j^+ b_j$$
$$+ \frac{S}{2} \sum_n^A \sum_\delta^A (\xi_{n,n+\delta} - \eta_{n,n+\delta})(a_n^+ b_{n+\delta}^+ + b_{n+\delta} a_n^+)$$
$$+ \frac{S}{2} \sum_n^A \sum_\delta^A (\xi_{n,n+\delta} + \eta_{n,n+\delta})(a_n b_{n+\delta} + a_n^+ b_{n+\delta}^+)$$
$$- \sum_j^B \sum_\delta^B J_{j,j+\delta} a_j^+ a_j b_{j+\delta}^+ b_{j+\delta}$$
$$- \frac{1}{8} \sum_n^A \sum_\delta^B (\xi_{n,n+\delta} - \eta_{n,n+\delta}) \left(a_n^+ a_n a_n b_{n+\delta}^+ + a_n b_{n+\delta}^+ b_{n+\delta} b_{n+\delta}\right.$$
$$\left. + a_n^+ a_{n+\delta} a_n^+ a_n + b_{n+\delta}^+ b_{n+\delta} a_n^+ b_{n+\delta}\right)$$
$$- \frac{1}{8} \sum_n^A \sum_\delta^A (\xi_{n,n+\delta} + \eta_{n,n+\delta}) \left(a_n^+ a_n a_n b_{n+\delta} + a_n b_{n+\delta} b_{n+\delta}^+ b_{n+\delta}\right.$$
$$\left. + a_n^+ b_{n+\delta} a_n^+ a_n + a_n b_{n+\delta}^+ b_{n+\delta}^+ b_{n+\delta}\right), \tag{10.94}$$

where the last four terms are anomalous terms resulting from magnon-magnon interactions. They can be neglected if we are interested only in effects of magnon-phonon interaction. We further assume that the wave function of the collective excitation state of the quasi-particles in the system is of the form

$$|\varphi(t)\rangle = \frac{1}{\lambda} \left[1 + \sum_n^A \varphi_{an}(t) a_n^+ + \sum_j^A \varphi_{bj}(t) b_j^+ \right] |0\rangle, \tag{10.95}$$

where $|0\rangle$ is the vacuum state (ground state). In the continuum approximation, *i.e.*,

$$J_{f,f+1} \approx J_0 - J_1 r_0 \frac{\partial u_f}{\partial x},$$
$$\xi_{f,f+1} \approx \xi_0 - \xi_1 r_0 \frac{\partial u_f}{\partial x},$$
$$\eta_{f,f+1} \approx \eta_0 - \eta_1 r_0 \frac{\partial u_f}{\partial x},$$
$$\varphi_{n,f\pm 1} \approx \varphi_{nf} \pm r_0 \frac{\partial}{\partial x} \varphi_{n'f} + \frac{1}{2} r_0^2 \frac{\partial^2}{\partial x^2} \varphi_{n'f} + \cdots, \quad (n = a, b; n' = b, a),$$

etc., we can get the following from the Heisenberg equations for φ_f,

$$\begin{aligned} i\hbar(\varphi_f)_t = 2SJ_0\varphi_f + S(\xi_0-\eta_0)\varphi_f + S(\xi_0+\eta_0)\varphi_f^* + \frac{S}{2}(\xi_0-\eta_0)r_0^2\varphi_{fxx} \\ +\frac{S}{2}(\xi_0+\eta_0)r_0^2\varphi_{fxx}^* - S(\xi_1-\eta_1)r_0u_{fx}\varphi_f \\ -S(\xi_1+\eta_1)r_0u_{fx}\varphi_f^* - 2J_1Sr_0u_{fx}\varphi_f, \end{aligned} \tag{10.96}$$

where

$$\varphi_f(x,t) = \varphi_{af}(x,t) + \varphi_{bf}(x,t).$$

Considering the symmetry of sublattices A and B, and the fact that the sublattices A and B are neighbors, from the Hamilton equation

$$-M(u_f)_{tt} = \frac{\partial}{\partial u_f}\langle\varphi(t)|H|\varphi(t)\rangle$$

where M is the mass of a lattice point (atom, for example), u_f its displacement, a classical quantity, we can get

$$-Mu_{ftt} \approx -K'r_0^2u_{fxx} + J_1Sr_0(|\varphi_f|^2)_x + \frac{1}{4}(\xi_1-\eta_1)Sr_0(|\varphi_f|^2)_x, \tag{10.97}$$

where $K' = mv_0^2/r_0^2$ is the force coefficient, u_f is defined as $u_f = u_{af}+u_{bf}$. We have also assumed that $M_a = M_b = M$. Equations (10.96) and (10.97) form a complete set of equations for the collective excitations in a Heisenberg antiferromagnetic system with magnon-phonon interactions.

For simplicity, we consider here only the anisotropic antiferromagnet with $\xi = \eta$ (other cases, of course, can be discussed in the same way). In this case, $J > \xi$ and $J < \xi$ correspond to the easy magnetic axis ($0z$) and the easy magnetic plane ($x0y$) in an antiferromagnet, respectively. Equation (10.96) reduces to

$$ih\varphi_t = 2J_0S\varphi + 2\xi_0S\varphi^* + S\xi_0r_0^2\varphi_{xx}^* - 2J_1Sr_0u_x\varphi - 2\xi_1Sr_0u_x\varphi^*. \tag{10.98}$$

Using (10.97) and its conjugate equation, and performing the transformation $\varphi_\pm = \varphi \pm \varphi^*$, we obtain

$$\varphi_{tt} - A_0\varphi_{xx} - B_0\varphi - C_0|\varphi|^2\varphi = 0, \tag{10.99}$$

where

$$A_0 = \frac{4\xi_0^2S^2r_0^2}{h^2} > 0,$$

$$B_0 = \frac{1}{\hbar^2}\left[4S^2(J_0^2-\xi_0^2) - 8S^2r_0(J_0J_1-\xi_0\xi_1)C\right],$$

$$C_0 = \frac{16J_1S^3r_0^2(J_0J_1-\xi_0\xi_1)}{\hbar^2(K'r_0^2-Mv^2)},$$

and C is an integral constant. It is clear that the magnon is still a soliton in nonlinear antiferromagnetic systems. It satisfies the ϕ^4-equation. Therefore, it can be described by the nonlinear quantum mechanics. In the case of $-\omega^2 < (A_0B_0)/(A_0 - v^2)$, there are non-topological soliton solutions to (10.99) if $C_0/(A_0 - v^2) > 0$, *i.e.*, if $J_0 > \xi_0$, and at the same time either (a) $J_0J_1 > \xi_0\xi_1$, $v < \min[\sqrt{K/m}r_0, 2\xi_0Sr_0/h]$, or (b) $J_0J_1 < \xi_0\xi_1$, $2\xi_0Sr_0/h < v < \sqrt{K/M}r_0$. The normalized solitary wave is then

$$\varphi = \sqrt{\frac{r_0}{W_s}}\,\mathrm{sech}\left(\frac{x - vt}{W_s}\right)e^{i(k'x-\omega t)}, \tag{10.100}$$

where

$$k' = \frac{v\omega}{A_0}, \quad W_s = \frac{4(A_0 - v^2)}{C_0r_0}, \quad \omega^2 = \frac{A_0B_0}{(A_0 - v^2)} - \frac{d^2A_0C_0^2}{16(A_0 - v^2)^2}.$$

Here v is velocity of the soliton. If $v^2 - A_0 > 0$, *i.e.*, $C_0/(A_0 - v^2) < 0$, equation (10.99) has the following topological soliton solution

$$\varphi = \sqrt{\frac{r_0}{2W_s}}\tanh\left(\frac{x - vt}{W_s}\right)e^{i(k'x-\omega t)}. \tag{10.101}$$

It can be seen from the above conditions that localized soliton can be excited by the magnon-phonon coupling only for the easy magnetic axis antiferromagnet. This was not observed before. No parallel has been obtained in ferromagnet. In the case being discussed, the coupling of the longitudinal lattice oscillations with the magnon results in a remarkable change in the transverse exchange integral of the antiferromagnet. It is the nonlinear interaction caused by the coupling that is vital to the formation of the soliton. In this case, the velocity of the soliton satisfies $v < \min[\sqrt{K'/M}r_0, 2\xi_0Sr_0/h]$.

We have so far only considered the collective excitation caused by magnon-phonon interaction. In fact, when magnon-magnon interaction in the system becomes too strong to be neglected, a new nonlinear interaction source will contribute to the collective excitation. The formation process and the properties of the collective excitation will change accordingly if this interaction is taken into consideration. The Hamiltonian of the system in this case is still given by (10.94), but the interaction term now includes direct interactions between neighboring magnons and other two-magnon effects, *i.e.*, influences of a magnon upon the transfer of other magnons and upon the magnon "resonance", *etc.*.

Pang *et al.* employed the following quasi-average field approximation to treat the effects of the anomalous correlation terms in (10.90) upon the soliton formation, and the quasi-particle energy in the collective excitation, that is

$$a_n^+a_na_nb_{n+\delta}^+ = \langle a_n^+a_n\rangle a_nb_{n+\delta}^+ + \langle a_nb_{n+\delta}^+\rangle a_n^+a_n - \langle a_n^+a_n\rangle\langle a_nb_{n+\delta}^+\rangle,$$
$$a_n^+a_na_n^+b_{n+\delta}^+ = \langle a_n^+a_n\rangle a_n^+a_{n+\delta}^+,$$
$$\cdots\cdots$$

Then the Hamiltonian of the system (10.94) becomes

$$
\begin{aligned}
H = E_0 &+ S\sum_n^A\sum_\delta^A J_{n,n+\delta}a_n^+a_n + S\sum_j^B\sum_\delta^B J_{j,j+\delta}b_j^+b_j \\
&+\frac{1}{2}S\sum_n^A\sum_\delta^A(\xi_{n,n+\delta}-\eta_{n,n+\delta})(a_nb_{n+\delta}^+ + a_n^+b_{n+\delta}) \\
&+\frac{1}{2}S\sum_n^A\sum_\delta^A(\xi_{n,n+\delta}+\eta_{n,n+\delta})(a_nb_{n+\delta} + a_n^+b_{n+\delta}^+) \\
&-\sum_n^A\sum_\delta^A J_{n,n+\delta}(\langle a_n^+a_n\rangle b_{n+\delta}^+b_{n+\delta} + \langle b_{n+\delta}^+b_{n+\delta}\rangle a_n^+a_n) \qquad (10.102)\\
-\frac{1}{8}\sum_n^A\sum_\delta^A&(\xi_{n,n+\delta}-\eta_{n,n+\delta})\left[(\langle a_n^+a_n\rangle + \langle b_{n+\delta}^+b_{n+\delta}\rangle)(a_nb_{n+\delta}^+ + a_n^+b_{n+\delta})+\right.\\
&\left.(\langle a_nb_{n+\delta}^+\rangle + \langle a_n^+b_{n+\delta}\rangle)(a_n^+a_n + b_{n+\delta}^+b_{n+\delta})\right]\\
-\frac{1}{8}\sum_n^A\sum_\delta^A&(\xi_{n,n+\delta}+\eta_{n,n+\delta})\left[(\langle a_n^+a_n\rangle + \langle b_{n+\delta}^+b_{n+\delta}\rangle)(a_nb_{n+\delta} + a_n^+a_{n+\delta}^+)\right],
\end{aligned}
$$

where

$$
\begin{aligned}
E_0 = T + V - J_0NS^2 &+ \sum_n^A\sum_\delta^A J_{n,n+\delta}\langle a_n^+a_n\rangle\langle b_{n+\delta}^+b_{n+\delta}\rangle + \qquad (10.103)\\
\frac{1}{8}\sum_n^A\sum_\delta^A&(\xi_{n,n+\delta}-\eta_{n,n+\delta})\left[(\langle a_n^+a_n\rangle + \langle b_{n+\delta}^+b_{n+\delta}\rangle)(\langle a_nb_{n+\delta}^+\rangle + \langle a_n^+b_{n+\delta}\rangle)\right].
\end{aligned}
$$

Similar to the derivations of (10.96) – (10.98) we can obtain

$$\varphi_{tt} - A\varphi_{xx} + B\varphi - g|\varphi|^2\varphi = 0, \qquad (10.104)$$

where

$$A = A_0, \quad B = B_0, \quad g = C_0 + \frac{8S(J_0^2 - v^2\xi_0^2)}{\hbar^2}.$$

Therefore, equation (10.104) has the same soliton solution as (10.100) or (10.101). Simply replacing C_0, A_0, and B_0 in (10.100) and (10.101) by g, A, and B, respectively, we get the solutions. Therefore, taking into account magnon-magnon interaction only changes the amplitude and velocity of the soliton and does not alter the fundamental nature of the collective excitation. The magnon-magnon interactions enhance the effects of the nonlinear interactions, thereby prompt formation of more stable solitons. This is because g is always greater than C_0 if solitons can exist. Furthermore, $J_0 \gg J_1$ and $\xi_0 \gg \xi_1$. We can see that in the presence of magnon-magnon interactions, they are only the soliton solution of the type (10.100) or (10.101) in the velocity range of $v^2 < A$. In this case

$$\omega^2 = \frac{AB}{A-v^2} - \frac{r_0^2 A g^2}{16(A-v^2)^2} = \frac{AB}{A-v^2} - \frac{r_0^2}{16}\frac{A}{(A-v^2)}\left[g + 8S(J_0^2 - v^2\xi_0^2)\right]^2.$$

Pang and Xu further studied nonlinear excitations of magnons in antiferromagnetic molecular crystal, such as $Ni(C_2H_8N_2)_2{-}NO_2(ClO_4)$ (NENP) and $Ni(C_3H_{10}N_2)_2NO_2(ClO_4)$ (NINO), and antiferromagnet with conservation of order parameters. It was found that the nonlinear excitations of magnons can still be described by (10.99) or (10.104). The only differences are the coefficients in these equations. Therefore, magnons in nonlinear antiferromagnetic systems obey the laws of nonlinear quantum mechanics.

It should be noted that even though magnon-phonon coupling and magnon-magnon interactions have same effects on the formation of solitons, there are differences between the two types of interactions. In the first place, the mechanism of localized nonlinear collective excitation caused by the first type is the breaking of kinetic symmetry, that is, it is caused by the interaction between magnon and lattice oscillation. In a steady state, the soliton and the localized deformation which depend on lattice oscillations propagate together with the same speed along the antiferromagnetic chain. The mechanism of excitation caused by the second type is the spontaneous breaking of symmetry brought about by the magnon-magnon interactions in the single-axis anisotropic antiferromagnet. However, the collective excitations resulting from both mechanisms have the same characteristics, *i.e.*, the structural anisotropy. As it was already mentioned earlier, the two mechanisms may cancel each other in isotropic antiferromagnetic chains, and no soliton can exist as in the case of ferromagnetic chains. It also indicates that soliton excitation of magnons in such systems is determined by the anisotropy of the system. As long as anisotropy exists in a given system, magnon-phonon coupling and nonlinear interactions between magnons will make the magnons "self-trapping" in a range of dimension $2W_s$ in the one-dimensional chains and a stable soliton is excited. When the anisotropy changes, the amplitude, the momentum, and the number of solitons all change accordingly.

Formation of solitons due to nonlinear interactions in anisotropic antiferromagnetic chain lead to many interesting physical phenomena. Indeed, anomalies have been observed in experiments. Attempts have been made to explain them using the magnetic soliton model, even though analytical expressions in place of (10.100) and (10.101) have not been obtained. (The readers are referred to work by Mikeska and Steiner 1991) For example, Boucher *et al.* used the soliton concept to explain the phenomenon of nuclear spin-lattice relaxation (NSLR) in antiferromagnetic chain $(CH_3)_4NMnCl_3$, even though theoretical expression for magnetic solitons has not been obtained. Through measurement, Boucher *et al.* obtained the ratio T_1^{-1} of NSLR of N^{15} in antiferromagnet, as a function of external field $\tilde{H}$ (2 kAm^{-1} $< \tilde{H} <$ 80 kAm^{-1}) and temperature T (2 K $\leq T \leq$ 4.2 K), and observed that T_1^{-1} diverged exponentially with $\tilde{H}/T$ at a certain temperature. With the analytical results for the soliton given above, we can explain the excitation and the behavior of the soliton in such systems, which in turn validates the theory.

It should also be pointed out that effects of external fields are not included in our discussion. If any such field is present and if it is in the direction along the easy magnetic axis, then its effect, due to the opposite magnetization directions of the two sublattices, will be equivalent to a periodic external field of period $2r_0$ which strengthens the discreteness of the lattice and thus invalidates the continuum approximation. However, if the direction of the external field is perpendicular to the antiferromagnetic spin direction, the continuum approximation will still be valid. For this reason, earlier experimental and theoretical studies were concentrated mainly on transverse fields, rather than longitudinal fields.

Bibliography

Aliotta, F. and Fontana, M. P. (1980). Optica Acta **27** (931) .

Antonchenko, V. Ya., Davydov, A. S. and Zolotaryuk, A. V. (1983). Phys. Stat. Sol.(b) **115** (631) .

Bell, R. P. (1973). The proton in chemistry, Chapman and Hall, London.

Bernal, J. D. and Fowler, R. H. (1933). J. Chem. Phys. **1** (515) .

Binder, K. (1984). Applications of the Monte Carlo Method, Springer-Verleg, Berlin.

Bjerrum, N. (1952). Science **115** (385) .

Bogoliubov, N. N. (1950). Ukraimian. Mat. J. **11** (1950) 276.

Bontis, T. (1992). Proton transfers in hydrogen bonded systems, Plenum Press, London.

Boucher, J. P., Pynn, R., Remoissewet, M., Regnault, L. P., Endo, Y. and Renard, J. P. (1980). Phys. Rev. Lett. **45** (486) .

Boucher, J. P., Pynn, R., Remoissewet, M., Regnault, L. P., Endo, Y. and Renard, J. P. (1990). Phys. Rev. Lett. **64** (1557) .

Braum, J. M. and Klvshar, Yu. S. (1990). Phys. Lett. A **149** (119) .

Braum, J. M. and Klvshar, Yu. S. (1991). Phys. Rev. B **43** (1060) .

Brown, D. W. and Ivic, Z. (1989). Phys. Rev. B **40** (9876) .

Brown, D. W., Lindenberg, K. and West, B. J. (1986). J. Chem. Phys. **84** (1574) .

Brown, D. W., Lindenberg, K. and West, B. J. (1987). J. Chem. Phys. **87** (6700) .

Brown, D. W., Lindenberg, K. and West, B. J. (1988). Phys. Rev. B **37** (2946) .

Brown, D. W., West, B. J. and Lindenberg, K. (1986). Phys. Rev. A **33** (4110) .

Buttiker, M. and Ladauer, R. (1981). Phys. Rev. A **23** (1397) .

Cao, Changliang (1993). Physics Sin. **22** (361) .

Chikazumi, S. (1981). Physics in high maginet fields, Springer-Verlag, Berlin.

Chochliouros, I. and Pouget, J. (1995). J. Phys. Condensed Matter **7** (8741) .

Currie, J. F., Krumhansl, J. A., Bishop, A. R. and Trullinger, S. E. (1980). Phys. Rev. B **22** (477) .

Davydov, A. S. (1982). Biology and quantum mechanics, Pergamon, New York.

Davydov, A. S. and Kislukha, N. I. (1973). Phys. Stat. Sol. (b) **59** (465) .

Davydov, A. S. and Kislukha, N. I. (1976). Sov. Phys.-JETP **44** (571) .

Derlin, J. J. (1990). Int. Rev. Phys. Chem. **9** (29) .

Desfontaines, H. and Peyrared, M. (1989). Phys. Lett. A **142** (128) .

Devreese, J. T. (1972). Polarons in ionic crystals and polarsecal counductors, NorthHolland, Amsterdam.

Dwicki, J. C.,*et al.* (1997). J. Am. Chem. Soc. **99** (7403) .

Eigen, M. and De Maeyer, L. (1958). Proc. Roy. Soc. London, Ser. A **247** (505) .

Eigen, M., de Maeyer, L. and Spatz, H. C. (1962). Physics of ice crystals, Coll. London.

Eisenberg, D. and Kanzmann, W. (1969). The structure and properties of water, Clarendon, Oxford.

Enderle, M.,*et al.* (2001). Phys. Rev. Lett. **87** (177203) .

Engelheart, H., Bullemer, B. and Riehl, N. (1969). in Physics of Ice, ed. N. Riehl, B. Bullemer and H. Engelheart, Plenun, New York.

Evans, M. W. (1982). J. Chem. Phys. **76** (5473 and 5480) ; ibid **77** (4632) .

Evans, M. W. (1983). J. Chem. Phys. **78** (925) .

Evans, M. W. (1987). J. Chem. Phys. **87** (6040) .

Fedyanin, V. and Yakushevich, L. (1976). JINR p17-96-27, Dubna.

Fedyanin, V. and Yakushevich, L. (1980). JINR p17-80-69, Dubna.

Fedyanin, V. and Yakushevich, L. (1981). Phys. Lett. A **85** (100) .

Fedyanin, V. and Yushankhai, Y. (1978). JINR p17-11996, Dubna.

Fedyanin, V. K. and Makhaankov, V. G. (1979). Phys. Scr. **20** (552) .

Fedyanin, V. S. and Yushankhai, V. (1978). Teor. Mat. Fiz **35** (240) .

Fisher, A. J., Hayes, W. and Wallance, D. S. (1989). J. Phys.: Condensed Matter **1** (5567) .

Fraggis, T., Pnevmatikos, St. and Economon, E. N. (1989). Phys. Lett. A **142** (361) .

Fröhich, H., Pelzer, H. and Zienau, S. (1950). Plilos. Mag. **41** (221) .

Fröhlich, H. (1954). Adv. Phys. **3** (325) .

Goldanskill, V. I., Krupyanskii, Yu. F. and Flerov, V. N. (1983). Dokl. Akad. Nauk (SSSR) **272** (978) .

Granicher, H. (1963). Phys. Kond. Materie l (1) .

Habbard, L. (1963). Proc. Roy. Soc. A **276** (238) .

Halding, J. and Londahl, P. S. (1988). Phys. Rev. A **37** (2608) .

Higashitani, K. (1992). J. Colloid and Interface Science **152** (125) .

Higashitani, K. (1993). J. Colloid and Interface Science **156** (90) .

Hobbs, P. V. (1974). Ice physics, Clarendon Press, Oxford.

Hochstrasser, D., Buttner, H., Defontaines, H. and Peyrared, M. (1988). Phys. Rev. A **38** (5332) .

Hochstrasser, D., Buttner, H., Defontaines, H. and Peyrared, M. (1989). J. Physique. Coll. **50** (3) .

Holstein, T. (1959). Ann. Phys. **8** (325) and 343.

Homilton, W. C. and Ibers, J. A. (1969). Hydrogen bonding in solids, Benjamin, New York.

Ivic, Z. and Brown, D. W. (1989). Phys. Rev. Lett. **63** (426) .

Jiang, Yijian, Jia Qingjiou, Zhang Peng Cheng and Xu Lu, (1992). J. Light Scattering **4** (7216) .

Joshi, K. M. and Kamat, P. V. (1965). J. Indian Chem. Soc. **43** (620) .

Karpman, V. I. and Maslov, E. M. (1977). Sov. Phys. JETP **46** (281) .

Karpman, V. I. and Maslov, E. M. (1978). Sov. Phys. JETP **48** (252) .

Kashimori, Y., Kikuchi, T. and Nishimoto, K. (1982). J. Chem. Phys. **77** (1904) .

Kawada, A., McGhie, A. R. and Labes, M. M. (1970). J. Chem. Phys. **52** (3121) .

Ke Laxin, B. N. (1982). Magnetization of water, Beijing, Measurement Press.

Klivshar, Yu. S. (1991). Phys. Rev. A **43** (3117) .

Koutselose, A. D. (1995). J. Chem. Phys. **102** (7216) .
Kristoforov, L. N. and Zolotaryuk, A. V. (1988). Phys. Stat. Sol. (b) **146** (487) .
Krumhansl, G. A. and Schrieffer, J. R. (1975). Phys. Rev. B **11** (3535) .
Kuper, C. G. and Whitfield, G. D. (1963). Polarons and excitons, Plenum Press, New York.
Kusalik, P. G. (1995). J. Chem. Phys. **103** (10174) .
Laedke, E. W., Spatschek, K. H., Wilkens, M. and Zolotaryuk, A. V. (1985). Phys. Rev. A **32** (1161) .
Landau, L. D. (1933). Phys. Z. Sowjetunion **3** (664) .
Li, Benyuan (1976). Physics Sin. **5** (240) .
Liemeza, J. (1976). Z. Phys. Chem. **99** (33) .
Lindner, U. and Fedyanin, V. (1978). Phys. Stat. Sol. (b) **89** (123) .
Lindner, U. and Fedyanin, V. (1978). Phys. Stat. Sol. (b) **95** (k83) .
Makhankov, V. G. (1981). Phys. Lett. A **81** (156) .
Makhankov, V. G. and Fedyanin, V. K. (1984). Phys. Rep. **104** (1) .
Makhankov, V. G., Makhaidiani, N. V. and Pashaev, O. K. (1981). Phys. Lett. A **81** (161) .
Marcheson, F. (1986). Phys. Rev. B **34** (6536) .
Mazenko, G. F. and Sahni, P. S. (1978). Phys. Rev. B **18** (6139) .
Mei, Y. P. and Yan, J. R. (1993). Phys. Lett. A **180** (259) .
Mei, Y. P., Yan, J. R., Yan, X. H. and You, J. Q. (1993). Phys. Rev. B **48** (575) .
Mikeska, H. J. (1979). J. Phys. C **11** (l29) .
Mikeska, H. J. and Steiner, M. (1991). Adv. Phys. **40** (p191-356) .
Mittal, R. and Howard, I. A. (1999). Physica D **125** (79) .
Mouritsen, O. G. (1978). Phys. Rev. B **18** (465) .
Mrevlislivil, G. M. (1979). Sov. Phys. USP. **22** (433) .
Mrevlislivil, G. M. (1979). Usp. Fiz. Nauk **128** (273) .
Muller, K. (1970). Z. Chem. **10** (216) .
Myneni, S.,*et al.* (2002). J. Phys. Condensed Matter **14** (L213) .
Nagle, J. F. and Morowitz, H. J. (1978). Proc. Natl. Acad. Sci. USA **75** (298) .
Nagle, J. F. and Nagle, T. S. (1983). J. Membr. Biol. **74** (1) .
Nagle, J. F., Mille, M. and Morowitz, H. J. (1980). J. Chem. Phys. **72** (3959) .
Nelmes, R. J. (1980). Ferroelectrics **24** (237) .
Nylund, E. S. and Tsironis, G. P. (1991). Phys. Rev. Lett. **66** (1886) .
Onsager, L. (1969). Science **166** (1359) .
Oraevskii, A. N. and M. Yu. Su, Dakov, (1987). Zh. Eksp. Teor. Fiz. **92** (1366) [Sov. Phys. JETP **65** (767)].
Pang, X. F. and Chen, X. R. (2002). Commun. Theor. Phys. **37** (715) .
Pang, Xiao-feng (1990). Acta Math. Phys. Sin. **10** (47) .
Pang, Xiao-feng (1993). Phys. Stat. Sol. (b) **189** (237) .
Pang, Xiao-feng (1994). Theory of nonlinear quantum mechanics, Chongqing Press, Chongqing, pp. 554-573 and pp. 574-583.
Pang, Xiao-feng (2000). Chin. Phys. **9** (86) .
Pang, Xiao-feng (2000). J. Shandong Normal Univ. Sin. (Nature) **15** (43) .
Pang, Xiao-feng (2002). Prog. Phys. Sin. **22** (214) .
Pang, Xiao-feng (2003). Phys. Stat. Sol. (b) **239** (862) .
Pang, Xiao-feng (2003). Soliton physics, Sichuan Sci. and Tech. Press, Chengdu, pp. 372-482 and pp. 714-771.
Pang, Xiao-feng and Feng, Yuan-ping, (2003). Chem. Phys. Lett. **373** (392) .
Pang, Xiao-feng and Feng, Yuan-ping, (2003). Chin. Phys. Lett. **20** (1662) .

Pang, Xiao-feng and Muller-Kersten, H. J. W. (2000). J. Phys.: Condens. Matter **12** (885) .
Pang, Xiao-feng and Zundel, G. (1997). Acta Phys. Sin. **46** (625) .
Pang, Xiao-feng and Zundel, G. (1998). Chin. Phys. **7** (70) .
Pan, Zhongcheng and Xun Chengxian (1985). Chin. Med. Phys. **7** (1985) 226.
Pauling, L. (1960). The nature of chemical bond, Cornell University, Ithaca.
Pekar, S. (1946). J. Phys. USSR. **10** (341) and 347.
Peyrard, M. (1995). Nonlinear excitation in biomolecules, Springer, Berlin.
Peyrard, M., Pnevmatikos, St. and Flytzanis, N. (1986). Physica D **19** (268) .
Peyrard, M., Pnevmatikos, St. and Flytzanis, N. (1987). Phys. Rev. A **36** (903) .
Pimentel, G. and McClellan, A. (1960). The hydrogen bond, Freeman, San Francisco.
Pnevmatikos, St. (1988). Phys. Rev. Lett. **60** (1534) .
Pnevmatikos, St., Flytzanis, N. and Bishop, A. R. (1987). J. Phys. C. Solid State Phys. **20** (2829) .
Pnevmatikos, St., Savin, A. V., Zolotariuk, A. V., Kivshar, Yu. S. and Velgakis, M. J. (1991). Phys. Rev. A **43** (5518) .
Pushkarov, D. and Pushkarov, Kh. (1977). J. Phys. (c) **10** (3711) .
Pushkarov, D. and Pushkarov, Kh. (1978). Phys. Stat. Sol. (b) **90** (361) .
Pushkarov, D. and Pushkarov, Kh. (1979). Phys. Stat. So1. (b) **81** (703) .
Pushkarov, D. and Pushkarov, Kh. (1979). Phys. Stat. Sol. (b) **93** (735) .
Pushkarov, D. and Pushkarov, Kh. (1984). Phys. Stat. Sol. (b) **123** (573) .
Revarat, Y. F.,*et al.* (1996). Phys. Rev. Lett. **77** (1861) .
Rodriges, K. and Fedyanin, V. (1981). JINR p. 17-81-169, Dubna; JINR p. 17-81-8, Dubna.
Ronnow, H. M.,*et al.* (2000). Phys. Rev. Lett. **84** (4469) .
Scheider, T. and Stoll, E. (1975). Phys. Rev. Lett. **35** (2961) .
Scheider, T. and Stoll, E. (1978). Phys. Rev. Lett. **41** (1429) .
Scheider, T. and Stoll, E. (1980). Phys. Rev. B **22** (5317) .
Schmidt, V. H., Drumeheller, J. E. and Howell, F. L. (1971). Phys. Rev. B **4** (4582) .
Schuster, P., Zundel, G. and Sandorfy, C. (1976). The hydrogen bond, recent developments in theory and experiments, North-Holland, Amsterdam.
Sergienko, A. I. (1987). Phys. Stat. Sol. (b) **144** (471) .
Sergienko, A. I. (1988). Sov. Phys. Solid State **30** (496) .
Sergienko, A. I. (1990). Sov. Phys-JETP **70** (710) .
Skrinjar, M. J., Kapor, D. W. and Stojanovic, S. D. (1988). Phys. Rev. A **38** (6402) .
Sokolov, N. D. (1955). Usp. Fiz. Nauk. **57** (205) .
Song, Dongning (1997). J. App. Math. and Mech. **18** (113) .
Tsironis, G. P. and Pnevnatikos, St. (1989). Phys. Rev. B **39** (7161) .
Walrafen, G. E. (1970). J. Chem. Phys. **52** (4276) .
Walrafen, G. E. (1986). J. Chem. Phys. **85** (6964) .
Wang, X. D., Brown, D. W., Lindenberg, K. and West, B. J. (1986). Phys. Rev. A **33** (4101) .
Weberpals, H. and Spatschek, K. H. (1987). Phys. Rev. A **36** (2946) .
Weiner, J. H. and Asker, A. (1970). Nature **226** (842) .
Whalley, E., Jones, S. J. and Grold, L. W. (1973). Physics and chemistry of ice, Roy. Soc Canada, Ottawa.
Whitfield, G. and Shaw, P. (1976). Phys. Rev. B **14** (3346) .
Xie, Wen Hui (1983). Magnetized water and its application, Science Press, Beijing.
Xu, C. T and Pang Xiao-feng (1996). Chin. J. Atom. Mol. Phys. **12** (508) .
Xu, C. T and Pang Xiao-feng (1998). Chin. J. Atom. Mol. Phys. **14** (219) .
Xu, C. T and Pang Xiao-feng (1999). Chin. J. Atom. Mol. Phys. **16** (506) .
Xu, C. T and Pang Xiao-feng (2000). Chin. J. Atom. Mol. Phys. **17** (99) and 513.

Yomosa, S. (1982). J. Phys. Soc. Japan **51** (3318) .
Yomosa, S. (1983). J. Phys. Soc. Japan **52** (1866) .
Zhang, Q., Romero-Rochin, V. and Silbey, R. (1988). Phys. Rev. A **38** (6409) .
Zolotariuk, A. V. (1986). Theor. Math. Fiz. **68** (415) .
Zolotariuk, A. V., Spatschek, K. H. and Laedke, E. W. (1984). Phys. Lett. A **101** (517) .
Zolotaryuk, A. V. and Pnevmatikos, S. (1990). Phys. Lett. A **143** (233) .
Zolotaryuk, A. V., Peyrard, M. and Spatschek, K. H. (2000). Phys. Rev. E **62** (5706) .
Zundel, G. (1972). Hydration and intermolecular interaction, Mir, Moscow.

Index